ANNUAL REVIEW OF PLANT PHYSIOLOGY AND PLANT MOLECULAR BIOLOGY

EDITORIAL COMMITTEE (1991)

Responsible for organization of Volume 42
(Editorial Committee, 1989)

ANNUAL REVIEW OF PLANT PHYSIOLOGY AND PLANT MOLECULAR BIOLOGY

VOLUME 42, 1991

WINSLOW R. BRIGGS, *Editor*

Carnegie Institution of Washington, Stanford, California

RUSSELL L. JONES, *Associate Editor*

University of California, Berkeley

VIRGINIA WALBOT, *Associate Editor*

Stanford University

ANNUAL REVIEWS INC.   4139 EL CAMINO WAY   PO BOX 10139   PALO ALTO, CALIFORNIA 94303-0897 USA

R ANNUAL REVIEWS INC.
Palo Alto, California, USA

International Standard Serial Number: 1040-2519
International Standard Book Number: 0-8243-0642-2
Library of Congress Catalog Card Number: A-51-1660

Annual Review and publication titles are registered trademarks of Annual Reviews
Inc.

Annual Reviews Inc. and the Editors of its publications assume no responsibility
for the statements expressed by the contributors to this *Review*.

Typesetting by Kachina Typesetting Inc., Tempe, Arizona; John Olson, President
Typesetting Coordinator, Janis Hoffman

PRINTED AND BOUND IN THE UNITED STATES OF AMERICA

ANNUAL REVIEWS INC. is a nonprofit scientific publisher established to promote the advancement of the sciences. Beginning in 1932 with the *Annual Review of Biochemistry*, the Company has pursued as its principal function the publication of high quality, reasonably priced *Annual Review* volumes. The volumes are organized by Editors and Editorial Committees who invite qualified authors to contribute critical articles reviewing significant developments within each major discipline. The Editor-in-Chief invites those interested in serving as future Editorial Committee members to communicate directly with him. Annual Reviews Inc. is administered by a Board of Directors, whose members serve without compensation.

For the convenience of readers, a detachable order form/envelope is bound into the back of this volume.

*Annual Review of Plant Physiology
and Plant Molecular Biology
Volume 42 (1991)*

CONTENTS

TISSUE, ORGAN, AND WHOLE PLANT EVENTS

ACCLIMATION AND ADAPTATION

INDEXES

RELATED ARTICLES OF INTEREST TO READERS

From the *Annual Review of Cell Biology*, Volume 6 (1990)

Chaperone Function: The Assembly of Ribulose Bisphosphate Carboxylase-
Oxygenase, *A. A. Gatenby and R. J. Ellis*
Biochemistry and Function of the Plastid Envelope, *R. Douce and J. Joyard*
The Structure and Insertion of Integral Proteins in Membranes, *S. J. Singer*
Calcium Channels, Stores, and Oscillations, *R. W. Tsien and R. Y. Tsien*

From the *Annual Review of Microbiology*, Volume 44 (1990)

Molecular Biology of Cellulose Degradation, *P. Béguin*

From the *Annual Review of Genetics*, Volume 24 (1990)

Tissue Specific Gene Expression in Plants, *J. W. Edwards and G. M. Coruzzi*
Gene-for-Gene Complementarity in Plant-Pathogen Interactions, *N. T. Keen*

From the *Annual Review of Phytopathology*, Volume 29 (1991)

Delignification by Wood Decay Fungi, *R. A. Blanchette*
Maintaining Genetic Diversity in Breeding for Resistance in Forest Trees,
    G. Namkoong
Plant Diseases and the Use of Wild Germplasm, *J. M. Lenné and D. Wood*
Environmentally Driven Cereal Crop Growth Models, *R. W. Rickman and
    B. Klepper*

From the *Annual Review of Biochemistry*, Volume 60 (1991)

Molecular Chaperones, *R. J. Ellis and S. M. van der Vies*
Eukaryotic DNA Polymerases, *T. S.-F. Wang*
Fidelity Mechanisms in DNA Replication, *H. Echols and M. F. Goodman*
Protein-Priming of DNA Replication, *M. Salas*
The Regulation of Histone Synthesis in the Cell Cycle, *M. A. Osley*
Phospholipid Transfer Proteins, *K. W. A. Wirtz*
Structure and Function of Signal-Transducing GTP-Binding Proteins, *Y. Kaziro,
    H. Itoh, T. Kozasa, M. Nakafuku, and T. Satoh*
Ribosomal RNA and Translation, *H. F. Noller*
Model Systems for the Study of Seven-Transmembrane-Segment Receptors,
    H. G. Dohlman, J. Thorner, M. G. Caron, and R. J. Lefkowitz
RNA Polymerase II, *R. A. Young*
RNA Antisense, *Y. Eguchi, T. Itoh, and J. Tomizawa*

ERRATA

We note with regret the following errors in Volume 41 of the *Annual Review of Plant Physiology and Plant Molecular Biology*.

page 21, line 1: The source line should read *"Annu. Rev. Plant Physiol. Plant Mol. Biol 1990. 41:21–53"*

page 29: The sentence beginning on line 14 should read: "Increasing turgor pressure under hypoosmotic conditions inhibited, while decreasing turgor pressure under hyperosmotic treatment stimulated, pump activity."

Érasmo Marré

Annu. Rev. Plant Physiol. Plant Mol. Biol. 1991. 42:1–20

SHORT STORY OF A PLANT PHYSIOLOGIST AND VARIATIONS ON THE THEME

Erasmo Marrè

Department of Biology, University of Milan, Italy

KEY WORDS: plant physiology, biology, vision of the world

CONTENTS

This is the use of memory: for liberation

T. S. Eliot, Little Gidding

The invitation to write this prefatory chapter is a great honor. However, what to write? Not much in my experience seems worth the telling.

While pondering my task, I happened to read the verse of T. S. Eliot cited above, which suggested an approach. I thus write this chapter with the aim of

1040-2519/91/0601-0001$02.00

obtaining, from a recollection and a reconsideration of my life, a better understanding of my actions and emotions, of their interrelationship, and of my interactions with others. I reflect, in other words, on how I have played my part in this world. If Eliot is right, these recollections of my history, seen as a whole and as objectively as possible, can lead to a liberation from futilities and self-attachment, and thus to a deeper understanding of the importance of events and persons in shaping my life.

Here I treat also of my activities and ideas in Plant Physiology. I feel no conflict between this "official" reportage and the more personal approach, for biology in general, and plant physiology in particular, have played a large part in my life and in my mental development.

BIOGRAPHICAL

Early Years

When very young—say between 5 and 15—I lived with few external restrictions in a small villa close to Genoa and the sea, a villa surrounded by gardens and orchards. Here I developed the wild side of my nature. Through fishing, through capturing and trying to tame small animals the seed of my interest in the functioning (the physiology!) of living beings began to germinate. Those who are strongly opposed to hunting forget that perhaps its deepest root is a kind of love: a longing to take into one's hands (to possess) the object of love. This distorted form of love can be a primitive stage of a less egoistic, more enlightened feeling for the beauty of nature and for our deep involvement in it. Fishing and hunting imply knowing the names and habits of fishes, birds, and squirrels; constructing traps stimulates technical ingenuity; taming the captured animals involves principles of both physiology and ethology. I gave up hunting at about the age of 20, when I realized that killing or caging did not give me what I sought—namely, an intimate contact with wild creatures.

I had access to a large and varied collection of books. My schoolwork took up only a small part of my time, and I dedicated much of the remainder to reading the classics: Shakespeare and Chesterton, de Kruif and Pascal, Fabre and Melville. This omnivorous reading increased in my high school years, when my family moved to Genoa. I became seriously interested both in literature and in the natural sciences. Thus I discovered the great theories and controversies related to biological evolution; I became familiar with the works of Lamarck, Darwin, Haeckel, and Huxley, and with the opposition of the Catholic Church. Seeking a bridge between biology, the human sciences, and religion, I studied Spencer, Bergson, and Theilard de Chardin.

In 1938 I entered the university as a student of medicine and immediately started laboratory work in anatomy and histology. Physiology became part of my studies almost immediately, inasmuch as my anatomy professor, Ganfini, was interested in the relationship between body weight and the cytological features of the hypophysis. This first contact with the problem of biological regulation informed my entire scientific life.

Having realized that I was interested in basic rather than in applied biology, and in dynamic rather than in descriptive biology, I switched from Medicine to Natural Sciences. Greatly influential in this decision was Ettore Remotti, professor of Comparative Anatomy and Physiology. Remotti was a man of modern scientific views who moved with equal ease and acuity in the fields of morphology and physiology. Remotti introduced me to the wonderful world of embryology, and my first published paper concerned the development of the embryo of a small viviparous fish and (of interest in the light of my later work) the mechanisms of nutrient exchange between mother and embryo.

The intensity of my dedication to laboratory work in those years left little time to other occupations. I was still reading very widely: philosophy (Spinoza, Kant) and novels (I discovered Poe, Dostoevski, Gogol). My social life was limited to contact with students attending the biological laboratories and to the membership of a Catholic association of students. In the latter organization I met two most enlightened priests, E. Guano and F. Costa, who had a profound influence upon me, helping to develop my feeling that a man's highest aim should be to understand human life in its wholeness. For some time I seriously considered entering a monastery. At about the same time, however, I was greatly attracted to a girl I met at the university. During vacation time, we mountaineered in the Alps and explored the Ligurian caves. In 1945 she agreed to become my wife; we still share the good and the bad, but that is another story.

The War

Three different tendencies (a passion for science, a fascination with philosophy, and love for a girl) coexisted in me in a largely unconscious but deep conflict, when the events of the war abruptly spurred me into action. The war until then had little affected my life; my feelings were definitely against nazism and fascism, but I had seen few opportunities to translate these feelings into action. However, July of 1943 saw the fall of Mussolini, and in October Italy surrendered to the Allies. The Germans invaded northern Italy, and the official government in southern Italy (occupied by the Allied armies) soon found itself in open conflict with its former ally. In northern Italy the fascists established a republican government, largely controlled by the Germans.

As a medical student I had been exempted from military service. However, in October 1943 I was informed that my call-up was imminent, and so, with three friends, I decided to leave Genoa for the south to join the Italian troops being organized to fight on the side of the Allies. After an adventurous trip through central Italy we crossed the line close to Termoli and, to cut a long story short, a month later we joined a group of some 30 people organized in Naples by the Italian National Liberation Committee in close connection with the American Vth Army. The purpose of this group was the dispatch to northern Italy of small teams of four men and a radio in order to establish contact and collaborate with the partisans operating against the fascists and the Germans.

After three months' training, including instruction in sabotage techniques and a parachutist course at Brindisi, I was parachuted into a partisan-occupied valley in the Alps, and moved to the mountains north of Genoa, in the area occupied by the partisans of the VI zone. This partisan group was politically controlled by the communists, but most of the military commanders were independent of political parties and were merely collaborating with the Communists in the fight against fascism. After some months of loyal and active collaboration with the partisans of the VI zone (organizing an information service for our base in South Italy, receiving air drops of weapons, sabotaging electrical lines, railways, etc), I was led by circumstances to assume a different role. A partisan brigade (the Arzani) had been dispersed by the Germans and its commander captured in December 1944. I was asked to take command of this brigade and to reorganize it. I spent the last four months of the war doing this, I think satisfactorily.

I dwell on this part of my story because I am convinced that these 18 months of "atypical military service" formed a crucial part of my education. Even if all my forces were concentrated on fighting fascism, my past experience conditioned my actions and decisions, and much of what I learned through action and experience was indelibly to influence my subsequent development. Thus, my earlier hunting, fishing, and mountain climbing made it easier than it would otherwise have been for me to move in the wilderness of the Apennine mountains; my training in experimental science helped me to see military problems in terms of final goals (survival, victory), to consider the means available before making plans, and, above all, to be aware of the difference between abstract plans and the concrete possibility of putting them into practice, when most of the parameters were unpredictable or uncertain. On the other hand, the human context in which I was acting was absolutely new to me. I had suddenly found myself part of an extremely mixed population in which communist idealists with years of imprisonment behind them, antifascist officers from the disbanded Italian army, liberal-minded´ intellectuals like myself (few indeed), socialist workers, and individualistic peasants moved to defend their rights all struggled to find a ground for unitary action under the pressure of circumstances. Understanding others and making myself understood in sometimes very difficult discussions, participating in actions, and confronting the emotions of such various people was an exceptionally important experience; it confirmed or corrected the vague concepts I had derived from books. As a result I formed a number of convictions that I still consider valid. First, although adherence to dogmatic interpretations of life may affect a person's objectivity and rationality, no one should be judged on the basis of ideology alone. Second, events (political and historical processes) are determined by internal, indefinite forces, arising from the depths of individual consciences, rather than by the conscious will and rationality of those holding official power (see Tolstoi's *War and Peace*). The behavior of each individual in a given moment is conditioned by impulses springing from the subconscious levels of personality, as shaped by genetics, environment, and previous choices. These impulses can be analyzed and modified at the conscious level where information of all kinds (i.e. culture in the widest sense) are integrated and utilized for practical action. Thus the development of rationality and knowledge serves as a filter for impulses, and both an artistic and a scientific education play a determinant role in this development. In this perspective, the contribution of the study of experimental sciences to the formation of personality can be ordered according to a hierarchy of values, rising from those studied in physics to those investigated in ethology and psychology, where a connection is established with subjects treated in the human sciences and arts.

I think I was born with a propensity to influence the events around me. This tendency developed further when I found myself the head of some 200 persons, with the power of decision in all matters concerning their military activity. Authority is responsibility first toward the objective justifying it and second toward the others working to accomplish this objective. What I rapidly realized was that authority is only legitimated—and accepted—when it is based on example and on the capacity to understand the needs of those who are subjected to it, to make good use of the abilities of others, and to win esteem rather than affection. Such qualities, developed while with the brigade, were to be useful long after the war, when I became the leader of a research group.

The war parenthesis ended in April 1945 with the surrender of the German army in Italy. My experience with political parties in the previous months caused me to decline invitations to enter politics. Instead I concentrated on my two main aims at that moment: to marry the girl I loved, and to find an acceptable job that would enable me to maintain a family. Of the few appointments offered to me, I chose an assistant professorship in Botany at the University of Genoa, which corresponded to my interests. At the beginning of 1946 I found myself married and fully involved in plant biology.

On the Way to Plant Physiology

At the end of the war the general situation of the biological sciences in Italy was far from satisfactory, for ancient as well as recent reasons. In spite of some great names (Malpighi, Redi, Spallanzani, Golgi—isolated peaks), the Italian cultural tradition after 1600 had favored speculative philosophy over experimental biology. The experimental investigation of the mechanisms of life was discouraged, the theory of evolution being considered an attack on religious dogmas. Thus most biological schools and studies were concerned more with description than with the elucidation of mechanism. This unfavorable situation had been worsened by fascism, with its idealistic roots and autarchic (cultural as well as economic) policy. As a consequence of these historical, cultural, and political conditions, the funds for experimental research (books, instruments, jobs) were scarce.

In 1945 the situation of experimental botany was particularly bad. But while most Italian botanists were then involved in descriptive work, in either taxonomy or morphology, some groups were developing more modern and dynamic approaches. In Florence, Chiarugi and collaborators, among whom was Francesco D'Amato, were carrying out good work in caryology, embryology, and cytogenetics, and some of their younger collaborators were initiating a fruitful genetic approach to plant biology. In Pavia, Ciferri was taking the first steps from plant pathology towards physiology, sending one of his best assistants, Felice Bertossi, abroad (to Gautheret's laboratory) to learn methods of tissue culture. Finally, and most important for my future work, in Padua a highly enlightened and dedicated scientist, Giuseppe Gola, was establishing the basis of a really modern Italian plant physiology, with regard both to problems and to methods. (As pointed out by Robert Hill, Gola was the first to recognize the role of a cytochrome in plant oxidation-reductions—in 1915.) The torch of plant physiology lit by Gola passed into the hands of some of his collaborators, in particular Sergio Tonzig, who in 1940 had moved to Milan, where he had started to set up a modern laboratory of plant physiology. The laboratory of Botany in Genoa, where I had found a job in 1946, was then ruled by Giuseppina Zanoni, an intelligent woman sincerely dedicated to biology. At that time, as the new full professor in Botany (having succeeded a strict taxonomist), she was just starting to organize a modern plant biology laboratory, building it up literally from its very foundations (the old building had been almost completely destroyed during the war).

For me this situation had advantages as well as disadvantages. Laboratory equipment and facilities were nonexistent, contacts with other botanists were scarce, and there was no established scientific tradition; but these disadvantages enabled me to start my scientific activity with a minimum of conditioning and a maximum of freedom in choosing a line of work.

Choice of Regulation as the Central Problem

A lucky coincidence of my interests and those of Professor Zanoni facilitated the choice of the field in which I could work. From the beginning of my studies I had been strongly attracted by the problem of hormonal correlations.

I was interested in the mechanisms that integrate the various life processes and allow each organism to behave as a unity, which I regarded as the most amazing and splendid mystery of life. Thus work on certain modest aspects of the problem of unity in a plant absorbed me.

My interest in hormonal correlations fitted in well with the views of Zanoni, who had for some years been conducting original research on the possible role of the sexual organs of plants in the control of differentiation, a role such organs were known to play in animals. While collaborating in her research on the specific actions of anthers and pistils, I concluded that these organs markedly influenced the growth of other plant parts only very early in development and that the removal not only of young anthers or ovaries but even of the growing apexes of shoots resulted in the disappearance of starch from the neighboring territories, an effect prevented by substituting auxin for these growth centers. Thus auxin—and possibly other plant hormones—could influence metabolism. This observation introduced me abruptly into the magic world of biochemistry, seen as a tool with which to investigate the mechanism of hormonal correlations. My new biochemical education was strongly stimulated by the sympathy and advice of Arturo Bonsignore, professor at Genoa and head of one of the best Italian schools of biochemistry, a physiologically minded biochemist of great imagination and technical skill. With his help, a lot of enthusiasm, and little knowledge, I set out to analyze my plant tissues for enzymes, substrates, and respiration. I like to remember, among my more instructive experiences in that period, the successful efforts I made to prepare the phosphate esters required to measure phosphorylase and other enzymes of carbohydrate metabolism.

This almost solitary fight against technical difficulties, this effort to enter a completely new world of information and methods, profoundly shaped my further development. Moreover, entering the world of biochemistry, then almost unknown to Italian botany, led to a series of friendly contacts with many other young Italian biochemists, among whom were Carlo Ricci and Sandro Pontremoli, two brilliant collaborators of Bonsignore's. Here I should also mention Enzo Boeri, whose talk in Siena on the interaction between large and small molecules opened my eyes to the control of protein activities. In the same years, circumstances led me to study the regulation of respiratory metabolism. As I was beginning to appreciate the importance of parameters such as temperature, pressure, and agitation with regard to quantitative O_2 and CO_2 measurements, a good friend of Zanoni's, Professor Vernoni, told her that I would benefit from a familiarity with what was already known! Thus, after some hurried reading of papers by Thunberg, Warburg, Krebs, and other luminaries of the field, I spent a week learning of the marvelous discoveries one can make by the correct use of a respirometer. My mentor in this field was Massimiliano Aloisi, now a leading figure in Italian pathology, to whom I feel bound by a warm and enduring gratitude.

The USA Experience

As I was still engaged in refining my biochemical and physiological tools with a view to attacking the problem of regulation, my research—until then almost autodidactic—was abruptly shifted to the open air and the international field of plant biology. In response to my request, Professor Zanoni arranged for me to study with Andrew Murneek, a distinguished and physiologically minded professor of Horticulture at Columbia, Missouri. Having obtained a Fulbright fellowship, I left for the United States in September 1951. On the way to Columbia, I stopped for a few days in Boston, where, in spite of my almost incomprehensible English, I had long talks with Kenneth Thimann—one of the most important events in my scientific life. These few hours revealed to me the extent of the difference between the Italian and the international scientific environment. In Thimann I found the closest approximation to my ideal of the scientist: a range of biological interests extending far beyond the boundaries of plant life; a capacity to apply the modern developments in biochemistry and biophysics to physiological problems; a sincere and intense interest in the development of science all over the world; the capacity to establish a warm, friendly relationship with younger people; a wide humanistic culture; and a faith in the value of scientific research accompanied by an awareness of the intrinsic limitations of this type of knowledge. I was prepared for this aspect of Thimann's personality by the quotation at the beginning of his book on plant hormones ("Peas, beans, oats and barley grow / But can you, or I, or anybody know / why peas, bean, oats and barley grow?"). Besides discussing and encouraging my plans and ideas with regard to regulation, Thimann gave me a precious overall picture of the enormous effort of renewal going on in plant biology, and of the impact on it of general biochemistry: People such as Kalckar, Lipmann, Krebs, Lynen, Ochoa, Kornberg, and many others were rapidly developing a far-reaching synthesis of the mechanistic analysis of processes and their physiological interpretation. As a consequence, plant physiology was then moving from the identification of forms and functions to the attempt to recognize and reconstruct the deterministic cause-effect chains of events mediating the life processes, the relationships between stimuli and responses, and (my chosen problem) the integration of the various processes in the unit of the organism.

Andrew Murneek, my Fulbright mentor, was a kind and intelligent man, interested in hormonal correlations in plants and, in particular, in the impact of hormones produced by the young flowers and fruits on photosynthesis, translocation, and productivity. It was easy to work out together a plan for my work at Columbia. Moving from Murneek's experience of auxin-induced parthenocarpy and my findings concerning the effects of growth centers on carbohydrate metabolism, I was able to confirm the hypothesis of an effect of growth centers on the activation of sugar metabolism and reserve accumulation in the neighboring tissues. In this, and on more general methodological issues, I was greatly helped by Nobel Prize winner Carl Cori, then teaching at Washington University, St. Louis, who kindly gave me some of his time, the opportunity to discuss my results, and some very precious advice. I also still remember with pleasure the long hours spent with biochemistry books, in the hard but exciting and rewarding effort to bridge the gap between my background and that of my young colleagues in the lab. With two of my fellow investigators I was to establish a lasting friendship and collaboration: Fred Teubner (who died prematurely some years later) and Bob Goodman, now a distinguished professor of Plant Pathology at the University of Missouri.

I was deeply immersed in my laboratory work and reading at Columbia when quite unexpectedly something happened that was to be of the utmost importance for my future. One of the leading Italian botanists, the already mentioned Sergio Tonzig, was then organizing his new laboratory in Milan. He had accepted the invitation of the Rockefeller Foundation to spend three months in the United States, to visit a series of modern botany laboratories, among which was Murneek's. I had never met Tonzig, and thought of him as one of the figures then ruling the universities in Italy by means of much politics and little science. But how wrong I was! The man immediately fascinated me with his subtle sense of humor and his passion for discussing the problems of Italian scientific research in general and the development of plant physiology in particular. I was also impressed by his willingness to discuss any problem, including those relating to his or my experimental work, on a plane of equality, despite the great difference in our age, experience, and academic status.

Tonzig amicably reproached me for the way I was spending my time in the United States, "stuck in a hole like a mouse who has found some cheese," instead of seeing that splendid country and meeting people who could give me new ideas. He persuaded me to accompany him on his visit to some plant physiology laboratories in California. The two weeks we spent together moving slowly west on Greyhound buses were very pleasant; short visits to famous beauty spots alternated with instructive visits to famous laboratories and discussions with some of the most eminent scientists in our field—from James Bonner to Went and Haagen Smit at Pasadena, Van Niel and Arnon at San Francisco, and many others. Tonzig's conversation touched on almost every subject: the *Divina Commedia* (Tonzig knew many passages by heart), the philosophy of science (plant sciences in particular), how to organize a modern plant physiology laboratory in Italy, and the importance of particular research areas.

Back at Columbia, much richer in information and ideas, I spent the time that remained completing my work on the influence of fertilization and of auxin-induced parthenocarpy on the levels of carbohydrate and phosphate esters and on the enzymes of the pentose phosphate pathway. Before returning to Italy, I spent some weeks in Thimann's laboratory at Harvard University. I was able to establish a precious and lasting contact with him, and friendly relations both with Carl Price, then beginning his attempt to clarify the mechanism of the primary action of auxin (a still unsolved problem), and with David Hackett, a promising plant biochemist who died tragically a few years later.

Contacts between European and American scientists are fundamentally important, not only because they advance experimental science but also because they help to develop the awareness that researchers are linked by common aspirations and interests irrespective of differences of nationality or age. The great hospitality offered by the United States to so many young people then fighting to find their way in an economically exhausted but intellectually still vital Europe was the main driving force for the development of a worldwide scientific community, in which discoveries and theories are exchanged, and friendships and collaborations formed without impediment. From the point of view of human progress (or cultural evolution) the development of this cooperation was more important than even the most brilliant achievements of science and technology.

Milan: a Research Team

Soon after my return to Genoa, I found myself in conflict with Professor Zanoni owing to my firm intention to maintain contacts and collaboration with other laboratories—the one directed by Tonzig in particular. Zanoni, like many other Italian botanists, had a restrictive concept of the identity of her laboratory and the importance of hierarchy. Ultimately I accepted Tonzig's invitation to move to his laboratory in Milan, where he offered me "nothing but every possible help to do good work," a promise he fulfilled far beyond any reasonable expectation. In a short time I had a reasonably well-equipped laboratory and, even more important, some young collaborators—the beginning of a small, fully independent research group specifically dedicated to plant physiology. Among the members of this group who made important contributions, were Giorgio Forti, Oreste Arrigoni, and, some time later, Renato Bianchetti. Our progress in the ensuing years was the fruit of a real and spontaneous unity of scientific purposes, our complementary tendencies and personalities, and our capacity to combine a lively and free discussion of the objectives to be pursued with unity of effort in carrying out the laboratory work. Thus Arrigoni (now Professor of Botany at Bari) contributed with his bright, sometimes poetic, imagination and enthusiasm, Forti with his outstanding capacity to sense the importance of new developments in the scientific world, and Bianchetti with his talent for lucid criticism and clear understanding of the key aspects of a problem. In the period from 1952 to 1959 the general philosophy of our research was substantially the one I had developed previously, namely to approach the problem of the regulation and integration of processes by seeking knowledge about the particular processes present in the organism, such as water relations, biosynthetic and energetic metabolism, transport of solutes, and so on.

The birth and growth of this group would not have been possible without the generous assistance of Tonzig, who showed a constant interest in our work, used his great authority in the Italian scientific world to obtain recognition for it, contributed to it with intelligent criticism and advice, and helped us to establish contacts with other laboratories, without ever restricting our freedom. As a result of Tonzig's efforts, in 1956 the Italian National Council established in our laboratory the first Center of Research in Plant Physiology. The institution of this Center (directed at first by Tonzig, then by me, and now by Giorgio Forti) was of decisive importance. It provided a stable budget, increased the number of research positions, and ensured a substantial scientific continuity in the group's work between 1953 and the present day.

Among our contributions of some importance in those years one can mention research on the role of ascorbic acid, on the extramitochondrial respiration pathways, on the metabolic changes associated with the transition from rest to growth, on the mechanism of thermoresistance, on the biochemistry of seed germination, and finally on photosynthesis.

Our work on ascorbic acid and on nonmitochondrial respiration sprang from the confluence between (a) Tonzig's observations of the effects of ascorbic acid on growth by elongation, plasma viscosity, and resistance to plasmolysis and (b) my hypothesis, developed while in the United States, that glucose-6-phosphate oxidative metabolism might be involved in auxin action and the regulation of growth. We started working on soluble dehydrogenases and oxidases involved in the transfer of electrons from glucose-6-phosphate to oxygen through pyridine coenzymes, glutathione, ascorbate, and polyphenols, as well as on the effect of auxin on the oxidation-reduction state of these factors. We also obtained evidence that auxin, as well as conditions breaking dormancy in resting tuber slices, reduced the activity of the pentose phosphate pathway and shifted the NADP, glutathione, and ascorbate systems toward reduction. Although far from definitive, these results pointed to an important role of this pathway in the regulation of cell activity. Work to substantiate this notion (mainly centered on the importance of thioredoxin) is still going on (or planned).

Our research in this region of metabolism immediately and quite naturally opened the way to the study of other aspects of respiration, in particular the possible involvement of the state of energization of the adenylate system in the action of hormones on growth and respiration. These studies, although they did not show such involvement definitively, were of great importance in increasing our direct experience with regard to respiratory and energetic metabolism and preparing the group for a new and exciting adventure, this time in the field of photosynthesis.

In 1952 Daniel Arnon had shown me how to prepare clean chloroplasts. Interested as we were in NADP function and curious about how this cofactor was manipulated in photosynthesis, we decided to look into this related problem. Working first with Servettaz, then with Forti, we were among the first to show that NADPH was indeed a link in the energy-yielding photosynthetic cycle. We obtained indications that a diaphorase and a cytochrome of the c type (cyt f) played roles in the cycle. (The nature of these roles was later independently worked out much more extensively by Giorgio Forti after his long and fruitful visit to André Jagendorf's laboratory.)

The Academy of Cats and the Birth of the Italian Society of Plant Physiology

In the same period, a considerable part of my time was devoted to social-political activity concerning the development of plant biology in Italian universities. The situation of young researchers in most of our botany laboratories (with very few exceptions, such as mine in Tonzig's laboratory) was then far from satisfactory. Old, politically powerful professors ruled the field in an autocratic way, severe and nondemocratic rules of hierarchy called for blind obedience on the part of assistants, and distances between the various academic ranks marked. Independent work was discouraged; most laboratories functioned as closed systems, and contacts among young researchers were discouraged or even prohibited. This situation seemed to me hardly tolerable. In particular, the hostility toward direct contact between young scientists belonging to different laboratories (regardless of the relations of their superiors) appeared to me an unbearable offense to dignity, and also a great impediment to the development of plant research in Italy. To these feelings and conclusions I was moved both by my experience during the partisan war and by what I had seen in the more liberal and cooperative scientific world of the United States. A few friends and I therefore started to establish a kind of mutual-help organization for young botanists, with the enthusiastic cooperation of several colleagues at various Italian laboratories. Notable among these were Felice Bertossi, an excellent plant physiologist at Pavia, Tullio Dolcher at Padova (who was to become my closest friend), Aldo Merola, a very good and complete botanist (ranging from taxonomy to ecology to physiology) at Naples, and Mario Orsenigo, a good plant pathologist at Padova; also included were Forti, Laudi, Arrigoni, Honsell, Pignatti, Meletti, Sarfatti, and Orio Ciferri. With these and some other younger researchers we gave birth to the "Academy of Cats" (cat signifying independence) with elaborate rules, the first of which was that no full professor could be admitted. This group operated efficiently (with periodic, semi-secret meetings) for some years, after which it was absorbed—with most of its original spirit—by the Italian Society of Plant Physiology (formed in 1961). Nearly all the members of the Academy of Cats are now full professors of either Plant Physiology or Botany and today play an important role in the development of plant sciences in Italy.

In 1959 Tonzig obtained for me a full professorship in Plant Physiology at Milan. This did not much alter the previous situation as far as relationships with Tonzig and his Institute of Botany were concerned: The old institute immediately offered hospitality and research facilities of all kinds to the new. My promotion, however, coincided with important changes in the organization of my research team. Most of my former collaborators had by this time reached a level of full independence in research. Some, including Arrigoni, Laudi, and Servettaz, now occupied stable positions in Botany with good prospects of advancement in their academic careers, and my closest collaborator, Giorgio Forti, then at André Jagendorf's laboratory at the Johns Hopkins University in the United States, had already started an independent line of work in photosynthesis. Only Renato Bianchetti (the youngest of my former associates) was still very efficiently cooperating in my research on the regulation of metabolism. I was therefore pleased to be allocated some research positions, which were soon occupied by three persons who were to be invaluable in carrying on my old line of work, modified according to the new circumstances and the advances of plant research in the world. These were Lilia Alberghina, Piera Lado, and Sergio Cocucci. I wish here to express my gratitude for their enthusiasm, spirit of collaboration, and dedication in the years they spent with me before moving on to independent positions, responsibilities, and areas of research.

The organization of the team coincided with certain changes in our research orientations. I had been asked by the *Annual Review of Plant Physiology* to write an article on "Phosphorylation in higher plants" and had been invited to give a lecture on respiratory metabolism in plants at the Botany Congress in Montreal. To prepare these papers I had to work hard to bridge the gaps in my biochemistry. Taking advantage of a grant from the Rockefeller Foundation, I spent three months visiting American laboratories, especially the laboratory of Harry Beevers, whose excellent work was then raising plant biochemistry to a level comparable to that of animal and microbial biochemistry. Harry was then engaged in the utilization of radioactive ^{14}C to determine the relative importance and physiological significance of the different metabolic pathways. It is difficult for me to find words to thank Harry for all the things I learned in those few weeks: from general ideas to specific techniques, from intellectual modesty to rigor in the interpretation of results.

Meanwhile the institution of full professorships in Plant Physiology in Italy had started a kind of autocatalytic process in this rather neglected area. The second chair in Plant Physiology had been given to Felice Bertossi, an intimate friend whom I have already mentioned as one of the founding members of the Academy of Cats. We, with most of the other members of that Academy, decided that the time had come to organize an Italian Society of Plant Physiology. This new, very democratic, and very active association was officially formed in 1961 and immediately began to promote contacts, collaborations, scientific meetings and various cultural activities. Particularly consequential was the initiation of three courses for researchers in Plant Physiology. These courses, organized by the Milan group with the active cooperation of Bertossi and of an excellent and open-minded geneticist, Giorgio Morpurgo, aimed to make our plant physiologists aware of the enormous advances that had been made in the fields of biochemistry, physiology, genetics, embryology, molecular biology, general physiology, chemistry, and physics. These courses, lasting two weeks and attended by some 40 young researchers, benefited from the contribution (as teachers and debaters) of most of the best scientists then available in Italy: from Falaschi to Tocchini Valentini for molecular biology; Monroy and Giudice for embryology; Scoffone for organic chemistry; Magni, Morpurgo, and Calef for genetics; Scarano and Guerritore for biochemistry; Ceccarelli for physics, and many others. Characteristic of these courses was the warm atmosphere of friendship created by the continuous informal contacts among relatively young people moved by a common purpose. Their influence on the development in Italy of a group of open-minded, modern plant physiologists was considerable.

The research activity of my new group in those years was initially concentrated on the relationships between growth stimulation and respiratory metabolism, with particular reference to the pentose phosphate pathway and to the NADP reduction state. By means of metabolite determinations and the ^{14}C techniques learned in Harry Beevers's laboratory, we were able to demonstrate that the auxin-induced stimulation of growth was associated with a marked decrease of activity of the pentose phosphate pathway and an increase of the levels of glucose-6-phosphate and reduced NADP, which tallied with our previous observations of the increase of the reduction state of glutathione and ascorbate in auxin-treated tissues.

From Protein Synthesis to Transport

This research was in full course when our attention was at least partially diverted to a new aspect of growth regulation, namely nucleic acid and protein synthesis. The need to include such an approach in our investigation of the problem of regulation was clearly suggested by certain dramatic successes in molecular biology—in particular, by the demonstration of specific mechanisms of regulation at the transcriptional level. Our entry into this field was efficiently catalyzed by the friendly, generous advice and collaboration of

Alberto Monroy, then carrying out excellent work on the changes in RNA and protein metabolism associated with sea urchin egg fertilization and embryonal development. Using Monroy's experimental approach as a guide, we started investigating the somewhat analogous transitions of macromolecular patterns in the three phases of seed development (namely growth and maturation, fall into dormancy and dehydration, and germination). Most of the forces of our laboratory came to be engaged full time in this new field. Among the main results obtained were the demonstration of the inducibility of isocitrate lyase and galactose kinase, of the reversible inactivation and reactivation of soluble and mitochondrial enzymes during the seed cycle, and, even more interesting, of the fundamental role of water availability (turgor?) in regulating protein synthesis.

We were slowly proceeding with these problems when, once again, circumstances (the most important, as usual, being contacts with clever people working in neighboring areas) modifed the main course of our research by introducing the physiology of transport as an important parameter of the general problem of regulation. Prompting us to start along this new course was the discovery of the fungal toxin fusicoccin and of its peculiar physiological action by two good friends: Antonio Graniti, who isolated the toxin and described its capacity to induce stomata opening and wilting, and Alessandro Ballio, who had worked out the structure of fusicoccin and had started investigating its mechanism of action. Collaborating with Ballio, we soon found that this toxin had a growth-promoting activity similar to that of auxin and could be used as a tool to elucidate the action of auxin. The excellent work of Cleland, Rayle, and Evans in the United States, and of Hager, Lüttge, and Haschke in Germany, had suggested that the effect of auxin on cell enlargement might depend on the stimulation of acid secretion. Using fusicoccin, we were able to demonstrate that growth stimulation was indeed correlated with acid secretion in the case of fusicoccin as well as in that of auxin. The investigation of the mechanism of acid secretion led us into the area of ion transport. In collaboration with Arnaldo Ferroni, a distinguished electrophysiologist at our University, we soon found that the fusicoccin- (and auxin-) stimulated acid secretion was associated with a hyperpolarization of transmembrane potential and with an increase of K^+ uptake, which suggested the operation of a proton pump. Because such work was completely outside our previous experience, our progress depended, once again, on the generous help and advice of several outstanding scientists then engaged in the biophysical and physiological aspects of ion transport in plants. Since 1968, this group of researchers had organized periodical workshops on transport. Characterized by a rare spirit of collaboration, this community of scientists played an extremely important role in the development of modern plant physiology and in the integration of physiology with biophysics, biochemistry, and molecular biology. I am still grateful to Noe Higinbotham for encouraging me to join this group. Thereafter, our work in plant transport proceeded in close contact with many "transportists," among whom I remember Jack Dainty, Clifford Slayman, Tom Hodges, Michael Pitman, Widmar Tanner, Ulrich Lüttge, John Raven, and Anton Novacky with particular gratitude.

Though we were now concerned with the mechanism of proton transport, our main problem remained that of the regulation of transport and the physiological consequences of changes in activities of transport processes. Here again precious help came from two distinguished scientists: Rainer Hertel of Freiburg, with whom we were able to give a first demonstration of a fusicoccin receptor in the plasma membrane, and Jean Guern, whose competence in the use of ^{31}P-NMR spectroscopy was of great value in establishing that fusicoccin-induced stimulation of the proton pump was indeed associated with an increase of cytosolic pH.

An interesting result of my effort to combine hormonology, biochemistry, and biophysics in the study of the integration of processes in plants was a workshop on "Hormonal regulation of transport" organized by our group at Pallanza in 1976, which was attended by many of the most representative scientists in these fields. The workshop was important in developing awareness of the complementarity of three fields: the physiology of transport, biochemistry, and hormonal regulation. It also gave rise to some unforeseen collaborations between people involved in quite different fields, such as between the electrophysiologist Noe Higinbotham and the hormonologist Bob Cleland.

The attention of our group since 1970 has been centered on proton transport; we have studied its regulation in intact cells and in membrane vesicles, its effects on intracellular pH, its consequences on metabolism, and its relationships with other proton transport systems such as the plasmalemma redox system; such work continues in the present. Some widening of our perspectives and programs has been prompted recently by advances in plant genetics and molecular biology as well as by the increasing tendency of Italian and international organizations to favor applied over basic research. Our recent programs thus tend towards the use and investigation of mutants in transport processes, with the twofold aim of obtaining further information about the mechanisms and regulation of transport, on the one hand, and of investigating the relationship between transport and productivity-limiting conditions (such as water, salt, and temperature stresses), on the other.

CONCLUSIONS

The recollection of one's past life tempts one to draw conclusions. Allow me here to reflect on certain of my firm beliefs concerning research, plant physiology, and the practice of experimental biology in relation to a general vision of the world.

Motivations for Research

At the root of my choice of research and teaching as a profession I find two distinct motivations: a thirst for self-fulfilment and a desire to enjoy the order and beauty of reality, and to share this enjoyment with others. While the impulse towards self-fulfilment has a well-defined physiological meaning (it helps human beings to develop by accepting challenges—e.g. climbing a mountain, solving a problem, overcoming laziness and discouragement) it also entails competitive confrontation with other members of the human community. On the other hand, the desire to enjoy the world and to communicate this pleasure gives rise to some of the highest manifestations of human nature, such as solidarity and friendship. The positive role of the spirit of competition is emphasized and encouraged by the official organization of science much more than the spirit of collaboration, possibly because of the political-social demand for practical results. I am deeply convinced, however, that the feeling of being associated in a common cause—which favors collaboration—is a much more important condition for scientific progress. Competition leads to the exploitation of our colleagues, considered merely as sources of information and ideas, whereas collaboration—when understood in the full meaning of the word—leads to solidarity and potentially to friendship, and is moved by the desire to help others and to share with others anything good achieved in our (or their) work. My conviction that friendly collaboration is far more favorable to scientific progress than competition is born of my experience. My scientific career was decisively influenced by collaborative contacts with other scientists, notably Thimann, Tonzig, Boeri, Monroy, Beevers, Higinbotham, Hertel, Guern, Tanner, Dainty, and many others. I am convinced that the positive contribution to science arising from their unselfish dedication in helping colleagues and students was even greater than that of their outstanding scientific discoveries.

The Political Trend towards Application and the Necessity of Integration of Plant Sciences

As in any other scientific area, the objective of research in plant physiology is twofold: knowledge for its own sake, and application to practical needs. Obviously enough, these two aspects are strictly interlinked: Good basic research produces tools for application, and application pressure stimulates research on basically central topics or problems.

Recently in Italy, and more or less all over the world, governments, under the pressure of expanding economic and social needs, have started to favor applied over basic research. Many of us feel that this political trend has something intrinsically wrong, demagogic, and illogical about it. Knowledge must precede application, and the progress of knowledge has its rules, among which intellectual freedom is the most important.

Some aspects of the trend toward greater support of applied science might be usefully exploited to increase awareness of the need to integrate the different scientific areas. The application of biological science—in our case of plant science—is a highly synthetic activity. It implicates a capacity to predict the behavior of an organism in given environments—behavior that results from the integration of all its activities. Knowledge of the mechanisms and knowledge of the integration of these processes are thus the two main bottlenecks that limit application. Our efforts to remove the "integration" bottleneck helps to identify the missing links in our knowledge about each mechanism, just as any increase in mechanistic knowledge contributes to an understanding of the integration.

From this point of view, the negative consequences of modern pressure in favor of application and technology are much greater in plant research than in animal and microbiological biology, where basic research is so advanced that the interactions among basic research areas and their impact on application have become obvious. In contrast, basic knowledge concerning plants is much poorer, and the need for closer interactions among the different areas (e.g. genetics, molecular biology, biophysics, systematics, plant physiology in its different branches, etc) has only recently become clear. An effort of interdisciplinary integration is urgently needed to enable us not only to understand better the phenomenon of life in its higher expressions, but also to establish a closer, more rational relationship between basic research and application.

Plant Physiology, Biology, and the Unity of Life

It seems there are two different ways to interpret one's role (significance) in life. One way is to accept—as a working hypothesis—that in the framework of a general design of the universe (difficult to rationalize and open only to intuitive understanding) each individual has a definite role, that individual roles aggregate to serve a primary purpose, and thus that all activities are organized to accomplish that purpose (just as the organization of physiological processes makes possible the survival and the functioning of an organism). Alternatively, one can see life as an incomprehensible puzzle, forget about the problem of unity, and allow oneself to behave according to one's impulses. Very near the end of a rather complex life, and having personally chosen the first alternative, I now wonder how much unity I have achieved, and how important it was to have dedicated so large a part of my time and thought to plant physiology.

Research has heavily influenced both my way of thinking and my vision of the world. Physiology is the study of the intimate nature and the dynamism of the most complex system existing in the universe, from its most microscopic to its most macroscopic manifestations. It is a study carried out by a well-defined method, the experimental method, which has been developed over hundreds of thousands of years, in a process that started with the appearance of *Homo sapiens,* gained rationality from the work of such men as Galileo, Newton, and Claude Bernard, and is still being refined today. Even with the limitations implicit in the circumstantial and hypothetical nature of scientific knowledge, the correct application of the experimental method remains necessary for the description and interpretation of the phenomena occurring in the world external to our consciousness, including the biological aspects of human history and behavior. A long-lasting dedication to the use of this method, as required for good work in physiology, tends to refine one's capacity to analyze, critically evaluate, and interpret any type of information received from the environment (including ideological, political, and historical information).

How has the study of biology influenced my vision of the world, and my attempt to organize it? My conviction that it has is difficult to explain, but I can identify some of its underlying aspects. The observation and analysis of life confronts us with the highest manifestation of complexity and order, thus nourishing our innate hunger for both truth and beauty (as C. R. Stocking so nicely put it in his prefatory chapter to Volume 35 of this *Annual Review*). This nourishment establishes an intimate and strong interaction between biology, as an experimental science, and all other forms of knowledge, including philosophical and artistic knowledge, which combine to form a single whole in the individual consciousness. In this regard, certain properties of physiology are of particular interest. The aim of physiology is to describe and understand the nature and integration of life processes. Even if we exclude a naive finalism from our effort to interpret reality, we are constantly led to recognize the fact that almost all the components of the biological machinery we investigate play an essential role in enabling it to achieve a final unitary result: the existence of life. As a consequence of this observation, the discovery of new processes is constantly followed by an effort to understand their physiological relevance. A similar situation is found when we move from organismic physiology to the analysis of the biological system as a whole, considering evolution as a process to be studied by the same method as we use for organisms, or parts of them. Can we identify a general design, and possibly a unitary aim, in the appearance of life and in its evolution to extreme manifestations of coordinated complexity? The recent attempts to reproduce in nonbiological systems the synthesis of biological energy-rich compounds and macromolecules, such as ATP, nucleic acids, and proteins, indicate that both physicochemical and historical restrictions may have played a decisive role in the origin of life, just as in the previous evolution of the universe. If this is the case, then the appearance of life is an improbable event privileged by the intrinsic nature of the matter-energy context. One might easily speculate that similar (although still unexplored) restrictions have channeled biological evolution in its progress towards coordinated complexity and consciousness. To accept this view implies the interpretation of evolution as the development of a design. It also implies that every element in the evolving system—myself included—has a definite role to play. Our responsibility is to accept our roles (insofar as we can understand them) in a way that conforms to the design.

I am well aware that a vision of the world based on the recognition of both a general design and a personal role implies a choice belonging to the domain of faith rather than to that of science. However, I am glad to avail myself of the present opportunity to state that in making this choice I see a confluence of my scientific experience with what knowledge I have been able to obtain from other sources. To paraphrase a famous sentence of Kant's, I like to say that two things in the world fill me with unlimited wonder: the evolution of the universe outside me, and the moral conscience within me. As a physiologist, I accept the challenge to seek a unitary interpretation of life.

Annu. Rev. Plant. Physiol. Plant Mol. Biol.. 1991. 42:21–53

SORTING OF PROTEINS IN THE SECRETORY SYSTEM

Maarten J. Chrispeels

Department of Biology and Center for Molecular Genetics, University of California, San Diego, La Jolla, California 92093-0116

CONTENTS

21

1040-2519/91/0601-0021$02.00

INTRODUCTION[1]

Every independent organelle in a eukaryotic cell contains a distinct set of proteins that determines the structure and biochemical activities of the organelle. The translation of all cellular proteins except for a small set of polypeptides synthesized in chloroplasts and mitochondria starts in the cytosol. At some point, either concurrent with translation or after the polypeptide has been released in the cytosol, a protein is targeted to one of more than 30 compartments in the cell. This intracellular protein traffic and the information necessary for targeting proteins to specific organelles are among the central issues in cell biology and have been studied intensively during the past five years. This research, carried out with mammalian, plant, and yeast *(Saccharamyces cerevisiae)* cells, has given rise to the following model of protein targeting. Each protein contains one or more specific targeting signals or domains within its primary amino acid sequence, and these domains interact with receptors. This interaction sends the protein to the organelle where it will subsequently function. For independent organelle systems, the receptor is the port of entry for the transported proteins, and the targeting signal initiates translocation of the protein across the limiting membrane. This is the case for proteins that enter the endoplasmic reticulum (ER), mitochondria, chloroplasts, and peroxisomes. Transmembrane domains can anchor proteins across a membrane, resulting in integral membrane proteins; multiple signal/anchor or transmembrane domains result in integral membrane proteins that span the lipid bilayer several times. Some signals retain proteins in specific locations. Such retention signals have been found on proteins that reside in the lumen of the ER and in the Golgi; they may also be present on small nuclear proteins. Finally, there are true sorting signals that direct proteins along branched transport pathways. Interaction between a sorting signal and a receptor ensures progress along one branch of the pathway only; disruption of that interaction after arrival allows the protein to remain in a specific compartment at the end of the pathway. The interaction between a transported protein and the receptor is sometimes mediated by a glycan on the protein that is being sorted, as happens for mammalian lysosomal hydrolases; however, the placement of these glycans along the polypeptide backbone and their exact modification in the Golgi to bring about a functional glycan group that interacts with the receptor are still a function of the primary amino acid sequence of the protein.

The organelles of of the secretory system include the ER (rough and

---

[1]Abbreviations: ER, endoplasmic reticulum; HRGP, hydroxyproline-rich glycoproteins; BiP, binding proteins; IgG, immunoglobulin G; PHA, phytohemagglutinin; GUS, β-glucuronidase; CaMV, cauliflower mosaic virus; CCV, clathrin-coated vesicles; CPY, carboxypeptidase Y; PrA, proteinase A; PR, pathogenesis related.

smooth), Golgi complex, the *trans*-Golgi network or partially coated reticulum, endosomes, secretory vesicles, vacuoles, the tonoplast and the plasma membrane, as well as a variety of transition or transport vesicles that mediate communication between and within these compartments. Transport and sorting of proteins through the secretory system have been studied since Palade (107) and his coworkers formulated the general outline of the pathway. Protein transport in the secretory system was reviewed in this series in 1976 by Chrispeels (21) and in 1985 by Akazawa & Hara-Nishimura (1). Recently, much new information has been obtained with plant cells, and this new information is summarized here. For recent reviews on protein sorting in other eukaryotic cells, see Pelham (110), Hurtley & Helenius (71), Kornfeld (86), Kornfeld & Mellman (87), and Rose & Doms (124). This review does not detail the evidence that vacuolar and extracellular proteins are transported in the secretory system, or the architecture of this system, but emphasizes mechanistic aspects of protein transport. The evidence that vacuolar and extracellular proteins are transported through the secretory pathway has been reviewed several times (23, 54, 78, 105), and the structure of the secretory system of plant cells has been reviewed by Harris (57).

MANY TYPES OF PROTEINS ARE TRANSPORTED IN THE SECRETORY SYSTEM OF PLANT CELLS

The secretory system delivers secretory proteins to the cell wall and the extracellular space as well as to the vacuole. Recent evidence indicates that the cell wall is a highly dynamic compartment with numerous proteins whose abundance is modulated depending on the stage of development or the external signals received. For an interesting description of this new point of view, see the review by Roberts (120). Vacuoles are no longer viewed as simply containing dilute solutions of ions and organic acids. Besides their complement of lysosomal acid hydrolases, vacuoles can store and later catabolize a great variety of proteins depending again primarily on the developmental stage of the cell or the signals that impinge on it. The study of protein transport in cell types with active secretory systems (storage parenchyma cells in cotyledons, aleurone cells, suspension-cultured cells) has been extremely useful, but it is necessary to emphasize the common occurrence of protein transport and sorting in many cell types, rather than its uniqueness in a few cell types. A few examples of recently discovered secretory proteins (vacuolar or extracellular) will suffice to illustrate this point. The cells of the papillae of the stigma or of the transmitting tissue in the style secrete large amounts of the S-allele glycoproteins that determine self-incompatibility (2, 79); when subjected to osmotic stress, cultured cells accumulate osmotin in their vacuoles (134); when subjected to nitrogen stress or continuously de-

podded, soybean plants accumulate vegetative storage protein in their vacuoles (138); many vegetative reserve tissues (tubers, roots, bark) accumulate
storage proteins in their vacuoles (135, 169); suspension-cultured carrot cells
secrete an auxin-regulated glycoprotein that is also abundantly secreted by
cells of dermal tissues (129). Particularly interesting is the response of cells to
pathogen invasion, which induces the synthesis of both vacuolar and extracellular forms of β-glucanase and chitinase (8, 157). Analysis of the amino
acid sequences of these proteins shows that these extracellular and vacuolar
forms are the products of different genes. Many more examples will be found
in the next few years as the techniques of gene cloning are combined with the
immunocytochemical localization of proteins whose amino acid sequences
have been determined. All these examples confirm the interpretation that
plant cells, like other eukaryotic cells, synthesize and secrete or transport to
their vacuoles a large variety of proteins, and that the abundance of these
proteins is highly regulated.

We presume that the transport of all the extracellular and vacuolar proteins
discussed above is mediated by the Golgi and involves all the components of
the secretory pathway. Can proteins bypass the Golgi apparatus? There is at
present, no good evidence that secretory proteins ever bypass the Golgi
apparatus during their transport; however, a direct ER-to-vacuole transport
may take place in the developing endosperm of certain cereals. During their
development, cereal grains synthesize and accumulate both alcohol-soluble
prolamines as well as globulins that resemble the storage proteins found in
other seeds. Both sets of proteins are synthesized on the rough ER and
accumulate in protein bodies. There is general agreement that the globulins
accumulate in protein storage vacuoles and that their transport is mediated by
the Golgi apparatus. However, the issue is not clear for the prolamines. In
maize and sorghum, two species that synthesize very little or no globulin,
prolamines aggregate in the ER and accumulate in ribosome-studded ER-
derived protein bodies (89, 153). In rice, prolamines also aggregate in the ER
and accumulate there, while the globulins (called glutelins in rice) are transported to vacuoles via the Golgi apparatus (88). Thus prolamines and glutelins
accumulate separately in distinct protein bodies. In wheat, barley, and oat,
prolamines first aggregate in the ER, but accumulate in vacuoles together with
globulins (83, 93, 108). The transport of the globulins is presumed to involve
the Golgi apparatus, but is this also the case for the prolamines in these
cereals? In wheat, prolamine aggregates are seen in vesicles associated with
the Golgi apparatus, and fusion of these vesicles with the tonoplast would
deliver the protein to the vacuole (83). In oat, on the other hand, prolamine
aggregates are seen in rough ER–derived 0.3 μm vesicles (93). Could such
vesicles be taken up directly into vacuoles by a process similar to autophagy?
Such a protein transport process would bypass the Golgi apparatus. In their
study (93), Lending et al failed to find Golgi structures in the oat endosperm

cells, and more work is needed to resolve the issue of the role of the Golgi in prolamine transport in endosperm cells.

In a recent study (133a) the subcellular fate of two closely related wheat prolamines, α- and γ-prolamine, was examined in *Xenopus* oöcytes that were injected with the mRNAs for these proteins. Most of the γ-gliadin was secreted from the oöcytes in a Golgi-mediated process, whereas most of the α-gliadin remained inside the oöcytes. The authors conclude that this different behavior of α- and γ-gliadins in oöcytes supports the hypothesis that there are also two separate transport routes in wheat endosperm. However, these results can also be explained by the failure of α-gliadin to become transport competent. For example, results obtained by A. Vitale (personal communication) with the legume storage protein phaseolin show that oligomer formation is a prerequisite for its secretion by *Xenopus* oöcytes. Oligomer formation depends on the amount of protein in the secretory system and can be modulated by changing the amount of mRNA that is injected.

A SIGNAL SEQUENCE IS NECESSARY AND SUFFICIENT FOR ENTRY INTO THE SECRETORY PATHWAY

The first step in the secretory process is the sequestration of proteins into the lumen of the ER or their incorporation into ER membranes. This sequestration occurs by virtue of the information contained in the signal sequence, an amino-terminal domain found on nearly all secretory proteins. Signal sequences of eukaryotic cells have the following properties in common: They are 13–30 amino acid residues long; they have a basic amino-terminal region, an uninterrupted stretch of 7 or 8 apolar, largely hydrophobic residues, and a more polar carboxyterminal region that defines the cleavage site (168). In 1975, Blobel & Dobberstein (7) proposed the now widely accepted signal hypothesis based on results obtained with a newly developed assay for protein translocation into microsomal membranes. Insertion of the protein into the lumen of the ER is achieved, according to this hypothesis, through the interaction of the signal peptide on a nascent polypeptide protruding from the ribosome with a signal recognition particle (an 11 S ribonucleoprotein) and a receptor or docking protein in the ER membrane. The signal peptide is cotranslationally cleaved from the nascent protein by an endopeptidase. An examination of the derived amino acid sequences of secreted and vacuolar plant proteins shows that all of them have an amino terminal domain with the properties of a eukaryotic signal sequence. (Possible exceptions are discussed below.)

Evidence obtained with transgenic plants shows that these signal sequences are both necessary and sufficient to direct proteins into the lumen of the ER. To examine whether the presence of a signal sequence would result in the insertion of a passenger protein into the lumen of the ER and its cotranslation-al glycosylation, Dorel et al (35) expressed a chimeric gene in transgenic tobacco and examined the glycosylation status of the resulting protein. The chimeric gene *phalb* consisted of the signal sequence of the vacuolar seed protein phytohemagglutinin (PHA), and the coding sequence of a cytosolic protein, the pea seed albumin PA2 (63). The chimeric gene was expressed with a seed-specific promoter in tobacco plants and tobacco seeds contained up to 0.7% PHALB protein. Immunoblot analyses with antibodies against PA2 showed the presence of four glycosylated polypeptides of PHALB, ranging in M_r from 29,000 to 32,000. These four polypeptides are glycoforms of a single polypeptide of M_r 27,000, and the heterogeneity was due to the presence of both high mannose and complex (Golgi-modified) glycans. These results showed that the signal peptide of PHA contains sufficient information to direct PHALB into the secretory system, causing it to be glycosylated. When the PA2 albumin (without a signal sequence) was expressed in tobacco cells, the protein did not enter the secretory pathway and was not glycosylated (D. Hunt and M. J. Chrispeels, unpublished). Thus, the information con-tained in the signal sequence is both necessary and sufficient for sequestration into the ER, and also for secretion (see below).

Similar experiments have also been carried out with chimeric gene con-structs consisting of the signal peptide of patatin, a potato tuber vacuolar protein, and β-glucuronidase (GUS) as the reporter protein (73), and with the signal peptide of an extracellular tobacco protein (pathogenesis-related pro-tein PR1) and various bacterial enzymes as the reporters (GUS, phosphi-nothricin acetyltransferase and neomycin phosphotransferase II) (30). Entry into the secretory pathway was demonstrated by showing that the proteins were glycosylated (GUS) and secreted by the transformed cells.

These experiments show not only that signal peptides from one plant species can function in another plant species, but also that there are no vacuole-specific and secretion-specific signal peptides. Indeed, signal pep-tides from vacuolar proteins (patatin, phytohemagglutin) allow reporter pro-teins to be secreted after entering the ER; the reverse (transport to the vacuole with a signal sequence from a secreted protein) has not yet been demonstrated because we do not yet know how to incorporate vacuolar targeting informa-tion in chimeric constructs. The signal sequences of plant proteins can also function in yeast cells and animal cells, emphasizing again the general lack of specificity of these domains. For example, phaseolin is secreted by *Xenopus* oocytes (163); and phytohemagglutinin is secreted by monkey COS cells (165) and transported to the vacuoles of yeast cells (148). A further lack of specificity in signal sequences is shown by the observation that bacterial chitinase with its own signal sequence is secreted, albeit inefficiently, by plant cells (97).

The presence of signal peptides on all proteins that enter the secretory system of plant cells implies that the translocation mechanism in plants is probably quite similar to the one found in mammalian cells, and may involve a signal recognition particle (6, 115). Tomato leaves contain a 7S RNA that shows 56% nucleotide identity with 7S RNA from human signal recognition particles, indicating again the enormous conservation in the secretory system (53). Whether posttranslational translocation of polypeptides into the lumen of the ER occurs in plants has not been determined.

Several anomalous observations concern the need for a functional signal sequence to enter the secretory system. In barley endosperm, chymotrypsin inhibitor-2 (CI-2) is a vacuolar protein of M_r 9380 (116). Yet, the cDNA has a stop codon in the signal sequence, and the initiation of translation occurs at a downstream ATG (172). mRNA for CI-2 is found associated with free polysomes as well as with membrane-bound polysomes. A related pro- teinaceous inhibitor CI-1 that is 45% homologous at the amino acid level lacks a signal peptide; however, this protein has not yet been localized at the subcellular level. If this protein is also found in vacuoles, then we must consider the possibility that a signal peptide is not necessary for the entry of a small protein into the secretory pathway or for its accumulation in vacuoles. Evidence from mammalian cells shows that a number of secreted proteins lack classical signal peptides (103).

We have little information on the topogenic sequences of integral mem- brane proteins that are incorporated into the membranes (ER, Golgi, tono- plast, plasma membrane) of the secretory system of plant cells. In mammalian cells, a hydrophobic domain is often located internally; it is not cleaved by the signal peptidase, but serves as a membrane anchor. Such a signal/anchor sequence can place the transmembrane protein in two different orientations: amino terminus outside or amino terminus inside. Many proteins have more than one signal/anchor, and span the membrane several times. So far, the amino acid sequences of only a few integral membrane proteins of the secretory system of plant cells have been determined: the plasma membrane H^+-ATPase (56) and a seed-specific tonoplast protein, called TIP (75). The hydropathy plots indicate that plasma membrane H^+-ATPase ($M_r = 100,000$) has 8 membrane-spanning segments, while TIP ($M_r = 27,000$) has 6. The derived amino acid sequences of these proteins do not have amino-terminal signal sequences, and we presume that insertion into the lipid bilayer starts with the first membrane-spanning fragment.

PROTEINS ACQUIRE TRANSPORT COMPETENCE IN THE LUMEN OF THE ENDOPLASMIC RETICULUM

Polypeptides that enter the ER undergo a number of cotranslational and posttranslational covalent modifications, acquire their secondary and tertiary structure, and form oligomers. As a result, they acquire transport competence, a still poorly defined property that determines whether proteins will efficiently proceed along the transport pathway or be retained and perhaps broken down. The proteins undergo at least two or more of the modifications discussed below, not necessarily in the order shown.

Cotranslational Removal of the Signal Sequence

Polypeptides isolated from the ER are smaller (by about 2000 Da) than the primary translation products that can be obtained by translation of mRNA in vitro. By separating polysomes from ER membranes with detergents and then completing polypeptide synthesis by run-off translation one can obtain polypeptides with and without signal sequence. These observations indicate that the removal of the signal peptide occurs cotranslationally. Integral membrane proteins generally do not have cleaved signal sequences (see above).

Cotranslational Attachment of High-Mannose Glycans to Specific Asparagine Residues (Asn X Thr/Ser)

Glycans with the structure $Glc_3Man_9Nac_2$ are assembled on dolichol lipid carriers (37) and transferred to specific asparagine residues on nascent polypeptides by oligosaccharyl transferase (38). That attachment of glycans occurs cotranslationally was shown for the vacuolar storage protein phaseolin by run-off translation of polypeptides already initiated on the rough ER. When run-off translation was done after dissolving the membranes with detergent, the completed polypeptides were found to carry one or two glycans (12). Attachment of glycans to specific Asn residues is documented by sequence analysis of glycopeptides obtained from trypsin digests of glycoproteins (143, 146). However, all available glycosylation sites are not necessarily utilized by the glycosylation machinery. Site-directed mutagenesis to replace either the Asn or the hydroxyamino acid (Ser or Thr) with other amino acids has been shown to abolish the glycan attachment site (137, 166, 171). Recent results show that novel glycosylation sites can be introduced by site-directed mutagenesis by positioning new Asn residues appropriately with respect to existing Ser or Thr residues, or by positioning new Ser or Thr residues appropriately with respect to existing Asn residues. When such modifications were made in the gene for phytohemagglutinin-L (PHA) and the modified genes introduced into plant cells, additional glycans were attached to the polypeptides of PHA (C. D. Dickinson and M. J. Chrispeels, unpublished).

Removal of Terminal Glucose Residues

Glucosidase I and II, two ER-resident enzymes (145), remove the 3 terminal Glc residues from the high-mannose glycans, yielding typical $Man_9GlcNac_2$ structures. When soluble proteins are isolated from the ER and denatured, their glycans are completely susceptible to degradation by α-mannosidase, yielding $ManGlcNac_2$, indicating that no terminal Glc residues are present to protect the glycan from degradation by α-mannosidase (44). There is only one documented case of a terminal Glc residue remaining on a high-mannose glycan of a mature protein: The vacuolar protein α-mannosidase that is abundantly present in jack bean cotyledons has a glycan with the structure $GlcMan_9GlcNac_2$ (144). The Glc residue is linked by an $\alpha,1 \rightarrow 3$ linkage to the same terminal Man residue that normally has 3 Glc residues in a $Glc_3Man_9GlcNac_2$ glycan, indicating that glucosidase II probably failed to remove the third Glc residue after transfer of the glycan to the nascent polypeptide.

Folding of the Polypeptide

Correct folding and assembly of secretory proteins is often necessary for their efficient exit from the ER and subsequent transport (71, 110). The requirement for assembly (formation of oligomers) is more stringent for membrane proteins, such as the vesicular stomatitis virus G-protein and influenza hemagglutinin, than for proteins that are soluble in the lumen of the ER. We presume that unfolded and/or unassembled proteins tend to form aggregates with themselves and with other proteins in the ER. Such aggregates may be unable to enter transport vesicles. Folding of proteins in vitro is a slow process, but in the lumen of the ER proteins achieve their final conformation within a few minutes. This efficiency of folding is due, at least in part, to ER-resident proteins that facilitate the folding process. The best characterized of these ER residents is BiP (Binding Protein; in mammalian cells also known as the glucose-regulated protein GRP78). BiP preferentially associates with abnormally folded proteins and incompletely assembled proteins; in this way, BiP contributes to the selectivity of transport (for a recent review, see 110).

Recent experiments indicate the BiP is also found in plant cells. A cDNA sequence that is highly homologous to the sequence of mammalian BiP has been found in a tomato cDNA library (A. Bennett, unpublished result) and a maize endosperm cDNA library (R. Boston, unpublished results). In maize endosperm, the abundance of BiP mRNA greatly increases when the cells are treated with tunicamycin, as is also found in mammalian cells. The inhibition of glycosylation by tunicamycin is thought to increase the amount of misfolded protein in the ER, and this in turn increases the need for BiP and the expression of BiP. Plant BiP has carboxyterminal HDEL, a typical ER-retention signal (see below) found on ER-resident proteins.

Formation of Disulfide Bonds

Many secretory proteins have disulfide bonds, the formation of which is catalyzed by the enzyme protein disulfide isomerase. This enzyme, which binds to unfolded proteins, catalyzes both the oxidation of thiols and disulfide exchange reactions. The importance of protein disulfide isomerase for correct folding is shown by experiments of Bulleid & Freedman (14), who studied the import of γ-gliadin by microsomes that were depleted (by extraction at pH 9.0) of their protein disulfide isomerase. Such depleted microsomes (from dog pancreas) imported γ-gliadin, but the formation of disulfide bonds was greatly inhibited. The γ-gliadin that was imported by microsomes that were reconstituted with protein disulfide isomerase was found to be disulfide bonded, demonstrating the role of the enzyme in disulfide bond formation. In mammalian cells, protein disulfide isomerase is an ER-lumenal enzyme. Whether this is also the case in plant cells remains to be demonstrated. The mammalian protein disulfide isomerase gene has now been cloned and the amino acid sequence of the protein determined. The protein exists both as a free monomer and as a subunit of 4-prolyhydroxylase. It also binds to the Asn-X-Ser/Thr acceptor sequence for N-linked glycosylation and appears to be essential for glycan transfer (48).

Modification of Amino Acid Residues

Most plant cells secrete extensins and other hydroxyproline-rich glycoproteins (HRGPs). The biosynthesis of these proteins involves the hydroxylation of prolyl residues by 4-prolyl hydroxylase followed by the glycosylation of many of the hydroxyprolyl residues with arabinosyl side chains (20). Recent evidence shows that 4-prolylhydroxylase is an ER-associated enzyme (3) while the glycosyl transferase(s) is(are) associated with the Golgi complex (46). It is not yet known whether the interesting protein homologies observed among 4-prolyl hydroxylase, protein disulfide isomerase, and glycosylation site binding protein of mammalian cells (48) are also present in the plant proteins.

Formation of Oligomers

Many vacuolar proteins are oligomers. For example, the cereal lectins such as wheat germ agglutinin are dimers, vicilin and other 7S storage proteins are trimers, many legume lectins (e.g. phytohemagglutinin) are tetramers, and legumin-type storage proteins are hexamers of two disulfide-bonded polypeptides. With the exception of legumin, which only forms trimers in the ER (26, 54), complete oligomer formation always occurs in the ER (24, 25, 141). Assembly of glycinin (a legumin-type storage protein) monomers into trimers and then into hexamers can occur in vitro (31, 32) and such an in vitro assembly system can be used to evaluate the effects of protein modification on oligomer formation (33). The assembly of glycinin trimers into hexamers requires the proteolytic cleavage of proglycinin into its basic and acidic subunits. Evidence from mammalian cells (9) indicates that BiP plays a role in oligomer assembly in vivo. In the ER, BiP is associated with unassembled heavy chains of IgG, as well as with IgG assembly intermediates, until the final light chain is added. If no light chains are present, then the heavy chains remain associated with BiP until they are degraded. Thus, BiP appears to play a role in both protein folding and oligomer assembly.

There is as yet no convincing evidence that BiP plays a similar role in the assembly of other multimeric proteins. However, in many instances the unassembled or partially assembled subunits of multimeric mammalian proteins do not leave the ER until correct oligomer formation has occurred (71). It is not yet known whether there is a similar mechanism for oligomeric plant vacuolar proteins. Recent experiments by A. Vitale (personal communication) show that unassembled monomers of phaseolin are not secreted by *Xenopus* oocytes and that oligomer formation is required for secretion.

CARBOXYTERMINAL KDEL AND HDEL FUNCTION AS ER RETENTION SIGNALS

Evidence from a number of laboratories shows that ER-resident proteins in mammalian and yeast cells have a carboxyterminal domain consisting of the four amino acids KDEL or HDEL (110). Recently, Inohara et al (72) obtained the amino acid sequence of the ER-associated auxin-binding protein of maize and found that this protein has a carboxyterminal KDEL sequence. This report has been confirmed by two groups (62, 154), and unpublished results from two other laboratories show that tomato BiP and maize BiP have carboxyterminal HDEL. It has recently been shown for mammalian and yeast cells that these tetrapeptides are ER retention signals: They are both necessary and sufficient for the retention of ER-resident proteins in the lumen of the ER (104, 111). We do not yet know what the mechanism of this ER retention is, but it seems unlikely that KDEL-containing proteins are permanently anchored to a membrane-bound receptor or a matrix in the lumen of the ER. The data of Ceriotti & Colman (19) on the effect of KDEL on the diffusion of BiP in *Xenopus* oocytes argue against this idea. Munro & Pelham (104) proposed that resident ER proteins escape from the ER and are continuously retrieved from the next compartment and returned to the ER. This proposal is supported by an analysis of the glycans of resident ER proteins such as glucosidase II and lysosomal proteins that have been modified with KDEL at their carboxytermini.

To test the possibility that carboxyterminal KDEL on a plant secretory protein can function as an ER retention signal in plant cells, we modified the cDNA of phytohemagglutinin (PHA) at the exact 3' end of the coding sequence, introduced the mutant gene into tobacco, and examined the subcellular location of the altered protein. Phytohemagglutinin, the seed lectin of the common bean *Phaseolus vulgaris,* is a vacuolar glycoprotein that is synthesized on the ER; its transport to the vacuole is mediated by the Golgi apparatus (22). Newly synthesized PHA-L (in the ER) has two high-mannose glycans that are attached to Asn^{12} and Asn^{60}. Transport through the Golgi complex results in the conversion of the glycan attached to Asn^{60} into a complex (endoglycosidase H–resistant) glycan (162). The specificity of this conversion is the same in tobacco as in bean (147). The presence of a complex glycan on PHA can normally be taken as evidence that the protein has left the ER and the *cis*-Golgi compartment and is now in or (more likely) beyond the *trans*-Golgi cisternae (86). To obtain a change exactly at the carboxyterminus of PHA, the nucleotide sequence was changed and slightly elongated so that the amino acid sequence would end in LNKDEL instead of LNQIL, the normal carboxyterminus of PHA-L. The genes for normal PHA-L and the mutant PHA-KDEL were introduced into tobacco with seed-specific promoters, and the subcellular location and posttranslational modification of the mutant protein were examined. A large proportion of the PHA-KDEL in mature tobacco seeds had two high-mannose glycans, indicating that it had not been modified in the Golgi complex. Immunocytochemical localization of the PHA-KDEL indicated substantial labeling of the ER and the nuclear envelope (61). Many studies have shown that the ER is continuous with the nuclear envelope, and it is therefore not unexpected that proteins that are in the lumen of the ER can also be found in the nuclear envelope, especially if transport out of the ER is prevented. Doms et al (34) obtained a similar result when transport out of the ER was inhibited with brefeldin A.

A considerable proportion (about half) of the PHA-KDEL was modified in the Golgi and proceeded to the protein storage vacuoles of the tobacco seeds. This partial retention of the PHA-KDEL in the ER could be due to a less than optimal display of the carboxyterminal tetrapeptide exactly at the end of the PHA polypeptide. Less than optimal display would cause poor recognition and inefficient recycling of PHA-KDEL back to the ER. To retain a protein in the lumen of the ER, it is probably not sufficient to have carboxyterminal KDEL. The other structural features of the protein must ensure the positioning of the tetrapeptide in such a way that it can be readily recognized by the receptor (161). Retardation of export rather than total ER retention was also recently observed for several mammalian secretory proteins that had been modified at the exact carboxyterminus to encode the sequence SEKDEL (173).

Recent experiments (T. J. V. Higgins, personal communications) show that carboxyterminal SEKDEL on pea vicilin greatly (100-fold) increased the accumulation of vicilin in the leaves of transgenic tobacco. Electron micrographs indicate the presence of a new type of "protein body" in the leaf mesophyll cells. These protein bodies, which contain the vicilin, may be derived from the ER and result from the high expression of the protein and its retention in the ER.

PROTEINS THAT ARE INCORRECTLY FOLDED OR LACK GLYCANS MAY BE BROKEN DOWN IN THE SECRETORY PATHWAY

Many secretory proteins have N-linked glycans; some, but not all, require these glycans for transport (reviewed in 106). In plant cells, for example, inhibition of glycan synthesis by tunicamycin does not prevent the transport out of the secretory system of the pea storage protein vicilin (25), of phytohemagglutinin (11), or of rice scutellum α-amylase (101). The transport of the glycosylated precursor of concanavalin A, on the other hand, is prevented by treatment of jack bean cotyledons with tunicamycin (40). It is of interest that the transport of the glycosylated precursor of barley lectin is more rapid when the pro-protein has no glycan (171). Although numerous functions have been suggested for N-linked glycans, the most clearly defined roles so far are in promoting correct protein folding and in protecting against proteolytic breakdown. These two phenomena may well be related, as incorrectly folded proteins may also be more susceptible to proteolysis.

Extensive work with the vesicular stomatitis virus G protein (reviewed in 124) showed that the unglycosylated polypeptides of this protein accumulate in the ER as large insoluble aggregates when cells are treated with tunicamycin. Mutant proteins that lacked the two glycosylation sites and had no glycans were poorly transported, and a single glycan was found to be sufficient for transport to the cell surface. Six of eight new glycosylation sites introduced at random positions in the polypeptide were glycosylated, and the glycans at two of these sites promoted transport. Further studies implied that the glycans enhanced transport because they promoted folding. As mentioned above, in plant cells tunicamycin treatment may also prevent the transport of unglycosylated polypeptides, as is the case for the precursor of concanavalin A, which normally has a single N-linked glycan on a domain that is lost during proteolytic processing.

A different effect of the inhibition of glycosylation is encountered in suspension-cultured plant cells treated with tunicamycin (70, 117). In these cells, treatment with tunicamycin causes a nearly complete inhibition of the accumulation of newly synthesized proteins in the culture medium and the cell wall. This result could be interpreted as an inhibition of synthesis and/or secretion. To understand the basis of this phenomenon, Faye & Chrispeels (42) studied the effect of tynicamycin on the synthesis and secretion of invertase in suspension-cultured carrot cells. They found that tunicamycin did not inhibit the synthesis of the unglycosylated invertase polypeptide, and pulse-chase experiments showed that the invertase polypeptides disappeared from the secretory system. However, no unglycosylated invertase accumulated in the cell wall or extracellular medium. These data are consistent with the interpretation that unglycosylated invertase is broken down along the secretory pathway. The absence of glycans may cause incorrect folding and/or may expose polypeptide regions that are sensitive to proteases present in the secretory system.

Breakdown as a result of incorrect folding was probably also the reason that a modified phaseolin did not accumulate in transgenic tobacco seeds. Hoffman et al (68) modified the β-phaseolin gene by introducing a 45-nucleotide piece that encoded 15 amino acids, 5 of which were methionine. The predicted secondary structure of the inserted peptide regions was α-helical, matching the structure of the peptide surrounding the insertion site. This high methionine (himet) β-phaseolin gene was introduced into tobacco with a seed-specific promoter and was expressed at the RNA level much the way control β-phaseolin was; however, little himet protein accumulated in tobacco seeds. Immunocytochemistry showed that himet phaseolin, unlike normal phaseolin, which accumulates in the matrix of the protein storage vacuoles of the seeds (51), was found only in the ER, Golgi cisternae, and Golgi-vesicles. This result indicates that himet phaseolin was probably broken down in the Golgi vesicles or just after entering the protein storage vacuoles. The three-dimensional structure of phaseolin has recently been deduced from X-ray crystallographic data, and it seems likely that the 15-amino-acid peptide that was introduced disrupted the tertiary structure of the protein (91).

MOVEMENT FROM THE ER TO THE GOLGI AND THEN TO THE PLASMA MEMBRANE IS BY BULK FLOW

Plant cells secrete proteins via the constitutive or unregulated pathway (15), meaning that secretory proteins are not stored in vesicles in the cytoplasm; rather, newly synthesized proteins continuously exit the cell, usually 20–40 min after they are synthesized. The available evidence indicates that each of the steps along the pathway of constitutive secretion (ER to Golgi, movement through the Golgi stacks, and movement from the Golgi to the plasma membrane via secretory vesicles) is a default step, defining overall the bulk-flow pathway through the secretory system (82). To be retained in the ER or the Golgi or to be sorted along one of the branch pathways at the *trans*-Golgi network (to regulated secretory vesicles or the lytic compartment), a protein needs additional positive topogenic information.

The best evidence for the existence of a default pathway in mammalian cells comes from the work of Wieland et al (170), who devised a bulk phase marker for the secretory pathway. This tripeptide (N-acetyl-Asn-Tyr-Thr-NH_2) passively crosses membranes and enters the ER where it is core glycosylated. The hydrophilic glycans inhibit exit through the ER membrane.

Much of the peptide is secreted, some acquiring complex glycans in the process, indicating passage through the Golgi apparatus. The secretion half-time of the N-acylglycotripeptides was 10–20 min. It is assumed that the tripeptide contains no transport or retention signals; it defines the bulk-flow rate of ER to cell surface movement and thereby the default pathway (82, 114, 125).

The same constructs that were used to demonstrate that signal sequences are necessary and sufficient for the entry of proteins in the ER (see above) have also been used to demonstrate that secretion is a bulk-flow pathway into plant cells. The three enzymes phosphinotricin acetyltransferase (PAT), neomycin phosphotransferase II (NPT II), and bacterial β-glucuronidase (GUS) were secreted by tobacco cells in which the chimeric genes were introduced by electroporation (30). The secretion index (the ratio of enzyme activity outside the cells to the activity inside the cell) differed for each enzyme. After 10 hr of incubation only 5% of the GUS activity, 20% of the NPT II activity, and about 50% of the PAT activity was outside the cells, indicating that secretion was relatively inefficient. The remainder of the enzyme activity was inside the cells and associated with the secretory system. These results show that secretion can occur independently of active sorting by nonspecific migration throughout the secretory system. The question that remains to be resolved is why secretory proteins are normally efficiently secreted whereas the proteins in these chimeric constructions were not.

That secretion occurs by a bulk-flow pathway is also shown by experiments in which a chimeric gene consisting of the CaMV 35S promoter, the nucleotide sequence corresponding to the signal peptide of a tobacco PR protein, and the coding sequence of human serum albumin (after deletion of the pro-domain) was expressed in tobacco. The resulting mature human serum albumin was secreted by the tobacco cells (133). If secretion occurs by bulk flow, then conversely vacuolar targeting should require specific information. To demonstrate that this is indeed the case, a chimeric gene consisting of the nucleotide sequence corresponding to the signal sequence of phytohemagglutinin and the coding sequence of a cytosolic seed protein (PA2) was expressed in tobacco (35). When this chimeric gene, *phalb*, was expressed with a seed-specific promoter the PHALB protein entered the secretory system (see Section 3) but did not accumulate in the protein storage vacuoles of the seeds. When *phalb* was expressed with the CaMV 35S promoter, PHALB was efficiently secreted by transformed suspension-cultured tobacco cells (D. Hunt and M. J. Chrispeels, *Plant Physiol.* 1991, in press).

Although regulated secretion has not yet been described for flowering plants, it occurs in the green alga *Chlamydomonas reinhardii* (13). In this alga, the adhesive interaction between mating-type-plus and mating-type-minus gametes induces a sexual signal leading to the release of lysin, a cell wall enzyme that causes wall release and degradation. The cell wall of *Chlamydomonas* consists almost entirely of HRGP molecules, and its removal allows the gamete protoplasts to fuse. The lysin is stored as inactive molecules of M_r 62,000, and sexual signaling converts these to active molecules of M_r 60,000 and causes them to be secreted from the cells.

ENDOSOMES, PARTIALLY COATED RETICULUM, CLATHRIN-COATED VESICLES

In mammalian cells, sorting of vacuolar proteins is thought to occur in the *trans*-Golgi network (TGN), a tubular reticulum in which proteins are segregated into different transport vesicles to be dispatched to their final destinations (52). The vesicles that transport proteins from the TGN to other compartments appear to be covered with a protein coat consisting of clathrin (28) and, in addition, a subset of assembly polypeptides characteristic for Golgi derived coated vesicles (123). Although clathrin-coated vesicles (CCV) function in receptor-mediated endocytosis primarily (50), they are also involved in the receptor-mediated transport of proteins to the lysosomes (17); and substantial evidence indicates that endocytic and lysosomal transport pathways meet in the TGN (158).

In plants, transport pathways are much less clear, but there is some evidence that the situation is similar to that in animal cells. The partially coated reticulum, an organelle consisting of tubular membranes bearing clathrin-like coats over parts of their cytoplasmic surface, was at first thought to be an independent structure (112, 113, 152). However, physical connections of the *trans*-Golgi in plant cells have been demonstrated (66) indicating that the partially coated reticulum resembles the TGN of animal cells. When protoplasts were allowed to take up electron-dense markers by endocytosis, these were found in the CCVs and in the partially coated reticulum (65, 74, 118, 151, 152), indicating that the partially coated reticulum is a part of the endocytic pathway in plants. Is there evidence that the CCVs and the partially coated reticulum are also part of the secretory pathway? Aleurone cells of barley secrete large amounts of α-amylase, and the protein has been localized by immunocytochemistry in the ER, the Golgi, and the partially coated reticulum (174). It is possible that α-amylase exits the partially coated reticulum via CCVs. Two recent studies show that precursors of vacuolar seed storage proteins and lectins are found in CCVs isolated from developing pea cotyledons (55, 121). These experiments can be interpreted as supporting the conclusion that CCVs participate in protein transport to vacuoles. However, it is not yet possible to eliminate the possibility that the presence of storage protein and lectin precursors in the preparation of CCVs resulted from contaminating ER and Golgi-derived membranes. The observation by Record &

Griffing (118) that the partially coated reticulum was not stained when they localized acid phosphatase in cells casts doubt on the participation of this organelle in protein transport. It is clear from these contradictory observations that further experimentation is needed to sort out the role of CCVs in protein transport.

SOME HIGH-MANNOSE GLYCANS ARE CONVERTED TO COMPLEX GLYCANS IN THE GOLGI APPARATUS

Mature vacuolar and extracellular glycoproteins have both high mannose and complex Asn-linked glycans. The conversion of a high-mannose glycan to a complex glycan takes place in the Golgi apparatus as a result of the sequential action of several glycosidases and glycosyl-transferases (for reviews, see 43, 81). The complex glycans found on plant glycoproteins differ substantially from those found on mammalian glycoproteins or the poly-mannosic glycans of yeast glycoproteins. The complex glycans of plants are characterized by an α, $1 \rightarrow 3$ fucose attached to the proximal N-acetyl glucosamine and a β, $1 \rightarrow 2$ xylose attached to the core mannose of the glycan. Glycans with similar structural features may be present in insects and molluscs (41). Hybrid glycans with structural features of both high mannose and complex glycans have not yet been found on plant glycoproteins.

The complex N-linked glycans on plant glycoproteins are resistant to digestion by endoglycosidase H. Because the enzymes that convert high mannose into complex glycans are associated with the Golgi apparatus (145), one may conclude that glycoproteins with complex glycans must be in, or have passed through, the Golgi apparatus. Such complex glycans can be readily detected by their resistance to endoglycosidase H or by interaction with antisera that react specifically with the complex glycans of plant glycoproteins (90, 100).

Why do some glycans remain in the high-mannose form while others are converted to complex glycans even on the same protein? For example, mature phytohemagglutinin has a high-mannose glycan attached to Asn^{12} and a complex glycan attached to Asn^{60} (143). When newly synthesized PHA is isolated from the ER it has high-mannose glycans in both positions; incubation of this ER-derived PHA with jack bean α-mannosidase resulted in the loss of most mannose residues from the glycan at Asn^{60} while the glycan at Asn^{12} remained unchanged (44). This result indicates that the glycan at Asn^{12} is not readily accessible to α-mannosidase and may not be accessible to the enzymes in the Golgi apparatus that need to act on it to convert it to a complex glycan. This lack of accessibility causes this particular high-mannose glycan to remain in that form as the protein moves through the Golgi apparatus.

The role of N-linked glycans is not well understood, but there is enough evidence to conclude that they do not contain vacuolar sorting information.

First, not all vacuolar proteins are glycoproteins. Although some vacuolar proteins are synthesized as glycosylated precursors and lose their glycans during posttranslational processing [e.g. concanavalin A (60), wheat germ agglutinin (98), and β-glucanase (132)], others have no glycans even as precursors in the ER (e.g. the 11S globulins legumin, glycinin, etc). These proteins are transported in the secretory system and properly targeted without ever being glycosylated. Second, tunicamycin, which inhibits N-linked glycosylation, does not cause missorting of normally glycosylated proteins (11). Finally, site-directed mutagenesis to eliminate the two glycosylation sites of phytohemagglutinin (166) and of patatin (137) and the single glycosylation site of the propolypeptide of barley lectin (171) did not cause the unglycosylated polypeptides to be mistargeted in a heterologous system. Together these data clearly indicate that the glycans of plant vacuolar proteins do not have targeting information; the situation in plants is therefore similar to the one in yeast (130).

Does it matter, as far as transport is concerned, whether glycans are in the high-mannose or the complex form? To investigate this question, Driouich et al (36) examined the effect of glycan processing inhibitors (38) on protein secretion by suspension-cultured cells. They found that inhibiting glycan processing did not prevent secretion of glycoproteins. Neither did the inhibition of glycan processing cause the polypeptides to be degraded, as happens when glycosylation is prevented by tunicamycin (42, 70, 117). A similar observation was made by Vitale et al (164), who studied the effect of 1-deoxymannojirimycin, an inhibitor of glycan processing, on the delivery of PHA and phaseolin to the protein storage vacuoles of bean cotyledons. They found no effect of the inhibitor on transport or stability.

An important feature of the complex glycans of plant glycoproteins is that they are highly immunogenic in rabbits and mice. When a glycoprotein with complex glycans is used as an antigen, many of the immunoglobulin molecules are directed to the complex glycans, resulting in an apparently nonspecific serum that contains a mixture of anti-glycan IgGs and anti-polypeptide IgGs. One can make these sera useful for further research by fractionating them into two sets of IgGs. This can be done by affinity chromatography using at the immunoadsorbant a covalently linked glycoprotein with complex glycans (e.g. horseradish peroxidase or pineapple stem bromelain).

THE PARADIGM OF VACUOLAR SORTING: MANNOSE-6-PHOSPHATE AND ITS RECEPTORS

In many mammalian cells, the correct targeting of newly synthesized lysosomal enzymes to lysosomes is dependent on their recognition by receptors for mannose-6-phosphate (Man-6-P) (167). The formation of the Man-6-P group occurs through the action of two enzymes that modify high-mannose glycans. A recent study of Pelham (109), who attached the ER-retention signal KDEL to the lysosomal enzyme cathepsin D, indicates that the first enzyme involved in phosphodiester formation occurs in a post-ER compartment (transition vesicles or salvage compartment) while the second enzyme, a phosphodiesterase, is probably located in the *cis*-Golgi.

The next step in the targeting of lysosomal hydrolases involves their binding to Man-6-P receptors in the Golgi. Two different receptors have been isolated: the cation-independent receptor (M_r 215,000) (96) and the cation-dependent receptor (M_r 46,000) (29). Both are glycoproteins with a single transmembrane domain and a short cytoplasmic portion. The small receptor has a single repeat of an extracellular domain that is repeated 15 times in the large receptor. Both receptors appear to function in the targeting of lysosomal enzymes, with the large cation-independent receptor as the dominant one. Immunocytochemical localization of the receptors shows that in the steady state most receptors reside in the *trans*-Golgi network and endosomes; only weak labeling occurs in the Golgi and the plasma membrane. No labeling is present over the lysosomal membranes (49). These observations suggest that the receptors recycle between the *trans*-Golgi and a pre-lysosomal, endosomal compartment. The low pH in the endosomes induces the discharge of the lysosomal enzymes and allows the receptors to recycle back to the Golgi (or to the plasma membrane as these receptors are also involved in endocytosis). These quite remarkable findings, discussed in detail in several reviews (87, 167), serve as a paradigm for the understanding of sorting of vacuolar protein in other eukaryotic cells. However, all information obtained so far indicates that sorting of vacuolar/lysosomal proteins in yeast, *Dictyostelium discoideum,* and plants occurs via a Man-6–independent mechanism. In yeast and plants, glycans are not necessary for correct vacuolar targeting (see above). Neither plants (47) nor *D. discoideum* (18) contains Man-6-P receptors. Furthermore, there is considerable evidence for all three systems that the sorting signal is contained within the polypeptide domain of vacuolar/ lysosomal proteins (76, 85, 119, 149, 150, 156).

VACUOLAR SORTING IN YEAST

When yeast *(Saccharomyces cerevisiae)* is viewed by phase-contrast microscopy, the vacuole is the most conspicuous organelle. Like the plant vacuole, it is an acidic, lytic compartment containing a H^+-ATPase in its limiting membrane and numerous acid hydrolases including proteinase A (PrA), proteinase B (PrB), carboxypeptidase Y (CPY), and membrane-bound α-mannosidase. The yeast vacuole also stores amino acids, phosphate, and inorganic ions.

Considerable progress has been made recently in understanding the sorting signals of yeast vacuolar enzymes. Transport of carboxypeptidase Y has been studied in the *sec* mutants, mutants that have specific blocks in the secretory pathway (139). CPY is synthesized as a preproprotein, and removal of the signal sequence in the ER exposes a propeptide which is cleaved after arrival of proCPY zymogen in the vacuole. Analysis of the extent of CPY processing in various *sec* mutants has shown that proCPY accumulates in *sec* mutants blocked in movement out of the ER and out of the Golgi, but not in mutants blocked in the transport of secretory vesicles to the plasma membrane. Thus, in yeast as in mammalian cells, the Golgi is a branch point in the secretory pathway.

Selection procedures such as screening for mutants defective in endocytic uptake of dyes (27) and in the accumulation of active CPY (77) have yielded yeast mutants defective in some aspect of vacuolar sorting. The observation that overproduction of vacuolar proteins leads to their secretion (128, 140) enabled the development of selection schemes for vacuolar sorting mutants. These schemes selected for the secretion of active carboxypeptidase Y enzyme (126) or for the secretion of a vacuolar CPY-invertase fusion protein (4, 5, 122). Although these mutants were selected for mislocalization of CPY or the CPY sorting determinant, they also mislocalize significant amounts of the soluble vacuolar glycoproteins proteinase A (PrA) and proteinase B (PrB). Some mutants also mislocalize the vacuolar membrane protein α-mannosidase (4). This result indicates that soluble vacuolar proteins share a common sorting pathway different from but overlapping the sorting of membrane proteins. The presumed sorting receptor for soluble proteins has not yet been identified in yeast.

As is the case with lysosomal enzyme sorting in mammalian cells, there may be independent mechanisms for the delivery of soluble and membrane proteins in yeast cells, and perhaps multiple mechanisms for each of these classes. This hypothesis is supported by the recent observation that a protein soluble in the vacuole is transported through the secretory pathway as an integral membrane protein (84).

That the sorting signal of CPY is contained within a polypeptide domain of the protein has been demonstrated by two methods: fusion of a nested set of carboxyterminal deletions to invertase (76) and a deletion analysis of wild-type CPY (156). These analyses identified the 111 amino acid propeptide as important for sorting. Actually, the NH$_2$-terminal 50 amino acids of the preprodomain (which includes a 20 residue signal sequence) are sufficient to sort invertase fusion protein to the vacuole while the first 30 amino acids of the preprodomain direct high levels of secretion. Deletion and point mutations in the CPY propeptide which direct secretion of CPY implicate the same region (155, 156). The analysis of mutations in a domain from amino acid

Lys$_{18}$ (in the signal sequence) to Leu$_{34}$ identified only four contiguous amino acids important for vacuolar sorting, QRPL$_{24}$ (155). It is noteworthy that deletions and fusions near, but not into, this four-amino-acid sequence can disrupt sorting, perhaps by interfering with the context of the signal (55, 76).

The sorting domain of proteinase A, another vacuolar hydrolase, has also been examined by gene fusions to invertase (85). Again a determinant sufficient for sorting invertase to the vacuole was localized to the 76-amino-acid PrA prepropeptide. It appears to be more COOH-terminal than the sorting domain in the CPY propeptide, and is located between amino acids 61 and 76. To show that this region is necessary for vacuolar sorting, large deletions were constructed in the proregion of wild-type PrA. These deletions resulted in very unstable proteins, but the small amount that was stable appeared to be correctly transported, even when the entire propeptide was deleted. Small deletions in the propeptide resulted in delayed kinetics of transport, but vacuolar transport remained directed. This could mean that there is a second signal in the mature portion of PrA. There is no significant homology between the CPY sorting domain and the PrA propeptide region.

Analysis of the protease B processing pathway in yeast raises additional questions (102). The 280-amino-acid propeptide of PrB is cleaved from the protein before it leaves the ER. It is improbable that any sorting decision is made at the level of the ER in yeast; therefore, either the sorting domain resides in the mature portion of the protein or the propeptide carries the domain and stays associated with the mature PrB. Analysis of the sorting signals of PrA and PrB shows that in these two proteins the sorting domain is unlikely to be a short linear amino acid sequence; thus, the work with CPY and the determination of a short sorting domain led to an initially simplistic view of the sorting process in yeast.

DEFINING THE SORTING SIGNAL OF A PLANT VACUOLAR PROTEIN IN YEAST

Because there is such a remarkable conservation of both general and specific features of the secretory pathway in eukaryotic cells, it might be possible to use yeast or mammalian cells to identify the targeting signal of a plant vacuolar protein. In mammalian cells, lysosomal (vacuolar) targeting is mediated by the Man-6-P sorting signal (see above), and plant vacuolar proteins are secreted by the bulk-flow pathway when expressed in mammalian cells (165), amphibian cells (163), or insect cells (16). Thus, animal cells do not provide a suitable model for the vacuolar sorting of plant proteins. What about yeast cells? Tague & Chrispeels (148) showed that when the gene for PHA-L is expressed under the control of the acid phosphatase promoter, the protein is transported to the yeast vacuole. This finding allowed them to use fusions of nested deletions of PHA with yeast invertase to define the vacuolar sorting domain of PHA. In these constructs, entry into the secretory system was assured by the presence of the signal sequence of PHA, the signal sequence of invertase having been removed by truncating the invertase gene. The 43 amino-terminal amino acids of mature PHA-L together with the 20-amino-acid signal sequence are sufficient to direct more than 90% of the invertase to the yeast vacuole. When only 33 amino acids of PHA were fused to invertase, 23% of the activity was secreted, and with only 24 amino acids of PHA, 80% of the invertase was secreted (150). These results show that the vacuolar sorting signal of PHA is located between amino acids 24 and 43 of the mature protein. Using internal deletions, the sorting domain was shown to be located between amino acids 33 and 44 of mature PHA-L (150).

This aminoterminal domain of PHA that targets invertase to the yeast vacuole contains sequence identity with many other plant vacuolar proteins (especially storage proteins and lectins) and with the yeast vacuolar enzyme CPY (150). This sequence, identified as $LQRD_{41}$ (numbering from the initiating methionine) in PHA-L, corresponds to $LQRP_{26}$ (numbering from the initiating methionine) in CPY. While the sequence is not strictly conserved in all legume lectins, the Gln residue (Q_{39} in PHA-L) is found in all legume lectins (39, 142). Most important is the fact that the LQRP domain of CPY contains the vacuolar targeting information (127, 156), and mutational analysis of this domain shows that the Gln residue in the $LQRP_{26}$ sequence is absolutely required for correct vacuolar sorting (156).

When fused to invertase, single amino acid changes in the putative vacuolar sorting domain of PHA at Leu_{38}, Gln_{39}, and Asn_{41} (all in the LQRD sequence) caused dramatic increases in the level of invertase secretion (and presumed decreases in vacuolar targeting) by the yeast cells. Similarly, replacement of QRD_{41} (from PHA) with EGN_{41} caused near total secretion of the invertase by the yeast cells (150). The effect of the small changes indicates that this region is recognized as a bona fide yeast vacuolar sorting sequence. However, when the QRD to EGN mutation was introduced into the longest PHA-invertase fusion (220 amino acids of mature PHA), the proportion of invertase that was secreted increased only from 7% to 28%. Introduction of this same mutation in full-length PHA did not give dramatic effects on the level of secretion of PHA when the mutant gene was expressed in yeast (150). This means that the amino-terminal region of 43 amino acids of PHA is sufficient for correct vacuolar targeting in yeast, but is not necessary. A second and perhaps independent signal may be present towards the middle of the polypeptide. This finding parallels similar observations made for the yeast vacuolar proteinases PrA and PrB (discussed above).

DEFINING VACUOLAR SORTING DOMAINS IN PLANTS

Several heterologous systems have now been used to demonstrate that vacuolar proteins are correctly targeted when their genes are expressed in another organism. Sengupta-Gopalan et al (131) reported that introduction of a phaseolin gene (phaseolin is the major seed storage protein of the common bean *Phaseolus vulgaris*) into tobacco resulted in the correct temporal and tissue-specific expression of the gene. Up to 1% of the total tobacco seed protein was found to be phaseolin. Immunocytochemical analysis of thin sections of the tobacco seeds by Greenwood & Chrispeels (51) showed that phaseolin was present in the amorphous matrix of the protein storage vacuoles in the embryos and endosperm of the tobacco seeds. Similarly, PHA (147) and pea vicilin (64) also accumulate in the amorphous matrix of the protein storage vacuoles in transgenic tobacco seeds, while the 15-kDa zein, when the gene is expressed with a phaseolin promoter, accumulates in the crystalloids of the protein storage vacuoles (67). Recent experiments show that patatin, the abundant vacuolar glycoprotein of potato tubers, is targeted to vacuoles in tobacco leaves (136) and that barley lectin, a vacuolar protein in barley embryos, is targeted to the vacuoles of tobacco leaves and roots (171). All these results show that the vacuolar sorting signal is part of the primary amino acid sequence of the protein and that correct targeting does not depend on the species or the organ or the cell type in which the protein is expressed. Three different approaches to defining such a sorting signal are described below.

Fusion Proteins

The approach described above for yeast—the use of nested carboxyterminal fusions of a vacuolar protein with a reporter protein (invertase)—has been tried in plants, but so far without success. A. von Schaewen and M. J. Chrispeels (unpublished results) introduced into *Arabidopsis thaliana* protoplasts the same PHA-invertase fusions that were expressed in yeast, after exchanging the yeast-specific promoter and 3' sequences for plant-specific sequences, and examined the subcellular location of invertase after three days of culture. Fusions with up to 73 amino acids of mature PHA allowed for the efficient secretion of invertase by the plant cells. Some of these constructs (with 43, 56, and 73 amino acids of mature PHA) gave efficient vacuolar transport of invertase in yeast. Thus, we can conclude that the yeast and *Arabidopsis thaliana* cells give different results using the same coding-region constructs. Furthermore, with the long constructs (166 or more amino acids of mature PHA) there were only low levels of invertase accumulation in the plant cells, although these same constructs gave good expression in the yeast cells. The approach that worked well in yeast (150) seems not to work as well in plants and has not yet yielded results to define the vacuolar sorting signal of phytohemagglutinin in plants.

Similar results have been obtained with pea legumin-yeast invertase fusions (G. Saalbach, personal communication). Targeting information (for the yeast vacuole) appears to be contained in both the N-terminal and C-terminal portions of the legumin protein.

Deletion of Propeptides

Recent experiments with sporamin (98a) and barley lectin (6a) are leading to the identification of the vacuolar sorting domain on these proteins. Sporamin, the vacuolar storage protein in sweet potato roots, is synthesized as a pre-proprotein with a short aminoterminal pro-domain (58, 59). Cleavage of this domain probably occurs in the vacuole. When sporamin is expressed in tobacco cells, the mature protein is correctly targeted to the vacuoles and processed to its mature form. However, when constructs that lack the coding sequence of the pro-domain are used, the mature protein is secreted by the tobacco cells. Similar results have also been obtained with barley lectin. Barley lectin (94), a homolog of wheat germ agglutinin (98), is synthesized as a preproprotein with a carboxyterminal pro-domain that carries an N-linked glycan. When a construct with the coding region of barley lectin is expressed in tobacco cells, the protein is correctly targeted to the vacuoles and processed (171). When the coding region of the pro-domain is deleted from the construct the mature protein is synthesized and secreted. Thus, in these two pro-teolytically processed proteins that are synthesized as proproteins the vacuolar sorting information appears to be contained in the pro-domains. Experiments to show that these domains are not only necessary but also sufficient for vacuolar sorting are now in progress.

Homologous Extracellular and Vacuolar Proteins

Insight into vacuolar sorting signals may come from comparing amino acid sequences of homologous extracellular and vacuolar proteins. Some enzymes such as β-glucanase, chitinase, and invertase have been shown to have both an extracellular and a vacuolar location. While these enzymes could be products of the same gene (as is the case for yeast cytosolic and extracellular invertase), they are more likely to be products of different genes (157). The most detailed study in this regard has been done with the so-called pathogenesis-related (PR) proteins of tobacco. When plants are infected by pathogens such as viruses, fungi, or bacteria they respond by the formation of a local necrosis. This is accompanied by the de novo synthesis of PR proteins, which can be subdivided into several classes of serologically related proteins (for reviews, see 8, 159). These PR proteins can be divided according to their isoelectric point into acidic and basic proteins. Acidic proteins often have immunologically related basic counterparts (69). The only PR proteins for which enzymatic activities are known are the two acidic chitinases (160) and their basic counterparts (92, 132) and the acidic β-glucanases and their basic counterparts (45, 80). Some of these PR proteins have been localized; in general, acidic proteins are found extracellularly while basic proteins are intracellular. It has been shown in some cases that this intracellular location is in the vacuole (10, 99). A comparison of the amino acid sequences of an acidic and a basic chitinase of tobacco shows that the higher molecular weight (5 kDa) of the basic chitinase is caused by the presence of a short N-terminal domain that is homologous to the chitin-binding domain of wheat germ agglutinin, a short insertion in the middle of the polypeptide, and carboxyterminal tail of 6 amino acids (95). Recent experiments (J. M. Neuhaus and T. Boller, personal communication) show that the carboxyterminal tail of basic tobacco chitinase contains significant vacuolar sorting information. When a construct with a normal basic tobacco chitinase cDNA is introduced into tobacco cells, the chitinase remains inside the cells. When the coding sequence for the carboxyterminal peptide of 7 amino acids is deleted, at least half of the chitinase is secreted. When this coding sequence is attached to the coding sequence of the secreted cucumber chitinase and the construct is expressed in tobacco cells, most of this chitinase remains inside the cells. The vacuolar location of this chitinase remains to be confirmed. Together, the most recent vacuolar targeting experiments indicate that vacuolar sorting information is not contained in a specific amino acid sequence, as there is no obvious sequence identity between the carboxyterminal predomain of wheat germ agglutinin, the aminoterminal prodomain of sporamin, and the 6 carboxyterminal amino acids of basic chitinase. The three-dimensional structure of these targeting domains may play a significant role in correct targeting. The carboxyterminal tail of wheat germ agglutinin forms an amphiphilic helix as does the carboxyterminal tail of barley β-glucanase (the vacuolar targeting domain of which is not yet known). However, the same is not true for the aminoterminal prodomain of sporamin or the short carboxyterminus of chitinase. Clearly, much work still needs to be done before we will understand the nature of vacuolar sorting information in plant cells.

ACKNOWLEDGMENTS

Research in the author's laboratory has been consistently supported by the National Science Foundation (Metabolic Biology or Cellular Biology), the United States Department of Agriculture (Competitive Research Grants Office), and the Department of Energy (Office of Basic Energy Sciences). The writing of this review was facilitated by many discussions with my colleagues at the University of California, especially Brian Tague. I thank David Robinson for his input into the section on coated vesicles, and Alessandro Vitale, Natasha Raikhel, Kenzo Nakamura, Thomas Boller, and T. J. V. Higgins for allowing me to quote their unpublished results. I am particularly grateful for the suggestions that Natasha Raikhel and Alessandro Vitale made on a near-final draft of this review. I thank Lynn Alkan for producing a pristine typescript.

Literature Cited

1. Akazawa, T., Hara-Nishimura, I. 1985. Topographic aspects of biosynthesis, extracellular secretion, and intracellular storage of proteins in plant cells. *Annu. Rev. Plant Physiol.* 36:441–72
2. Anderson, M. A., McFadden, G. I., Bernatzky, R., Atkinson, A., Orpin, T., et al. 1989. Sequence variability of three alleles of the self-incompatibility gene of *Nicotiana alata. Plant Cell* 1:483–91
3. Andreae, M., Blankenstein, P., Zhang, Y., Robinson, D. G. 1988. Towards the subcellular localization of plant prolyl hydroxylase. *Eur. J. Cell Biol.* 47:181–92
4. Bankaitis, V. A., Johnson, L. M., Emr, S. D. 1986. Isolation of yeast mutants defective in protein targeting to the vacuole. *Proc. Natl. Acad. Sci. USA* 83:9075–79
5. Banta, L. M., Robinson, J. S., Klionsky, D. J., Emr, S. D. 1988. Organelle assembly in yeast: characterization of yeast mutants defective in vacuolar biogenesis and protein sorting. *J. Cell Biol.* 107:1369–83
6. Bassüner, R., Wobus, U., Rapoport, T. A. 1984. Signal recognition particle triggers the translocation of storage globulin polypeptides from field bean (*Vicia faba* L) across mammalian ER membrane. *FEBS Lett.* 166:314–20
6a. Bednarek, S. Y., Wilkins, T. A., Dombrowski, J. E., Raikhel, N. Y. 1990. A carboxyterminal propeptide is necessary for proper sorting of barley lectin to vacuoles in tobacco. *Plant Cell* 2:1145–55
7. Blobel, G., Dobberstein, B. 1975. Transfer of proteins across membranes. I. Presence of proteolytically processed and unprocessed nascent immunoglobulin light chains on membrane-bound ribosomes of murine myeloma. *J. Cell Biol.* 67:835–51
8. Bol, J. F., Linthorst, H. J. M., Cornelissen, B. J. C. 1990. Plant pathogenesis-related proteins induced by virus infection. *Annu. Rev. Phytopathol.* 28:113–38
9. Bole, D. G., Hendershot, L. M.,

Kearny, J. F. 1986. Posttranslational association of immunoglobulin heavy chain binding protein with nascent heavy chains in non-secreting and secreting hybridomas. *J. Cell Biol.* 1558–66
10. Boller, T., Vögeli, U. 1984. Vacuolar localization of ethylene-induced chitinase in bean leaves. *Plant Physiol.* 74:442–44
11. Bollini, R., Ceriotti, A., Daminati, M. G., Vitale, A. 1985. Glycosylation is not needed for the intracellular transport of phytohemagglutinin in developing *Phaseolus vulgaris* cotyledons and for the maintenance of its biological activities. *Physiol. Plant.* 65:15–22
12. Bollini, R., Vitale, A., Chrispeels, M. J. 1983. In vitro and in vivo processing of seed reserve protein in the endoplasmic reticulum: evidence for two glycosylation steps. *J. Cell Biol.* 96:999–1007
13. Buchanan, M. J., Iman, S. H., Eskue, W. A., Snell, W. J. 1989. Activation of the cell wall degrading protease, lysine during sexual signalling in *Chlamydomonas:* the enzyme is stored as an inactive, higher relative molecular mass precursor in the periplasm. *J. Cell Biol.* 108:199–207
14. Bulleid, N. J., Freedman, R. B. 1988. Defective co-translational formation of disulphide-isomerase-deficient microsomes. *Nature* 335:649–51
15. Burgess, T. L., Kelly, R. B. 1987. Constitutive and regulated secretion of proteins. *Annu. Rev. Cell Biol.* 3:243–93
16. Bustos, M. M., Luckow, V. A., Griffing, L. R., Summers, M. D., Hall, T. C. 1988. Expression, glycosylation and secretion of phaseolin in a baculovirus system. *Plant Mol. Biol.* 10:475–88
17. Campbell, C. H., Rome, L. H. 1983. Coated vesicles from rat liver and calf brain contain lysosomal enzymes bound to mannose 6-phosphate receptors. *J. Biol. Chem.* 258:13347–352
18. Cardelli, J. A., Golumbeski, G. S., Dimond, R. L. 1986. Lysosomal enzymes in *Dictyostelium discoideum* are transported at distinctly different rates. *J. Cell Biol.* 102:1264–70

19. Ceriotti, A., Colman, A. 1988. Binding to membrane proteins within the endoplasmic reticulum cannot explain the retention of the glucose-regulated protein GRP 78 in *Xenopus* oöcytes. *EMBO J.* 7:633–38

20. Chrispeels, M. J. 1970. Biosynthesis of cell wall protein: sequential hydroxylation of proline, glycosylation of hydroxyproline and secretion of the glycoprotein. *Biochem. Biophys. Res. Commun.* 39:732–37

21. Chrispeels, M. J. 1976. Biosynthesis, intracellular transport and secretion of extracellular molecules. *Annu. Rev. Plant Physiol.* 27:19–38

22. Chrispeels, M. J. 1983. The Golgi apparatus mediates the transport of phytohemagglutinin to the protein bodies in bean cotyledons. *Planta* 158:140–51

23. Chrispeels, M. J. 1985. The role of the Golgi apparatus in the transport and posttranslational modification of vacuolar (protein body) proteins. *Oxford Surv. Plant Mol. Cell Biol.* 2:43–68

24. Chrispeels, M. J., Bollini, R. 1982. Characteristics of membrane-bound lectin in developing *Phaseolus vulgaris* cotyledons. *Plant Physiol.* 70:1425–28

25. Chrispeels, M. J., Higgins, T. J. V., Craig, S., Spencer, D. 1982. Role of the endoplasmic reticulum in the synthesis of reserve proteins and the kinetics of their transport to protein bodies in developing pea cotyledons. *J. Cell Biol.* 93:5–14

26. Chrispeels, M. J., Higgins, T. J. V., Spencer, D. 1982. Assembly of storage protein oligomers in the endoplasmic reticulum and processing of the polypeptides in the protein bodies of developing pea cotyledons. *J. Cell Biol.* 93:306–13

27. Chvatchko, Y., Howald, I., Riezman, H. 1986. Two yeast mutants defective in endocytosis are defective in pheromone response. *Cell* 46:355–64

28. Crowther, R. A., Finch, J. T., Pearse, B. M. F. 1976. On the structure of coated vesicles. *J. Mol. Biol.* 103:785–98

29. Dahms, N. M., Lobel, P., Breitmeyer, J., Chirgwin, J. M., Kornfeld, S. 1987. 46 kd mannose 6-phosphate receptor: cloning, expression, and homology to the 215 kd mannose 6-phosphate receptor. *Cell* 50:181–92

30. Denecke, J., Botterman, J., Deblaere, R. 1990. Protein secretion in plant cells can occur via a default pathway. *Plant Cell* 2:51–59

31. Dickinson, C. D., Fleoner, L. A., Lilley, G. G., Nielsen, N. C. 1987. Self-assembly of proglycinin and hybrid proglycinin synthesized in vitro from cDNA. *Proc. Natl. Acad. Sci. USA* 84:5525–29

32. Dickinson, C. D., Hussein, E. H. A., Nielsen, N. C. 1989. Role of posttranslational cleavage in glycinin assembly. *Plant Cell* 1:459–69

33. Dickinson, C. D., Scott, M. P., Hussein, E. H. A., Argos, P., Nielsen, N. C. 1990. Effect of structural modifications on the assembly of a glycinin subunit. *Plant Cell* 2:403–13

34. Doms, R. W., Russ, G., Yewdell, J. W. 1989. Brefeldin A redistributes resident and itinerant Golgi proteins to the endoplasmic reticulum. *J. Cell Biol.* 109:61–72

35. Dorel, C., Voelker, T. A., Herman, E. M., Chrispeels, M. J. 1989. Transport of proteins to the plant vacuole is not by bulk flow through the secretory system, and requires positive sorting information. *J. Cell Biol.* 108:327–37

36. Driouich, A., Gonnet, P., Makkie, M., Laine, A.-C., Faye, L. 1989. The role of high-mannose and complex asparagine-linked glycans in the secretion and stability of glycoproteins. *Planta* 180:96–104

37. Elbein, A. D. 1979. The role of the lipid-linked saccharides in the biosynthesis of complex carbohydrates. *Annu. Rev. Plant Physiol.* 30:239–72

38. Elbein, A. D. 1988. Glycoprotein processing and glycoprotein processing inhibitors. *Plant Physiol.* 87:291–95

39. Etzler, M. E. 1985. Plant lectins: molecular and biological aspects. *Annu. Rev. Plant Physiol.* 36:209–34

40. Faye, L., Chrispeels, M. J. 1987. Transport and processing of the glycosylated precursor of concanavalin A in jackbean. *Planta* 170:217–24

41. Faye, L., Chrispeels, M. J. 1988. Common antigenic determinants in the glycoproteins of plants, molluscs and insects. *Glycoconjugate J.* 5:245–56

42. Faye, L., Chrispeels, M. J. 1989. Apparent inhibition of β-fructosidase secretion by tunicamycin may be explained by breakdown of the unglycosylated protein during secretion. *Plant Physiol.* 89:845–51

43. Faye, L., Johnson, K. D., Sturm, A., Chrispeels, M. J. 1989. Structure, biosynthesis, and function of asparagine-linked glycans on plant glycoproteins. *Physiol. Plant* 75:309–14

44. Faye, L., Sturm, A., Bollini, R., Vitale, A., Chrispeels, M. J. 1986. The position of the oligosaccharide side-chains of phytohemagglutinin and their accessibility to glycosidases determines their sub-

sequent processing in the Golgi. *Eur. J. Biochem.* 158:655–61

45. Felix, G., Meins, F. 1985. Purification, immunoassay and characterization of an abundant, cytokinin-regulated polypeptide in culture tobacco tissues. Evidence the protein is a β-1,3-glucanase. *Planta* 164:424–28

46. Gardiner, M., Chrispeels, M. J. 1975. Involvement of the Golgi apparatus in the synthesis and secretion of hydroxyproline-rich cell wall glycoproteins. *Plant Physiol.* 55:536–41

47. Gaudreault, P., Beevers, L. 1984. Protein bodies and vacuoles as lysosomes. *Plant Physiol.* 76:228–32

48. Geetha-Habib, M., Noiva, R., Kaplan, H. A., Lennarz, W. J. 1988. Glycosylation site binding protein, a component of oligosaccharyl transferase, is highly similar to three other 57 kd luminal proteins of the ER. *Cell* 54:1053–60

49. Geuze, H. J., Stoorvogel, W., Strous, G. J., Slot, J. W., Bleekemolen, J. E., Mellman, I. 1988. Sorting of mannose 6-phosphate receptors and lysosomal membrane proteins in endocytic vesicles. *J. Cell Biol.* 107:2491–2501

50. Goldstein, J. L., Brown, M. S., Anderson, R. G. W., Russell, D. W., Schneider, W. T. 1985. Receptor-mediated endocytosis: concepts emerging from the LDL receptor system. *Annu. Rev. Cell Biol.* 1:1–39

51. Greenwood, J. S., Chrispeels, M. J. 1985. Correct targeting of the bean storage protein phaseolin in the seeds of transformed tobacco. *Plant Physiol.* 79: 65–71

52. Griffiths, G., Simons, K. 1986. The *trans* Golgi network: sorting at the exit site of the Golgi complex. *Science* 234: 438–43

53. Haas, B., Klanner, A., Ramm, K., Sänger, H. L. 1988. The 7S RNA from tomato leaf tissue resembles a signal recognition particle RNA and exhibits a remarkable sequence complementarity to viroids. *EMBO J.* 7:4063–74

54. Hara-Nishimura, I., Nishimura, M., Akazawa, T. 1985. Biosynthesis and intracellular transport of 11S globulin in developing pumpkin cotyledons. *Plant Physiol.* 77:747–52

55. Harley, S. M., Beevers, L. 1989. Coated vesicles are involved in the transport of storage proteins during seed development in *Pisum sativum* L. *Plant Physiol.* 91:674–78

56. Harper, J. F., Surowy, T. K., Sussman, M. R. 1989. Molecular cloning and sequence of cDNA encoding the plasma membrane proton pump (H$^+$-ATPase) of

Arabidopsis thaliana. Proc. Natl. Acad. Sci. USA 86:1234–38

57. Harris, N. 1986. Organization of the endomembrane system. *Annu. Rev. Plant Physiol.* 37:73–92

58. Hattori, T., Ichihara, S., Nakamura, K. 1987. Processing of a plant vacuolar protein precursor *in vitro Eur. J. Biochem.* 166:533–38

59. Hattori, T., Nakagawa, T., Maeshima, M., Nakamura, K., Asahi, T. 1985. Molecular cloning and nucleotide sequence of cDNA for sporamin, the major soluble protein of sweet potato tuberous roots. *Plant Mol. Biol.* 5:313–20

60. Herman, E. M., Shannon, L. M., Chrispeels, M. J. 1985. Concanavalin A is synthesized as a glycoprotein precursor. *Planta* 165:23–29

61. Herman, E. M., Tague, B. W., Hoffman, L. M., Kjemtrup, S. E., Chrispeels, M. J. 1990. Retention of phytohemagglutinin with carboxyterminal KDEL on the nuclear envelope and the endoplasmic reticulum. *Planta*. In press

62. Hesse, T., Feldwisch, J., Balshüsemann, D., Bauw, G., Puype, M., et al. 1989. Molecular cloning and structural analysis of a gene from *Zea Mays* (L.) coding for a putative receptor for the plant hormone auxin. *EMBO J.* 8:2453–61

63. Higgins, T. J. V., Beach, L. R., Spencer, D., Chandler, P., Randall, P. S., et al. 1987. cDNA and protein sequence of a major pea seed albumin (PA2: *Mr* 26,000). *Plant Mol. Biol.* 8:37–47

64. Higgins, T. J. V., Newbigin, E. J., Spencer, D., Llewellyn, D. J., Craig, S. 1989. The sequence of a pea vicilin gene and its expression in transgenic tobacco plants. *Plant Mol. Biol.* 11:683–95

65. Hillmer, S., Depta, H., Robinson, D. G. 1986. Confirmation of endocytosis in higher plant protoplasts using lectin-gold conjugates. *Eur. J. Cell Biol.* 41:142–49

66. Hillmer, S., Freundt, H., Robinson, D. G. 1988. The partially coated reticulum and its relationship to the Golgi apparatus in higher plant cells. *Eur. J. Cell. Biol.* 47:206–12

67. Hoffman, L. M., Donaldson, D. D., Bookland, R., Rashka, K., Herman, E. M. 1987. Synthesis and protein body deposition of maize 15-kd zein in transgenic tobacco seeds. *EMBO J.* 6: 3213–21

68. Hoffman, L. M., Donaldson, D. D., Herman, E. M. 1988. A modified storage protein is synthesized, processed and degraded in the seeds of transgenic plants. *Plant Mol. Biol.* 11:717–29

69. Hooft van Huijsduijnen, R. A. M., Kauffmann, S., Brederode, F. T., Cornelissen, B. J. C., Legrand, M., et al. 1987. Homology between chitinases that are induced by TMV infection of tobacco. *Plant Mol. Biol.* 9:411–20

70. Hori, H., Elbein, A. D. 1981. Tunicamcyin inhibits protein glycosylation in suspension-cultured soybean cells. *Plant Physiol.* 67:882–86

71. Hurtley, S. M., Helenius, A. 1989. Protein oligomerization in the endoplasmic reticulum. *Annu. Rev. Cell Biol.* 5:277–307

72. Inohara, N., Shimomura, S., Fukui, T., Futai, M. 1989. Auxin-binding protein located in the endoplasmic reticulum of maize shoots: molecular cloning and complete primary structure. *Proc. Natl. Acad. Sci. USA* 86:3564–68

73. Iturriaga, G., Jefferson, R. A., Bevan, M. V. 1989. Endoplasmic reticulum targeting and glycosylation of hybrid proteins in transgenic tobacco. *Plant Cell* 1:381–90

74. Joachim, S., Robinson, D. G. 1984. Endocytosis of cationic ferritin by bean leaf protoplasts. *Eur. J. Cell Biol.* 34:212–16

75. Johnson, K. D., Höfte, H., Chrispeels, M. J. 1990. An intrinsic tonoplast protein of protein storage vacuoles in seeds is structurally related to a bacterial solute transporter (Glpf). *Plant Cell* 2:525–32

76. Johnson, L. M., Bankaitis, V. A., Emr, S. D. 1987. Distinct sequence determinants direct intracellular sorting and modification of a yeast vacuolar protease. *Cell* 48:875–85

77. Jones, E. W. 1977. Proteinase mutants of *Saccharomyces cerevisiae*. *Genetics* 85:23–33

78. Jones, R. L., Robinson, D. G. 1989. Protein secretion in plants. *New Phytol.* 111:567–97

79. Kandasamy, M. K., Paolillo, D. J., Faraday, C. D., Nasrallah, J. B., Nasrallah, M. E. 1989. The S-locus specific glycoproteins of *Brassica* accumulate in the cell wall of developing stigma papillae. *Dev. Biol.* 134:462–72

80. Kauffmann, S., Legrand, M., Geoffroy, P., Fritig, B. 1987. Biological function of "pathogenesis-related" proteins: four PR proteins of tobacco have 1,3-β-glucanase activity. *EMBO J.* 6:3209–12

81. Kaushal, G. P., Elbein, A. D. 1989. Glycoprotein processing enzymes of plants. *Methods Enzymol.* 179:452–75

82. Kelly, R. B. 1985. Pathways of protein secretion in eukaryotes. *Science* 230:25–32

83. Kim, W. T., Franceschi, V. R., Krishnan, H. B., Okita, T. W. 1988. Formation of wheat protein bodies: involvement of the Golgi apparatus in gliadin transport. *Planta* 176:173–82

84. Klionsky, D. J., Emr, S. D. 1989. Membrane protein sorting: biosynthesis, transport and processing of yeast vacuolar alkaline phosphatase. *EMBO J.* 8:2241–50

85. Klionsky, D. J., Banta, L. M., Emr, S. D. 1988. Intracellular sorting and processing of a yeast vacuolar hydrolase: proteinase A propeptide contains vacuolar targeting information. *Mol. Cell Biol.* 8:2105–16

86. Kornfeld, S. 1987. Trafficking of lysosomal enzymes. *FASEB J.* 1:462–68

87. Kornfeld, S., Mellman, I. 1989. The biogenesis of lysosomes. *Annu. Rev. Cell Biol.* 5:483–525

88. Krishnan, H. B., Franceschi, V. R., Okita, T. W. 1986. Immunochemical studies on the role of the Golgi complex in protein body formation in rice seeds. *Planta* 169:471–80

89. Larkins, B. A., Hurkman, W. J. 1978. Synthesis and deposition of zein in protein bodies of maize endosperm. *Plant Physiol.* 62:256–63

90. Laurière, M., Laurière, C., Chrispeels, M. J., Johnson, K. D., Sturm, A. 1989. Characterization of a xylose-specific antiserum that reacts with the complex asparagine-linked glycans of extracellular and vacuolar glycoproteins. *Plant Physiol.* 90:1182–88

91. Lawrence, M. C., Suzuki, E., Varghese, J. N., Davis, P. C., Donkelaar, A. V., et al. 1990. The three-dimensional structure of the seed storage protein phaseolin at 3 Å resolution. *EMBO J.* 9:9–15

92. Legrand, M., Kauffmann, S., Geoffroy, P., Fritig, B. 1987. Biological function of pathogenesis-related proteins: four tobacco pathogenesis-related proteins are chitinases. *Proc. Natl. Acad. Sci. USA* 84:6750–54

93. Lending, C. R., Chesnut, R. S., Shaw, K. L., Larkins, B. A. 1989. Immunolocalization of avenin and globulin storage proteins in developing endosperm of *Avena sativa* L. *Planta* 178:315–24

94. Lerner, D. R., Raikhel, N. V. 1989. Cloning and characterization of root-specific barley lectin. *Plant Physiol.* 91:124–29

95. Linthorst, J. H. M., Van Loon, L. C., van Rossum, C. M. A., Mayer, A., Bol, J. F., et al. 1990. Analysis of acid and basic chitinases from tobacco and petunia and their constitutive expression in

transgenic tobacco. *Mol. Plant Microbe Interact.* 3:252–58

96. Lobel, P., Dahms, N. M., Breitmeyer, J., Chirgwin, J. M., Kornfeld, S. 1987. Cloning of the bovine 215-kDa cation-independent mannose 6-phosphate receptor. *Proc. Natl. Acad. Sci. USA* 84:2233–37

97. Lund, P., Lee, R. Y., Dunsmuir, P. 1989. Bacterial chitinase is modified and secreted in transgenic tobacco. *Plant Physiol.* 91:130–35

98. Mansfield, M. A., Peumans, W. J., Raikhel, N. V. 1988. Wheat-germ agglutinin is synthesized as glycosylated precursor. *Planta* 173:482–89

98a. Matsuoka, M., NaKamura, K. 1991. Propeptide of a precursor to a plant vacuolar protein required for vacuolar targeting. *Plant Cell* 2:941–50

99. Mauch, F., Staehelin, L. A. 1989. Functional implications of the subcellular localization of ethylene-induced chitinase and β-1,3-glucanase in bean leaves. *Plant Cell* 1:447–57

100. McManus, M. T., McKeating, J., Secher, D. S., Osborne, D. J. Ashford, D., et al. 1988. Identification of a monoclonal antibody to abscission tissue that recognises xylose/fucose-containing N-linked oligosaccharides from higher plants. *Planta* 175:506–12

101. Miyata, S., Akazawa, T. 1982. Enzymic mechanism of starch breakdown in germinating rice seeds. 12. Biosynthesis in α-amylase in relation to protein glycosylation. *Plant Physiol.* 70:147–53

102. Moehle, C. M., Dixon, C. K., Jones, R. B. 1989. Processing pathway for protease B of *Saccharomyces cerevisiae*. *J. Cell Biol.* 108:309–24

103. Muesch, A., Hartmann, E., Rohde, K., Rubartelli, A., Sitia, R., Rapoport, T. A. 1990. A novel pathway for secretory proteins? *Trends Biochem. Sci.* 15:86–88

104. Munro, S., Pelham, H. R. B. 1987. A C-terminal signal prevents secretion of luminal ER proteins. *Cell* 48:988–97

105. Müntz, K. 1989. Intracellular protein sorting and the formation of protein reserves in storage tissue cells of plant seeds. *Biochem. Physiol. Pflanz.* 185:315–35

106. Olden, K., Parent, J. B., White, S. L. 1982. Carbohydrate moieties of glycoproteins: a re-evaluation of their function. *Biochim. Biophys. Acta* 650:209–32

107. Palade, G. E. 1975. Intracellular aspects of the processing of protein synthesis. *Science* 189:347–58

108. Parker, M. L., Hawes, C. R. 1982. The Golgi apparatus in developing endosperm of wheat (*Triticum aestivum* L.). *Planta* 154:277–83

109. Pelham, H. R. B. 1988. Evidence that luminal ER proteins are sorted from secreted proteins in a post-ER compartment. *EMBO J.* 7:913–18

110. Pelham, H. R. B. 1989. Control of protein exit from the endoplasmic reticulum. *Annu. Rev. Cell Biol.* 5:1–23

111. Pelham, H. R. B., Hardwick, K. G., Lewis, M. J. 1988. Sorting of soluble ER proteins in yeast. *EMBO J.* 7:1757–62

112. Pesacreta, T. C., Lucas, W. J. 1984. The plasma membrane coat and a coated vesicle-associated reticulum of membranes: their structures and possible interrelationship in *Chara corallina*. *J. Cell Biol.* 98:1537–45

113. Pesacreta, T. C., Lucas, W. J. 1985. Presence of a partially coated reticulum in angiosperms. *Protoplasma* 125:173–84

114. Pfeffer, S. R., Rothman, J. E. 1987. Biosynthetic protein transport and sorting by the endoplasmic reticulum and Golgi. *Annu. Rev. Biochem.* 56:829–52

115. Prehn, S., Wiedemann, M., Rapoport, T. A., Zwieb, C. 1987. Protein translocation across wheat germ microsomal membranes requires an SRP-like component. *EMBO J.* 6:2093–97

116. Rasmussen, U., Munck, L., Ullrich, S. E. 1990. Immunogold localization of chymotrypsin inhibitor-2, a lysine-rich protein, in developing endosperm. *Planta* 180:272–77

117. Ravi, K., Hu, C., Reddi, P. S., Huystee, R. B. V. 1986. Effect of tunicamycin on peroxidase release by cultured peanut suspension cells. *J. Exp. Bot.* 37:1708–15

118. Record, R. D., Griffing, L. R. 1988. Convergence of the endocytic and lysosomal pathways in soybean protoplasts. *Planta* 176:425–32

119. Richardson, J. M., Wyochik, N. A., Ebert, E. L., Dimond, R. L., Cardelli, J. A. 1989. Inhibition of early but not late proteolytic processing events leads to the missorting and oversecretion of precursor forms of lysosomal enzymes in *Dictyostelium discoideum*. *J. Cell Biol.* 106:2097–2107

120. Roberts, K. 1989. The plant extracellular matrix. *Curr. Opin. Cell Biol.* 1:1020–27

121. Robinson, D. G., Balusek, K., Freundt, H. 1989. Legumin antibodies recognize polypeptides in coated vesicles isolated from developing pea cotyledons. *Protoplasma* 150:79–82

122. Robinson, J. S., Klionsky, D. J., Banta, L. M., Emr, S. D. 1988. Protein sorting in *Saccharomyces cerevisiae:* isolation of mutants defective in the delivery and processing of multiple vacuolar hydrolases. *Mol. Cell. Biol.* 8:4936–48

123. Robinson, M. S. 1987. 100 kD coated vesicle proteins: molecular heterogeneity and intracellular distribution studied with monoclonal antibodies. *J. Cell Biol.* 204:887–95

124. Rose, J. K., Doms, R. W. 1988. Regulation of protein export from the endoplasmic reticulum. *Annu. Rev. Cell Biol.* 4:257–88

125. Rothman, J. E. 1987. Protein sorting by selective retention in the endoplasmic reticulum and Golgi stack. *Cell* 50:521–22

126. Rothman, J. H., Stevens, T. H. 1986. Protein sorting in yeast: mutants defective in vacuole biogenesis mislocalize vacuolar proteins into the late secretory pathway. *Cell* 47:1041–51

127. Rothman, J. H., Howald, I., Stevens, T. H. 1989. Characterization of genes required for protein sorting and vacuolar function in the yeast *Saccharomyces cerevisiae*. *EMBO J.* 8:2057–65

128. Rothman, J. H., Hunter, C. P., Valls, L. A., Stevens, T. H. 1986. Overproduction-induced mislocalization of a yeast vacuolar protein allows isolation of its structural gene. *Proc. Natl. Acad. Sci. USA* 83:3248–52

129. Satoh, S., Fujii, T. 1988. Purification of GP57, and auxin regulated extra-cellular glycoprotein of carrots, and its immunocytochemical localization in dermal tissues. *Planta* 175:363–73

130. Schwaiger, H., Hasilik, A., von Figura, K., Wiemken, A., Tanner, W. 1982. Carbohydrate-free carboxypeptidase Y is transferred into the lysosome-like yeast vacuole. *Biochem. Biophys. Res. Commun.* 104:950–56

131. Sengupta-Gopalan, C., Reichert, N. A., Barker, R. F., Hall, T. C., Kemp, J. D. 1985. Developmentally regulated expression in the bean β-phaseolin gene in tobacco seed. *Proc. Natl. Acad. Sci. USA* 82:3320–24

132. Shinshi, H., Wenzler, H., Neuhaus, J., Felix, G., Hofsteenge, J. F., Meins, J. 1988. Evidence for N- and C-terminal processing of a plant defense-related enzyme: primary structure of tobacco pre-pro-β-1, 3-glucanase. *Proc. Natl. Acad. Sci. USA* 85:5541–45

133. Sijmons, P. C., Dekker, B. M. M., Schrammeijer, B., Verwoerd, T. C., van den Elzen, P. J. M., Hoekema, A. 1990. Production of correctly processed human serum albumin in transgenic plants. *Bio/Technology* 8:217–21

133a. Simon, R., Altschuler, Y., Rubin, R., Galili, G. 1990. Two closely related wheat storage proteins follow a markedly different subcellular route in *Xenopus laevis* oöcytes. *Proc. Natl. Acad. Sci. USA* 88:834–38

134. Singh, N. K., Bracker, C. A., Hasegawa, P. M., Handa, A. K., Buckel, S., et al. 1987. Characterization of osmotin. *Plant Physiol.* 85:529–36

135. Sonnewald, U., Studer, D., Rocha-Sosa, M., Willmitzer, L. 1989. Immunocytochemical localization of patatin, the major glycoprotein in potato (*Solanum tuberosom* L.). *Planta* 178:176–83

136. Sonnewald, U., Sturm, A., Chrispeels, M. J., Willmitzer, L. 1989. Targeting and glycosylation of patatin, the major potato tuber protein in leaves of transgenic tobacco. *Planta* 179:171–80

137. Sonnewald, U., von Schaewen, A., Willmitzer, L. 1990. Expression of mutant patatin protein in transgenic tobacco: role of glycans and intracellular location. *Plant Cell* 2:345–55

138. Staswick, P. E. 1990. Novel regulation of vegetative storage protein genes. *Plant Cell* 2:1–6

139. Stevens, T. H., Esmon, B., Schekman, R. 1982. Early stages in the yeast secretory pathway are required for transport of carboxypeptidase Y to the vacuole. *Cell* 30:439–48

140. Stevens, T. H., Rothman, J. H., Payne, G. S., Schekman, R. 1986. Gene dosage-dependent secretion of yeast vacuolar carboxypeptidase Y. *J. Cell Biol.* 102:1551–57

141. Stinissen, H. M., Peumans, W. J., Chrispeels, M. J. 1985. Subcellular site of lectin synthesis in developing rice embryos. *EMBO J.* 3:1979–85

142. Strosberg, A. D., Buffard, D., Lauwereys, M., Foriers, A. 1986. Legume lectins: a large family of homologous proteins. In *The Lectins*, ed. I. E. Liener, N. Sharon, I. J. Goldstein, pp. 249–64. San Diego:Academic

143. Sturm, A., Chrispeels, M. J. 1986. The high mannose oligosaccharide of phytohemagglutinin is attached to Asparagine 12 and the modified oligosaccharide to Asparagine 60. *Plant Physiol.* 81:320–22

144. Sturm, A., Chrispeels, M. J., Wieruszeski, J. M., Strecker, G., Montreuil, J. 1987. Structural analysis of the N-linked oligosaccharides from jack bean α-mannosidase. *Proc. 9th Int. Symp. Glycoconjugates*, Abstr. B28

145. Sturm, A., Johnson, K. D., Szumilo, T., Elbein, A. D., Chrispeels, M. J. 1987. Subcellular localization of glycosidases and glycosyltransferases involved in the processing of N-linked oligosaccharides. *Plant. Physiol.* 85:741–45

146. Sturm, A., van Kuik, J. A., Vliegenthart, J. F. G., Chrispeels, M. J. 1987. Structure, position, and biosynthesis of the high mannose and the complex oligosaccharide side chains of the bean storage protein phaseolin. *J. Biol. Chem.* 262:13392–403

147. Sturm, A., Voelker, T. A., Herman, E. M., Chrispeels, M. J. 1988. Correct glycosylation, Golgi-processing, and targeting to protein bodies of the vacuolar protein phytohemagglutinin in transgenic tobacco. *Planta* 175:170–83

148. Tague, B. W., Chrispeels, M. J. 1987. The plant vacuolar protein, phytohemagglutinin, is transported to the vacuole of transgenic yeast. *J. Cell Biol.* 105:1971–79

149. Tague, B. W., Chrispeels, M. J. 1989. Identification of the targeting domain of the vacuolar glycoprotein phytohemagglutinin in transgenic yeast and tobacco. *J. Cell Biochem. Suppl.* 13D:230

150. Tague, B. W., Dickinson, C. D., Chrispeels, M. J. 1990. A short domain of the plant vacuolar protein phytohemagglutinin targets invertase to the yeast vacuole. *Plant Cell* 2:533–46

151. Tanchak, M. A., Griffing, L. R., Mersey, B. G., Fowke, L. C. 1984. Endocytosis of cationized ferritin by coated vesicles of soybean protoplasts. *Planta* 162:481–86

152. Tanchak, M. A., Rennie, P. J., Fowke, L. C. 1988. Ultrastructure of the partially coated reticulum and dictyosomes during endocytosis by soybean protoplasts. *Planta* 175:433–41

153. Taylor, J. R. N., Schussler, L., Liebenberg, N. v. d. W. 1984. Protein body formation in starchy endosperm of developing *Sorghum bicolor* (L.) Moench seeds. *S. Afr. J. Bot.* 51:35–40

154. Tillmann, U., Viola, G., Kayser, B., Siemeister, G., Hesse, T., et al. 1989. cDNA clones of the auxin-binding protein from corn coleoptiles (*Zea mays* L.): isolation and characterization by immunogold methods. *EMBO J.* 8:2463–67

155. Valls, L. A. 1988. *A mutational analysis of a yeast vacuolar sorting determinant.* PhD thesis. Univ. Oregon

156. Valls, L. A., Hunter, C. P., Rothman, J. H., Stevens, T. H. 1987. Protein sorting in yeast: the localization determinant of yeast vacuolar carboxypeptidase Y resides in the propeptide. *Cell* 48:887–97

157. Van den Bulcke, M., Bauw, G., Castresana, C., Montagu, M. V., Vandekerckhove, J. 1989. Characterization of vacuolar and extracellular $\beta(1,3)$-glucanases of tobacco: evidence for a strictly compartmentalized plant defense system. *Proc. Natl. Acad. Sci. USA* 86:2673–77

158. Van Deurs, B., Sandvig, K., Petersen, O. W., Olsnes, S., Simons, K., Griffiths, G. 1988. Estimation of the amount of internalized ricin that reaches the trans-Golgi network. *J. Cell Biol.* 106:253–67

159. Van Loon, L. C. 1989. Stress proteins in infected plants. In *Plant-Microbe Interactions: Molecular and Genetic Perspectives*, ed. T. Kosuge, E. W. Nester, 3:198–237. New York: McGraw-Hill

160. Van Loon, L. C., Gerritsen, Y. A. M., Ritter, C. E. 1987. Identification, purification and characterization of pathogenesis-related proteins from virus-infected Samsun NN tobacco leaves. *Plant Mol. Biol.* 9:593–609

161. Vaux, D., Tooze, J., Fuller, S. 1990. Identification by anti-idiotype antibodies of an intracellular membrane protein that recognizes a mammalian endoplasmic reticulum retention signal. *Nature* 345:495–502

162. Vitale, A., Ceriotti, A., Bollini, R., Chrispeels, M. J. 1984. Biosynthesis and processing of phytohemagglutinin in developing bean cotyledons. *Eur. J. Biochem.* 141:97–104

163. Vitale, A., Sturm, A., Bollini, R. 1986. Regulation of processing of a plant glycoprotein in the Golgi complex: a comparative study using *Xenopus* oocytes. *Planta* 169:108–16

164. Vitale, A., Zoppè, M., Bollini, R. 1988. Mannose analog 1-deoxymannojirimycin inhibits the Golgi-mediated processing of bean storage glycoproteins. *Plant Physiol.* 89:1079–84

165. Voelker, T. A., Florkiewicz, R. Z., Chrispeels, M. J. 1986. Secretion of phytohemagglutinin by monkey COS cells. *Eur. J. Cell Biol.* 42:218–23

166. Voelker, T. A., Herman, E. M., Chrispeels, M. J. 1989. In vitro mutated phytohemagglutinin genes expressed in tobacco seeds: role of glycans in protein targeting and stability. *Plant Cell* 1:95–104

167. von Figura, K., Hasilik, A. 1986. Lysosomal enzymes and their receptors. *Annu. Rev. Biochem.* 55:167–93

168. von Heijne, G. 1986. Towards a com-

parative anatomy of N-terminal topogenic protein sequences. *J. Mol. Biol.* 189:239–42

169. Wetzel, S., Demmers, C., Greenwood, J. S. 1989. Seasonally fluctuating bark proteins are a potential form of nitrogen storage in three temperate hardwoods. *Planta* 178:275–81

170. Wieland, F. T., Gleason, M. L., Serafini, T. A., Rothman, J. E. 1987. The rate of bulk flow from the endoplasmic reticulum to the cell surface. *Cell* 50:289–300

171. Wilkins, T. A., Bednarek, S. Y., Raikhel, N. V. 1990. Role of propeptide glycan in post-translational processing and transport of barley lectin to vacuoles in transgenic tobacco. *Plant Cell* 2:301–13

172. Williamson, M. S., Forde, J., Buxton, B., Kreis, M. 1987. Nucleotide sequence of barley chymotrypsin inhibitor-2 (CI-2) and its expression in normal and high-lysine barley. *Eur. J. Biochem.* 165:99–106

173. Zagouras, P., Rose, J. K. 1989. Carboxyterminal SEKDEL sequences retard but do not retain two secretory proteins in the endoplasmic reticulum. *J. Cell Biol.* 109:2633–52

174. Zingen-Sell, I., Hillmer, S., Robinson, D. G., Jones, R. L. 1990. Localization of α-amylase isoenzymes within the endomembrane system of barley aleurone. *Protoplasma* 154:16–24

Annu. Rev. Plant Physiol. Plant Mol. Biol. 1991. 42:55–76

ROOT SIGNALS AND THE REGULATION OF GROWTH AND DEVELOPMENT OF PLANTS IN DRYING SOIL

W. J. Davies and Jianhua Zhang

Division of Biological Sciences, Institute of Environmental and Biological Sciences, University of Lancaster, Bailrigg, Lancashire LA1 4YQ, United Kingdom

KEY WORDS: abscisic acid (ABA), stomata, leaf growth, xylem, drought, root signals

CONTENTS

Man will occasionally stumble across the truth but most of the time he will pick himself up and continue on.

Winston Churchill

INTRODUCTION

Much early writing on the effects of drought on plant growth and crop production emphasized the damaging effects of water deficits on many aspects of shoot growth and functioning. It is now clear that well before this type of damage becomes apparent, soil drying exerts many regulatory effects, including reduced rates of cell division and cell growth previously considered to be examples of damage to plants. Much of this regulation of growth and development in response to soil drying involves gene action (10, 21, 83). It is likely that in the next few years we will greatly increase our understanding of the molecular biology of drought responses, but such progress will be of little benefit without increased understanding of the biochemistry and physiology of droughted plants.

Here we address the factors regulating the development of the whole plant growing in drying soil. We extend certain arguments outlined in an earlier *Annual Review* chapter (89), but differ from most reviews in this area by proposing an important role for the root in sensing the amount of water available in the soil.

CHEMICAL AND HYDRAULIC SIGNALING

Soil Drying and Leaf Water Status

For many years now it has been generally accepted that as the soil dries, water uptake is reduced and leaf water status declines. It is argued that many shoot processes are influenced by this perturbation in shoot water status; and the influential writings of Kramer (49), Slayter (98, 99), and others have led us to ascribe a central regulatory role to the variation in shoot water status. This is the case even though the literature contains many data suggesting that leaf water status does not always play a central role in the regulation of drought responses. Until recently these data have been overlooked or interpreted to fit prevailing theory.

We do not deny that soil drying changes leaf water relationships or that these changes influence the biochemistry, physiology, growth, and development of the shoot. Such responses occur commonly. We are concerned here with shoot responses to soil drying that occur in the absence of any detectable change in leaf water status. Because such regulation almost certainly also occurs in parallel with the hydraulic effects of soil drying, it is important to understand its mechanistic basis. At the very least we must be clear that use of the term "water stress" is not limited to situations where water relations variables are modified.

Leaf water potential is the most commonly used indicator of shoot water status, and particular potentials are routinely associated with particular degrees of stress (33). Not only is this practice dubious, but even when we consider plant turgors as well as water potentials we may have problems: Plants in drying soil (and plants subjected to other environmental perturbations as well), can exhibit high, undisturbed turgor and yet show greatly reduced growth rates (e.g. 64, 68, 81, 105). Similarly, unwatered plants can have water potentials identical to or higher than those of well-watered plants but stomata that are almost closed (2, 60, 61, 66). Variables such as leaf conductance and extension rate may be more useful indicators of plant stress than the more commonly used variables of leaf water relations. Either way, it is necessary to address the question of why stomata and leaf growth rates show sensitive responses to soil drying.

Regulation of Plant Development: a Role for the Root?

INDIRECT EVIDENCE   Shoot growth and the physiology of plants in drying soil can be modified as a function of soil drying, even when shoot water relations are not perturbed. In cases where leaf water status does change, variation in shoot physiology can often be linked more closely to changes in soil water status than to changes in leaf water status (108). Plants must thus "sense" the drying of the soil around the root and communicate this information to the shoot by some means other than a reduction in the flux of water to the shoots. It is possible to explain some of the results cited here by suggesting that leaf growth rate and stomatal functioning are finely attuned to the water relationships of the plant (7). Boyer & Nonami (7a) suggest that such drying may result in the collapse of the water potential gradient between the xylem and the growing cells. This change would prevent water movement into the growing zone, limiting growth of plants with high leaf water potentials. Nevertheless, there are several reasons to look beyond such an explanation. For example, Jones and coworkers (46) found that over a period of several weeks, xylem water potentials at midday were higher in unwatered apple trees than in irrigated controls. This status resulted from reduced stomatal conductances and is apparently an example of overcontrol of water potential. The association of low conductance with high water potential in this apple crop argues for the regulation of leaf water status by stomata (43) rather than the more commonly held (e.g. 107) converse.

Why do plants in unwatered soil show low conductances when their shoot water potentials are high? One explanation is that the stomata do not open because plants can sense the availability of water in the soil and regulate stomatal behavior accordingly, whatever the water status of their shoots. Jones (42) and Cowan (15) have suggested that such detection involves the transfer of chemical information from roots to shoots. Recent evidence suggests that root signals not only influence stomatal behavior (and therefore carbon gain) but also regulate leaf initiation, leaf expansion, and other developmental processes. It remains to be shown that a single signaling system is involved in the regulation of all these processes.

DIRECT EVIDENCE   In an investigation of the leaf growth characteristics of clonal apple trees (25), root systems of individual plants were split between two containers, with approximately half the system in each container. With-

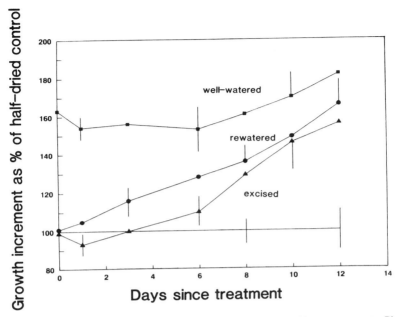

Figure 1   Daily increment in leaf area by apple trees with roots divided between two pots. Plants were either watered onto both halves of the root system (■) or onto one half of the root system only. Results are for plants that have received the above treatments for 24 days and are expressed as percentages of the growth of the plants that remained with half their roots in dry soil(−). Other groups of plants were either rewatered (●) or had the roots in dry soil excised (▲). All points are 3-day pooled means (n = 6) with standard errors shown. [From 25]

holding water from half of the root system reduced the rate of leaf area development by nearly 50% over a three-week period, in the absence of any significant leaf water deficit. Both the expansion of individual leaves and the initiation of new leaves were reduced by soil drying. As might be expected, rewatering restored the rates of leaf expansion and leaf initiation to the rates shown by plants that had remained well watered throughout the experimental period. The same result was achieved by excising from the plant the roots in dry soil (Figure 1). It is difficult to see how this treatment could have resulted in increased water supply to the leaves or restoration of the water potential gradient between the xylem and the growing leaves. A logical hypothesis is that the roots in drying soil chemically inhibited the initiation and expansion of leaves, even when shoots were well supplied with water, and that removal of these roots removed this inhibitory influence.

Split-root experiments with other species have resulted in reports of restricted leaf conductance in plants with no apparent water deficit (6, 124). In a recent paper, Saab & Sharp (88) report that drying part of the root system of

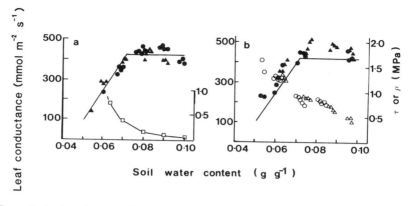

Figure 2   Leaf conductance of wheat plants as a function of changing soil water status: (*a*) plants whose water potentials were uncontrolled during the monotonic drying of the soil. Soil water suction (□) is also shown, (*b*) plants that were kept constantly at full turgor during the drying of the soil by application of a balancing pressure to the roots (○△). Results are shown for two plants. The solid line in *b* is a summary of the relationship shown in *a*. [From 24]

maize plants significantly reduced leaf growth rate but had no substantial effect on leaf conductance.

A different approach, one that allows destruction of the nexus between soil drying and a reduction in water supply to the leaves, has been taken by Passioura and colleagues, who pressurized the roots of wheat and sunflower plants by just enough to counterbalance the increase in soil suction as the soil dried (78). In this way, the shoot water relationships of unwatered plants were maintained at the value recorded in leaves before the soil was allowed to dry, but the roots of these plants were still in contact with drying soil. Stomata of pressurized plants showed an increasing restriction of conductance as soil dried, even though leaf water potentials did not decline (24; Figure 2). The relationship between leaf conductance and soil water content was identical in pressurized and unpressurized plants. More recently, Passioura (77) has shown that root signals may also control leaf expansion by wheat seedlings in drying soil.

Why Is a Root Signal Needed to Regulate Development?

We have seen above that leaf water potential can be higher for plants in moderately dry soil than for plants in wet soil. This finding alone argues against the general relevance of leaf water potential as a regulator of plant development. Leaf water potentials can vary from minute to minute (for example under cloudy conditions), further complicating the picture. With control systems it is normally appropriate to have a signal that varies over a time scale similar to that of the process being controlled (45), and it is therefore clear that rapid changes in leaf water potential are not an appropriate control signal for plant development. In addition, the water relations of shoot meristems of many plants may be buffered against abrupt changes in the water relations of the laminae (e.g. 68, 105). It might be suggested that some "measure" of soil water availability would be needed for effective regulation of development in drying soil.

Jones (42, 44, 45) has argued that plants are more likely to develop successfully if they make maximal use of the water available in the soil. Behavior that results in the conservation of water (i.e. any maximizing of water use efficiency) (15, 75) may place an individual at a disadvantage to a competitor that does not conserve water. Jones's analysis of water use patterns in unpredictable environments shows that an optimistic or water-spending approach was close to ideal as long as it included a mechanism that prevented death as available water started to run out. Such a mechanism might involve restricted development of transpiring area and restricted stomatal opening, but these responses would need to be keyed to a measure of the amount of soil water available to the plant.

How Can Plants "Measure" Water Availability?

CHANGES IN ROOT WATER STATUS   Particularly for large plants rooted in large volumes of soil, water availability may vary substantially from one part of the root system to another. For example, a tree with most of its roots in the upper 10 cm or so of soil will dry this soil layer substantially while plenty of water may still be available deeper in the profile. Information about differential water availability will be useful in "fixing" a plant's developmental pattern, and thus plants would benefit from a mechanism that integrates information from all roots. A bulk shoot or stem-base water potential would not be adequate, as it would be biased towards the water status of the wettest part of the soil (45). Several recent papers (80, 81, 94, 96, 97) consider the factors regulating the water relations of root tips. Some of this work suggests that the tips of individual roots can show water relations somewhat different from those of the bulk of the root system in other parts of the soil profile. Little information is available on water relations profiles along roots, or even on root-to-root variation in water relations; but if only a few roots start to dry the soil, these may dehydrate; as a result, they may generate a chemical response that moves to the shoot and exerts an influence, even when the flux of water is not reduced. The dehydration of more and more roots to a greater and greater extent would contribute to a signal of increasing intensity. It is well known that in drying soil, root cortex often dehydrates and dies, leaving the turgid root tip connected to the rest of the plant by the stele. The influence of this dehydration on root signaling should be investigated.

Kramer (50) has criticized the idea that roots can dehydrate substantially, arguing that because leaves lose water to the atmosphere they will always dehydrate before roots and will therefore be a more sensitive indicator of any change in soil water availability. In addition, he notes that roots may osmoregulate to sustain turgor in drying soil and for this reason also may be poor sensors of soil drying. While there is no doubt that water can be lost from leaves before it is pulled out of roots, the important point would seem to be that surface soil can dry very substantially (76) and that a root in this soil layer may therefore dehydrate substantially. The leaves can be well supplied with water from other roots in wet soil and may therefore show no dehydration relative to leaves of well watered plants. Dehydration of these shallow roots would be expected to influence metabolism in the root tips greatly and thereby provide some chemical indication of soil drying under circumstances where the supply of water to the shoots would not be a sensitive indicator of changes in the soil. Direct water loss to the atmosphere from surface soil may even mean that roots dry to water potentials lower than those the leaves ever experience (76).

Effective osmoregulation has been reported for seminal roots and primary roots of maize seedlings at lower water potential (93–95, 113). Zhang & Davies (119) suggest that secondary and tertiary roots of maize may be much less effective at turgor maintenance as soil dries, and emphasize that the root system cannot be treated as a homogeneous mass. Water relations and metabolic activity may vary substantially from one part of the root system to another. Certain classes of roots may be better sensors of soil drying than other classes. We need more information on this very important point.

RESPONSES TO CHANGES IN MECHANICAL IMPEDANCE AND OTHER CHANGES IN THE SOIL   The sensing mechanism proposed above could involve all of the live roots on a plant that can contribute information to the shoot, probably mostly through the xylem stream. Another possibility is that the plant may "measure" the increase in mechanical impedance as the soil dries. This would be accomplished as growing roots encounter increased soil strength at reduced soil water content. Masle & Passioura (63) have tested the importance of such changes by varying the mechanical impedance of soil around roots of wheat plants (by varying both water content and bulk density). Plants grew less well in soil with a high impedance, and at a given high mechanical impedance it was not possible to distinguish between the effects of lower water content and high bulk density. Thus a reduction in soil water content may be sensed through a change in soil strength. Masle & Passioura were able to rule out a limiting supply of nutrients and water and a reduced carbon supply as explanations for the restricted growth of wheat plants in soils with a high mechanical impedance. They speculated that shoot growth was reduced primarily in response to some hormonal message induced in the roots as they encountered soil of high mechanical impedance.

An estimate of changing soil water status might also be provided by a change in the ionic status in soil around the root. Such a change might be reflected in a reduced supply of particular nutrients to the leaves. For example, Shaner & Boyer (92) showed that variation in nitrogen supply to leaves could regulate enzyme activity in plants in drying soil. It seems doubtful, however, that such a mechanism would have the sensitivity required to explain certain observations. For example, Neales et al (72) have noted that a very small proportion of a root system can be isolated and dehydrated independently of the soil around the root mass. The effect of this procedure on the stomata can be substantial. Thus the responses are sensitive; and dehydration of part of the root system can have an important role in root-to-shoot communication, whether or not the reaction of roots to other changes in the soil is important. The possibility that nutrient deficiency may also stimulate the production of other chemical growth regulators has also been discussed (11a).

Localized salt concentrations may substantially increase as soil dries. Hartung & Radin (29) have suggested that this increase may generate strong ion imbalance in the xylem, which could provide information on soil water status to the shoots.

NATURE OF THE CHEMICAL INFORMATION CONVEYED FROM ROOTS TO SHOOTS

Rapid responses to dehydration of only a small proportion of the root system (72) suggest that this type of chemical signaling is not simply a reflection of a general adjustment in root functioning. It appears that small changes in the turgor, volume, or even pressure on membranes of only a few roots may be all that is necessary to modify shoot growth and physiology.

Results of the split-root experiment described above (25) suggest that the signaling between root and shoot of the effects of soil drying must be primarily of a "positive" nature (40)—i.e. an increase in the supply of some physiologically active substance. Removal of the drought-induced growth limitation by excision of partially dehydrated roots would not be expected if the principal message was either "negative" or "accumulative"—i.e. the reduced supply of something normally transmitted from roots to shoots, or the accumulation in shoots of something normally transmitted to roots. This is not to say that such messages are not ever important. All three types of signaling are considered here.

Positive Signals

Analysis of the composition of xylem sap from unwatered plants shows effects of soil drying on cations, anions, pH, buffer capacity, amino acids, and plant hormones (91). Concentrations of most components decline as the soil around the roots is allowed to dry. One clear exception is the concentration of abscisic acid (ABA), which increases substantially following only mild soil drying.

SOIL DRYING AND ABA SYNTHESIS IN ROOTS   Despite early reports to the contrary, many species synthesize ABA in their roots (14, 35, 54, 87, 110). The relationship between root water status and root ABA content described by Zhang & Davies (118) suggests that accumulation of ABA by roots may be a sensitive measurement of root water status.

Leaves also synthesize abscisic acid; even leaves of well-watered plants can contain a lot of ABA. If ABA arriving in the leaves via the xylem stream conveys information about the water status of the soil, the shoot must be able to differentiate at least between root-sourced and leaf-sourced ABA and possibly also between ABA arriving in the xylem stream and xylem-derived ABA that arrived earlier in the day. Shoots may well be able to do this, since at least in the well-watered plant in the light, much ABA is effectively sequestered within the chloroplasts, which act as anion traps (16, 31). Unless leaf turgor falls, leaf-sourced ABA should remain quite separate from any ABA arriving in the transpiration stream. A proportion of the transpiration stream arrives directly at sites of evaporation in the epidermal cell walls adjacent to guard cells (65), which are also thought to be the sites of action for ABA on the guard cells (26). The epidermis may also be a site for the regulation of leaf growth (18). The transpiration stream may therefore provide a direct link between the point in the root tip where the soil drying is sensed and locations in the leaf where plant water balance and growth can be regulated. It is not clear how much of the xylem-derived ABA ends up in the epidermis, but we note below that root signals can only operate if ABA does not dam up at the sites of water loss. This suggests that the leaf has an effective means of metabolizing or compartmentalizing ABA (31).

Strong evidence that ABA is synthesized in increased quantities in roots in drying soil is now provided by several different studies. (*a*) In split-root experiments, the half of the root system in drying soil contains substantially increased concentrations of ABA compared to the other half of the root system in wet soil (124). (*b*) A small proportion of the root system separated from the main soil mass and allowed to dehydrate partially in air contains higher concentrations of ABA within a few hours (72). (*c*) Comprehensive analysis of ABA contents of roots in different parts of the soil profile shows significant differences in concentration through the profile. These differences reflect the water status of the soil surrounding the individual roots (119). It appears that as bulk soil water content around the roots falls towards −0.2 to −0.3 MPa, partial dehydration of roots stimulates increased ABA production.

ABA IN THE XYLEM AND THE CONTROL OF STOMATAL BEHAVIOR AND LEAF GROWTH   If roots are loaded with ABA, increased concentrations of this compound can be detected in the leaves of transpiring plants soon after lights have been switched on. Covering leaves with tin foil prevents transpiration, and enhanced ABA concentrations are not detected, suggesting that this compound moves from roots to shoots primarily through the xylem stream (118).

Xylem sap from well-watered maize and sunflower plants contains an ABA concentration around 10 nM; in plants subjected to comparatively mild soil drying this concentration can rise by one or two orders of magnitude (91, 120, 121), even when the leaf water status is not perturbed by soil drying. In other plants, increases in xylem ABA concentration of 25 (29) and 50 fold (71) have been reported as responses to mild soil drying. Substantial increases in concentration can be detected in the xylem before the ABA concentration in the leaf is seen to increase (121), suggesting that xylem ABA concentration may be a more sensitive indicator of the effect of soil drying than is an estimate of the ABA concentration in the leaf. Other species may show much smaller increases in ABA concentration (31, 100), and we must therefore ask how large the concentration increases must be to be "read" by the shoots. Bioassay results (101) suggest that increases in concentration of less than an order of magnitude produce only small responses, but some plants are particularly sensitive to the chemical signal (see below).

It is well known that ABA fed into the xylem can substantially affect leaf gas exchange (e.g. 51, 56). Loveys (55) and others have reported on the effect of xylem-derived ABA on the stomatal behavior of leaves of a variety of species. For both maize and sunflower plants during a period when both temperature and photon flux density varied substantially, the correlation between limitation in stomatal conductance and extra ABA in the xylem sap resulting from soil drying was highly significant. This single relationship held up for two separate drying cycles (120, 121). Feeding ABA to well-watered plants resulted in an almost identical relationship between conductance and xylem ABA concentration, which provides good correlative evidence that the drought-induced restriction in stomatal conductance results from ABA accumulation in the xylem stream. More quantitative evidence for this contention is provided by an experiment (123) where xylem sap collected from unwatered maize plants was passed through an immunoaffinity column composed of ABA antibodies (48). This treatment removed nearly all the ABA from the sap and also nearly all of the antitranspirant activity, as determined by both an epidermal strip bioassay (74) and a transpiration bioassay (71)

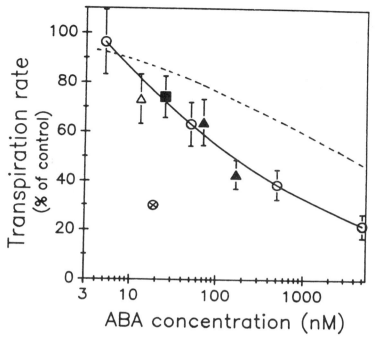

Figure 3 Transpiration of detached wheat leaves as a function of concentration of ABA in assay solutions. The solutions were synthetic ABA in artificial xylem sap (○), xylem sap from well-watered maize plants (■), xylem sap from unwatered maize plants (▲), and xylem sap from unwatered maize plants but with ABA removed by immunoaffinity column (△) [from 123]. The dotted line [from 70] shows the dose response of detached wheat leaves to synthetic ABA solutions. The crossed point is a typical result for xylem sap extracted from unwatered wheat plants and fed to detached wheat leaves. Data are expressed as percentages of the rate of water loss from control leaves fed with artificial xylem sap only (no ABA). Data from reference 123 are means of 5 observations. The bars are double standard deviations divided by the mean transpiration rate of the control leaves.

(Figure 3). This experiment suggests that ABA is the major stomatally active component in the xylem sap of maize plants.

In wheat, on the other hand, even the ABA concentration in the sap of unwatered plants is relatively low (71). When this sap was fed through the midribs of isolated wheat leaves, the reduction in transpiration was very much greater than would have been expected from the application of a comparable concentration of ABA in artificial xylem sap (Figure 3) (70). Removal of ABA from the sap using an immunoaffinity column did not remove the antitranspirant activity, leading Munns & King to suggest that the xylem sap of wheat must contain an as-yet-unidentified compound with antitranspirant activity. This possibility is worthy of further investigation.

Although many workers have described the inhibition of leaf growth by applied ABA (34, 84, 106, 109), the physiological relevance of these results has been questioned because it has also been shown that younger, rapidly growing leaves contain higher concentrations of ABA than do mature leaves (115, 116). As noted above, however, much ABA in turgid leaves may be sequestered within the chloroplasts and may therefore have little immediate influence on growth and development. Zhang & Davies (122) have recently shown an excellent relationship between the ABA concentration in the xylem and the growth rate of leaves of unwatered maize and sunflower plants, despite the fact that turgor is maintained in these leaves as the soil dries. As in the experiments reported above, these relationships can be duplicated by applying ABA to well-watered plants, again suggesting that ABA may play an important role in the regulation of growth of unwatered plants. Bensen et al (3) have shown that an increase in endogenous ABA in the stems of un-watered soybean seedlings coincides with a decrease in wall extensibility and tissue conductance (8, 73). Others have shown that ABA may reduce the growth rate of leaves by influencing the properties of cell walls (53, 109). Applied ABA does not, however, cause the same changes in mRNA in soybean stems that are caused by withholding water (17), although there is evidence that ABA and soil drying can cause similar changes in polypeptide and mRNA accumulation in more mature tissues such as leaves (9).

It seems likely that even in plants where root signals play a dominant role in the early stages of soil drying, continued soil drying may eventually result in some shoot dehydration. We have found that with sunflower plants (120), stomata of older leaves may lose some sensitivity to ABA (1), and the turgor of these leaves then declines before the turgor of younger leaves. Presumably, these older leaves contribute ABA to the transpiration stream, and this ABA helps to sustain turgor in younger leaves. The net import of ABA to younger leaves is one of the reasons these leaves contain such high concentrations (116). This leaf-to-leaf signaling may be important in controlling turgor of young leaves, which have only a limited capacity for ABA synthesis (13, 57, 117).

Claes et al (12) have identified a gene in rice that is expressed in response to a number of stresses, including drought and ABA applied to roots. The induction of the gene was seen clearly in the sheath tissue, a part of the plant that is not an obvious site of dehydration. This type of response indicates a chemical influence on development via a change in gene expression. Work with heterophyllus species (23, 114) has shown the very substantial effect that ABA can have on leaf development.

FIELD STUDIES   Most of the comparatively few reports of diurnal variation in ABA content in shoots of field-grown plants evidence little relationship between ABA and stomatal behavior (e.g. 11, 82). These studies often measured bulk leaf ABA content, while stomatal behavior may be more closely linked to xylem ABA concentration.

In a recent field study by Tardieu et al (104), maize plants were grown on two field plots with contrasting soil structures. Soil compaction reduced stomatal conductance even though plants on compacted soil showed water relations comparable to those of plants on uncompacted soils. Plants on compacted soils contained more ABA than plants on uncompacted soils, an observation that seemed to correlate with a limitation on water uptake in compacted soil. There was a reasonable correlation between extra ABA in xylem sap and the relative restriction in conductance, and the diurnal variation in conductance could to some extent be explained by diurnal variation in xylem ABA concentration. Similar results have been obtained for grape vines in the field in Australia by Loveys (55). Despite apparent daily correlations, Loveys et al (59) have found no conclusive relation between xylem ABA and conductance assessed on a seasonal basis. Nevertheless, Wartinger et al (111) report that xylem ABA concentration may have set the range of maximum conductance for almond trees during the first part of a soil drying cycle in the field in Israel. At lower soil water contents, conductance was apparently determined by other factors. These results suggest that xylem ABA may have a role in controlling stomatal behavior of plants in the field, but more work is needed to elucidate these responses.

Negative and Accumulative Signals

Even when we put aside the evidence from experiments such as those of Gowing et al (25), it is hard to argue that negative or accumulative signals can act alone in root-to-shoot communication of the effects of soil drying. We have seen, above, that chemical information from only a few roots in drying soil is enough to influence stomatal behavior and leaf growth. If we assume that transport from these roots is completely stopped, this would mean a reduction of only a few per cent in the transport of a putative substance normally moving from roots to shoots. It is not clear that the physiology and growth of the shoot could "read" such a small change with sufficient precision for negative signals to work in wet soil. In addition, Jones (45) has pointed out that because the magnitude of a negative signal would vary with root volume, whether or not the roots were stressed, such a signal by itself would not have the required information content. There are several possible candidates as negative signals, but the most-discussed must be the transport of cytokinin from root to shoot (38, 89).

After early work by Went & Bonner (112), Kulaeva (52), and Kende (47) demonstrating much cytokinin activity in xylem sap, Itai & Vaadia (39) showed that drought stress reduced the production and transport of cytokinin from sunflower roots. There is no doubt that cytokinins applied externally to plants can increase stomatal apertures and transpiration of many plants (36, 37), although these responses seem to be best developed only in older leaves (5). These two sets of observations suggest the possibility that soil drying may influence stomatal behavior via a reduced supply of cytokinin from roots. Despite the reservations aired above, there is some evidence that patterns of stomatal behavior of sugarcane can be interpreted as a function of the delivery of chemical promoters of stomatal opening (zeatin riboside and potassium) (67). It seems possible, then, that in drying soil a reduction in cytokinin supply, perhaps acting in concert with other signals, could reduce stomatal conductance.

High concentrations of cytokinin can override the effects of ABA on stomata (4, 85, 86); one could therefore argue that as the soil dries, reduction in cytokinin supply would amplify shoot responses to an increasing concentration of ABA. Schurr & Gollan (91) have clearly demonstrated a role for inorganic ions in this respect and have shown that the sensitivity of stomata of sunflower to ABA in the xylem sap can vary enormously, largely as a function of a reduced supply of nitrate and calcium through the xylem of plants in drying soil. Radin and coworkers (85, 86) have also emphasized the influence of the ionic status of the soil on stomatal sensitivity to ABA. It thus seems possible that positive and negative signals may combine and interact in their effects on shoot processes, perhaps with positive signals dominating in plants in moist soil and negative signals increasing in importance as soil dries further.

Jackson & Hall (41) have argued that the increase in the concentration of ABA in leaves of flooded plants is an accumulative signal. This would be the case if this ABA is shoot-sourced and would normally have been transported from leaves to roots in the phloem. A reduction in phloem transport may not be surprising following total inundation of the root system and a substantial change in root system functioning. It seems less likely in drying soil, where root activity is often sustained and may even increase (62). This type of response may be important when soil drying becomes very severe and root growth and activity are severely limited.

Other Signals and Some Problems

STRONG ION DIFFERENCES AND pH   We have noted above that even leaves of well-watered plants contain enough ABA to influence stomatal behavior if it is released from anion traps to sites of action on the guard cells. It is generally held that ABA will be released from the mesophyll to the apoplast of the leaf if for any reason the pH of the apoplast is increased (30). Schurr & Gollan (91) and Hartung & Radin (29) report data which suggest that the pH of the xylem sap will increase as the soil and the roots dry, but the response is not very sensitive to soil drying. Xylem sap from sunflower may be well buffered in the pH range 6.5–7.5 (91), but apoplastic sap of cotton leaves has a low buffer capacity in the pH range 6.3–7.5 (30). Transmission of a pH change from root to shoots could redistribute ABA in the leaf and influence stomatal behavior independently of any change in shoot water status. The drought-induced pH change and the increase in ABA content in the xylem sap can be seen to have an additive effect.

Stewart (102, 103) has demonstrated that tiny imbalances between strong cations and anions (strong ion differences—SID) can control the pH of an aqueous solution. Schurr & Gollan (91) found an excess of anions in the xylem sap of well-watered plants and an excess of cations in unwatered plants. Hartung & Radin (29) propose that such a strong SID can alkalize the apoplastic fluid and may serve as a signal to the shoots—a signal that will redistribute any ABA compartmentalized within the leaf.

Hartung and coworkers have modeled the redistribution of ABA throughout the leaf as a function of changing pH (31). This is an important area of considerable uncertainty that needs further investigation if we are to understand the dynamics of stomatal behavior in droughted plants. The stomata and presumably also the rate of leaf growth will ultimately respond to the amount of ABA in the apoplast. This response will be a function of the flux into the leaf (supply from roots × transpiration rate), the compartmentation, and any biosynthesis, metabolism, or conjugation that is taking place under stressed or unstressed conditions. Tight relationships between ABA concentration in the xylem and physiological response perhaps suggest that mechanisms exist that can keep apoplastic ABA concentration relatively low. One of these may be rapid redistribution because of the high ABA permeability of the plasmalemma (31), and inasmuch as protoplasts of barley are unable to metabolize ABA (58) it may be that at least some ABA metabolism takes place in the apoplast. Mathematical models (31) have suggested that the epidermis and the phloem play an important role in ABA redistribution under water deficit, but we need more information on the regulation of apoplastic ABA concentration in leaves with high and low turgor. Large amounts of ABA arrive in the leaf via the xylem. These must be rapidly dealt with; otherwise stomata would remain permanently closed. Clearly, these mechanisms are an integral part of any root-to-shoot signaling system.

FLUXES OF ABA   Jones (45) has questioned how ABA that is synthesized in increased quantities by root tips in dry soil (which are not contributing to transpiration) can be transported to the epidermis in higher amounts than in control plants. The answer is presumably that these roots are still contributing to the transpiration stream and that even very dry roots release ABA as they rehydrate and dehydrate again after a dark period. Schurr & Gollan (91) have calculated fluxes of ABA from roots in moist and dry soil and have shown that even though the transpiration stream from roots in dry soil is much reduced, the increase in ABA concentration is so large that the input of ABA to the leaf still increases dramatically as the soil dries.

If we are to assess critically the involvement of root-sourced ABA in the control of growth and physiology, it is important that our sampling procedure for xylem sap does not in itself change the concentration of xylem components. Munns (70) has drawn attention to this problem and suggested that destructive sampling of plants may concentrate growth regulators into a reduced volume of xylem sap. In practice, such small volumes of sap are required by immunological tests that it has proved possible to sample sap from the volume present in the vessels before destructive sampling (122). Rather surprisingly, sequential sampling from exuding cut stumps showed that ABA concentrations in the sap were relatively stable, a dilution rather than a concentration occurring with time.

It has proved difficult to collect xylem sap and feed it back to cut leaves. This is an important way of testing for antitranspirant activity (71); but we have found that particulate matter in the sap results in blockage of the xylem, and stomatal closure is stimulated by loss of turgor (123). It is important to rule out these effects in assessments of the importance of chemical control of stomatal behavior.

ABA IN THE SOIL   At least some of the extra ABA found in the xylem of unwatered plants may come from the soil. This possibility has received little attention, but it may cause problems in an assessment of root responses to soil drying. Muller et al (69) have noted that soil fungi (22) and algae (32) may synthesize ABA and release it to the rhizosphere. Root particles can also release ABA into the soil (27). These possibilities were confirmed by Hartung et al (28), who analyzed water from desert soils. The ABA may be produced in increasing amounts as the soil dries; it may be taken up directly by the plant and exported but may also influence root development (69) and activity. Plant roots may acidify their immediate environment (28), an action that would aid in the uptake of undissociated ABA into roots.

CONCLUSIONS

During the last three or four years, some of the changing concepts in plant water relations considered above have stimulated spirited arguments (7, 20, 50, 76, 90). Kramer (50) remarks that we have spent 50 years of progress in our understanding of plant water relations by shifting emphasis away from the soil to the shoot of the plant. Our most recent evidence suggests that in doing so we may have overlooked some important mechanisms that allow the plant

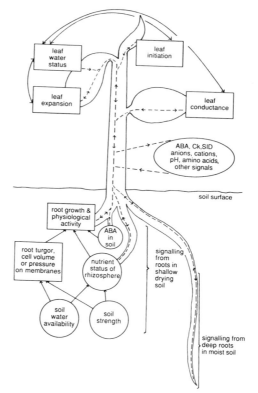

Figure 4 Diagrammatic representation of factors influencing the generation of chemical information (dotted lines) in roots in drying soil, the transfer of this information to leaves, and its
effects on shoot processes. Soil effects are shown as circles and plant physiological and
developmental processes are shown as rectangles. Hydraulic signaling is not shown, but it is
assumed that reduced water uptake will influence leaf water status, which will in turn have both
hydraulic (solid lines) and chemical effects on shoot processes. It is assumed that changes in
shoot processes will exert feedback effects on the roots, and therefore chemical information is
also shown moving from shoots to roots. This flux also represents accumulative signaling. For a
discussion of each of these effects and of the identity of possible chemical signals, see the text.

to regulate development as a function of soil water status. Our current
thinking in this area is summarized in Figure 4. It is clear that root signals of
the type described above may exert relatively dynamic control over stomatal
behavior. A major role for such signals, however, may be in the regulation of
development over an extended period. Thus the signals will not be less
effective in large trees with long transport pathways—trees that may have
relatively slow rates of xylem transport.

We have not suggested here that the water status of the shoot is never relevant to shoot functioning. Rather, we hold that the effects of shoot water status can be overridden and certainly modified by signals arising in the roots. Current evidence (20) suggests that these signals are chemical in nature, but we should keep an open mind to such other possibilities as action potentials, which have been involved in the coordination of whole plant responses to other stresses such as wounding, pathogens, or cold (19, 45, 79).

ACKNOWLEDGEMENTS

We are grateful to the Agricultural and Food Research Council for their financial support for some of the research reported in this review. We thank our colleagues for encouragement throughout the work and for helpful comments on the manuscript. WJD is grateful to Hans Meidner, Theodore Kozlowski, and Paul Kramer for early inspiration.

Literature Cited

1. Atkinson, C. J., Davies, W. J., Mansfield, T. A. 1989. Changes in stomatal conductance of intact aging wheat leaves in response to abscisic acid. *J. Exp. Bot.* 40:1021–28
2. Bates, L. M., Hall, A. E. 1982. Diurnal and seasonal responses of stomatal conductance for cowpea plants subjected to different levels of environmental drought. *Oecologia* 54:304–8
3. Bensen, R. J., Boyer, J. S., Mullet, J. E. 1988. Water deficit-induced changes in abscisic acid content, growth, polysomes, and translatable RNA in soybean hypocotyls. *Plant Physiol.* 88:289–94
4. Blackman, P. G., Davies, W. J. 1983. The effects of cytokinins and ABA on stomatal behaviour of maize and *Commelina. J. Exp. Bot.* 34:1619–26
5. Blackman, P. G., Davies, W. J. 1984. Age-related changes in stomatal response to cytokinins and ABA. *Ann. Bot.* 54:121–25
6. Blackman, P. G., Davies, W. J. 1985. Root to shoot communication in maize plants of the effects of soil drying. *J. Exp. Bot.* 36:39–48
7. Boyer, J. S. 1989. Water potential and plant metabolism: comments on Dr. P. J. Kramer's article, "Changing concepts regarding plant water relations". *Plant Cell Environ.* 12:213–16
7a. Boyer, J. S., Nonami, H. 1990. Xylem hydraulics, turgor and wall properties during growth. See Ref. 20, pp. 45–52
8. Bozarth, C. S., Mullet, J. E., Boyer, J. S. 1987. Cell wall proteins at low water potentials. *Plant Physiol.* 85:261–67
9. Bray, E. A. 1988. Drought- and ABA-induced changes in polypeptide and mRNA accumulation in tomato leaves. *Plant Physiol.* 88:1210–14
10. Bray, E. A. 1989. Gene expression during environmental stress and its regulation by abscisic acid. *Plant Growth Regulat. Soc. Am. Q.* 17:112–16
11. Burschka, C., Tenhunen, J. D., Hartung, W. 1983. Diurnal variations in abscisic acid content and stomatal response to applied abscisic acid in leaves of irrigated and non irrigated *Arbutus unedo* plants under naturally fluctuating environmental conditions. *Oecologia* 58:128–31
11a. Chapin, F. S. 1990. Effect of nutrient deficiency on plant growth: evidence for a centralized stress-response system. See Ref. 20, pp. 135–48
12. Claes, B., Dekeyser, R., Villarroel, R., Van den Bulcke, M., Bauw, G., et al. 1990. Characterisation of a rice gene showing organ-specific expression in response to salt stress and drought. *Plant Cell* 2:19–27
13. Cornish, K., Zeevaart, J. A. D. 1984. Abscisic acid metabolism in relation to water stress and leaf age in *Xanthium strumarium. Plant Physiol.* 76:1029–35
14. Cornish, K., Zeevaart, J. A. D. 1985. Abscisic acid accumulation by roots of *Xanthium strumarium* L. and *Lycopersicon esculentum* Mill in relation to water stress. *Plant Physiol.* 79:653–58
15. Cowan, I. R. 1982. Regulation of water use in relation to carbon gain in higher plants. In *Physiological Plant Ecology II*, ed. O. L. Lange, P. S. Novel, C. B. Osmond, H. Ziegler, pp. 589–614. Berlin: Springer-Verlag
16. Cowan, I. R., Raven, J. A., Hartung,

W., Farquhar, G. D. 1982. A possible role for abscisic acid in coupling stomatal conductance and photosynthetic carbon metabolism in leaves. *Aust. J. Plant Physiol.* 9:489–98

17. Creelman, R. A., Mason, H. S., Bensen, R. J., Boyer, J. S., Mullet, J. E. 1990. Water deficit and ABA cause differential inhibition of shoot vs. root growth in soybean seedlings. Analysis of growth, osmoregulation and gene expression. *Plant Physiol.* 92:205–14

18. Dale, J. E. 1988. The control of leaf expansion. *Annu. Rev. Plant Physiol. Plant Mol. Biol.* 39:267–95

19. Davies, E. 1987. Action potentials as multifunctional signals in plants: a unifying hypothesis to explain apparently disparate wound responses. *Plant Cell Environ.* 10:623–31

20. Davies, W. J., Jeffcoat, B., eds. 1990. *Importance of Root to Shoot Communication in the Response to Environmental Stress.* Monogr. 21. Bristol: Br. Soc. Plant Growth Regul. 398 pp.

21. Davies, W. J., Mansfield, T. A., Hetherington, A. M. 1990. Sensing of soil water status and the regulation of plant growth and development. *Plant Cell Environ.* 13:709–19

22. Dorffling, K., Peterson, W., Sprecher, E., Urvasch, I., Peterson, H. 1984. Abscisic acid in phytopathogenic fungi of the genera *Botrytis, Ceratocystis, Fusarium* and *Rhizoctona. Z. Naturforsch. Teil C* 39:683–84

23. Goliber, T. E. 1989. Regulation of leaf development in heterophyllous aquatic plants. See Ref. 86a, pp. 184–206

24. Gollan, T., Passioura, J. B., Munns, R. 1986. Soil water status affects the stomatal conductance of fully turgid wheat and sunflower leaves. *Aust. J. Plant Physiol.* 13:459–64

25. Gowing, D. J., Davies, W. J., Jones, H. G. 1990. A positive root-sourced signal as an indicator of soil drying in apple, *Malus* x *domestica* Borkh. *J. Exp. Bot.* 41:1535–40

26. Hartung, W. 1983. The site of action of abscisic acid at the guard cell plasmalemma of *Valerianella locusta. Plant Cell Environ.* 6:427–28

27. Hartung, W., Abou-Mandour, A. A. 1980. Abscisic acid in root cultures of *Phaseolus coccineus* L., *Z. Pflanzenphysiol.* 97:265–69

28. Hartung, W., Heilmeier, H., Wartinger, A., Kettemann, I., Schulze, E.-D. 1990. Ionic content and abscisic acid relations of *Anastatica hierochuntica* L. under arid conditions. *Isr. J. Bot.* 39: 373–82

29. Hartung, W., Radin, J. W. 1989. Abscisic acid in the mesophyll apoplast and the root xylem sap of water-stressed plants: the significance of pH gradients. See Ref. 86a, pp. 110–24

30. Hartung, W., Radin, J. W., Hendrix, D. L. 1988. Abscisic acid movement into the apoplastic solution of water-stressed cotton leaves: role of apoplastic pH. *Plant Physiol.* 86:908–13

31. Hartung, W., Slovik, S., Baier, M. 1990. pH Changes and redistribution of abscisic acid within the leaf under stress. See Ref. 20, pp. 215–36

32. Hirsch, R., Hartung, W., Gimmler, H. 1989. Abscisic acid content of algae under stress. *Bot. Acta* 102:326–34

33. Hsiao, T. C. 1973. Plant responses to water stress. *Annu. Rev. Plant Physiol.* 24:519–70

34. Huber, W., Sankhla, N. 1974. Abscisic acid-kinetin interaction in growth and activities of enzymes of amino-acid metabolism in *Pennisetum tyhoides* seedlings. *Z. Pflanzenphysiol.* 73:160–66

35. Hubick, K. T., Taylor, J. S., Reid, D. M. 1986. The effect of drought on levels of ABA, cytokinin, gibberellins and ethylene in aeroponically grown sunflower plants. *J. Plant Growth Regul.* 4:139–51

36. Incoll, L., Jewer, P.C. 1987. Cytokinins and stomata. In *Stomatal Function*, ed. E. Zeiger, G. D. Farquhar, I. R. Cowan, pp. 281–92. Stanford: Stanford Univ. Press

37. Incoll, L., Jewer, P.C. 1987. Cytokinins and the water relations of whole plants. In *Cytokinins: Plant Hormones in Search of a Role.* Monogr. 14, ed. R. Horgan, B. Jeffcoat, pp. 85–97. Bristol: Br. Plant Growth Regul. Group

38. Incoll, L., Ray, J. P., Jewer, P. C. 1990. Do cytokinins act as root to shoot signals? See Ref. 20, pp. 185–99

39. Itai, C., Vaadia, Y. 1965. Kinetin-like activity in root exudate of water stressed sunflower plants. *Physiol. Plant.* 18:941–44

40. Jackson, M. B. 1990. Communication between the roots and shoots of flooded plants. See Ref. 20, pp. 115–34

41. Jackson, M. B., Hall, K. C. 1987. Early stomatal closure in waterlogged pea plants is mediated by abscisic acid in the absence of foliar water deficits. *Plant Cell Environ.* 10:121–30

42. Jones, H. G. 1980. Interaction and integration of adaptive responses to water stress: the implications of an unpredictable environ. In *Adaptation of Plants to Water and High Temperature Stress*, ed.

N. C. Turner, P. J. Kramer, pp. 353–65. New York: Wiley

43. Jones, H. G. 1985. Physiological mechanisms involved in the control of leaf water status: implications for the estimation of tree water status. *Acta Hortic.* 171:291–96

44. Jones, H. G. 1987. Breeding for stomatal behaviour. In *Stomatal Function,* ed. E. Zeiger, G. D. Farquhar, I. R. Cowan, pp. 431–43. Stanford: Stanford Univ. Press

45. Jones, H. G. 1990. Control of growth and stomatal behaviour at the whole plant level: effects of soil drying. See Ref. 20, pp. 81–93

46. Jones, H. G., Luton, M. T., Higgs, K. H., Hamer, P. J. C. 1983. Experimental control of water status in an apple orchard. *J. Hortic. Sci.* 58:301–16

47. Kende, H. 1965. Kinetin-like factors in the root exudate of sunflowers. *Proc. Natl. Acad. Sci. USA* 53:1302–7

48. Knox, J. P., Galfre, G. 1986. Use of monoclonal antibodies to separate the enantiomers of abscisic acid. *Anal. Biochem.* 155:92–94

49. Kramer, P. J. 1969. *Plant and Soil Water Relationships: A Modern Synthesis.* New York/London/Toronto: McGraw-Hill. 482 pp.

50. Kramer, P. J. 1988. Changing concepts regarding plant water relations. *Plant Cell Environ.* 11:565–68

51. Kriedemann, P. E., Loveys, B. R., Fuller, G. L., Leopold, A. C. 1972. Abscisic acid and stomatal regulation. *Plant Physiol.* 49:842–47

52. Kulaeva, O. N. 1962. The effect of roots on leaf metabolism in relation to the action of kinetin on leaves. *Sov. Plant Physiol. (Engl. Transl. Fiziol. Rast.)* 9:182–89

53. Kutschera, U., Schopfer, P. 1986. *In vivo* measurement of cell-wall extensibility in maize coleoptiles: effect of auxin and abscisic acid. *Planta* 169:437–47

54. Lachno, D. R., Baker, D. A. 1986. Stress induction of abscisic acid in maize roots. *Physiol. Plant.* 68:215–21

55. Loveys, B. R. 1984. Diurnal changes in water relations and abscisic acid in field-grown *Vitis vinifera* cultivars. III. The influence of xylem-derived abscisic acid on leaf gas exchange. *New Phytol.* 98:563–73

56. Loveys, B. R., During, H. 1984. Diurnal changes in water relations and abscisic acid in field-grown *Vitis vinifera* cultivars. II. Abscisic acid changes under semi-arid conditions. *New Phytol.* 97:37–47

57. Loveys, B. R., Milborrow, B. V. 1984. Metabolism of abscisic acid. In *The Biosynthesis and Metabolism of Plant Hormones,* ed. A. Crozier, J. R. Hillmann. *Soc. Exp. Biol. Semin. Ser.* 23: 71–104. Cambridge: Cambridge Univ. Press

58. Loveys, B. R., Robinson, S. P. 1987. Abscisic acid synthesis and metabolism in barley leaves and protoplasts. *Plant Sci.* 49:23–30

59. Loveys, B. R., Robinson, S. P., Downton, W. J. S. 1987. Seasonal and diurnal changes in abscisic acid and water relations of apricot leaves. *New Phytol.* 107:15–27

60. Ludlow, M. M., Sommer, K. J., Flower, D. J., Ferraris, R., So, H. B. 1989. Influence of root signals resulting from soil dehydration and high soil strength on the growth of crop plants. See Ref. 86a, pp. 81–99

61. Ludlow, M. M., Sommer, K. J., Muchow, R. C. 1990. Agricultural implications of root signals. See Ref. 20, pp. 251–67

62. Malik, R. S., Dhankar, J. S., Turner, N. C. 1979. Influence of soil water deficits on root growth of cotton seedlings. *Plant Soil* 53:109–15

63. Masle, A., Passioura, J. B. 1987. Effect of soil strength on the growth of young wheat plants. *Aust. J. Plant Physiol.* 14:643–56

64. Matsuda, K., Riazi, A. 1981. Stress-induced osmotic adjustment in growing regions of barley leaves. *Plant Physiol.* 68:571–76

65. Meidner, H. 1975. Water supply, evaporation and vapour diffusion in leaves. *J. Exp. Bot.* 26:666–73

66. Meinzer, F. C., Grantz, D. A. 1990. Stomatal and hydraulic conductance in growing sugarcane: stomatal adjustment to water transport capacity. *Plant Cell Environ.* 13:383–88

67. Meinzer, F. C., Grantz, D. A., Smit, B. 1990. Stomatal behaviour in growing sugarcane in relation to factors in the transpiration stream. See Ref. 20, pp. 315–16

68. Michelena, V. A., Boyer, J. S. 1982. Complete turgor maintenance at low water potentials in the elongating regions of maize leaves. *Plant Physiol.* 69:1145–49

69. Muller, M., Deigele, C., Ziegler, H. 1989. Hormonal interactions in the rhizosphere of maize (*Zea mays* L.) and their effects on plant development. *Z. Pflanzenernaehr. Bodenkd.* 152:247–54

70. Munns, R. 1990. Chemical signals mov-

ing from roots to shoots: the case against ABA. See Ref. 20, pp. 175–84

71. Munns, R., King, R. W. 1988. Abscisic acid is not the only stomatal inhibitor in the transpiration stream. *Plant Physiol* 88:703–8

72. Neales, T. F., Masia, A., Zhang, J., Davies, W. J. 1989. The effects of partially drying part of the root system of *Helianthus annuus* on the abscisic acid content of the roots, xylem sap and leaves. *J. Exp. Bot.* 40:1113–20

73. Nonami, H., Boyer, J. S. 1988. Wall extensibility at low water potentials. *Plant Physiol.* 86:S34

74. Ogunkanmi, A. B., Tucker, D. J., Mansfield, T. A. 1973. An improved bio-assay for abscisic acid and other antitranspirants. *New Phytol.* 72:277–82

75. Passioura, J. B. 1983. Roots and drought resistance. *Agric. Water Manage.* 7:265–80

76. Passioura, J. B. 1988. Response to Dr. P. J. Kramer's article, "Changing concepts regarding plant water relations". *Plant Cell Environ.* 11:569–71

77. Passioura, J. B. 1988. Root signals control leaf expansion in wheat seedlings growing in drying soil. *Aust. J. Plant Physiol.* 15:687–93

78. Passioura, J. B., Munns, R. 1984. Hydraulic resistance of plants. II. Effects of rooting medium, and time of day, in barley and lupin. *Aust. J. Plant Physiol.* 11:341–50

79. Pickard, B. G. 1973. Action potentials in higher plants. *Bot. Rev.* 39:172–201

80. Pritchard, J., Barlow, P. W., Adam, J. S., Tomos, A. D. 1990. Biophysics of the inhibition of the growth of maize roots by lowered temperature. *Plant Physiol.* 93:222–30

81. Pritchard, J., Wyn Jones, R. G., Tomos, A. D. 1990. Measurement of yield threshold and cell wall extensibility of intact wheat roots under different ionic, osmotic and temperature treatments. *J. Exp. Bot.* 41:669–75

82. Quarrie, S. A. 1983. Genetic differences in abscisic acid physiology and their potential uses in agriculture. In *Abscisic Acid*, ed. F. T. Addicott, pp. 356–420. New York: Praeger

83. Quarrie, S. A. 1990. Water stress proteins and abscisic acid. See Ref. 20, pp. 13–28

84. Quarrie, S. A., Jones, H. G. 1977. Effects of abscisic acid and water stress on development and morphology in wheat. *J. Exp. Bot.* 28:182–203

85. Radin, J. W. 1984. Stomatal responses to water stress and to abscisic acid in phosphorus-deficient cotton plants. *Plant Physiol.* 76:392–94

86. Radin, J. W., Parker, L. L., Quinn, G. 1982. Water relations of cotton plants under nitrogen deficiency. V. Environmental control of abscisic acid accumulation and stomatal sensitivity to abscisic acid. *Plant Physiol.* 70:1066–70

86a. Randall, D. D., Blevins, D. G., eds. 1989. *Current Topics in Plant Biochemistry and Physiology 1989*, Vol. 8. Columbia: Univ. Missouri-Columbia

87. Robertson, J. M., Pharis, R. P., Huang, Y. Y., Reid, D. M., Yeung, E. C. 1985. Drought-induced increases in abscisic acid levels in the root apex of sunflower. *Plant Physiol.* 79:1086–89

88. Saab, I. N., Sharp, R. E. 1989. Non-hydraulic signals from maize roots in drying soil: inhibition of leaf elongation but not stomatal conductance. *Planta* 179:466–74

89. Schulze, E.-D. 1986. Carbon dioxide and water vapour exchange in response to drought in the soil. *Annu. Rev. Plant Physiol.* 37:247–74

90. Schulze, E.-D., Steudle, E., Gollan, T., Schurr, U. 1988. Response to Dr. P. J. Kramer's article, "Changing concepts regarding plant water relations". *Plant Cell Environ.* 11:573–76

91. Schurr, U., Gollan, T. 1990. Composition of xylem sap of plants experiencing root water stress—a descriptive study. See Ref. 20, pp. 201–14

92. Shaner, D. L., Boyer, J. S. 1976. Nitrate reductase activity in maize leaves. II. Regulation by nitrate flux at low leaf water potential. *Plant Physiol.* 58:555–59

93. Sharp, R. E. 1990. Comparative sensitivity of root and shoot growth and physiology to low water potentials. See Ref. 20, pp. 29–44

94. Sharp, R. E., Davies, W. J. 1979. Solute regulation and growth by roots and shoots of water stressed maize plants. *Planta* 147:43–49

95. Sharp, R. E., Davies, W. J. 1989. Regulation of growth and development of plants growing with a restricted supply of water. In *Plants under Stress*, ed. H. G. Jones, T. J. Flowers, M. B. Jones. *Soc. Exp. Biol. Semin. Ser.* 39:71–93. Cambridge: Cambridge Univ. Press

96. Sharp, R. E., Hsiao, T. C., Silk, W. K. 1990. Growth of the maize primary root at low water potentials. II. The role of growth and deposition of hexose and potassium in osmotic adjustment. *Plant Physiol.* 93:1337–46

97. Sharp, R. E., Silk, W. K., Hsiao, T. C.

1988. Growth of the maize primary root at low water potentials. I. Spatial distribution of expansive growth. *Plant Physiol.* 87:50–57

98. Slatyer, R. O. 1967. *Plant Water Relationships.* New York: Academic

99. Slatyer, R. O., Taylor, S. A. 1960. Terminology in plant and soil-water relations. *Nature* 187:992

100. Smith, P. G., Dale, J. E. 1988. The effects of root cooling and excision treatments on the growth of primary leaves of *Phaseolus vulgaris* L. Rapid and reversible increases in abscisic acid content. *New Phytol.* 110:293–300

101. Snaith, P. J., Mansfield, T. A. 1982. Stomatal sensitivity to abscisic acid: can it be defined? *Plant Cell Environ.* 5:309–11

102. Stewart, P. A. 1981. *How to Understand Acid-Base. A Quantitative Acid-Base Primer for Biology and Medicine.* London: Edward Arnold

103. Stewart, P. A. 1983. Modern quantitative acid-base chemistry. *Can. J. Physiol. Pharmacol.* 61:1444–61

104. Tardieu, F., Katerji, N., Bethenod, O., Zhang, J., Davies, W. J. 1991. Maize stomatal conductance in the field: its relationship with soil and plant water potentials, mechanical constraints and root messages. *Plant Cell Environ.* In press

105. Thomas, A., Tomos, A. D., Stoddart, J. L., Thomas, H., Pollock, C. J. 1989. Cell expansion rate, temperature and turgor pressure in growing *Lolium temulentum* leaves. *New Phytol.* 112:1–5

106. Thomas, T. H., Wareing, P. F., Robinson, P. M. 1965. Action of the sycamore 'dormin' as a gibberellin antagonist. *Nature* 205:1279

107. Turner, N. C. 1974. Stomatal response to light and water under field conditions. *R. Soc. NZ Bull.* 12:423–32

108. Turner, N. C., Schulze, E.-D., Gollan, T. 1985. The responses of stomata and leaf gas exchange to vapour pressure deficit and soil water content. II. In the mesophytic herbaceous species *Helianthus annuus. Oecologia* 65:348–55

109. Van Volkenburgh, E., Davies, W. J. 1983. Inhibition of light stimulated leaf expansion by abscisic acid. *J. Exp. Bot.* 34:835–45

110. Walton, D. C., Harrison, M. A., Cote, P. 1976. The effects of water stress on abscisic acid levels and metabolism in roots of *Phaseolus vulgaris* L. and other plants. *Planta* 131:141–44

111. Wartinger, A., Heilmeier, H., Hartung, W., Schulze, E.-D. 1991. Daily and seasonal courses of leaf conductance and abscisic acid in the xylem sap of almond trees *(Prunus dulcis)* under desert conditions. *New Phytol.* In press

112. Went, F. W., Bonner, D. M. 1943. Growth factors controlling stem growth in darkness. *Arch. Biochem.* 1:439–52

113. Westgate, M. E., Boyer, J. S. 1985. Osmotic adjustment and the inhibition of leaf, root, stem, and silk growth at low water potentials in maize. *Planta* 164:540–49

114. Young, J. P., Dengler, N. G., Horton, R. F. 1987. Heterophylly in *Ranunculus flabellaris:* the effect of abscisic acid on leaf anatomy. *Ann. Bot.* 69:117–25

115. Zeevaart, J. A. D. 1977. Sites of ABA synthesis and metabolism in *Ricinus communis* L. *Plant Physiol.* 59:788–91

116. Zeevaart, J. A. D., Boyer, G. L. 1984. Accumulation and transport of abscisic acid and its metabolites in *Ricinus and Xanthium. Plant Physiol.* 74:934–39

117. Zeevaart, J. A. D., Creelman, R. A. 1988. Metabolism and physiology of abscisic acid. *Annu. Rev. Plant Physiol. Plant Mol. Biol.* 39:439–73

118. Zhang, J., Davies, W. J. 1987. Increased synthesis of ABA in partially dehydrated root tips and ABA transport from roots to leaves. *J. Exp. Bot.* 38:2015–23

119. Zhang, J., Davies, W. J. 1989. Abscisic acid produced in dehydrating roots may enable the plant to measure the water status of the soil. *Plant Cell Environ.* 12:73–81

120. Zhang, J., Davies, W. J. 1989. Sequential responses of whole plant water relations towards prolonged soil drying and the mediation by xylem sap ABA concentrations in the regulation of stomatal behaviour of sunflower plants. *New Phytol.* 113:167–74

121. Zhang, J., Davies, W. J. 1990. Changes in the concentration of ABA in xylem sap as a function of changing soil water status will account for changes in leaf conductance. *Plant Cell Environ.* 13:277–85

122. Zhang, J., Davies, W. J. 1990. Does ABA in the xylem control the rate of leaf growth in soil-dried maize and sunflower plants. *J. Exp. Bot.* 41:1125–32

123. Zhang, J., Davies, W. J. 1991. Antitranspirant activity in the xylem sap of maize plants. *J. Exp. Bot.* In press

124. Zhang, J., Schurr, U., Davies, W. J. 1987. Control of stomatal behaviour by abscisic acid which apparently originates in roots. *J. Exp. Bot.* 38:1174–81

Annu. Rev. Plant Physiol. Plant Mol. Biol. 1991. 42:77–101

FRUCTAN METABOLISM IN GRASSES AND CEREALS

C. J. Pollock and A. J. Cairns

Environmental Biology Department, AFRC Institute of Grassland and Environmental Research, Welsh Plant Breeding Station, Plas Gogerddan, Aberystwyth, Dyfed SY23 3EB, Wales, UK

KEY WORDS: sucrose, oligosaccharide, glycosyl transferase, vacuole, assimilate partitioning

CONTENTS

INTRODUCTION

Although starch is the major chemical species harvested from the grains of temperate cereals, for much of the growing season it forms only a minor component of the total reserve carbohydrate within the plant. Prior to an-

1040-2519/91/0601-0077$02.00

thesis, the leaves, stems, and apices accumulate soluble carbohydrates, often to extremely high concentrations. These carbohydrates consist of a mixture of monosaccharides, sucrose, and polymers of fructose (fructans).[1] Fructans have been recognized as a distinct class of carbohydrates for over 150 years (88), and their chemistry, biochemistry, and physiology have been studied intermittently over that period. Prior to 1969, a number of reviews were published covering the chemistry and physiology of fructans extracted from both monocots and dicots (2, 27, 44, 63, 90). Interest in the biochemistry, enzymology, and cell biology of these compounds has led to a substantial body of newer work summarized in recent reviews (41, 66, 69, 75, 78, 83). However, much of the work described in these reviews was performed on the fructans accumulated by members of the *Asterales* and *Liliales* rather than by the Gramineae.

Our review concentrates upon the specific aspects of fructan metabolism associated with their occurrence in grasses and cereals; it emphasizes recent studies. The distinctions between the ways fructan is metabolized in different genera are highlighted, and current areas of uncertainty are discussed critically. The methodology associated with studies of sucrose and fructan metabolism has also been reviewed recently (3, 82) and is not covered in detail here.

OCCURRENCE OF FRUCTANS WITHIN THE GRAMINEAE

Although chemotaxonomic surveys of plant distribution are widespread, comparatively few studies within the Gramineae have concentrated upon their carbohydrates. Investigators using a limited range of species have suggested that fructans were restricted to C3 or cool-season grasses but that not all C3 grasses contained fructans (8, 71, 103, 104). It was also proposed, based on these and other more restricted observations, that fructan accumulation in vegetative organs was associated with high concentrations of sucrose and low concentrations of starch (76, 123). In a recent more detailed survey, Chatterton et al (17) examined 185 grass accessions grown under controlled environmental conditions using two temperature regimes. Fructans were absent from C4 species (or were only present at very low concentrations), nor did C3 grasses from the genera *Danthonia, Stipa, Oryzopsis,* and *Phragmities,* thought to be of Gondwanic origin (6, 21), accumulate much fructan. This research demonstrated that fructan and starch were not necessarily alternative forms of carbohydrate storage in grasses and that fructan accumulation was not always associated with high total sucrose content. For example, *Agropyron cristatum,* when grown at low temperatures, was observed to contain 30% fructan, 13% sucrose, and 26% starch (expressed on a structural dry weight

---

[1]Here we adopt the recommendations on nomenclature of the Fructan Working Group (113).

basis). Fractionation by thin-layer chromatography (TLC) of fructans from a range of Gramineae indicated some similarities in distribution at the generic, sub-tribal, and tribal levels (105).

Precise analytical methods are required to distinguish among sucrose, low-DP (degree of polymerization) fructo-oligosaccharides, and other soluble oligosaccharides such as the raffinose series (78). Errors in attribution are likely to be more frequent when the soluble carbohydrate contents are low and when modern chromatographic techniques have not been employed (76). Such methodological problems may be the cause of differences among reports concerning the minor occurrence of fructans in specific genera (8, 103). Recently Smouter & Simpson (105) reported that representatives of tropical genera accumulate some fructans under conditions where export of sucrose from leaves is blocked, suggesting that the physiological state of the tissue at harvest may also affect determinations of occurrence.

STRUCTURES OF FRUCTANS FROM GRASSES AND CEREALS

Early studies on fructans from a variety of sources showed that they were acid-labile, water-soluble, polydisperse polymers that yielded only fructose and a trace of glucose following acid hydrolysis (63, 66). Development of paper chromatography and the techniques of methylation analysis (44) led to progressively better understanding of fructan structures, summarized in the review by Edelman & Jefford (27) concerning the fructans of *Helianthus tuberosus,* a member of the Asterales. In this species, the native polymer consisted predominantly of an homologous series of linear β-1,2-linked oligofructosides connected to a terminal sucrosyl moiety (50). The maximum size of the linear chains appeared to be about 35 hexose residues and, by virtue of the linkage pattern, the polymer was nonreducing. With minor exceptions (16) later studies have confirmed this structural identification, but recent work has also demonstrated that fructans with this simplicity and regularity of structure are rare, particularly among those extracted from temperate Gramineae. Initial studies (summarized in 44) showed that the predominant linkage pattern between adjacent fructose residues in the higher-molecular-weight fructans from grasses was β-2,6. Because of the preponderance of alternative linkage patterns, fructans from members of the Asterales (mainly 2,1 linked) are often referred to as inulins and those from grasses (predominantly 2,6 linked) as levans (by analogy with bacterial fructans; 78).

Estimates of mean molecular size have shown large differences among the fructans from different species of temperate grasses (Table 1). In species such as *Dactylis glomerata* and *Phleum pratense* that accumulate fructans of high

Table 1 Variation in mean molecular size of fructans from different species[a]

Species	Mean molecular size (kD)	Method employed
Gramineae		
Phleum pratense	42	size-exclusion chromatography
Phleum pratense	46	end-group analysis
Lolium perenne	12	size-exclusion chromatography
Festuca pratensis	12	size-exclusion chromatography
Dactylis glomerata	20	size-exclusion chromatography
Bromus inermis	5	end-group analysis
Liliales		
Allium sativum	9	size-exclusion chromatography
Asterales		
Helianthus tuberosus	5	end-group analysis

[a] Data summarized from reference 78.

molecular weight, it is often difficult to detect the full range of oligosaccharides intermediate in size between sucrose and the high-molecular-weight polymer (72, 114). Results obtained by the use of anion exchange chromatography coupled with pulsed amperometric detection suggest that such oligosaccharides are present, although in low concentrations (19). However, genera such as *Lolium, Festuca, Hordeum,* and *Triticum,* with smaller fructans, do contain appreciable amounts of oligosaccharide material (73, 78, 104). More detailed recent analyses of such oligosaccharides using high-performance liquid-chromatography (HPLC) and high-resolution TLC have revealed considerable additional complexity, and a number of different isomeric oligosaccharides have been observed (5, 13, 47, 105, 107, 117).

   Three isomers of monofructosyl sucrose have been described (1, 4, 39), and it is thought that these form the simplest members of three distinct homologous linear series of fructans. The series based on 1-kestose [O-α-D-glucopyranosyl-(1-2)-O-β-D-fructofuranosyl-(1-2)-D-fructofuranoside] predominate in the fructans of the Asterales, but 1-kestose has also been isolated from many grasses (5, 13, 16, 117) and appears to be the most rapidly labeled fructan formed after exposure of leaves of *Lolium temulentum* to[14]CO_2 (73).6-Kestose[O-α-D-glucopyranosyl-(1-2)-O-β-D-fructofuranosyl-(6-2)-D-fructofuranoside] can be considered the parent structure of the 2,6-linked fructans which are the principal component of the high-molecular-weight fructans of many temperate grasses (7, 16, 56, 58, 79, 90, 115). The third trisaccharide, neo-kestose [O-β-D-fructofuranosyl-(2-6)-O-α-D-glucopyranosyl-(1-2)-D-fructofuranoside] contains an included glucose residue, giving rise to the possibility of chain elongation in both directions and with different linkage patterns. This class of fructans predominates in *Asparagus officianalis* (100). Branched fructans have also been isolated from grasses. They are common in cereal leaves, stems, and grains (65, 90). Such fructans probably have backbones based upon both the 1-kestose and 6-kestose series (66).

The development of high-resolution TLC (13, 55, 105, 107, 117) has allowed the heterogeneity of the oligosaccharides from different species to be visualized. Figure 1 shows the patterns observed when leaf fructans extracted from five different species of temperate grasses are separated, together with the oligosaccharides extracted from the tubers of *H. tuberosus*. The simplicity of the pattern in fructans from *H. tuberosus* is immediately apparent, and it has been shown that the oligosaccharide mobilities observe the mathematical relationships expected of an homologous series (35). The grass fructans, however, show much more complexity, together with substantial differences among species. This complexity of oligosaccharide structure in grasses has been confirmed by other studies using TLC (105, 107), nuclear magnetic resonance spectroscopy (31), and anion exchange chromatography (19). Variation in the relative abundances of different oligosaccharides is clearly visible, as are differences in whether the three trisaccharide isomers are present or absent. However, 1-kestose is present in all samples. Isomeric oligofructosides have recently been separated using reverse-phase HPLC (52); and a combination of size-exclusion chromatography, reverse-phase HPLC, and TLC has been used to separate the individual fructans from leaves of *L. temulentum* up to the pentasaccharide level. Methylation analysis, followed by separation and analysis of partially methylated alditol acetates by gas chromatography/mass spectrometry has permitted structures to be assigned to each of the isomers. The results are summarized in Table 2. Members of all the series described above are present in different amounts, and all become labeled throughout the molecule if $^{13}CO_2$ is administered during the course of fructan accumulation (I. M. Sims, R. Horgan, C. J. Pollock, unpublished observations). A range of complementary methods have shown the high-molecular-weight fructans produced in *Lolium*, *Festuca*, and *Dactylis* to be predominantly 2,6-linked linear molecules (16, 56, 79, 115). It must be assumed, therefore, that the proportion of such linkages in the oligosaccharide pool increases with increasing DP, but this has not yet been demonstrated unequivocally. Anion exchange chromatography, allied with pulsed amperometric detection, will permit resolution of individual oligosaccharides up to at least DP30 (19); this resolution, coupled with micro-methods of methylation analysis, should allow this problem to be addressed.

The significance of this complexity, and the marked species variation within it, is obscure, particularly when it is contrasted with the simplicity of

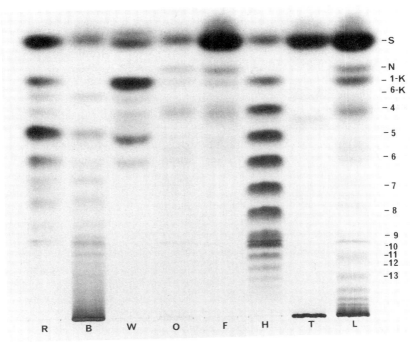

Figure 1   Thin-layer chromatographic separations of fructans from leaves of various grass species and from tubers of *H. tuberosus*. Techniques for extraction, purification, and separation are as described in Reference 13. Vertical symbols indicate mobilities of individual compounds as follows: S: Sucrose; N: Neokestose; 1-K: 1-Kestose; 6-K: 6-kestose; 4 to 13: individual isomeric oligosaccharides of appropriate DP from *H. tuberosus*. Horizontal symbols correspond to plant species as follows: R: rye; B: barley; W: wheat; O: oats; F: fescue; H: *H. tuberosus;* T: timothy; L: *L. temulentum*.

fructan structures observed in *H. tuberosus* and in other Asterales (51). It would be of considerable interest to use such sensitive indicators of structural heterogeneity as chemotaxonomic aids within the Gramineae and to compare the data thus obtained with other taxonomic observations.

Little is known of the interrelationships between fructan structure and the physical properties of the molecules, either as solids or in solution. Crystal structures have been assigned to 1- and 6-kestose (30, 53), and a range of helical structures for longer chains are feasible in solution (33, 62). Levans favor a left-handed conformation whereas inulins may assume a range of both left- and right-handed conformations with only small amounts of steric strain (34). The effects of a low percentage of branching on such structures have not been elucidated. Empirical observations suggest that the high degree of water

Table 2 Defined chemical structures separated and characterized from the pentasaccharide fraction of excised illuminated leaves of *L. temulentum*[a]

Structure	Parent trisaccharide		Number of isomers
G-1,(2-F-6)$_3$,2-F	6-kestose		1
G-1,(2-F-1)$_3$,2-F	1-kestose		1
F-2,(1-F-2)$_2$,6-G-1,2-F	neokestose		1
F-2,6-G-1,(2-F-1)$_2$,2-F	neokestose	2,1 linked	1
F-2,1-F-2,6-G-1,2-F-1,2-F	neokestose		1
F-2,(6-F-2)$_2$,6-G-1,2-F	neokestose		1
F-2,6-G-1,(2-F-6)$_2$,2-F	neokestose	2,6 linked	1
F-2,6-F-2,6-G-1,2-F-6,2-F	neokestose		1
F-G-(F)$_2$-F	neokestose	unresolved	
F-(F$_2$)-G-F	neokestose	(mixed 2,1/	5
F-F-G-F-F	neokestose	2,6 linked)	
Total			11

[a] Leaves were harvested 18 hr after excision. Neutral soluble carbohydrates were fractionated by size-exclusion chromatography and the pentasaccharide fraction further separated by reverse-phase HPLC. Purified oligosaccharides were methylated and analyzed by gas chromatography/mass spectrometry (I. M. Sims, R. Horgan, C. Pollock, unpublished).

solubility observed, particularly among levans, would be characteristic of a flexible polymer chain with a relatively low frequency of interchain hydrogen bonding.

PHYSIOLOGY OF FRUCTAN ACCUMULATION IN GRAMINEAE

Fructans function as accessible reserve carbohydrates. Fructans can be stored at different sites within the plant, and very different patterns of synthesis and degradation occur under the influence of a range of external and internal factors. The principle distinction that can be made is between storage (*a*) in primary heterotrophic organs, such as leaf sheaths, shoots, roots, stems, and grains, and (*b*) in autotrophic organs such as leaves and some floral tissues (76, 78). Storage in leaves is intimately connected with the synthesis and export of sucrose, and consequently with chloroplast carbon metabolism; in heterotrophic sinks, on the other hand, synthesis is from imported carbon and thus is less affected by short-term environmental fluctuations (78).

Seasonal Accumulation in Stems, Leaf Sheaths, and Roots

Because of the large mass of tissue involved, accumulation in cereal and grass stems during flower development is the most readily detected physiological correlate of fructan metabolism. Final fructan concentrations can be as high as 30% of the dry weight (2, 80, 104), with a gradient of accumulation between the apex and the base (80, 104). Accumulation continues during stem growth, flowering, and anthesis; fructan contents then fall during the later stages of grain filling (2, 9, 11). Disappearance of fructan is almost complete in cereal stems (2, 9) and in summer-dormant pasture grasses where the above-ground portions die (4a), but high concentrations of fructan persist in the lower stem internodes of perennial temperate grasses where nodal buds sprout to form vegetative tillers in the fall (80). The fate of the fructan mobilized from stems during grain filling is still debated (111). It is probable that some, at least, can be used to sustain grain growth during periods where flag leaf photosynthesis is limited (9, 11, 40, 57, 111). The woody nature of cereal and grass stems makes them poor subjects for biochemical analysis; the few studies of the regulation of fructan metabolism in such tissues have been concerned principally with fructan breakdown (68, 102, 124), although enzymological measurements have been made recently during fructan accumulation in wheat stems, and both genotypic and environmental differences have been observed (24). Long-term labeling experiments, possibly using ^{13}C rather than ^{14}C, could provide more detailed information on the timing of synthesis and breakdown in the various tissues of the stem, but the production of specific antibodies and cDNA sequences for enzymes of fructan metabolism, purified from leaves or from other less recalcitrant tissues, probably offers the best opportunity for characterizing these important processes at the biochemical and molecular level. In perennial species accumulation (predominantly in the leaf sheaths) continues throughout the fall and early winter (64, 80). Fructan concentrations decline, with a transient accumulation of sucrose, prior to the onset of rapid spring growth (80). Fructans can be isolated from the roots of temperate grasses (22, 64, 108). Although few studies have been done of the seasonal patterns of synthesis and degradation in these organs, we know that accumulation continues throughout the winter in roots of *Lolium* species. The highest mean molecular weights of root fructans were, however, observed during the summer (64).

Long-Term Storage in Reproductive Tissues

Mature seeds of grasses contain small amounts of fructan (61). During the initial stages of grain filling, the concentration of fructan is much higher (29, 92). Ho & Gifford (45) claimed that this accumulation occurred in the endospermic sap of wheat grains, but more recent evidence suggests that the compartmentation depends upon the stage of kernel development (92). Observations of Housley & Daughtry (46) suggest that active fructan synthesis occurs early in grain development but declines in significance as starch synthesis becomes established, leaving a static fructan pool to form a declining proportion of the total grain weight. The physiological significance of this process is not clear, although it has been suggested that fructan synthesis would help to maintain a concentration gradient of sucrose between phloem and sink tissue while minimizing the accumulation of osmotically active solutes (78).

Accumulation in Growth Zones

Grass and cereal stubbles often show high concentrations of soluble carbohydrates (104). While mature leaf sheaths contain fructan (see above), high concentrations may also accumulate in the basal parts of growing leaves which will be enclosed by mature sheaths (49, 74, 107). In a series of detailed studies (91, 93–95), Schnyder et al observed a spatial gradient of fructan content within the extension zone. This gradient, which generally correlated with the rate of water uptake and the relative elongation rate within the extension zone, was stable under specified experimental conditions, although irradiance levels and diurnal transitions altered the pattern of carbohydrate deposition and utilization. Because the leaves of grasses and cereals grow from a basal intercalary meristem, individual cells pass through the extension zone with transit times in the order of 24 hr. The occurrence of a stable fructan gradient in the presence of moving cells indicates that the fructan contents of individual cells must be changing rapidly, and the data have been used to calculate the magnitude of the fluxes occurring during these processes. Rates of net fructan deposition as high as 6 μg mg^{-1} tissue water hr^{-1} were observed. To a rough approximation, this is equivalent to 6 mg g^{-1} fresh weight hr^{-1}, which compares closely with net accumulation rates of 5 mg g^{-1} fresh weight hr^{-1} observed in excised leaves of *L. temulentum* (48). Both these processes are much faster than the net rate of fructan accumulation in long-term sinks such as flowering stems of grasses and cereals (< 1 mg g^{-1} fresh weight day^{-1}; 80). The metabolic control of fructan metabolism within the extension zone is likely to be of considerable significance in determining overall patterns of leaf growth and cell wall deposition. Unfortunately the masses of tissue involved are very small, and suitable techniques for measuring the relevant enzyme activities have not been developed (78). The availability of specific antisera or cDNA sequences would permit further study of this interesting and potentially important manifestation of fructan metabolism. The role of fructan accumulation in this system is not immediately clear. Fructan synthesis will again enhance the gradient of sucrose concentration between phloem and young cells and also minimize the osmotic consequences of carbon storage (78), but it would appear that the same problems are not encountered by C4 grasses such as *Zea mays* or *Sorghum bicolor,* where fructans cannot be detected in the extension zone (C. J. Pollock, unpublished observations).

Fructan Metabolism in Leaves

PATTERNS OF ACCUMULATION   Field measurements of the fructan contents of mature leaf blades generally yield lower concentrations than those present in sheaths and stems (80). Interest in leaf fructan metabolism has, however, increased following (a) studies on the effects of low temperature on fructan accumulation in vegetative tillers of perennial grasses (75, 81) and (b) the observation that net rates of carbohydrate accumulation can be drastically increased by excising and illuminating leaves of grasses and cereals (48, 117). These experiments appear to confirm the suggestions of Archbold and others (2, 10, 75) that fructan accumulation is stimulated by treatments that lead to a reduction in demand for fixed carbon (such as chilling, withholding mineral nutrients, applying growth retardants, etc) or by treatments that lead to an increased supply of carbon (such as increased leaf area, photoperiod, irradiance, CO_2 concentration etc). The principle virtue of the excised grass leaf, however, has been to establish a convenient model system in which both the biosynthesis of fructan and the concomitant alteration in the patterns of gene expression can be studied easily and reproducibly. All species of temperate grasses hitherto tested will accumulate large amounts of fructans if hydrated excised leaves are illuminated, and these systems generally produce fructans similar in size, distribution, and oligosaccharide composition to the fructans that accumulate in attached leaves as a result of other, less destructive, treatments (C. J. Pollock, A. J. Cairns, unpublished observations).

If leaves are illuminated under controlled conditions, rates of total soluble carbohydrate accumulation are constant for up to 72 hr following excision (48). Initially this accumulation is almost entirely as sucrose, but after a lag period, trisaccharides accumulate as (subsequently) do higher-order oligosaccharides and high-molecular-weight fructans (Figure 2). Total tissue carbohydrate contents can reach 60–70% of dry weight, with up to 60% of this as fructan; but sucrose concentrations stabilize after a few hours of incubation. In all cases examined in detail, using either mass or radioactivity fixed from $^{14}CO_2$ as a measure of fructan synthesis, 1-kestose was the primary trisaccharide intermediate formed (13, 47, 117). This is also true in experiments on attached leaves held at low temperature (20, 73). The kinetics of movement of both mass and radioactivity into higher-molecular-weight fructans generally fit the precursor-product relationship suggested by Edelman & Jefford (27). Under this model, chain elongation from sucrose occurs by successive attachment of fructose residues, each oligosaccharide being, therefore, the immediate precursor of the next largest. Suggestions have been made, however, that some oligofructans that accumulate during net fructan synthesis may arise by specific breakdown of higher-order forms (87).

That sucrose is the major intermediate was confirmed when it was demon-

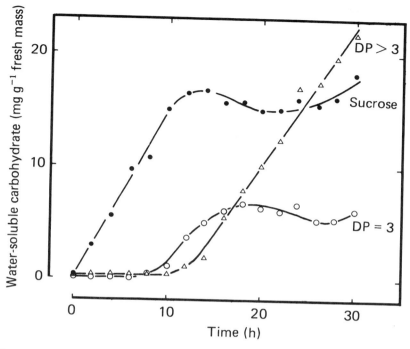

Figure 2 Accumulation of water-soluble carbohydrate fractions by excised leaves of *L. temulentum* after various periods of illumination (reproduced from 13 with permission). DP=3 refers to total trisaccharides; DP>3 refers to total fructo-oligosaccharides with a degree of polymerization greater than 3.

strated that fructan synthesis could be induced in darkness by feeding excised leaves with sucrose (48, 119). The process is delayed if fructose is administered rather than sucrose; under these circumstances, the initiation of fructan accumulation is preceded by the synthesis of sucrose from the fed hexose (A. J. Cairns, C. J. Pollock, unpublished observations). In these studies, sugar was administered by allowing the leaves to transpire while the bases were immersed in a sugar solution. Similar techniques have been used to administer specific inhibitors of protein synthesis or nucleic acid metabolism (14, 119). Inhibitors effectively blocked the conversion of sucrose into fructan without affecting the net rate of carbohydrate accumulation. This observation suggests that the primary processes of CO_2 fixation and sucrose biosynthesis were unaffected but indicates that fructan synthesis itself was dependent upon altered patterns of gene expression. These alterations could be associated either with changes in the relevant synthetic enzymes or, perhaps, in the amounts of some proteinaceous modulator (14).

SITES OF ACCUMULATION   Early microscopic studies suggested that the vacuole was the most likely site of fructan accumulation in plants (27, 76). More recently, this has been confirmed by Weimken and coworkers for leaf tissue of *H. vulgare* and tuber tissue of *H. tuberosus* (32, 117). In these studies, significant amounts of sucrose were shown to occur both in the vacuole and elsewhere, whereas trisaccharides and higher-order fructans were largely confined to the vacuole. It should be borne in mind that current cell fractionation techniques are insufficiently precise to determine whether small discrepancies in fructan recovery during vacuole release are due to leakiness of the vacuoles or to the natural occurrence of a small proportion of oligosaccharides outside the vacuole. However, isolated vacuoles from leaf mesophyll cells of *H. vulgare* appear not to be permeable in vitro to 1-kestose (E. Martinoia, U. Heck, unpublished observations).

It is tacitly assumed that, within the leaf, fructans accumulate predominantly in mesophyll cells. However, mechanically isolated epidermal tissue from the upper surface of leaves of *L. temulentum* contains significant quantities of fructan (J. E. Ashton, C. J. Pollock, unpublished observations). In addition, non-uniformity of distribution of fructans along the length of the leaf has been observed (120), and there is independent evidence to suggest that heterogeneity exists within the photosynthetic tissue of cereal leaves with respect to carbohydrate storage (122). Such non-uniform patterns of distribution have generally been ignored when metabolite determinations have been made (77), and it is important that, in future studies on regulatory pathways, more attention should be paid to this inhomogeneity of distribution.

BIOSYNTHESIS AND DEGRADATION OF FRUCTAN

The Proposed Enzymic Mechanism

An integrated hypothesis describing the enzymology of fructan metabolism in *H. tuberosus* was proposed by Edelman & Jefford (27); it has formed the basis for most subsequent studies, including the few concerning temperate grasses and cereals. The essence of Edelman & Jefford's proposals was that the reactions of synthesis are catalyzed by a number of specific fructosyl transferases, with sucrose acting directly as the primary fructosyl donor (76, 78, 83). The overall stoichiometry of the reaction involving sucrose can be written as $n(G-F) \rightarrow G-F-(F)_{n-1}+n-1(G)$; but it was suggested that this reaction is actually catalyzed by the concerted action of at least two distinct enzymes: (*i*) sucrose-sucrose fructosyl transferase (SST, EC 2.4.1.99), $G-F + G-F \rightarrow G-F-F + G$, and (*ii*) fructan-fructan fructosyl transferase (FFT, EC 2.4.1.100), $G-F-(F)_n + G-F-(F)_m \rightleftarrows G-F-(F)_{n+1} + G-F-(F)_{m-1}$.

The donor and acceptor specificities of FFT were considered to be such that the continuous production of 1-kestose by SST generated fructosyl donors for

FFT and hence permitted the progressive chain elongation of acceptor fructan molecules. Tentative alternative suggestions involving fructosyl transfer from sugar nucleotide fructoses (37, 116) have not received significant experimental support (72). Edelman & Jefford (27) suggested that breakdown of fructan occurs via the sequential removal of terminal fructose residues by a specific β-fructofuranosidase (fructan exohydrolase, FEH, EC 3.2.1.26), $G-F-(F)_n \rightarrow G-F-(F)_{n-1} + F$.

In a series of papers (25–27, 96) Edelman and coworkers, using growing, dormant, and sprouting tubers of *H. tuberosus*, measured, characterized, and partially purified the activities described above. Because they lacked today's techniques for assessing purity by the use of polyacrylamide gel electrophoresis (PAGE) followed by silver-staining (67), their data may not relate to the properties of a single protein species.

Implicit in the above scheme is the fact that the activities measured were specific for the β-2,1-linked fructan series. In *A. officinalis*, where fructans of the neokestose series also accumulated, Shiomi (98) detected an additional enzyme activity, catalyzing reactions of the type:

$$G\text{-}F\text{-}F + \overset{*}{G}\text{-}\overset{*}{F} \rightleftharpoons G\text{-}F + F\text{-}\overset{*}{G}\text{-}\overset{*}{F}.$$

$$\text{isokestose} \quad \text{sucrose} \quad \text{sucrose} \quad \text{neokestose}$$

A recent review of these findings (100) suggests that the action of this enzyme, together with ones having properties similar to the SST and FFT from *H. tuberosus*, can synthesize in vitro fructans of both the isokestose and neokestose series. All these enzymes have been characterized and at least partially purified. In this case, native polyacrylamide gels stained with amido black were used as a criterion of purity (98, 99, 101).

The Fructan Enzymes of Grasses and Cereals

No complete purification of proteins equivalent to those described above has been reported from members of the Gramineae. In the main, studies have been concerned with measuring changes in maximum extractable activities and comparing these, under various treatment regimes, with the rates of fructan accumulation in vitro. Unfortunately, the problem of artefactual fructosyl transfer by other enzymes assayed at high substrate concentrations has been largely ignored, and thus it is often difficult to draw valid conclusions from these studies.

SUCROSE-SUCROSE FRUCTOSYL TRANSFERASE   The synthesis in vitro of isokestose by extracts of temperate grasses has been recorded on a number of occasions (14, 47–49, 74, 117). The enzyme responsible for this activity has not, to our knowledge, been purified to homogeneity or indeed to a state where it is wholly free from invertase activity (sucrase,β-fructofuranosidase, EC3.2.1.26). In contrast, preparations of SST from *A. officinalis* (100) and from *H. tuberosus* (96; based only upon preliminary observations) are reported not to hydrolyze sucrose. Measurable SST activity in grasses has been shown to increase following chilling (74), excision and illumination (14, 48, 117), and the administration of exogenous sugars (121). Conditions that led to an increase in the sucrose concentration and the initiation of fructan synthesis were accompanied by an increase in SST activity (14, 117), although feeding trehalose to detached barley leaves in the dark was associated with an increase in SST activity without the trehalose itself being transformed into fructan (121). Treatments that prevented the initiation of fructan accumulation (subjection to darkness or administration of inhibitors of gene expression) were associated either with a decline in SST activity (121) or with the prevention of any detectable increase (14). Partial purification of SST activity from barley protoplasts using ion exchange media yielded two forms, one that had a high SST-to-invertase ratio when measured at 100 mM sucrose and which synthesized mainly 1-kestose, and a second that had a high invertase-to-SST ratio and synthesized mainly 6-kestose, the second most abundant trisaccharide found in barley leaves in vivo (119). The purity of these preparations was not determined using PAGE/silver-staining, and kinetic characterization of the material was not reported. In particular, the effects of sucrose concentration, assay pH, or temperature upon the relative activity of sucrose cleavage and isokestose formation were not reported.

In parallel studies using enzymes extracted from detached illuminated leaves of *L. temulentum*, it was demonstrated that partially purified material (where both invertase and SST activities were detectable) would synthesize 1- and 6-kestose. 6-Kestose was not observed in significant amounts in vivo in illuminated excised leaves until late in the process of accumulation, whereas neokestose, which was not synthesised in vitro, did accumulate in vivo (15). Increase in total trisaccharide synthesis was linear with increasing sucrose concentration in the range 10–700 mM. No saturation was observed even at grossly nonphysiological sucrose concentrations. The specific synthesis of 1-kestose had an apparent K_m of 380 mM. In addition, treatment of excised illuminated leaves with the supposedly inactive L isomer of the protein synthesis inhibitor 2-(4-methyl-2,6-dinitroanilino)N-methyl propionamide(L-MDMP) indicated that fructan accumulation could be induced in the absence of any detectable rise in extractable SST activity when measured at 600 mM sucrose (15). Similar discrepancies between in vivo synthetic rates and SST activities measured in vitro were observed in chilled leaves of barley, where fructan accumulated markedly in the lower leaf segments despite this tissue's possessing low extractable SST activity (120).

FRUCTAN-FRUCTAN FRUCTOSYL TRANSFERASE   There has been no full purification from grasses of proteins capable, alone or in concert with other proteins, of the synthesis in vitro of the full range of higher-order fructans found in vivo in any particular species. Crude extracts, however, can catalyze some chain-elongation reactions, ranging from the ability to synthesize small amounts of short-chain oligomers (15, 47) to the extensive synthesis of high-DP material (114). The latter observation was based upon a particulate preparation from young seedlings of *Phleum pratense* that showed a pH optimum of 7.0. The properties of this preparation contrasted markedly with data in previous reports concerning the partially purified FFT from tubers of *H. tuberosus,* which had an acidic pH optimum and was freely soluble (27). Both of these observations are, of course, consistent with a vacuolar location for this activity (117).

In an attempt to overcome the limitations of chemical assays, we developed an assay involving the measurement of enzyme-dependent radioactivity transferred from $U^{14}C$ sucrose to existing nonradioactive fructan acceptors (12). When measured using crude extracts of excised illuminated leaves of *L. temulentum,* this reaction proceeded at physiological sucrose concentrations (15–30 mM) and resulted in uniform labeling of the full range of oligo- and polyfructans observed in vivo. The activity was not detected in leaves where fructan synthesis had not been induced, and it could be blocked by D-MDMP but not by the inactive L-isomer (see above; A. J. Cairns, unpublished observation). Use of anion exchange chromatography and pulsed amperometric detection (82) to separate and quantitate the mass of product formed at low sucrose concentrations in the absence of acceptor fructan showed that crude extracts from induced leaves do not synthesize significant quantities of oligosaccharides with a DP exceeding 5 (C. J. Pollock, D. B. Gibeault, unpublished observations).

Studies by Edelman & Dickerson (25) showed that FFT would catalyze reversible fructosyl transfer in the absence of net accumulation. Thus incubation of radioactive sucrose with nonradioactive 1-kestose in the presence of such enzyme preparations yields radioactive 1-kestose according to the reaction:

$$\overset{*\ *}{G\text{-}F} + G\text{-}F\text{-}F \rightleftharpoons \overset{*\ *}{G\text{-}F\text{-}F} + G\text{-}F.$$

Similar reversible transfer of fructose residues from higher-molecular-weight donors would lead to progressive appearance of label in such material, still in the absence of net synthesis. Methylation analysis, followed by gas chromatographic separation of partially methylated alditol acetates and measurement of radioactivity using flow-through-gas-phase ^{14}C counting, can be used to measure the relative specific activities of the different glycosyl moieties within a carbohydrate molecule (97). Isokestose produced by the action of SST should have equal specific activity for both the terminal and the linkage fructose moiety, whereas, if this compound is produced by the FFT reaction described above, a substantial reduction in the relative specific activity of the terminal residue would be expected. Administration of $^{14}CO_2$ to excised and illuminated leaves of *L. temulentum* midway through the induction period yielded fructan preparations showing a marked reduction in the relative specific activity of the terminal fructose moiety. This specific activity gradually increased with time until it became equivalent to the specific activity of the other two glycosyl moieties. This is consistent with the occurrence of FFT-mediated isokestose synthesis in vivo against a background of net accumulation (T. L. Housley, D. B. Gibeault, N. C. Carpita, I. M. Sims, C. J. Pollock, unpublished observations).

No comparisons have been made between FFT preparations of different species in terms of the relative affinities or different oligosaccharide donors or acceptors, so it is not known whether the species-dependent differences in oligosaccharide structure (see above) reflect differences in the properties of the synthetic enzymes or in the occurrence of some kind of partial degradation following synthesis (87).

FRUCTAN EXO-HYDROLASE   This enzyme has been partially purified from stem bases of *Dactylis glomerata*. The final preparation showed a single band on native PAGE when stained by amido black (125). Polyclonal antibodies raised against this preparation gave a single cross-reacting band when blotted against PAGE separations of crude proteins, and the density of this band altered in parallel with changes in extractable enzyme activity. These changes were consistent with the suggestions that activity rose during times of net fructan mobilization (e.g. following defoliation) and fell again during fructan accumulation (124). The enzyme has been shown to cleave the terminal fructose moiety from fructan, to have an acidic pH optimum, and, at least in barley, to be located in the vacuole (118). However, a partially purified preparation from barley has been shown to be active against inulin fructans but inactive against bacterial levan containing a 2,6 backbone and 2,1 branches (42). This would suggest either that more than one FEH is required in order to degrade the full range of structures observed in vivo or that the particular barley enzyme that was purified may be involved in hydrolytic modifications to fructans during synthesis (87). This preparation was apparently inactive against sucrose. This inactivity was not observed in a partially characterized preparation from *L. temulentum*, where multiple isoforms were detected by activity stains following starch gel electrophoresis. These forms showed different relative affinities for sucrose and high-molecular-weight fructan, and their abundance changed during the initiation of fructan degradation or re-synthesis. In all cases, however, the partially purified preparations were active both against sucrose and fructan (R. J. Simpson, R. P. Walker, C. J. Pollock, unpublished observations). Sucrose appeared to be a strong inhibitor of FEH activity, an observation made previously for the enzyme extracted from *H. tuberosus* (26) but not for the enzyme extracted from *Taraxacum officinale* (89).

Fructan Enzymology in Grasses: Problems and Progress

All the reactions of fructan metabolism described above can be written in terms of the group transfer reaction:

$$R\text{-}O\text{-}\overset{*}{R} + \overset{+}{R} + -O\text{-}\dot{R} \leftrightharpoons R\text{-}O\text{-}\dot{R} + \overset{+}{R} + -O\text{-}\overset{*}{R}.$$

The identity of R can vary; it may be a hydrogen atom, a single glycosyl moiety, or the residue of a polymer chain. Under various circumstances, therefore, the nature of the substrate provided in an assay in vitro might determine whether the catalytic function was attributed to the presence of FFT, SST, FEH, or invertase. Indeed, it has been known for some time that invertase itself can catalyze fructosyl transfer at high sucrose concentrations (1, 4, 39, 52, 77, 112) and can yield all three isomers of monofructosyl sucrose. We have measured transferase activity in a range of purified invertase preparations derived from plants and microorganisms that do not accumulate fructans naturally, and such activity appears to be a property of all preparations (A. J. Cairns, C. J. Pollock, unpublished observations). Purified invertase can also hydrolyze at least some fructans (112) and can synthesize fructans of DP greater than 3 (52). If a single protein from an organism that does not accumulate fructan can catalyze all the reactions required for fructan synthesis and degradation, how can we assign a sensible function in vitro to crude or partially purified proteins from fructan accumulators that, where the matter has been studied in detail, are usually capable of catalyzing more than one such reaction?

Changes in the relative rates of the different activities during physiological treatments can be correlated with observed changes in the patterns of fructan metabolism in vivo (14, 47, 48, 117). These observations support the physiological relevance of some of the initial measurements. They do not, however, permit unequivocal assignment of function to purified proteins unless these have been shown to be monofunctional. We would propose that such demonstrations are rare, requiring, as they do, characterization under a wide range of substrate and competitor concentrations, pH values, incubation times, temperatures, etc. Differences in pH or temperature optima between different activities, which have been used as support for the presence of different enzymes in crude extracts (18, 117), can also be observed when the transferase and hydrolase activities of purified yeast invertase are compared (A. J. Cairns, unpublished observations). This suggests that these criteria, too, are not always valid, a suggestion borne out by observations that similar enzymes from prokaryotic microorganisms may possess distinct sites for hydrolase and transferase activities on the same protein (126). We suggest that an unambiguous assignment of function will depend upon fulfilling the following conditions:

1. Demonstration of protein purity using the best current methods (e.g. two-dimensional denaturing PAGE, followed by silver staining).
2. Full kinetic characterization, including measurement of all possible competing and alternative activities over an appropriate range of substrate concentrations, pH, and temperature values. (Note: the chemical activity of water is an independent variable that is generally ignored, even though water is a potential substrate; 70).
3. Determination of developmental changes in activity and demonstration of their correspondence to alterations in the rate of fructan metabolism in vivo.
4. Demonstration that the products formed in vitro resemble, in structure if not in relative abundance, those observed in vivo.

Faced with the rigid application of all these criteria, it might be concluded that no enzyme of fructan metabolism has been properly purified from any higher plant. While this is probably an exaggeration (see above), we feel that much more caution is required in the interpretation of existing data than has sometimes been evident.

THE CONSEQUENCES OF FRUCTAN ACCUMULATION FOR THE REGULATION OF CARBON METABOLISM

In Source Leaves

Although the enzymological details may be unclear (see above), there is convincing evidence that sucrose is the major substrate for fructan metabolism (27, 75, 76, 83). Thus the regulation of fructan metabolism must be considered in terms of the production and utilization of sucrose and of the recycling of hexose formed either during fructosyl transfer from sucrose to the growing chain or during fructan breakdown. The apparent insensitivity of sucrose synthesis to increasing carbohydrate content in leaves of grasses (13, 48, 117) contrasts with the situation in plants such as spinach where, under conditions of reduced export, an increased proportion of the total photosynthetic carbon is accumulated as starch (110). The regulatory properties of cytoplasmic fructose-1,6 bisphosphatase (FBPase; EC3.1.3.11) and the changes in fructose-2,6 bisphosphate (F2,6 BP) concentrations that follow excision and illumination of *L. temulentum* leaves (B. E. Collis, C. J. Pollock, unpublished observations; 77) are both consistent with Stitt et al's (109) model for the regulation of substrate supply to sucrose phosphate synthase (SPS; EC 2.4.1.14) in high starch leaves. Thus the reasons for the apparent insensitivity of sucrose synthesis in grass leaves to such changes are obscure. Measurement of whole-leaf F2,6 BP concentrations may, however, mask changes in particular tissues that might still permit significant fluxes through cytoplasmic FBPase into sucrose to occur during fructan accumulation. Some reports suggest that metabolic compartmentation of recently fixed sucrose occurs (48), but this may be due to tissue-level rather than cell-level compartmentation (120, 122). In leaves, the transport of sucrose into the vacuole is thought not to occur against a concentration gradient (54); it is difficult, therefore, to consider how this step might contribute to the regulation of the balance between vacuolar and cytoplasmic carbon metabolism.

Free hexose rarely accumulates to any significant degree during fructan synthesis (117). Under these circumstances, it must be presumed that resynthesis of sucrose occurs from the hexose released. Measurement of maximal catalytic activities of the various enzymes required for this process suggests that there are sufficient activities to account for the reutilization of 1 mole of glucose for each 1 mole of fructose transferred into fructan. Such reutilization would, however, have the effect of uncoupling the flux through cytoplasmic FBPase from that through SPS (B. E. Collis, C. J. Pollock, unpublished observations). Estimates of the magnitude of such fluxes suggest that the flux through SPS may be up to 50% higher than that through cytoplasmic FBPase.

During fructan utilization there is also restricted hexose accumulation in grass tissues (121), and thus a similar resynthesis of sucrose from free fructose must be presumed to occur. Since this usually occurs in vivo when demand for fixed carbon exceeds supply, it is less easy to determine the extent to which SPS activity is increased relative to FBPase. It is important to note that the utilization of both starch and fructan to maintain export from leaves under conditions of reduced photosynthesis will require the resynthesis of sucrose. Only the direct mobilization of vacuolar sucrose can proceed without metabolic transformations. It seems likely that this step is the initial change that triggers the mobilization of both starch (38) and fructan (121) in leaves of temperate grasses.

In Sink Tissues

Few studies have been done on the integration of fructan metabolism and cytoplasmic carbon metabolism in sink tissues. It is believed that sucrose uptake into vacuoles in sink tissues is an active process (60), so regulation at the level of sucrose transport becomes a possibility. Recycling of hexose is assumed to occur because of the relatively low tissue contents, but the magnitude of such processes has not been measured. In tubers of *H. tuberosus*, chilling induces a net depolymerization of fructan with no net loss of carbohydrate. Under these conditions resynthesis of short-chain fructan acceptors would be required, although the control of this process has never been studied in detail (78). It is not known whether such cold-induced depolymerization occurs in graminaceous storage tissue.

CONCLUDING REMARKS

The Selective Advantages of Fructan Metabolism

A number of suggestions have been made concerning the selective advantage of fructan as a storage carbohydrate. These are generally associated with some aspect of the osmotic behavior of fructan and have been discussed in some detail in recent reviews and papers (28, 41, 76, 78, 83). There is no generally compelling evidence for any single specific factor that might confer selective advantage upon fructan accumulators, and the habitual proximity of accumulators and nonaccumulators within any particular flora means that it is difficult to define environmental situations for which the ability to accumulate fructans is a requirement (41). On the other hand, the abundance of the compounds within at least two highly advanced genera argues against selective neutrality (59). In leaves of grasses, at least, the concentration of soluble carbohydrates within fructan accumulators can far exceed those of starch, suggesting that the utilization of the vacuole as a storage compartment would fit such plants for exploiting marginal environments where periods of positive carbon balance were frequently interspersed with ones where net mobilization of reserves was required to sustain growth. Such situations do indeed occur frequently during the perennial life cycle of northern temperate grasses (78).

The Experimental Advantages of Fructan Metabolism

Recent emphasis on the metabolism of fructans in leaves and its integration with the synthesis and export of sucrose reflects current advances in understanding the regulation of photosynthetic carbon metabolism (78, 84, 109). Vacuolar fructan accumulation also provides a useful tool with which to study the role of sucrose in inducing altered patterns of gene expression (14, 121) together with the synthesis and processing of vacuolar proteins (117, 118) and their turnover in a lytic compartment (121). The successful purification of any of the enzymes of fructan metabolism induced during illumination of excised grass and cereal leaves will permit, via the raising of antisera or the isolation and cloning of the relevant genetic sequences, the mechanism of targeting and turnover to be studied in an active system where the balance between synthesis and degradation can easily be altered by varying the overall carbohydrate status of the tissue (121). Such advances will also facilitate studies of fructan metabolism in previously recalcitrant experimental systems such as apices, developing seeds, and woody tissues. We hope that, based upon substantial recent interest in these distinctive and significant metabolic events, progress in the areas described above will be rapid. Studies of this kind will help identify the selective advantages of a range of compounds that are abundant in the plant kingdom but that have suffered experimentally in comparison with starch because of their limited agricultural significance. Recent production of fructans from sucrose on an industrial scale as a food additive (43), together with studies on the utilization of natural fructans as a renewable chemical feedstock (36), should also result in increased interest in these distinctive compounds.

ACKNOWLEDGMENTS

We acknowledge the helpful advice and access to unpublished material offered by Ian Sims, Ana Winters, Tom Housley, Dave Gibeaut, Nick Carpita, Enrico Martinoia, Doug Randall, and Richard Simpson. The Institute of Grassland and Environmental Research is supported financially by the Agricultural and Food Research Council, UK. This work was carried out during the tenure of NATO Grant CRG0706/87 (to CJP).

Literature Cited

1. Albon, N., Bell, D. J., Blanchard, P. H., Gross, D., Rundell, J. T. 1953. Kestose, a trisaccharide formed from sucrose by yeast invertase. *J. Chem. Soc.*, pp. 24–27
2. Archbold, H. K. 1940. Fructosans in the monocotyledons, a review. *New Phytol.* 39:185–219
3. Avigad, G. 1990. Disaccharides. See Ref. 23, pp. 111–88
4. Bacon, J. S. D., Bell, D. J. 1953. A new trisaccharide produced from sucrose by mold invertase. *J. Chem. Soc.*, pp. 2528–30
4a. Ballard, R. A., Simpson, R. J., Pearce, G. R. 1990. Losses of the digestible components of annual ryegrass (*Lolium rigidum* Gaudin) during senescence. *Aust. J. Agric. Res.* 41:719–31
5. Bancal, P., Gaudillere, J. P. 1989. Oligofructan separation and quantification by high performance liquid chromatography. Application to *Asparagus officinalis* and *Triticum aestivum*. *Plant Physiol. Biochem.* 27:1–6
6. Barkworth, M. E., Everett, J. 1986. Evolution in the stipeae: identification and relationships of its monophyletic taxa. See Ref. 106, pp. 251–64
7. Bell, D. J., Palmer, A. 1952. Structural studies on inulin from *Inula helenicum* and on levan from *Dactylis glomerata* and *Lolium italicum*. *J. Chem. Soc.*, pp. 763–70

8. Bender, M. M., Smith, D. 1973. Classification of starch and fructosan-accumulating grasses as C-3 or C-4 species by carbon isotope analysis. *J. Br. Grassl. Soc.* 28:97–100
9. Blacklow, W. M., Darbyshire, B., Pheloung, P. 1984. Fructans polymerised and depolymerised in the internodes of winter wheat as grain-filling progressed. *Plant Sci. Lett.* 36:213–18
10. Borland, A. M., Farrar, J. F. 1987. The influence of low temperature on diel patterns of carbohydrate metabolism in leaves of *Poa annua* L. and *Poa x jemtlandica* (Almq.) Richt. *New Phytol.* 105:255–63
11. Borrell, A. K., Incoll, L. D., Simpson, R. J., Dalling, M. J. 1989. Partitioning of dry matter and the deposition and use of stem reserves in a semi-dwarf wheat crop. *Ann. Bot.* 63:527–39
12. Cairns, A. J. 1989. Fructan biosynthesis in excised leaves of *Lolium temulentum* L. IV. Cell-free ^{14}C labelling of specific oligofructans at low sucrose concentration. *New Phytol.* 112:465–73
13. Cairns, A. J., Pollock, C. J. 1988. Fructan biosynthesis in excised leaves of *Lolium temulentum* L. I. Chromatographic characterisation of oligofructans and their labelling patterns following ^{14}CO$_2$ feeding. *New Phytol.* 109:399–405
14. Cairns, A. J., Pollock, C. J. 1988. Fruc-

tan biosynthesis in excised leaves of *Lolium temulentum* L. II. Changes in fructosyl transferase activity following excision and application of inhibitors of gene expression. *New Phytol.* 109:407–13

15. Cairns, A. J., Winters, A., Pollock, C. J. 1989. Fructan biosynthesis in excised leaves of *Lolium temulentum* L. III. A comparison of the *in vitro* properties of fructosyl transferase activities with the characteristics of *in vivo* fructan accumulation. *New Phytol.* 112:343–52

16. Carpita, N. C., Kanabus, J., Housley, T. L. 1989. Linkage structure of fructans and fructan oligomers from *Triticum aestivum* and *Festuca arundinacea* leaves. *J. Plant Physiol.* 134:162–68

17. Chatterton, N. J., Harrison, P. A., Bennett, J. H., Asay, K. H. 1989. Carbohydrate partitioning in 185 accessions of Gramineae grown under warm and cool temperatures. *J. Plant Physiol.* 134:169–79

18. Chatterton, N. J., Harrison, P. A., Thornley, W. R., Bennett, J. H. 1988. Characterization of sucrose:sucrose fructosyl transferase from crested wheatgrass. *New Phytol.* 109:29–33

19. Chatterton, N. J., Harrison, P. A., Thornley, W. R., Draper, E. A. 1990. Oligosaccharides in foliage of *Agropyron, Bromus, Dactylis, Festuca, Lolium* and *Phleum. New Phytol.* 114:167–71

20. Chatterton, N. J., Thornley, W. R., Harrison, P. A., Bennett, J. H. 1988. Dynamics of fructan and sucrose biosynthesis in crested wheatgrass. *Plant Cell Physiol.* 29:1103–8

21. Concert, H. J. 1986. Current concepts in the systematics of the Arundinoideae. See Ref. 106, pp. 239–50

22. Cooper, R. J., Street, J. R., Henderlong, P. R., Koski, A. J. 1988. An analysis of the carbohydrate status of mefluidide-treated annual bluegrass. *Agron. J.* 80:410–14

23. Dey, P. M., ed. 1990. *Methods in Plant Biochemistry*, Vol. 2, *Carbohydrates*. London/San Diego: Academic

24. Dubois, D., Winzeler, M., Nösberger, J. 1990. Fructan accumulation and sucrose:sucrose fructosyl transferase activity in stems of spring wheat genotypes. *Crop Sci.* 30:315–19

25. Edelman, J., Dickerson, A. G. 1966. The metabolism of fructose polymers in plants. Transfructosylation in tubers of *Helianthus tuberosus* L. *Biochem. J.* 98:787–94

26. Edelman, J., Jefford, T. G. 1964. The metabolism of fructose polymers in plants. 4. β-fructofuranosidases of tubers of *Helianthus tuberosus* L. *Biochem. J.* 93:148–61

27. Edelman, J., Jefford, T. G. 1968. The mechanism of fructosan metabolism in higher plants as exemplified in *Helianthus tuberosus* L. *New Phytol.* 67:517–31

28. Englmaier, P. 1987. Carbohydrate metabolism of salt-tolerant fructan grasses as exemplified with *Puccinellia peisonis. Biochem. Physiol. Pflanz.* 182:165–82

29. Escalada, J. A., Moss, D. N. 1976. Changes in the non-structural carbohydrate fractions of developing spring wheat kernels. *Crop Sci.* 16:627–31

30. Ferretti, V., Bertolasi, V., Gilli, G. 1984. Structure of 6-kestose monohydrate, $C_{18}H_{31}O_{16}H_2O$. *Acta Crystallogr. C* 40:531–35

31. Forsythe, K. L., Feather, M. S., Gracz, H., Wong, T. C. 1990. Detection of kestoses and kestose-related oligosaccharides in extracts of *Festuca arundinacea, Dactylis glomerata* L. and *Asparagus officinalis* L. root cultures and invertase by ^{13}C and ^1H nuclear magnetic resonance spectroscopy. *Plant Physiol.* 92:1014–20

32. Frehner, M., Keller, F., Wiemken, A. 1984. Fructan metabolism in *Helianthus tuberosus:* compartmentation in protoplasts and vacuoles isolated from tubers. *J. Plant Physiol.* 116:197–208

33. French, A. D. 1988. Accessible conformation of the β-D-(2→1)- and -(2→6)-linked D-fructans inulin and levan. *Carbohydr. Res.* 176:17–29

34. French, A. D. 1989. Chemical and physical properties of fructans. *J. Plant Physiol.* 134:125–36

35. French, D., Wild, G. M. 1953. Correlation of carbohydrate structure with papergram mobility. *J. Am. Chem. Soc.* 75:2612–16

36. Fuchs, A. 1987. Potentials for non-food use of fructose and inulin *Staerke* 39:335–43

37. Gonzalez, N. S., Pontis, H. G. 1963. Uridine diphosphate fructose and uridine diphosphate acetyl-galactosamine from dahlia tubers. *Biochim. Biophys. Acta* 69:179–81

38. Gordon, A. J., Ryle, G. J. A., Webb, G. 1980. The relationship between starch and sucrose during "dark" export from leaves of uniculm barley. *J. Exp. Bot.* 31:845–50

39. Gross, D., Blanchard, P. H., Bell, D. J. 1954. Neokestose, a trisaccharide formed from sucrose by yeast invertase. *J. Chem. Soc.*, pp. 1727–30

40. Hendrix, J. E., Linden, J. C., Smith,

D. H., Ross, C.W., Park, I. K. 1986. Relationship of pre-anthesis fructan metabolism to grain numbers in winter wheat (*Triticum aestivum* L.). *Aust. J. Plant Physiol.* 13:391–98

41. Hendry, G. 1987. The ecological significance of fructan in a contemporary flora. *New Phytol.* 106:201–16

42. Henson, C. A. 1989. Purification and properties of barley stem fructan exohydrolase. *J. Plant Physiol.* 134:186–91

43. Hidaka, H., Hirayama, M., Sumi, N. 1988. A fructo-oligosaccharide-producing enzyme from *Aspergillus niger* ATCC 20611. *Agric. Biol. Chem.* 52:1181–83

44. Hirst, E. L. 1957. Some aspects of the chemistry of the fructosans. *Proc. Chem. Soc.,* pp. 193–204

45. Ho, L. C., Gifford, R. M. 1984. Accumulation and conversion of sugars by developing wheat grains. V. The endosperm apoplast and apoplastic transport. *J. Exp. Bot.* 35:58–73

46. Housley, T. L., Daughtry, C. S. T. 1987. Fructan content and fructosyl transferase activity during wheat seed growth. *Plant Physiol.* 83:4–7

47. Housley, T. L., Kanabus, J., Carpita, N. C. 1989. Fructan synthesis in wheat leaf blades. *J. Plant Physiol.* 134:192–95

48. Housley, T. L., Pollock, C. J. 1985. Photosynthesis and carbohydrate metabolism in detached leaves of *Lolium temulentum* L. *New Phytol.* 99:499–507

49. Housley, T. L., Volenec, J. J. 1988. Fructan content and synthesis in leaf tissues of *Festuca arundinacea*. *Plant Physiol.* 86:1247–51

50. Howarth, W. N., Learner, A. 1928. Polysaccharides. Part 1. The structure of inulin. *J. Chem. Soc.,* pp. 619–25

51. Incoll, L. D., Bonnett, G. D., Gott, B. 1989. Fructans in the underground storage organs of some Australian plants used for food by Aborigines. *J. Plant Physiol.* 134:196–202

52. Ivin, P. C., Clarke, M. L. 1987. Isolation of kestoses and nystose from enzyme digests by high performance liquid chromatography. *J. Chromatogr.* 408:393–98

53. Jeffrey, G. A., Park, Y. J. 1972. The crystal and molecular structure of 1-kestose. *Acta Crystallogr. B* 28:257–67

54. Kaiser, G., Heber, U. 1984. Sucrose transport into vacuoles isolated from barley mesophyll protoplasts. *Planta* 161:562–68

55. Kanaya, K. I., Chiba, S., Shimomura, T. 1978. Thin-layer chromatography of linear oligosaccharides. *Agr. Biol. Chem.* 42:1947–48

56. Kühbauch, W. 1974. Fructosangelhalt, -polymerisationsgrad und -struktur in verschiedenen Pflanzenteilen von Leischgras. *Z. Pflanzenphysiol.* 74:121–29

57. Kühbauch, W., Thome, U. 1989. Nonstructural carbohydrates of wheat stems as influenced by sink-source manipulations. *J. Plant Physiol.* 134:243–50

58. Laidlaw, R. A., Reid, J. S. G. 1951. Studies on fructosans. III. A fructosan from *Lolium perenne*. *J. Chem. Soc.,* pp. 1830–34

59. Lewis, D. H. 1984. Occurrence and distribution of storage carbohydrates in vascular plants. In *Storage Carbohydrates in Vascular Plants*, ed. D. H. Lewis, pp. 1–52. Cambridge: Cambridge Univ. Press

60. Lucas, W. J., Madore, M. A. 1988. Recent advances in sugar transport. See Ref. 85, pp. 35–84

61. MacLeod, A. M., McCorquodale, H. 1958. Water-soluble carbohydrates of seeds of the Gramineae. *New Phytol.* 57:168–82

62. Marchessault, R., Bleha, T., Deslandes, Y., Revol, J. F. 1980. Conformation and crystalline structure of (2-1)-β-D-fructofuranan (inulin). *Can. J. Chem.* 58:2415–22

63. McDonald, E. J. 1946. The polyfructosans and difructose anhydrides. *Adv. Carbohydr. Chem.* 2:253–77

64. McGrath, D. 1988. Seasonal variation in the water-soluble carbohydrates of perennial and Italian ryegrass under cutting conditions. *Irish J. Agric. Res.* 27:131–39

65. Medcalf, D. G., Cheung, P. W. 1971. Composition and structure of glucofructans from durum wheat flour. *Cereal Chem.* 48:1–8

66. Meier, H., Reid, J. S. G. 1982. Reserve polysaccharides other than starch in higher plants. In *Enclyclopedia of Plant Physiology, New Ser.,* ed. F. A. Loewus, W. Tanner, 13:418–71. Berlin: Springer-Verlag

67. Merril, C. R. 1990. Gel staining techniques. *Methods Enzymol.* 182:477–88

68. Mino, Y., Maeda, K. 1974. Metabolism of sucrose and phlein in the haplocorm of timothy. *J. Jpn. Grassl. Soc.* 20:6–10

69. Nelson, C. J., Smith, D. 1986. Fructans: their nature and occurrence. *Curr. Top. Plant Biochem. Physiol.* 5:1–16

70. Nelson, J. M., Schubert, M. P. 1928. Water concentration and the rate of hydrolysis of sucrose by invertase. *J. Am. Chem. Soc.* 50:2188–93

71. Ojima, K., Isawa, T. 1968. The varia-

tion of carbohydrates in various species of grasses and legumes. *Can. J. Bot.* 46:1507–11

72. Pollock, C. J. 1979. Pathway of fructosan synthesis in leaf bases of *Dactylis gomerata Phytochemistry* 18:777–79
73. Pollock, C. J. 1982. Oligosaccharide intermediates of fructan synthesis in *Lolium temulentum*. *Phytochemistry* 21:2461–65
74. Pollock, C. J. 1984. Sucrose accumulation and the initiation of fructan biosynthesis in *Lolium temulentum* L. *New Phytol.* 96:527–34
75. Pollock, C. J. 1986. Environmental effects on sucrose and fructan metabolism. See Ref. 86, pp. 32–46
76. Pollock, C. J. 1986. Fructans and the metabolism of sucrose in higher plants. *New Phytol.* 104:1–24
77. Pollock, C. J., Cairns, A. J., Collis, B. E., Walker, R. P. 1989. Direct effects of low temperature upon components of fructan metabolism in leaves of *Lolium temulentum* L. *J. Plant Physiol.* 134:203–8
78. Pollock, C. J., Chatterton, N. J. 1988. Fructans. See Ref. 85, pp. 109–40
79. Pollock, C. J., Hall, M. A., Roberts, D. P. 1979. Structural analysis of fructose polymers by gas-liquid chromatography and gel filtration. *J. Chromatogr.* 171:411–15
80. Pollock, C. J., Jones, T. 1979. Seasonal patterns of fructan metabolism in forage grasses. *New Phytol.* 83:8–15
81. Pollock, C. J., Ruggles, P. A. 1976. Cold-induced fructan synthesis in leaves of *Dactylis glomerata. Phytochemistry* 15:1643–46
82. Pontis, H. G. 1990. Fructans. See Ref. 23, pp. 353–69
83. Pontis, H. G., del Campillo, E. 1985. Fructans. In *Biochemistry of Storage Carbohydrates in Green Plants*, ed. P. M. Dey, R. A. Dixon, pp. 205–27. New York: Academic
84. Preiss, J. 1988, Biosynthesis of starch and its regulation. See Ref. 85, pp. 181–254
85. Preiss, J., ed. 1988. *The Biochemistry of Plants, a Comprehensive Treatise*, Vol. 14. *Carbohydrates*. San Diego/London: Academic
86. Randall, D. D., Miles, C. D., Nelson, C. J., Blevins, D. G., Miernyk, J. A., eds. 1986. *Current Topics in Plant Biochemistry and Physiology*, Vol. 5, Columbia: Univ. Missouri
87. Rocher, J.-P. 1967. Les levanes de *Lolium italicum* synthése dans les organes végétatifs. *Physiol. Veg.* 5:71–80

88. Rose, N. 1804. Ueber eine eigenthumliche vegetabilische Substanz. *Neues Allgem. Chem.* 3:217–19
89. Rutherford, P. P., Deacon, A. C. 1972. β-fructofuranosidases from roots of dandelion (*Taraxacum officinale* Weber). *Biochem. J.* 126:569–73
90. Schlubach, H. H. 1961. Der Kohlenhydratstoffweschel im Roggen und Weizen. In *Progress in the Chemistry of Organic Natural Products*, ed. L. Zechmeister, 19:291–316. Berlin: Springer-Verlag
91. Schnyder, H. 1986. Carbohydrate metabolism in the growth zone of tall fescue leaf blades. See Ref. 86, pp. 47–58
92. Schnyder, H., Ehses, U., Bestajovsky, J., Merhoff, R., Kühbauch, W. 1988. Fructan in wheat kernels during growth and compartmentation in the endosperm and pericarp. *J. Plant Physiol.* 132:333–38
93. Schnyder, H., Nelson, C. J. 1987. Growth rates and carbohydrate fluxes within the elongation zone of tall fescue leaf blades. *Plant Physiol.* 85:548–53
94. Schnyder, H., Nelson, C. J. 1989. Growth rates and assimilate partitioning in the elongation zone of tall fescue leaf blades at high and low irradiance. *Plant Physiol.* 90:1201–6
95. Schnyder, H., Nelson, C. J., Spollen, W. G. 1988. Diurnal growth of tall fescue leaf blades. II. Dry matter partitioning and carbohydrate metabolism in the elongation zone and adjacent expanded tissue. *Plant Physiol.* 86:1077–83
96. Scott, R. W., Jefford, T. D., Edelman, J. 1966. Sucrose fructosyl transferase from higher plant tissues. *Biochem. J.* 100:23 p
97. Shea, E. M., Gibeaut, D. M., Carpita, N. C. 1989. Structural analysis of the cell walls regenerated by carrot protoplasts. *Planta* 179:293–308
98. Shiomi, N. 1981. Purification and characterization of 6G-fructosyltransferase from the roots of asparagus (*Asparagus officinalis* L.). *Carbohydr. Res.* 96:281–92
99. Shiomi, N. 1982. Purification and characterization of 1-F-fructosyltransferase from the roots of asparagus (*Asparagus officinalis* L.) *Carbohydr. Res.* 99:157–69
100. Shiomi, N. 1989. Properties of fructosyltransferases involved in the synthesis of fructan in liliaceous plants. *J. Plant Physiol.* 134:151–55
101. Shiomi, N., Izawa, M. 1980. Purification and characterization of sucrose:sucrose 1-fructosyltransferase from the roots of asparagus (*Asparagus offici-*

nalis L.). *Agric. Biol. Chem.* 44:603–14

102. Smith, A. E. 1976. β-Fructofuranosidase and invertase activity in tall fescue culm bases. *J. Argic. Food Chem.* 24:476–78

103. Smith, D. 1968. Classification of several native North American grasses as starch or fructosan accumulators in relation to taxonomy. *J. Br. Grassl. Soc.* 23:306–9

104. Smith, D. 1973. The nonstructural carbohydrates. In *The Biochemistry of Herbage*, ed. G. W. Butler, R. W. Bailey, 2:105–55. New York: Academic

105. Smouter, H., Simpson, R. J. 1989. Occurrence of fructans in the Gramineae (Poaceae). *New Phytol.* 111:359–68

106. Soderstrom, T. R., Hilu, K. W., Campbell, C. S., Barkworth, M. E., eds. 1986. *Grass Systematics and Evolution.* Washington: Smithsonian Inst. Press

107. Spollen, W. G., Nelson, C. J. 1988. Characterization of fructan from mature leaf blades and elongation zones of developing leaf blades of wheat, tall fescue and timothy. *Plant Physiol.* 88:1349–53

108. Steen, E., Larsson, K. 1986. Carbohydrates in roots and rhizomes of perennial grasses. *New Phytol.* 104:339–46

109. Stitt, M., Huber, S., Kerr, P. 1987. Control of photosynthetic sucrose formation. In *The Biochemistry of Plants. A Comprehensive Treatise*, Vol. 10, *Photosynthesis*, ed. M. D. Hatch, N. K. Boardman, pp. 327–409. San Diego/London: Academic

110. Stitt, M., Kurzel, B., Heldt, H. W. 1984. Control of photosynthetic sucrose synthesis. II. Partitioning between sucrose and starch. *Plant Physiol.* 75:554–60

111. Stoy, V. 1970. The storage and remobilization of carbohydrates in cereals. In *Crop Physiology and Cereal Breeding*, ed. J. H. J. Spiertz, Th. Kramer, pp. 55–59. Wageningen: Centre Agric. Publ Doc.

112. Straathof, A. J. J., Kieboom, A. P. G., van Bekkum, H. 1986. Invertase-catalysed fructosyl transfer in concentrated solutions of sucrose. *Carbohydr. Res.* 146:154–59

113. Suzuki, M. 1989. Nomenclature of fructans. *Fructan Newsl.* 1:5

114. Suzuki, M., Pollock, C. J. 1986. Extraction and characterisation of the enzymes of fructan biosynthesis in timothy (*Phleum pratense*). *Can. J. Bot.* 64:1884–87

115. Tomasic, J., Jennings, H. J., Glaudemans, C. P. J. 1978. Evidence for a single type of linkage in a fructofuranan from *Lolium perenne. Carbohydr. Res.* 62:127–33

116. Umemara, Y., Nakamura, M., Funahashi, S. 1967. Isolation and characterization of uridine diphosphate fructose from tubers of Jerusalem artichoke (*Helianthus tuberosus* L.). *Arch. Biochem. Biophys.* 199:240–52

117. Wagner, W., Keller, F., Wiemken, A. 1983. Fructan metabolism in cereals: induction in leaves and compartmentation in protoplasts and vacuoles. *Z. Pflanzenphysiol.* 112:359–72

118. Wagner, W., Wiemken, A. 1986. Properties and subcellular localisation of fructan hydrolase in the leaves of barley (*Hordeum vulgare* L. cv. Gerbel). *J. Plant Physiol.* 123:429–39

119. Wagner, W., Wiemken, A. 1987. Enzymology of fructan synthesis in grasses. Properties of sucrose-sucrose fructosyl transferase in barley leaves (*Hordeum vulgare* L. cv. Gerbel). *Plant Physiol.* 85:706–10

120. Wagner, W., Wiemken, A. 1989. Fructan metabolism in expanded primary leaves of barley (*Hordeum vulgare* L. cv. Gerbel): Change upon ageing and spatial organisation along the leaf blade. *J. Plant Physiol.* 134:237–42

121. Wagner, W., Wiemken, A., Matile, Ph. 1986. Regulation of fructan metabolism in leaves of barley (*Hordeum vulgare* L. cv. Gerbel). *Plant Physiol.* 81:444–47

122. Williams, M. L, Farrar, J. F., Pollock, C. J. 1989. Cell specialisation within the parenchymatous bundle sheath of barley. *Plant Cell Environ.* 12:909–18

123. Wille, F. 1917. Anatomisch-physiologische Untersuchungen am Gramineenrhizom. *Beih. Bot. Zentralbl.* 33:1–70

124. Yamamoto, S., Mino, Y. 1985. Partial purification and properties of pheinase induced in stem base of orchard grass after defoliation. *Plant Physiol.* 78:591–95

125. Yamamoto, S., Mino, Y. 1989. Mechanism of phleinase induction in the stem base of orchard grass after defoliation. *J. Plant Physiol.* 134:258–60

126. Yamashita, Y., Hanada, N., Itoh-Andoh, M., Takehara, T. 1989. Evidence for the presence of two distinct sites of sucrose hydrolysis and glucosyl transfer activities on 1,3-α-D-glucan synthase of *Streptococcus mutans. FEBS Lett.* 243:343–46

Annu. Rev. Plant Physiol. Plant Mol. Biol. 1991. 42:103–28

pH AND IONIC CONDITIONS IN THE APOPLAST

C. Grignon and H. Sentenac

Biochimie et Physiologie Végétales, Ecole Nationale Supérieure Agronomique, Institut National de la Recherche Agronomique, Centre National de la Recherche
Scientifique URA 573, 34060 Montpellier Cedex 1, France

KEY WORDS: cell wall fixed charges, Donnan equilibrium, uronic acid, ion exchange,
              unstirred layer

CONTENTS

In higher plants, the apoplast occupies 5% or less of the tissue volume of
aerial organs (11, 20, 36, 71, 120, 168) and of root cortexes (179). The
structural and ionic characteristics of this small compartment are important for
the physiology of the plant because (*a*) they determine the ionic composition
of the medium that bathes the cell membrane, (*b*) they control extracellular

transports of solutes, and (c) they affect mechanical and osmotic phenomena involved in cell growth.

In this chapter we restrict our comments to the advances, since the review by Haynes (74), concerning pH and ionic conditions in the apoplast. We do not treat the root cation exchange capacity (CEC) and the transport of Ca^{2+} in the xylem and in nonvascularized tissues because we do not know significantly more about them now than when they were last reviewed (29, 53, 74).

IONIC PROPERTIES OF THE APOPLAST

The Apoplast as an Ion Exchanger

Cell walls contain high concentrations of uronic acids. All measurements of ionic and electrical properties of cell walls indicate that the main charges are those of dissociated weak acids, with pK values similar to that of polygalacturonic acid (about 3) (2, 83, 110, 111, 142, 149, 155, 165). Thus, cations are accumulated in the cell walls, and anions are excluded. Most of the accumulated cations are reversibly retained in the cell walls, either as free hydrated ions, or immobilized by various mechanisms but easily exchangeable. Cell walls may also contain multivalent cations precipitated (188) or irreversibly bound to polyuronic acids and other ligands (194). Removal of Ca^{2+} from apoplast leads to solubilization of pectin (77, 80, 150) and thus to a decrease in CEC of cell walls (9, 67, 80). This is because a fraction of the immobilized Ca^{2+} ions participate in Ca-bridges between pectin chains (61). The binding of Ca^{2+} induces a special conformation of pectin (97), which probably results in the cooperative formation of cross-linking binding sites, different from those responsible for the CEC (67). These sites, which stabilize the matrix pectin, could be ligand cavities selective for Ca^{2+} (69).

The simplest description of the distribution of the exchangeable ions in the apoplast is the Donnan model: The cell walls are viewed as a solution of "indiffusible" anions restricted to the free space (21) by a fictitious membrane permeable to all other ions and statistically neutralized by diffusible cations. Increasing the salt concentration in the medium results in a screening of the electrostatic field of the fixed charges (ionic strength effect) and in a decrease in both the accumulation ratio of cations and the exclusion ratio of anions. The predictions of this model are generally qualitatively correct but often quantitatively less satisfactory. For instance, the observed magnitude of anion exclusion may be less than predicted. This may be due to specific binding of anions on basic amino acids of extracellular proteins, but also to overestimation of the activity of dissociated acid groups (111, 142). Also, one often observes higher cation binding than expected and, in some cases, differences in binding selectivity between cations with the same valency, a phenomenon unexplained by the Donnan model. These discrepancies between observations and the predictions of the Donnan model indicate that some cations can be tightly associated to the indiffusible anions, either chemically (as well known for H^+, and Ca^{2+} in chelates), electrostatically (105), or structurally (as for Ca^{2+} in the egg box model) (69). These mechanisms form neutral ion pairs by masking the charges. They differ from the statistical neutralization of fixed anions by free cations (the ideal Donnan system), in which the charges remain unmasked. Charge-masking associations may be incorporated in the Donnan model, either as specific binding described by the mass action law, with ad hoc specific affinity constants (155), or as electrostatic condensation (49, 111, 142).

The interaction of masking and screening effects may result in complex behavior of cell walls, such as apparent values of affinity constants (including pK) different from the intrinsic ones due to local accumulation of H^+ and cations (e.g. 142, 155). Since the electrostatic accumulation ratio decreases as ionic strength increases, the apparent affinity constants generally vary with salt concentration in titration experiments. This may simulate the existence of binding sites with different affinities in systems containing only one kind of site (15). Also, high concentrations of univalent salts may paradoxically increase the concentration of divalent cations in the cell walls by decreasing that of H^+, thus diminishing the protonation of the fixed charges (48, 191).

The ionic composition of the medium inside the cell walls depends on the concentration of the fixed charges. This parameter is difficult to assess because of uncertainties about the volume occupied by the Donnan phase (22, 141, 142). The volume of the apoplast accessible to solutes is classically (21) divided between water free space (WFS) and Donnan free space (DFS). WFS is defined as the volume of the free space in which, the electrical interactions of the free ions and the fixed charges being negligible, both anions and cations are present in equivalent amounts as freely diffusible ions. DFS corresponds to the regions in which the free ions are in the range of the electrostatic field of the ionized fixed charges. There is no defined frontier between DFS and WFS because the electrostatic interactions vanish progressively with distance from the charge-bearing polymers, in a manner depending (a) on the concentration of the ionized charges (determined by the structural characteristics of the polymer, by the pH, and by the concentrations of the cations capable of charge-masking association) and (b) on the ionic strength of the medium. The distinction between WFS and DFS may be considered a convenient representation of the fact that the charges are immobilized on structural polymers rather than being free in an homogenous phase (an implicit assumption of the ideal Donnan model). Indeed, treatment of the cell walls as charged surfaces, as described by the Gouy-Chapman model (45), may give results as satisfactory as the Donnan model, provided that the electrical diffuse layers overlap. This condition is probably satisfied only at low ionic strength, where the Debye length is comparable to the width of the micropores in the walls. Except in lignified materials (191), the DFS volume changes when the swelling state of the cell wall material is modified by changes (*a*) in electrostatic interactions between the charged polymers and (*b*) in osmotic gradients between the medium and the free space (111, 142, 143). Reciprocally, stretching the cell walls modifies the electrostatic interactions (68).

The amounts of ions exchanged between the cell walls and the medium may be used for thermodynamic formulations of selectivity coefficients based on the law of mass action, independently of the cell wall volume (2, 22, 48, 180). But in practice these coefficients are of limited value in explaining phenomena (141). When directly measured after blotting or centrifuging samples of isolated cell walls, the water content ranges from 1 to 6 ml·g^{-1} DW, with most of the estimates centered between 1 and 2 ml·g^{-1} DW (22, 111, 141, 143, 165, 181, 191). On the other hand, many experimental data on ion exchanges may be fitted with the predictions of theoretical (Donnan) models, when the DFS volume is considered as an adjustable parameter. In this case, the values chosen must be of the same magnitude as the direct estimates of the water content (1–2 ml·g^{-1} DW) (125, 142, 155, 161, 196). Thus, in many cases, the whole free space may be regarded as a DFS, the volume of which depends on the theoretical assumptions made about the mechanisms of ion accumulation. Other estimates of the fixed charge concentration have been obtained without using DFS volume, by measuring the electrical potential recorded by a microelectrode just touching the cell surface. The activity of the ionized fixed charges may be calculated in simple situations (149, 165, 166), assuming that the potential difference (PD) between the walls and the bulk medium is equal to the Donnan potential. However, the meaning of the recorded potential is not clear, because it depends on the actual structure of the polyelectrolyte phase at the tip of the microelectrode, which is not well defined. Thus, the estimates of the fixed charge activity obtained from electrical measurements may not be better than the other ones. The main advantage of the contact electrodes is the possibility of probing the macro-heterogeneities of the fixed charge distribution at the cell or tissue level (see below).

The mean concentrations of the fixed charges in the cell walls, calculated from CEC and DFS volume, or from electrical potential, with living tissues or isolated cell walls, range from about 0.1 M to 1.5 M. The highest values have been found in hydrophytes (143, 181), and the lowest ones in monocotyledons (83, 125, 156). Most of the values for terrestrial dicotyledons range from 0.2 M in parenchyma cells of the bean petiole (166), to 1 M in legume roots (155) and tomato xylem (191). Some of these values represent charge densities averaged over whole tissues and organs, and may mask large inhomogeneities. Starrach & Mayer (166) mapped the distribution of the apoplastic fixed charges in bean pulvinus. In both parenchyma and collenchyma cells they observed that the density of these charges was three times as high at the external surface of the cell walls as at the surface facing the cell lumen, in accordance with the histochemical data on pectin distribution and cation-binding capacity (145, 183). The walls of the motor cells displayed an especially high density of fixed charges, again in accordance with histochemical data.

The Apoplast as a Diffusion Barrier

Unstirred layers (ULs) are the regions in the medium where the rate of laminar flow is restricted by interactions with a solid surface. ULs may be conveniently modeled as regions in which solutes move only by diffusion, a process much slower than convection over macroscopic distances. Their thicknesses vary from 10 μm (phytoplankton in a well-stirred solution) to 1000 μm (*Chara corallina* in stagnant solution) (160, 186). In addition to the unstirred layers adjacent to the plant cells, there are unstirrable regions inside the micropores of the cell walls. In the absence of suberin or lignin incrustation, the micropores are large enough to allow the diffusive transport of water, ions, and even small proteins. Their diameter has been determined to be in the range 3–8 nm (6, 26, 27, 68, 159) and may be enlarged by mild treatment with pectinase, which suggests that the sieving properties of the walls are determined by pectin organization (6).

An UL can be looked upon as a membrane in series with the actual membrane, with a permeability coefficient determined by its thickness and by the diffusion coefficient of the molecule in aqueous solution. Simple calculations show that the UL permeability may be of the same order of magnitude as actual membrane permeability for rapidly permeating solutes. Hence, net fluxes between the cell and the medium may create significant concentration gradients inside the UL, and the transport of solutes may be rate-controlled by the UL. These phenomena strongly influence measurements of solute uptake by plant cells. The most well-documented effects are related to measurements of the kinetic parameters K_M and J_{max}. Apparent K_M values could differ from the true values by an order of magnitude or more because of the UL effect, as shown for phosphate uptake by *Anacystis nidulans* (compare 52 and 109). In the presence of ULs, the linear transformations commonly used to extract K_M and J_{max} introduce systematic errors in both these parameters. ULs cause convex curvature of Eadie-Hofstee transformations of kinetic data for single mediated transport systems, while multiple transport systems produce concave curvature in this plot, and these opposing effects can cancel out, leading to an erroneous apparently linear relation (128). In tissues with irregular surface, the UL thickness is larger over depressions than over protuberances. The theoretical analysis of this situation (174, 175) shows that a single transport system evenly distributed along the irregular surface can masquerade as multiple transport systems with differing K_M values. These effects of variable UL thickness may concern ion uptake by roots if the cells within the cortex are directly involved in ion absorption (46, 85). In summary, the naive application of the Lineweaver-Burk or Eadie-Hofstee plots to estimate kinetic parameters of transport systems that are influenced by ULs in the medium and the apoplast can lead to serious errors.

Another source of error due to ULs in the analysis of kinetic data comes from the surface pH shift maintained by active H^+ excretion coupled to rate-limiting H^+ diffusion through the ULs. Such a surface pH shift has been shown in the ULs adjacent to (102, 116) and inside (157, 173) the cell walls. The magnitude of the surface pH shift is dependent on the stirring rate, on the buffering power in the medium (101, 129, 157, 173), and on the pK value of the buffer (129). Since the CO_2/H_2CO_3 system can buffer the medium, the surface pH shift also depends on the partial pressure of CO_2 (178). Any modification in the external solution that changes the magnitude of the surface pH shift is likely to affect any process dependent on the H^+ electrochemical potential at the membrane surface. For instance, increasing the buffering power, or the HCO_3^- concentration in the absorption solution at constant bulk pH, decreased the uptake of hexose, NO_3^-, and $H_2PO_4^-$ by corn roots (70, 173, 178). In *Chara*, HCO_3^- uptake is usually estimated by measuring the rate of O_2 release with an O_2 electrode or the rate of $(^{14}C)HCO_3^-$ incorporation. The usual difference in stirring conditions between the two procedures leads to a 10–20-fold difference in UL thickness, and the resulting discrepancies between the measures of HCO_3^- uptake may be the source of misinterpretations (99). The activities of all extracellular, pH-sensitive enzymes (28, 61) are also likely to be affected by the steady-state surface pH-shift in ULs (see below).

Ion Diffusion in the Apoplast as Affected by Fixed Charges

The transport of ions through the cell wall ULs is classically thought to be influenced by the wall fixed charges. Diffusion in cell walls may be approached by considering them as a rigid, macroreticular ion exchanger consisting in a solid framework plus an interstitial pore phase. Diffusion in the pore phase is slower than in the medium for three reasons (75). First, pores occupy only a fraction of the total cross section of the cell walls. Second, the diffusion path is tortuous. Third, the actual mobility of the solutes in the pores is reduced by mechanical frictions and electrostatic interactions with the pore walls. The effective diffusion coefficients of solutes in ion exchangers integrate these three kinds of retarding effects. As a rule, the retardation is stronger for counter-ions than for co-ions, and stronger for polyvalent than for monovalent counter-ions, probably because of electrostatic attraction of counter-ions by the fixed ionic groups. Classical estimates of diffusion coefficients in cell walls are one order of magnitude lower than in free solution for univalent ions (124, 187), and still lower for multivalent ones (177).

There is no simple predictive relation between the effective diffusion coefficient of a given ion in an ion exchanger and the strength of the electrostatic interactions between the fixed charges and this ion (75, 177). Thus, the theoretical models of the cell wall ionic atmosphere at thermodynamic equilibrium (45, 49, 142, 155) cannot be used to predict how the electrostatic interactions affect ion movement in the apoplast. Studies in this field are scarce (66, 177). These complex phenomena seem to be responsible for the differences observed between rates of lateral escape of ions from the xylem (192, 193). They have not been explicitly taken into account in studies on ion uptake by plant cells, and one does not know whether they have any physiological importance.

Electrostatic Interactions Between the Fixed Charges and the Membrane Transporters

Membrane transport may be expected to be sensitive to the electrical properties of the cell walls for two reasons. First, the electrostatic field of the fixed charges may determine the ion concentrations near the active sites of the transporters (although concentrated in the middle lamellae, pectin is also present in the cell wall matrix, near the plasmalemma; see 126, 150). Of course, this could occur only if the mean distance between these sites and the fixed charges is not larger than the width of the electrical diffuse layer. Second, the diffusion of the ions across the walls, from the medium to the membrane and vice versa, may be controlled by the fixed charges (see above). Again, this could have significant consequences only if there is no short circuit by channels lined with neutral materials (as the ones hypothesized to support the old concept of WFS). The theoretical consequences of the surface electrostatic interactions on ion transport kinetics have been analyzed by Borst-Pauwels and his colleagues (8, 14, 146). The predicted apparent changes in maximal rates of uptake and in affinity constants, and departures from Michaelian kinetics, have been experimentally verified on yeast (13, 15, 172). The fixed charges responsible for these effects may belong to the walls, or to the membrane components. In yeast, the kinetic characteristics of cation uptake were not modified when walls were removed by digestion (62), which suggests that only the surface charge of the membrane is responsible for the electrostatic control mentioned above. In corn roots, both Ca^{2+} and La^{3+} (thought to be restricted in the apoplast) increased the apparent K_M of the high-affinity K^+ transport system (95), by the amount that would be theoretically expected if La^{3+} or Ca^{2+} binding had depolarized the cell surface 25–30 mV. This estimate falls well within the predictions of the model of Sentenac

& Grignon (155) applied to the electrochemical parameters of the cell walls of corn roots (156). There are also indications that the concentration of anions near the cell membrane is lowered by electrostatic repulsion by the cell wall. It has long been known that phosphate (Pi) uptake is stimulated by polyvalent cations. Since the importance of the stimulation depends only on the valency of the cation (58), and occurs without any lag (154), it is probably mediated by electrostatic interactions at the cell surface. In corn roots, the observed kinetics of phosphate influx versus pH differs from that predicted from the $H_2PO_4^-$ concentration in the medium, but is strikingly similar to that predicted from the concentration in the walls, as affected by the fixed charges (156). Thus, the cell wall Donnan potential seems to determine the effective concentration of the substrate of the high-affinity Pi transport system, as proposed by Franklin (58). In principle, the validity of this conclusion could be checked by comparing the transports in intact tissues and protoplasts. Unfortunately, the available data on this point (93, 96) have not been obtained in conditions suitable for evaluating the effects of the cell walls, because the high ionic strength and buffering power of the medium were likely to have attenuated the electrostatic interactions within the walls, as well as the surface pH shift due to H^+ excretion. The large depolarization often observed in protoplasts has been attributed to the loss of contact between the cell walls and the plasma membrane during plasmolysis (149). However, there is no indication that electrostatic interactions with the cell walls are the underlying causes of the important differences in membrane properties between cells and protoplasts (5, 33).

Apoplastic pH (pH_{cw})

Most of the data on the extracellular pH in roots are not really estimates of pH_{cw}, but of the pH in the ULs external to the root surface. They have been obtained with pH indicators in agar (107, 113, 189) or in porous beads stuck to the root surface (123), and by pH microelectrodes localized a few micrometers away from the epidermal surface (86, 116). Indirect estimates of the pH near the plasmalemma have been derived from measurements of weak acid influx (157, 173). In aerial organs, four kinds of methods have been used to estimate pH_{cw}. One is the measurement of the pH of sap extracted under pressure (1, 72), or by centrifugation (104), or of solution percolated on abraded leaf surface (47, 65). Another method uses H^+-selective electrodes placed in small volumes of solution equilibrated with the apoplast, in small holes cut in tissues (82, 88, 91, 92, 167), or simply appressed to tissue surface (3, 31, 182). In some cases, microelectrodes could be inserted in the apoplast of selected cells (17, 50). Spectral changes of fluorescent probes infiltrated in the apoplast have also been used for qualitative (24, 25, 50) or quantitative (122) determinations of pH_{cw}. Finally, pH_{cw} values have been estimated by incubating tissues in solutions with different pHs and determinating the "equilibrium" pH that corresponds to zero H^+ net flux (134, 184). No systematic bias is evident when comparing these methods, perhaps because of the variety of materials and of experimental conditions. The lowest reported values are just above pH 4, and the highest just above pH 7, but the majority lie between pH 5 and pH 6.5. As a whole, reported pH_{cw} values are higher for dicotyledons than for monocotyledons (coleoptiles, essentially), and higher for angiosperms than for gymnosperms (122).

Although the cell walls present a significant buffering capacity (122), pH_{cw} is dependent upon many factors. The apoplast is acidified in the light (182), because H^+ extrusion is stimulated by photosynthesis (106), and acidification is also evident in growing tissues (79, 103) (see below). Inhibition of growth by hydric stress is accompanied by an increase in pH_{cw} (72, 182). Nitrogen nutrition seems also to be a determinant of leaf pH_{cw}, NO_3^- assimilation being probably responsible for relatively high values (134). Finally, large differences in pH_{cw} exist between the various cells and tissues within a leaf. Tissues involved in assimilate transport (3, 24, 25), stomata (50), and motor cells of pulvini (88, 91, 92, 167) display acidified or alkalinized apoplasts, according to their physiological functions and states.

Solutes in the Apoplast

As discussed above, ion concentrations in the cortical apoplast of roots depend on those in the medium, and on the electrostatic characteristics of the fixed charges. The depletion in K^+, NH_4^+, and NO_3^- of the root ULs observed in diluted media (76, 86, 116) indicate that they also depend on the intensity of membrane transport. In the apoplast of aerial organs, the solute concentrations are determined by the balance of import via the xylem, absorption by cells, and export by the phloem. The importance of the xylem import for this balance is shown by the significant decrease in apoplastic solute concentration that is observed upon cessation of transpiration (12, 78, 118). The mean concentration of solutes is normally lower in the leaf apoplast than in the incoming xylem sap (12, 78), and it is buffered by cell transport against imposed changes (12). Such findings point to the importance of the rate of absorption in determining apoplast concentrations.

In growing tissues, as in stomata and in motor cells of pulvini, relatively high apoplastic concentrations of osmotically active solutes have been observed (36) and/or inferred from the water relations of the tissues (37, 38, 94, 108, 168). Inorganic electrolytes have been reported to account for about 25% of the apoplastic solutes (36), most of them being K^+ salts (59, 94, 168). Apoplasmic K^+ in aerial organs has been estimated either on extracted apoplastic sap (72) or in situ, with K^+-selective microelectrodes (12, 16, 87, 98), by the "equilibrium" method (190), and by compartmental analysis of washout curves (59). Electron microprobes have also been used (23, 60), but the results are likely to be affected by cytoplasmic K^+. As for H^+, three orders of magnitude separate the lowest estimates of the activity of K^+ in the apoplast from the highest ones, in conditions thought to be representative of normal functioning of various aerial organs. Most of the values lie between 2 and 5 mM and 100 mM, with the exception of those measured by Blatt (12) with a K^+-selective microelectrode inserted in drops of solution equilibrated with the apoplast of attached leaves of various plants (20–50 μM in nontranspiring leaves, and 150 μM in transpiring ones). This variability may not be totally accounted for by physiological differences, and part of it may originate from the experimental procedures (see below).

PHYSIOLOGICAL CONSEQUENCES

The Apoplast as an Ion Reservoir

Since the relative volume of the apoplast is very small, large variations of solute concentration may be obtained from small changes in membrane net fluxes. Thus, transporting solutes into or out of the apoplast could be the easiest way for rapid regulation of the turgor pressure (37, 39). Such a reservoir function of the apoplast is compatible with the high CEC of the apoplast, which facilitates the storage of cations and permits their mobilization upon acidification by the active extrusion of H^+ from the cell. Since H^+ ions, in contrast to K^+, are capable of masking the fixed charges, the H^+/K^+ exchange in the apoplast has probably two main consequences. First, the disappearance of part of the fixed charges augments the K^+ electrochemical potential, and this can favor K^+ uptake. Second, the osmotic potential in the walls is augmented because the bound H^+ ions are not osmotically active. The reservoir function of the apoplast is best established for stomata and motor cells of pulvini, in which rapid, massive exchanges of K^+ occur between specialized cells and their cell walls. The opening of stomata is accompanied by a transfer of K^+ from epidermal cells to guard cells, and closure is accompanied by the reverse transfer. Since guard cells are symplastically isolated (119), these changes in internal contents result from transports from, or into, the apoplast. In the open stomata of *Commelina* (16), the apoplastic K^+ activity varies from 3–5 mM in the guard cells, to 20 mM in epidermal cells. Dark-induced closure results in inverted gradients, with accumulation of K^+ (100 mM) at the surface of guard cells and depletion in K^+ of epidermal cell apoplast. The finding of high concentration of weak acid groups bound to the cell walls of *Commelina* stomata prompted Saftner & Raschke (149) to propose that the protonation of these groups during cell wall acidification could displace K^+ and make it available for uptake by the guard cells. This hypothesis has been strengthened by the demonstration of pH gradients in the apoplast around the stomata (17). In epidermal strips of *Commelina* (50), the steady-state pH in the apoplast of the different cells of the stomatal complex and epidermis is neutral when stomata are closed. After opening has been triggered by light, a burst of H^+ ions, originating from guard cells, decreases transiently their pH_{cw} by two pH units, and an acidification wave progress outwards (50). In parallel, an extracellular electrical current, attributed to H^+ ions extruded by guard cells, flows in the apoplast (18). During stomatal movements, 3–4 pmol K^+ are rapidly exchanged between the individual guard cell and its surrounding apoplast (132). From published data on cell volumes (64), and assuming that the volume of the free space is about 5% of the cell volume, one may estimate the volume of the apoplast of the guard cells plus the adjacent cells as 4–8 pL. This volume may contain 1–4 pmol fixed anionic charges (149). These figures indicate that the peristomatal apoplast has the capacity to act as a reservoir, to provide a significant part of the K^+ ions used for stomatal aperture.

The nastic movements of the leaflets of various Leguminosae are due to opposing changes in turgor of two sets of cells, in the extensor and flexor regions. These changes partly result from large variations in K^+ distribution between flexor and extensor, associated with variations in H^+ net transport between the cells and the medium or the apoplast (151). Compartmental analysis of K^+-washout curves from tangentially cut sections of the extensor and the flexor regions of the pulvinus of *Phaseolus* was used by Freundling et al (59) to determine apoplastic K^+. In both these tissues, the lowest apoplastic levels of K^+ were observed in states corresponding to swollen cells, and to highest levels in intracellular K^+. On the other hand, when the cells shrank, there was a two-fold increase in extracellular K^+ concentration, accompanied by a decrease in intracellular K^+. On the basis of these figures, the authors concluded that 30–40% of the K^+ gained or lost by the cells during the leaf movements was shuttled between the protoplasts and their near apoplast (the other 60–70% being exchanged with more distant tissues). More dynamic measurements with K^+-selective microelectrodes confirmed that the changes in K^+ activity were opposite in the apoplast and the protoplasts of the motor cells of *Samanea* (98). Two minutes after closure had been dark-triggered, the K^+ concentration began to increase from 15 mM in the apoplast of the (shrinking) extensor cells, to a stationary value of 75–100 mM. In parallel, the K^+ level decreased in the flexor apoplast. A different picture had previously been derived from X-ray microanalysis in freeze-dried sections of pulvini, with K^+ simultaneously accumulated in the apoplast and the protoplast of expanded cells (23, 60). But, even if the K^+ flows were driven by membrane transport across the apoplastic barriers that separate the flexor and the extensor (152), the results could not be understood because the concentration gradients described in the apoplastic part of the pathway should oppose (rather than drive) the ion flows from extensor to flexor during closure. Probably, the apoplast was contaminated by cytoplasmic K^+.

Potassium accumulation in specific regions of the pulvinus apoplast is facilitated by the high concentration of negative fixed charges (165), especially in the walls of the motor cells (166). As in stomata, the apoplastic fixed charges in pulvinus are weak acids (59), and their protonation upon H^+ excretion by the cells probably mobilizes K^+. Reversible changes in pH_{cw} accompanying spontaneous or triggered pulvinar movements, leading to pH_{cw} gradients exceeding 2 pH units across the pulvini, have been recorded by several authors using pH microelectrodes (88, 91, 92, 167). By comparing the changes in pH_{cw} that accompany swelling of motor cells in the extensor cell walls (167), and the titration curve for the fixed charges in these walls (59), one can calculate that during acidification half the anionic groups become protonated. The corresponding amount of fixed charges masked by protonation is close to the amount of K^+ that disappeared from the apoplast, as determined by washout curves (59).

In summary, the presence of fixed weak acid groups in the apoplast allows large accumulations of K^+ as well as binding of large amounts of H^+ when the apoplast is low in K^+ and mildly acidified. The storing capacities of the cell walls for both H^+ and K^+ are qualitatively and quantitatively adequate for the apoplast to act as a reservoir of ions during rapid regulation of the cell turgor.

H^+-Cotransports

The pH of the cell wall solution in roots or in submerged cells of hydrophytes may differ from that of the surrounding solution because of electrostatic H^+ accumulation in the DFS, and the restriction of H^+ diffusion through the ULs. For instance, H^+ extrusion may shift pH_{cw} to a value lower than that which would be expected at equilibrium from the Donnan potential alone. The magnitude of the pH drop due to electrostatic interactions can be more than one pH unit in dilute solutions, as calculated for cell walls equilibrated with a solution at pH 5.5 (155). Discrepancies between the effects of pH on weak acid (AH) uptake and theoretical titration curves have been ascribed to such an apoplasmic pH drop in studies dealing with IAA or abscisic acid uptake, the pH_{cw} drop being thought to increase the local concentration of the permeant AH form (51, 148). However, this reasoning is unsound (133) because equilibrium H^+ accumulation is necessarily accompanied by A^- exclusion, so that AH concentration remains unaffected. Similarly, the drop in pH_{cw} due to the Donnan effect has been proposed to facilitate carbon uptake by *Chara*, by speeding up the rate of HCO_3^- to CO_2 conversion (130, 160). However, the opposite electrostatic effects on HCO_3^- and H^+ concentrations would cancel out one another (54). In other words, cell walls at equilibrium with the surrounding medium cannot modify the electrochemical potential of the solutes that enter them from the external solution. Thus, the Donnan potential cannot directly participate to the energetic coupling of H^+-cotransports. The situation will be different when the pH and HCO_3^-/CO_2 concentrations in the apoplast are maintained out of equilibrium by active H^+ excretion and respiratory flux. In this case, the surface pH shift does increase H^+ electrochemical potential in cell walls (157). This phenomenon has been shown to play a part in the energetic coupling of ion uptake by corn roots (173, 178) and of HCO_3^- uptake by *Chara* cells (99, 160, 186). In corn roots, increasing the bulk pH from 6 to 8 would be expected to collapse the proton motive force. However, since the surface pH shift due to H^+ excretion has been estimated to be about 1 pH unit at pH 6 and 2.5 pH units at pH 8 (173), the proton motive force seems to be maintained, and drives H^+ recirculation across the cell membrane and associated cotransports (173). For instance, $H_2PO_4^-$ influx was not inhibited upon variation of bulk pH from pH 6 to pH 8, except when the surface pH shift had been suppressed, either by inhibiting the proton pump, or by buffering the medium (173, 178). In *Chara*, H^+ ions are known to participate in the transport of HCO_3^- (99, 130, 186). The surface of *Chara* internode cells exhibits distinct acid and alkaline bands, the former being specially involved in HCO_3^-/CO_2 uptake. The surface pH shift that characterizes the acid bands is likely to play a part in the energetic coupling of HCO_3^- transport (99, 186).

The magnitude of the pH shift induced by H^+ excretion is likely to depend both on H^+ transport rate and on cell wall resistance to H^+ diffusion. Several known anatomical differentiations seem to combine conditions for high-rate H^+ excretion with ULs highly resistant to H^+ diffusion. This combination could locally maintain pH_{cw} at a low value, independent of the bulk pH, and favorable to optimal functioning of H^+-cotransport. This may be one of the functions of the labyrinthine invaginations of the transfer cell walls of *Elodea* and *Potamogeton* (131), of swollen root tips and rhizodermal transfer cells of Fe-stressed plants (144), of proteoid roots (63), of the sealed apoplast at the symbiont interface in ectomycorrhiza (4), and of the so-called charasomes in *Chara* cells (130). Such an hypothesis has already been put forward in discussions dealing with mechanisms for nutrient transport between fungus and host in mycorrhizal roots (30, 162) but is better supported by experimental data for charasomes. These structures are complex invaginations of the plasma membrane of some charophytes, giving rise to a large increase in membrane surface area (55). Cytochemical studies have shown that the charasome membrane has considerable ATPase activity. It is surrounded by an extensive periplasmic space, distinct from the cell wall (56, 57) and filled with a particulate substance (121). It has been proposed that these features increase the diffusive resistance to H^+ efflux (130). Charasomes are specifically associated with the acid regions (55, 130); their formation is induced by akaline pH and stimulated at low HCO_3^- availability. In these conditions, there is a clear relationship between the density of charasomes in the acid bands and the capacity of the cell to transport HCO_3^- (130). From this whole set of observations, it seems reasonable to accept the hypothesis that charasomes are sites of high rates of H^+ excretion and accumulation at the membrane surface, capable of improving the energetic coupling of HCO_3^- transport. It is worthwhile noting that charasomes are probably involved in Cl^- uptake too (57).

Longitudinal pH profiles measured in the ULs of *Chara* (101) indicate that pH gradients of about 3 pH units can occur between alkaline and acid regions, within less than 1 mm. The corresponding pH gradients within the cell walls and along the membrane surface must be even greater. In roots, the estimates of the surface pH shift due to H^+ excretion in well-stirred medium, obtained by using weak acid influx as a pH probe (157), are minimum values, because they average the possible radial heterogeneities (158) and the longitudinal heterogeneities well demonstrated at the surface of roots (107, 113, 123, 189).

Assimilate transports in leaves are expected to depend on pH_{cw}, because these solutes are absorbed by H^+ cotransport systems. Gibberellic acid GA_3 has opposite effects on net H^+ transports by mesophyll and phloem, and it has been proposed that pH_{cw} is differentially regulated by GA_3, so that sucrose efflux occurs in the mesophyll and sucrose influx in the phloem (3, 43). Canny (24, 25) supplied leaves with sulforhodamine G (SR) via the transpiration stream and observed that this dye penetrated selectively some tissues. Since SR was considered a weak acid, highly permeant when protonated, its uptake was attributed to local acidification of the apoplast. The data collected on numerous species of legumes and trees reveal acidification of the apoplast of the cells that are transiently involved in recovery of xylem-born assimilates and storage of nitrogen (paraveinal mesophyll in early states of plant development), and of cells permanently involved in transport of assimilates into exporting vascular systems (bundle sheath). In view of the importance of the question, it would be interesting to assess rigorously the validity of SR as a probe of pH_{cw}.

Extracellular Enzyme Activity

As discussed above for the transporters, enzymes sequestered in the cell walls may be in the action range of the electrostatic fields generated by the fixed charges of cell walls and membrane surface. If the substrates and/or effectors are ionized, the apparent kinetic behavior of the immobilized enzymes will be dependent on the electrostatic properties of the cell walls. Further complications of the behavior of extracellular enzymes may result from the effects of the local pH on their binding to the cell walls. This may control their kinetic parameters and their stability, as for example the acidic peroxydases (147). The effects of the electrostatic repulsion of anionic substrate by the cell wall fixed charges are illustrated by the case of an extracellular phosphatase of sycamore and soybean cells, which displays Michaelian kinetics when purified in a soluble form and apparent cooperativity when bound to the cell walls (117). This cooperativity is enhanced when the electrostatic charge of the walls is augmented by removal of Ca^{2+} (41) and disappears if the electrostatic field is screened by high ionic strength (40, 42), reflecting the electrostatic interaction of the p-nitrophenyl phosphate by the cell wall fixed charges. The complex interplay of diffusion and electric phenomena, near a charged surface bearing an enzyme with ionized substrate and product, may create instabilities in substrate and product concentrations and propagate them along the surface (138, 139). How these phenomena could be involved in enzymic control of cell wall extension is illustrated by a model applied to soybean cell growth (140). The growth process is viewed as involving an extension step mediated by sliding of cellulose microfibrils, followed by partial hydrolysis of uronic acid methyl esters. These two steps are antagonist, with regard to the electrostatic characteristics of the walls: The charge density of the walls is diminished during the extension step, whereas it is increased by the following demethylation. In the pH 5–pH 8 range, the enzymes involved in the extension step are activated by decreasing pH (170), but the pectin methyl esterase is inhibited (112). The theoretical analysis of this system predicts reciprocal controls of the activities of the two sets of enzymes, mediated by the changes in pH_{cw} that result from the opposite changes in charge density during the cell wall extension, and the demethylation step (112, 115).

Apoplastic pH and Acid-Growth

The classical acid-growth theory states that IAA acidifies the apoplast by stimulating active H^+ excretion and that this results in activation of various processes that participate in cell wall loosening and cell extension (170). When care is taken to ensure a good contact between H^+ electrodes and the cell walls, good correspondence is observed between the apoplast acidification and the onset of IAA-stimulated elongation (31, 79). One fact in experimental support of this theory is that exogenous H^+ ions may partly mimic the effects of IAA on cell wall loosening associated with cell extension. The pH of the external buffers that reproduce the IAA-stimulated elongation rate could be considered representative of the natural pH_{cw} (184). But the meaning of such estimates is obscured by two main uncertainties, which have fueled the current controversy about the validity of the acid-growth theory. First, there is no agreement about the value of pH_{cw} that really triggers cell wall extension in natural conditions. Second, it is not certain whether pH mediates the effects of IAA, or IAA and pH act additively on cell wall extension.

In cell walls placed under tension, acid-induced extension has two components: an initial burst that decays exponentially with time, plus a constant rate extension persisting for several hours, and likely responsible for in vivo elongation (32). This long-term extension was stimulated when the pH was < 6, and the maximum response was obtained below pH 4 in isolated cell walls of cucumber hypocotyl and oat coleoptile (32, 35). On the other hand, Kutschera & Schopfer (89), using segments of living corn coleoptiles, in conditions that allowed the control of the apoplastic pH by external buffers, observed that the growth rate immediately after a pH jump was significantly augmented only at pH values lower than 4.5, and reached a maximum at pH 2.5. Analogous results have been obtained by Schopfer (153) with oat coleoptile, in which a pH of 3.5–4.0 was needed to mimic the IAA-stimulated elongation. In these two studies, the segments were incubated in water before use, and the apoplast pH equilibrated to approximately pH 5. On the other hand, authors who observed acid-induced growth at pH 5–6 (32, 79, 135) equilibrated the apoplast with strong neutral buffers before acid stimulation. Indeed, such pretreatment, in place of preincubation in water, rendered the elongation of corn coleoptiles much more sensitive to acid stimulation, since the maximum rate was obtained at pH 4 instead of pH 2.5 (89). Schopfer (153) showed that the growth stimulation of peeled oat coleoptiles at pH 5–6 observed by Rayle (135) is likely to correspond to spontaneous relaxation of the cell walls, after peeling eliminated the constraints due to the rigid epidermis. This spontaneous burst extension was inhibited during the pretreatment at neutral pH but took place when the tissue was transferred to buffers of lower pH. However, this phenomenon may only explain the initial burst of elongation, and not the sustained linear growth described by Cleland et al (32) and Cosgrove (35).

In summary, reproducing the effect of IAA on cell wall extension with acid buffers may necessitate high H^+ ion concentrations, at least one order of magnitude greater than that found (at equilibrium) in the WFS of tissue stimulated by IAA alone. In theory, part of this discrepancy could be accounted for by differences in electrostatic accumulation of H^+ ions in the DFS, and in surface pH shift in the ULs, between the treatments with IAA and those with buffers. However, the discrepancy has been observed even when pH_{cw} during IAA treatment was controlled by buffering the medium (135), or was independently determined by the zero H^+ net flux method (184). Finally, Kutschera & Schopfer (89) observed that addition of IAA at various pH values resulted in a constant augmentation in the initial elongation rate of abraded corn coleoptile, even in pH conditions that did not permit acid-induced growth. Thus, comparison of acid-induced, and IAA-induced cell elongation does not provide a mean for estimating pH_{cw} in growing tissues, perhaps because H^+ ions are not the only mediators of IAA action on the cell wall extension. Interestingly, the predictions of the acid-growth theory concerning the relation between growth and pH_{cw} seem valid in the case of FC-induced growth, at least for coleoptiles (90, 184).

Cell Expansion and Water Transport

The current controversy about the limiting factors of cell expansion is a consequence of our poor knowledge of the osmotic conditions in the apoplast. Growing tissues have low water potentials, typically −0.3 MPa. Assuming that this value is representative of the driving force for water movement leads to the conclusion that the water transport into the expanding cells is limited, not by the driving force, but by the resistance offered to water movement through the tissues (19). However, different conclusions may be reached if the osmotic properties of the solutes present in the apoplast are taken into account (34). A conventional view, which ignores the role of apoplast solutes, predicts that the water potential and the turgor pressure should be increased if the cellular growth rate is artificially reduced, or if the access of water to growing tissue is facilitated by infiltration under vacuum. However these predictions were not confirmed in growing pea internodes (37). Also, direct measurement of the turgor pressure of individual cells showed that the driving force for water movement through the cortex was very small. The conclusion was that the resistance to water flow was much lower than estimated from the water flow and water potential, and that the origin of the error was the presence of osmotically active solutes in the cell walls. Indeed, if most of the decrease in water potential occurs in the cell wall, rather than across the cell membrane, it would not be a very efficient way to drive water because the reflection coefficient is very low in the walls. This interpretation is supported by various data on the presence of significant amounts of osmotically active solutes in the apoplast of leaves and of the growing region of stems (36, 73, 94, 176). But other data support a totally different picture. Nonami & Boyer (118) pressurized the roots of pea and soybean seedlings, to force exudation of solution at the surface of the elongating region of the stems, at a rate approximately equal to that of the growth of cell volume. The osmotic potentials they measured were much smaller than those estimated by Cosgrove & Cleland (37). Independent determinations of the water potential and the hydrostatic potential of the apoplast confirmed that the osmotic potential was negligible. The conclusion was that the movement of water from the xylem to the growing cells was driven by large gradients in water potential, due to the hydrostatic tension developed by the expansion of cells, and transmitted to the apoplast, and that the limiting factor was the resistance to water movement in enlarging tissues. In the same study, Nonami & Boyer showed that the experimental procedures used by Cosgrove & Cleland (36) could augment artefactually the concentration of solutes in the apoplastic sap. It seems that solute concentration in the apoplast of aerial organs may undergo rapid variations upon slight changes in the balance between import and export rates by cells and xylem. This would explain the discrepancies between the results obtained on the same plants with different methods for collecting the apoplastic sap.

Apoplastic Ca^{2+} and Mechanical Properties of the Cell Walls

In some fruits, ripening is accompanied by tissue softening, which is associated with disorganization of the middle lamella and degradation of pectin by polygalacturonases (7, 126). On the other hand, exogenously applied Ca^{2+} and high pH treatment increase tissue firmness (84) and tensile strength (169). These results support the hypothesis that removal of bound Ca^{2+} by exchange for excreted H^+ is involved in the cell wall softening (84). However, the expected increase in H^+ extrusion and Ca^{2+} displacement have not been demonstrated (53). Many studies on cell growth also suggest that the load-bearing bonds that oppose the plastic deformation of walls are Ca^{2+} bridges. Most of these data concern cell walls from aerial axes of dicotyledonous plants rendered freely accessible to Ca^{2+} or Ca^{2+} chelators from the medium (81, 114, 127, 163, 185). Calcium and other cations inhibit elongation of some stems and bind to their isolated cell walls with the same selectivity sequence (10, 171). Only a minority of apoplastic Ca^{2+} ions seem to be associated with load-bearing bonds, because the plastic extensibility of the walls is increased greatly when the wall Ca^{2+} falls below some threshold (81, 185). Calcium associated to such bonds is the most difficult to extract, casting some doubt on whether in vivo acidification of the apoplast removes sufficient amounts of Ca^{2+} from the cell walls to reach the critical threshold. However, since the fraction of apoplasmic Ca^{2+} corresponding to this threshold is close to the fraction of uronic acids associated with Ca^{2+} in soybean "native" cell walls (195), it is possible that in natural conditions only a minute amount of Ca^{2+} needs to be removed to permit the deformation of the walls. Furthermore, since prolonged acidic treatments following removal of Ca^{2+} by chelators results in further augmentation of the plastic extensibility of boiled soybean cell walls, it is possible that acid-labile links, different from the Ca^{2+} bridges, exist in the walls (185).

The data on monocotyledonous plants (mostly from *Avena*) do not accord with those from the dicotyledonous species. In *Avena*, acidic treatments increased the cell wall extensibility only in "native" preparations. Boiling suppressed this response, suggesting that the acid-induced loosening is mediated by enzymes. Also, treatments known to extract Ca^{2+} were ineffective in increasing the plastic extensibility, and the inhibition of cell elongation by exogenous Ca^{2+} seemed to result from the inhibitory effect of this cation on enzymes responsible for cell wall loosening (137). Rayle (136) recently showed that EGTA may increase the plastic extensibility of frozen-thawed cell walls of *Avena* coleoptile but that this is due to the acidification caused by H^+ ions displaced by Ca^{2+} from EGTA. Displacement of the cell wall Ca^{2+} without a loosening effect was possible in conditions preventing acidification. Thus, the load-bearing bonds may differ between dicotylodenous plants and *Avena*. In the former, they may be Ca^{2+} bridges and possibly H^+ labile bonds. In the latter, their rupture necessitates attack by enzymes that are activated by H^+ and inhibited by Ca^{2+}

CONCLUSION

During the last ten years, following the review of Haynes (74), the focus of studies on the apoplast has shifted from roots to aerial parts, and from selectivity of cation exchanges to apoplastic pH, K^+, and osmotically active solutes. One of the reasons for this change is that it has not been possible to progress towards a mechanistic explanation of the well-documented correlations between the root CEC and the global mineral composition of the plants (74). On the other hand, the spreading of the chemiosmotic concepts and the development of the acid-growth theory have prompted many workers to pay attention to pH conditions at the cell surface. Finally, interesting models for the study of the role of the apoplast in cell transports have been provided by work on the rapid, massive ion transports characteristic of the motor cells. The emerging view is that, in aerial organs, and even in roots, the pH, ionic, and osmotic conditions in the apoplast can be controlled by the membrane transports, and reciprocally, can affect cell function. Since in many experimental situations, these conditions are different from those known in the bulk medium, they must be explicitly taken into account in the interpretations of the results. Unfortunately, much of our knowledge of these matters is based on circumstantial evidence. This is because preservation of the steady-state conditions in the apoplast is difficult, owing to the small relative size of this compartment, and direct measurements of these conditions are hampered by its complex structure. These difficulties have stimulated the elaboration of many theoretical models of the apoplast ionic atmosphere, which are still of limited value in explaining the effect of the apoplast on the cell biology. In this context, progress may be expected from the development of methods for measuring local concentrations, such as ion-selective microelectrodes and optical probes.

ACKNOWLEDGMENTS

We gratefully thank David Clarkson, Danièle Lacotte, and Jean-Claude Lacotte for their kindly help.

Literature Cited

1. Ackerson, R. C. 1982. Synthesis and movement of abscissic acid in water—stressed cotton leaves. *Plant Physiol.* 69:609–13

2. Allan, D. L., Jarrell, W. M. 1989. Proton and copper adsorption to maize and soybean root cell walls. *Plant Physiol.* 89:823–32

3. Aloni, B., Daie, J., Wyse, R. E. 1988. Regulation of apoplastic pH in source leaves of *Vicia faba* by gibberellic acid. 1988. *Plant Physiol.* 88:367–69

4. Ashford, A. E., Allaway, W. G., Peterson, C. A., Cairney, J. W. G. 1989. Nutrient transfer and the fungus-root interface. *Aust. J. Plant Physiol.* 16:85–97

5. Barrier-Brygoo, H., Ephritikhine, G., Klämbt, D., Ghislain, M., Guern, J. 1989. Functional evidence for an auxin receptor at the plasmalemma of tobacco mesophyll protoplasts. *Proc. Natl. Acad. Sci. USA* 86:891–95

6. Baron-Epel, O., Gharayal, P. K., Schindler, M. 1988. Pectin as mediators of wall porosity in soybean cells. *Planta* 175:389–95

7. Bartley, I. M., Knee, M., Casimir, M. A. F. 1982. Fruit softening. I. Changes in cell wall composition and endopolygalacturonase activity in ripening pears. *J. Exp. Bot.* 33:1248–55

8. Barts, P. W. J. A., Borst-Pauwels, G. W. F. H. 1985. Effects of membrane potential and surface potential on the kinetics of solute transport. *Biochim. Biophys. Acta* 813:51–60

9. Baydoun, E. A. H., Brett, C. T. 1984. The effect of pH on the binding of calcium to pea epicotyl cell walls and its implications for the control of cell extension. *J. Exp. Bot.* 35:1820–31

10. Baydoun, E. A. H., Brett, C. T. 1988. Properties and possible physiological significance of cell wall calcium binding in etiolated pea epicotyls. *J. Exp. Bot.* 39:199–208

11. Berlin, J., Quisenberry, J. E., Bailey, F., Woodworth, M., McMichael, B. L. 1982. Effect of water stress on cotton leaves. I. An electron microscopic stereological study of the palisade cells. *Plant Physiol.* 70:238–43

12. Blatt, M. R. 1985. Extracellular potassium activity in attached leaves and its relation to stomatal function. *J. Exp. Bot.* 36:240–51

13. Borst-Pauwels, G. W. F. H. 1981. Ion transport in yeast. *Biochim. Biophys. Acta* 650:88–127

14. Borst-Pauwels, G. W. F. H., Severens, P. P. J. 1984. Effect of the surface potential upon ion selectivity found in competitive inhibition of divalent cation uptake. A theoretical approach. *Physiol. Plant.* 60:86–91

15. Borst-Pauwels, G. W. F. H., Theuvenet, A. P. R. 1984. Apparent saturation kinetics of divalent cation uptake in yeast caused by a reduction in the surface potential. *Biochim. Biophys. Acta* 771:171–76

16. Bowling, D. J. F. 1987. Measurement of the apoplastic activity of K^+ and Cl^- in the leaf epidermis of *Commelina communis* in relation to stomatal activity. *J. Exp. Bot.* 38:1351–55

17. Bowling, D. J. F., Edwards, A. 1984. pH gradients in the stomatal complex of *Tradescantia virginiana. J. Exp. Bot.* 35:1641–45

18. Bowling, D. J. F., Edwards, M. C., Gow, N. A. R. 1986. Electrical currents at the leaf surface of *Commelina communis* and their relationship to stomatal activity. *J. Exp. Bot.* 37:876–82

19. Boyer, J. S. 1985. Water transport. *Annu. Rev. Plant Physiol.* 36:473–516

20. Briarty, L. G. 1980. Stereological analysis of cotyledon cell development in *Phaseolus.* I. Analysis of a cell model. *J. Exp. Bot.* 31:1379–86

21. Briggs, G. E., Robertson, R. N. 1957. Apparent free space. *Annu. Rev. Plant Physiol.* 8:11–30

22. Bush, D. S., McColl, J. G. 1987. Mass-action expressions of ion exchange applied to Ca^{2+}, H^+, K^+ and Mg^{2+} sorption on isolated cell walls of leaves from *Brassica oleracea. Plant Physiol.* 85: 247–60

23. Campbell, N. A., Satter, R. L., Garber, R. C. 1981. Apoplastic transport of ions in the motor organ of *Samanea. Proc. Natl. Acad. Sci. USA* 78:2981–84

24. Canny, M. J. 1987. Locating active proton extrusion in leaves. *Plant Cell Environ.* 10:271–74

25. Canny, M. J. 1988. Bundle sheath tissues of legume leaves as a site of recovery of solutes from the transpiration stream. *Physiol. Plant.* 73:457–64

26. Carpita, N. C. 1982. Limiting diameter of pores and the surface structure of plant cell walls. *Science* 218:813–14

27. Carpita, N. C., Sabularse, D., Montezinos, D., Delmer, D. P. 1979. Determination of the pore size of cell walls of living plant cells. *Science* 205:1144–47

28. Cassab, G. I., Varner, J. E. 1988. Cell wall proteins. *Annu. Rev. Plant Physiol. Plant Mol. Biol.* 39:321–53

29. Clarkson, D. T. 1984. Calcium transport between tissues and its distribution in the plant. *Plant Cell Environ.* 7:449–56

30. Clarkson, D. T. 1985. Factors affecting mineral nutrient acquisition by plants. *Annu. Rev. Plant Physiol.* 36:77–115

31. Cleland, R. E. 1976. Fusicoccin-induced growth and hydrogen ion excretion of *Avena* coleoptiles: relation to auxin responses. *Planta* 128:20–6

32. Cleland, R. E., Cosgrove, D., Tepfer, M. 1987. Long-term acid-induced wall extension in an *in-vitro* system. *Planta* 170:379–85

33. Cornel, D., Grignon, C., Rona, J.-P., Heller, R. 1983. Measurement of intracellular potassium activity in protoplasts of *Acer pseudoplatanus:* origin of their electropositivity. *Physiol. Plant.* 57:203–9

34. Cosgrove, D. J. 1986. Biophysical control of plant cell growth. *Annu. Rev. Plant Physiol.* 37:377–405

35. Cosgrove, D. J. 1989. Characterization of long-term extension of isolated cell walls from growing cucumber hypocotyls. *Planta* 177:121–30

36. Cosgrove, D. J., Cleland, R. E. 1983. Solutes in the free space of growing stem. *Plant Physiol.* 72:326–31

37. Cosgrove, D. J., Cleland, R. E. 1983. Osmotic properties of pea internodes in relation to growth and auxin action. *Plant Physiol.* 72:332–38

38. Cosgrove, D. J., Steudle, E. 1981. Water relations of growing pea epicotyl segments. *Planta* 153:343–50

39. Cram, W. J. 1980. See Ref. 164, pp. 3–13

40. Crasnier, M., Giordani, R. 1985. Elution of acid phosphatase from sycamore cell walls. *Plant Sci.* 40:35–41

41. Crasnier, M., Moustacas, A. M., Ricard, J. 1985. Electrostatic effects and calcium ion concentration as modulators of acid phosphatase bound to plant cell walls. *Eur. J. Biochem.* 151:187–90

42. Crasnier, M., Noat, G., Ricard, J. 1980. Purification and molecular properties of acid phosphatase from sycamore cell walls. *Plant Cell Environ.* 3:217–24

43. Daie, J. 1987. Interaction of cell turgor and hormones on sucrose uptake rates in isolated phloem of celery. *Plant Physiol.* 84:1033–37

44. Dainty, J., De Michelis, M. I., Marrè, E., Rasi-Caldogno, F., ed. 1989. *Plant Membrane Transport: The Current Position.* Amsterdam/New York/Oxford: Elsevier. 712 pp.

45. Dainty, J., Hope, A. B. 1961. The electric double layer and the Donnan equilibrium in relation to plant cell walls. *Aust. J. Biol. Sci.* 14:541–51

46. Dalton, F. N. 1984. Dual pattern of potassium transport in plant cells: a physical artifact of a single uptake mechanism. *J. Exp. Bot.* 35:1723–32

47. Delrot, S., Faucher, M., Bonnemain, J.-L., Bomort, J. 1983. Nycthemeral changes in intracellular and apoplastic sugars in *Vicia faba* leaves. *Physiol. Vég.* 21:459–67

48. Demarty, M., Morvan, C., Thellier, M. 1978. Exchange properties of isolated cell walls of *Lemna minor* L. *Plant Physiol.* 62:477–81

49. Demarty, M., Ripoll, C., Thellier, M. 1980. See Ref. 164, pp. 33–44

50. Edwards, M. C., Smith, G. N., Bowling, D. J. F. 1988. Guard cells extrude protons prior to stomatal opening—a study using fluorescence microscopy and pH microelectrodes. *J. Exp. Bot.* 39:1541–47

51. Everat-Bourbouloux, A., Delrot, S., Bonnemain, J. L. 1984. Propriétés de l'absorption et distribution de l'acide abscissique dans les tissus caulinaires de *Vicia faba. Physiol. Vég.* 22:37–46

52. Falkner, G., Horner, F., Simonis, W. 1980. The regulation of the energy-dependent phosphate uptake by the blue-green alga *Anacystis nidulans. Planta* 149:138–43

53. Ferguson, I. B. 1984. Calcium in plant senescence and fruit ripening. *Plant Cell Environ.* 7:477–89

54. Ferrier, J. M. 1980. Apparent bicarbonate uptake and possible plasmalemma proton efflux in *Chara corallina. Plant Physiol.* 66:1198–99

55. Franceschi, V. R., Lucas, W. J. 1980. Structure and possible function(s) of charasomes: complex plasmalemma–cell wall elaborations present in some characean species. *Protoplasma* 104:253–71

56. Franceschi, V. R., Lucas, W. J. 1981. The charasome periplasmic space. *Protoplasma* 107:269–84

57. Franceschi, V. R., Lucas, W. J. 1982. The relationship of the charasome to chloride uptake in *Chara corallina:* physiological and histochemical investigations. *Planta* 154:525–37

58. Franklin, R. E. 1969. Effect of adsorbed cations on phosphorus uptake by excised roots. *Plant Physiol.* 44:697–700

59. Freundling, C., Starrach, N., Flach, D., Gradmann, D., Mayer, W.-E. 1988. Cell walls as reservoirs of potassium ions for reversible volume changes of pulvinar motor cells during rhythmic leaf movements. *Planta* 175:193–203

60. Fromm, J., Eschrich, W. 1988. Transport processes in stimulated and nonstimulated leaves of *Mimosa pudica*. III. Displacement of ions during seismonastic leaf movements. *Trees* 2:65–72

61. Fry, S. C. 1986. Cross-linking of matrix polymers in the growing cell walls of angiosperms. *Annu. Rev. Plant Physiol.* 37:165–86

62. Gage, R. A., Wijngaarden, W., Theuvenet, A. P. R., Borst-Pauwels, G. W. F. H. 1985. Inhibition of Rb$^+$ uptake in yeast by Ca^{2+} is caused by a reduction in the surface potential and not in the Donnan potential of the cell wall. *Biochim. Biophys. Acta* 812:1–8

63. Gardner, W. K., Parbery, D. G., Barber, D. A. 1982. The acquisition of phosphorus by *Lupinus albus* L. I. Some characteristics of the soil/root interface. *Plant Soil* 68:19–32

64. Garrec, J.-P., Vavasseur, A., Michalowicz, G., Laffray, D. 1983. Stomatal movements and repartition of the elements K, Cl, Na, P, Ca, Mg and S in the stomatal complexes of *Vicia faba* and *Commelina communis*. Electron probe studies. *Z. Pflanzenphysiol.* 112:35–42

65. Geiger, D. R., Sovonick, S. A., Shock, T. L., Fellows, R. J. 1984. Role of free space in translocation in sugar beet. *Plant Physiol.* 54:892–98

66. Gillet, C., Lefebvre, J. 1978. Ionic diffusion through the *Nitella* cell wall in the presence of calcium. *J. Exp. Bot.* 29:1155–59

67. Gillet, C., Van Cutsem, P., Voue, M. 1989. Correlation between the weight loss induced by alkaline ions and the cationic exchange capacity of the *Nitella* cell wall. *J. Exp. Bot.* 40:129–33

68. Gogarten, J. P. 1988. Physical properties of the cell wall of photoautotrophic suspension cells from *Chenopodium rubrum* L. *Planta* 174:333–39

69. Grant, G. T., Morris, E. R., Rees, D. A., Smith, P. J. C., Thom, D. 1973. Biological interactions between polysaccharides and divalent cations: the egg-box model. *FEBS Lett.* 32:195–98

70. Grunwaldt, G., Ehwald, R. 1983. Specificity and reversibility of the bicarbonate effect on hexose uptake by maize root tips. *J. Exp. Bot.* 34:1347–57

71. Hajibagheri, M. A., Hall, J. L., Flowers, T. J. 1984. Stereological analysis of leaf cells of the halophyte *Suaeda maritima* (L.) Dum. *J. Exp. Bot.* 35:1547–57

72. Hartung, W., Radin, J. W., Hendrix, D. L. 1988. Abscissic acid movement into the apoplastic solution of water-stressed cotton leaves. *Plant Physiol.* 86:908–13

73. Harvey, D. M. R., Hall, J. L., Flowers, T. J., Kent, B. 1981. Quantitative ion localization within *Sueda maritima* leaf mesophyll cells. *Planta* 151:555–60

74. Haynes, R. J. 1980. Ion exchange properties of roots and ionic interactions within the root apoplasm: their role in ion accumulation by plants. *Bot. Rev.* 46:75–99

75. Helfferich, F. 1962. *Ion Exchange*. New York: McGraw-Hill. 624 pp.

76. Henriksen, G. H., Bloom, A. J., Spanswick, R. M. 1990. Measurement of net fluxes of ammonium and nitrate at the surface of barley roots using ion-selective microelectrodes. *Plant Physiol.* 93:271–80

77. Homblé, F., Ritcher, C., Dainty, J. 1989. Leakage of pectins from the cell wall of *Chara corallina* in the absence of divalent cations. *Plant Physiol. Biochem.* 27:463–68

78. Jachetta, J. J., Appleby, A. P., Boersma, L. 1986. Use of the pressure vessel to measure concentrations of solutes in apoplastic and membrane-filtered symplastic sap in sunflower leaves. *Plant Physiol.* 82:995–99

79. Jacobs, M., Ray, P. M. 1976. Rapid auxin-induced decrease in free space pH and its relationship to auxin-induced growth in maize and pea. *Plant Physiol.* 58:210–13

80. Jarvis, M. C. 1982. The proportion of calcium-bound pectin in plant cell walls. *Planta* 154:344–46

81. Jarvis, M. C., Logan, A. S., Duncan, H. J. 1984. Tensile characteristics of collenchyma cell walls at different calcium contents. *Physiol. Plant.* 61:81–86

82. Johnson, K. A., Jacobs, M. 1983. Measurement of free space pH in intact plants using a pH microelectrode. *Plant Physiol.* 72:S80

83. Keller, E. R. J., Dahse, I., Müller, E. 1980. Ion exchange properties of the corn coleoptile cell wall space. I. Concentration potentials and fixed charge activity. *Biochem. Physiol. Pflanz.* 175:643–52

84. Knee, M. 1982. Fruit softening. III. Requirement for oxygen and pH effects. *J. Exp. Bot.* 33:1263–69

85. Kochian, L. V., Lucas, W. J. 1983. Potassium transport in corn roots. II. The significance of root periphery. *Plant Physiol.* 73:208–15

86. Kochian, L. V., Shaff, J. E., Lucas, W. J. 1989. High affinity K$^+$ uptake in maize roots. A lack of coupling with H$^+$ efflux. *Plant Physiol.* 91:1202–11

87. Kumon, K., Tsurumi, S. 1984. Ion efflux from pulvinar cells during slow downward movement of the petiole of

Mimosa pudica L. induced by photo-stimulation. *J. Plant Physiol.* 115:439–43

88. Kumon, K., Suda, S. 1985. Changes in extracellular pH of the motor cells of *Mimosa pudica* L. during movement. *Plant Cell Physiol.* 26:375–77

89. Kutschera, U., Schopfer, P. 1985. Evidence against the acid-growth theory of auxin action. *Planta* 163:483–93

90. Kutschera, U., Schopfer, P. 1985. Evidence for the acid-growth theory of fusicoccin action. *Planta* 163:494–99

91. Lee, Y., Satter, R. L. 1988. Effects of temperature on H^+ uptake and release during circadian rhythmic movements of excised *Samanea* motor organs. *Plant Physiol.* 86:352–54

92. Lee, Y., Satter, R. L. 1989. Effects of white, blue, red light and darkness on pH of the apoplast in the *Samanea* pulvinus. *Planta* 178:31–40

93. Lefebvre, D. D., Clarkson, D. T. 1987. High affinity phosphate absorption is independent of the root cell wall in *Pisum sativum*. *Can. J. Bot.* 65:1504–8

94. Leigh, R. A., Tomos, A. D. 1983. An attempt to use isolated vacuoles to determine the distribution of sodium and potassium in cells of storage roots of red beet *(Beta vulgaris)*. *Planta* 159:469–75

95. Leonard, R. T., Nagahashi, G., Thomson, W. W. 1975. Effect of lanthanum on ion absorption in corn roots. *Plant Physiol.* 55:542–46

96. Lin, W. 1979. Potassium and phosphate uptake in corn roots. Further evidence for an electrogenic H^+/K^+ exchanger and an OH^-/Pi antiporter. *Plant Physiol* 63:952–55

97. Liners, F., Letesson, J.-J., Didembourg, C., Van Cutsem, P. 1989. Monoclonal antibodies against pectin. Recognition of a conformation induced by calcium. *Plant Physiol.* 91:1419–24

98. Lowen, C. Z., Satter, R. L. 1989. Light-promoted changes in apoplastic K^+ activity in the *Samanea* pulvinus, monitored with liquid membrane microelectrodes. *Planta* 179:421–27

99. Lucas, W. J. 1985. See Ref. 100, pp. 229–54

100. Lucas, W. J., Berry, J. A., ed. 1985. *Inorganic Carbon Uptake by Aquatic Photosynthetic Organisms*. Rockville: Am. Soc. Plant Physiol. 494 pp.

101. Lucas, W. J., Keifer, D. W., Sanders, D. 1983. Bicarbonate transport in *Chara corallina*: evidence for cotransport of HCO_3^- with H^+. *J. Membr. Biol.* 73:263–74

102. Lucas, W. J., Smith, F. A. 1973. The formation of alkaline and acid regions at the surface of *Chara corallina* cells. *J. Exp. Bot.* 24:1–14

103. Lürsen, K. 1978. The surface pH of *Avena* coleoptiles and its correlation with growth. *Plant Sci. Lett.* 13:309–13

104. Madore, M., Webb, J. A. 1981. Leaf free space analysis and vein loading in *Cucurbita pepo*. *Can. J. Bot.* 59:2550–87

105. Manning, G. S. 1969. Limiting laws and counterion condensation in polyelectrolyte solutions. I. Colligative properties. *J. Chem. Phys.* 51:924–33

106. Marrè, M. T., Albergoni, F. G., Moroni, A., Marrè, E. 1989. Light-induced activation of electrogenic H^+ extrusion and K^+ uptake in *Elodea densa* depends on photosynthesis and is mediated by the plasma membrane ATPase. *J. Exp. Bot.* 40:343–52

107. Marschner, H., Römheld, V., Ossenberg-Neuhaus, H. 1982. Rapid method for measuring changes in pH and reducing processes along roots of intact plants. *Z. Pflanzenphysiol.* 105:407–16

108. Mayer, W.-E., Flach, D., Raju, M. V. S., Starrach, N., Wiech, E. 1985. Mechanics of circadian pulvini movements in *Phaseolus coccineus* L. *Planta* 163:381–90

109. Mierle, G. 1985. A method for estimating the diffusion resistance of the unstirred layers of microorganisms. *Biochim. Biophys. Acta* 812:827–34

110. Morvan, C., Demarty, M., Thellier, M. 1979. Titration of isolated cell walls of *Lemna minor* L. *Plant Physiol.* 63:1117–22

111. Morvan, C., Demarty, M., Thellier, M. 1985. Propriétés physicochimiques des parois de *Lemna minor*: échanges entre les ions calcium et lithium. *Physiol. Vég.* 23:333–44

112. Moustacas, A.-M., Nari, J., Diamantidis, G., Noat, G., Crasnier, M., Borel, M., Ricard, J. 1986. Electrostatic effects and the dynamics of enzyme reactions at the surface of plant cells. 2. The role of pectin methyl esterase in the modulation of electrostatic effects in soybean cell walls. *Eur. J. Biochem.* 155:191–98

113. Mulkey, T. J., Evans, M. L. 1981. Geotropism in corn roots: evidence for its mediation by differential acid efflux. *Science* 212:70–71

114. Nakajima, N., Morikawa, H., Igasashi, S., Senda, M. 1981. Differential effects of calcium and magnesium on mechanical properties of pea stem cell walls. *Plant Cell Physiol.* 22:1305–15

115. Nari, J., Noat, G., Diamantidis, G., Woudstra, M., Ricard, J. 1986. Electrostatic effects and the dynamics of enzyme reactions at the surface of plant cells. 3. Interplay between limited cell wall autolysis, pectin methyl esterase activity, and electrostatic effects in soybean cell walls. *Eur. J. Biochem.* 155: 199–202

116. Newman, I. A., Kochian, L. V., Grusak, M. A., Lucas, W. J. 1987. Fluxes of H^+ and K^+ in corn roots. Characterization and stoichiometries using ion-selective microelectrodes. *Plant Physiol.* 84:1177–84

117. Noat, G., Crasnier, M., Ricard, J. 1980. Ionic control of acid phosphatase activity in plant cell walls. *Plant Cell Environ.* 3:225–29

118. Nonami, H., Boyer, J. S. 1987. Origin of growth-induced water potential. Solute concentration is low in apoplast of enlarging tissues. *Plant Physiol.* 83: 596–601

119. Palevitz, B. A., Hepler, P. K. 1985. Changes in dye coupling of stomatal cells of *Allium* and *Commelina* demonstrated by micro-injection of lucifer yellow. *Planta* 164:473–79

120. Parkhurst, D. F. 1982. Stereological methods for measuring internal leaf structure variables. *Am. J. Bot.* 69:31–39

121. Pesacreta, T. C., Lucas, W. J. 1984. Plasma membrane coat and a coated vesicle–associated reticulum of membranes: their structure and possible interrelationship in *Chara corallina*. *J. Cell Biol.* 98:1537–45

122. Pfanz, H., Dietz, K.-J. 1987. A fluorescent method for the determination of the apoplastic proton concentration in intact leaf tissues. *J. Plant Physiol.* 29:41–48

123. Pilet, P. E., Versel, J. M., Mayor, G. 1983. Growth distribution and surface pH patterns along maize roots. *Planta* 158:398–402

124. Pitman, M. G. 1982. Transport across plant roots. *Q. Rev. Biophys.* 15:481–54

125. Pitman, M. G., Lüttge, U., Kramer, D., Ball, E. 1974. Free space characteristics of barley leaf slices. *Aust. J. Plant Physiol.* 1:65–75

126. Platt-Aloia, K. A., Thomson, W. W., Young, R. E. 1980. Ultrastructural changes in the walls of ripening avocados: transmission, scanning and freeze fracture microscopy. *Bot. Gaz.* 141: 366–73

127. Prat, R., Gueissaz, M. B., Goldberg, R. 1984. Effects of Ca^{2+} and Mg^{2+} on elongation and H^+ secretion of *Vigna*

radiata hypocotyl sections. *Plant Cell Physiol.* 25:1459–67

128. Preston, R. L. 1982. Effects of unstirred layers on the kinetics of carrier mediated solute transport by two systems. *Biochim. Biophys. Acta* 668:422–28

129. Price, G. D., Badger, M. R. 1985. Inhibition by proton buffers of photosynthetic utilization of bicarbonate in *Chara corallina*. *Aust. J. Plant Physiol.* 12: 257–67

130. Price, G. D., Badger, M. R., Bassett, M. E., Whitecross, M. I. 1985. Involvement of plasmalemmasomes and carbonic anhydrase in photosynthetic utilization of bicarbonate in *Chara corallina*. *Aust. J. Plant Physiol.* 12:241–56

131. Prins, H. B. A., Snel, J. F. H., Zanstra, P. E., Helder, R. J. 1982. The mechanism of bicarbonate assimilation by the polar leaves of *Potamogeton* and *Elodea*. CO_2 concentration at the leaf surface. *Plant Cell Environ.* 1:231–38

132. Raschke, K., Hedrich, R., Reckmann, U., Schroeder, J. I. 1988. Exploring biophysical and biochemical components of the osmotic motor that drives stomatal movement. *Bot. Acta* 101:283–94

133. Raven, J. A. 1975. Transport of indolaecetic acid in plant cells in relation to pH and electrical potential gradients, and its significance for polar IAA transport. *New Phytol.* 74:163–72

134. Raven, J. A., Farquhar, G. D. 1989. See Ref. 44, pp. 607–10

135. Rayle, D. L. 1973. Auxin-induced hydrogen-ion secretion in *Avena* coleoptiles and its implications. *Planta* 114:63–73

136. Rayle, D. L. 1989. Calcium bridges are not load-bearing cell walls bonds in *Avena* coleoptiles. *Planta* 178:92–95

137. Rayle, D. L., Cleland, R. E. 1977. Reevaluation of the effects of calcium ions on auxin-induced elongation. *Plant Physiol.* 60:709–12

138. Ricard, J., Noat, G. 1984. Enzyme reactions at the surface of the living cell. I. Electric repulsion of charged ligands and recognition of signals from the external milieu. *J. Theor. Biol.* 109:555–69

139. Ricard, J., Noat, G. 1984. Enzyme reactions at the surface of the living cell. II. Destabilization in the membranes and conduction of the signals. *J. Theor. Biol.* 109:571–80

140. Ricard, J., Noat, G. 1986. Electrostatic effects and the dynamics of enzyme reactions at the surface of plant cells. 1. A theory of the ionic control of a complex multienzyme system. *Eur. J. Biochem.* 155:183–90

141. Ritcher, C., Dainty, J. 1989. Ion be-

havior in plant cell walls. I. Characterization of the *Sphagnum russowii* cell wall ion exchanger. *Can. J. Bot.* 67:451–59

142. Ritcher, C., Dainty, J. 1989. Ion behavior in plant cell walls. II. Measurement of the Donnan free space, anion-exclusion space, anion exchange capacity, and cation-exchange capacity in delignified *Sphagnum russowii* cell wall. *Can. J. Bot.* 67:460–65

143. Ritchie, R. J., Larkum, A. W. D. 1982. Cation exchange properties of the cell walls of *Enteromorpha intestinalis* L. Link. (Ulvales, Chlorophyta). *J. Exp. Bot.* 132:125–39

144. Römheld, V., Marschner, H. 1981. Iron deficiency stress induced morphological and physiological changes in root tips of sunflower. *Physiol. Plant.* 53:354–60

145. Roland, J.-C., Bessoles, M. 1968. Mise en évidence du calcium dans les cellules du collenchyme. *CR Acad. Sci. Paris Sér. D* 267:589–92

146. Roomans, G. M., Borst-Pauwels, G. W. F. H. 1978. Co-transport of anions and neutral solutes with cations across charged biological membranes. Effect of surface potential on uptake kinetics. *J. Theor. Biol.* 73:453–58

147. Ros Barceló, A., Pedreño, M. A., Sabater, F., Muñoz, R. 1989. Control by pH of cell wall peroxydase activity involved in lignification. *Plant Cell Environ.* 30:237–41

148. Rubery, P. H., Sheldrake, A. R. 1973. Effect of pH and surface charge on cell uptake of auxin. *Nature New Biol.* 244:285–88

149. Saftner, R. A., Raschke, K. 1981. Electrical potentials in stomatal complexes. *Plant Physiol.* 76:1124–32

150. Sasaki, K., Nagahashi, G. 1989. Autolysis-like release of pectic polysaccharides from regions of cell walls other than the middle lamella. *Plant Cell Physiol.* 30:1159–69

151. Satter, R. L., Galston, A. W. 1981. Mechanisms of control of leaf movements. *Annu. Rev. Plant Physiol.* 32: 83–110

152. Satter, R. L., Garber, R. C., Khairallah, L., Cheng, Y.-S. 1982. Elemental analysis of freeze-dried thin sections of *Samanea* motor organs: barriers to ion diffusion through the apolast. *J. Cell Biol.* 95:893–902

153. Schopfer, P. 1989. pH-Dependence of extension growth in *Avena* coleoptiles and its implications for the mechanism of auxin action. *Plant Physiol* 90:202–7

154. Sentenac, H. 1988. *Relation structure/-fonction dans la racine: effets du squelette pariétal sur les transports membranaires.* Thèse de Doctorat d'Etat. Univ. Montpellier II, 173 pp.

155. Sentenac, H., Grignon, C. 1981. A model for predicting ionic equilibrium concentrations in cell walls. *Plant Physiol.* 68:415–19

156. Sentenac, H., Grignon, C. 1985. Effect of pH on orthophosphate uptake by corn roots. *Plant Physiol.* 1985:136–41

157. Sentenac, H., Grignon, C. 1987. Effect of H^+ excretion on the surface pH of corn root cells evaluated by using weak acid influx as a pH probe. *Plant Physiol.* 84:1367–72

158. Sentenac, H., Thibaud, J.-B., Grignon, C. 1989. See Ref. 44, pp. 611–14

159. Shepherd, V. A., Goodwin, P. B. 1989. The porosity of permeabilised *Chara* cells. *Aust. J. Plant Physiol.* 16:231–39

160. Smith, F. A. 1985. See Ref. 100, pp. 1–15

161. Smith, F. A., Fox, A. L. 1975. The free space of *Citrus* leaf slides. *Aust. J. Plant Physiol.* 2:441–46

162. Smith, F. A., Smith, S. E. 1989. Membrane transport at the biotrophic interface: an overview. *Aust. J. Plant Physiol* 16:33–43

163. Soll, H. J., Böttger, M. 1982. The mechanism of proton-induced increase in wall extensibility. *Plant Sci. Lett.* 24:163–71

164. Spanswick, R. M., Lucas, W. J., ed. 1980. *Plant Membrane Transport: Current Conceptual Issues.* Amsterdam/ New York/Oxford: Elsevier/North-Holland: Biomedical Press. 670 pp.

165. Starrach, N., Flach, D., Mayer, W.-E. 1985. Activity of fixed negative charges of isolated extensor cell walls of the laminar pulvinus of primary leaves of *Phaseolus. J. Plant Physiol.* 120:441–55

166. Starrach, N., Mayer, W.-E. 1986. Unequal distribution of fixed negative charges in isolated cell walls of various tissues in primary leaves of *Phaseolus. J. Plant Physiol.* 126:213–22

167. Starrach, N., Mayer, W.-E. 1989. Changes in the apoplastic pH and K^+ concentration in the *Phaseolus* pulvinus *in situ* in relation to rhythmic leaf movements. *J. Exp. Bot.* 40:865–73

168. Steudle, E., Smith, J. A. C., Lüttge, U. 1980. Water-relation parameters of individual mesophyll cells of *Kalanchoë daigremontiana. Plant Physiol.* 66: 1155–63

169. Stow, J. 1989. The involvement of calcium ions in maintenance of apple fruit tissue structure. *J. Exp. Bot.* 40:1053–57

170. Taiz, L. 1984. Plant cell extension: regulation of cell wall mechanical properties. *Annu. Rev. Plant Physiol.* 35: 585–657

171. Tepfer, M., Taylor, I. E. P. 1981. The interaction of divalent cations with pectic substances and their influence in acid-induced cell wall loosening. *Can. J. Bot.* 59:1522–25

172. Theuvenet, A. P. R., Borst-Pauwels, G. W. F. H. 1983. Effect of surface potential on Rb⁺ uptake in yeast. The effect of pH. *Biochim. Biophys. Acta* 734:62–69

173. Thibaud, J.-B., Davidian, J.-C., Sentenac, H., Soler, A., Grignon, C. 1988. H⁺ cotransports in roots as related to the surface pH shift induced by H⁺ excretion. *Plant Physiol.* 88:1469–73

174. Thomson, A. B. R. 1979. Limitations of the Eadie-Hofstee plot to estimate kinetic parameters of intestinal transport in the presence of an unstirred water layer. *J. Membr. Biol.* 47:39–57

175. Thomson, A. B. R., Dietschy, J. M. 1980. Experimental demonstration of the effect of unstirred water layer on the kinetic constants of the membrane transport of D-glucose in rabbit jejunum. *J. Membr. Biol.* 54:221–29

176. Tomos, A. D., Steudle, E., Zimmermann, U., Schulze, E.-D. 1981. Water relations of leaf epidermal cells of *Tradescantia virginiana. Plant Physiol.* 68:1135–43

177. Touchard, P., Demarty, M., Ripoll, C., Morvan, C., Thellier, M. 1989. Estimation of ionic mobilities in flax cell walls. In *Plant Membrane Transport: The Current Position,* ed. J. Dainty, M. I. De Michelis, E. Marrè, F. Rasi-Caldogno, pp. 603–6. Amsterdam: Elsevier. 712 pp.

178. Toulon, V., Sentenac, H., Thibaud, J.-B., Soler, A., Clarkson, D. T., Grignon, C. 1989. Effect of HCO₃⁻ concentration in the absorption solution on the energetic coupling of H⁺-cotransports in roots of *Zea mays* L. *Planta* 179:235–41

179. Vakhmistrov, D. B. 1967. Localization of the free space in barley roots. *Fiziol. Rasten.* 14:397–404

180. Van Cutsem, P., Gillet, C. 1982. Activity coefficients and selectivity values of Cu⁺⁺, Zn⁺⁺ and Ca⁺⁺ ions adsorbed in the *Nitella flexilis* L. cell wall during triangular ion exchanges. *J. Exp. Bot.* 33:847–53

181. Van Cutsem, P., Gillet, C. 1983. Proton-metal cation exchange in the cell wall of *Nitella flexilis. Plant Physiol.* 73:865–67

182. Van Volkenburgh, E., Boyer, J. S. 1985. Inhibitory effect of water deficit on maize leaf elongation. *Plant Physiol.* 77:190–94

183. Varner, J. E., Taylor, R. 1989. New ways to look at the architecture of plant cell walls. *Plant Physiol.* 91:31–33

184. Vesper, M. J. 1985. Use of a pH-response curve for growth to predict apparent wall pH in elongating segments of maize coleoptiles and sunflower hypocotyls. *Planta* 166:96–104

185. Virk, S. S., Cleland, R. E. 1988. Calcium and the mechanical properties of soybean hypocotyl cell walls: possible role of calcium in cell-wall loosening. *Planta* 176:60–67

186. Walker, N. A. 1985. See Ref. 100, pp. 31–37

187. Walker, N. A., Pitman, M. G., 1976. Measurement of fluxes across membranes. In *Transport in Plants. Encyclopedia of Plant Physiology,* ed. U. Lüttge, M. G. Pitman, 2A:93–126. Berlin: Springer-Verlag. 490 pp.

188. Watt, W. M., Morrell, C. K., Smith, D. L., Steer, M. W. 1987. Cystolith development and structure in *Pilea cadieri Urticaceae. Ann. Bot.* 60:71–84

189. Weisenseel, M. H., Dorn, A., Jaffe, L. F. 1979. Natural H⁺ currents traverse growing roots and root hairs of barley (*Hordeum vulgare* L.). *Plant Physiol.* 64:512–18

190. Widders, I. E., Lorenz, O. A. 1983. Effect of leaf age on potassium efflux and net flux in tomato leaf slices. *Ann. Bot.* 52:499–506

191. Wolterbeek, H. T. 1987. Cation exchange in isolated xylem cell walls of tomato. I. Cd²⁺ and Rb⁺ exchange in adsorption experiments. *Plant Cell Environ.* 10:39–44

192. Wolterbeek, H. T. 1987. Relationships between adsorption, chemical state and fluxes of cadmium applied as Cd(NO₃)₂ in isolated xylem cell walls of tomato. *J. Exp. Bot.* 38:419–32

193. Wolterbeek, H. T., Van Luipen, J., De Bruin, M. 1984. Non-steady state xylem transport of fifteen elements into the tomato leaf as measured by gamma-ray spectroscopy: a model. *Physiol. Plant.* 61:599–606

194. Wuytack, R., Gillet, C. 1978. Nature des liaisons de l'ion calcium dans la paroi de *Nitella flexilis. Can. J. Bot.* 56:1439–43

195. Yamaoka, T., Chiba, N. 1983. Changes in the coagulating ability of pectin during the growth of soybean hypocotyls. *Plant Cell Physiol.* 24:1281–90

196. Zid, E., Grignon, C. 1985. Sodium-calcium interactions in leaves of *Citrus aurantium* grown in the presence of NaCl. *Physiol. Vég.* 23:895–903

Annu. Rev. Plant Physiol. Plant Mol. Biol. 1991. 42:129–44

METABOLITE TRANSLOCATORS OF THE CHLOROPLAST ENVELOPE

Ulf-Ingo Flügge

Lehrstuhl für Botanik I der Universität Würzburg, Mittlerer Dallenbergweg 64, D-8700 Würzburg, Germany

Hans Walter Heldt

Institut für Biochemie der Pflanze, Untere Karspüle 2, D-3400 Göttingen, Germany

CONTENTS

INTRODUCTION

Chloroplasts are enclosed by two membranes; the outer and the inner envelope. Owing to the presence of porins in the outer envelope, this membrane is nonspecifically permeable to substances up to a molecular mass of about 10 kDa (17). The inner envelope membrane, however, is the actual permeability

1040-2519/91/0601-0129$02.00

barrier between the chloroplast stroma and the cytosol. It is the site at which different metabolite translocators coordinate the metabolism in both compartments (e.g. photosynthesis, photorespiration, biosyntheses of sucrose, starch, amino acids). The energy-transducing thylakoid membranes, located within the chloroplasts, are distinct from the envelope membranes. Stroma and thylakoid membranes each account for about 50% of the total chloroplast protein whereas the two envelope membranes represent less than 1%.

When the last extensive review on the function of the chloroplast envelope appeared in this series 10 years ago (30) the basic properties of several metabolite translocators of the inner envelope membrane had been described. Here we focus on aspects of the transport of metabolites across the chloroplast envelope and give an account of the progress achieved in this field in the last decade. The number of laboratories involved in this research has been small, perhaps because transport measurements with intact chloroplasts and work with isolated membrane transport proteins are technically difficult. For a more extensive treatise on other functional aspects of the plastid envelope the reader is referred to a recent review by Douce & Joyard (9).

THE PHOSPHATE-TRIOSE PHOSPHATE-3-PHOSPHOGLYCERATE TRANSLOCATOR

Transport in C_3 Plants

This translocator, for convenience called simply the phosphate translocator, mediates the export of the fixed carbon (in form of triose phosphates) from the chloroplasts into the cytosol, where it is converted into other substances—e.g. sucrose or amino acids. The inorganic phosphate released during biosynthetic processes is shuttled back via the phosphate translocator into the chloroplasts for the formation of ATP catalyzed by the thylakoid ATPase (see Figure 1). This transport process may therefore be regarded as the main transport function of the chloroplast envelope (21). It was investigated first in chloroplasts and later, after isolation of the purified translocator, in artificial, reconstituted phospholipid vesicles. The phosphate translocator is the major protein component of the inner envelope membrane, accounting for as much as 10–15% of the total protein of this membrane. In chloroplasts from the C_3 plants spinach and pea, the translocator, catalyzing a strict counterexchange, accepts at its binding site either P_i or a phosphate molecule attached to the end of a three-carbon chain, such as dihydroxyacetone phosphate, glyceraldehyde 3-phosphate, and glycerol 3-phosphate (15, 21). Three carbon compounds in which the phosphate is attached to carbon atom 2, such as 2-phosphoglycerate or phosphoenolpyruvate, show very little interaction with the C_3 translocator (Table 1).

Transport mediated by the phosphate translocator is affected by illumina-

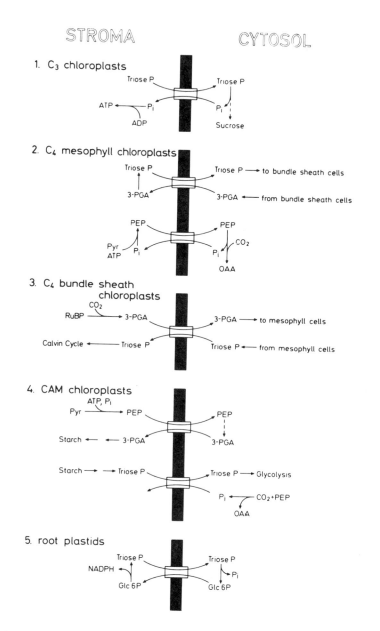

Figure 1 Physiological functions of the phosphate translocator in plastids

Table 1 Specificity of phosphate translocators in different plastids

	K_M(phosphate)[mM]	K_i[mM]				
		DHAP[a]	3-PGA	PEP	2-PGA	Glc6P
C$_3$-type chloroplasts						
a. spinach	0.32	0.20	0.36	4.1	3.8	≫12
b. pea	0.41	0.49	0.47	4.6	4.5	≫12
C$_4$-type chloroplasts						
maize mesophyll[b]	0.045	0.084	0.053	0.086	0.073	nd
Plastids from pea roots[c]	0.18	0.11	0.31	0.20	nd	0.33

[a] DHAP, dihydroxyacetone phosphate; PGA, phosphoglycerate; PEP, phosphoenolpyruvate; Glc6P, glucose 6-phosphate
[b] Data from Ref. 27
[c] S. Borchert and H. W. Heldt, unpublished data

tion. For example, in chloroplasts kept in the dark, inorganic phosphate, triose phosphates and 3-phosphoglycerate are transported with about similar K_m and V_{max}, whereas under illumination the outward transport of 3-phosphoglycerate is restricted. This effect of light is linked to the H$^+$ gradient across the envelope, resulting from the light-driven proton transport into the thylakoids. Because the phosphate translocator transports twice-negatively charged anions (15), and 3-phosphoglycerate (unlike inorganic phosphate and triose phosphate) carries at physiological pH three negative charges, an exchange of 3-phosphoglycerate with any of the other substrates implies an uncoupler-sensitive net movement of protons across the envelope (21). A linkage of proton transfer with a phosphate/3-phosphoglycerate exchange has been demonstrated with intact spinach chloroplasts (15) and with vesicles from envelope membranes from spinach and pea that had been loaded with the fluorescent pH indicator pyranine (37). These findings suggest that an alkalization of the chloroplast stroma as observed upon illumination would restrict the outward transport of 3-phosphoglycerate, since it would involve a transfer of a proton against a concentration gradient, whereas the inward transport of 3-phosphoglycerate should be enhanced. This hypothesis could be verified by studies of the isolated phosphate translocator from spinach chloroplasts reconstituted into liposomes, where it could be demonstrated that, in addition to the direct effect of the pH on the changes of the K_m values, the application of a transmembrane pH gradient similar to that found in illuminated chloroplasts resulted in a restriction of 3-phosphoglycerate transport (20).

Transport in C$_4$ and CAM Plants

C$_4$ plants contain two different CO$_2$-fixing cell types: mesophyll and bundle sheath cells. In C$_4$ mesophyll cells, the phosphoenolpyruvate formed by the pyruvate phosphate dikinase in the stroma is transferred into the cytosol to be carboxylated to oxaloacetate, and the remaining phosphate reenters the stroma (see Figure 1). Since the bundle sheath cells of some C_4 plants—e.g. maize—are deficient in photosystem II activity, a triose phosphate/3-phosphoglycerate shuttle is additionally required for the transport of 3-phosphoglycerate from the bundle sheath chloroplasts into the mesophyll chloroplasts, where it is reduced to triose phosphate and shuttled back into the bundle sheath chloroplasts (see Figure 1).

In CAM (crassulacean acid metabolism) plants, phosphoenolpyruvate, synthesized during the day via the pyruvate P_i dikinase inside the chloroplasts, serves as a precursor for starch biosynthesis. Since the activities of enolase and phosphoglyceromutase are considered to be exclusively cytoplasmatic (71), phosphoenolpyruvate has to be exported first into the cytosol, where it is converted to 3-phosphoglycerate. The latter is then imported back into the chloroplasts for the subsequent biosynthesis of starch. During the night period, triose phosphates formed during phosphorolytic mobilization of starch can be exported from the chloroplasts in exchange with inorganic phosphate (see Figure 1).

Specific transport of phosphoenolpyruvate with inorganic phosphate and also with triose phosphate and 3-phosphoglycerate was found in mesophyll chloroplasts from the C_4 plants *Digitaria sanguinalis* (38, 56), maize (7, 27) and *Panicum miliaceum* (43), and also in the facultative CAM plant *Mesembryanthemum crystallinum* (41).

The measurement of the counterexchange of substrates (7, 27, 38), the observed competitive inhibition of the transport of 3-phosphoglycerate, triose phosphate and inorganic phosphate by phosphoenolpyruvate (27, 28, 59), and the finding that the inhibition of the phosphate translocator by 4,4' diisothiocyanostilbene-2,2'-disulfonic acid (DIDS) could be protected by the addition of phosphoenolpyruvate (as well as inorganic phosphate and 3-phosphoglycerate) (57) indicated that all these substances were transported by the same translocator. A comparison between the substrate specificities of the phosphate translocators from C_4 (maize) mesophyll chloroplasts and those from C_3 plants (spinach, pea) is shown in Table 1. Bundle sheath chloroplasts from the C_4 plant *Panicum miliaceum* also transport phosphoenolpyruvate in addition to inorganic phosphate, 3-phosphoglycerate, and triose phosphates, although the affinity for phosphoenolpyruvate and 3-phosphoglycerate of the bundle sheath translocator is lower than that of the mesophyll translocator (43). These results indicate that leaves of C_4 plants possess two different phosphate translocators.

Transport in Root Plastids

A major function of root plastids is the reduction of nitrite, for which the necessary redox equivalents are probably provided by the oxidative pentose phosphate pathway (6). Because pea root plastids have no fructose 1,6-bisphosphatase activity, the functioning of the oxidative pentose phosphate pathway would involve an uptake of hexose phosphate and a release of triose phosphate. An uptake of hexose phosphate would also be required for starch biosynthesis. Experiments with intact pea root plastids showed that glucose 6-phosphate was a better candidate than glucose 1-phosphate for oxidation by the oxidative pentose phosphate pathway (6), indicating that glucose 6-phosphate was the hexose phosphate species taken up by the plastids.

A method of preparing intact plastids from pea roots (12) made the study of transport across the plastid envelope possible. The existence of a phosphate translocator in these plastids was first indicated by measurements of [^{32}P]-labeled phosphate uptake, which was reduced in the presence of triose phosphates and 3-phosphoglycerate (13). A subsequent systematic study of this transport demonstrated that these plastids possessed a phosphate translocator transporting in a counterexchange mode not only inorganic phosphate, triose phosphate, 3-phosphoglycerate and phosphoenolpyruvate but also glucose 6-phosphate (5; S. Borchert and H. W. Heldt, in preparation; see Table 1). The transport of glucose 6-phosphate is quite specific, glucose 1-phosphate showing only a weak interaction with the translocator (5). It appears that the root plastid phosphate translocator exhibits the broadest substrate specificity of the phosphate translocators dealt with in the preceding experiments. Whereas the C_3 chloroplast translocator accepts only phosphate and C_3 compounds with the phosphate attached to the C_3 atom (i.e. triose phosphate and 3-phosphoglycerate), and the C_4 mesophyll chloroplast translocator also transports C_3 compounds with the phosphate attached to the C_2 atom (i.e. phosphoenolpyruvate and 2-phosphoglycerate), the root plastid phosphate translocator transports all these substances and, in addition, glucose 6-phosphate (see Table 1).

Like the phosphate translocator in chloroplasts, the root plastid phosphate translocator is inhibited by DIDS and pyridoxal-5'-phosphate. Transport of glucose 6-phosphate and of inorganic phosphate has been also found in chloroplasts from the alga *Codeum fragile* (60). It remains to be elucidated whether this transport process is related to the root plastid system.

Properties of the Phosphate Translocator Protein

The phosphate translocator has been identified in spinach chloroplasts. Its molecular mass has been estimated by specific reaction with pyridoxal-5'-phosphate or trinitrobenzenesulfonic acid followed by labeling with [^{3}H]NaBH$_4$ to be 29 kDa as revealed by sodium dodecyl sulfate/polyacrylamide gel electrophoresis (SDS-PAGE) (23). The same method yielded for the phosphate translocator in maize mesophyll chloroplasts a molecular mass of 30 kDa (65). Specific labeling of the chloroplast phosphate translocators was also achieved by treatment of the intact chloroplasts with tritiated dihydro-4,4'-diisothiocyanostilbene-2,2'-disulfonic acid ([³H]-H_2DIDS), yielding for the spinach and pea phosphate translocator a molecular mass of 29 kDa, and for the maize mesophyll phosphate translocator a molecular mass of 29–30 kDa (27, 58).

The phosphate translocator has been isolated in the presence of detergents (23). Hydrodynamic studies of the isolated spinach translocator-detergent micelle revealed that the translocator protein had in its functional state a molecular mass of about 62 kDa (16), thus consisting of two identical subunits. The hydrodynamic studies further indicated that the translocator protein had a prolate ellipsoidal shape, the semimajor axis being 6.6 nm long (16). The phosphate translocator is large enough to protrude from both sides of the inner envelope membrane, thus rendering the translocator accessible to its substrates. The hydrodynamic properties of the phosphate translocator have been confirmed by measuring the rotational mobility of the eosin-5-isothiocyanate (EITC)-labeled translocator reconstituted in liposomes (67).

As is the case for most of the chloroplast envelope membrane proteins, the phosphate translocator protein is coded for by nuclear genes and synthesized in the cytosol as a higher-molecular-weight precursor protein with a N-terminal extension, the transit peptide (22); it is subsequently transported into the chloroplasts, where it is processed to its mature size by a specific protease (19). The entire cDNA sequence of the spinach phosphate translocator precursor protein has recently been elucidated (19). It contains, besides noncoding regions at the 5' and 3' ends, a 1212-bp coding region encoding the entire 404 amino acid residues of the precursor protein (M_r 44,234). Using an AvaII fragment of the spinach translocator as a hybridization probe a cDNA clone coding for the phosphate translocator from pea chloroplasts has also recently been isolated and sequenced (70). This clone contains a 1206-bp coding region corresponding to a polypeptide of 402 amino acid residues (M_r 43,671). Determination of the N-terminal sequence of the mature translocator proteins from both spinach and pea chloroplasts showed that the transit peptides consist of 80- and 72-amino-acid residues, respectively. Both transit peptides contain a positively charged amphiphilic α-helix. This structure had been identified as an essential feature of mitochondrial transit peptides but was found to be absent in transit peptides of higher-plant chloroplast proteins destined for the stroma or the thylakoids (66). However, such an amphiphilic α-helix appears to be sufficient to direct an inner envelope polypeptide to its membrane (10).

The mature part of the spinach translocator protein contains 324 amino acid residues (M_r 35,603) and that of the pea translocator contains 330 amino acid residues (M_r 35,957). The molecular masses of both translocators are significantly higher than those determined by SDS-PAGE (about 29 kDa). This probably results from the hydrophobic nature of the translocator, because both translocators contain a high percentage of nonpolar amino acid residues that might bring about an unusually high binding of detergent. As deduced from hydrophobicity distribution analyses both proteins contain six to seven hydrophobic stretches about 20–23 amino acid residues long that can form α-helical structures spanning the envelope membrane and thus anchoring the translocator within the membrane. Some of these α-helixes have an amphiphilic character and might form a hydrophilic translocation channel through which the substrates could be transported across the membrane. Earlier experiments had shown that a lysine and an arginine residue are involved in the binding of the twice-negatively charged substrates to the translocator, with only one binding site being present in the dimeric transloca-tor protein (23, 27). Presumably, this substrate binding site is exposed to only one side of the membrane at a time, thereby catalyzing the counterexchange of substrates. It may be noted that a lysine and an arginine residue (Lys-273 and Arg-274 in the spinach translocator) are the only charged amino acid residues contained in that part of the protein embedded within the membrane. We speculate that these two cationic residues are involved in the binding of the negatively charged substrates.

Both mature proteins contain about 87% identical amino acid residues and, in addition, 5% conservative substitutions; but their amino acid sequences do not share any structural features with known mitochondrial or bacterial trans-port proteins, and even a search of the EMBL databank reveals no homology with any other known nucleotide or amino acid sequence. Thus, plastid translocators may represent a class of transport proteins distinct from those of other cells.

Characterization of the phosphate translocators in chloroplasts and plastids of C_3 plants, in chloroplasts of CAM plants, and in the mesophyll and bundle sheath chloroplasts of C_4 plants indicated that in different plastids contained in a single plant species, different isoforms of phosphate translocators occur, probably encoded by different genes. Work is in progress to identify the DNA sequences coding for these translocators and to elucidate how the different functions of the various phosphate translocators are linked to structural fea-tures of the protein.

DICARBOXYLATE TRANSPORT

In spinach chloroplasts a number of dicarboxylates such as L-malate, suc-cinate, 2-oxoglutarate, aspartate, and glutamate were initially found to be transported in a counterexchange mode across the inner envelope membrane (40). A transport of dicarboxylates was also observed in chloroplasts from pea (50) and maize mesophyll cells (8). The study of dicarboxylate transport in spinach chloroplasts had indicated that the transport might be mediated by at least two different translocators with overlapping specificities (40). This suggestion was supported by the observation that pea leaf chloroplasts exhibited two kinds of aspartate transport, differing in their half-saturation by aspartate (69). Moreover, studies of the effect of dicarboxylates on 2-oxoglutarate uptake suggested that, for transport, the binding of 2-oxoglutarate occurred at a binding site different from that of other transported dicarboxylates (11, 51, 75). An uptake of 2-oxoglutarate into, and a release of glutamate from, C_3 chloroplasts is required for the photorespiratory NH_3 reassimiliation via the glutamine synthetase/glutamate synthase pathway (2, 74). A detailed study of the kinetics of the uptake of these two substances in spinach chloroplasts demonstrated that two transport processes are involved, both catalyzing an exchange with malate: an uptake of 2-oxoglutarate in exchange with stromal malate on the 2-oxoglutarate translocator (which does not transport glutamate) and a release of glutamate in exchange with external malate occurring on the dicarboxylate translocator. These processes result in a net 2-oxoglutarate/glutamate exchange with no net malate transport (25, 76). Mutants of *Arabidopsis thaliana* deficient in dicarboxylate transport activities are not viable under conditions that require photorespiration. The putative polypeptide associated with this mutation has been identified in *A. thaliana* (62), probably representing the glutamate-malate translocator. In the *Arabidopsis* mutants mentioned here, the transport of glutamine into the chloroplasts was not affected (61), suggesting the existence of a separate glutamine translocator. The existence of a glutamine translocator was substantiated by transport measurements with chloroplasts from spinach and oat. This specific glutamine translocator does not transport any other dicarboxylates except L-glutamate (77), whereas otherwise glutamine transport occurs at low rates and in exchange with other dicarboxylates, probably mediated by the dicarboxylate (i.e. the glutamate-malate) translocator (3). It is proposed that this very active and specific glutamine translocator plays a role during photorespiratory NH_3 refixation by catalyzing the uptake of glutamine formed by the cytosolic isoform of glutamine synthase into chloroplasts, where it is transformed to glutamate by the glutamate synthase located exclusively in the stroma.

The dicarboxylate translocator can also transport oxaloacetate with about the same affinity as malate (8, 40). For a physiological role, however, such a transport of oxaloacetate seems of minor importance. Due to the malate dehydrogenase equilibrium [K_{equil} at pH 7 = $3 \cdot 10^5$ (63)], the cellular concentration of malate is expected to be orders of magnitude higher than that of oxaloacetate, making the transport of oxaloacetate and that of malate by the same translocator virtually impossible (26). On the other hand, polarographic studies of oxaloacetate reduction by illuminated chloroplasts suggested that oxaloacetate can be efficiently taken up into chloroplasts at low concentrations even in the presence of malate (1). Measurements of the uptake of

[14]C-labeled oxaloacetate into chloroplasts from spinach and maize mesophyll cells demonstrated that these chloroplasts indeed contained a highly specific oxaloacetate translocator, with very low competition by an even 1000-fold excess of other dicarboxylates (29). This translocator appears to play an important physiological role in C_4 metabolism, mediating the uptake of oxaloacetate formed by phosphoenolpyruvate carboxylase in the cytosol into the mesophyll chloroplasts, where it is reduced to malate. In C_3 plants, this translocator is involved in the functioning of the oxaloacetate-malate shuttle that enables the transfer of redox equivalents from the chloroplasts into the cytosol (32).

TRANSPORT OF MONOCARBOXLATES

The Glycolate/Glycerate Translocator

Due to the oxygenase activity of the ribulose bisphosphate carboxylase, phosphoglycolate as a byproduct of C_3 photosynthesis has to be reconverted into 3-phosphoglycerate, which is finally fed back into the Calvin cycle. The photorespiratory cycle requires the participation of chloroplasts, peroxisomes, and mitochondria and involves several transport steps of intermediates across the different organelle membranes. Two molecules of glycolate resulting from the hydrolysis of phosphoglycolate in the stroma are exported from the chloroplasts in exchange for one molecule of glycerate, which is subsequently phosphorylated via the glycerate kinase, yielding 3-phosphoglycerate. Both glycolate and glycerate rapidly permeate the chloroplast inner envelope membrane. Whereas initial uptake studies suggested that these compounds were nonspecifically permeating the envelope by diffusion of the protonated acids (14, 64), measurements of the uptake of radioactively labeled glycolate (33, 34) and glycerate (52-54) clearly demonstrated that the uptake of these substances is translocator mediated. The transport of both substances showed about the same sensitivity to inhibitors like N-ethylmaleimide or p-chloromercuribenzene sulfonic acid. It turned out that both substances are transported by the same translocator (36), having about the same affinities for glycolate, glyoxylate, D-glycerate and D-lactate. L-lactate, L-glycerate, and 2-hydroxymonocarboxylates with more than three carbon atoms are only poorly transported (35). The activity of the glycolate-glycerate translocator appears to be high enough to account for in vivo photorespiratory carbon fluxes (34). Such a translocator has also been found in mesophyll and bundle sheath chloroplasts from the C_4 plant *Panicum miliaceum* (45). Glycerate and the other transported carboxylates can be accumulated in the stroma by means of a light-generated pH gradient across the envelope (34, 54) which can be explained by an H^+/carboxylate symport or a macroscopically identical exchange of the carboxylate for hydroxyl ions.

The Pyruvate Translocator

Pyruvate is not transported by spinach chloroplasts but is readily taken up by chloroplasts of C_4 plants. In these plants, pyruvate has to be transported into the mesophyll chloroplasts, where it is converted to the primary CO_2 acceptor, phosphoenolpyruvate. A specific translocator was first found in mesophyll chloroplasts from *Digitaria sanguinalis*, a C_4 plant of the $NADP^+$-malic enzyme type (39), and also in pea chloroplasts an uptake of pyruvate has been demonstrated (50). In *Digitaria*, pyruvate transport was found to be inhibited by the sulfhydryl reagent mersalyl and by pyruvate analogs such as phenylpyruvate and α-cyano-4-hydroxycinnamic acid (39). Transport in *Digitaria* was studied only in the dark where the internal concentrations of pyruvate were always much lower than in the external medium. A light-dependent and uncoupler-sensitive active transport of pyruvate was first demonstrated in maize mesophyll chloroplasts (24). Apparently, the accumulation of pyruvate is driven by a light-dependent cation gradient and is involved in driving the C_4 dicarboxylate/pyruvate shuttle of C_4 photosynthesis. Pyruvate uptake was also found in other C_4 plants—e.g. in mesophyll and bundle sheath chloroplasts of the C_4 plant *Panicum miliaceum* (NAD^+-malic enzyme type) (46). Interestingly enough, a light-dependent accumulation was found only in mesophyll chloroplasts but not in bundle sheath chloroplasts. A correlation of pyruvate uptake and the stromal pH suggested that active pyruvate uptake was primarily driven by the pH gradient across the envelope (47). Testing the hypothesis of an H^+/pyruvate symport showed that pyruvate uptake into darkened mesophyll chloroplasts from *Panicum miliaceum* could be driven by an artificially applied Na^+ gradient, while the H^+ gradient itself gave negative results, suggesting that pyruvate transport can be driven by a Na^+/pyruvate symport (48). Similar results were obtained with mesophyll chloroplasts from the C_4 plants *Urochloa panicoides* and *Panicum maximum* (phosphoenolpyruvate carboxykinase types), whereas in the C_4 plants *Zea mays* and *Sorghum bicolor* ($NADP^+$-malic enzyme types) an effect of Na^+ on pyruvate transport was not found. We suggest that in C_4 plants of both the NAD^+ malic enzyme type and the phosphoenolpyruvate carboxykinase type, pyruvate uptake is driven by a Na^+ gradient, whereas in C_4 plants of the $NADP^+$ malic enzyme type the driving force for active pyruvate transport is provided by an H^+ gradient (44, 49).

ATP/ADP TRANSPORT

In chloroplasts from pea leaves, ATP exchanged not only with ADP and inorganic pyrophosphate, as known from the chloroplast ATP/ADP translocator (30, 31) but also with phosphoenolpyruvate (72). ATP transport in chloroplasts is not affected by atractyloside, a specific inhibitor of the mitochondrial

ATP/ADP translocator but could effectively be inhibited by a derivative of ATP, adenosine 2',3'-dialdehyde 5'-triphosphate (18). Bongkrekic acid, another inhibitor of the mitochondrial ATP/ADP translocator, was reported to inhibit transport of ATP (and of phosphoenolpyruvate) in pea chloroplasts (73) but had no effect on the ATP translocator from spinach chloroplasts (18). In spinach chloroplasts, the V_{max} of the ATP/ADP translocator is lower by about one order of magnitude than that of the phosphate translocator (18, 31), although the activity was shown to depend on the developmental state of the chloroplasts. In young pea leaves the activity of the ATP translocator was found to be much higher than in mature leaves (55). Presumably, this translocator serves to supply chloroplasts with ATP generated by mitochondrial oxidative phosphorylation during biogenesis and, in mature leaves, during the night period only. A role of the ATP translocator in protein translocation into chloroplasts has also been demonstrated (18). Although the present evidence suggests that the chloroplast ATP/ADP translocator is different from the mitochondrial one, it has been shown recently that inner envelope membranes from sycamore amyloplasts (but not from chloroplasts) may contain a 32-kDa polypeptide that cross-reacts with an antiserum directed against the ATP/ADP translocator from mitochondria of *Neurospora crassa* (42). However, final experimental proof for the presence of a mitochondrial-like ATP/ADP translocator in amyloplast envelope membranes is still lacking.

TRANSPORT OF OTHER COMPOUNDS

In spinach chloroplasts a rapid uptake of ascorbate has been observed. Substrate saturation of the uptake, and the inhibition by didehydroascorbic acid and the sulfhydryl reagent p-chloromercuriphenyl sulfonic acid, indicate that the ascorbate is taken up by a specific translocator (4).

CONCLUDING REMARKS

In this review we have summarized our still limited knowledge about different metabolite translocators of the chloroplast envelope. The most significant advance in this field has been the recent cloning and sequencing of the first envelope membrane protein, the chloroplast phosphate translocator. The availability of the cloned gene has now opened the way to insights into structural and functional relationships. The molecular identities of all the other envelope translocators remain to be elucidated. Unfortunately, no specifically and covalently reacting inhibitors for these translocators are known that would allow their identification. Therefore one has to rely on unspecifically reacting amino acid reagents that are proved to be inhibitory, and one has to correlate the transport inhibition parameters with those of the incorporation of the labeled inhibitor into a particular membrane polypeptide. Another approach would be to select a detergent that does not affect the functional activity of the translocator and then to purify the protein whereby its activity is detected by reconstitution into artificial phospholipid vesicles. Much progress is needed before we shall obtain a physical descripton of the molecular mechanism of chloroplast metabolite transport, but further efforts to identify and structurally analyze other envelope membrane translocators will contribute to a deeper understanding of chloroplast metabolite transport.

ACKNOWLEDGMENT

Work in the authors' laboratories has been supported by the Deutsche Forschungsgemeinschaft.

Literature Cited

1. Anderson, J. W., House, C. M. 1979. Polarographic study of dicarboxylic-acid-dependent export of reducing equivalents from illuminated chloroplasts. *Plant Physiol.* 64:1064-69
2. Anderson, J. W., Walker, D. A. 1984. Ammonia assimilation and oxygen evolution by a reconstituted chloroplast system in the presence of 2-oxoglutarate and glutamate. *Planta* 159:247-53
3. Barber, D. J., Thurman, D. A. 1978. Transport of glutamine into isolated pea chloroplasts. *Plant Cell Environment* 1:297-303
4. Beck, E., Burkert, A., Hofmann, H. 1983. Uptake of ascorbate by intact spinach spinacia-oleracea cultivar vital chloroplasts. *Plant Physiol.* 73:41-45
5. Borchert, S., Große, H., Heldt, H. W. 1989. Specific transport of phosphate, glucose 6-phosphate, dihydroxyacetone phosphate and 3-phosphoglycerate into amyloplasts from pea roots. *FEBS Lett.* 253:183-86
6. Bowsher, C. G., Hucklesby, D. B., Emes, M. J. 1989. Nitrite reduction and carbohydrate metabolism in plastids purified from roots of *Pisum sativum* L. *Planta* 177:359-66
7. Day, D. A., Hatch, M. D. 1981. Transport of 3-phosphoglyceric acid, phosphoenolpyruvate, and inorganic phosphate in maize mesophyll chloroplasts, and the effect of 3-phosphoglyceric acid on malate and phosphoenolpyruvate production. *Arch. Biochem. Biophys.* 211:743-49
8. Day, D. A., Hatch, M. D. 1981. Dicarboxylate transport in maize mesophyll chloroplasts. *Arch. Biochem. Bio phys.* 211:738-42
9. Douce, R., Joyard, J. 1990. Biochemistry and function of the plastid envelope. *Annu. Rev. Cell Biol.* 6:173-216
10. Dreses-Werringloer, U., Fischer, K., Wachter, E., Link, T. A., Flügge, U.-I. 1990. cDNA sequence and deduced amino acid sequence of the precursor of the 37 kDa inner envelope membrane polypeptide from spinach chloroplasts: Its transit peptide contains an amphiphilic α-helix as the only detectable structural element. *Eur. J. Biochem.* In press
11. Dry, I. B., Wiskich, J. T. 1983. Characterization of dicarboxylate stimulation of ammonia, glutamine and 2-oxoglutarate-dependent O_2 evolution in isolated pea chloroplasts. *Plant Physiol.* 72:291-96
12. Emes, M. J., England, S. 1986. Purification of plastids from higher plant roots. *Planta* 168:161-66
13. Emes, J. J., Traska, A. 1987. Uptake of inorganic phosphate by plastids purified from the roots of *Pisum sativum* L. *J. Exp. Bot.* 38:1781-88
14. Enser, U., Heber, U. 1980. Metabolic regulation by pH gradients: Inhibition of photosynthesis by indirect proton transfer across the chloroplast envelope. *Biochim. Biophys. Acta* 592:577-91
15. Fliege, R., Flügge, U.-I., Werdan, K., Heldt, H. W. 1978. Specific transport of inorganic phosphate, 3-phosphoglycerate and triosephosphates across the inner membrane of the envelope in spinach chloroplasts. *Biochim. Biophys. Acta* 502:232-47
16. Flügge, U.-I. 1985. Hydrodynamic properties of the Triton X-100 solubilized chloroplast phosphate translocator. *Biochim. Biophys. Acta* 815:299-305

17. Flügge, U. I., Benz, R. 1984. Pore forming activity in the outer membrane of the chloroplast envelope. *FEBS Lett.* 169:85-89

18. Flügge, U. I., Hinz, G. 1986. Energy dependence of protein translocation into chloroplasts. *Eur. J. Biochem.* 160:563-70

19. Flügge, U.-I., Fischer, K., Gross, A., Sebald, W., Lottspeich, F., Eckerskorn, C. 1989. The triose phosphate-3-phosphoglycerate-phosphate translocator from spinach chloroplasts: nucleotide sequence of a full-length cDNA clone and import of the *in vitro* synthesized precursor protein into chloroplasts. *EMBO J.* 8:39-46

20. Flügge, U.-I., Gerber, J., Heldt, H. W. 1983. Regulation of the reconstituted chloroplast phosphate translocator by a proton gradient. *Biochim. Biophys. Acta* 725:229-37

21. Flügge, U.-I., Heldt, H. W. 1984. The phosphate-triose phosphate-phosphoglycerate translocator of chloroplasts. *Trends Biochem. Sci.* 9:530-33

22. Flügge, U.-I., Wessel, D. 1984. Cell-free synthesis of putative precursors for envelope membrane polypeptides of spinach chloroplasts. *FEBS Lett.* 168:255-59

23. Flügge, U.-I., Heldt, H. W. 1986. Chloroplast phosphate–triose phosphate–phosphoglycerate translocator: its identification, isolation and reconstitution. *Methods Enzymol.* 125:716-30

24. Flügge, U.-I., Stitt, M., Heldt, H. W. 1985. Light driven uptake of pyruvate into mesophyll chloroplasts from maize. *FEBS Lett.* 183:335-39

25. Flügge, U.-I., Woo, K. C., Heldt, H. W. 1988. Characteristics of 2-oxoglutarate and glutamate transport in spinach chloroplasts. *Planta* 174:534-41

26. Giersch, C. 1982. Capacity of the malate/oxaloacetate shuttle for transfer of reducing equivalents across the envelope of leaf chloroplasts. *Arch. Biochem. Biophys.* 219:379-87

27. Gross, A., Brückner, G., Heldt, H. W., Flügge, U.-I. 1990. Comparison of the kinetic properties, inhibition and labeling of the phosphate translocators from maize and spinach mesophyll chloroplasts. *Planta* 180:262-71

28. Hallberg, M., Larsson, C. 1983. Highly purified intact chloroplasts from mesophyll protoplasts of the 4 carbon pathway plant *Digitaria sanguinalis*: inhibition of phosphoglycerate reduction by orthophosphate and by phosphoenolpyruvate. *Physiol. Plant.* 57:330-38

29. Hatch, M. D., Dröscher, L., Heldt, H.

W. 1984. A specific translocator for oxaloacetate transport in chloroplasts. *FEBS Lett.* 178:15-19

30. Heber, U., Heldt, H. W. 1981. The chloroplast envelope: structure, function and role in leaf metabolism. *Annu. Rev. Plant Physiol.* 32:139-68

31. Heldt, H. W. 1969. Adenine nucleotide translocation in spinach chloroplasts. *FEBS Lett.* 5:11-14

32. Heldt, H. W., Heineke, D., Heupel, R., Krömer, S., Riens, B. 1990. Transfer of redox equivalents between subcellular compartments of a leaf cell. In *Proceedings of the Eighth International Congress on Photosynthesis, Stockholm*, ed. M. Baltscheffsky, 4:15.1-15.7. New York: Academic

33. Howitz, K. T., McCarty, R. E. 1983. Evidence for a glycolate transporter in the envelope of pea chloroplasts. *FEBS Lett.* 154:339-42

34. Howitz, K. T., McCarty, R. E. 1985. Kinetic characteristics of the chloroplast envelope glycolate transporter. *Biochemistry* 24:2645-52

35. Howitz, K. T., McCarty, R. E. 1985. Substrate specificity of the pea chloroplast glycolate transporter. *Biochemistry* 24:3645-50

36. Howitz, K. T., McCarty, R. E. 1986. D-glycerate transport by the pea chloroplast glycolate carrier. Studies on ^{14}C D-glycerate uptake and D-glycerate dependent O_2 evolution. *Plant Physiol.* 890:390-95

37. Howitz, K. T., McCarty, R. E. 1988. Measurement of H^+ linked transport activities in pyranine loaded chloroplast inner envelope vesicles. *Plant Physiol.* 86:999-1001

38. Huber, S. C., Edwards, G. E. 1977. Transport in C_4 mesophyll chloroplasts. Evidence for an exchange of inorganic phosphate and phosphoenolpyruvate. *Biochim. Biophys. Acta* 462:603-12

39. Huber, S. C., Edwards, G. E. 1977. Transport in C_4 mesophyll chloroplasts. Characterization of the pyruvate carrier. *Biochim. Biophys. Acta* 462:583-602

40. Lehner, K., Heldt, H. W. 1978. Dicarboxylate transport across the inner membrane of the chloroplast envelope. *Biochim. Biophys. Acta* 501:531-44

41. Neuhaus, H. E., Holtum, J. A. M., Latzko, E. 1988. Transport of phosphoenolpyruvate by chloroplasts from *Mesembryanthemum crystallinum* L. exhibiting crassulacean acid metabolism. *Plant Physiol.* 87:64-68

42. Ngernprasirtsiri, J., Takabe, T., Akazawa, T. 1989. Immunochemical analysis shows that an ATP/ADP translocator is

associated with the inner envelope membrane of amyloplasts from *Acer pseudoplatanus* L. *Plant Physiol.* 89:1024-27

43. Ohnishi, J., Flügge, U.-I., Heldt, H. W. 1990. Phosphate translocator of mesophyll and bundle sheath chloroplasts of a C₄ plant, *Panicum miliaceum* L. Identification and kinetic characterization. *Plant Physiol.* 91:1507-11

44. Ohnishi, J., Flügge, U.-I., Heldt, H. W., Kanai, R. 1990. Involvement of Na⁺ in active uptake of pyruvate in mesophyll chloroplasts of some C₄ plants: Na⁺/pyruvate transport. *Plant Physiol.* In press

45. Ohnishi, J., Kanai, R. 1988. Glycerate uptake into mesophyll and bundle sheath chloroplasts of a C₄ plant, *Panicum miliaceum*. *J. Plant Physiol.* 133:119-21

46. Ohnishi, J. I., Kanai, R. 1987. Pyruvate uptake by mesophyll and bundle sheath chloroplasts of a C₄ plant *Panicum miliaceum*. *Plant Cell Physiol.* 28:1-10

47. Ohnishi, J. I., Kanai, R. 1987. Light dependent uptake of pyruvate by mesophyll chloroplasts of a C₄ plant *Panicum miliaceum* L. *Plant Cell Physiol.* 28:243-51

48. Ohnishi, J. I., Kanai, R. 1987. Na⁺ induced uptake of pyruvate into mesophyll chloroplasts of a C₄ plant *Panicum miliaceum*. *FEBS Lett.* 219:347-50

49. Ohnishi, J.-I., Kanai, R. 1990. Pyruvate uptake induced by a pH jump in mesophyll chloroplasts of maize and sorghum, NADP-malic enzyme type C₄ species. *FEBS Lett.* 269:122-24

50. Proudlove, M. O., Thurman, D. A. 1981. The uptake of 2-oxoglutarate and pyruvate by isolated pea chloroplasts. *New Phytol.* 88:255-64

51. Proudlove, M. O., Thurman, D. A., Salisbury, J. 1984. Kinetic studies on the transport of 2-oxoglutarate and L-malate in isolated pea chloroplasts. *New Phytol.* 96:1-8

52. Robinson, S. P. 1982. Transport of glycerate across the envelope membrane of isolated spinach chloroplasts. *Plant Physiol.* 70:1032-38

53. Robinson, S. P. 1982. Light stimulates glycerate uptake by spinach chloroplasts. *Biochim. Biophys. Acta* 106:1027-34

54. Robinson, S. P. 1984. Lack of ATP requirement for light stimulation of glycerate transport into intact isolated chloroplasts. *Plant Physiol.* 75:425-30

55. Robinson, S. P., Wiskich, J. T. 1977. Pyrophosphate inhibition of carbon dioxide fixation in isolated pea chloro-

plasts by uptake in exchange for endogenous adenine nucleotides. *Plant Physiol.* 59:422-27

56. Rumpho, M. E., Edwards, G. E. 1984. Inhibition of 3-phosphoglycerate-dependent O₂ evolution by phosphoenolpyruvate in C₄ mesophyll chloroplasts of *Digitaria sanguinalis* (L) *Scop.* *Plant Physiol.* 76:711-18

57. Rumpho, M. E., Edwards, G. E. 1985. Characterization of 4,4'-diisothiocyano-2,2'-disulfonic acid stilbene inhibition of 3-phosphoglycerate-dependent O₂ evolution in isolated chloroplasts. Evidence for a common binding site on the C₄ phosphate translocator for 3-phosphoglycerate, phosphoenolpyruvate, and inorganic phosphate. *Plant Physiol.* 78:537-44

58. Rumpho, M. E., Edwards, G. E., Yousif, A. E., Keegstra, K. 1988. Specific labelling of the phosphate translocator in C₃ and C₄ mesophyll chloroplasts by tritiated dihydro-DIDS (1,2-ditritio-1,2-[2,2'-disulfo-4,4'-diisothiocyano]-diphenylethane). *Plant Physiol.* 86:1193-98

59. Rumpho, M. E., Wessinger, M. E., Edwards, G. E. 1987. Influence of organic-phosphates on 3-phosphoglycerate dependent O₂ evolution in C₃ and C₄ mesophyll chloroplasts. *Plant Cell Physiol.* 28:805-14

60. Rutter, J. C., Cobb, A. H. 1983. Translocation of orthophosphate and glucose-6-phosphate in *Codeum fragile* chloroplasts. *New Phytol.* 95:559-68

61. Somerville, S. C., Ogren, W. L. 1983. An *Arabidopsis thaliana* mutant defective in chloroplast dicarboxylate transport. *Proc. Natl. Acad. Sci. USA* 80:1290-94

62. Somerville, S. C., Somerville, C. R. 1985. A mutant of *Arabidopsis* deficient in chloroplast dicarboxylate transport is missing an envelope protein. *Plant Sci. Lett.* 37:217-20

63. Stern, J. R., Ochoa, S., Lynen, F. 1952. Enzymatic synthesis of citric acid. V. Reaction of acetyl coenzyme A. *J. Biol. Chem.* 198:313-20

64. Takabe, T., Akazawa, T. 1981. Structure and function of chloroplast proteins. Mechanism of glycolate transport in spinach leaf chloroplasts. *Plant Physiol.* 68:1093-97

65. Thompson, A. G., Brailsford, M. A., Beechey, R. B. 1987. Identification of the phosphate translocator from maize mesophyll chloroplasts. *Biochem. Biophys. Res. Commun.* 143:164-69

66. von Heijne, G., Steppuhn, J., Herrmann, R. G. 1989. Domain structure of

mitochondrial and chloroplast targeting peptides. *Eur. J. Biochem.* 180:535-45

67. Wagner, R., Apley, E. C., Gross, A., Flügge, U.-I. 1989. The rotational diffusion of chloroplast phosphate translocator and of lipid molecules in bilayer membranes. *Eur. J. Biochem.* 182:165-73

68. Wallsgrove, R. M., Kendall, A. C., Hall, N. P., Turner, J. C., Lea, P. J. 1986. Carbon and nitrogen metabolism in a barley mutant with impaired chloroplast dicarboxylate transport. *Planta* 168:324-29

69. Werner-Washburne, M., Keegstra, K. 1985. L-Aspartate transport into pea chloroplasts. *Plant Physiol.* 78:221-27

70. Willey, D. L., Fischer, K., Wachter, E., Link, T. A., Flügge, U.-I. 1990. Molecular cloning and structural analysis of the phosphate translocator from pea chloroplasts and its comparison to the spinach phosphate translocator. *Planta.* In press

71. Winter, K., Forster, J. G., Edwards, G. E., Holtum, J. A. M. 1982. Intracellular localization of enzymes of carbon metabolism in *Mesembryanthemum crystallinum* exhibiting C_3 photosynthetic characteristics or performing crassulacean acid metabolism. *Plant Physiol.* 69:300-7

72. Woldegiorgis, G., Voss, S., Shrago, E., Werner-Washburne, M., Keegstra, K. 1983. An adenine nucleotide-phospho-enolpyruvate counter-transport system in C_3 and C_4 plant chloroplasts. *Biochem. Biophys. Res. Commun.* 116:945-51

73. Woldegiorgis, G., Voss, S., Shrage, E., Werner-Washburne, M., Keegstra, K. 1985. Adenine nucleotide translocase-dependent anion transport in pea chloroplasts. *Biochim. Biophys. Acta* 810:340-45

74. Woo, K. C., Osmond, C. B. (1982). Stimulation of ammonia and 2-oxoglutarate-dependent O_2 evolution in isolated chloroplasts by dicarboxylates and the role of the chloroplast in photorespiratory nitrogen recycling. *Plant Physiol.* 69:591-96

75. Woo, K. C., Flügge, U.-I., Heldt, H. W. 1984. Regulation of 2-oxoglutarate and dicarboxylate transport in spinach chloroplasts by ammonia in the light. In *Proceedings of the Sixth International Congress on Photosynthesis, Brussels,* ed. V. Sybesma, 3:685-88. The Hague: W. Junk

76. Woo, K. C., Flügge, U.-I., Heldt, H. W. 1987. A two-translocator model for the transport of 2-oxoglutarate and glutamate in chloroplasts during ammonia assimilation in the light. *Plant Physiol.* 84:624-32

77. Yu, J., Woo, K. C. 1988. Glutamine transport and the role of the glutamine translocator in chloroplasts. *Plant Physiol.* 88:1048-54

Annu. Rev. Plant Physiol. Plant Mol. Biol. 1991. 42:145–88

PLANT LIPOXYGENASE: STRUCTURE AND FUNCTION

James N. Siedow

Department of Botany, Duke University, Durham, North Carolina 27706

KEY WORDS: lipid peroxidation, jasmonic acid, plant stress response, lipoxygenase pathway

CONTENTS

INTRODUCTION

The presence of an enzyme activity in plants, termed "lipoxidase," that could catalyze the oxidation of fatty acids was first reported almost 60 years ago (5). A second activity, "carotene oxidase," which was associated with the degradation of carotenoids, was found to be due to the same enzyme (154), and the name "lipoxygenase" has superseded both of these earlier terms when referring to this enzyme activity. Lipoxygenase (linoleate:oxygen oxidoreductase, EC 1.13.11.12) from soybean (*Glycine max*) seeds, among the first enzymes to be crystallized (156), was originally reported to contain neither a metal cofactor nor a prosthetic group of any kind. This result, even at the time considered surprising for an enzyme catalyzing an oxygenation reaction, has since been shown to be incorrect, possibly on both counts. More recent work has established that plant lipoxygenases are members of a class of nonheme iron-containing dioxygenases that catalyze the addition of molecular oxygen to fatty acids containing a *cis,cis*-1,4-pentadiene system to give an unsaturated fatty acid hydroperoxide (Figure 1). Commercially, lipoxygenase was an important enzyme almost before its discovery (7). Small amounts of soybean flour have long been added to wheat flour to bleach carotenoids and modify the rheological properties of the resulting dough (7). Because many products of the lipoxygenase reaction (or derivatives thereof) are aromatic, the presence of lipoxygenase activity in many foodstuffs can affect their properties, particularly during long-term storage, in both desirable and undesirable ways. The commercial importance of lipoxygenase (not a major concern in this review) has been treated in previous accounts (7, 47).

Although lipoxygenase was one of the first enzymes to be crystallized, research on plant lipoxygenases showed only limited progress until the early 1970s when Axelrod and coworkers first separated and purified individual lipoxygenase isozymes from soybean seeds (7). This work stimulated further studies that served to characterize the mechanistic features of the lipoxygenase catalytic reaction (63, 166), but comparable progress toward understanding the physiological function of plant lipoxygenase has not been made. Little is known with certainty about the role of lipoxygenase in any plant cell. In animals, on the other hand, where the discovery of lipoxygenase activity is of much more recent vintage (110, 138), it is now well established that the products of several different mammalian lipoxygenases are the primary metabolites on pathways that lead to the formation of important regulatory molecules in inflammatory responses, leukotrienes, and lipoxins (135, 139). However, the potential for gaining a better understanding of lipoxygenase in plants has been advanced considerably through the recent application of

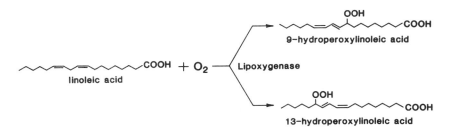

Figure 1   The primary reaction catalyzed by lipoxygenase using linoleic acid (*cis,cis*-9,12-octadecadienoic acid) as a substrate indicating the two possible reaction products, 9- and 13-hydroperoxylinoleic acid.

molecular technologies to the study of lipoxygenase, which has provided new insights into the nature of the lipoxygenase protein as well as the tools needed to facilitate further study of possible role(s) of lipoxygenase in plant cells. The similarity between the primary amino acid sequences of the soybean cotyledon lipoxygenases and lipoxygenases found both in other organs of the soybean and in other plant species has facilitated recent studies of lipoxygenase behavior outside of the confines of the soybean seed.

Plant lipoxygenase has been the subject of numerous reviews (7, 63, 165, 166), including some relatively recent ones (73, 103, 139, 174). Because much of the earlier literature on plant lipoxygenases focuses on the mechanistic features of the catalytic reaction, I make no attempt to reiterate that material except as is needed to put the more recent work in context. My goals are to summarize the present status of research on the structure of lipoxygenase in the light of recent molecular characterizations of lipoxygenase primary structures and to indicate where research on lipoxygenase is going as researchers begin to use biochemical, molecular, and genetic technologies in combination to elucidate the role(s) this enzyme plays in plants.

BIOCHEMICAL AND MOLECULAR PROPERTIES

General Features

Lipoxygenase activity has been reported in a wide range of organisms including more than 60 species of higher plants (47, 174), eukaryotic algae (186), baker's yeast and other fungi (72, 137), and a cyanobacterium (10). The latter paper is the only report of lipoxygenase activity in a prokaryote (although the authors refer to their organism as "the green alga *Oscillatoria*"). It would be of interest to have a clearer picture of how widely distributed lipoxygenase is among prokaryotic organisms, because bacteria are known to contain several different nonheme iron dioxygenases (e.g. catechol 1,2-dioxygenase, protocatechuate 3,4-dioxygenase) that show some physicochemical similarities to plant lipoxygenases (127). As noted, lipoxygenase activity is also found in a variety of mammalian cell types (135). Within a given plant, the level of lipoxygenase activity present in any tissue can vary markedly not only among different organs of the plant, but also between different developmental stages of a particular tissue type (see below).

Axelrod (7) expected lipoxygenase to be ubiquitous in plants and suggested that failure to measure lipoxygenase activity in a given species reflected the insensitivity of the detection method used rather than absence of enzyme activity. This point is well taken with respect to the assays of lipoxygenase activity commonly utilized (8, 70). However, the availability of antibodies against soybean cotyledon lipoxygenases has provided extremely sensitive probes to detect lipoxygenase protein levels in soybean and other legume tissues using Western immunoblots (121) and ELISAs (179). These assays should leave no question about the presence or absence of lipoxygenase protein in a given tissue at a particular stage of development. Further, molecular probes based upon conserved regions within the nucleotide coding sequence of the lipoxygenase gene make possible the development of very sensitive assays using Northern blotting techniques (2) to ascertain the presence of lipoxygenase transcripts within nonleguminous plant tissues.

The general reaction catalyzed by all lipoxygenases (Figure 1) involves the addition of molecular oxygen across the *cis,cis*-1,4-pentadiene system found in such commonly occurring unsaturated fatty acids as linoleic (18:2, an 18-carbon carboxylic acid containing double bonds at carbons 9 and 12), linolenic (18:3), and arachidonic (20:4) acids. The resulting hydroperoxide product contains a set of *cis,trans* double bonds that are now in conjugation as a result of a double bond migration during the catalytic cycle. Note that, in theory, the oxygen can add to either end of the pentadiene system. In the case of linoleic acid, this leads to two possible products, the 9- and 13-hydroperoxy linoleic acids (Figure 1). Lipoxygenase activity is most commonly measured either by taking advantage of the appearance of an absorption band at 234 nm ($E = 25$ mM^1 cm^1) associated with the newly formed conjugated double bond system in the hydroperoxide product or by following the course of the reaction by measuring the uptake of molecular oxygen with a Clark-type oxygen electrode (8, 70). Although numerous polyunsaturated fatty acids or fatty acid–containing molecules can serve as lipoxygenase substrates, linoleic acid is, by far, the most widely utilized substrate experimentally. For looking at unpurified mixtures containing more than one isozyme, Funk et al (61) have developed a procedure for separating the individual isozymes on an isoelectric focusing gel and visualizing the isozymes in the gel using previously developed activity stains (167).

Legume seeds contain particularly high levels of lipoxygenase activity. As a result, much of the characterization of lipoxygenase has been associated with the enzyme isolated from soybean cotyledons (7). Prior to the early 1970s, soybean seed lipoxygenase had generally been regarded as a single entity. Axelrod and coworkers changed this when they separated soybean cotyledon lipoxygenase into four distinct isozymes that were originally designated lipoxygenases-1, -2, -3a, and -3b (22, 23). However, there are marked similarities between lipoxygenases-3a and -3b (23); and in naturally occurring mutants that lack lipoxygenase-3 activity (88, 89), both lipoxygenase-3a and -3b activities are missing. It therefore appears that any differences between lipoxygenases-3a and -3b result from posttranslational modification and do not reflect their originating from separate genetic loci. This has led most workers to associate only three isozymes (lipoxygenases-1, -2, -3) with soybean cotyledons (74). All three isozymes are globular, water-soluble proteins that consist of a single polypeptide having a molecular weight of roughly 96,000. The three isozymes differ with respect to their isoelectric points, showing values of 5.68, 6.25, and 6.15 for lipoxygenases-1, -2, and -3, respectively (8). The large charge difference between lipoxygenase-1 and lipoxygenases-2 and -3 allows for relatively easy separation of lipoxygenase-1 from the other two isozymes using anion exchange chromatography (8, 22). Separation of lipoxygenases-2 and -3 initially proved difficult (8), but more recent applications of high performance liquid chromatography (128) and chromatofocusing (61) have provided relatively facile routes for the separation and purification of these two isozymes. While lipoxygenase was originally reported to require no metal cofactor for activity (156), analyses of the purified soybean isozymes indicated the presence of one atom of iron per 96 kDa protein (19, 123), which was later shown to be essential for catalysis (29, 124).

The major differences among the three soybean cotyledon lipoxygenase isozymes can be seen at the level of their reactivities. Lipoxygenase-1 has a pH optimum for activity centered around 9.0, while lipoxygenase-2 shows a sharp pH maximum around pH 6.5 and lipoxygenase-3 displays a broad optimum centered around pH 7 (8). Lipoxygenase-1 shows a marked preference for charged fatty acids and, as such, shows little reactivity with fatty acids that are esterified (i.e. as they would be found in a membrane). Lipoxygenases-2 and -3 are more reactive toward neutral fatty acids but will react with free fatty acids, particularly at pHs below 7.0. Related to this, lipoxygenase-1 can effectively utilize the sulfate ester of linoleic acid as a substrate and does so with a greater reactivity at pH 6.8 than 9.0 (12). Lipoxygenases-2 and -3 are unable to utilize the sulfate analog of linoleate as a substrate (12). The apparent ability of charge to so markedly affect isozyme reactivity toward a particular substrate represents an important distinction between lipoxygenase-1 and isozymes -2 and -3. As structural studies begin to elucidate the amino acids associated with the lipoxygenase active site, it may turn out that nothing more substantial than an appropriately positioned positive charge in the active site of lipoxygenase-1 accounts for the marked differential in pH optimum between lipoxygenase-1 and the other two soybean cotyledon isozymes.

The product of the reaction of linoleic acid with soybean cotyledon lipoxygenase-1 is almost exclusively 13-hydroperoxy linoleic acid (7, 8). With lipoxygenases-2 and -3, roughly equal amounts of the 9- and 13-hydroperoxy products are obtained. In addition to the primary reaction products, lipoxygenases have long been known to catalyze the appearance of secondary reactions, which bring about the formation of such products as dimers and oxodienoic acids (66). These secondary reactions serve as the source of the carotenoid oxidase activity seen in the early studies of lipoxygenase and are responsible for the pigment bleaching properties of soybean flour (7). Lipoxygenase-3 is the most active isozyme with respect to secondary reactions and catalyzes them under both aerobic and anaerobic conditions (129). Lipoxygenase-1 shows these secondary reactions only at either low oxygen tensions or under anaerobic conditions while lipoxygenase-2 shows little secondary product formation under any condition (129). An additional reaction that has been attributed to lipoxygenase is the formation of singlet oxygen, although the literature on this topic has been somewhat controversial (39). More recently, Kanofsky & Axelrod (83) used infrared chemiluminescence to demonstrate conclusively that singlet oxygen could be generated during the lipoxygenase reaction, but it required the presence of 13-hydroperoxylinoleic acid. In addition, singlet oxygen production showed the same pattern among the soybean seed isozymes as outlined above for the secondary reactions; lipoxygenase-3 formed singlet oxygen under any condition, lipoxygenase-2 was essentially inactive, and lipoxygenase-1 only showed appreciable singlet oxygen formation at low oxygen concentrations.

Following the separation and characterization of the three soybean cotyledon isozymes, a screening of soybean germplasm collections led to the discovery of lines that lacked the activities associated with lipoxygenases-1, -2 and -3 (75, 88, 89). Analysis of seeds derived from these lipoxygenase null-lines using an SDS gel system that effectively separates the three isozymes and immunodisc electrofocusing with an anti-lipoxygenase-2 antibody that cross-reacts with all three isozymes indicated that each of the three lipoxygenase null-lines lacked their respective protein component (88). More recently, Park & Polacco (115) reported the presence of low levels of the protein and activity in the null-lines of lipoxygenases-1 and -2 but not in the line that lacked lipoxygenase-3. This suggests that there may be some low level of leakiness in the former two lines. Studies using probes specific for each of the three isozymes in Northern blots of total RNA obtained from each of the null-lines have further established that the lines lacking lipoxygenases-1 and -3 accumulate little, if any, of their respective mRNAs (150). These data indicate that the mutations in these two lines affect either transcription or some aspect of transcript stability. Northern blotting using a probe specific for lipoxygenase-2 indicated that the line that lacked lipoxygenase-2 showed a considerable level of transcript formation. However, this same lipoxygenase probe also showed a markedly reduced apparent transcript level in the line lacking lipoxygenase-1 (150). All three lipoxygenase null-lines are inherited as simple Mendelian recessive traits (27, 76, 89). Further, lipoxygenases-1 and -2 are closely linked to one another, but each segregates independently of lipoxygenase-3 (27). Crosses between the separate null-lines have led to the development of double recessives lacking lipoxygenases-1 and -3 (89) and lipoxygenases-2 and -3 (27). No report has appeared of the development of any line lacking both lipoxygenases-1 and -2. This could be the result of the close linkage between the two loci or the possibility that loss of both of these cotyledon isozymes is a lethal trait. None of the individual lipoxygenase null-lines (or any of the combinations) have shown any obvious phenotypic manifestations beyond the loss of their particular isozyme activity (27, 75, 88).

Characterization of the separate soybean seed isozymes in the 1970s also led to the development of a nomenclature that separated lipoxygenases into "type 1" or "type 2" isozymes depending primarily upon whether a given isozyme shows a pH optimum for activity at high or neutral pH, respectively (63). Using this system, most plant lipoxygenases reported to date are type 2 enzymes; type 1 enzymes are uncommon, but not unheard of, outside of soybean cotyledons. When this scheme was first put forward, it was designed to eliminate ambiguities associated with the use of such terms as "alkaline," "acid," "neutral," or "ester" enzymes (63). However, the properties originally associated with the type 2 isozymes included more than simply the pH optimum for reactivity. Type 2 lipoxygenases were also characterized as showing a 50:50 ratio of 9- and 13-hydroperoxy products using linoleic acid as a substrate and to have high levels of carotenoid co-oxidation activity. The latter, as noted, is primarily associated with the ability of a given isozyme to carry out secondary product formation. The two categories were also developed at a time when the status of lipoxygenase-3 as a distinct species was still uncertain (63).

In retrospect, it appears that the characteristics originally outlined for type 2 enzymes were a combination of properties associated with lipoxygenases-2 and -3 and do not represent sufficient criteria for separating those two isozymes from lipoxygenase-1. For example, lipoxygenase-2 (unlike lipoxygenase-3) shows little carotenoid co-oxidation activity, while lipoxygenase-1 can show that activity at low oxygen tensions (129). Further, as noted above, the differential in pH optima may ultimately be reduced to the presence of a single charged species within the active site rather than to any larger structural features of the lipoxygenase protein. As the primary structures of more plant lipoxygenases become available, a clearer delineation of lipoxygenase subtypes should be possible that will enable the development of a more structure-based classification scheme. At this point, application of the type 1/2 nomenclature can be misleading, and it would be good to see its use discontinued.

Reaction Mechanism

As noted previously, a considerable amount of the work on plant lipoxygenases in the 1970s and 1980s was devoted to elucidating the reaction mechanism (139, 165, 166), and there is now a reasonable consensus regarding the major features by which soybean cotyledon lipoxygenase-1 catalysis takes place. In resting lipoxygenase-1, the single atom of nonheme iron at the active site exists in a high-spin Fe(II) state (37, 146). The resting enzyme is essentially colorless, shows no electron paramagnetic resonance (EPR) spectroscopic features, and is not catalytically active. The active form of lipoxygenase-1 can be generated from the resting enzyme by reaction with one equivalent of hydroperoxide product which oxidizes the active site iron to a high-spin Fe(III) state (29, 123). This form of the enzyme has a characteristic pale yellow color and an EPR signal in the g-6 region, characteristic of high-spin Fe(III). A second form of oxidized lipoxygenase-1 can be generated in the presence of a stoichiometric excess of product. This form has a distinctly purple color and exhibits an additional EPR signal around g-4.3 (29). The latter signal is relatively intense but accounts for only a small amount (10%) of the total integrated Fe(III) EPR signal (147). The purple form of the enzyme is unstable and will decay to the yellow species within a matter of minutes at room temperature and within several hours if held on ice (28, 165). An initial report of EPR spectroscopy performed on oxidized soybean lipoxygenase-2 indicated that the spectrum was predominantly of the g-4.3 type with very little of the g-6 signal present (49). More recently, Draheim et al (35) carried out EPR studies on two soybean lipoxygenase isozymes (apparently corresponding to lipoxygenases-2 and -3) and found a behavior qualitatively similar to that seen with lipoxygenase-1. At this point it seems well established that the resting, inactive form of all three soybean cotyledon isozymes exists in the Fe(II) state, and that it is activated in the presence of added product by conversion to the Fe(III) state. The catalytic significance of the purple form of the enzyme is, at present, less obvious.

In addition to EPR spectroscopy, several other spectroscopic techniques have been brought to bear on soybean lipoxygenase-1 in an attempt to determine the nature of the ligands associated with the iron atom in the active site. Extended x-ray absorption fine structure (EXAFS) studies of lipoxygenase-1 showed a best fit to a simulation in which the active site iron contained six ligands, four imidazole nitrogens and two oxygens, the latter possibly coming from carboxylate residues (108). Mössbauer spectroscopy of [^{57}Fe]lipoxygenase-1, labeled by culturing immature soybean seeds on an ^{57}Fe-containing medium (59), indicated that the iron atom is octahedrally coordinated, the six coordinating ligands being some combination of N- and/or O-containing species (37). A roughly octahedral six-coordinate structure for the iron-ligand system in resting state lipoxygenase-1 was also indicated by magnetic circular dichroism spectroscopy (177).

Once resting lipoxygenase-1 has been oxidized to the active yellow form, it can carry out the enzymic reaction. Neither the resting state nor the oxidized, activated lipoxygenase-1 binds molecular oxygen directly, so catalysis is thought to be initiated by the binding of an appropriate unsaturated substrate at the active site. Lipoxygenases are always described as reacting with polyunsaturated fatty acids containing a *cis,cis*-1,4-pentadiene system. This uniformity appears to be related to the fact that naturally occurring fatty acids nearly always exist in the *cis,cis* configuration. Funk et al (58) have demonstrated that both the *cis,trans* and the *trans,cis* isomers of the naturally occurring linoleic acid (*cis,cis*-9,12-octadecadienoic acid) can serve as substrates for soybean cotyledon lipoxygenase-1, suggesting that the active site is relatively flexible in its ability to accommodate substrate. Analysis of the resulting products indicated that substrate can apparently bind to the active site in either of two possible orientations (58).

Following binding of fatty acid substrate, the catalytic mechanism was originally thought to involve the stereospecific abstraction of a hydrogen atom from the methylene group at the center of the pentadiene system (C-11 in linoleic acid, Figure 1) (42). The presence of a marked kinetic isotope effect with linoleic acid labeled with deuterium at C-11 established this hydrogen abstraction as the rate-determining step for the overall reaction (41). This hydrogen abstraction was at first associated with a direct reduction of the Fe(III), but little consideration was given to what ultimately happened to the accompanying proton (63). More recently, studies using mechanism-based lipoxygenase inhibitors have led to a revision of this scheme, the initial event in catalysis being the extraction of a methylene proton by an unspecified basic group in the active site (24). This proton extraction facilitates an electrophilic attack by the Fe(III) atom on one of the double bonds in the pentadiene system, which leads to the formation of Fe(II) and an organic radical whose unpaired electron is delocalized over the entire pentadienyl system. This reaction sequence could also take place in a concerted fashion (24). However, there is no consensus even about this mechanism. Corey and coworkers have developed an alternative mechanism: Following proton extraction, the Fe(III) covalently bonds with a terminal carbon on the resulting pentadienyl carbanion and does not formally undergo reduction (25, 26). Until recently, the data in the literature were not sufficiently unambiguous to distinguish readily between these two possibilities. However, Funk et al (60) have recently used the combination of EPR and Mössbauer spectroscopy to demonstrate that the iron atom in the yellow, Fe(III) form of lipoxygenase-1 is reduced to the Fe(II) state upon addition of substrate (linoleic acid) to the enzyme under anaerobic conditions. This result is inconsistent with the formation of an organoiron intermediate. Continuing with the redox-based mechanism, oxygen now reacts with the pentadienyl radical to give a hydroperoxy radical that is subsequently reduced to hydroperoxide product by the Fe(II), restoring the enzyme to its initial Fe(III) state. Presumably steric constraints within the enzyme act to direct the stereospecific addition of oxygen that leads predominantly to the 13-hydroperoxy product seen in the case of lipoxygenase-1.

The secondary reactions catalyzed by lipoxygenase have been the subject of considerable discussion in previous reviews (7, 63, 166, 174). Because so much of the mechanistic work on lipoxygenase has been carried out with the soybean cotyledon lipoxygenase-1, formation of secondary products has often been referred to as the "anaerobic reaction" of lipoxygenase (63, 174), reflecting the fact that secondary products only appear with lipoxygenase-1 under conditions of low oxygen tension. Mechanistically considered, the initial abstraction of a hydrogen atom from the methylene carbon of the substrate (by whichever of the routes outlined above) leads to formation of an intermediate fatty acid radical-containing species (L$^{\cdot}$). Under anaerobic conditions, there will be no oxygen available to react with the bound radical, so eventually some of the radicals will dissociate from the enzyme. Being free in solution, this reactive species can now undergo any of the standard secondary reactions associated with organic free radicals (e.g. dimer formation). The appearance of additional secondary products, such as oxodienoic acids and pentane, in the presence of low levels of oxygen can be attributed to the formation of free hydroperoxy radical–containing species (LOO$^{\cdot}$). This happens either via dissociation of the hydroperoxy radical intermediate from the enzyme in an abortive catalytic cycle or through reaction of the dissociated L$^{\cdot}$ species with oxygen in solution to produce LOO$^{\cdot}$. Oxodienoic acid and pentane can then result from the spontaneous decomposition of the hydroperoxy radical (166). At ambient atmospheric oxygen concentrations, the normal catalytic cycle of lipoxygenase-1 is apparently rapid and efficient enough that dissociation of the intermediate radical species (L$^{\cdot}$ or LOO$^{\cdot}$) rarely takes place. The ability of soybean cotyledon lipoxygenase-3 to catalyze the formation of secondary products under both aerobic and anaerobic conditions suggests that the enzyme-bound radical intermediates formed during its reaction sequence are more readily able to dissociate from the enzyme, even at high oxygen concentrations, than those formed by lipoxygenase-1.

One reason for belaboring the lipoxygenase secondary reactions involves the fact that the co-oxidation reactions of lipoxygenase, which include its ability to bleach pigments such as carotenoids and chlorophylls, are now believed to be associated with secondary product formation and not the result of direct oxidation of pigment by the fatty acid hydroperoxide product (174). While this has been understood by workers in the area for some time, it was not always made apparent in earlier reviews, where the two reactions were often treated in separate sections (7, 63, 166). In addition, the relationship between the co-oxidation reactions and secondary product formation (129), coupled with the variability of the latter reaction among different lipoxygenase isozymes (8), provides a good rationale for avoiding the use of reactions such as carotenoid bleaching to measure lipoxygenase activity (70).

Finally, a potentially important observation with respect to the lipoxyge-nase reaction mechanism is the recent report that purified soybean cotyledon lipoxygenase-1 contains covalently bound pyrroloquinoline quinone (PQQ, methoxatin) as an organic cofactor (162). If correct, this report would bring lipoxygenase from the point where its supposed lack of either a metal or an organic cofactor made it unique among oxidative enzymes of this general type (155), to the point where it contains both. PQQ has previously been associ-ated with several different copper-containing enzymes, where it plays a direct role in catalysis by providing ligands to the active-site metal, which is believed to undergo redox changes during the course of catalysis (36, 64).

The presence of PQQ in lipoxygenase-1 was established by treating the enzyme with phenylhydrazine, which had previously been shown to be an inhibitor (67). Reaction of lipoxygenase with phenylhydrazine led to the appearance of a broad-spectrum species with an absorbance maximum cen-tered around 350 nm. The latter spectrum was attributed to formation of a PQQ-phenylhydrazine adduct that had been used previously as a diagnostic for the presence of PQQ in several copper-containing proteins (163, 164). Verification of the PQQ adduct was carried out using HPLC after the adduct had been released from the enzyme by extensive proteolysis with pronase E (162). The latter result indicated that, as with other PQQ-containing enzymes (64), the quinone moiety was covalently bound to the protein. A stoichiome-try of 0.88 moles of PQQ per mole of enzyme was reported which coincided with a value of 0.91 moles of iron per mole of enzyme in the same preparation (162).

Van der Meer & Duine (162) suggested that PQQ could serve in the active site of lipoxygenase as a terdentate ligand to the iron atom, binding through the nitrogen of the PQQ pyridine ring and oxygen atoms from the carboxylate group at C(3) and an OH group present at C(5) in hydrated PQQ. However, at this stage we have no direct evidence to support an interaction between PQQ and iron in lipoxygenase. Furthermore, PQQ is a difficult cofactor to quantify reliably (64); the accuracy of previous reports of enzymes containing PQQ remains controversial (175). After I had drafted this review, Veldink et al (164a) reported that they were unable to demonstrate the presence of any PQQ in purified soybean lipoxygenase-1 using either the procedure described by van der Meer & Duine (162) or several alternative assays of PQQ. At present, the concept that PQQ is a lipoxygenase cofactor should be viewed with considerable skepticism.

Structure Studies

Given the wide diversity of plants in which lipoxygenase activity has been reported (47, 174), it is somewhat surprising that the number of species from which the enzyme has been well characterized is limited. Nonetheless, where plant lipoxygenases have been characterized to some extent [plants as widely divergent as rice, soybean, cotton and sunflower (47, 174)], the enzyme consists of a protein whose molecular weight is in the vicinity of 95,000. Exceptions have been reported; molecular weights ranging from 72,000 to 108,000 have been published for lipoxygenases isolated from pea (63). However, the cDNA sequence for one of the pea lipoxygenase isozymes (admittedly there are several) indicates a molecular weight of about 97,000 for the resulting protein, which shares an appreciable amino acid sequence homology with the three soybean cotyledon isozymes (40). The sample size is still small, but the present evidence indicates that there will not be a collection of widely divergent lipoxygenase proteins, at least with respect to polypeptide size, but rather a group of related proteins all of which are similar to the well-characterized isozymes from soybean cotyledons.

Beyond the fact of having similar molecular weights, the first real indication of relatedness among the three soybean cotyledon isozymes was derived from immunological studies using polyclonal antibodies generated against the individual purified isozymes. These results have been discussed in detail in recent reviews (103, 174). To summarize, among soybeans, antibodies raised against purified cotyledon lipoxygenase-2 generally show a broad cross-reactivity both with the other two cotyledon isozymes and with proteins found in other organs of the plant (2, 121). However, comparisons using a variety of immunological detection procedures suggested that the lipoxygenase isozymes located in the soybean hypocotyl, leaves, and roots, while immunologically related to cotyledon lipoxygenase-2, were nonetheless distinct isozymes. Cross-reactivity is most readily observed through the use of Western immunoblots (121), and some of the apparent lack of cross-reactivity reported in the past (168, 179) may be related as much to differences in the nature of the immunoassay used as to any specific differences in the antibodies used. This may be why polyclonal antibodies raised against purified cotyledon lipoxygenase-1 have not been found to be broadly cross-reactive (168, 179). Given the marked amino acid sequence homology among soybean cotyledon lipoxygenases-1, -2, and -3 (see below), this result is surprising. It is further tempered by the recent report of an antilipoxygenase-1 that is cross-reactive with other soybean lipoxygenase isozymes on immunoblots of isoelectric focusing gels (115). Only one report of an antibody against purified soybean cotyledon lipoxygenase-3 has appeared, and its reactivity on an ELISA was restricted to that isozyme; no cross-reactivity was seen with either of the other two isozymes (179). No mention was made in this study of whether cross-reactivity had been tested using immunoblots.

Looking beyond soybean, only a limited number of studies have compared the cross-reactivity of other lipoxygenases with antibodies against the soybean cotyledon isozymes. While reactivity cannot be demonstrated with all antibodies in all types of immunoassays, examples of cross-reactivity between antibodies against the soybean seed isozymes and lipoxygenases from other legumes do exist (18, 168). Only two studies have looked for lipoxygenase antibody cross-reactivity outside of the legumes, giving somewhat mixed results. Trop et al (159) saw inhibition of lipoxygenase activity in potato tuber and eggplant fruit by antibodies raised against soybean lipoxygenase, while Vernooy-Gerritsen et al (168) saw no cross-reactivity with extracts from maize, wheat, flax, eggplant, and potato in immunodiffusion assays using an antibody against highly purified soybean cotyledon lipoxygenase-2. The marked differences between these two studies, both with respect to the nature of the purified enzymes used to generate the antibodies in the first place and with respect to the immunological assays, does not allow one to draw conclusions at present about the immunological relatedness of soybean cotyledon lipoxygenases and the lipoxygenases found in other plant families. Given the widespread availability of lipoxygenase-2 antibodies and the proven effectiveness and sensitivity of "Western" immunoblots in establishing cross-reactivity (121), it would be useful to see a broad survey of the extent of cross-reactivity of soybean cotyledon antilipoxygenase-2 with a range of different plant species. Even though the ease of using the polymerase-catalyzed chain reaction to clone homologous genes from different species has reduced the utility of cross-reactive antibodies for this purpose, there will still be instances where access to a cross-reactive antibody will facilitate cloning a particular sequence. It would be useful to know how applicable the soybean cotyledon lipoxygenase antibodies might be to such a contingency.

Beyond plants, Pages et al (114) reported raising antibodies in rabbits against Sigma lipoxygenase. They found that the resulting serum could partially inhibit mouse macrophage lipoxygenase activity. They further demonstrated that in *Drosophila*, the male reproductive system and the digestive tracts of both males and females reacted with the antiserum in immunocytochemical localization studies (114). This study makes no mention of having additionally purified the Sigma lipoxygenase fraction before using it to generate the antibodies. The omission is problematic because antibodies against proteins other than lipoxygenase could be responsible for the observed immunofluorescence in the fruit fly sections. On the other hand, the inhibition of the mouse lipoxygenase activity is an intriguing observation that deserves further study since the primary sequence similarity between soybean and mammalian lipoxygenases is limited to a few regions within the protein (see below).

Without question, the most important recent advance toward understanding lipoxygenase structure involves the cloning and sequencing of the cDNAs for a number of different lipoxygenases, including all three of the soybean cotyledon isozymes. Start et al (150) first reported the cloning of three lipoxygenase cDNAs from soybean seeds in a project that was reputedly designed to isolate the gene for urease. One clone, pAL-134, was found to produce an antigen in an expression system that cross-reacted with an antibody against soybean lipoxygenase-1. Two additional clones, pLX-65 and pLX-10, were obtained by screening a size-selected cDNA library with pAL-134, but none of these three clones represented a full-length lipoxygenase cDNA. Northern blots of each of these clones against total RNA isolated from soybean cultivars lacking each of the three seed isozymes confirmed that pAL-134 was the cDNA for lipoxygenase-1 and further indicated that pLX-10 was associated with lipoxygenase-3. The third clone, pLX-65 was described as being "lipoxygenase-1-like," given its marked sequence similarity with pAL-134 (150). Since then, the sequences of full-length cDNAs for all three soybean cotyledon isozymes have been reported (141, 142, 183), as has the complete sequence for a lipoxygenase from pea seed (40) and the partial sequence of a putative lipoxygenase from tobacco (14). To add to the procession, the explosive growth in research on mammalian lipoxygenases has led to the cloning and sequencing of cDNAs for human and rat 5-lipoxygenases (9, 32, 105), human and rabbit 15-lipoxygenases (53, 145), and a porcine 12-lipoxygenase (184). It should be noted that mammalian lipoxygenases are designated according to the position of addition of the -OOH moiety in the primary hydroperoxide product using arachidonic acid (5,8,11,14-all-*cis*-eicosatetraenoic acid) as a substrate.

Not unexpectedly, the derived amino acid sequences of the three soybean cotyledon isozymes are markedly similar. Lipoxygenase-1 contains 838 amino acids and gives a deduced molecular weight of 94,038 (142). The corresponding values for lipoxygenases-2 and -3 are 865 ($M_r = 97,053$) and 857 ($M_r = 96,541$), respectively (141, 183). Comparison of these nucleotide sequences with the three clones originally reported by Start et al (150) confirms the original assignments of lipoxygenases-1 (pAL-134) and -3 (pLX-10), but the lipoxygenase-1-like clone, pLX-65, is actually associated with lipoxygenase-2. When the amino acid sequences of all three isozymes are aligned to maximize sequence homologies, with gaps put in to accommodate differences in total amino acids (141, 183), lipoxygenases-1 and -2 show an 81% sequence identity. Comparisons of lipoxygenases-2 and -3 and lipoxygenases-1 and -3 give values of 74% and 70%, respectively. Yenofsky et al (183) also isolated and sequenced a genomic DNA for soybean lipoxygenase-3. When compared with the sequence for the corresponding cDNA, the gene for lipoxygenase-3 was found to contain nine exons and eight introns. Given that the genomic DNA has been isolated for lipoxygenase-3, it would be interesting to know how the nucleotide sequence in the lipoxygenase-3 null-lines (89) differs from that of the wild-type, since at least one of the null-lines appears to result from a failure to accumulate the lipoxygenase-3 mRNA (150).

Most of the size differential between lipoxygenase-1 and the other two cotyledon isozymes is associated with a region containing the N-terminal 55 amino acids of lipoxygenases-2 and -3. Most of these amino acids are missing in lipoxygenase-1. Within the gene for lipoxygenase-3, these residues are all located in exon 1. The derived amino acid sequence for the pea seed lipoxygenase contains 861 amino acids ($M_r = 97,628$) (40). Optimized alignment of the pea amino acid sequence with those of the soybean lipoxygenases indicates a 69% amino acid identity with lipoxygenase-1 but an 86% identity with lipoxygenase-3. The latter value is increased to 93% if conservative substitutions are included. Ealing & Casey (40) pointed out that the N-terminal region of the pea lipoxygenase sequence contains a stretch of amino acids (35–44) having an 80% identity with the soybean lipoxygenase-3, while the comparable region of lipoxygenase-1 was missing six of these amino acids. This run of amino acids is within the N-terminal region referred to above. The putative tobacco lipoxygenase amino acid sequence was derived from a clone originally screened for cross-hybridization with the cDNA for soybean lipoxygenase-3 (14). Compared to the three soybean cotyledon isozymes, this clone is incomplete, missing a portion of the N-terminal coding region, but the remainder of the sequence is very similar to the sequences of all three cotyledon isozymes.

Studies of lipoxygenases isolated from mammalian sources have reported molecular weights between 61,000 for the rabbit reticulocyte 15-lipoxygenase and 100,000 for an enzyme from human platelets (139). However, the amino acid sequences derived from the mammalian lipoxygenase cDNAs indicate that there are 661 amino acids ($M_r = 74,600$) in the human and rabbit 15-lipoxygenases (53, 145) and 662 amino acids ($M_r = 74,911$) in the porcine 12-lipoxygenase (184), while the human 5-lipoxygenase has 673 amino acids ($M_r = 77,830$) (105). Comparison with the larger soybean isozymes shows that most of the size differential is associated with the fact that the N-terminal 110 (lipoxygenase-1) to 140 (lipoxygenase-2) amino acids found in the soybean lipoxygenases have no apparent counterparts in the mammalian enzymes. Based upon the gene structure for soybean lipoxygenase-3 (183), the sequence of the mammalian enzymes begins roughly in the middle of exon 2. Aligning the C-terminal portion of the plant lipoxygenases with the mammalian enzymes to maximize sequence comparisons, the human 5-lipoxygenase shows only 22% sequence identity with soybean lipoxygenase-3, which increases to about 45% if conservative substitutions are considered (53). However, even when the human 5- and 15-lipoxygenases are aligned and compared with each other, they show values of only 39% (identity) and

61% (similarity), respectively (145). The 12- and 15-lipoxygenases are more closely related, showing 89% identity (184). Among the mammalian enzymes, only the gene for the human 5-lipoxygenase has been isolated to date (57). It contains 14 exons divided by 13 introns and shows no obvious structural similarity to the soybean lipoxygenase-3 gene.

Recognizing the danger of drawing too many conclusions from sequence comparisons in the absence of a known structure, there are nonetheless some regions within the different lipoxygenase sequences worth pointing out. There are numerous examples distributed throughout the sequences where runs of 5–7 amino acids may differ among the three soybean cotyledon isozymes, but there are only two large regions where the sequences differ markedly. The first is the earlier-noted region of dissimilarity at the N-terminus, which is missing in lipoxygenase-1 and is not particularly highly conserved between lipoxygenases-2 and -3. The second region is a stretch of roughly 35 amino acids (spanning amino acids 396–431 in lipoxygenase-2) where lipoxygenase-3 shows a clear divergence from lipoxygenases-1 and -2. There is only about a 50% identity between lipoxygenases-1 and -3 throughout this region, while lipoxygenases-1 and -2 are 80% identical over the same stretch of amino acids. Yenofsky et al (183) noted that these amino acids essentially represent the region coded for within exon 6 of the lipoxygenase-3 gene. Not surprisingly, the pea lipoxygenase sequence is almost identical to that of soybean lipoxygenase-3 in this region (40).

It is more difficult to identify regions of similarity when the overall proteins display 70–80+% sequence identity in the first place, but several have been noted. With the exception of tobacco, the C-terminal 9 amino acids of all the plant lipoxygenases sequenced to date are identical (RGIPNSISI). Tobacco is unusual in that it not only shows no sequence similarity with these 9 amino acids but also adds an additional 20 amino acids onto the C-terminus (14). Three other regions of sequence similarity have drawn more attention. In all three soybean cotyledon isozymes, a group of six histidines appear within a 37-amino-acid span (522–559, L-2) that itself is buried in a somewhat larger region of high sequence homology (Figure 2A). Of interest is the observation that five of these six histidines are conserved in all of the mammalian lipoxygenases, and the larger region within which they are located represents a region of considerable sequence relatedness between the plant and animal enzymes (Figure 2A). For example, there is roughly 50% sequence identity between soybean lipoxygenase-3 and human 5-lipoxygenase over this region (53). Within this region both the pea and tobacco lipoxygenase sequences are highly conserved, relative to the soybean enzymes, but the pea sequence contains all six histidines while tobacco lacks one of six (H545, L-2). However, the missing histidine in tobacco is the same one that is not found in the mammalian lipoxygenases (Figure 2A). No obvious sequence similarities between any plant lipoxygenase and other known enzymes have been reported, but it has been pointed out (32) that a 15-amino-acid run in the human (32) and rat (9) 5-lipoxygenases (368–382, Hu5-LOX) shares some sequence relatedness (50% similarity) to the interface binding domains of human lipoprotein lipase (178) and rat hepatic lipase (92). While this sequence lies within the region under discussion here, no comparable similarity between the plant sequences in this region and the lipase binding domains is apparent. A second highly conserved region includes a short run of 13 amino acids (714–726, L-2) that are completely conserved among the four legume enzymes (ASALHAAVNFGQY) and differ by only one amino acid in the tobacco sequence, where the initial alanine residue is a threonine. Among the mammalian lipoxygenases, the human and rat 5-lipoxygenases show only the single substitution of a glutamine for the one leucine in the sequence (9, 32). In the 15- and 12-lipoxygenase sequences, this region shows no obvious similarity to the plant enzymes (53, 145, 184) except that all lipoxygenases sequenced to date retain the histidine found within this region. Curiously, the mammalian 15- and 12-lipoxygenases show a zinc-finger, metal-binding motif [C-(X)$_3$-C-(X)$_3$-H-(X)$_3$-H] in this same region (184) that is not seen in either the plant enzymes or the mammalian 5-lipoxygenases. There is a third region (369–395, L-2) where the amino acid sequence is 93% identical between lipoxygenases-1 and -2 and over 81% identical between lipoxygenases-1 and -3 (Figure 2B). This region is also highly conserved in the pea (40) and tobacco (14) enzymes. The same region shows a greater than 70% sequence identity when we compare the mammalian 5- and 15-lipoxygenases, and a 44% identity (60% similarity) between soybean lipoxygenase-3 and either the 5- or 15-lipoxygenases (53) (Figure 2B). One can carry the analysis of sequence similarities on ad nauseam, but the regions outlined above maintain a clearly recognizable primary sequence similarity across two kingdoms. Based upon past experience with other, better characterized enzymes, this suggests that these are regions to focus on as more structural and mechanistic information is developed about the lipoxygenase reaction in both plants and animals.

With the deluge of primary sequences that has resulted from our ability to clone and sequence nucleic acids easily, a cottage industry has grown up around trying to glean information about higher-level protein structure based solely upon primary sequence data. Hydropathy plots have routinely been applied to the analysis of membrane-bound proteins (45). Most plant lipoxygenases are certainly not membrane-bound proteins; but such plots do reflect, at some level, local structural features even within soluble proteins (20). This may be particularly true of an enzyme whose primary substrates are quite hydrophobic. Figure 3 shows the hydropathy profiles calculated for soybean cotyledon lipoxygenases-1 and -2 and human 5-lipoxygenase using the pro-

A.

```
                                              *           *           *
LOX-1   479   A T S   K A Y V I V N D S C Y H Q L M S H W L N T H A A M E P F I V
LOX-2   507   W L L A - A Y - V V N - S C Y - Q L M S - W - N T - A V I - P F I I
LOX-3   498   W L L A - A Y - V V N - S C Y - Q L V S - W - N T - A V V - P F I I
P-LOX   502   W L L A - A Y - I V N - S C Y - Q L V S - W - N T - A V V - P F V I
5LOX    347   W L L A - I W - R S S - F H V - Q T I T - L - R T - L V S - V F G I
15LOX   340   W L L A - C W - R S S - F Q L - E L Q S - L - R G - L M A - V I V V
                            #                   #     #                         #
```

```
                                        *                 *
LOX-1   512   A T H R H L S V L H P I Y K L L T P H Y R N N M N I N A L A R Q S L
LOX-2   541   - T N - H - S A L - - - Y - - L T P - Y - D T M N - - A L - - Q S -
LOX-3   532   - T N - H - S V V - - - Y - - L T P - Y - D T M N - - G L - - L S -
P-LOX   536   - T N - H - S C L - - - Y - - L Y P - Y - D T M N - - S L - - L S -
5LOX    381   - M Y - Q - P A V - - - F - - L V A - V - F T I A - - T K - - E Q -
15LOX   374   - T M - C - P S I - - - F - - I I P - L - Y T L E - - V R - - T G -
                            #             #     #
```

B.

```
LOX-1   340   W M T D E E F A R E M I A G V N P C V I R G L E E F P
LOX-2   369   - M T - E E - A R E M V A - V - - V C I R G L Q E F -
LOX-3   358   - M T - E E - A R E M L A - V - - N L I R C L K E F -
P-LOX   362   - M T - E E - A R E M L A - V - - N V I C C L Q E F -
5LOX    226   - Q E - L M - G Y Q F L N - C - - V L I R R C T E L -
15LOX   222   - K E - A L - G Y Q F L N - A - - V V L R R S A H L -
                        #             #               #
```

Figure 2 Comparison of the amino acid sequences of plant and mammalian lipoxygenases within two highly conserved regions. The amino acid sequences for soybean lipoxygenase-1 (LOX-1) (142), soybean lipoxygenase-2 (LOX-2) (141), soybean lipoxygenase-3 (LOX-3) (183), pea seed lipoxygenase (P-LOX) (40), human 5-lipoxygenase (5LOX) (105) and human 15-lipoxygenase (15LOX) (145) were aligned for maximum sequence overlap by combining the fits developed by Shibata et al (141) and Fleming et al (53). —, indicates amino acid identity among all sequences; #, indicates sites of nonidentity that nevertheless have conservative amino acid substitutions among all sequences; blank spaces are gaps inserted to maximize sequence alignment; *, indicates histidine residues that are conserved among all sequences. See text for details.

cedure of Kyte & Doolittle (94). Comparable profiles are obtained for soybean lipoxygenase-3 and the sequence derived from pea. The most obvious difference between the two plant enzymes lies at the N-terminus where lipoxygenase-2 contains roughly 30 additional amino acids (Figure 3, region 1'). This gives rise to a more broadly hydrophobic region in lipoxygenase-2 (and lipoxygenase-3) than that seen at the N-terminus of lipoxygenase-1. After this initial difference, the hydropathy profiles for the two plant enzymes track each other quite well. This is not surprising given the marked similarity in their amino acid sequences. Even taking into account that the human 5-lipoxygenase has roughly 200 fewer amino acids than the plant enzymes and that 110 (L-1) to 140 (L-2) of the missing amino acids are located at the N-terminus of the plant enzymes, the overall hydropathy profile for the

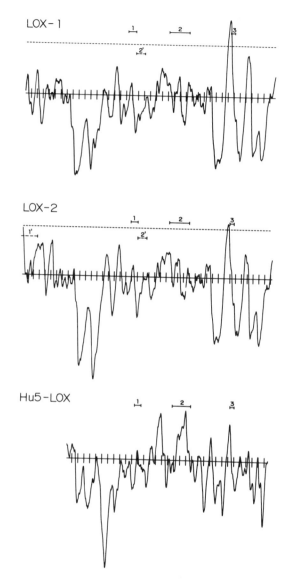

Figure 3 Hydropathy plots of soybean lipoxygenase-1 (LOX-1), soybean lipoxygenase-2 (LOX-2), and human 5-lipoxygenase (Hu5-LOX), aligned for maximum sequence homology. Hydropathy was calculated using the amino acid hydrophobicity data base of Kyte & Doolittle (94) with a sliding interval of 19 residues. The distance between tick marks on the abscissa corresponds to 20 amino acids. Regions of high sequence homology (1, 2, 3) and sequence dissimilarity (1', 2') are described in the text.

mammalian enzyme shows only a limited relatedness to that of the plant enzyme, depending upon how willing one is to align strongly hydrophobic and/or hydrophilic regions.

Of more interest is a comparison of the three regions of sequence similarity discussed above. As has been pointed out (32, 74), the largest stretch of sequence similarity (Figure 2A) contains a hydrophobic region in the central portion of the sequence (Figure 3, region 2). This "spike" of hydrophobicity is smallest in soybean lipoxygenase-1 and is more highly pronounced in the mammalian enzymes than in the plant ones. More intriguing is the conserved 13-amino-acid sequence in the C-terminal region (714–726, L-2). In the hydropathy profiles of all lipoxygenases sequenced to date, this stretch coincides with residues that contribute to the second half of a major peak of hydrophobicity located in the C-terminal third of the molecule (Figure 3, region 3). The third region of sequence homology (369–395, L-2) also shows a similar, but less pronounced, hydropathy profile among all lipoxygenases, being centered around a small hydrophobic region that interrupts two larger hydrophilic domains (Figure 3, region 1). It is interesting that the hydrophilic domain to the C-terminal side of this conserved region in the soybean lipoxygenases consists of the 35-amino-acid run that makes up the major region of sequence dissimilarity between lipoxygenase-3 and the other two cotyledon isozymes (Figure 3, region 2'). Application of standard secondary structure algorithms (21) to these regions of sequence similarity did not show any obvious pattern for any region, or part thereof, when compared across all the known lipoxygenase sequences (J. N. Siedow, unpublished observation).

Several additional features of lipoxygenase structure can be considered in the light of information provided in the primary sequences. Histidine has been found to be a nitrogenous ligand to iron in a variety of nonheme iron centers, including hemerythrin (140), ribonucleotide reductase (109), and the bacterial photosynthetic reaction center (30), among others. Given the presence of six invariant histidine residues among all known lipoxygenase sequences (distributed among two of the three highly conserved regions between the plant and mammalian enzymes), the suggestion has been made that one or more of these histidines might function as ligands to the lipoxygenase active site iron atom (53, 141). As noted above, EXAFS and Mössbauer spectroscopic studies have indicated that histidine imidazoles probably contribute four of the six ligands to the iron in the active site of soybean cotyledon lipoxygenase-1 (37, 108). However, Funk et al (56) recently reported the synthesis of human 5-lipoxygenase in insect cells using a baculovirus expression system that allowed them to carry out site-directed mutagenesis on the enzyme. They changed two of the invariant histidine residues in this region (H362 or H372; equivalent to H522 and H532 in L-2) to serine and observed that neither modification affected the ability of the resulting transformed cells to convert arachidonic acid to the appropriate hydroperoxy product (56). Thus whatever the evolutionary driving force that has maintained these two histidines, it does not appear to be related to their serving as essential ligands to the iron in the lipoxygenase active site. In addition to histidine, the acidic side chains of aspartate and glutamate have been found to serve as iron ligands in many nonheme iron proteins (17, 30, 109). Within the conserved region containing the five invariant histidines (Figure 2A), two acidic residues (D518/E536, L-2) are unchanged throughout all known sequences, and their possible role as ligands to the active site iron should not be overlooked. Considering other potential iron ligands, the purple form of lipoxygenase is similar to a blue spectral species of the enzyme protocatechuate dioxygenase, which has been shown by resonance Raman spectroscopy to result from a charge-transfer complex between a tyrosine and the active site iron in the latter enzyme (17). With this in mind, Shibata et al (141) pointed out that tyrosine residues were located adjacent to two of the five invariant histidines located in the large region of sequence similarity (Figure 2A) and suggested that they might also serve as ligands to the iron atom, at least in the Fe(III) state. Hildebrand et al (74) noted that neither of these tyrosines is conserved in the putative tobacco sequence nor in any of the mammalian sequences; and one is adjacent to His-522 (L-2), which seems to have been ruled out as a possible iron ligand (56). However, given the metastable nature of the purple form of the soybean enzyme as well as the absence of a demonstration that these other lipoxygenases can even form such a spectral species, it is difficult at this time to make a cross-species comparison using this particular property of soybean cotyledon lipoxygenases.

Appearance of the primary sequences has resolved one long-standing controversy about the soybean cotyledon lipoxygenases — the number of cysteine residues associated with each isozyme. The values are 4, 5, and 7 for lipoxygenases-1, -2, and -3, respectively (141, 142, 183), and there is no indication that any of these form disulfide bridges, at least in lipoxygenases-1 (148) or -2 (141). In an earlier study, Spaapen et al (148) found that treatment of soybean lipoxygenase-1 with methylmercuric halides led to inhibition of the enzyme (but only to a maximum of 50%), a decrease in the stereospecificity of hydroperoxide product formation, and an increase in secondary product formation under ambient oxygen concentrations —characteristics more commonly associated with lipoxygenase-3 (129). Among the three soybean cotyledon isozymes, three cysteines are conserved in all the sequences (C157/C520/C707, L-2), and one of these (C520) appears within the region containing the five invariant histidines (Figure 2A). However, the latter cysteine is not conserved in the tobacco sequence (14), and none of these cysteines is maintained in any of the mammalian sequences (53). The fact that mercurial treatment does not completely eliminate lipoxygenase-1 activity indicates that the affected cysteine is not an essential part of the catalytic cycle, and the altered catalytic properties of the enzyme could just as well be associated with steric factors deriving from interaction between the mercurial-modified enzyme and substrate. Further, Feiters et al (50) found that methylmercuric iodide did not affect the activity of soybean "type-2" lipoxygenase, but that one less sulfhydryl group underwent modification with the type-2 lipoxygenase than with lipoxygenase-1. At this point, there is no compelling reason for thinking that any of the sulfhydryl groups in the soybean cotyledon lipoxygenases play a particularly important role in catalysis. Finally, several studies have implicated the presence of one or more tryptophan residues within the active site of soybean lipoxygenase-1 (51, 90). Shibata et al (141) found that 12 tryptophan residues were conserved among the three soybean isozymes. Within the region containing the five invariant histidines, there is one conserved tryptophan (W528, L-2) that is maintained among all the plant sequences, but it is not found in any of the animal lipoxygenases (Figure 2A). Among the several tryptophans that are conserved in all the plant sequences obtained to date, two (W369/W646, L-2) are also conserved in the mammalian sequences, and one of these (W369) appears at the beginning of the most N-terminal of the three regions showing high sequence homology among all lipoxygenases (Figure 2B).

More work is needed to put on a sounder footing much of the above speculation about residues of potential importance in the lipoxygenase catalytic cycle. However, the presence of so many primary sequences will facilitate isolating homologous sequences from other sources, permitting the kinds of further sequence comparisons that have provided insights into other proteins in the absence of a crystallographic structure (80). Further, reports of preliminary X-ray analyses of crystalline soybean lipoxygenases-1 (16, 152) and -2 (149) have appeared recently, holding out the hope that the three-dimensional structure of a lipoxygenase will be determined in the not-too-distant future.

PHYSIOLOGICAL ROLE

For all the advances that have been made in understanding the structure of the lipoxygenase protein and its mechanism of action, the larger question of what physiological role(s) this seemingly ubiquitous enzyme plays in the functioning of the plant cell remains unanswered. Recent work from several laboratories has opened up intriguing avenues for further investigation. My goal in this section is not to attempt a compilation of the results of all recent reports involving physiological characterizations of lipoxygenase action, but rather to summarize results from the areas I think are most likely to elucidate the role of lipoxygenase in higher plants. Recall that while the role of any lipoxygenase in higher plants has yet to be established conclusively, functions for the various lipoxygenase activities found in mammals have become steadily more apparent. The 15-lipoxygenase activity found in red blood cells was the first mammalian enzyme to be well characterized, and it appears to be involved in the degradation of mitochondrial membranes during the maturation of reticulocytes (139). However, the more common role of lipoxygenases in mammalian cells is in the synthesis of hydroperoxy adducts of arachidonic acid, which serve as precursors for the biosynthesis of several important regulatory molecules (the leukotrienes and lipoxins) involved in mediating inflammatory responses in different tissues (135). Lipoxygenases that serve this function include the leukocyte 5-lipoxygenase, the 12-lipoxygenase in platelets and the 15-lipoxygenase in epithelial cells (135, 139). More recently, a role has been suggested for an 8-lipoxygenase in the induction of phorbol ester production and subsequent tumor growth in epithelial cells (52). It is not obvious how the inflammatory response in animals correlates with any specific process in plants, but some of the plant lipoxygenases may play a role in the synthesis of regulatory molecules (see below). However unrelated the topic may seem, it remains important for workers in the plant area to keep abreast of the rapidly expanding developments in the world of animal lipoxygenases.

While a function of any lipoxygenase in plants has yet to be established, two roles attributed to lipoxygenase in the past can be safely eliminated from consideration. Many plant tissues, when subjected to a variety of stresses (wounding, chilling temperatures, desiccation) or are undergoing senescence, respond by producing the hormone ethylene (182). Prior to the elucidation of the pathway of ethylene biosynthesis from S-adenosylmethionine through 1-aminocyclopropane-1-carboxylic acid (ACC) (182), it had been suggested that, at least under some conditions, ethylene might be synthesized through the spontaneous decomposition of fatty acid hydroperoxide intermediates via an ill-defined free radical mechanism (100). Formation of the fatty acid hydroperoxides could, of course, involve lipoxygenase. Characterization of the ACC pathway for ethylene biosynthesis appeared to eliminate any role for lipoxygenase. However, the possibility that lipoxygenase might be involved in the ACC-mediated formation of ethylene arose with reports of an in vitro system in which lipoxygenase and linoleic acid brought about the synthesis of ethylene in the presence of added ACC (15, 82). Several nonphysiological features associated with in vitro systems for synthesizing ethylene from ACC have been cited, and they need to be taken into account before accepting the physiological relevance of any given cell-free system (153, 182). With this in mind, Pirrung (122) demonstrated that however similar the features of the model system were to those of ethylene formation in vivo, there were sufficient differences to warrant ruling out the lipoxygenase-mediated reaction (15) as a source of ethylene in vivo. Kende (86) has reviewed the status of nonstandard routes of ethylene formation and concluded that none of them represent viable alternatives; in vivo, ethylene formation proceeds solely via the conversion of ACC to ethylene catalyzed by the ethylene-forming enzyme.

A second postulated role for lipoxygenase was as a component of the cyanide-resistant, alternative electron transfer pathway commonly found in plant mitochondria (143). Goldstein et al (68) first reported a correlation between (a) the level of alternative pathway in isolated wheat mitochondria following Percoll density-gradient purification and (b) lipoxygenase activity. This relationship has not held up in numerous subsequent studies, which found normal levels of the alternative pathway in isolated mitochondria having little or no measurable lipoxygenase activity (38, 144). In addition, differential inhibitor sensitivities for the alternative pathway and lipoxygenase have been established (107, 119). Subsequently, Rustin et al (131), responding to difficulties encountered in attempts to isolate and purify the alternative oxidase, put forward a model system in which the oxidation of ubiquinol by the alternative pathway proceeded not by way of a specific quinol oxidase but via a complex cycle that involved the interconversion of fatty acid and fatty acid hydroperoxy radicals ($L\cdot/LOO\cdot$). The role of lipoxygenase in this system was to bring about formation of the initiating hydroperoxy radicals (131). The operation of such a cycle in vivo is difficult to prove or disprove conclusively. For most workers in the field this model's explanatory power is insufficient to warrant displacing the concept of a specific ubiquinol oxidase (143). Further, Elthon and colleagues (43, 44) have since developed antibodies against a specific protein that represents a component of the alternative oxidase, and these antibodies have been used to isolate a cDNA that encodes this protein (130). This clear demonstration of a definable alternative oxidase protein has eliminated any need to search for other explanations of the alternative pathway activity in plant mitochondria and, it is hoped, will forever put to rest the phoenix-like suggestion that lipoxygenase is involved.

Other attempts to establish a role for lipoxygenase in plant tissues have focused on a few not entirely unrelated areas. As outlined in previous reviews (73, 74, 103, 174), investigators have sought lipoxygenase involvement in (a) plant growth and development, (b) the biosynthesis of regulatory molecules, (c) plant senescence, and (d) plant responses to wounding and pathogen infection.

Growth and Development

Studies looking for possible roles of lipoxygenase in plant growth and development have, until recently, been limited primarily to correlations between the appearance of lipoxygenase activity and the temporal course of a specific developmental sequence. The high levels of lipoxygenase activity found in legume seedlings have made this the tissue of choice in a number of these studies. In maturing soybean seeds the levels of all three of the seed-associated lipoxygenase isozymes continuously increase for 5–10 days before the seeds reach physiological maturity, with some variation among the different isozymes and the particular soybean genotype being measured (77). This increase in activity is associated with the de novo synthesis of lipoxygenase protein (59), and Funk and coworkers have taken experimental advantage of this fact to incorporate ^{59}Fe (59) and ^{57}Fe (37) into lipoxygenase in cultured soybean seeds.

Following germination, Hildebrand & Hymowitz (77) observed a massive decrease in the measured lipoxygenase-2+3 activity over the first 24 hr followed by an activity increase that peaked between days five and six, depending upon the cultivar. Over this same time period, lipoxygenase-1 simply showed a continuously decreasing activity (77). Studies that have separated cotyledon lipoxygenase activity from that of the rest of the growing seedling show a somewhat different pattern. A continuous loss of both the protein and activity of all three cotyledon lipoxygenase isozymes is seen over a 5–9-day period following imbibition (115, 120). However, Park & Polacco (115) also reported a marked increase in lipoxygenase activity in the expanding axis (hypocotyl/radicle) following germination. Consistent with earlier immunological studies (121) the lipoxygenase appearing in the axis tissue differed from the isozymes found in the cotyledon (115). This was based on the fact that the two species appearing in the hypocotyl/radicle had more acidic isoelectric points that any of the three cotyledon isozymes on isoelectric focusing gels, and that the appearance of the hypocotyl/radicle lipoxygenases was unchanged in lipoxygenase null lines lacking each of the individual cotyledon isozymes (115). These results suggest that the reported increase in lipoxygenase-2+3 activity following germination (77) may not, in fact, be associated with either cotyledon lipoxygenase-2 or -3 but rather is due to the appearance of distinct new isozymes in the emerging hypocotyl/radicle. Labeling studies with [^{35}S]methionine indicated that the axis lipoxygenase species resulted from de novo synthesis, and that there was no apparent synthesis of any of the three cotyledon isozymes following germination (115). A similar observation of the appearance of new, more acidic, lipoxygenase isozymes in developing soybean seedlings has been reported by Hildebrand et al (79). However, Hildebrand et al also noted that total lipoxygenase-2+3 activity in cotyledons from either the lipoxygenase-2 or -3 null lines was unchanged relative to that seen in the wild type. This could indicate that a compensatory regulation takes place in these mutant lines to increase the expression of one cotyledon lipoxygenase isozyme when another activity is missing.

At this time, it seems that distribution of the three well-characterized soybean seed lipoxygenase isozymes is limited to the cotyledons and that synthesis of these enzymes is restricted to the period of seed maturation. The accumulated data on the behavior of the cotyledon isozymes following germination (2, 77, 115, 120), coupled with the observation that there are no known deleterious phenotypic manifestations associated with the absence of any of the three cotyledon isozymes in the lipoxygenase null lines (27, 75, 88), suggests that earlier speculation to the effect that lipoxygenase-1 functions as a seed storage protein (77, 120) can be expanded to include cotyledon lipoxygenases-2 and -3 as well. The developmental expression of all three cotyledon isozymes is consistent with that of other known seed storage proteins (103, 120), and it now seems clear that the new lipoxygenase activities that appear in developing soybean seedlings following germination are not derived from the same genetic loci as any of the three cotyledon isozymes (2, 77, 115, 121). On the other hand, the fact that high levels of lipoxygenase activity are found in rapidly growing hypocotyl and radicle tissue following germination (2, 115) suggests that these newly synthesized lipoxygenases may play an important role in some aspect of seedling growth.

Studies of the behavior of lipoxygenase in the early stages of seedling growth outside of soybean are few, but they have the advantage of being unencumbered by the presence of a high background activity caused by the large amount of cotyledon lipoxygenase present in soybean seeds. Increases in lipoxygenase activity soon after germination and during the early stages of seedling growth have been reported for a number of species, including rice (113), wheat (71), mustard (112), lupin (11), and cucumber (104). A common feature in all these studies is that the increase in lipoxygenase activity that appears in the developing seedling within the first 50–72 hr after germination is followed by an equally substantial loss of lipoxygenase activity over the subsequent 48 hr. In rice for example, total seedling lipoxygenase activity increased 20-fold three days after germination, and most of the activity appeared in the developing shoot (113). Based upon elution patterns using DEAE chromatography, the lipoxygenase that appeared following germination was shown to be a distinctly different isozyme from that which made up the major component of the lipoxygenase activity in ungerminated seedlings. Further, the increase in lipoxygenase activity could be inhibited by addition of cycloheximide, suggesting that protein synthesis is required for its appearance. By day five, total shoot lipoxygenase activity had decreased almost 80% from its level on day three. Not all plants show the pattern described above. In corn and sunflower, lipoxygenase activity also increases following germination, but the maximum isn't reached until four days after germination (171). Lipoxygenase activity peaks at day six in watermelon seedlings (169). In dwarf pea (6), there is an initial decline in total seedling lipoxygenase, probably associated with degradation of cotyledon activity, followed by a slow increase in activity that plateaus after about two weeks.

The effects of light on the appearance of lipoxygenase activity have not been addressed in most of these seedling studies, which almost uniformly utilize etiolated seedlings. There are indications that light can affect the expression of lipoxygenase during seedling development, but the results vary considerably among different species. No effect of light on the expression of lipoxygenase in developing sunflower seedlings was observed (171). The effects on corn are minor: Maximum lipoxygenase activity appeared six days after germination instead of four (171). However, in dark-grown mustard seedlings, the sharp rise in lipoxygenase activity two days after germination is prevented by treatment with light in a process that appears to involve phytochrome (112). This light treatment also inhibits hypocotyl elongation, pointing up a general correlation that has appeared between the observed levels of lipoxygenase activity and rapidly elongating tissues. Outside of the cotyledons, the highest levels of lipoxygenase activity (per mg soluble protein) in soybeans are associated with expanding axis tissue (2). However, much higher lipoxygenase activity is seen in young, expanding leaves than in fully expanded ones (2). A similar correlation has been noted between hypocotyl elongation rates and lipoxygenase activity in winter rape seedlings (93). Further, treatments that inhibit hypocotyl elongation in winter rape, such as light or cold, reduce seedling lipoxygenase levels, while stimulation of hypocotyl elongation by added gibberellic acid increased lipoxygenase activity above that of untreated controls (93). The lipoxygenase measurements in the latter study were made on whole seedlings. Given the current recognition of a differential distribution of individual lipoxygenase isozymes among different plant tissues (2, 104, 113, 115), it would be interesting to know whether the changes in lipoxygenase observed upon inhibition or stimulation of hypocotyl growth in rape seedlings (93) are predominantly associated with lipoxygenase activity specifically localized in the hypocotyl. Most studies correlating growth rates and lipoxygenase activity have used tissues that would best be described as shoot-associated. However, relatively high levels of lipoxygenase activity have been reported in developing soybean radicle tissue (79, 115), while very low activities are seen in mature soybean roots (2), suggesting that the same general correlation might hold for the below-ground portion of the plant as well.

To summarize, there appears to be a correlation between the amount of lipoxygenase activity found in a given plant tissue and its rate of elongation, the highest levels of lipoxygenase being found in rapidly growing tissues. What role lipoxygenase, or more specifically the products of the lipoxygenase reaction, might play in such tissues is not presently understood. However, new tools available to workers in the field (i.e. molecular probes) will allow, for example, both Northern blotting (2, 34, 115) and in situ hybridization to be applied in studies of the developmental expression of specific lipoxygenase mRNAs in different soybean tissues. While this will not, in and of itself, provide an understanding of the role of lipoxygenase in plant growth, it should lead to a more refined picture than presently exists of the tissue specificity of the developmental expression of different soybean lipoxygenase isozymes. At the same time, we must better characterize the individual lipoxygenase isozymes that appear in different tissues at different stages of plant development. These isozymes must be understood with respect to not only their primary protein structure and their temporal and spatial patterns of expression but also such biochemical properties as their primary substrates and the nature of their reaction products. This combination of approaches has proved successful in delineating the different roles of the numerous mammalian lipoxygenases (135, 139). It is hoped that similar efforts applied to plant lipoxygenases will ultimately disclose what role(s) lipoxygenase plays, if any, in plant growth and development.

Synthesis of Regulatory Molecules

In seeking a role for lipoxygenase, an obvious question arises. Of what use are the fatty acid hydroperoxide products that result from the lipoxygenase reaction? The primary fatty acid hydroperoxide products of the lipoxygenase reaction are reactive molecules capable of harming biological systems, as evidenced by their participation in the degradation of reticulocyte mitochondria (139). In attempting to ascertain a possible role for lipoxygenase in the synthesis of regulatory molecules in plants, it seems most logical to consider not the lipoxygenase products themselves but rather potential plant metabolites that can be derived from these fatty acid hydroperoxides. This is certainly the case in mammals where many lipoxygenase reaction products serve as precursors for the synthesis of important molecules involved in regulating specific cellular responses (135). Two major routes for metabolizing fatty acid hydroperoxides exist in higher plants (Figure 4). These have been referred to collectively as the "lipoxygenase pathway" (174). Because these have been discussed in excellent reviews by Vick & Zimmerman (174) and Hildebrand (73), I undertake only a brief overview of the pathway here. In one branch of the pathway the enzyme hydroperoxide lyase acts on the lipoxygenase product, 13-hydroperoxylinolenic acid, resulting in the formation of hexanal and 12-oxo-*cis*-9-dodecenoic acid (Figure 4). The latter compound can undergo isomerization to the more stable 12-oxo-*trans*-10-dodecenoic acid, also known as traumatin or "wound hormone" (174). This compound has been reported to mimic the physiological effects seen upon wounding of plant tissues, inducing cell division and subsequent callus formation in a cucumber hypocotyl assay (185). A related compound, *trans*-

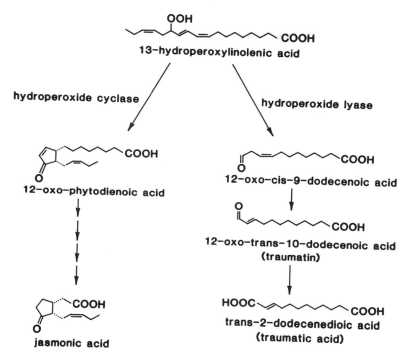

Figure 4 The "lipoxygenase pathway" for the biosynthesis of jasmonic acid, traumatin, and traumatic acid from the lipoxygenase product, 13-hydroperoxylinolenic acid.

2-dodecenedioic acid, or traumatic acid, was identified in the 1930s as the active component of traumatin in a similar callus forming assay (13, 46). Traumatin can be converted to traumatic acid by oxidation of the aldehyde moiety on the former, which readily occurs nonenzymically (Figure 4) (174). At present it appears that both of these compounds could be involved in plant responses to wounding (65, 174), and they clearly result from the action of lipoxygenase in plants. Further, there is evidence for a lipoxygenase response to wounding in plants (see below). The exciting possibility that traumatic acid and/or traumatin are primary agents of signal transduction in plant wound responses needs clarification. Less certain is how traumatic acid or traumatin might be associated with any role played by lipoxygenase in rapidly growing tissues.

The second pathway in plants for the metabolism of lipoxygenase products (Figure 4) is initiated by an enzyme, hydroperoxide cyclase, which reacts with 13-hydroperoxylinolenic acid to give 8-[2-(*cis*-2'-pentenyl)-3-oxo-*cis*-4-cyclopentenyl] octanoic acid, commonly referred to as 12-oxo-phytodienoic acid (170, 174). The latter compound, in turn, serves as a precursor for the formation of jasmonic acid (Figure 4) (172). Jasmonic acid and its commonly occurring methyl ester were originally reported to have growth-inhibitory properties in plants that mimic those seen with abscisic acid. Jasmonic acid was therefore suggested to act as a promoter of plant senescence (1, 160). Physiological effects reported for jasmonic acid include inhibition of soybean callus growth in the standard cytokinin bioassay (161), stimulation of ethylene production in ripe fruits (136), induction of specific proteins in senescent barley leaves (176), and a general inhibition of growth in rice seedlings (181), among others (174). More recently, it has been shown that jasmonic acid can stimulate the accumulation of vegetative storage proteins when added to fully expanded soybean leaves or soybean cell cultures (4). In the cultured cells, appearance of the storage protein is correlated with induction of its mRNA following jasmonic acid addition (151). These storage proteins are thought to represent temporary and readily mobilizable storage forms of nitrogen in vegetative tissues (151).

Most studies on jasmonic acid to date have involved effects associated with its exogenous application, and clear demonstration of a specific role for endogenous jasmonic acid in plants is still needed. However, an intriguing observation related to jasmonic acid and the role of lipoxygenase is the fact that the greatest amounts of endogenous jasmonic acid are most commonly seen in young, actively growing tissues, not senescent ones (91, 106). Not surprisingly, the same is true of the enzyme activities associated with jasmonic acid biosynthesis (173). While this correlation does not prove anything per se, it is consistent with the observation that the greatest lipoxygenase activities most commonly appear in actively growing tissues (2, 93, 115). Unclear (to me, at least) is why the very high levels of lipoxygenase measured in young tissues would be required for synthesis of what appears to be a regulatory molecule. Nevertheless, as the role of jasmonic acid in plant processes becomes clearer, a role for lipoxygenase will likewise appear.

Senescence

That lipoxygenase plays a role in plant senescence is probably the oldest of the currently extant theories about this enzyme's function. The reasons for this are obvious: Senescence is essentially a degradative process, which includes the loss of membrane integrity (157). The most prevalent source of potential lipoxygenase substrate is found among the polyunsaturated fatty acid side chains present in membrane lipids. Likewise, the products of the lipoxygenase reaction are themselves relatively reactive and can result in further membrane deterioration and increased permeability (95), among other possibilities (74, 103). As noted previously, there is good evidence that the 15-lipoxygenase in mammalian reticulocytes plays a role in mitochondrial degradation during red-blood-cell maturation (139). An additional, enticing feature of a role for lipoxygenase in senescence is the ability of the enzyme and its products to generate free radicals (63, 166), a factor that appears to be de rigeur in discussions of plant senescence (96, 157). Thus if soybean cotyledon lipoxygenases play any role beyond that of storage proteins during the early stages of seedling growth it might be to catalyze the disruption of storage cell membranes, enhancing their permeability and subsequently facilitating the movement of large amounts of needed metabolites to the growing embryo (174). The possible role of lipoxygenase in senescence has been discussed at length in recent reviews (103, 174) and is only summarized here.

Several studies have indicated that an increase in lipoxygenase activity is a common feature of senescing plant tissues (69, 98, 99, 117) and that treatments believed to delay the onset of senescence, such as the addition of cytokinins (69, 98) or antioxidants (97, 98), reduce the level of endogenous lipoxygenase activity relative to untreated controls. However, some of these studies induced senescence in tissues by detaching them from the intact plant (69). Thus any observed change in lipoxygenase could be a wound-induced response (see below) not related to senescence at all. Even in detached tissues there is no common lipoxygenase response, whatever its origin; Kar & Feierabend (84) saw a marked decline in lipoxygenase activity in detached, senescing wheat and rye leaves that was not affected by added kinetin. In addition, Peterman & Siedow (120) followed the developmental sequence associated with the senescence of soybean cotyledons that remained attached to the plant and observed that lipoxygenase levels were almost undetectable in older cotyledons and remained so throughout the course of senescence. If senescent cotyledons were subsequently "rejuvenated" by removal of the epicotyl, lipoxygenase activity was induced (120). Because in many respects cotyledon rejuvenation represents a reversal of the senescence process, this induction was taken as an additional indication that lipoxygenase plays no obvious role in soybean cotyledon senescence. Finally, inhibiting lipoxygenase activity does not affect the amount of lipid peroxidation observed in senescing *Avena* or *Rumex* leaves (31).

Thompson and coworkers have pointed out that there need not be high levels of lipoxygenase activity in order to initiate senescence-associated lipid peroxidation because the process will be self-propagating once started (158). In conjunction with this idea, they have described a microsomal membrane-bound lipoxygenase activity that is capable of generating superoxide anion radicals (101) and whose activity is seen to increase in aging, etiolated *Phaseolus vulgaris* cotyledons (116) and senescing carnation petals (54, 158). This intriguing membrane-bound lipoxygenase activity is unusual relative to other, better-characterized plant lipoxygenases, which are almost uniformly soluble proteins. If there were a lipoxygenase activity associated with senescence, it might be expected to be membrane bound to more readily attack the membranes that will ultimately be degraded (158), but clearly a better characterization of this particular activity is needed if its role in senescence is to be conclusively demonstrated.

At present, the available evidence argues against a role for any soluble lipoxygenase in plant senescence. The consistent finding of high lipoxygenase activities in actively growing plant tissues contrasts markedly with the senescence literature, which displays a mixed set of observations. The most clear-cut potential role for lipoxygenase in senescence — that of providing precursors for the synthesis of jasmonic acid (174) —has nothing to do directly with membrane degradation. Further, we await a clear demonstration that endogenous jasmonic acid even plays a role in senescence. The appearance of jasmonic acid in young, actively growing plant tissues (91, 106) suggests that promoting senescence may not rank among its primary roles. At present there are more viable possibilities for lipoxygenase function in plants than a role in senescence.

Wound and Other Stress Responses

That plant lipoxygenases might play a role in plant responses to wounding and other related stresses has long been suggested (62). In recent years this research area has seen remarkable growth. A role for plant lipoxygenase as a response to external stresses would provide a functional homology with one of its major roles in animals, where the lipoxygenase-mediated formation of leukotrienes and lipoxins serves as a signal to initiate stress responses within the organism (135). The induction of lipoxygenase in plants in response to wounding by an external biological agent such as a bacterial, fungal, or viral pathogen could be advantageous for several reasons (74). A burst of lipoxygenase activity accompanying physical wounding of the cell would lead to the production of fatty acid hydroperoxides, various free radical-containing species, and their accompanying decomposition products which, in turn, could help stimulate the breakdown of cellular membranes leading to a localized cell death commonly associated with the "hypersensitive reaction" (33, 125). Enhanced lipid peroxidation has been demonstrated to be an important step in the development of the hypersensitive reaction in cucumber cotyledons following infection with *Pseudomonas syringae* pv. *pisi* (87). Hydroperoxy radicals could also help promote the lignification process that accompanies the hypersensitive reaction, although this role is more commonly assigned to peroxidase-derived radicals (125). Playing a more direct role, lipoxygenase reaction products might act as agents toxic to the invading pathogen (74). Finally, a rationale for the induction of lipoxygenase in nearby tissues might be to increase the strength of the ensuing reponse if those tissues are subsequently invaded.

Initial studies of the response of lipoxygenase to pathogen infection indicated that lipoxygenase activity in potato tubers increased for a period of up to four weeks following fungal infection with *Phoma exigua* (62). Similarly, lipoxygenase activity increased up to seven-fold in tobacco leaves over a period of 11 days following infection with a powdery mildew, *Erysiphe cichoracearum* (101). Numerous similar reports have appeared since these initial observations with increases in lipoxygenase activity upon pathogen infection being seen as soon as 6 hr after inoculation (118). In addition, increases in lipoxygenase activity can be induced by addition of either the pathogen or an elicitor (55, 111, 118) suggesting that lipoxygenase induction represents a standard component of the plant's defense response mechanism (33) and does not simply result from mechanical wounding of the tissue during the infection process. For example, lipoxygenase activity in tomato leaves responded similarly to glycoprotein nonspecific elicitor or a specific elicitor derived from *Cladosporium fulvum*-infected tomatoes (118). Lipoxygenase activity increased and reached a maximum 12 hr after inoculation and declined to the control value after 24 hr. Both elicitors produced necrotic lesions on the leaves, stimulated electrolyte leakage over a 3-hr period following inoculation, and promoted enhanced lipid peroxidation after 12 hr. Curiously, addition of a lipoxygenase inhibitor, piroxicam, to the leaves reduced the amount of electrolyte leakage seen upon addition of the nonspecific elicitor but did not affect the leakage associated with the specific elicitor. None of the other responses were affected by piroxicam, including necrotic lesion formation. Given the lack of knowledge of how specific an inhibitor piroxicam is in plants, it is difficult to draw many meaningful conclusions from these observations. Another aspect of the lipoxygenase response to pathogen infection is the observation that resistant and susceptible plant lines show differential responses. Lipoxygenase activity increased in *Nicotiana tabacum* cv. Xanthi nc leaves upon infection with tobacco mosaic virus, but was unchanged in a susceptible tobacco variety (132). These results paralleled the hypersensitive reaction that appeared in the former but not in the latter. In resistant oat lines, the induction of two new lipoxygenase isozymes accounted for an increase in activity observed upon infection with crown rust, *Puccinia coronata*, while no change in lipoxygenase activity was seen following infection of susceptible lines (180). An additional system worth mentioning involves an observed increase in the susceptibility of postharvest avocado fruits to infection by the fungus *Colletotrichum gloeosporioides* which actually correlates with an increase in lipoxygenase activity in the fruit peel (85, 126). There appears to be a naturally occurring antifungal compound in the peel, *cis,cis*-1-acetoxy-2-hydroxy-4-oxo-heneicosa-12,15-diene, which is degraded by the action of lipoxygenase. In addition, the peel contains a potent inhibitor of lipoxygenase, epicatechin. As the avocado fruit ripens, the epicatechin level decreases, increasing the endogenous lipoxygenase activity which, in turn, breaks down the antifungal compound. Lipoxygenase protein levels remain roughly constant during ripening as judged immunologically using an ELISA (85). Ultimately, the amount of endogenous antifungal compound becomes sufficiently reduced to allow the development of disease symptoms. Knowing that lipoxygenase is responsible for the breakdown of the antifungal compound should facilitate the development of a postharvest treatment for avocados that minimizes infection. Numerous antioxidants are known to inhibit lipoxygenase activity (47) and many could safely be used to treat avocados.

Results such as those cited above are still at a relatively simple correlative stage of comparing a lipoxygenase activity response with pathogen infection. Additional experimental manipulations are now required to establish whether a mechanistic relationship exists between the observed increase in lipoxygenase activity and pathogen resistance/susceptibility. Host-pathogen interactions are exceedingly complex, involving numerous enzymes and a host of regulatory features (178). There is no reason to think that lipoxygenase represents the "switch" that determines resistance or susceptibility; more likely, its response, or lack thereof, will prove to be linked to a more global regulatory element that also controls the response of numerous other cellular components to pathogen infection. However, there seems little doubt that lipoxygenase is involved in plant responses to pathogen infection, and the future should see some exciting developments as a combination of biochemical and molecular approaches are used to address its specific role in the overall plant defense response.

Studies of the response of lipoxygenase to mechanical wounding are not as well developed as those of the response to pathogen invasion, but some do exist. Galliard (62) originally reported that lipoxygenase activity increased over a 7-day period in potato tubers in response to mechanical wounding. Wounding of yam tubers led to a transient increase in lipoxygenase activity after 24 hr followed by a decay to the control level over a subsequent 3–5-day period (81). Hildebrand et al (74, 78) have looked at the effects of mechanical wounding compared to mite infestation on lipoxygenase activity in soybean leaves. An increase in both lipoxygenase enzyme activity and gene expression was observed with soybean leaves that had been wounded daily for a week, but the variability in the activity profile was large. A similar variability was seen with the mite infestation (78), but an earlier report did note that a larger response of lipoxygenase to infestation was observed in a resistant cultivar than in a susceptible one (74).

At this point we have too few data to determine whether there is a general lipoxygenase response to mechanical wounding comparable to that commonly seen following pathogen infection. However, the lipoxygenase response to wounding is not as great as that of other, well-established wound responses

(133). The transduction pathway for signaling a wound or insect attack in solanaceous species is rapidly being elucidated (48, 134), and until recently there was no indication of the involvement of any lipoxygenase-derived product. However, Farmer & Ryan (48a) have demonstrated that the induction of proteinase inhibitors I and II in tomato can be brought about by either application of methyl jasmonate to leaves or by the isopiestic movement of the volatile ester from an adjacent cotton-tipped dowel placed in the same chamber as the plants. These proteinase inhibitors are known to be synthesized as a response to wounding or pathogen infection in members of the Solanaceae (133, 134). Treatment with methyl jasmonate was also shown to induce the expression of proteinase inhibitors in tobacco and alfalfa (48a). These results suggest not only that jasmonic acid might act along the signal transduction pathway within a given plant for the activation of defensive genes in response to pathogen attack or mechanical wounding, but also that methyl jasmonate may actually serve as a signal transducing agent between separate plants. In addition, the vegetative storage proteins induced in response to jasmonic acid treatment have also been reported to be induced upon mechanical wounding (151); and Anderson (3) had earlier suggested that jasmonic acid may serve as a mobile signal for plant stress based upon its originating as a lipoxygenase metabolite. However, we do not know whether, in a given plant cell, production of regulatory levels of jasmonic acid would require an increase in the endogenous level of lipoxygenase. The behavior of neither lipoxygenase nor endogenous jasmonic acid (or methyl jasmonate) is sufficiently well characterized with respect to any wound response to ascertain what relationship might exist between the two species following an appropriate stimulus. Clearly, more work is needed to establish conclusively the extent to which traumatin, traumatic acid, and/or jasmonic acid are involved as signal transducers in any wound response in higher plants. The establishment of a clearcut role for any of these species in a wound response would establish a concomitant role for lipoxygenase.

CONCLUDING REMARKS

The past 20 years have seen major advances in our understanding of the protein structure and the mechanism of the lipoxygenase reaction catalyzed by the isozymes isolated from soybean cotyledons. This aspect of our understanding of lipoxygenase should continue to develop rapidly. The appearance of diffractable crystals (16, 149, 152) of soybean cotyledon lipoxygenases promises that a three-dimensional structure of at least one of these isozymes will be available soon. If we can couple elucidation of lipoxygenase structure with a system for expressing genetically engineered lipoxygenase gene sequences (with modified amino acids in the region of the active site and elsewhere), we should be able to understand the mechanism of the lipoxygenase reaction as well as we do that of other well-characterized enzymes. The baculovirus system that is able to express mammalian 5-lipoxygenase (56) could prove amenable to expression of the plant enzymes. Likewise, the question of whether lipoxygenases are actually found in cyanobacteria is worth further exploration. Cyanobacteria are currently proving convenient for the expression of site-directed mutations of photosynthetic electron transfer proteins. It may not be too far-fetched to suggest that within the next five years we will thoroughly understand the structure of soybean cotyledon lipoxygenase and, by analogy, the structures of other closely related plant isozymes that appear to be involved in various aspects of plant growth and development.

Elucidating the specific physiological roles played by lipoxygenase in plants is another, more daunting, challenge. At this time, understanding the role of lipoxygenase seems inextricably linked to better understanding the roles of such potential regulatory molecules as traumatic acid, traumatin, and jasmonic acid. This is where efforts in the functional realm should be focused in the near future. With respect to lipoxygenase itself, beyond screening for mutants lacking the different cotyledon isozymes (75, 88, 89), little attempt has been made to generate lipoxygenase mutants to aid the study of lipoxygenase function. This may not be surprising given the lack of an obvious screen, but as the structure and patterns of expression of more non-cotyledon isozymes become better characterized, we should be able to develop mutants of one or more of these different isozymes. Such progress, coupled with a thorough biochemical characterization of the reactants and products associated with these isozymes, should lead to an understanding of the role they play in the life of the plant.

This is an exciting time in lipoxygenase research. We can now do more than look for correlations between enzyme activity and a particular developmental pattern or general physiological response. A series of events as interesting as that provided by the formation of leukotrienes in the mammalian inflammatory response may well exist as part of the plant defense response to pathogen invasion, as could a seminal role for lipoxygenase in initiating this response. It will remain for future research to elucidate such a response and demonstrate how it functions in higher-plant cells.

ACKNOWLEDGMENTS

I thank David Hildebrand and Max Funk for communicating results prior to publication, David Saravitz for helpful comments and discussion on the function of lipoxygenase, and Cyril Kaspi without whose assistance this manuscript would never have been completed on time.

Literature Cited

1. Aldridge, D. C., Galt, S., Giles, D., Turner, W. B. 1971. Metabolites of *Lasidioplodia theobromae. J. Chem. Soc.*, pp. 1623-27
2. Altschuler, M., Collins, G. B., Hildebrand, D. F. 1989. Developmental expression of lipoxygenases in soybeans. *Plant Sci.* 63:151-58
3. Anderson, J. M. 1989. Membrane derived fatty acids as precursors to second messengers. In *Second Messengers in Plant Growth and Development*, ed. W. F. Boss, D. J. Morre, pp. 181-212. New York: Allen R. Liss
4. Anderson, J. M., Spilatro, S. R., Klauer, S. F., Franceschi, V. R. 1989. Jasmonic acid dependent increase in the level of vegetative storage protein in soybean. *Plant Sci.* 62:45-52
5. Andre, E., Hou, K.-W. 1932. The presence of a lipoid oxidase in soybean, *Glycine soya*, Lieb. *C. R. Acad. Sci. (Paris)* 194:645-47
6. Anstis, P. J. P., Friend, J. 1974. The isozyme distribution of etiolated pea seedling lipoxygenase. *Planta* 115:329-35
7. Axelrod, B. 1974. Lipoxygenases. *ACS Adv. Chem. Ser.* 136:324-48
8. Axelrod, B., Cheesbrough, T. M., Laasko, S. 1981. Lipoxygenase from soybeans. *Methods Enzymol.* 71:441-51
9. Balcarek, J. M., Theisen, T. W., Cook, M. N., Varrichio, A., Hwang, S.-M., Strohsacker, M. W., Crooke, S. T. 1988. Isolation and characterization of a cDNA encoding rat 5-lipoxygenase. *J. Biol. Chem.* 263:13937-41
10. Beneytout, J.-L., Andrianarison, R.-H., Rakotoarisoa, Z., Tixier, M. 1989. Properties of a lipoxygenase in green algae (*Oscillatoria* sp.). *Plant Physiol.* 91:367-72
11. Beneytout, J. L., Najid, A., Tixier, M. 1988. Changes in lipoxygenase activity during seedling development of *Lupinus albus. Plant Sci.* 58:35-41
12. Bild, G. S., Ramadoss, C. S., Axelrod, B. 1977. Effect of substrate polarity on the activity of soybean lipoxygenase isozymes. *Lipids* 12:732-35
13. Bonner, J., English, J. 1938. A chemical and physiological study of traumatin, a plant wound hormone. *Plant Physiol.* 13:331-48
14. Bookjans, G., Altschuler, M., Brockman, J., Yenofsky, R., Polacco, J., et al. 1988. Molecular biological studies of plant lipoxygenases. In *Proceedings of the World Conference on Biotechnology for the Fat and Oil Industry*, ed. T. H. Applewhite, pp. 301-4. Urbana: Am. Oil Chem. Soc. Press
15. Bousquet, J. P., Thimann, K. V. 1984. Lipid peroxidation forms ethylene from 1-aminocyclopropane-1-carboxylic acid and may operate in leaf senescence. *Proc. Natl. Acad. Sci. USA* 81:1724-27
16. Boyington, J. C., Gaffney, B. J., Amzel, L. M. 1990. Crystallization and preliminary X-ray analysis of soybean lipoxygenase-1, a non-heme iron-containing dioxygenase. *J. Biol. Chem.* 265:12771-73
17. Bradley, F. C., Lindstedt, S., Lipscomb, J. D., Que, L., Row, A. L., Rundgren, M. 1986. 4-hydroxyphenylpyruvate dioxygenase is an iron-tyrosinate protein. *J. Biol. Chem.* 261:11693-96
18. Casey, R., Domoney, C., Nielsen, N. C. 1985. Isolation of a cDNA clone for pea (*Pisum sativum*) seed lipoxygenase. *Biochem. J.* 232:79-85
19. Chan, H. W. S. 1973. Soy-bean lipoxygenase: an iron-containing dioxygenase. *Biochim. Biophys. Acta* 327:32-35
20. Chothia, C., Finkelstein, A. V. 1990. The classification and origins of protein folding patterns. *Annu. Rev. Biochem.* 59:1007-39
21. Chou, P. Y., Fasman, G. D. 1978. Emperical predictions of protein conformation. *Annu. Rev. Biochem.* 47:251-76
22. Christopher, J. P., Pistorius, E. K., Axelrod, B. 1970. Isolation of an isozyme of soybean lipoxygenase. *Biochim. Biophys. Acta* 198:12-19
23. Christopher, J. P., Pistorius, E. K., Axelrod, B. 1972. Isolation of a third isozyme of soybean lipoxygenase. *Biochim. Biophys. Acta* 284:54-62
24. Corey, E. J. 1986. Mechanism of enzymic lipoxygenation of arachidonic acid. Key role of organoiron intermediates. In *Stereochemistry of Organic and Bioorganic Transformations*, ed. W. Bartmann, K. B. Sharpless, pp. 1-12. Weinheim, F.R.G.: VCH Publishers
25. Corey, E. J., d'Alarcao, M., Matsuda, S. P. T. 1986. A new irreversible inhibitor of soybean lipoxygenase; relevance to mechanism. *Tetrahedron Lett.* 27:3585-88
26. Corey, E. J., Nagata, R. 1987. Evidence in favor of an organoiron-mediated pathway for lipoxygenation of fatty acids by

soybean lipoxygenase. *J. Am. Chem. Soc.* 109:8107-8

27. Davies, C. S., Nielsen, N. C. 1986. Genetic analysis of a null-allele for lipoxygenase-2 in soybean. *Crop Sci.* 26:460-63

28. deGroot, J. J. M. C., Garssen, G. J., Veldink, G. A., Vliegenthart, J. F. G., Boldingh, J. 1975. On the interaction of soybean lipoxygenase-1 and 13-L-hydroperoxylinoleic acid, involving yellow and purple coloured enzyme species. *FEBS Lett.* 56:50-54

29. deGroot, J. J. M. C., Vliegenthart, J. F. G., Boldingh, J., Weber, R., Van Gelder, B. F. 1975. Demonstration by EPR spectroscopy of a functional role of iron in soybean lipoxygenase-1. *Biochim. Biophys. Acta* 377:71-79

30. Deisenhofer, J., Epp, O., Miki, K., Huber, R., Michel, H. 1985. Structure of the protein subunits in the photosynthetic reaction center of *Rhodopseudomonas viridis* at 3A resolution. *Nature* 318:618-24

31. Dhindsa, R. S., Plumb-Dhindsa, P. L., Reid, D. M. 1982. Leaf senescence and lipid peroxidation: Effects of some phytohormones and scavengers of free radicals and singlet oxygen. *Physiol. Plant.* 56:453-57

32. Dixon, R. A. F., Jones, R. E., Diehl, R. E., Bennett, C. D., Kargman, S., Rouzer, C. A. 1988. Cloning of the cDNA for human 5-lipoxygenase. *Proc. Natl. Acad. Sci. USA* 85:416-20

33. Dixon, R. A., Lamb, C. J. 1990. Molecular communication in interactions between plants and microbial pathogens. *Annu. Rev. Plant Physiol. Plant Mol. Biol.* 41:339-67

34. Domoney, C., Firmin, J. L., Sidebottom, C., Ealing, P. M., Slabas, A., Casey, R. 1990. Lipoxygenase heterogeneity in *Pisum sativum. Planta* 181:35-43

35. Draheim, J. E., Carroll, R. T., McNemar, T. B., Dunham, W. R., Sands, R. H., Funk, M. O. 1989. Lipoxygenase isozymes: a spectroscopic and structural characterization of soybean seed enzymes. *Arch. Biochem. Biophys.* 269:208-18

36. Duine, J. A., Jongejan, J. A. 1989. Quinoproteins, enzymes with pyrroloquinoline quinone as cofactor. *Annu. Rev. Biochem.* 58:403-26

37. Dunham, W. R., Carroll, R. T., Thompson, J. F., Sands, R. H., Funk, M. O. 1990. The initial characterization of the iron environment in lipoxygenase by Mössbauer spectroscopy. *Eur. J. Biochem.* 190:611-17

38. Dupont, J., Rustin, P., Lance, C. 1982. Interaction between mitochondrial cytochromes and linoleic acid hydroperoxide: possible confusion with lipoxygenase and alternative pathway. *Plant Physiol.* 69:1308-14

39. Duran, N. 1982. Singlet oxygen in biological processes. In *Chemical and Biological Generation of Excited States*, ed. A. Walderman, G. Cilento, pp. 345-69. New York: Academic

40. Ealing, P. M., Casey, R. 1988. Complete amino acid sequence of a pea (*Pisum sativum*) seed lipoxygenase predicted from a near full-length cDNA. *Biochem. J.* 253:915-18

41. Egmond, M. R., Veldink, G. A., Vliegenthart, J. F. G., Boldingh, J. 1973. C-11 H-abstraction from linoleic acid, the rate-limiting step in lipoxygenase catalysis. *Biochem. Biophys. Res. Commun.* 54:1178-84

42. Egmond, M. R., Vliegenthart, J. F. G., Boldingh, J. 1972. Stereospecificity of the hydrogen abstraction at carbon atom n-8 in the oxygenation of linoleic acid by lipoxygenases from corn germs and soya beans. *Biochem. Biophys. Res. Commun.* 48:1055-60

43. Elthon, T. E., McIntosh, L. 1987. Identification of the alternative terminal oxidase of higher plant mitochondria. *Proc. Natl. Acad. Sci. USA* 84:8399-403

44. Elthon, T. E., Nickels, R. L., McIntosh, L. 1989. Monoclonal antibodies to the alternative oxidase of higher plant mitochondria. *Plant Physiol.* 89:1311-17

45. Engelman, D. M., Steitz, T. A., Goldman, A. 1986. Identifying nonpolar transbilayer helices in amino acid sequences of membrane proteins. *Annu. Rev. Biophys. Biophys. Chem.* 15:321-53

46. English, J., Bonner, J., Haagen-Smit, A. J. 1939. Structure and synthesis of a plant wound hormone. *Science* 90:329

47. Eskin, N. A. M., Grossman, S., Pinsky, A. 1977. Biochemistry of lipoxygenase in relation to food quality. *Crit. Rev. Food Sci. Nutr.* 9:1-41

48. Farmer, E. E., Pearce, G., Ryan, C. A. 1989. In vitro phosphorylation of plant plasma membrane proteins in response to the proteinase inhibitor inducing factor. *Proc. Natl. Acad. Sci. USA* 86:1539-42

48a. Farmer, E. E., Ryan, C. A. 1990. Interplant communication: airborne methyl jasmonate induces synthesis of proteinase inhibitors in plant leaves. *Proc. Natl. Acad. Sci. USA* 87:7713-16

49. Feiters, M. C., Aasa, R., Malmström,

B. G., Veldink, G. A., Vliegenthart, J. F. G. 1986. Spectroscopic studies on the interactions between lipoxygenase-2 and its product hydroperoxides. *Biochim. Biophys. Acta* 873:182-89

50. Feiters, M. C., Veldink, G. A., Vliegenthart, J. F. G. 1986. Heterogeneity of soybean lipoxygenase-2. *Biochim. Biophys. Acta* 870:367-71

51. Finazzi-Agro, A., Avigliano, L., Veldink, G. A., Vliegenthart, J. F. G., Boldingh, J. 1973. The influence of oxygen on the fluorescence of lipoxygenase. *Biochim. Biophys. Acta* 326:462-70

52. Fisher, S. M., Baldwin, J. K., Jasheway, D. W., Patrick, K. E., Cameron, G. S. 1988. Phorbol ester induction of 8-lipoxygenase in inbred SENCAR (SSIN) but not C57BL/6J mice correlated with hyperplasia, edema, and oxidant generation but not ornithine decarboxylase induction. *Cancer Res.* 48:658-64

53. Fleming, J., Thiele, B. J., Chester, J., O'Prey, J., Janetzki, S., et al. 1989. The complete sequence of the rabbit erythroid cell-specific 15-lipoxygenase mRNA: comparison of the predicted amino acid sequence of the erythrocyte lipoxygenase with other lipoxygenases. *Gene* 79:181-88

54. Fobel, M., Lynch, D. V., Thompson, J. E. 1987. Membrane deterioration in senescing carnation flowers: coordinated effects of phospholipid degradation and the action of membraneous lipoxygenase. *Plant Physiol.* 85:204-11

55. Fournier, J., Pelissier, B., Esquerre-Tugaye, M. T. 1986. Induction d'une activite lipoxygenase dans les cellules de tabac (Nicotiana tabacum) en culture, par des eliciteurs d'ethylene de *Phytophthora parasitica* var. nicotianae. *C. R. Acad. Sci. Paris* 303:651-54

56. Funk, C. D., Gunne, H., Steiner, H., Izumi, T., Samuelsson, B. 1989. Native and mutant 5-lipoxygenase expression in a baculovirus/insect cell system. *Proc. Natl. Acad. Sci. USA* 86:2592-96

57. Funk, C. D., Hoshiko, S., Matsumoto, T., Radmark, O., Samuelsson, B. 1989. Characterization of the human 5-lipoxygenase gene. *Proc. Natl. Acad. Sci. USA* 86:2587-91

58. Funk, M. O., Andre, J. C., Otsuki, T. 1987. Oxygenation of trans polyunsaturated fatty acids by lipoxygenase reveals steric features of the catalytic mechanism. *Biochemistry* 26:6880-84

59. Funk, M. O., Carroll, R. T., Thompson, J. F., Dunham, W. R. 1986. The lipoxygenases in developing soybean seeds, their characterization and synthesis in vitro. *Plant Physiol.* 82:1139-44

60. Funk, M. O., Carroll, R. T., Thompson, J. F., Sands, R. H., Dunham, W. R. 1990. Role of iron in lipoxygenase catalysis. *J. Am. Chem. Soc.* 112:5375-76

61. Funk, M. O., Whitney, M. A., Hausknecht, E. C., O'Brien, E. M. 1985. Resolution of the isozymes of soybean lipoxygenase using isoelectric focusing and chromatofocusing. *Anal. Biochem.* 146:246-51

62. Galliard, T. 1978. Lipolytik and lipogenase enzymes in plants and their action in wounded tissues. In *Biochemistry of Wounded Plant Tissues*, ed. G. Kahl, pp. 155-201. Berlin: Walter de Gruyter

63. Galliard, T., Chan, H. W.-S. 1980. Lipoxygenases. In *The Biochemistry of Plants*, ed. P. K. Stumpf, E. E. Conn, 4:131-61. New York: Academic

64. Gallop, P. M., Paz, M. A., Fluckiger, R., Kagan, H. M. 1989. PQQ, the elusive coenzyme. *Trends Biochem. Sci.* 14:343-46

65. Gardner, H. W. 1979. Lipid hydroperoxide reactivity with proteins and amino acids: a review. *J. Agric. Food Chem.* 27:220-29

66. Garssen, G. J., Vliegenthart, J. F. G., Boldingh, J. 1971. An aerobic reaction between lipoxygenase, linoleic acid and its hydroperoxides. *Biochem. J.* 122:327-32

67. Gibian, M. J., Singh, K. 1986. Irreversible inhibition of soybean lipoxygenase by phenyldiazine, autoxidizing phenylhydrazine and related materials. *Biochim. Biophys. Acta* 878:79-92

68. Goldstein, A. H., Anderson, J. O., McDaniel, R. G. 1981. Cyanide-insensitive and cyanide-sensitive O_2 uptake in wheat. *Plant Physiol.* 67:594-96

69. Grossman, S., Leshem, Y. Y. 1978. Lowering endogenous lipoxygenase activity in *Pisum sativum* foliage by cytokinin as related to senescence. *Physiol. Plant.* 43:359-62

70. Grossman, S., Zakut, R. 1979. Determination of the activity of lipoxygenase (lipoxidase). In *Methods of Biochemical Analysis*, ed. D. Glick, 3:303-29. New York: Wiley Interscience

71. Guss, P. L., Macko, V., Richardson, T., Stahmann, M. A. 1968. Lipoxidase in early growth of wheat. *Plant Cell Physiol.* 9:415-22

72. Hamberg, M. 1986. Isolation and structure of lipoxygenase from *Saprolegnia parasitica*. *Biochim. Biophys. Acta* 876:688-92

73. Hildebrand, D. F. 1989. Lipoxygenases. *Physiol. Plant.* 76:249-53
74. Hildebrand, D. F., Hamilton-Kemp, T. R., Legg, C. S., Bookjans, G. 1988. Plant lipoxygenases: occurrence, properties and possible functions. *Curr. Top. Plant Biochem. Physiol.* 7:201-19
75. Hildebrand, D. F., Hymowitz, T. 1981. Two soybean genotypes lacking lipoxygenase-1. *J. Am. Oil Chem. Soc.* 58: 583-86
76. Hildebrand, D. F., Hymowitz, T. 1982. Inheritance of lipoxygenase-1 activity in soybean seeds. *Crop Sci.* 22:851-53
77. Hildebrand, D. F., Hymowitz, T. 1983. Lipoxygenase activities in developing and germinating soybean seeds with and without lipoxygenase-1. *Bot. Gaz.* 144: 212-16
78. Hildebrand, D. F., Rodriguez, J. G., Legg, C. S., Brown, G. C., Bookjans, G. 1989. The effects of wounding and mite infestation on soybean leaf lipoxygenase levels. *Z. Naturforsch. Teil C* 44:655-59
79. Hildebrand, D. F., Snyder, K. M., Hamilton-Kemp, T. R., Bookjans, G., Legg, C. S., Andersen, R. A. 1989. Expression of lipoxygenase isozymes in soybean tissues. In *Biological Role of Plant Lipids,* ed. P. A. Biacs, K. Gruiz, T. Kremmer, pp. 51-56. New York: Plenum
80. Howell, N. 1989. Evolutionary conservation of protein regions in the protonmotive cytochrome *b* and their possible roles in redox catalysis. *J. Mol. Evol.* 29:157-69
81. Ikediobi, C. O., Chelvarajan, R. L., Ukoha, A. I. 1989. Biochemical aspects of wound healing in yams (*Dioscorea* spp). *J. Sci. Food Agric.* 48:131-39
82. Kacperska, A., Kubacka-Zebalska, M. 1985. Is lipoxygenase involved in the formation of ethylene from ACC? *Physiol. Plant.* 64:333-38
83. Kanofsky, J. R., Axelrod, B. 1986. Singlet oxygen production by soybean lipoxygenase isozymes. *J. Biol. Chem.* 261:1099-1104
84. Kar, M., Feierabend, J. 1984. Metabolism of activated oxygen in detached wheat and rye leaves and its relevance to the initiation of senescence. *Planta* 160:385-91
85. Karni, L., Prusky, D., Kobiler, I., Bar-Shira, E., Kobiler, D. 1989. Involvement of epicatechin in the regulation of lipoxygenase activity during activation of quiescent *Colletotrichum gloeosporioides* infections of ripening avocado fruits. *Physiol. Mol. Plant Pathol..* 35:367-74
86. Kende, H. 1989. The enzymes of ethylene biosynthesis. *Plant Physiol.* 91:1-4
87. Keppler, L. D., Novacky, A. 1986. Involvement of membrane lipid peroxidation in the development of a bacterially induced hypersensitive reaction. *Phytopathology* 76:104-8
88. Kitamura, K. 1984. Biochemical characterization of lipoxygenase-lacking mutants, L-1-less, L-2-less, and L-3-less soybeans. *Agric. Biol. Chem.* 48:2339-46
89. Kitamura, K., Davies, C. S., Kaizuma, N., Nielsen, N. C. 1983. Genetic analysis of a null-allele for lipoxygenase-3 in soybean seeds. *Crop Sci.* 23:924-27
90. Klein, B. P., Cohen, B.-S., Grossman, S., King, D., Malovany, H., Pinsky, A. 1985. Effect of modification of soybean lipoxygenase-1 with N-bromosuccinimide on linoleate oxidation, pigment bleaching and carbonyl production. *Phytochemistry* 24:1903-6
91. Knofel, H. D., Bruckner, C., Kramell, R., Sembdner, G., Schreiber, K. 1984. A radioimmunoassay for jasmonic acid. *Biochem. Physiol. Pflanzen.* 179:317-25
92. Komaromy, M. C., Schotz, M. C. 1987. Cloning of rat hepatic lipase cDNA: evidence for a lipase gene family. *Proc. Natl. Acad. Sci. USA* 84:1526-30
93. Kubacka-Zebalska, M., Kacperska-Palacz, A. 1980. Lipoxygenase, an enzyme involved in plant growth? *Physiol. Veg.* 18:339-47
94. Kyte, J., Doolittle, R. F. 1982. A simple method for displaying the hydrophobic character of a protein. *J. Mol. Biol.* 157:105-32
95. Leibowitz, M. E., Johnson, M. C. 1971. Relation of lipid peroxidation to loss of cations trapped in liposomes. *J. Lipid Res.* 12:662-70
96. Leshem, Y. Y. 1988. Plant senescence processes and free radicals. *Free Radic. Biol. Med.* 5:39-49
97. Leshem, Y. Y., Barness, G. 1982. Lipoxygenase as affected by free radical metabolism: senescence retardation by the xanthine oxidase inhibitor allopurinol. In *Biochemistry and Metabolism of Plant Lipids,* ed. J. F. G. M. Wintermans, P. J. C. Kuiper, pp. 275-78. Amsterdam: Elsevier
98. Leshem, Y. Y., Grossman, S., Frimer, A., Ziv, J. 1979. Endogenous lipoxygenase control and lipid-associated free radical scavenging as modes of cytokinin action in plant senescence retardation. In *Advances in the Biochemistry and Physiology of Plant Lipids,* ed. L.

A. Appelqvist, C. Liljenberg, pp. 193-98. Amsterdam: Elsevier

99. Leshem, Y. Y., Wurzburger, J., Grossman, S., Frimer, A. A. 1981. Cytokinin interaction with free radical metabolism and senescence: effects on endogenous lipoxygenase and purine oxidation. *Physiol. Plant.* 53:9-12

100. Lieberman, M. 1979. Biosynthesis and action of ethylene. *Annu. Rev. Plant Physiol.* 30:533-91

101. Lupu, R., Grossman, S., Cohen, Y. 1980. The involvement of lipoxygenase and antioxidants on pathogenesis of powdery mildew on tobacco plants. *Physiol. Plant Pathol.* 16:241-48

102. Lynch, D. V., Thompson, J. E. 1984. Lipoxygenase-mediated production of superoxide anion in senescing plant tissue. *FEBS Lett.* 173:251-54

103. Mack, A. J., Peterman, T. K., Siedow, J. N. 1987. Lipoxygenase isozymes in higher plants: biochemical properties and physiological role. *Isozymes: Curr. Top. Biol. Med. Res.* 13:127-54

104. Matsui, K., Kajiwara, T., Hayashi, K., Hatanaka, A. 1988. Tissue specific heterogeneity of lipoxygenase in cucumber seedlings. *Agric. Biol. Chem.* 52:3219-21

105. Matsumoto, T., Funk, C. D., Radmark, O., Hoog, J.-O., Jornvall, H., Samuelsson, B. 1988. Molecular cloning and amino acid sequence of human 5-lipoxygenase. *Proc. Natl. Acad. Sci. USA* 85:26-30

106. Meyer, A., Miersch, O., Buttner, C., Dathe, W., Sembdner, G. 1984. Occurrence of the plant growth regulator jasmonic acid in plants. *J. Plant Growth Regul.* 3:1-8

107. Miller, M. G., Obendorf, R. L. 1981. Use of tetraethylthiuram disulfide to discriminate between alternative respiration and lipoxygenase. *Plant Physiol.* 67:962-64

108. Navaratnam, S., Feiters, M. C., Al-Hakim, M., Allen, J. C., Veldink, G. A., Vliegenthart, J. F. G. 1988. Iron environment in soybean lipoxygenase-1. *Biochim. Biophys. Acta* 956:70-76

109. Nordlund, P., Sjoberg, B.-M., Eklund, H. 1990. Three-dimensional structure of the free radical protein of ribonucleotide reductase. *Nature* 345:593-98

110. Nugteren, D. H. 1975. Arachidonate lipoxygenase in blood platelets. *Biochim. Biophys. Acta* 380:299-307

111. Ocampo, C. A., Moerschbacher, B., Grambow, H. J. 1986. Increased lipoxygenase activity is involved in the hypersensitive response of wheat leaf cells infected with avirulent rust fungi or treated with fungal elicitor. *Z. Naturforsch. Teil C* 41:559-63

112. Oelze-Karow, H., Schopfer, P., Mohr, H. 1970. Phytochrome-mediated repression of enzyme synthesis (lipoxygenase: a threshold phenomenon). *Proc. Natl. Acad. Sci. USA* 65:51-57

113. Ohta, H., Ida, S., Mikami, B., Morita, Y. 1986. Changes in lipoxygenase components of rice seedlings during germination. *Plant Cell Physiol.* 27:911-18

114. Pages, M., Rosello, J., Casas, J., Gelpi, E., Gualde, N., Rigaud, M. 1986. Cyclooxygenase and lipoxygenase-like activity in *Drosophilia melanogaster*. *Prostaglandins* 32:729-40

115. Park, T. K., Polacco, J. C. 1989. Distinct lipoxygenase species appear in the hypocotyl/radicle of germinating soybean. *Plant Physiol.* 90:285-90

116. Pauls, K. P., Thompson, J. E. 1984. Evidence for the accumulation of peroxidized lipids in membranes of senescing cotyledons. *Plant Physiol.* 75:1152-57

117. Peary, J. S., Prince, T. A. 1990. Floral lipoxygenase: activity during senescence and inhibition by phenidone. *J. Am. Soc. Hort. Sci.* 115:455-57

118. Peever, T. L., Higgins, V. J. 1989. Electrolyte leakage, lipoxygenase, and lipid peroxidation induced in tomato leaf tissue by specific and nonspecific elicitors from *Cladosporium fulvum*. *Plant Physiol.* 90:867-75

119. Peterman, T. K., Siedow, J. N. 1983. Structural features required for inhibition of lipoxygenase-2 by propyl gallate. *Plant Physiol.* 71:55-58

120. Peterman, T. K., Siedow, J. N. 1985. Behavior of lipoxygenase during establishment, senescence, and rejuvenation of soybean cotyledons. *Plant Physiol.* 78:690-95

121. Peterman, T. K., Siedow, J. N. 1985. Immunological comparison of lipoxygenase isozymes-1 and -2 with soybean seedling lipoxygenases. *Arch. Biochem. Biophys.* 238:476-83

122. Pirrung, M. C. 1986. Mechanism of a lipoxygenase model for ethylene biosynthesis. *Biochemistry* 25:114-19

123. Pistorius, E. K., Axelrod, B. 1974. Iron, an essential component of lipoxygenase. *J. Biol. Chem.* 249:3183-86

124. Pistorius, E. K., Axelrod, B., Palmer, G. 1976. Evidence for participation of iron in lipoxygenase reaction from optical and electron spin resonance studies. *J. Biol. Chem.* 251:7144-48

125. Ponz, F., Breuning, G. 1986. Mechanisms of resistance to plant viruses. *Annu. Rev. Phytopathol.* 24:355-81

126. Prusky, D., Kobiler, I., Jacoby, B., Sims, J. J., Midland, S. L. 1985. Inhibitors of avocado lipoxygenase: their possible relationship with the latency of *Colletotrichum gloeosporioides*. *Physiol. Plant Pathol.* 27:269-79

127. Que, L. 1980. Non-heme iron dioxygenases. *Struct. Bond.* 40:39-72

128. Ramadoss, C. S., Axelrod, B. 1982. High performance liquid chromatographic separation of lipoxygenase isozymes in crude soybean extracts. *Anal. Biochem.* 127:25-31

129. Ramadoss, C. S., Pistorius, E. K., Axelrod, B. 1978. Coupled oxidation of carotene by lipoxygenase requires two isozymes. *Arch. Biochem. Biophys.* 190:549-52

130. Rhoads, D. M., McIntosh, L. 1991. Isolation and characterization of a cDNA clone encoding an alternative oxidase protein of *Sauromatum guttatum* (Shott). *Proc. Natl. Acad. Sci. USA.* In press

131. Rustin, P., DuPont, J., Lance, C. 1983. Involvement of lipid peroxy radicals in the cyanide-resistant electron transport pathway. *Physiol. Veg.* 22:643-63

132. Ruzicska, P., Gombos, Z., Farkas, G. L. 1983. Modification of the fatty acid composition of phospholipids during the hypersensitive reaction in tobacco. *Virology* 128:60-64

133. Ryan, C. A. 1987. Oligosaccharide signaling in plants. *Annu. Rev. Cell Biol.* 3:295-317

134. Ryan, C. A. 1988. Oligosaccharides as recognition signals for the expression of defensive genes in plants. *Biochemistry* 27:8879-83

135. Samuelsson, B., Dahlen, S.-E., Lindgren, J. A., Rouzer, C. A., Serhan, C. N. 1987. Leucotrienes and lipoxins: structures, biosynthesis and biological effects. *Science* 237:1171-76

136. Saniewski, M., Czapski, J. 1985. Stimulatory effect of methyl jasmonate on the ethylene production in tomato fruits. *Experientia* 41:256-57

137. Schechter, G., Grossman, S. 1983. Lipoxygenase from baker's yeast: purification and properties. *Int. J. Biochem.* 15:1295-304

138. Schewe, T., Halangk, W., Hiebsch, C., Rapoport, S. M. 1975. A lipoxygenase in rabbit reticulocytes which attacks phospholipids and intact mitochondria. *FEBS Lett.* 60:149-52

139. Schewe, T., Rapoport, S. M., Kuhn, H. 1986. Enzymology and physiology of reticulocyte lipoxygenase: comparison with other lipoxygenases. *Adv. Enzymol. Mol. Biol.* 58:191-272

140. Sheriff, S., Hendrickson, W. A., Smith, J. L. 1987. Structure of myohemerythrin in the azidomet state at 1.7/1.3 A resolution. *J. Mol. Biol.* 197:273-96

141. Shibata, D., Steczko, J., Dixon, J. E., Andrews, P. C., Hermodson, M., Axelrod, B. 1988. Primary structure of soybean lipoxygenase-2. *J. Biol. Chem.* 263:6816-21

142. Shibata, D., Steczko, J., Dixon, J. E., Hermodson, M., Yazdanparast, R., Axelrod, B. 1987. Primary structure of soybean lipoxygenase-1. *J. Biol. Chem.* 262:10080-5

143. Siedow, J. N., Berthold, D. A. 1986. The alternative oxidase: A cyanide-resistant respiratory pathway in higher plants. *Physiol. Plant.* 66:569-73

144. Siedow, J. N., Girvin, M. E. 1980. Alternative respiratory pathway: its role in seed respiration and its inhibition by propyl gallate. *Plant Physiol.* 65:669-74

145. Sigal, E., Craik, C. S., Highland, E., Grunberger, D., Costello, L. L., et al. 1988. Molecular cloning and primary structure of human 15-lipoxygenase. *Biochem. Biophys. Res. Commun.* 157:457-64

146. Slappendel, S., Malmstrom, B. G., Petersson, L., Ehrenberg, A., Veldink, G. A., Vliegenthart, J. F. G. 1982. On the spin and valence state of iron in native soybean lipoxygenase-1. *Biochem. Biophys. Res. Commun.* 108:673-77

147. Slappendel, S., Veldink, G. A., Vliegenthart, J. F. G., Aasa, R., Malmstrom, B. G. 1981. EPR spectroscopy of soybean lipoxygenase-1. Description and quantification of the high-spin Fe(III) signals. *Biochim. Biophys. Acta* 667:77-86

148. Spaapen, L. J. M., Verhagen, J., Veldink, G. A., Vliegenthart, J. F. G. 1980. The effect of modification of sulfhydryl groups in soybean lipoxygenase-1. *Biochim. Biophys. Acta* 618:153-62

149. Stallings, W. C., Kroa, B. A., Carroll, R. T., Metzger, A. L., Funk, M. O. 1990. Crystallization and preliminary X-ray characterization of a soybean seed lipoxygenase. *J. Mol. Biol.* 211:685-87

150. Start, W. G., Ma, Y., Polacco, J. C., Hildebrand, D. F., Freyer, G. A., Altschuler, M. 1986. Two soybean seed lipoxygenase nulls accumulate reduced levels of lipoxygenase transcripts. *Plant Mol. Biol.* 7:11-23

151. Staswick, P. E. 1990. Novel regulation of vegetative storage protein genes. *Plant Cell* 2:1-6

152. Steczko, J., Muchmore, C. R., Smith, J. L., Axelrod, B. 1990. Crystallization and preliminary X-ray investigation of lipoxygenase-1 from soybeans. *J. Biol. Chem.* 265:11352-54

153. Stegink, S. J., Siedow, J. N. 1986. Ethylene production from 1-aminocyclopropane-1-carboxylic acid in vitro: a mechanism for explaining ethylene production by a cell-free preparation from pea epicotyls. *Physiol. Plant.* 66:625-31

154. Sumner, J. B., Sumner, R. J. 1940. The coupled oxidation of carotene and fat by carotene oxidase. *J. Biol. Chem.* 134: 531-33

155. Tappel, A. L. 1961. Biocatalysts: lipoxidase and hematin compounds. In *Autoxidation and Antioxidants*, ed. W. O. Lundberg, pp. 325-66. New York: Interscience

156. Theorell, H., Holman, R. T., Akeson, A. 1947. Crystalline lipoxidase. *Acta Chem. Scand.* 1:571-76

157. Thomas, H., Stoddart, J. L. 1980. Leaf senescence. *Annu. Rev. Plant Physiol.* 31:83-111

158. Thompson, J. E., Paliyath, G., Brown, J. H., Duxbury, C. L. 1987. The involvement of activated oxygen in membrane deterioration during senescence. In *Plant Senescence*, ed. W. W. Thompson, E. A. Nothnagel, R. C. Huffaker, pp. 146-55. Rockville: Am. Soc. *Plant Physiol.*

159. Trop, M., Grossman, S., Veg, Z. 1974. The antigenicity of lipoxygenase from various plant sources. *Ann. Bot.* 38:783-94

160. Ueda, J., Kato, J. 1980. Isolation and identification of a senescence-promoting substance from wormwood (*Artemesia absinthium* L.). *Plant Physiol.* 66:246-49

161. Ueda, J., Kato, J. 1982. Identification of jasmonic acid and abscisic acid as senescence-promoting substances from *Cleyera ochnacea*. *Agric. Biol. Chem.* 46:1975-76

162. van der Meer, R. A., Duine, J. A. 1988. Pyrroloquinoline quinone (PQQ) is the organic cofactor in soybean lipoxygenase-1. *FEBS Lett.* 235:194-200

163. van der Meer, R. A., Jongejan, J. A., Duine, J. A. 1987. Phenylhydrazine as a probe for cofactor identification in amine oxidoreductases. *FEBS Lett.* 221:299-304

164. van der Meer, R. A., Jongejan, J. A., Duine, J. A. 1988. Dopamine β-hydroxylase from bovine medulla contains covalently-bound pyrroloquinoline quinone. *FEBS Lett.* 231:303-7

164a. Veldink, G. A., Boelens, H., Maccarrone, M., van der Lecq, F., Vliegenthart, J. F. G., et al. 1990. Soybean lipoxygenase-1 is not a quinoprotein. *FEBS Lett.* 270:135-38

165. Veldink, G. A., Vliegenthart, J. F. G. 1985. Lipoxygenases, nonheme iron-containing enzymes. *Adv. Inorg. Biochem.* 6:139-61

166. Veldink, G. A., Vliegenthart, J. F. G., Boldingh, J. 1977. Plant lipoxygenases. *Prog. Chem. Fats Other Lipids* 15:131-66

167. Verhue, W. M., Francke, A. 1972. The heterogeneity of soyabean lipoxygenase. *Biochim. Biophys. Acta* 284:43-53

168. Vernooy-Gerritsen, M., Veldink, G. A., Vliegenthart, J. F. G. 1982. Specificities of antisera directed against soybean lipoxygenases-1 and -2 and purification of lipoxygenase-2 by affinity chromatography. *Biochim. Biophys. Acta* 708:330-34

169. Vick, B. A., Zimmerman, D. C. 1976. Lipoxygenase and hydroperoxide lyase in germinating watermelon seedlings. *Plant Physiol.* 57:780-88

170. Vick, B. A., Zimmerman, D. C. 1979. Substrate specificity for the synthesis of cyclic fatty acids by a flaxseed extract. *Plant Physiol.* 63:490-94

171. Vick, B. A., Zimmerman, D. C. 1982. Levels of oxygenated fatty acids in young corn and sunflower plants. *Plant Physiol.* 69:1103-8

172. Vick, B. A., Zimmerman, D. C. 1983. The biosynthesis of jasmonic acid: a physiological role for plant lipoxygenase. *Biochem. Biophys. Res. Commun.* 111:470-77

173. Vick, B. A., Zimmerman, D. C. 1984. Biosynthesis of jasmonic acid by several plant species. *Plant Physiol.* 75:458-61

174. Vick, B. A., Zimmerman, D. C. 1987. Oxidative systems for modification of fatty acids: the lipoxygenase pathway. In *The Biochemistry of Plants*, ed. P. K. Stumpf, E. E. Conn, 9:53-90. Orlando: Academic

175. Wang, N., Southan, C., DeWolf, W. E., Wells, T. N. C., Kruse, L. I., Leatherbarrow, R. J. 1990. Bovine dopamine-hydroxylase, primary structure determined by cDNA cloning and amino acid sequencing. *Biochemistry* 29:6466-74

176. Weidhase, R. A., Kramell, H., Lehmann, J., Liebisch, H., Lerbs, W., Parthier, B. 1987. Methyl jasmonate-induced changes in the polypeptide pattern of senescing barley leaf segments. *Plant Sci.* 51:177-86

177. Whittaker, J. W., Solomon, E. I. 1988. Spectroscopic studies on ferrous non-heme iron active sites: magnetic circular dichroism of mononuclear Fe sites in superoxide dismutase and lipoxygenase. *J. Am. Chem. Soc.* 110:5329-39

178. Wion, K. L., Kirchgessner, T. G., Lusis, A. L., Schotz, M. C., Lawn, R. M. 1987. Human lipoprotein lipase complementary DNA sequence. *Science* 235:1638-41

179. Yabuuchi, S., Lister, R. M., Axelrod, B., Wilcox, J. R., Nielsen, N. C. 1982. Enzyme-linked immunosorbent assay for the determination of lipoxygenase isozymes in soybean. *Crop Sci.* 22:333-37

180. Yamamoto, H., Tani, T. 1986. Possible involvement of lipoxygenase in the mechanism of resistance of oats *Avena sativa* to *Puccinia coronata* pv *avenae*. *Jpn. J. Phytopathol.* 116:329-37

181. Yamane, H., Sugawara, J., Suzuki, Y., Shimamura, E., Takahashi, N. 1980. Synthesis of jasmonic acid related com-pounds and their structure-activity rela-tionships on the growth of rice seed-lings. *Agric. Biol. Chem.* 44:2857-64

182. Yang, S. F., Hoffman, N. E. 1984. Ethylene biosynthesis and its regulation in higher plants. *Annu. Rev. Plant Phys-iol.* 35:155-89

183. Yenofsky, R. L., Fine, M., Liu, C. 1988. Isolation and characterization of a soybean (*Glycine max*) lipoxygenase-3 gene. *Mol. Gen. Genet.* 211:215-22

184. Yoshimoto, T., Suzuki, H., Yamamoto, S., Takai, T., Yokoyama, C., Tanabe, T. 1990. Cloning and sequence analysis of the cDNA for arachidonate 12-lipoxygenase of porcine leukocytes. *Proc.Natl. Acad. Sci. USA* 87:2142-46

185. Zimmerman, D. C., Coudron, C. A. 1979. Identification of traumatin, a wound hormone, as 12-oxo-trans-10-dodecenoic acid. *Plant Physiol.* 63:536-41

186. Zimmerman, D. C., Vick, B. A. 1973. Lipoxygenase in *Chlorella pyrenoidosa*. *Lipids* 8:264-66

Annu. Rev. Plant Physiol. Plant Mol. Biol. 1991. 42:189–204

ISOLATION AND CHARACTERIZATION OF SPERM CELLS IN FLOWERING PLANTS

Scott D. Russell

Department of Botany and Microbiology, University of Oklahoma, Norman, Oklahoma 73019-0245

KEY WORDS: angiosperm, gametogenesis, fertilization, male cytoplasmic inheritance

CONTENTS

INTRODUCTION

Considering the crucial role of the male gamete in double fertilization, it is remarkable that research on the physiology of sperm cells in flowering plants got under way only recently. According to Mendelian genetics, offspring receive one nucleus of the male gamete and one from the female gamete. The effect of DNA-containing cytoplasmic organelles in the male gamete (namely

189

mitochondria and plastids) on the cytoplasmic organization of the offspring depends on the presence and quantity of these organelles in the sperm.

This review treats the function, isolation, and characterization of sperm cells, therefore summarizing research conducted mainly since 1984. Literature concerning pollen development (26), physiology (31, 32), incompatibility (9, 45), and the isolation and characterization of generative cells (30, 49) has been reviewed elsewhere.

DEVELOPMENTAL CONTEXT OF SPERM CELLS

The sperm of angiosperms are cellular descendants of meiotic divisions occurring in the anther of the flower. Meiosis in each meiocyte forms four genetically dissimilar microspores. Each microspore undergoes a mitotic division to form a large vegetative cell and a smaller, reproductive generative cell. Together these cells form the pollen grain or, strictly speaking, the microgametophyte. Subsequently, the generative cell migrates into the vegetative cell and becomes an elongated, spindle-shaped, or even filiform cell, depending on the species; each generative cell retains its own plasma membrane and is, in turn, enveloped by the vegetative cell membrane (26). The generative cell undergoes a second mitotic division to form the two sperm.

Depending on the timing of generative cell mitosis the two sperm cells may be formed before germination of the pollen (tricellular) or after pollen germination (bicellular). Of 243 families surveyed (11), bicellular pollen was present in 137 families (56%), tricellular pollen in 55 families (23%), and both types of pollen in 51 families (21%). Bicellular pollen is typically regarded as the ancestral condition in most families, tricellular pollen as the derived condition (11). Most of the research reviewed here concerns sperm cells isolated from pollen grains in tricellular species. Bicellular species must be cultured in order to grow pollen tubes, trigger mitosis, and obtain sperm cells.

Structural Characterization

Flowering plant sperm are structurally simple, apparently nonmotile cells that contain a normal complement of cellular organelles, including heritable cytoplasmic organelles such as mitochondria and, in some plants, plastids (10, 19). In contrast to animal species, competition in angiosperms is a property of the gametophyte (pollen grain and tube) and not the gamete; once this competition is completed, *both* sperm cells contribute to the formation of the embryo during double fertilization: One sperm cell fuses with the egg to produce the embryo, the other fuses with the central cell to produce the nutritive endosperm.

The sperm cells are typically connected to one another and physically

Table 1 Summary of the fertilization characteristics of the sperm cells in angiosperms

Pollen type and species	Method[a]	Morphology[b]	Functional sperm[c]	Sample	Reference
Tricellular pollen:					
Monocotyledons:					
Hordeum vulgaris	Recon	I	—	n = 5	40
Zea mays	Recon	D	B+ (3, 55)	n = 1, 1, 1	37, 56, 57
Dicotyledons:					
Brassica campestris	Recon	MtD	—	n = 5	36
Brassica oleracea	Recon	MtD	—	n = 3	36
Euphorbia dulcis	Recon	MtD	—	n = 1	43
Gerbera jamesonii	Recon	MtD	—	n = 1	48
Plumbago zeylanica	Recon	MtPD	P+ (60)	n = 11	59
Spinacia oleracea	Recon	MtD	—	n = 7	78
Bicellular pollen:					
Monocotyledons:					
Gladiolus gandavensis	ImAn	D	—	n = 6	66
Dicotyledons:					
Nicotiana tabacum	Recon	I	—	n = 9	H.-S. Yu, personal communication
Petunia hybrida	Recon	I	—	n = 1	77
Rhododendron spp.	ImAn	D	—	n = 7	66

[a] ImAn: image analysis of nuclear fluorochromatic patterns; Recon: serial TEM and 3-D reconstruction
[b] I: isomorphic; D: dimorphic (condition of organellar DNA unknown); MtD: mitochondrial dimorphism; MtPD: mitochondrial and plastid dimorphism
[c] B+: sperm cell containing B-chromosomes; P+: sperm cell containing most plastids

associated with the vegetative nucleus, transported within the pollen tube as a "linked unit" (64) termed the "male germ unit" (14). The linkage of the sperm cells increases the effectiveness of gamete delivery, reduces heterofertilization, and ensures nearly simultaneous transmission of the two sperm cells into the receptive embryo sac (64). This association is present in most species of flowering plants studied to date (24) or is formed soon after pollen germination (41).

Sperm Dimorphism and Preferential Fertilization

Although the association between one sperm cell and the vegetative nucleus imposes a degree of polarity, strong cellular differences may also occur between the two sperm, including differences in the content of heritable organelles termed *cytoplasmic heterospermy*, and differences in the nucleus termed *nuclear heterospermy* (60). Although cytoplasmic heterospermy may be relatively common in flowering plants (Table 1), nuclear differences have been reported only in *Zea mays*, in which B-chromosomes frequently undergo nondisjunction during generative cell mitosis (3). The most common form of cytoplasmic dimorphism is mitochondrial inequality: Sperm cells associated with the vegetative nucleus usually contain more mitochondria than the other cell. The most extreme form examined to date is found in *Plumbago zeylanica,* where both plastid and mitochondrial content differ (10, 59).

Preferential fertilization, in which one sperm cell has a greater likelihood of fusing with the egg, may also occur in species where the sperm cells differ (Table 1). In *Plumbago,* fusion between the egg and the plastid-rich sperm cell occurs in 94% of the cases examined (60), selectively transmitting male plastids into the zygote (58). Transmission of male mitochondria into both the egg and central cell, however, occurs at a nearly constant 1:1000 ratio of male:female mitochondria (62), because the differences in mitochondrial content in the egg and central cell match those of the dimorphic sperm cells. Whether this ratio represents the maximum permissible dose of sperm mitochondria to prevent sperm transmission of mitochondrial DNA or the minimum needed to make recombination of mitochondrial DNA possible is unclear (62).

Preferential fertilization (termed meiotic drive by geneticists) also occurs in *Zea,* in which the sperm cell containing one set of extra B-chromosomes fuses with the egg about 65% of the time (2, 3, 55). Rather than simply providing a marker for identifying the sperm cell destined to fuse with the egg, however, B-chromosomes appear to confer a selective advantage to the sperm cell that contains them (2). It is interesting that preferentiality may be eliminated by introducing excessive B-chromosomes into the sperm or by introducing a specific B-chromosome (TB-9b) into the egg. This observation suggests that maternal discrimination is active only under relatively strict conditions (2).

Maternal control of fertilization is also suggested in a mutant line of barley in which seeds are produced containing normal endosperm but haploid embryos (38). One sperm cell apparently fuses with the central cell regardless of pollen source whereas the other remains unfused (38). Research on the mechanism of fertilization in normal lines of barley indicates that the zygote of plants with uniparental, maternal cytoplasmic inheritance are *cytoplasmically restrictive:* Male cytoplasmic organelles may be shed by the separation or pinching off of cellular folds during maturation of the sperm (40) and the sperm cytoplasm excluded from the egg, remaining outside of the cell (39). However, the central cell in the same plant may be *cytoplasmically permissive,* allowing sperm organelles to be transmitted into the central cell (39). Plants with biparental inheritance appear to be cytoplasmically permissive during the entire double fertilization process (58, 60).

ISOLATION OF SPERM CELLS

Sperm cell isolation was first reported in detail by Cass, who placed pollen grains of barley in a Brewbaker-Kwack (BK) pollen germination medium (1) with 20% sucrose (4), obtaining release of the sperm cells by osmotic shock and observing the behavior of the living cells using Nomarski differential interference contrast microscopy. Over 30 min, the originally paired sperm cells separated, lost their spindle shape, and became ellipsoidal to spheroidal. Alternative media, such as filtrates of ovular material, were also examined but there were no specific benefits or effects on sperm cell appearance or behavior (4).

Technical Details and Protocols

Numerous isolation protocols have emerged (Table 2), including some modified for collection of sperm cells en masse (Table 3). Sperm cell release has been conducted using two general techniques: osmotic shock, or direct physical separation by grinding or applied pressure. The pollen cytoplasm and gametes are usually incubated for an additional 10–20 min to allow the gametes to separate from vegetative cell cytoplasm, and then pollen cell walls are removed by filtration. A nylon mesh two thirds to one half the average diameter of the pollen efficiently captures the pollen walls. En masse isolation of sperm cells involves either centrifugation on a discontinuous gradient or filtration using a polycarbonate (e.g. Nuclepore) filter. Details of the procedures used and results are given in Table 2.

Isolation of sperm cells using osmotic shock relies on the greater sensitivity of the vegetative cell membrane to changes in the milieu. In some cases, the shock of entering an aqueous medium from a nearly desiccated state is sufficient to rupture pollen grains regardless of the osmotic concentration used (29), but sperm cells may be surprisingly intact (61). After prehydration (for example, exposure in a humid chamber), pollen is more gradually introduced to a hydrated state, possibly enhancing viability of the isolated sperm cells (67). Pollen grains may then be subjected to an abrupt change in osmotic concentration (80) or pH (54) or lysed in other ways.

The isolation of sperm cells can also be accomplished by using physical stress to break the pollen plasma membrane. Methods used include glass tissue homogenizers (6, 67, 68) or the pressure of a glass roller applied to a smooth glass surface (71–73). Clearance height in the homogenizer is not critical, because the resistant exines will set the minimum. Isotonic or slightly hypertonic media are best for the isolation of the cells (68).

The most frequently used media for sperm isolation are modifications of the BK medium (1)—e.g. 100 ppm H_3BO_3, 300 ppm $Ca(NO_3)_2 \cdot 4H_2O$, 200 ppm $MgSO_4 \cdot 7H_2O$, 100 ppm KNO_3, and 10% sucrose. Other successfully applied media include the Roberts medium, which is a Tris-buffered modification of BK medium (52); a modified K3 Kao protoplast culture medium containing BK macronutrients, plus micronutrients, coenzymes, vitamins, and hormones (44); and the RY-2 protoplast medium used for plantlet regeneration (79).

Table 2 Summary of methods used for isolating sperm cells in angiosperms and evaluating the isolated cells

Pollen type and species	Medium[a]	Procedure[b]	Evaluation technique[c]	Reference
Tricellular Pollen:				
Monocotyledons:				
Hordeum vulgare	BKS 30	OS	DIC	4
Lolium perenne	RY-2 20	OS/G	FCR+, DAPI+, PCM	75
Secale cereale	BKS +[d]	GH	FCR+, Ho+	80
Triticum aestivum	BKS 30	OS/G	FCR+, DAPI+, EtB+	35
Triticum aestivum	BKS 20	OS	FCR+, DAPI+	69
Zea mays	NA	OS	DAPI+, PCM	6
Zea mays	BKS 15	OS/G	FCR+, PCM, TEM	35
Zea mays	BKS 15	OS	FCR+, DAPI+, TEM, SEM	15
Zea mays	BKS 15	OS	FCR+, DAPI+, PCM	34
Zea mays	BKS 15 + V[e]	OS	FCR+, PCM	53
Zea mays	BKS 15	OS	EvB−, DIC, PCM, TEM, ABB, CBB	5
Zea mays	BKS/K3 30/15%[f]	2-step OS/G	FCR+, Ho+	80
Dicotyledons:				
Ambrosia sp.	BKS	OS	FCR+, EtB+, PCM	33
Artemesia sp.	BKS	OS	FCR+, EtB+, PCM	33
Bellis sp.	BKS	OS	FCR+, EtB+, PCM	33
Beta vulgaris	20 +[g]	OS	FCR±, PI−, Ho, DIC	46
Brassica campestris	NA	GH	Ho+, PCM	6
Brassica campestris	M-T[h]	GH	Ho+, PCM	23
Brassica napus	BKS 12.5	GH	FCR+, PCM	33
Brassica napus	RM+[d]	GH	FCR+, Ho+	80
Brassica oleracea	BKS 15	OS/G	FCR+, DAPI+, EtB+	35
Gerbera jamesonii	M-S+[e]	GH	FCR−, SEM, TEM	67

Species	Medium	Treatment	Method	Ref.
Gerbera jamesonii	S 1 M[i]	GH	EvB−, DAPI+, DIC, SEM	68
Impatiens sp.	S 20% +[g]	OS		46
Plumbago zeylanica	BKS 5-50	OS	DIC	64
Plumbago zeylanica	20	OS	EvB−, FCR−, Ho+, DIC	61
Senecio sp.	BKS	OS	FCR+, EtB+, PCM	33
Spinacia oleracea	BKS 25	P	FCR+, PCM	71, 72
Spinacia oleracea	BKS 25	P	FCR+, PCM, TEM	73
Taraxacum sp.	BKS	OS	FCR+, EtB+, PCM	33
Bicellular Pollen:				
Monocotyledons:				
Lilium longiflorum	BKS 10[j]	Enz[j], GH	My+, PCM	67
Gladiolus gandavensis	semi-vivo[k]	OS[k] or Enz[l]	DAPI+, EtB, Ho, SEM	66
Dicotyledons:				
Rhododendron spp.	semi-vivo[k]	OS[k] or Enz[l]	DAPI+, EtB, Ho, SEM	66

[a] BKS: Brewbaker-Kwack pollen growth medium, plus sucrose (in %) (1); K3: modified Kao protoplast culture medium (44); M-S: mannitol-sucrose; M-T: mannitol-Tris; RM: Roberts medium (52); S: sucrose (in %); RY-2: medium (79); V: vitamins

[b] G: grinding; GH: glass homogenizer; OS: osmotic shock; P: pressure

[c] +: positive (dye incorporated); −: negative (dye excluded); ABB: Aniline blue black; CBB: Coomassie brilliant blue; DAPI: 4-6-diamidino-2-phenylindole (8); DIC: differential interference contrast microscopy; EtB: ethidium bromide (22); EvB: Evans blue (17); FCR: fluorochromatic reaction using fluorescein diacetate (21); Ho: Hoechst 33258 (22); My: mythramycin (8); PCM: phase contrast microscopy; PI: propidium iodide (25)

[d] 5% sucrose, 8% sorbitol, 0.3% potassium dextran sulfate, 10 μg ml^{-1} fluorescein diacetate

[e] Vitamins used: 0.5 mg ml^{-1} nicotinic acid, 0.5 mg ml^{-1} pyridoxine, 0.1 mg ml^{-1} thiamine HCl, 2 mg ml^{-1} glycine

[f] Sperm stored in 5% sucrose, 8% sorbitol, 0.5% potassium dextran sulfate, 10 μg ml^{-1} fluoroscein diacetate; K3: macronutrients

[g] 0.4 M mannitol, 10 mM Tris pH 7.5, 3 mM $CaCl_2$, 1.5 mM $MgCl_2$, 10 mM NaCl

[h] 0.4 M mannitol contains 150 mg l^{-1} $CaCl_2 \cdot 2H_2O$; isolation medium contains 30% sucrose, 150 mg l^{-1} $CaCl_2 \cdot 2H_2O$, and 100 mg l^{-1} H_3BO_3

[i] 1 M sucrose, 2.1 mM $Ca(NO_3)_2$, 1.6 mM H_3BO_3; pollen 20 mg 0.2 ml^{-1}; grinding medium: 1 M mannitol, 0.2 M sucrose, 10 mM HEPES buffer (pH 7.2), 0.3-1.0% bovine serum albumin, 0.3% polyvinylpyrrolidone, 10 mM cysteine, 2.1 mM $Ca(NO_3)_2$

[j] Pollen tubes grown in vitro; enzymes: 2% cellulysin, 11 units mg^{-1} pectinase for 2 hr at end of in vitro pollen tube growth

[k] Styles implanted in BKS 12 plus 0.6% agar for 24 hr in the dark until pollen tubes emerge [8 hr Gladiolus, 24 hr Rhododendron], then incubated in drop of 5 or 7.5% sucrose in medium for OS, or in enzymes[l]

[l] Enzymes: 0.5% macerozyme R-10 (Serva), 1% cellulase (Onozuka R-10, Serva) in BKS medium

Table 3 Summary of methods for collecting sperm cells en masse.

Pollen type and species	Centrifugation[a]	Method[a]	Concentration[b]	Viability[c]	Half-Life[d]	Yield[e]	Reference
Tricellular Pollen:							
Monocotyledon:							
Lolium perenne	15,000 × g 30 min	P (0*/30%)	NA	PCM+	overnight[f]	2%	75
Triticum aestivum	NA	Sor (20/40%)	NA	most	15 min	NA	69
Zea mays	NA	NDC (1.07 g ml⁻¹) Su*	NA	NA	NA	NA	6
Zea mays	9,000 × g 40 min	P (15*/40%)[g]	3 × 10⁶	90% FCR	20 hr	20%	15
Zea mays	55 × g 3 min	BKS (15/*/30)[h]	1.5 × 10⁶	>50% EvB	3 hr	30%	5
Dicotyledons:							
Beta vulgaris	3,000 rpm 15 min	P (20/30*/50%)	7.3 × 10⁶	30% F-PI	NA	NA	46
Brassica campestris	NA	NDC (1.22 g ml⁻¹) Pt*	NA	NA	NA	NA	6
Brassica campestris	800 × g 10 min P*	—	NA	NA	NA	NA	23
Gerbera jamesonii	850 × g P*	—	2 × 10⁵	NA	NA	NA	67
Gerbera jamesonii	850 × g 10 min P*	—	NA	NA	NA	NA	68
Plumbago zeylanica	8,000 × g 8 min	S (30%) Pt*	8.8 × 10⁶	most EvB	8 hr	60%	61
Spinacia oleracea	13,000 × g 30 min	P (*/10/30/50%)	4 × 10⁶	90% F	18 hr	5–10%	71
Spinacia oleracea	13,000 × g 40 min	P (20%*)	1 × 10⁶	90% F	30 hr[i]	5–10%	72, 73
Bicellular Pollen:							
Monocotyledon:							
Gladiolus gandavensis	omitted	—	NA	NA	NA	60%	66
Dicotyledon:							
Rhododendron spp.	omitted	—	NA	NA	NA	NA	66

[a] Media for centrifugation: BKS: Brewbaker-Kwack pollen growth medium, plus sucrose (in %); Sor: sorbitol gradient (discontinuous gradient, in %); P: Percoll gradient (discontinuous gradient, in %); NA: not available; S: sucrose pad (in %); Pt: pellet or Su: supernatant layer; —: medium replaced by a 1 μm polycarbonate filter. Recovery location for sperm cells is indicated by asterisk (*) next to appropriate layer.

[b] Sperm cells per milliliter

[c] Percentage viability as indicated by: EvB: Evans blue (−); FCR: fluorescein diacetate (+); F-PI: FCR (+) propidium iodide (−)

[d] Viability half-life

[e] Percentage recovery compared to the number of pollen grains treated × 2 sperm per pollen grain

Because few of the cited reports use parallel methodology, it is difficult to conclude which, if any, of the media is ideal for most angiosperms; but the life expectancy of isolated sperm cells seems strongly improved by complex media (Tables 2 and 3) and particularly by the addition of vitamins (53). As protocols are refined, a distinction will likely be made between isolation media (designed to optimize recovery) and storage or culture media (designed to extend sperm cell viability).

Assessment of Sperm Cell Quality

Sperm cells have been assessed for (a) structural intactness, (b) membrane integrity, (c) enzyme activity, and (d) respiratory activity. All of these aspects are appropriate; perhaps most crucial would be an assay of the cell's ability to effect fertilization, which is not currently feasible.

Techniques to assay structural intactness include transmission (6, 15, 42, 68, 73, 74) and scanning electron microscopy (66, 68), which can be used to assess membrane structure and cell shape; intramembrane particle distribution can be inferred from freeze-fracture replicas (74). Integrity of membranes can be assessed using the exclusion of polar dyes such as Evans blue (17) or propidium iodide (25), both of which are rapidly absorbed into dead cells (Table 2). Enzyme activity is readily demonstrated using the so-called fluorochromatic reaction (21) in which fluorescein diactate is cleaved by cellular esterase. Fluorescein diacetate is easily absorbed by living cells but becomes fluorescent only if esters are cleaved from the fluorescein molecule and the cell is intact enough to retain the polar dye. Respiratory activity has been demonstrated using an assay for ATP involving cleavage by luciferin yielding measurable luminescence. The results of this method roughly correspond to those using fluorescein diacetate (54) but ATP is still evident even when the cells are no longer viable according to the previous method.

During isolation, the sperm cells commonly assume an ellipsoidal to spheroidal shape and lose their microtubules (5, 15, 73). A comparable loss of microtubules has been described in detail using anti-tubulin immunofluorescence of freshly isolated generative cells in bicellular plants (70). These changes appear comparable to those observed in isolated generative cells (70). An exception to this is the isolation of sperm cells in *Gerbera*, in which the native shape is retained (68). Spheroidal sperm cells are not typically seen in young growing pollen tubes in vivo, but a significant rounding of these cells seems characteristic as pollen tubes near the ovule (38, 58, 64, 65). Although spheroidal sperm cells obtained through isolation are artifactual, this feature is common in somatic plant protoplasts, typically without impairing cell quality.

Isolated sperm cells lose the plasma membrane of the vegetative cell that usually surrounds them in vivo (26). This exposes the true surface of the sperm cells (15, 64) and makes it possible to characterize the plasma membrane. In vivo, the same modification of the surface of the sperm cell occurs once sperm cells are discharged into the embryo sac (38, 58, 65) and is postulated to be a prerequisite for sperm fusion in angiosperms (58).

A significant remaining question about the condition of isolated sperm cells is whether they are completely mature in the pollen grain or whether essential gene products remain to be transcribed or translated during pollen tube growth prior to fertilization (66). Given the vast range of times between pollination and gametic fusion in different angiosperms (20 min to several months), these plants likely vary in this respect.

PHYSIOLOGICAL CHARACTERIZATION OF SPERM CELLS

Immunological Characterization

Hybridoma antibodies have been elicited to an inoculum of intact sperm cells of *Brassica* (23) and *Plumbago* (47) to generate antibody libraries to surface epitopes. Intact cells were used because of the small number of sperm cells available and the technical problems of preparing isolated plasma membranes. Depending on how long sperm cells remain intact in the inoculum, whole cells would be likely to elicit some antibodies to surface compounds. Initial attempts were partially successful, but techniques for retaining integrity of the sperm cells during inoculation and for increasing the sensitivity of screening for surface-specific epitopes will be needed before it is possible to produce a useful monoclonal antibody library for recognition studies.

The only detailed report on antibodies elicited to isolated sperm, in *Plumbago* (47), indicates the feasibility of this approach. Sperm antigens are effective in eliciting antibody-producing hybridoma lines to a wide spectrum of different epitopes in which pollen wall compounds, including allergens, are not immunodominant. The secreting lines were specific for epitopes in the sperm-enriched (23%), cytoplasmic-particulate (4%), and water-soluble fractions (8%). Antibody-producing lines were mainly IgM (61%), reflecting the high glycoprotein content of the cells (47). Identification of a recognition factor from this approach is complicated by the need to develop an appropriate functional assay. Further research in this area is warranted if specific recognition compounds occur as in pollen tube–pistil interactions (9, 45).

Polypeptide/Protein Characterization

Biochemical characterization of the proteins of male gametes using SDS-PAGE has been conducted on a small number of species: *Brassica, Gerbera*

(27), *Plumbago* (18), and *Zea* (P. Roeckel, C. Dumas, personal communication). These results indicate that unique polypeptide bands are readily identified in sperm fractions of higher plants. In order to reduce this effect, presence/absence comparisons with possible contaminants were conducted in *Plumbago* (18).

Two-dimensional gel electrophoresis (IEF/SDS-PAGE) was conducted to compare polypeptide heterogeneity in *P. zeylanica* (18). The sperm-enriched fraction and two selected contaminant fractions were examined. Cytoplasmic particulates pelleted at 100,000 × g constituted the first fraction; water soluble molecules collected from the supernatant formed the second fraction. The sperm-enriched fraction contained the most polypeptide spots (515 from M_r 33,000 to 205,000). The cytoplasmic-particulate fraction yielded 427 spots, and the water-soluble fraction had 285. One quarter of the spots were found in all fractions, about half were found in two fractions, and one quarter were found in a single fraction. Of the latter group, 51.9% of the polypeptides were unique to the sperm-enriched fraction, 40.4% to the water-soluble fraction, and 2% to the cytoplasmic-particulate fraction. Polypeptides of sufficient molecular weight to be involved in recognition (M_r 70,000 to >200,000) were present in the sperm-enriched fraction (18). Some of the differences between polypeptides undoubtedly result from generational differences of gene expression between the sporophyte and gametophyte (32). Although pollen wall proteins are synthesized by the sporophyte, gametophytic gene products may control gametic recognition, fusion, and fertilization.

No studies of developmental polypeptide changes in the male gametic line are currently available. Polypeptide changes in developing pollen and microspores of *Zea* (16), *Triticum* (76), and *Brassica* (13) indicate that developmental changes do occur. In *Triticum,* 11 new bands were detected in SDS-PAGE and IEF gels and 4 bands were lost during pollen maturation, indicating a small number of stage-specific changes. Changes during anther development in *Zea* (12) reflect similar stage-specific changes in the proteins and activities of specific enzymes. With further research, considerable progress in identifying sperm-specific polypeptides may be expected.

Elemental Characterization

Energy-dispersive X-ray analysis of pollen components indicates that sperm cells contain significant quantities of carbon, oxygen, phosphorus, calcium, and potassium, with smaller amounts of magnesium, silicon, manganese, and detectable levels of chromium. Although mostly similar, the dimorphic sperm of *Plumbago* exhibit small differences in Kα lines of calcium, potassium, and phosphorus (63). Other studies of pollen have indicated similar elemental composition in the pollen tube tip (50). The surfaces of pollen grains in different species, however, display significant elemental differences (7).

Molecular Characterization

Conflicting information is available regarding the synthesis of RNA in sperm cells. In the tricellular pollen of *Secale cereale*, [5-^{3}H]uridine was incorporated into both the sperm and vegetative nuclei during in vitro pollen tube growth (20). The types of RNA synthesized by the sperm nuclei were not determined. However, in the bicellular pollen of *Hyoscyamus niger*, exposure to short pulses of [5-^{3}H]uridine did not detect incorporation (51). Determination of whether RNA synthesis is related to the timing of sperm cell formation will require additional research.

Translational products within sperm cells have been obtained by means of in vitro translation of isolated RNA species using [^{35}S]methionine, combined with electrophoresis and radioautography (P. Roeckel and C. Dumas, personal communication). These studies demonstrate that translatable mRNA pools exist for a number of polypeptides in the sperm cells of corn; their identity has not yet been fully determined. The use of stage-specific promoters (e.g. 73a) and reporter genes in combination with particle-gun transformation to elicit transgenic plants seems to be a promising approach to understanding the developmental processes underlying sperm differentiation. Adaptation and application of existing molecular biological methods to sperm cell biology are critically needed.

CONCLUSIONS AND OVERVIEW

Sperm cells have only recently been isolated and used in basic and applied science. Living sperm cells have been isolated in many species of flowering plants; now key methodological improvements will be required if we are to extend the useful lifespan of these cells. Further research involving immunological characterization, cell sorting, cell culture, radiolabeling, gene expression, and molecular biology of the sperm cells will require basic knowledge about the maintenance and storage of sperm cells.

After these technical problems are solved, sperm cells could play a unique role in the developing field of reproductive-cell engineering. Because the cytoplasm of some sperm cells lacks plastids, it could be used to produce fusion hybrids with selected cytoplasmic constitutions. Sperm cells are naturally produced, haploid cells that may be a logical substitute for somatic protoplasts in specific instances or in germ plasm storage. Recently, Kranz and coworkers have combined the isolated egg and sperm cells of *Zea* using electrofusion to produce a microcallus (28). As problems coping with gametic cells are resolved, in vitro fusion of male and female gametes of higher plants ex ovulo may help us to model and thus better understand fertilization in angiosperms.

Literature Cited

1. Brewbaker, J. L., Kwack, B. H. 1963. The essential role of calcium ion in pollen germination and pollen tube growth. *Am. J. Bot.* 50:859–65
2. Carlson, W. R. 1969. Factors affecting preferential fertilization in maize. *Genetics* 62:543–54
3. Carlson, W. R. 1986. The B-chromosome of maize. *Crit. Rev. Plant Sci.* 3:201–26
4. Cass, D. D. 1973. An ultrastructural and Nomarski-interference study of the sperms of barley. *Can. J. Bot.* 51:601–5
5. Cass, D. D., Fabi, G. C. 1988. Structure and properties of sperm cells isolated from the pollen of *Zea mays*. *Can. J. Bot.* 66:819–25
6. Cass, D. D., Hough, T., Knox, B., McConchie, C., Singh, M. 1986. Isolation of sperms from pollen of corn and oilseed rape. *Plant Physiol.* 80(Suppl.): 130
7. Cerceau-Larrival, M.-T., Derouet, L. 1988. Relation possible entre les éléments inorganiques détectés par Spectrométrie X à Sélection d'Energie et l'allergénicité des pollens. *Ann. Sci. Nat. Bot. (Paris) Ser. 13* 9:133–52
8. Coleman, A. W., Goff, L. J. 1985. Applications of fluorochromes to pollen biology. 1. Mithramycin and 4-6-diamidino-2-phenylindole (DAPI) as vital stains and for quantitation of nuclear DNA. *Stain Technol.* 60:145–54
9. Cornish, E. C., Anderson, M. A., Clarke, A. E. 1988. Molecular aspects of fertilization in flowering plants. *Annu. Rev. Cell Biol.* 4:209–28
10. Corriveau, J. L., Coleman, A. W. 1988. Rapid screening method to detect potential biparental inheritance of plastid DNA and results for over 200 angiosperm species. *Am. J. Bot.* 75:1443–58
11. Davis, G. L. 1966. *Systematic Embryology of the Angiosperms.* New York: Wiley
12. Delvallee, I., Dumas, C. 1987. Anther development in *Zea mays*: changes in protein, peroxydase, and esterase patterns. *J. Plant Physiol.* 132:210–17
13. Detchepare, S., Heizmann, P., Dumas, C. 1989. Changes in protein patterns and

protein synthesis during anther development in *Brassica oleracea*. *J. Plant Physiol.* 135:129–37
14. Dumas, C., Knox, R. B., McConchie, C. A., Russell, S. D. 1984. Emerging physiological concepts in fertilization. *What's New in Plant Physiol.* 15:17–20
15. Dupuis, I., Roeckel, P., Matthys-Rochon, E., Dumas, C. 1987. Procedure to isolate viable sperm cells from corn (*Zea mays* L.) pollen grains. *Plant Physiol.* 85:876–78
16. Frova, C., Binelli, G., Ottaviano, E. 1987. Isozyme and HSP gene expression during male gametophyte development in maize. *Isozymes: Current Topics in Biological and Medical Research* ed. M. C. Rattazzi, J. G. Scandalios, G. S. Whitt, 15:97–120. New York: Alan R. Liss
17. Gaff, D. F., Okang'o-Ogola. 1971. The use of nonpermeating pigments for testing the survival of cells. *J. Exp. Bot.* 22:756–58
18. Geltz, N. R., Russell, S. D. 1988. Two-dimensional electrophoretic studies of the proteins and polypeptides in mature pollen grains and the male germ unit of *Plumbago zeylanica*. *Plant Physiol.* 88:764–69
19. Hagemann, R., Schröder, M.-B. 1989. The cytological basis of the plastid inheritance in angiosperms. *Protoplasma* 152:57–64
20. Haskell, D. W., Rogers, O. M. 1985. RNA synthesis by vegetative and sperm nuclei of trinucleate pollen. *Cytologia* 50:805–9
21. Heslop-Harrison, J., Heslop-Harrison, Y., Shivanna, K. R. 1984. The evaluation of pollen quality and a further appraisal of the fluorochromatic (FCR) test procedure. *Theor. Appl. Genet.* 67: 367–75
22. Hough, T., Bernhardt, P., Knox, R. B., Williams, E. G. 1985. Applications of fluorochromes to pollen biology. 2. The DNA probes ethidium bromide and Hoechst 33258 in conjunction with the callose-specific aniline blue fluorochrome. *Stain Technol.* 60:155–62
23. Hough, T., Singh, M. B., Smart, I. B., Knox, R. B. 1986. Immunofluorescent

screening of monoclonal antibodies to surface antigens of animal and plant cells bound to polycarbonate membranes. *J. Immunol. Methodol.* 92:103–8

24. Hu, S.-Y. 1990. Male germ unit and sperm heteromorphism: the current status. *Acta Bot. Sinica* 32:230–40

25. Jones, K. H., Senft, J. A. 1985. An improved method to determine cell viability by simultaneous staining with fluorescein diacetate and propidium iodide. *J. Histochem. Cytochem.* 33:77–79

26. Knox, R. B., Singh, M. B. 1987. New perspectives in pollen biology and fertilization. *Ann. Bot.* 60 (Suppl. 4):15–37

27. Knox, R. B., Southworth, D. A., Singh, M. B. 1988. Sperm cell determinants and control of fertilization in plants. In *Eukaryote Cell Recognition: Concepts and Model Systems,* ed. G. P. Chapman, C. C. Ainsworth, C. J. Chatham, pp. 175–93. Cambridge: Cambridge Univ. Press

28. Kranz, E., Bautor, J., Lörz, H. 1990. In vitro fertilization of single, isolated gametes, transmission of cytoplasmic organelles and cell reconstitution of maize (*Zea mays* L.). In *Progress in Plant Cellular and Molecular Biology,* ed. H. H. J. Nijamp, L. H. W. van der Plas, and J. van Aartrijk, pp. 252–57. Dordrecht: Kluwer

29. Lidforss, B. 1896. Zur Biologie des Pollens. *Jahrb. Wiss. Biol.* 29:1–38

30. Mascarenhas, J. P. 1975. The biochemistry of angiosperm pollen development. *Bot. Rev.* 41:259–314

31. Mascarenhas, J. P. 1989. The male gametophyte of flowering plants. *Plant Cell* 1:657–64

32. Mascarenhas, J. P. 1990. Gene activity during pollen development. *Annu. Rev. Plant Physiol. Plant Mol. Biol.* 41:317–38

33. Matthys-Rochon, E., Dumas, C. 1988. The male germ unit: retrospect and prospects. In *Plant Sperm Cells as Tools for Biotechnology,* ed. H. J. Wilms, C. J. Keijzer, pp. 51–60. Wageningen: PUDOC

34. Matthys-Rochon, E., Detchepare, S., Wagner, V. T., Roeckel, P., Dumas, C. 1988. Isolation and characterization of viable sperm cells from tricellular pollen grains. In *Sexual Reproduction in Higher Plants,* ed. M. Cresti, P. Gori, E. Pacini, pp. 245–50. Berlin: Springer-Verlag

35. Matthys-Rochon, E., Vergne, P., Detchepare, S., Dumas, C. 1987. Male

germ unit isolation from three tricellular pollen species: *Brassica oleracea, Zea mays,* and *Triticum aestivum. Plant Physiol.* 83:464–66

36. McConchie, C. A., Russell, S. D., Dumas, C., Knox, R. B. 1987. Quantitative cytology of the mature sperm cells of *Brassica campestris* and *B. oleracea. Planta* 170:446–52

37. McConchie, C. A., Hough, T., Knox, R. B. 1987. Ultrastructural analysis of the sperm cells of maize, *Zea mays. Protoplasma* 139:9–19

38. Mogensen, H. L. 1982. Double fertilization in barley and the cytological explanation for haploid embryo formation, embryoless caryopses, and ovule abortion. *Carlsberg Res. Commun.* 47:313–54

39. Mogensen, H. L. 1988. Exclusion of male mitochondria and plastids during syngamy as a basis for maternal inheritance. *Proc. Natl. Acad. Sci. USA* 85:2594–97

40. Mogensen, H. L., Rusche, M. L. 1985. Quantitative analysis of barley sperm: occurrence and mechanism of cytoplasm and organelle reduction and the question of sperm dimorphism. *Protoplasma* 128:1–13

41. Mogensen, H. L., Wagner, V. T. 1987. Associations among components of the male germ unit following in vivo pollination in barley. *Protoplasma* 138:161–72

42. Mogensen, H. L., Wagner, V. T., Dumas, C. 1990. Quantitative, three-dimensional ultrastructure of isolated sperm cells of *Zea mays* L. *Protoplasma* 153:136–40

43. Murgia, M., Wilms, H. L. 1988. Three dimensional image and mitochondrial distribution in sperm cells of *Euphorbia dulcis.* See Ref. 33, pp. 75–79

44. Nagy, J. I., Maliga, P. 1976. Callus induction and plant regeneration from mesophyll protoplasts of *Nicotiana sylvestris. Z. Pflanzenphysiol.* 78:453–55

45. Nasrallah, M. E., Nasrallah, J. B. 1988. Cell-specific expression of the S-gene in *Brassica* and *Nicotiana.* See Ref. 27, pp. 39–45

46. Nielsen, J. E., Olesen, P. 1988. Isolation of sperm cells from the trinucleate pollen of sugar beet (*Beta vulgaris*). See Ref. 33, pp. 111–22

47. Pennell, R. I., Geltz, N. R., Koren, E., Russell, S. D. 1987. Production and partial characterization of hybridoma antibodies elicited to the sperm of *Plumbago zeylanica. Bot. Gaz.* 148:401–06

48. Proovost, E., Southworth, D., Knox, R. B. 1988. Three dimensional reconstruc-

tion of sperm cells and vegetative nucleus in pollen of *Gerbera jamesonii.* See Ref. 33, pp. 69–73

49. Raghavan, V. 1987. Developmental strategies of the angiosperm pollen: a biochemical perspective. *Cell Differ.* 21:213–26

50. Reiss, H.-D., Herth, W., Schnepf, E. 1983. The tip-to-base calcium gradient in pollen tubes of *Lilium longiflorum* measured by proton-induced X-ray emission (PIXE). *Protoplasma* 115: 153–59

51. Reynolds, T. L., Raghavan, V. 1982. An autoradiographic study of RNA synthesis during maturation and germination of pollen grains of *Hyoscyamus niger. Protoplasma* 111:177–82

52. Roberts, I. N., Gaude, T. C., Harrod, G., Dickinson, H. G. 1983. Pollen-stigma interactions in *Brassica oleracea;* a new pollen germination medium and its use in elucidating the mechanism of self incompatibility. *Theor. Appl. Genet.* 65:231–38

53. Roeckel, P., Dupuis, I., Detchepare, S., Matthys-Rochon, E., Dumas, C. 1988. Isolation and viability of sperm cells from corn (*Zea mays*) and kale (*Brassica oleracea*) pollen grains. See Ref. 33, pp. 105–9

54. Roeckel, P., Matthys-Rochon, E., Dumas, C. 1990. Pollen and isolated sperm cell quality in *Zea mays.* In *Characterization of Male Transmission Units in Higher Plants,* ed. B. Barnabás, K. Liszt, pp. 41–48. Budapest: Agric. Res. Inst., Hungarian Acad. Sci.

55. Roman, H. 1948. Directed fertilization in maize. *Proc. Natl. Acad. Sci. USA* 34:36–42

56. Rusche, M. L. 1988. The male germ unit of *Zea mays* in the mature pollen grain. See Ref. 33, pp. 61–67

57. Rusche, M. L., Mogensen, H. L. 1988. The male germ unit of *Zea mays:* quantitative ultrastructure and three-dimensional analysis. See Ref. 34, pp. 221–26

58. Russell, S. D. 1983. Fertilization in *Plumbago zeylanica:* gametic fusion and the fate of the male cytoplasm. *Am. J. Bot.* 70:416–34

59. Russell, S. D. 1984. Ultrastructure of the sperm of *Plumbago szeylanica:* 2. Quantitative cytology and three-dimensional reconstruction. *Planta* 162: 385–91

60. Russell, S. D. 1985. Preferential fertilization in *Plumbago:* ultrastructural evidence for gamete-level recognition in an angiosperm. *Proc. Natl. Acad. Sci. USA* 82:6129–32

61. Russell, S. D. 1986. A method for the isolation of sperm cells in *Plumbago zeylanica. Plant Physiol.* 81:317–19

62. Russell, S. D. 1987. Quantitative cytology of the egg and central cell of *Plumbago zeylanica* and its impact on cytoplasmic inheritance patterns. *Theor. Appl. Genet.* 74:693–99

63. Russell, S. D. 1990. Energy-dispersive X-ray microanalysis of cellular components in the pollen of an angiosperm, *Plumbago zeylanica.* In *Proceedings of the XII International Congress for Electron Microscopy,* pp. 804–5. San Francisco Press: San Francisco

64. Russell, S. D., Cass, D. D. 1981. Ultrastructure of the sperm of *Plumbago zeylanica.* 1. Cytology and association with the vegetative nucleus. *Protoplasma* 107:85–107

65. Russell, S. D., Rougier, M., Dumas, C. 1990. Organization of the early postfertilization megagametophyte of *Populus deltoides.* 1. Ultrastructure and implications for male cytoplasmic transmission. *Protoplasma* 155:153–65

66. Shivanna, K. R., Xu, H., Taylor, P., Knox, R. B. 1988. Isolation of sperms from the pollen tubes of flowering plants during fertilization. *Plant Physiol.* 87:647–50

67. Southworth, D., Knox, R. B. 1988. Methods for the isolation of sperm cells from pollen. See Ref. 33, pp. 87–95

68. Southworth, D., Knox, R. B. 1989. Flowering plant sperm cells: isolation from pollen of *Gerbera jamesonii* (Asteraceae). *Plant Sci.* 60:273–77

69. Szakács, E., Barnabás, B. 1990. Sperm cell isolation from wheat (*Triticum aestivum* L.) pollen. See Rev. 54, pp. 37–40

70. Tanaka, I., Nakamura, S., Miki-Hirosige, H. 1989. Structural features of isolated generative cells and their protoplasts from pollen of some liliaceous plants. *Gamete Res.* 24:361–74

71. Theunis, C. H., van Went, J. L., Wilms, H. J. 1988. A technique to isolate sperm cells of mature spinach pollen. See Ref. 34, pp. 233–38

72. Theunis, C. H., van Went, J. L. 1989. Isolation of sperm cells from mature pollen grains of *Spinacia oleracea. Sexual Plant Reprod.* 2:97–102

73. Theunis, C. H., van Went, J. L. 1990. Isolated sperm cells of *Spinacia oleracea* L. See Ref. 54, pp. 25–29

73a. Ursin, V. M., Yamaguchi, J., McCormick, S. 1989. Gametophytic and sporophytic expression of anther-specific genes in developing tomato anthers. *Plant Cell* 1:727–36

74. van Aelst, A. C., Theunis, C. H., van

Went, J. L. 1990. Freeze-fracture studies on isolated sperm cells of *Spinacia oleracea*. *Protoplasma* 153:204–7

75. van der Maas, H. M., Zaal, M. A. C. M. 1990. Sperm cell isolation from pollen of perennial ryegrass (*Lolium perenne* L.). See Ref. 54, pp. 31–35

76. Vergne, P., Dumas, C. 1988. Isolation of viable wheat gametophytes of different stages of development and variations in their protein patterns. *Plant Physiol.* 88:969–72

77. Wagner, V. T., Mogensen, H. L. 1987. The male germ unit in the pollen and pollen tubes of *Petunia hybrida:* ultrastructural quantitative and three dimensional features. *Protoplasma* 143: 93–100

78. Wilms, H. J. 1986. Dimorphic sperm cells in the pollen grain of *Spinacia*. In *Biology of Reproduction and Cell Motility in Plants and Animals,* ed. M. Cresti, R. Dallai, pp. 193–98. Siena: Univ. Siena

79. Yamada, Y., Yang, Z.-Q., Tang, D.-T. 1986. Plant regeneration from protoplast-derived callus of rice. *Plant Cell Rep.* 5:85–88

80. Yang, H.-Y., Zhou, C. 1989. Isolation of viable sperms from pollen of *Brassica napus, Zea mays* and *Secale cereale*. *Chinese J. Bot.* 1:80–84

Annu. Rev. Plant Physiol. Plant Mol. Biol. 1991. 42:205–225

GENE TRANSFER TO PLANTS:
Assessment of Published Approaches and Results

I. Potrykus

Institute of Plant Sciences, Swiss Federal Institute of Technology (ETH), CH-8092 Zürich, Switzerland

KEY WORDS: Assessment of Gene Transfer Techniques

CONTENTS

1040-2519/91/0601-0205$02.00

INTRODUCTION

Since the first transgenic plants were regenerated (30) the area of gene transfer to plants has seen exciting progress. A review on this area could discuss many topics: which genes have been transferred to which plants, how much we have learned about plant genes and their regulation, how elegantly transgenic plants and reporter genes have been applied to study plant development, how efficiently transient expression systems have been used to study gene regulation, how much progress has been achieved in the application of gene technology to the improvement of crop plants, and so on. These and other topics deserve discussion, and many reviews are available (e.g. 15, 22, 25, 30, 42, 47–49, 51, 65, 90). The present chapter focuses on a critical assessment of gene transfer methods, a problem little discussed so far (68, 69). Numerous laboratories have transferred genes to a wide collection of plants, including not only those considered easy experimental models [such as tobacco *(Nicotiana tabacum)* and petunia *(Petunia hybrida)]* but also those considered recalcitrant [such as grain legumes—e.g. soybean *(Glycine max)*—or cereals—e.g. rice *(Oryza sativa)* and maize *(Zea mays)]*. Certainly, for someone interested in studying gene regulation in *Arabidopsis thaliana* gene transfer is no longer a technical problem. One can choose between several established methods. Either *Agrobacterium*-mediated gene transfer to root explants and seeds, or direct gene transfer to protoplasts can be expected to function, although one might encounter difficulties with some genotypes or with the seed approach. One could even use the biolistic approach, or microinjection. Someone interested in studying tissue specificity of promoter constructs or the functionality of homologous or heterologous genes in differentiated tissues could apply biolistics to plant tissues with a fair chance of success. In many cases, however, where the experimenter is not free to choose the experimental organism, and where reproducible production of transgenic plants of a given genotype of a given species is required, as is the case with most applied projects or with projects where gene technology should be applied to complex and genetically well-defined marker strains, gene transfer can still be a serious experimental problem. The quest is still open for a method that will allow routine and efficient gene transfer into all desired genotypes of any plant species.

For some time there was good reason to believe that *Agrobacterium tumefaciens* was the vector system with the capacity for gene transfer to any plant species and variety. As this is not the case (I will propose a hypothesis to explain why), numerous alternative approaches have been tested; some have been successful, including "direct gene transfer to protoplasts" (65), "biolistics" (77), and "microinjection into proembryos" (61). Numerous others have yielded interesting but probably artifactual data. Because *Agrobacterium* and the other "successful" methods have inherent limitations it may be unrealistic to hope that a general method will be possible at all; a wide variety of methods might be the solution for future gene transfer problems. Therefore it seems worthwhile to discuss even approaches that have not been successful so far. The present assessment of gene transfer technology focuses on problems. The achievements and perspectives have been highlighted (often too strongly) in the original publications and previous reviews. I focus on cereals as an example of a group of plants difficult to transform. Discussions of cereal problems probably also apply to other plant species difficult to transform.

Because gene transfer problems started with *Agrobacterium*, I start by discussing why *Agrobacterium* may not function with cereals (and other plant species). I focus on integrative transformation leading to stable transgenic plants. It will not be possible to include a careful discussion of the exciting achievements of transient expression systems (25, 32).

My assessment here is based on a rigid definition of "proof of integrative transformation." This is mandatory because many data in the literature have been misinterpreted on the basis of *indicative* evidence. Those who disagree with the view that indicative evidence can be misleading will not agree with my assessment. This assessment is also based on an interpretation of biological parameters affecting gene transfer, and I make several statements for which no solid experimental data are available. Acceptance or refusal of these statements will not alter the assessment of the available data but may influence interpretation of the future potential of the various approaches. The literature and public presentations in the area of gene transfer methods are confusing; for someone not personally involved, they are probably difficult to understand. I hope my assessment will help to clarify the situation. The assessment rests on two components: It accepts integrative transformation only if definite proof is available, and it tries to understand the biology behind the various gene transfer approaches. It is to be understood as a working hypothesis that can (so far) consistently explain the success or failure of the various approaches. It has also been efficient in predicting failure of novel approaches over many years. Though in agreement with all available data on gene transfer, it is a provocative hypothesis to be falsified or verified by future experimentation.

BASIC CONSIDERATIONS

Proof of Integrative Transformation

Based on indicative evidence, there are numerous methods to produce transgenic plants, including extremely simple ones. Because no transgenic plants could be recovered after use of most of these methods, however, indicative evidence must be untrustworthy. Indeed many researchers have obviously been misled by artifacts, and it is good advice to demand suitable proof that transgenic plants have been produced. Neither genetic, phenotypic, nor physical data alone are acceptable. Proof of integrative transformation requires: 1. Controls for treatment and analysis; 2. a tight correlation between treatment and predicted results; 3. a tight correlation between physical (e.g. Southern blot) and phenotypic (e.g. enzyme assay) data; 4. *complete* Southern analysis containing (*a*) the predicted signals in high-molecular-weight DNA, including hybrid fragments between host DNA and foreign gene, and the presence of the complete gene, and (*b*) evidence for the absence of contaminating DNA fragments or identification of such fragments; 5. data that allow discrimination between false positives and correct transformants in the evaluation of the phenotypic evidence; 6. correlation of the physical and phenotypic evidence with transmission to sexual offspring; and 7. molecular and genetic analysis of offspring populations.

Biology of Gene Transfer Protocols

Certain biological parameters affect the delivery of foreign genes to cells and the fate of these genes in the cells. Consideration of these parameters may help us both to understand the problems of some approaches and to design better experiments in future.

1. Not all plant cells are totipotent.
2. Plants differ in their capacity to respond to triggers: They are *competent* for specific triggers.
3. Transgenic plants can be regenerated only from cells competent both for regeneration (in a broad sense) and integrative transformation (simultaneously or sequentially).
4. Plant tissues are mixed populations of cells with competence for many different responses. Considering the states of competence essential for recovery of transgenic plants, the following situation must be considered: (*a*) A very small (and varying) minority of cells in plant tissues will be *competent for both transformation and regeneration*. (*b*) Other cells will be competent for transformation *or* regeneration. (*c*) A larger fraction of cells will be *potentially competent,* which means that given the correct treatment they will have the potential to shift to the competent state. (*d*) A variable portion of the cells will not even be potentially competent but will be *noncompetent*.
5. The relative composition of cell populations in tissues is determined by the species, the genotype, the type of organ, the developmental state of the organ and tissue areas within the organ, and even by the individual history of the experimental plant.

6. The most effective trigger for shifting cells potentially competent for regeneration into the competent state is mechanical (and enzymatic?) wounding. The "wound response" (45) is probably the biological basis for proliferation and regeneration from somatic cells.

7. Plant species differ in their wound response, as do different cells of the same plant. Graminaceous plant species, especially the cereals including maize (and probably grain legumes), have only a rudimentary wound response.

8. For some genotypes it is possible to cause cells competent for regeneration to proliferate under experimental conditions that maintain this state ("embryogenic suspensions"; 86). Such cell cultures contain cells competent for regeneration and (after protoplasting?) competent for integrative transformation.

9. Plant cell walls are efficient barriers and traps for DNA molecules of the size of a functional gene. A recent publication (16) seems to contradict this statement. However, I present an alternative hypothesis for the interpretation of the data in the section below on free passage of functional genes across cell walls.

10. So far, genes can be transported into walled plant cells only with the help of *Agrobacterium,* viruses, microinjection, and biolistics.

11. Production of transgenic plants requires efficient gene transfer into cells competent (simultaneous or consecutive) for integrative transformation and regeneration.

12. Competence for integrative transformation has obviously little relation to competence for transient expression.

13. Nonviral DNA can integrate into the host genome; its presence in the cells does not guarantee integration.

14. Nonviral DNA does not travel from cell to cell; it is restricted to the cell to which it has been delivered.

15. Viral DNA does not integrate into the host genome even if present at very high copy numbers.

16. Viral DNA (and RNA) moves from cell to cell and can spread systemically throughout the plant; it is, however, excluded from the meristems and the "germ line."

Of course, statements on biological phenomena never fit 100%.

The above statements include the possibility of rare exceptions. If the reader can agree with the definition of proof and the statements concerning the biological parameters, the assessment of the literature on gene transfer to plants is relatively easy and straightforward, even in the evaluation of the future potential of the various approaches. The present assessment is thus in agreement with the published data and with data presented at public meetings.

It incorporates discussions of an EMBO workshop on gene transfer to plants organized by me in the fall of 1988 (70).[1]

ASSESSMENT OF GENE TRANSFER APPROACHES TO PLANTS

Agrobacterium-Mediated Gene Transfer

Agrobacterium tumefaciens and *A. rhizogenes* provide excellent vector systems for the production of transgenic plants. We witness an ever-increasing list of novel transgenic species. Numerous excellent reviews (e.g. 22, 42, 48) and laboratory manuals (e.g. 20, 31) have been published. Molecular analysis of the events leading finally to the transfer of the T-DNA to host plant cells is close to completion, and elegant protocols have been worked out for using this biological vector in basic and applied research. The transfer process requires a coordinated interaction between the bacterium and the plant host. Unfortunately, though we know much about the contribution of the bacterial partner, we know little about the interaction from the plant cell—the latter being far more difficult to study. As long as it is possible to work with plant systems that cooperate, this is not a serious problem; but numerous plants (especially economically important ones) do not cooperate. In these cases more knowledge about the biological basis of the plant's participation would be welcome. Decades ago A. Braun (8) collected valuable information on the importance of a state of competence for tumor transformation in the plant cell. Along the same lines A. Binns (6, 7) is now reviewing our knowledge of the biology of "host range limitations." Lack of competence in their cells is probably the reason some plants can be transformed (*a*) only with great difficulty and after screening of numerous plant genotypes and *Agrobacterium* strains (e.g. 41), (*b*) during a short developmental window of specific organs, or (*c*) not at all.

The common denominator may be found in the phenomenon of "wound response" (45). Plants and tissues differ in their wound response. Only plants with a pronounced wound response develop larger populations of wound-adjacent competent cells for regeneration and transformation. Plants that have been recalcitrant to transformation with *Agrobacterium* probably do not express the appropriate wound response. This is probably the reason for the complete failure to transform cereal plants with *Agrobacterium,* despite the enormous effort so far invested in this approach. Most experiments have not been published because they failed; some promising data have been presented

[1]As to the participants of this workshop, I am grateful to my coworkers, Drs. S. K. Datta, G. Neuhaus, J. Paszkowski, M. W. Saul, C. Sautter, and G. Spangenberg, for discussions and for information they collected at meetings and contributed from the literature.

at international meetings, but because no proof has been presented to date, the data are considered artifactual.

Transformation of "monocots" (e.g. 10, 43, 78) is of no importance in this context: It is not because they are monocots that cereals are difficult to transform but because they do not have the proper wound response. Monocots with wound response (e.g. *Asparagus*) are as easy to transform as dicots with wound response; and dicots without proper wound response (e.g. grain legumes?) are probably as difficult to transform as cereals. The report on transformation of maize seedlings (34) does not present proof and did not lead to transgenic offspring; the data are therefore considered artifacts.

Why is it probably impossible to transform cereal plants with *Agrobacterium?* Wounding of differentiated cereal tissues does not lead to wound response–induced dedifferentiation of wound-adjacent cells and accumulation of competent cells. Instead wound-adjacent cells accumulate phenols and die. Although *Agrobacterium* obviously transfers T-DNA into wound-adjacent cells efficiently in cereals (see the section below on agroinfection), even integration of this T-DNA cannot lead to transgenic clones and plants. It is not as easy to understand why even experiments with meristems (e.g. leaf base and split shoot tip), which can form proliferating cultures in vitro, did not yield transgenic clones and plants. A possible explanation may lie in the fact that cereal cell cultures are not the consequence of proliferating wound meristems but rather are based on adventitious or axillary meristems (46). Wounding plus in vitro culture does not lead to many competent cells but to a few meristem initials that proliferate as meristems. And meristematic cells may not be competent for transformation (see the section below on microinjection).

Of the many attempts to transform cereal cell cultures with *Agrobacterium*, (which should be possible), only one has been reported successful (73). Unfortunately, insufficient data are presented to constitute proof as defined above. From three independent lines of experimentation three different types of indicative evidence are presented, each of which could be considered an artifact. If co-incubation of rice suspension cultures with *Agrobacterium* could indeed lead to transgenic clones it should be relatively easy to produce conclusive data. The key problem in *Agrobacterium*-mediated transformation of cereals (and other recalcitrant plants?) probably lies neither with *Agrobacterium* (it transfers its T-DNA to cereals) nor with the host range (cereals are probably included) but rather with the availability and accessibility of cells competent for integrative transformation and regeneration.

Agroinfection

Viral DNA integrated into the T-DNA of the Ti-plasmid can be delivered into plant cells with the normal *Agrobacterium* T-DNA transfer process (29, 35).

The consequences in plants *with* wound response are the following: The viral genome enters the plant cell as part of the T-DNA. It is released to form a functional virus that replicates and spreads systemically. It may not be necessary for the T-DNA to integrate in order to release the virus. However, T-DNA *can* integrate; thus agroinfection can lead to integration of viral DNA in competent wound-adjacent cells and consequently to transgenic plants containing integrated viral DNA. This is similar to normal T-DNA transformation.

Agroinfection with maize streak virus in maize has led to systemic spread of the virus (37). Showing for the first time that T-DNA transfer to cereals is possible, this finding caused considerable excitement; it was taken as evidence that the well-established *Agrobacterium* vector system could be used for genetic engineering of cereals. Later it was even demonstrated that the efficiency of such T-DNA transfer is comparable to that in dicot species (36).

Does agroinfection, then, have potential for the production of trangenic cereal? Of course, the virus is released into wound-adjacent cells in cereals, whence it can spread systemically. Unfortunately, the spreading virus does not integrate, even if it reaches one of the rare competent cells in the plant. At the wound site, however, the T-DNA and the virus face the problem discussed above: Even if integration were to occur it would have no consequences because the wound-adjacent cells die. Therefore agroinfection has no better chance of yielding transgenic cereals than does *Agrobacterium* infection alone. Nevertheless, it does have considerable potential for studies in virus biology, because it can transfer deletion mutants and even single viral genes. It also has the merit of having shown, by its amplifying effect, that *Agrobacterium* interacts with cereals.

Viral Vectors

In 1984 an experiment was published (9) that we hoped would initiate important contributions of viral vectors to the genetic engineering of plants. A small bacterial antibiotic resistance gene integrated into a deletion mutant of a DNA virus spread systemically throughout infected plants and made them resistant to the antibiotic; but so far hopes that multicopy amplification and systemic spread of engineered viruses could be exploited to produce large quantities of genes or gene products have not been fulfilled. Viral genomes are obviously so compact that they do not easily tolerate foreign genes (27).

The discovery that RNA viruses can be reverse-transcribed to yield cDNA clones that again are infective opened up the possibility of applying genetic engineering technology to the far larger group of RNA viruses (1, 2); but because viral DNA does not integrate into the host genome, and is excluded from the meristems and offspring, it is difficult to envisage how viral vectors could contribute to the production of transgenic plants. The difficulty might be overcome if a transposable element were part of a systemic virus, and if the transposable element carried a gene of interest, and if the transposable element could be induced to excise and integrate.

Protoplasts and Direct Gene Transfer

Protoplasts (isolated plant cells without cell wall) are ideal for gene transfer.

1. The freely accessible plasmalemma guarantees that genes can reach and enter each and every protoplast at DNA concentrations that can be regulated experimentally.
2. The enzymatic (or mechanical) isolation procedures, however, induce wound response; this shifts potentially competent cells into the competent state, thus increasing the proportion of cells competent for regeneration and (?) integrative transformation.
3. The foreign genes reach every competent cell, thus increasing the chance for recovery of transgenic plants from a given population.
4. Gene transfer does not require any biological vector; DNA uptake is a physical process, thus circumventing any possible hostrange problem.

We developed "direct gene transfer" as an alternative to use of *Agrobacterium* because of the foreseeable difficulties with cereals (67). DNA uptake can be promoted by various treatments including polyethylene glycol and/or electroporation (26, 60, 80). Integrative transformation can be very efficient (60) and leads to stable inheritance of predominantly single-gene loci of the foreign gene (71). Cotransformation efficiently transfers nonselectable genes (79). Homologous recombination enables gene targeting (66). A barrier to gene transfer has so far not been detected: Virtually every protoplast system has proven transformable, though with different efficiencies.

So, is there no problem with the recovery of transgenic plants from protoplasts? Unfortunately, there are severe problems, all related to plant regeneration from protoplasts. Although exciting progress in this respect has been made (75) and further progress can be foreseen, plant regeneration from protoplasts will probably always be a delicate process (with exceptions); it will probably also depend upon parameters not under experimental control [e.g. species and genotype-dependent competence for wound response and regeneration (72)]. Transgenic plants have been recovered recently from protoplasts from important crops that could not be transformed with *Agrobacterium:* Several laboratories reported on transgenic Japonica-type rice (81, 84, 92, 93), one on Indica-type rice (14); following production of sterile transgenic maize (74), a very recent report described recovery of numerous fertile transgenic maize plants (19). Because plant regeneration from protoplasts in wheat (87) and plantlet regeneration in barley have also been reported (91), and because other graminaceous species were regenerated earlier, further transgenic graminaceous monocots will probably follow. However, these successes are far from promising routine application of gene technology to grasses. All these experimental systems depend on the establishment of an embryogenic cell culture system (86) that cannot be established easily for every desired commercial variety.

Biolistics or Particle Gun

Acceleration of heavy microparticles coated with DNA has been developed into a technique that carries genes into virtually every type of cell and tissue (50, 76). No gene transfer approach since the early *Agrobacterium*-mediated gene transfer experiments has met with so much enthusiasm, and in no other gene transfer approach has there been a comparable investment in experimentation and manpower. Some expected that this technique would solve all gene transfer problems. Indeed the biolistic approach has advantages and potential for general applicability:

1. It is easy to handle.
2. One shot can lead to multiple hits (transfer of genes into many cells).
3. Cells survive the intrusion of one (?) particle.
4. The genes coated on the particle have biological activity.
5. Target cells can be as different as pollen, cell culture cells, cells in differentiated tissues and meristems.
6. They can be located at the surface or in deeper layers of organs.
7. The method depends on physical parameters only, and so on.

Thus the method allows the transport of genes into many cells at nearly any desired position in a plant without too much manual effort. The enormous investment into this technique has paid off, and transgenic plants have been recovered that would have been difficult to produce by other methods. (One wonders whether with a similar investment other methods might not have been made successful, too.)

The first transgenic soybean plants were reported simultaneously via *Agrobacterium* vectoring (41) and biolistics (58); however, biolistics has by now become far more successful than use of *Agrobacterium* in this crop (12). The real breakthrough for biolistics came with the recovery of fertile transgenic maize in three independent laboratories (23, 33; S. Jayne, J. Suttie, M. Koziel, G. Pace et al, personal communication), and other laboratories may be close to a similar success. This success with maize will guarantee further investment in this technique. However, an interesting question remains: Why, given the advantages listed above, is this technique so inefficient in yielding stable integrative events, especially in experimental systems as ideal as those used (i.e. embryogenic suspensions)? If one compares (*a*) the number of fertile transgenic maize plants recovered from biolistic treatment in large-scale experiments using embryogenic suspensions with (*b*) those recovered from a comparable small-scale experiment on direct gene transfer to protoplasts isolated from such embryogenic suspensions, then the low yield from the biolistic experiments is surprising. In my opinion, plants difficult to transform using *Agrobacterium* probably have very few competent cells; the particle has to reach these rare cells by a random hit, and the DNA has to integrate into the genome of these cells. Considering the low conversion rate of transient events (hits) to stable integrative events in biolistic systems, integrative transformation in recalcitrant transformation systems must be expected to be rare. As I see it, the real advantage of the biolistic technique to date lies in its application in *transient* gene expression studies in differentiated tissues (32). In this area the technique has little competition.

Microinjection into Zygotic and Microspore-Derived Proembryos

Microinjection uses microcapillaries and microscopic devices to deliver DNA into defined cells in such a way that the injected cell survives and can proliferate (61). This technique has produced transgenic clones from protoplasts (59) and transgenic chimeras from microspore-derived proembryos in oilseed rape (62). Like biolistics, microinjection definitely delivers DNA into walled plant cells. In comparison with biolistics, microinjection has disadvantages: Only one cell receives DNA per injection, and handling requires more skill and instrumentation. It also has advantages:

1. The quantity of DNA delivered can be optimized.
2. The experimenter can decide into which cell to deliver DNA.
3. Delivery is precise and predictable, even into the cell nucleus, and is under visual control.
4. Cells of small structures (e.g. microspores and few-celled proembryos, which are not available in the large quantities required for the biolistic technique, can be precisely targeted.
5. Defined microinjected cells can be microcultured.
6. In combination with protocols for the culture of zygotic proembryos, microinjection should offer an approach to transformation open for every species and variety with sexual propagation.

On the assumption that few-celled zygotic proembryos contain competent cells, our group has established plant regeneration from isolated zygotic proembryos of maize *(Zea mays)*, wheat *(Triticum aestivum)*, rice *(Oryza sativa)*, barley *(Hordeum vulgare)*, soybean *(Glycine max)*, cotton *(Gossy-*

pium hybrid), sunflower *(Helianthus annuus)*, tobacco *(Nicotiana tabacum)*, and *Arabidopsis thaliana*. Following multiple microinjections with proven marker genes, putative primary transgenic chimeras and sexual offspring have been analyzed for the presence of the foreign genes (G. Neuhaus, G. Spangenberg, S. K. Datta, personal communication). So far we have only indicative evidence for transgenic chimeras. As we have no proof yet for transgenic offspring, these may be artifacts. Gene transfer to structures consisting of more than one cell can at best produce transgenic chimeras. Therefore two interpretations of the data are possible to date: (*a*) Larger experiments will increase the chance for transmission of the transgene to the offspring (as exemplified with the biolistic approach), or (*b*) meristematic cells have little competence for integrative transformation. To test the second hypothesis an experiment has been performed using the well-established gene transfer system of *Agrobacterium* and zygotic proembryos of *N. tabacum* var SR1, a well-documented model for *Agrobacterium*-mediated transformation. No transgenic tissue could be detected in the primary regenerants and in the sexual offspring (G. Neuhaus, A. Matzke, M. Matzke, personal communication). The microinjection data from proembryos indicated above, the data from the experiment just described, the difficulties in transforming embryogenic cultures and meristems with the biolistic approach [transgenic soybeans (58) are not from meristematic cells but from differentiated cells below the meristem developing to adventitious shoot buds], the negative experience using *Agrobacterium* with meristematic cells in general, and the well known phenomenon of virus exclusion from meristems—all this may point to a biological problem well worth studying: Do meristematic (embryonic) cells have a mechanism that prevents integration of incoming nucleic acids?

Four of the approaches discussed so far have provided proof for the production of transgenic plants (*Agrobacterium* vectoring, direct gene transfer to protoplasts, biolistics, and microinjection); two have not (agroinfection and viral vectors). One key difference between the former group of approaches and the following one is the concern about the cell wall. *Agrobacterium* uses a still-unexplained biological mechanism to overcome the cell wall; viral vectors use amplification of rare events; protoplasts have stripped off their cell wall; and biolistics and microinjection apply brute force. All of the following (and *so far unsuccessful*) approaches hope (with the exception of the microlaser technique) for a free movement of genes across the plant cell wall.

Free Passage of Functional Genes across Cell Walls?

The experimental design of the gene transfer approaches discussed below requires free movement of large DNA molecules across one, several, or many cell walls to achieve transformation of target cells. The feasibility of this movement will determine the potential of most of these following approaches, of which none has yet yielded a single proven transgenic plant. In this context a recent publication (16) on transient expression of marker genes in rice (*Oryza sativa*) tissues following electroporation of tissue slices from seedlings is of crucial importance. Using GUS and NPTII genes under constitutive and tissue-specific expression signals and electroporation with cross sections of seedlings, the authors describe data on transient expression of foreign genes *within* the rice tissue that convince even the critical reader. If the interpretation of the data in this publication is correct, then functional genes actively travel, without problems, across many cell walls. It is especially striking how evenly GUS gene expression was distributed over the entire cross section of leaf whorls and how well organ-specific expression fits the expectation. The assumption is that the genes reached every cell in the tissue. However, owing to (*a*) hundreds of experiments that had failed to transport genes across cell walls (including electroporation with other experimental systems), (*b*) the theoretical problem of envisaging how entire genes might cross the physical barrier of the cell walls, (*c*) the fact that cellulose fibrils are efficient adsorbers of DNA, (*d*) the knowledge that electroporation does not move DNA but just opens pores in membranes, and (*e*) the fact that there is not even a gradient in the experimental data, the interpretation of the data is difficult to understand; the data may fit an alternative hypothesis better.

I propose the following interpretation, which to my understanding fits all the data presented and agrees with the previous experience with DNA and cell walls. In the experimental system used, DNA was transferred *not into walled cells* but into *protoplasts within opened cells* at the cut surface of the tissue. (Cutting of plasmolyzed tissues with elongated cells was a standard technique in the late 19th and early 20th century to collect protoplasts.) If this were the case, the GUS stain would diffuse to neighboring cells and would be transported over longer distances along vascular bundles. Histological examination at different distances from the treated surface would mimic constitutive or organ-specific expression. The response of the system to the optimization of the electroporation protocol would be (and was) identical to the response of protoplast populations. At present, without further experimentation, it cannot be decided which interpretation of the data is correct. At present I continue in the view that passage of functional genes across plant cell walls does not normally occur.

Incubation in DNA of Dry Seeds or Embryos

Whereas in the previous experimental protocol one could envisage even recovery of transgenic cell clones (from protoplasts within opened cells), this is difficult to conceive in the following approach (82, 83) unless DNA moves freely into and between cells. The authors describe experiments in which every precaution was taken to avoid the experimental pitfalls of earlier reports (52, 53). Incubation of dry seeds and embryos (from cereals and grain legumes) in viral or nonviral DNA yielded interesting evidence for gene expression and recombination. Although the experiments included convincing controls and clearly demonstrated the presence and expression of defined marker genes as well as the replication of engineered viral DNA, they do not provide proof of integrative transformation. The authors' conclusion that the data demonstrate uptake of foreign DNA into the cells of the embryo, and that this approach therefore has potential for the recovery of transgenic cereals and grain legumes, is probably too optimistic. On the assumption that the cell wall is no problem, these experiments should (but do not) yield transgenic plants. On the assumption that the cell wall is a problem, it is difficult to envisage how the DNA might reach those meristematic cells that will form the new plant. Alternatively, the data could be explained without the necessity of massive DNA transport across cell walls: The dry embryos were split off from the endosperm, creating a giant wound across the scutellum. Because the tissues are dry, the cell contents do not leak out. Incubation in DNA solutions could create a micro-environment in the open cells that enables in vitro transcription, translation, and replication. Further experiments will have to show which of the possible hypotheses is correct. I do not see, however, much potential here for the production of transgenic cereals.

Incubation in DNA of Turgescent Tissue or Cells

Over a period of more than 20 years, seedlings, organs, tissue explants, cell cultures, and cells have been brought in direct contact with DNA and defined marker genes with the hope of transformation. Included have been experimental designs involving open plasmodesmata or loosening of cell wall structures, as well as treatments assuring that potentially competent cells were present at high frequencies. However, even in experiments that would have detected extremely rare events, not a single case of integrative transformation has been proven. Experiments relying on the passage of functional genes across plant cell walls obviously have little chance for success. Plant cell walls are not only efficient barriers, but they are also efficient traps for DNA molecules. It would be very surprising if DNA can cross cell walls efficiently (16).

Pollen Transformation

This approach, under experimental challenge since the early 1970s, is based on the hope that DNA can be taken up into germinating pollen and can either integrate into the sperm nuclei or reach the zygote with the pollen tube (39). Indeed, this would be the ideal method for gene transfer into plants. Although surprising phenotypes have been recovered that could be interpreted as indicative evidence for gene transfer (18, 39, 63), in no case so far has proof been provided. As numerous large-scale experiments in experienced laboratories and with defined marker genes have only given negative results, this approach does not seem a promising one. Not only the cell wall but also external and internal nucleases and the heterochromatic state of the acceptor DNA present problems for this technique. The latter problems may be overcome with the approach of "in vitro maturation" (5), where immature microspores are treated with DNA, matured to pollen, and used for pollination.

Pollen Tube Pathway

If it were possible to deliver DNA to the zygote via open pollen tubes, this approach would be very attractive. Unfortunately, a recent publication (56) providing phenotypic and molecular data on transgenic rice plants does not present proof. The Southern data show neither integration into high-molecular-weight DNA nor defined hybrid fragments, and can easily be understood as artifacts; the dot blot technique used is prone to artifacts, and the enzyme data are not reliable because cereals have a rich record in false positives with the assay used. Data from a recent poster presentation (38) do not clarify the situation but require complicated additional assumptions for explanation. It is also difficult to understand how the DNA applied to the cut pistil might reach the zygote: The pollen tubes are not open pipes but are sealed off with callose plugs; the DNA will be trapped in cell wall material; there are probably nucleases in the pollen tube and also in the synergids. However, the approach is attractive enough to be worthy of rigorous testing. Considering the possibility of transformation of contaminating endophytes might help to exclude this possible source of artifacts in future experiments. So far, transgenic plants have not been recovered from this approach.

Macroinjection

Use of injection needles with diameters greater than cell diameters destroys the cells into which DNA is delivered. DNA integration into cells would, therefore, require the DNA to move into wound-adjacent cells. This is not possible via plasmodesmata because of the size of the DNA molecules and because plasmodesmata are sealed off immediately upon wounding; nor is it possible across the cell walls. In the experiment that produced not only phenotypic changes in the offspring but also exciting molecular data (17), the DNA would have had to travel through many cell layers. A marker gene was injected into the stem below the immature floral meristem of rye (Secale cereale) to reach the sporogenic tissue. Hybridization with the marker gene and enzyme assays with sexual offspring having survived selection yielded strong indicative evidence, but no proof of transformation. Unfortunately, it has so far not been possible either to reproduce these data in several large-scale experiments with other cereals or to establish proof with the original material. This approach probably has little chance of success.

Electroporation

Discharge of a capacitor across cell populations leads to transient openings in the plasmalemma. This electroporation facilitates entry of DNA molecules into cells if the DNA is in direct contact with the membrane. For protoplast systems, electroporation is one of several standard techniques for routine and efficient transformation (24–26, 80). Since in numerous important plant species regeneration is possible from cell cultures and tissue explants but not (yet) from protoplasts, it has been important to test whether electroporation could transfer genes into walled cells (54). A great variety of experimental systems have been challenged, including germinating pollen tubes, suspension cultures, and tissue explants. There were interesting phenotypic changes, but no proof of transformation was obtained. The most interesting case (16), transient expression in rice, has been discussed above.

Electrophoresis

In contrast to electroporation, electrophoresis can be expected to *transport* DNA. A series of experiments has been performed using shoot meristems of barley seeds to test whether electrophoresis of DNA across tissues could transport genes into cells (4). The data obtained included radioactively labeled cell walls (after use of radiolabeled DNA), positive GUS assays, and a protein band on SDS-PAGE with GUS mobility. There was no proof for integrative transformation and the data can be interpreted as artifacts. It might be worthwhile to test this idea with a simpler experimental system that can provide clearcut answers. One assumes, however, that even electrophoresis cannot overcome the cell wall barrier.

Liposome Fusion

Fusion of DNA-containing liposomes is an established technique for the production of transgenic plants (11). It has no obvious advantage over simpler methods of direct gene transfer and is not much in use. DNA-containing liposomes have also been applied to various tissues, cell cultures (28), and pollen tubes (3) with the rationale that liposomes might help in transporting the DNA via plasmodesmata or directly across cell walls. It has been shown that liposomes can carry small dye molecules into cells within tissues via fusion with the plasmalemma, but there is no proof of transport and integration of marker genes. As plasmodesmata are sealed off immediately upon wounding, this route is not open to even very small liposomes; impregnation of cell walls with phospholipids does not seem to change their barrier function.

Liposome Injection

Microinjection into differentiated cells can easily deposit the DNA into the vacuole, where it is degraded. Microinjection of liposomes into the vacuole, however, can lead to fusion with the tonoplast, thus releasing the content of the liposome into the cytoplasm, as demonstrated with cytoplasm-activated fluorescent dyes (57). It was an elegant idea to exploit this situation for transformation of vacuolated cells (55). Unfortunately, activity of DNA delivered by this method has yet to be shown. Though elegant, this method probably has no advantage over straightforward microinjection, especially for the transformation of recalcitrant plants, because those probably have to be regenerated from meristematic cells which do not contain a large central vacuole.

Microlaser

A microlaser beam focused into the light path of a microscope can be used to burn holes into cell walls and membranes (89). It was hoped that incubation of perforated cells in DNA solutions could serve as a basis for vector-independent gene transfer into walled cells (88). There are no conclusive data available on DNA uptake, and there are problems with DNA adsorption to cell wall material even before it could be taken up. As microinjection and biolistics definitely transport DNA into walled plant cells, the microlaser could offer advantages only in very specific cases where those techniques would not be applicable.

SUMMARY

This review is selective; it does not attempt an encyclopedic overview of the many publications and presentations in the area. However, I hope it is complete in discussing all the various approaches to integrative transformation. I here interpret the available data on the basis of a rigorous definition of proof and on the assumption that DNA does not freely cross cell walls.

Of the numerous approaches to integrative transformation *Agrobacterium*-mediated gene transfer and direct gene transfer to protoplasts are routine and efficient methods; further optimization may lead to better efficiency in integrative transformation for the biolistic process; microinjection probably has more potential than has been realized so far. Compared to the enormous investment in biolistics, microinjection has had only marginal support; experiments have focused exclusively on meristematic cells, which may not even be competent for integrative transformation.

Future developments in gene transfer will result from protocols effective with meristematic cells. Transformation of meristematic cells is probably the bottleneck through which production of transgenic plants from recalcitrant species and varieties (those lacking a proper wound response) must proceed. There is a series of important questions to be studied, if we are interested in establishing a solid ground for further progress: What constitutes the type of wound response that provides the basis for the formation of wound meristems? What makes a cell competent for dedifferentiation, proliferation, and regeneration? What makes a cell competent for integrative transformation? Why are dedifferentiating cells so much easier to transform than proliferating cells? Do meristematic or embryonic cells indeed have a mechanism that prevents integrative transformation? Are viruses "excluded" from meristems because they cannot enter or because they cannot replicate? Development of more efficient protocols for routine transformation of recalcitrant species and varieties is less a technical than a biological problem. Efficient methods to transfer genes into plant cell already exist. *Agrobacterium* use is probably as efficient as biolistics, microinjection, and direct gene transfer. (I cannot share the optimism of those who hope for a free passage of DNA across plant cell walls.) First of all, however, we must learn *into which cells* to deliver the genes.

Literature Cited

1. Ahlquist, P., Pacha, R. F. 1990. Gene amplification and expression by RNA viruses and potential for further application to plant gene transfer. *Physiol. Plant.* 79:163–67
2. Ahlquist, P., French, R., Bujarski, J. J. 1987. Molecular studies of Brome mosaic virus using infectious transcripts from clones cDNA. *Adv. Virus Res.* 32:214–42
3. Ahokas, H. 1987. Transfection by DNA-associated liposomes evidenced at pea pollination. *Hereditas* 106:129–38
4. Ahokas, H. 1989. Transfection of germinating barley seed electrophoretically with exogenous DNA. *Theor. Appl. Genet.* 77:469–72
5. Alwen, A., Eller, N., Kastler, M., Benito-Moreno, R. M., Heberle-Bors, E. 1990. Potential of in vitro pollen maturation for gene transfer. *Physiol. Plant.* 79:194–96
6. Binns, A. N. 1990. *Agrobacterium*-mediated gene delivery and the biology of host range limitations. *Physiol. Plant.* 79:135–39
7. Binns, A. N., Thomashow, M. F. 1988. Cell biology of *Agrobacterium* infection and transformation of plants. *Annu. Rev. Microbiol.* 42:575–606
8. Braun, A. C. 1952. Conditioning of the host cell as a factor in the transformation process in crown gall. *Growth* 16:65–74

9. Brisson, N., Paszkowski, J., Penswick, J. R., Gronenborn, B., Potrykus, I., Hohn, T. 1984. Expression of a bacterial gene in plants by using a viral vector. *Nature* 310:511–14
10. Bytebier, B., Deboeck, F., DeGreve, H., VanMontagu, M., Hernalsteens, J. P. 1987. T-DNA organization in tumor cultures and transgenic plants of the monocotyledon *Asparagus officinalis*. *Proc. Natl. Acad. Sci. USA* 84:5345–49
11. Caboche, M. 1990. Liposome-mediated transfer of nucleic acids into plant cells. *Physiol. Plant.* 79:173–76
12. Christou, P., McCabe, D. E., Martinell, B. J., Swain, W. F. 1990. Soybean genetic transformation—commercial production of transgenic plants. *Trends Biotechnol.* 8:145–51
13. Deleted in proof
14. Datta, S. K., Peterhans, A., Datta, K., Potrykus, I. 1990. Genetically engineered fertile Indica-rice recovered from protoplasts. *BioTechnology* 8:736–40
15. Davey, M. R., Rech, E. L., Mulligan, B. J. 1989. Direct DNA transfer to plant cells. *Plant Mol. Biol.* 13:273–85
16. Dekeyser, R. A., Claes, B., De Rycke, R. M. U., Habets, M. E., Van Montagu, M. C., Caplan, A. B. 1990. Transient gene expression in intact and orga-

nized rice tissue. *The Plant Cell* 2:591–602

17. De la Pena, A., Lörz, H., Schell, J. 1987. Transgenic plants obtained by injecting DNA into young floral tillers. *Nature* 325:274–76

18. DeWet, J. M. J., Bergquist, R. R., Harlan, J. R., Brink, D. E., Cohan, C. E., et al. 1985. Exogenous DNA transfer in maiz *(Zea mays)* using DNA-treated pollen. In *Experimental Manipulation of Ovule Tissue*, ed. G. P. Chapman, S. H. Mantell, W. Daniels, pp. 197–209. London: Longman

19. Donn, G., Nilges, M., Morocz, S. 1990. Stable transformation of maize with a chimaeric, modified phosphinotrycin-acyltransferase gene from *Streptomyces viridochromogenes*. In *Progress in Plant Cellular and Molecular Biology*, ed. H. J. J. Nijkamp, L. H. W. Van der Plas, J. Aartrijk, Abstr. A2-38, p. 53. Dordrecht: Kluwer

20. Draper, J., Scott, R., Armitage, P., Walden, R., eds. 1988. *Plant Genetic Transformation and Gene Expression. A Laboratory Manual*. Oxford: Blackwell Scientific Publ.

21. Feldmann, K. A., Marks, M. D. 1987. *Agrobacterium*-mediated transformation of germinating seeds of *Arabidopsis thaliana: a* non-tissue culture approach. *Mol. Gen. Genet.* 208:1–9

22. Fraley, R. T., Rogers, S. G., Horsch, R. B. 1986. Genetic transformation in higher plants. *CRC Crit. Rev. Plant Sci.* 4:1–46

23. Fromm, M. E., Morrish, F., Armstrong, C., Williams, R., et al. 1990. Inheritance and expression of chimeric genes in the progeny of transgenic maize plants. *BioTechnology* 8:833–39

24. Fromm, M., Callis, J., Taylor, L. P., Walbot, V. 1987. Electroporation of DNA and RNA into plant protoplasts. *Methods Enzymol.* 153:351–66

25. Fromm, M., Walbot, V. 1987. Transient expression of DNA in plant cells. In *Plant Gene Research: Plant DNA Infectious Agents*, ed. T. Hohn, J. Schell, pp. 304–10. Wien/New York: Springer

26. Fromm, M. E., Taylor, L. P., Walbot, V. 1986. Stable transformation of maize after electroporation. *Nature* 319:791–93

27. Fütterer, J., Bonneville, J. M., Hohn, T. 1990. Cauliflower mosaic virus as a gene expression vector for plants. *Physiol. Plant.* 79:154–57

28. Gad, A. E., Rosenberg, N., Altmann, A. 1990. Liposome-mediated gene delivery into plant cells. *Physiol. Plant.* 79:177–83

29. Gardner, R., Chonoles, K., Owens, R. 1986. Potato spindle tuber viroid infections mediated by the Ti-plasmid of *Agrobacterium tumefaciens*. *Plant Mol. Biol.* 6:221–28

30. Gasser, C. S., Fraley, R. T. 1989. Genetically engineering plants for crop improvement. *Science* 244:1293–99

31. Gelvin, S. B., Schilperoort, R. A., Verma, D. P. S. 1989/1990. *Plant Molecular Biology Manual*. Dordrecht: Kluwer

32. Goff, S. A., Klein, T. M., Roth, B. A., Fromm, M. E., Cone, K. C., Radicella, J. P., Chandler, V. 1990. Transactivation of anthocyanin biosynthesis genes following transfer of B regulatory genes into maize tissues. *EMBO J.* 9:2517–22

33. Gordon-Kamm, W. J., Spencer, T. M., Mangano, M. L., Adams, T. R., Daines, R. J., et al. 1990. Transformation of maize cells and regeneration of fertile transgenic plants. *The Plant Cell* 2:603–18

34. Graves, A. C. F., Goldmann, S. L. 1986. The transformation of *Zea mays* seedlings via the *Agrobacterium tumefaciens*: detection of T-DNA-specified enzyme activities. *Plant Mol. Biol.* 7:43–50

35. Grimsley, N. H. 1990. Agroinfection. *Physiol. Plant.* 79:147–53

36. Grimsley, N. H., Hohn, B., Ramos, C., Kado, C., Rogowsky, P. 1989. DNA transfer from *Agrobacterium* to *Zea mays* or *Brassica* by agroinfection is dependent on bacterial virulence functions. *Mol. Gen. Genet.* 217:309–16

37. Grimsley, N. H., Hohn, T., Davies, J. W., Hohn, B. 1987. *Agrobacterium*-mediated delivery of infectious maize streak virus into maize plants. *Nature* 325:177–79

38. Hensgens, L. A. M., Meijer, E. G. M., Van Os-Ruygrok, P. E., Rueb, S., dePater, B., Schilperoort, R. A. 1990. Rice transformation, vectors, expression of transferred genes and rice genes. *Abstr. 4th Annu. Meet. Rockefeller Found. Int. Prog. Rice Biotechnol.*, IRRI, Manila

39. Hess, D. 1987. Pollen based techniques in genetic manipulation. *Int. Rev. Cytol.* 107:169–90

40. Hess, D. 1969. Versuche zur Transformation höherer Pflanzen: Induktion und konstante Weitergabe der Anthocyansynthese bei *Petunia hybrida*. *Z. Pflanzenphysiol.* 60:348–58

41. Hinchee, M. A. W., Connor-Ward, D. V., Newell, C. A., McDonnell, R. E., Sato, S. S., et al. 1988. Production of transgenic soybean plants using *Agro-*

bacterium-mediated DNA transfer. *BioTechnology* 6:915–22

42. Hooykaas, P. J. J. 1989. Transformation of plant cells via *Agrobacterium*. *Plant Mol. Biol.* 13:327–36

43. Hooykaas-VanSlogteren, G. M. S., Hooykaas, P. J. J., Schilperoort, R. A. 1984. Expression of Ti-plasmid genes in monocotyledonous plants infected with *Agrobacterium* tumefaciens. *Nature* 311:763–64

44. Horn, M. E., Shillito, R. D., Conger, B. V., Harms, C. T. 1988. Transgenic plants of orchardgrass (*Dactylis glomerata* L.) from protoplasts. *Plant Cell Rep.* 7:469–72

45. Kahl, G. 1982. Molecular biology of wound healing: the conditioning phenomenon. In *Molecular Biology of Plant Tumors,* ed. G. Kahl, J. Schell, pp. 211–67. New York: Academic

46. King, P. J., Potrykus, I., Thomas, E. 1978. In vitro genetics of cereals: problems and perspectives. *Physiol. Vég.* 16:381–99

47. Klee, H., Horsch, R., Rogers, S. 1987. *Agrobacterium*-mediated plant transformation and its further applications to plant biology. *Annu. Rev. Plant Physiol.* 38:467–86

48. Klee, H. J., Rogers, S. G. 1989. Plant genetic vectors and transformation: plant transformation systems based on the use of *Agrobacterium tumefaciens*. See Ref. 78a, pp. 2–25

49. Klein, T. M., Goff, S. A., Roth, B. A., Fromm, M. E. 1990. Applications of the particle gun in plant biology. In *Progress in Plant Cellular and Molecular Biology,* ed. H. J. J. Nijkamp, L. H. W. Van der Plas, J. Van Aartrijk, pp. 56–66. Dordrecht: Kluwer

50. Klein, T. M., Fromm, M. E., Gradziel, T., Sanford, J. C. 1988. Factors influencing gene delivery into *Zea mays* cells by high velocity microprojectiles. *BioTechnology* 6:923–26

51. Klein, T. M., Roth, B. A., Fromm, E. M. 1989. Advances in direct gene transfer into cereals. In *Genetic Engineering, Principles and Methods,* ed. J. K. Setlov, 11:13–31. New York/London: Plenum

52. Ledoux, L., Huart, R. 1969. Fate of exogenous desoxyribonucleic acid in barley seedlings. *J. Mol. Biol.* 43:243–62

53. Ledoux, L., Huart, R., Jacobs, M. 1974. DNA-mediated genetic correction of thiaminless *Arabidopsis thaliana*. *Nature* 249:17–21

54. Lindsey, K., Jones, M. G. K. 1990.

Electroporation of cells. *Physiol. Plant* 79:168–72

55. Lucas, W. J., Lansing, A., de Wet, J. R., Walbot, V. 1990. Introduction of foreign DNA into walled plant cells via liposomes injected into the vacuole: a preliminary study. *Physiol. Plant.* 79: 184–89

56. Luo, Z., Wu, R. 1988. A simple method for transformation of rice via the pollentube pathway. *Plant Mol. Biol. Rep.* 6:165–74

57. Madore, M. A., Oross, J. W., Lucas, W. J. 1986. Symplastic transport in *Ipomea tricolor* source leaves: demonstration of functional symplastic connections from mesophyll to minor veins by a novel dye tracer method. *Plant Physiol.* 82:432–42

58. McCabe, D. E., Swain, W. E., Martinell, B. J., Christou, P. 1988. Stable transformation of soybean (*Glycine max*) by particle acceleration. *BioTechnology* 6:923–26

59. Miki, L. A., Reich, T. J., Iyer, V. N. 1987. Microinjection: an experimental tool for studying and modifying plant cells. In *Plant Gene Research: Plant DNA Infectious Agent,* ed. T. Hohn, J. Schell, pp. 249–66. Wien/New York: Springer

60. Negrutiu, I., Shillito, R. D., Potrykus, I., Biasini, G., Sala, F. 1987. Hybrid genes in the analysis of transformation conditions. I. Setting up a simple method for direct gene transfer to protoplasts. *Plant Mol. Biol.* 8:363–73

61. Neuhaus, G., Spangenberg, G. 1990. Plant transformation by microinjection techniques. *Physiol. Plant.* 79:213–17

62. Neuhaus, G., Spangenberg, G., Scheid, O. M., Schweiger, H. G. 1987. Transgenic rape seed plants obtained by microinjection of DNA into microspore-derived proembryoids. *Theor. Appl. Genet.* 75:30–36

63. Ohta, Y. 1986. High efficiency genetic transformation of maize by a mixture of pollen and exogenous DNA. *Proc. Natl. Acad. Sci. USA* 83:715–19

64. Deleted in proof

65. Paszkowski, J., Saul, M. W., Potrykus, I. 1989. Plant gene vectors and genetic transformation: DNA-mediated direct gene transfer to plants. See Ref. 78a, pp. 52–68

66. Paszkowski, J., Baur, M., Bogucki, A., Potrykus, I. 1988. Gene targeting in plants. *EMBO J.* 7:4021–27

67. Paszkowski, J., Shillito, R. D., Saul, M. W., Mandak, V., Hohn, T., et al.

1984. Direct gene transfer to plants. *EMBO J.* 3:2712–22

68. Potrykus, I. 1990. Gene transfer to plants: assessment and perspectives. *Physiol. Plant.* 79:125–34

69. Potrykus, I. 1990. Gene transfer to cereals: an assessment. *BioTechnology* 8: 535–42

70. Potrykus, I., ed. 1990. Gene transfer to plants: a critical assessment. Proceedings of an EMBO workshop. *Physiol. Plant.* 79:123–220

71. Potrykus, I., Paszkowski, J., Saul, M. W., Petruska, J., Shillito, R. D. 1985. Molecular and general genetics of a hybrid foreign gene introduced into tobacco by direct gene transfer. *Mol. Gen. Genet.* 199:167–77

72. Potrykus, I., Shillito, R. D. 1986. Protoplasts: isolation, culture and plant regeneration. *Methods Enzymol.* 118: 549–78

73. Raineri, D. M., Bottino, P., Gordon, M. P., Nester, E. W. 1990. *Agrobacterium*-mediated transformation of rice (*Oryza sativa* L.). *BioTechnology* 8:33–38

74. Rhodes, C. A., Pierce, D. A., Mettler, I. J., Mascarenhas, D., Detmer, J. J. 1989. Genetically transformed maize plants from protoplasts. *Science* 240: 204–7

75. Roest, S., Gilissen, L. J. W. 1989. Plant regeneration from protoplasts: a literature review. *Acta Bot. Neerl.* 38:1–23

76. Sanford, J. C. 1990. Biolistic plant transformation. *Physiol. Plant.* 79:206–9

77. Sanford, J. C. 1988. The biolistic process. *Trends Biotechnol.* 6:299–302

78. Schäfer, W., Görz, A., Kahl, G. 1987. T-DNA integration and expression in a monocot crop plant after induction of *Agrobacterium. Nature* 327:529–32

78a. Schell, J., Vasil, I. K., eds. 1989. *Cell Culture and Somatic Cell Genetics of Plants*, Vol. 6: *Molecular Biology of Plant Nuclear Genes.* San Diego: Academic

79. Schocher, R. J., Shillito, R. D., Saul, M. W., Paszkowski, J., Potrykus, I. 1986. Co-transformation of unlinked foreign genes into plants by direct gene transfer. *BioTechnology* 4:1093–96

80. Shillito, R. D., Saul, M. W., Paszkowski, J., Müller, M., Potrykus, I. 1985. High frequency direct gene transfer to plants. *BioTechnology* 3:1099–1103

81. Shimamoto, K., Terada, R., Izawa, T., Fujimoto, H. 1989. Fertile rice plants regenerated from transformed protoplasts. *Nature* 338:274–76

82. Töpfer, R., Gronenborn, B., Schaefer, S., Schell, J., Steinbiss, H. H. 1990. Expression of engineered wheat dwarf virus in seed-derived embryos. *Physiol. Plant.* 79:158–62

83. Töpfer, R., Gronenborn, B., Schell, J., Steinbiss, H. H. 1989. Uptake and transient expression of chimeric genes in seed-derived embryos. *The Plant Cell* 1:133–39

84. Toriyama, K., Arimoto, Y., Uchimiya, H., Hinata, K. 1988. Transgenic rice plants after direct gene transfer into protoplasts. *BioTechnology* 6:1072–74

85. Valvekens, D., Van Montagu, M., Van Lijsebettens, M. 1988. *Agrobacterium-tumefaciens*-mediated transformation of *Arabidopsis thaliana* root explants by using kanamycin selection. *Proc. Natl. Acad. Sci. USA* 85:5536–40

86. Vasil, I. K. 1987. Developing cell and tissue culture systems for the improvement of cereal and grass crops. *J. Plant Physiol.* 128:193–218

87. Vasil, V., Redway, F., Vasil, I. K. 1990. Regeneration of plants from embryogenic suspension culture protoplasts of wheat (*Triticum aestivum* L.). *BioTechnology* 8:429–34

88. Weber, G., Monajembashi, S., Greulich, K. O., Wolfrum, J. 1988. Injection of DNA into plant cells with a UV laser microbeam. *Naturwissenschaften* 75:35–36

89. Weber, G., Monajembashi, S., Wolfrum, J., Greulich, K. O. 1990. Genetic changes induced in higher plant cells by a laser microbeam. *Physiol. Plant.* 79:190–93

90. Weising, K., Schell, J., Kahl, G. 1988. Foreign genes in plants: transfer, structure, expression, and applications. *Annu. Rev. Genet.* 22:421–77

91. Yan, Q., Zhang, X., Shi, J., Li, J. 1990. Green plant regeneration from protoplasts of barley (*Hordeum vulgare* L.). *Kexue Tongbao* 35:1–6

92. Zhang, W., Wu, R. 1988. Efficient regeneration of transgenic rice plants from rice protoplasts and correctly regulated expression of foreign genes in the plants. *Theor. Appl. Genet.* 76:835–40

93. Zhang, H. M., Yang, H., Rech, E. L., Golds, T. J., Davis, A. S., et al. 1988. Transgenic rice plants produced by electroporation-mediated plasmid uptake intoprotoplasts. *Plant Cell Rep.* 7:379–84

Annu. Rev. Plant Physiol. Plant Mol. Biol. 1991. 42:227–40

THIONINS

Holger Bohlmann and Klaus Apel

Institut für Pflanzenwissenschaften der ETH-Zürich, Universitätsstraße 2, CH-8092 Zürich, Switzerland

KEY WORDS: Thionin, toxin, plant resistance, sulfur-rich proteins

CONTENTS

In higher plants a large number of toxic compounds exist whose physiological roles within the plant are only rarely understood. Among these, thionins form a well-defined group of low-molecular-weight polypeptides whose toxic effects on bacteria (20, 62), fungi (5, 18), yeast (27, 62), and animal and plant cells (19, 33, 39, 50) have been known for quite a while. Only recently has evidence for a possible biological function emerged from studies of a newly discovered group of thionins, the leaf thionins of barley (4, 23). In the present review we'll summarize work on thionins with special emphasis on leaf thionins. A review dealing mainly with seed-specific thionins was published recently (22).

The Distribution of Thionins within Higher Plants

In 1885 Jago & Jago suggested the existence of a substance lethal to brewer's yeast in wheat flower (cited in 41). They observed that when brewer's yeast was used to ferment dough in bread making, production of CO_2 was depressed. In 1942 Balls et al extracted the toxic substance, identified it as a

1040-2519/91/0601-0227$02.00

low-molecular-weight polypeptide, and named it purothionin (2). Subsequent work demonstrated that hexaploid wheat produces three genetic variants, the α_1-, α_2-, and β-purothionins, each of about M_r 5000 (9, 21, 47). The complete amino acid sequences of these variants have been determined (31, 32, 37, 40, 41). The polypeptides contain 45 amino acid residues. The 8 cysteine residues and the basic amino acids are distributed throughout the molecule (Figure 1). Similar proteins have been extracted from the endosperm of many other members of the *Aegilops-Triticum* group (21, 30) and from other Gramineae such as barley (28, 45, 46, 48), oat (3), maize (30), and rye (26). The term "thionin" was chosen for all these proteins to indicate their close similarity (26).

Related proteins have been identified also in several dicotyledonous plants. The best-known example of this group of thionins is the viscotoxin, which has been extracted from leaves and stems of European and American mistletoes (52–56, 75). The poisonous properties of mistletoe have been known since ancient times. Extracts from mistletoe have been used against a variety of diseases, and such extracts are still in use as constituents of herbal remedies (61). Some of their effects are a toxic manifestation attributable to a mixture of small basic proteins, first isolated by Winterfeld & Bijl (75) and given the name viscotoxin. Viscotoxins have three disulfide bonds in common locations corresponding to three out of four locations of purothionin (Figure 1). A closely related thionin has been isolated from seeds of *Pyrularia pubera,* a parasitic plant of the same order as the mistletoes (Santalales) but of a different family (70). Its structure is intermediate between those of the viscotoxins and the cereal grain thionins. Like the cereal thionins the *Pyrularia* thionin has eight cysteine residues covalently linked by four disulfide bridges (70). Although the viscotoxins contain only three disulfide bonds, they have slightly more sequence homology with the *Pyrularia* protein than do the other thionins (70).

Finally, another thionin-like protein has been identified in seeds of the crucifer *Crambe abyssinica* (68). Dubbed crambin, it consists of a single chain of 46 amino acids, and its three disulfide bonds are found at positions corresponding to those of viscotoxin (64). Crambin is very hydrophobic and, in contrast to other thionins, has no positive net charge (64). While other thionins are toxic to a large number of organisms, crambin exhibits no such biological activity (64, 67).

Figure 1 shows a comparison of amino acid sequences of the different thionins (determined primarily by protein sequencing). The sequences of the viscotoxins from *Viscum album* have been corrected for the last 4 amino acid residues at the C-termini of the polypeptides. As discussed by Vernon et al (70), it seems likely that this part of the published amino acid sequence is

```
Position:           1          10         20          30         40        47

Purothionin Alpha1  K S C C R S T L G R N C Y N L C R A R G - - A Q K L C A G V C R C K I S S G L S C P K G F P K   (32, 40, 41)
Purothionin Alpha2  K S C C R R T T L G R N C Y N L C R S R G - - A Q K L C S T V C R C K L T S G L S C P K G F P K   (32)
Purothionin Beta    K S C C K S T L G R N C Y N L C R A R G - - A Q K L C A N V C R C K L T S G L S C P K D F P K   (37, 40, 41)
Avenothionin Alpha  K S C C R N T L G R N C Y N L C R S R G - - A P K L C A T V C R C K I S S G L S C P K D F P K   (3)
Avenothionin Beta   K S C C K N T L G R N C Y N L C R A R G - - A P K L C S T V C R C K L T S S G L S C P K D F P K   (3)
Hordothionin Alpha  K S C C R S T L G R N C Y N L C R V R G - - A Q K L C A G V C R C K L T S S G K C P T G F P K   (50, 46)
Hordothionin Beta   K S C C R S T L G R N C Y N L C R V R G - - A Q K L C A N A C R C K L T S G L K C P S S F P K   (28)

DB4                 K S C C K D T L A R N C Y N T C H F A G G - S R P V C A G A C R C K I I S G P K C P S D Y P K   (4)
DC4                 K S C C K D T L A R N C Y N T C R F A G G - S R P V C A G A C R C K I I S G P K C P S D Y P K   (4)
DG3                 K S C C K N T T G R N C Y N A C R F A G G - S R P V C A T A C G C K I I S G P T C P R D Y P K   (4)

Pyrularia Toxin     K S C C R N T W A R N C Y N V C R L P G T I S R E I C A K K C D C K I I S G T T C P S D Y P K   (70)

Viscotoxin A2       K S C C P N T T G R N I Y N T C R F G G G - S R E V C A S L S G C K I I S A S T C P S D Y P K   (43)
Viscotoxin A3       K S C C P N T T G R N I Y N A C R L T G A - P R P T C A K L S G C K I I S G S T C P S D Y P K   (60)
Viscotoxin B        K S C C P N T T G R N I Y N T C R L G G G - S R E R C A S L S G C K I I S A S T C P S D Y P K   (58)
Viscotoxin 1-PS     K S C C P N T T G R N I Y N T C R F G G G - S R E V C A R I S G C K I I S A S T C P S D Y P K   (57)
Phoratoxin A        K S C C P T T A R N I Y N T C R F G G G - S R P V C A K L S G C K I I S G T K C D S G W N H   (38, 65)
Phoratoxin B        K S C C P T T A R N I Y N T C R F G G G - S R P I C A K L S G C K I I S G T K C D S G W N H   (38, 65)
Ligatoxin           K S C C P S T T A R N I Y N T C R L T G T - S R P T C A S L S G C K I I S G S T C D S G W N H   (66)
Denclatoxin         K S C C P T T A A R N G Y N I C R L P G T - P R P V C A A L S G C K I I S G T G C P P G Y R H   (59)

Crambin 1           T T C C P S I V A R S N F N V C R L P G T - P E A L C A T Y T G C I I I P G A T C P G D Y A N   (64, 69)
Crambin 2           T T C C P S I V A R S N F N V C R L P G T - S E A I C A T Y T G C I I I P G A T C P G D Y A N   (64, 69)
```

Figure 1 Comparison of the amino acid sequences of seed-specific thionins of cereals (purothionin, hordothionin, avenothionin), leaf thionins of barley (DB4, DC4, DG3), *Pyrularia* toxin, viscotoxin and related thionins, and crambin.

incorrect. Recently obtained sequence data of cDNAs for viscotoxins A3 and B confirm this view (G. Schrader, unpublished results).

A striking feature of all thionins known so far is the conservation of 6 or 8 cysteine residues. In all thionins except crambin, the first and last 2 amino acid residues have been conserved as well. It remains to be seen whether or not the C-terminal sequences of ligatoxin, denclatoxin, and phoratoxin need correction as in the case of the viscotoxins. In all thionins but crambin studied so far, amino acids in positions 7, 9–11, 13, 14, 17, 20, 28, 34, 37, and 42 are highly conserved.

Along with the homologies in the primary structure of the thionins goes a homology in their three-dimensional structures. The three-dimensional structure has been determined for several thionins. The most intensively studied thionin in this regard is crambin (e.g. 7, 8, 11, 25, 74), because it can easily be crystallized (63). Other thionins for which the three-dimensional structure has been analyzed include the purothionins (e.g. 12, 13, 35, 73), hordothionins (35), viscotoxins (36, 73), and phoratoxin (14, 36). These studies show thionin to be a compact, L-shaped molecule. The long arm of the L is formed by two alpha-helixes, the short arm by two short antiparallel beta-sheets and the last (approximately 10) C-terminal amino acid residues (Figure 2).

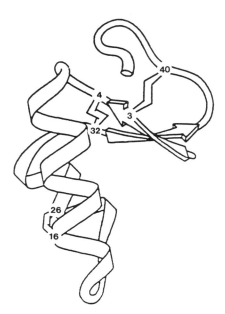

Figure 2   Schematic drawing of the backbone of crambin (73). Arrows depict β-strands. The positions of cysteine residues have been labeled by numbers.

This three-dimensional structure is stabilized by 3 or 4 disulfide bridges. These account for the stability of thionins. Viscotoxins, for example, can be heated at 100° C for 30 min without loss of their toxic properties (56). Disulfide bridges have been found in all thionins investigated in this regard— for instance in viscotoxin (e.g. 44), crambin (64), and purothionins (24). The bridges are formed between the cysteine residues in positions 3–39, 4–31, 12–29 (these are missing in crambin and mistletoe toxins), and 16–25.

Thionins have an amphipathic structure that stems primarily from the two amphipathic alpha helixes: for example, amino acids 10–19 and 23–28 in α_1-purothionin (13), 7–19 and 23–30 in crambin (11), and 7–19 and 23–29 in phoratoxin (14). Hydrophobic residues are mainly at the outer surface of the long arm of the L, whereas hydrophilic residues are mainly found at the inner surface of the L and at the outer surface of the corner of the L (12, 25, 64).

The ubiquity of thionin homologs throughout the plant world raises questions about the evolutionary and genetic connections among the species from which they are extracted (Figure 3). Since thionins have been found primarily

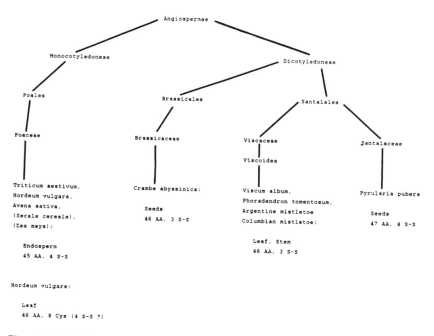

Figure 3   The phylogenetic relationship of higher plants from which thionins have been isolated and identified by sequence analysis.

in species that have some agricultural or pharmaceutical importance and thus have been analyzed very intensively, it seems likely that upon further search thionin-like proteins may be found in other species as well. In agreement with this prediction, thionin-like proteins have been isolated from several higher-plant species that belong to widely separate families—e.g. Solanaceae (tomato), Juglandaceae (walnut), Anacardiaceae (mango), and Caricaceae (melon tree) (17). The identification of these proteins as putative thionin-like proteins was based on their low molecular weight, their heat stability, their positive net charge, and their high content of sulfhydryl groups. Even though these features are similar to those of most thionins, this identification is tentative and has to be verified more rigorously by sequence analysis of the purified proteins.

The Toxicity of Thionins

A main characteristic of all thionins except crambin is their toxic effect on different biological systems. Shortly after the discovery of purothionins their bactericidal and fungicidal properties (62), as well as their toxicity to several small mammals when injected intraperitoneally or intravenously (16), were reported. On the other hand, oral administration of up to 229 mg kg^{-1} body weight had no effect on guinea pigs (16). Viscotoxins are also toxic when administered intraperitoneally (LD 50 in mice, 0.5 mg kg^{-1} bodyweight) or intravenously (LD 50 in cats, 0.1 mg kg^{-1} bodyweight) but are not toxic when eaten (56). A toxicity has also been demonstrated in insects; when injected into the hemocoel, endosperm thionins killed *Manduca sexta* (tobacco hornworm) larvae (34). After the first report in 1942, toxicity of thionins has repeatedly been demonstrated in yeast (e.g. 27, 62) and bacteria (e.g. 20, 62).

Besides the toxicity on whole organisms, cytotoxicity in various animal cell cultures has been reported by several groups. Nakanishi et al (39) showed that purothionin had the most pronounced toxic effect on mammalian cells in the S-phase. Kashimoto et al (33) demonstrated leakage of cytoplasmic (but not mitochondrial) enzymes and ions from bovine adrenal medullary cells treated with purothionin. Carrasco et al (10) found differing sensitivities of different mammalian cell lines for thionins (purothionin, hordothionin, viscotoxin). They suggested that thionins modify membrane permeability. The mechanism by which these toxic effects are exerted is not known. Interestingly, the only thionin known so far that is not toxic, is the seed protein crambin (64, 67). Crambin is not basic, as are all other toxic thionins, and has no tyrosine residue in position 13. Wada et al (72) have shown that chemical modification of the amino groups in purothionin led to the loss of toxicity, and modification of the tyrosine residue led to a considerably reduced activity. An essential role of this tyrosine residue in *Pyrularia* thionin has recently been reported by

Evans et al (19). Their data suggest that the toxic action of *Pyrularia* thionin in mammalian cells is mediated by Ca^{2+} influx (perhaps via ion channels), membrane depolarization, and activation of phospholipase A2 (19). These data suggest that the primary action of thionins might be the alteration of ion channels, while the amphipathic structure of thionins indicates that the toxicity is exerted by a direct interaction with lipid membranes. The hydrophobic face of the thionins could interact with the hydrophobic aliphatic chains of the phospholipids, while the positively charged basic amino acids could interact with the negatively charged phosphates of the phospholipids. These salt bridges cannot be formed with crambin, which has only two basic residues (Arg 10 and Arg 17), both blocked through intramolecular bonds (73).

The Biological Function of Thionins

The function of most thionins within the plant is not known with any degree of confidence. Several physiological roles have been assigned to these proteins. Crambin, *Pyrularia* toxin, and the endosperm thionin of cereals, all of which are localized within the seeds, may function as storage proteins. Most thionins show a broad range of toxic effects exerted at the level of cell membranes of bacteria, fungi, and higher organisms (see above). In the case of the seed-specific thionins of cereals, it had been suggested that in addition to their possible role as storage proteins they could act as a general-purpose defense toxin and might protect the starchy endosperm against invasive bacteria and fungi (20). However, besides the toxicity of thionins in vitro there is no experimental evidence yet to support this assumed function of seed-specific thionins in cereals. It has been demonstrated that purothionin can be reduced in vitro by the thioredoxin system of wheat seeds (29, 71). The sulfhydryl form of purothionin reductively activates fructose-1,6-biphosphatase. It was proposed that purothionin could function as a secondary thiol messenger (71). In a similar reaction, purothionin blocked DNA synthesis by inhibiting the activity of ribonucleotide reductase with reduced thioredoxin as hydrogen donor (29). This inhibition of DNA synthesis has been considered a possible cause of the known toxic effect of purothionin on mammalian cells that undergo active chromosome duplication (39). However, the physiological relevance of these activities for the plant cell is difficult to assess. Most thionins are basic proteins. Thus, it is not surprising that these proteins interact strongly with DNA. The interaction of viscotoxins with DNA results in the protection of the DNA double helix against thermal denaturation (76). The in vivo relevance of this phenomenon has not been investigated yet.

Leaf Thionins of Barley

Recently, a new group of thionins has been detected in barley leaves (4, 23). Subsequent studies of this novel thionin group have provided several independent lines of evidence suggesting very strongly that leaf thionins may play an important role during the defense against pathogens (5, 18). The existence of leaf thionins has been predicted from cDNA sequences that represent transcripts encoding a M_r 15,000 thionin precursor polypeptide (4, 23). These transcripts are highly abundant in leaves of dark-grown barley seedlings. During illumination, however, their concentration rapidly declines (49). Mature thionins could be detected in the cell wall and the central vacuole of barley leaf cells (5, 50). Thionins were detectable in all cell walls of etiolated barley leaves, but by far the highest concentration was present in the outer cell wall of the epidermal cell layer (49). The toxicity of leaf thionins isolated from cell walls and vacuoles was tested with two phytopathogenic fungi and with isolated protoplasts of tobacco leaves. The two leaf thionin fractions strongly inhibited the growth of both fungi and the protoplasts (5, 50).

The toxicity in vitro of the leaf thionins and their predominance in the outer cell wall of the epidermal cell layer of etiolated barley leaves indicate that these thionins may be part of a resistance mechanism of barley plants against pathogens. The only result difficult to reconcile with such a hypothesis is the rapid decline in thionin mRNA concentration once the seedling is exposed to light (49). The fact that barley seedlings accumulate large amounts of thionins in the dark is not per se in conflict with their being part of a resistance mechanism. It is known that during the early stage of plant development barley seedlings are exposed to a large number of soil-borne pathogens that are ready to invade the seedling (49). Under these circumstances it would not seem surprising that etiolated seedlings synthesize defense-related proteins in large amounts. In more mature, nonstressed plants that are fully adapted to a normal day-night cycle, thionin mRNAs are hardly detectable. However, when exposed to the fungal pathogen *Erisyphe graminis* (powdery mildew), plants rapidly began to reaccumulate thionin mRNAs (5). Roughly two days after the plants had been inoculated with fungal spores, the thionin mRNA content reached its maximum and began to decline afterwards (5).

At the same time, marked changes in the distribution of thionins at the site of infection occurred (18). One of the first host reactions to powdery mildew is the formation of the papilla, a locally restricted cell wall apposition that begins to form in the epidermal cell below the penetration site of the fungal appressorium (1). In susceptible barley leaves the papilla was devoid of thionins, and in cell walls close to the infection site the concentration of thionins was drastically reduced. In resistant leaves, however, thionins accumulate plentifully in the papilla; and in the cell wall close to the infection site the thionin concentration remains high (18). These results strongly suggest that leaf thionins play an important part in the defense mechanism of barley plants.

This interpretation is in agreement with results on the organization of leaf thionin genes. Leaf thionins of barley are distinct from other thionins in being encoded by a complex multigene family composed of 50–100 genes per haploid genome (5). Most of these genes are not identical and seem to encode different thionin variants (4, 5). The polymorphism of leaf thionin genes is in good agreement with their presumed role as defense factors. Such a polymorphism is common among toxins, being almost an inevitable consequence of the interplay between predator and prey (e.g. 42). If leaf thionins do contribute to the plant's resistance against microbial attack, the continuous variability of pathogens would likely drive the evolution of new thionin variants. One would expect such thionin variants to be functionally different. Even though the functional specialization of leaf thionin variants has not yet been tested, there is evidence from work with other thionins to support functional differentiation. For instance, α- and β-purothionins differ markedly in their biological activity (20). Thus minor differences in the amino acid sequences of thionin isopeptides may change the specificity of these toxins.

Open Questions

We have no proof yet that leaf thionins are indeed part of a resistance mechanism; nor has such a function been proven for other putative defense-related plant proteins (6). One way of testing the proposed role of leaf thionins as a resistance factor would be to express thionin genes in a transgenic plant and observe whether the protein affects the resistance of this plant to its own pathogens. Such experiments are under way in several laboratories, but final results have not yet been published. One difficulty that one may encounter in such an experiment with transgenic plants stems from the toxicity of thionins. It is not known yet how plants cope with the presence of these toxic compounds. It has been suggested that the acidic part of the thionin precursor might neutralize a toxic effect of the mature, basic thionin (50). If this is so one would expect such an acidic domain to be present not only in the precursors of leaf thionins but also in those of other toxic thionins. Indeed, precursor polypeptides of hordothionin and viscotoxin A3, whose amino acid sequences have been deduced from cDNA sequences, contain an acidic portion similar to that of the barley leaf thionin precursor (Figure 4; G. Schrader, K. Apel, in preparation). The charge differences between the basic thionin and the acidic domain of the precursor polypeptide would favor a close interaction between them that could prevent the correct folding of the bioactive mature thionin. However, this proposal does not explain why the plant can tolerate the large amounts of mature thionins that accumulate within the central vacuole of the cell (50).

So far the leaf thionins have been considered as putative general-purpose defense toxins that may be directed against invading pathogens, but another

```
Precursor DB4 (Hordeum vulgare)          Leaf          M A P S K S I K S V
Precursor HTH1 (Hordeum vulgare)         Endosperm     M G L - K G - - - V
Precursor Viscotoxin A3 (Viscum album)   Leaf + Stem   M E V V R G - S S L


V I C V L I L G L V L E Q V Q V E G▼K S C C K D T L A R N C Y N T C H F A
M V C L L I L G L V L E Q V Q V E G K S C C R S T L G R N C Y N L C R V R
V L L V L L L G A L L V S N V E E S K S C C P N T T G R N I Y N A C R L T


G G S R P V C A G A C R C K I I S G P K C P S D Y P K▼L N L L P E S G E P
G A Q K - L C A G V C R C K L T S S G K C P T G F P K L A L V S N S D E P
G A P R P T C A K L S G C K I I S G S T C P S D Y P K F - - - - - - - -


D V T Q Y C T I G C R N S V C D - N M D N V F R G Q E M K F D M G L C S N
D T V K Y C N L G C R A S N C D Y M V N A A A D D E E M K L Y L E N C G D
- - - - Y C T M G C E S S Q C - - - A T N S - - - - N G D A E A V R C K T


A C A R F C N D G A V I Q S V E A   (4)
A C V N F C N G D A G L T S L T A   (51)
A C S D L C Q F V - - - - - D D A   (G.Schrader, unpublished)
```

Figure 4 Comparison of the derived amino acid sequences of cDNAs encoding the precursor of a leaf thionin (DB4), α-hordothionin (HTH1), and viscotoxin A3. The black triangles mark the beginning and the end of the three thionin sequences.

possible function cannot yet be ruled out. As mentioned above, isolated leaf thionins damage not only fungi but also plant protoplasts (50). Thus, if bioactive thionins are released from the cell wall and the central vacuole, they may both block the growth of the pathogen and affect the plant cell. One of the first defense reactions of plants is the hypersensitive reaction. Cells close to the infection site collapse and form a necrotic area, which prevents spreading of pathogens (15). So far it is not known how this suicidal activity in plant cells is regulated. Leaf thionins may be involved in triggering such a defense reaction.

Conclusions

At the moment open questions still outweigh known facts about the function of thionins. Thionins in evolutionary disparate species such as *Viscum album*, *Crambe abyssinica*, and *Triticum aestivum* bear striking structural similarities. It appears likely that thionins have evolved from a common ancestor and that they play an important role in the growth and development of these plants, but so far only little is known about their physiological role. Several independent lines of evidence suggest that leaf thionins of barley are part of a defense mechanism. It remains to be seen whether other members of the thionin group function similarly in other plant species.

ACKNOWLEDGMENTS

We are indebted to members of our group and to colleagues who over the years have contributed to a better understanding of what thionins might do within the barley leaf. Work in the author's laboratory would have been impossible without continuous support from the Deutsche Forschungsgemeinschaft.

Literature Cited

1. Aist, J. R. 1983. Structural responses as resistance mechanisms. In *The Dynamics of Host Defence,* ed. J. A. Bailey, B. J. Deverall, pp. 33–70. New York: Academic

2. Balls, A. K., Hale, W. S., Harris, T. H. 1942. A crystalline protein obtained from a lipoprotein of wheat flour. *Cereal Chem.* 19:279–88

3. Bekes, F., Lasztity, R. 1981. Isolation and determination of amino acid sequence of avenothionin, a new purothionin analogue from oat. *Cereal Chem.* 58:360–61

4. Bohlmann, H., Apel, K. 1987. Isolation and characterization of cDNAs coding for leaf-specific thionins closely related to the endosperm-specific hordothionin of barley (*Hordeum vulgare* L.). *Mol. Gen. Genet.* 207:446–54

5. Bohlmann, H., Clausen, S., Behnke, S., Giese, H., Hiller, C., et al. 1988. Leaf-specific thionins of barley—a novel class of cell wall proteins toxic to plant-pathogenic fungi and possibly involved in the defense mechanism of plants. *EMBO J.* 7:1559–65

6. Boller, T. 1987. Hydrolytic enzymes in plant disease resistance. In *Plant-Microbe Interactions,* ed. T. Kosuge, E. W. Nester, 2:385–413. New York/London: Macmillan

7. Brünger, A. T., Campbell, R. L., Clore, G. M., Gronenborn, A. M., Karplus, M., et al. 1987. Solution of a protein crystal structure with a model obtained from NMR interproton distance restraints. *Science* 235:1049–53

8. Brünger, A. T., Clore, G. M., Gronenborn, A. M., Karplus, M. 1986. Three-dimensional structure of proteins determined by molecular dynamics with interproton distance restraints: application to crambin. *Proc. Natl. Acad. Sci. USA* 83:3801–5

9. Carbonero, P., Garcia-Olmedo, F. 1969. Purothionin in *Aegilops-Triticum* spp. *Experientia* 25:1110–11

10. Carrasco, L., Vazquez, D., Hernandez-Lucas, C., Carbonero, P., Garcia-Olmedo, F. 1981. Thionins: plant peptides that modify membrane permeability in cultured mammalian cells. *Eur. J. Biochem.* 116:185–89

11. Clore, G. M., Brünger, A. T., Karplus, M., Gronenborn, A. M. 1986. Application of molecular dynamics with interproton distance restraints to three-dimensional protein structure determination. A model study of crambin. *J. Mol. Biol.* 191:523–51

12. Clore, G. M., Nilges, M., Sukumaran, D. K., Brünger, A. T., Karplus, M., Gronenborn, A. M. 1986. The three-dimensional structure of α 1-purothionin in solution: combined use of nuclear magnetic resonance, distance geometry and restrained molecular dynamics. *EMBO J.* 5:2729–35

13. Clore, G. M., Sukumaran, D. K., Gronenborn, A. M., Teeter, M. M., Whitlow, M., Jones, B. L. 1987. Nuclear magnetic resonance study of the solution structure of α 1-purothionin. Sequential resonance assignment, secondary structure and low resolution tertiary structure. *J. Mol. Biol.* 193:571–78

14. Clore, G. M., Sukumaran, D. K., Nilges, M., Gronenborn, A. M. 1987. Three-dimensional structure of phoratoxin in solution: combined use of nuclear magnetic resonance, distance geometry, and restrained molecular dynamics. *Biochemistry* 26:1732–45

15. Collinge, D. B., Slusarenko, A. J. 1987. Plant gene expression in response to pathogens. *Plant Mol. Biol.* 9:938–410

16. Coulson, E. J., Harris, T. H., Axelrod, B. 1942. Effect on small laboratory animals of the injection of the crystalline hydrochloride of a sulfur protein from wheat flour. *Cereal Chem.* 19:301–7

17. Daley, L. S., Theriot, L. J. 1987. Electrophoretic analysis, redox activity, and other characteristics of proteins sim-

ilar to purothionins from tomato (Lyco-persicum esculenta), mango (Mangifera indica), papaya (Carica papaya), and walnut (Juglans regia). J. Agric. Food Chem. 35:680–87

18. Ebrahim-Nesbat, F., Behnke, S., Kleinhofs, A., Apel, K. 1989. Cultivar-related differences in the distribution of cell-wall-bound thionins in compatible and incompatible interactions between barley and powdery mildew. Planta 179:203–10

19. Evans, J., Wang, Y., Shaw, K.-P., Vernon, L. P. 1989. Cellular responses to Pyrularia thionin are mediated by Ca^{2+} influx and phospholipase A2 activation and are inhibited by thionin tyrosine iodination. Proc. Natl. Acad. Sci. USA 86:5849–53

20. Fernandez de Caleya, R., Gonzalez-Pascual, B., Garcia-Olmedo, F., Carbonero, P. 1972. Susceptibility of phytopathogenic bacteria to wheat purothionins in vitro. Appl. Microbiol. 23:998–1000

21. Fernandez de Caleya, R., Hernandez-Lucas, C., Carbonero, P., Garcia-Olmedo, F. 1976. Gene expression in alloploids: genetic control of lipopur-othionins in wheat. Genetics 83:687–99

22. Garcia-Olmedo, F., Rodriguez-Palenzuela, P., Hernandez-Lucas, C., Ponz, F., Marana, C., et al. 1989. The thionins: a protein family that includes purothionins, viscotoxins and crambins. Oxford Surv. Plant Mol. Cell Biol. 6:31–60

23. Gausing, K. 1987. Thionin genes specifically expressed in barley leaves. Planta 171:241–46

24. Hase, T., Matsubara, H., Yoshizumi, H. 1978. Disulfide bonds of purothi-onin, a lethal toxin for yeasts. J. Biochem. 83:1671–78

25. Hendrickson, W. A., Teeter, M. M. 1981. Structure of the hydrophobic protein crambin determined directly from the anomalous scattering of sulphur. Nature 290:107–13

26. Hernandez-Lucas, C., Carbonero, P., Garcia-Olmedo, F. 1978. Identification and purification of a purothionin homo-logue from rye (Secale cereale L.). J. Agric. Food Chem. 26:794–96

27. Hernandez-Lucas, C., Fernandez de Caleya, R., Carbonero, P. 1974. Inhibition of brewer's yeasts by wheat purothi-onins. Appl. Microbiol. 28:165–68

28. Hernandez-Lucas, C., Royo, J., Paz-Ares, J., Ponz, F., Garcia-Olmedo, F., Carbonero, P. 1986. Polyadenylation site heterogeneity in mRNA encoding

the precursor of the barley toxin β-hordothionin. FEBS Lett. 200:103–6

29. Johnson, T. C., Wada, K., Buchanan, B. B., Holmgren, A. 1987. Reduction of purothionin by the wheat seed thiore-doxin system. Plant Physiol. 85:446–51

30. Jones, B. L., Cooper, D. B. 1980. Purification and characterization of a corn (Zea mays) protein similar to purothionins. J. Agric. Food Chem. 28:904–8

31. Jones, B. L., Lookhart, G. L., Mak, A. S., Cooper, D. B. 1982. Sequences of purothionins and their inheritance in diploid, tetraploid, and hexaploid wheats. J. Hered. 73:143–44

32. Jones, B. L., Mak, A. S. 1977. Amino acid sequences of the two α-purothionins of hexaploid wheat. Cereal Chem. 54:511–23

33. Kashimoto, T., Sakakibara, R., Huynh, Q. K., Wada, H., Yoshizumi, H. 1979. The effect of purothionin on bovine adrenal medullary cells. Res. Commun. Chem. Pathol. Pharmacol. 26:221–24

34. Kramer, K. J., Klassen, L. W., Jones, B. L., Speirs, R. D., Kammer, A. E. 1979. Toxicity of purothionin and its homologues to the tobacco hornworm, Manduca sexta (L.) (Lepidoptera: Sphingidae). Toxicol. Appl. Pharmacol. 48:179–83

35. Lecomte, J. T. J., Jones, B. L., Llinas, M. 1982. Proton magnetic resonance studies of barley and wheat thionins: structural homology with crambin. Biochemistry 21:4843–49

36. Lecomte, J. T. J., Kaplan, D., Llinas, M., Thunberg, E., Samuelsson, G. 1987. Proton magnetic resonance characterization of phoratoxins and homologous proteins related to crambin. Biochemistry 26:1187–94

37. Mak, A. S., Jones, B. L. 1976. The amino acid sequence of wheat β-purothionin. Can. J. Biochem. 54:835–42

38. Mellstrand, S. T., Samuelsson, G. 1974. Phoratoxin, a toxic protein from the mistletoe Phoradendron tomentosum subsp. macrophyllum (Loranthaceae). The amino acid sequence. Acta Pharmacol. Suecica 11:347–60

39. Nakanishi, T., Yoshizumi, H., Tahara, S., Hakura, A., Toyoshima, K. 1979. Cytotoxicity of purothionin-A on various animal cells. Gann 70:323–26

40. Ohtani, S., Okada, T., Kagamiyama, H., Yoshizumi, H. 1975. The amino acid sequence of purothionin A, a lethal toxic protein for brewer's yeasts from wheat. Agric. Biol. Chem. 39:2269–70

41. Ohtani, S., Okada, T., Yoshizumi, H., Kagamiyama, H. 1977. Complete primary structures of two subunits of purothionin A, a lethal protein for brewer's yeast from wheat flour. *J. Biochem.* 82:753–67

42. Olivera, B. M., Rivier, J., Clark, C., Ramilo, C. A., Corpuz, G. P., et al. 1990. Diversity of *Conus* neuropeptides. *Science* 249:257–63

43. Olson, T., Samuelsson, G. 1972. The amino acid sequence of viscotoxin A2 from the European mistletoe (*Viscum album* L., Loranthaceae). *Acta Chem. Scand.* 26:585–95

44. Olson, T., Samuelsson, G. 1974. The disulphide bonds of viscotoxin A2 from the European mistletoe (*Viscum album* L. Loranthaceae). *Acta Pharmacol. Suecica* 11:381–386

45. Ozaki, Y., Wada, K., Hase., T., Matsubara, H., Nakanishi, T., Yoshizumi, H. 1980. Amino acid sequence of a purothionin homolog from barley flour. *J. Biochem.* 87:549–55

46. Ponz, F., Paz-Ares, J., Hernandez-Lucas, C., Garcia-Olmedo, F., Carbonero, P. 1986. Cloning and nucleotide sequence of a cDNA encoding the precursor of the barley toxin α-hordothionin. *Eur. J. Biochem.* 156:131–35

47. Redman, D. G., Fisher, N. 1968. Fractionation and comparison of purothionin and globulin components of wheat. *J. Sci. Food Agric.* 19:651–55

48. Redman, D. G., Fisher, N. 1969. Purothionin analogues from barley flour. *J. Sci. Food Agric.* 20:427–32

49. Reimann-Philipp, U., Behnke, S., Batschauer, A., Schäfer, E., Apel, K. 1989. The effect of light on the biosynthesis of leaf-specific thionins in barley, *Hordeum vulgare*. *Eur. J. Biochem.* 182:283–89

50. Reimann-Philipp, U., Schrader, G., Martinoia, E., Barkholt, V., Apel, K. 1989. Intracellular thionins of barley. A second group of leaf thionins closely related to but distinct from cell wall-bound thionins. *J. Biol. Chem.* 264:8978–84

51. Rodriguez-Palenzuela, P., Pintor-Toro, J. A., Carbonero, P., Garcia-Olmedo, F. 1988. Nucleotide sequence and endosperm-specific expression of the structural gene for the toxin α-hordothionin in barley (*Hordeum vulgare* L.). *Gene* 70:271–81

52. Samuelsson, G. 1958. Phytochemical and pharmacological studies on *Viscum album* L. I. Viscotoxin, its isolation and properties. *Swed. J. Pharmacol.* 62:169–89

53. Samuelsson, G. 1961. Phytochemical and pharmacological studies on *Viscum album* L. V. Further improvements in the isolation methods for viscotoxin. Studies on viscotoxin from *Viscum album* growing on *Tilia cordata* Mill. *Swed. J. Pharmacol.* 65:481–94

54. Samuelsson, G. 1966. Screening of plants of the family Loranthaceae for toxic proteins. *Acta Pharmacol. Suecica* 3:353–62

55. Samuelsson, G. 1969. Screening of the plants of the family Loranthaceae for toxic proteins. Part II. *Acta Pharmacol. Suecica* 6:441–46

56. Samuelsson, G. 1974. Mistletoe toxins. *Syst. Zool.* 22:566–69

57. Samuelsson, G., Jayawardene, A. L. 1974. Isolation and characterization of viscotoxin 1Ps from *Viscum album* L. ssp. *austriacum* (Wiesb.) Vollmann, growing on *Pinus silvestris*. *Acta Pharmacol. Suecica* 11:175–84

58. Samuelsson, G., Pettersson, B. 1971. The amino acid sequence of viscotoxin B from the European mistletoe (*Viscum album* L., Loranthaceae). *Eur. J. Biochem.* 21:86–89

59. Samuelsson, G., Pettersson, B. 1977. Toxic proteins from the mistletoe *Dendrophtora clavata*. II. The amino acid sequence of denclatoxin B. *Acta Pharmacol. Suecica* 14:245–54

60. Samuelsson, G., Seger, L., Olson, T. 1968. The amino acid sequence of oxidized viscotoxin A3 from the European mistletoe (*Viscum album* L. Loranthaceae). *Acta Chem. Scand.* 22:2624–42

61. Selawry, O. S., Vester, F., Mai, W., Schwartz, M. R. 1961. Zur Kenntnis der Inhaltsstoffe von *Viscum album*. II. Tumorhemmende Inhaltsstoffe. *Hoppe-Seyler's Z. Physiol. Chem.* 324:262–81

62. Stuart, L. S., Harris, T. H. 1942. Bactericidal and fungicidal properties of a crystalline protein isolated from unbleached wheat flour. *Cereal Chem.* 19:288–300

63. Teeter, M. M., Hendrickson, W. A. 1979. Highly ordered crystals of the plant seed protein crambin. *J. Mol. Biol.* 127:219–23

64. Teeter, M. M., Mazer, J. A., L'Italien, J. J. 1981. Primary structure of the hydrophobic plant protein crambin. *Biochemistry* 20:5437–43

65. Thunberg, E. 1983. Phoratoxin B, a toxic protein from the mistletoe *Phoradendron tomentosum* subsp. *macrophyllum*. *Acta Pharmacol. Suecica* 20:115–22

66. Thunberg, E., Samuelsson, G. 1982. The amino acid sequence of ligatoxin A,

from the mistletoe *Phoradendron liga*. *Acta Pharmacol. Suecica* 19:447–56

67. Van Etten, C. H., Gagne, W. E., Robbins, D. J., Booth, A. N., Daxenbichler, M. E., Wolff, I. A. 1969. Biological evaluation of *Crambe* seed meals and derived products by rat feeding. *Cereal Chem.* 46:145–55

68. Van Etten, C. H., Nielsen, H. C., Peters, J. E. 1965. A crystalline polypeptide from the seed of *Crambe abyssinica*. *Phytochemistry* 4:467–73

69. Vermeulen, J. A. W. H., Lamerichs, R. M. J. N., Berliner, L. J., De Marco, A., Llinas, M., et al. 1987. ^{1}H-NMR characterization of two crambin species. *FEBS Lett.* 219:426–30

70. Vernon, L. P., Evett, G. E., Zeikus, R. D., Gray, W. R. 1985. A toxic thionin from *Pyrularia pubera*: purification, properties, and amino acid sequence. *Arch. Biochem. Biophys.* 238:18–29

71. Wada, K., Buchanan, B. B. 1981. Purothionin: a seed protein with thioredoxin activity. *FEBS Lett.* 124:237–40

72. Wada, K., Ozaki, Y., Matsubara, H., Yoshizumi, H. 1982. Studies on purothionin by chemical modifications. *J. Biochem.* 91:257–63

73. Whitlow, M., Teeter, M. M. 1985. Energy minimization for tertiary structure prediction of homologous proteins: α_1-purothionin and viscotoxin A3 models from crambin. *J. Biochem. Struct. Dynam.* 2:831–48

74. Williams, R. W., Teeter, M. M. 1984. Raman spectroscopy of homologous plant toxins: crambin and α1- and β-purothionin secondary structures, disulfide conformation, and tyrosine environment. *Biochemistry* 23:6796–6802

75. Winterfeld, V. K., Bijl, L. H. 1949. Viscotoxin, ein neuer Inhaltsstoff der Mistel (*Viscum album* L.). *Liebigs Ann. Chem.* 561:107–15

76. Woynarowski, J. M., Konopa, J. 1980. Interaction between DNA and viscotoxins. Cytotoxic basic polypeptides from *Viscum album* L. *Hoppe-Seyler's Z. Physiol. Chem.* 361:1535–45

Annu. Rev. Plant Physiol. Plant Mol. Biol. 1991. 42:241–79

THE ROLE OF HOMEOTIC GENES IN FLOWER DEVELOPMENT AND EVOLUTION

Enrico S. Coen

John Innes Institute, AFRC Institute of Plant Science Research, Norwich NR4 7UH, United Kingdom

KEY WORDS: *Antirrhinum majus, Arabidopsis thaliana*, inflorescence, heterochronic, *Peloria*

CONTENTS

1040-2519/91/0601-0241$02.00

INTRODUCTION

Historical Background

The term homeosis was coined in 1894 by Bateson (7) and defined as "the assumption by one member of a meristic series, of the form or characters proper to other members of the series." However, the importance of homeosis for the study of development had already been appreciated as early as 1790 by Goethe (35), who was an accomplished botanist as well as a great poet. He argued that plants are constructed from a series of equivalent (homologous) organs and that in the course of development there is a progressive transformation (metamorphosis) in the form assumed by the organs, starting with the leaf and culminating in the parts of the flower. He proposed that the mechanism underlying the progressive metamorphosis was the sequential purification of an initially crude sap. In support of these ideas he described several homeotic transformations (retrograde metamorphoses), abnormalities in which one organ assumed the form normally associated with a different organ.

In spite of this remarkable study, mainstream plant developmental biology has left the area of homeosis largely unexplored (69), partly because there has been no easy way to address experimentally the detailed mechanisms involved. Studies on homeosis in plants have therefore been mainly restricted to the domains of morphology and genetics. In the 19th and early 20th centuries, several plant morphologists compiled catalogs of the many cases of plant teratologies (63, 70, 78, 107). The arrival of classical genetics also stimulated some limited studies on the inheritance of plant abnormalities, and these showed that single genes (homeotic genes) could control homeotic phenotypes (11, 20, 86, 102).

The advent of molecular genetic techniques has caused a resurgence of interest in homeotic genes. Dramatic progress has been made in the study of these genes in *Drosophila,* and similar studies are now being carried out in plants (27, 94, 108). Here the background and current state of our knowledge of homeotic genes in plants and their relationship to development and evolution are described. Rather than to give an exhaustive list of cases of homeosis, the aim is to describe the major principles they illustrate and to highlight some important problems and areas that need to be addressed in the future.

Features of Plant Development

Most of our current notions of how homeotic genes act come from studies on segmentation in *Drosophila* (33, 49, 71). In certain respects there are many similarities between homeotic genes in plants and those in *Drosophila*. In both cases, the genes switch the developmental pathways of homologous structures and many of the genes appear to interact in a combinatorial manner. However, there are also some important differences that reflect the distinct modes of plant and fruit fly growth. Organs are produced throughout a plant's life history from meristems by sequential addition, giving plants the potential for indeterminate growth. In contrast, the segments of *Drosophila* are produced synchronously by subdivision of an egg of finite size. The additive and sequential plant growth pattern is reflected in the homeotic phenotypes, which can result in premature developmental arrest or continuous reiteration of a developmental program. Consequently, as well as being homeotic, some such mutations can be considered to be heterochronic because they change the relative timing of developmental events (2).

Plant cells are immobile and surrounded by rigid cell walls. The form of an organ is therefore dictated by the pattern of cell divisions, sizes, and shapes (61). In contrast, animal cells can be mobile and lack cell walls, so that overall form depends on a complex interaction between specialized skeletal or supporting structures and the other tissues in the body. The properties of plant cells provide great potential for modification of form through relatively simple changes in basic cell behavior. Taken together, these aspects of plant growth allow for an enormous diversity in the number, identity, and arrangement of organs both within one individual and in different species. The study of homeotic genes in plants therefore provides an excellent opportunity to study fundamental processes involved in development and evolution.

EXPERIMENTAL SYSTEMS

Antirrhinum *and* Arabidopsis *as Model Systems*

Many different plants species have been studied and characterized genetically [e.g. maize (23, 90), barley (73), wheat (89), tomato (83), pea (14), *Arabidopsis* (55), and *Antirrhinum* (39, 99)], providing a valuable resource for comparative studies of gene action. At present, the molecular and genetic analysis of floral homeotic genes is most advanced in two species: *Antirrhinum majus* and *Arabidopsis thaliana*. *Antirrhinum* has several advantages: well-characterized transposable elements, large flowers that are easy to emasculate and cross, and ease of vegetative propagation (26, 39). *Arabidopsis* has different merits: a small genome size, a rapid generation time, ease of transformation, and small plant volume (67). These two species are particularly interesting from a comparative view because their flower and inflorescence morphologies show some distinct features. In this review I concentrate on principles derived from studies on these species, also discussing their general applicability to other species.

Description of Wild Types

Following germination of wild-type *Antirrhinum* seed, initial growth of the plant above soil level is generated by a vegetative apical meristem that usually produces a pair of opposite (180°) leaves at each node in a decussate phyllotaxis (i.e. each pair is at a right angle to the previous one). After a period of vegetative growth, the apex undergoes a transition to become an inflorescence meristem (Figure 1). In comparison with vegetative growth, the inflorescence meristem produces much smaller leaves (bracts) in a spiral arrangement, generally with a single bract at each node separated by short internodes. In the axils of these bracts, floral meristems are initiated that produce four concentric whorls of organs separated by very short "internodes" so that consecutive whorls appear adjacent to each other. The aerial parts of the plant may therefore be considered to be generated by three homologous meristems—vegetative, inflorescence, and floral—that differ in their respective phyllotaxis (decussate, spiral, whorled), lateral organs (leaves, bracts, floral organs), and internode length (long, short, very short).

Arabidopsis aerial parts are also generated from three meristems, but some of their properties are different from those of *Antirrhinum*. The vegetative meristem produces a rosette of leaves separated by short internodes in a spiral phyllotaxis (Figure 1). The inflorescence meristem initially produces several small leaves, called cauline leaves, separated by relatively large internodes; in the axils of the cauline leaves, further inflorescence meristems arise. Above the cauline leaves, the inflorescence meristem produces floral meristems laterally in a spiral arrangement. The floral meristems of *Arabidopsis*, in common with those of other crucifers, are not subtended by bracts (87). As in *Antirrhinum*, the floral meristems produce a series of concentric whorls of organs separated by very short "internodes."

Both species are considered to have four whorls of floral organs, numbered starting from the outside of the flower; thus whorl 1 is outermost and whorl 4 is central. Organs within a whorl are referred to as upper or lower, depending on their position relative to the bract and/or stem (bract is lower, stem is upper, Figure 1). The whorl primordia appear sequentially: whorl 1 first, then whorls 2 and 3 almost simultaneously, and finally whorl 4 (5, 92).

Antirrhinum flowers are zygomorphic: they can be divided into two halves by a single longitudinal plane that passes through the vertical axis (Figures 1 and 2). Whorl 1 comprises five sepals, the lowest two being alternate to the bract. The corolla occupies whorl 2 and consists of five petals; these are

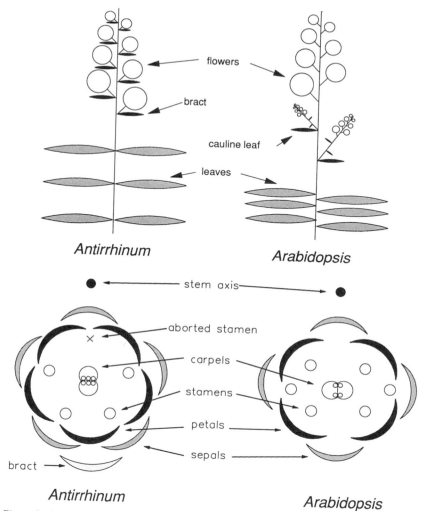

Figure 1   Schematic illustrations of wild-type *Antirrhinum* and *Arabidopsis*. Upper illustrations show whole plants; lateral shoots arising in the axils of leaves are omitted for clarity. Lower illustrations show floral diagrams.

united for part of their length to form a tube that terminates in five lobes (i.e. the corolla is sympetalous). The shape of the two upper lobes is very distinct from that of the lower three, and the junction between the upper and lower lobes forms a hinge region that allows the flower to be opened by insects (Figure 2). In whorl 3, five stamen primordia are initiated that alternate with the petals, but the uppermost primordium fails to develop fully and yields an

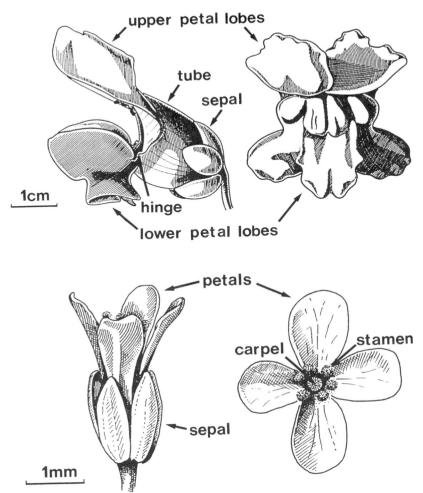

Figure 2   Flowers of *Antirrhinum (upper)* and *Arabidopsis (lower)*. Flowers are shown in side view *(left)* or face view *(right)*. The *Antirrhinum* flower shown in side view is opened slightly to illustrate the hinge between upper and lower petals. The face view of *Arabidopsis* is at a 45° orientation relative to the floral diagram in Figure 2. *Antirrhinum* drawings were adapted from Weberling (106); *Arabidopsis* drawings were by Keith Roberts.

aborted or rudimentary structure. Consequently, the adult flower has only four stamens, the upper two being shorter than the lower pair. Whorl 4 is occupied by two united carpels forming a gynoecium with a bilocular ovary.

Arabidopsis flowers are actinomorphic: They can be divided into two equal halves by more than one longitudinal plane passing through the flower center

(Figures 1 and 2). Whorl 1 contains four sepals and whorl 2 contains four separate equal petals, which alternate with the sepals. Whorl 3 contains six stamens, and their arrangement results in the flower's having only two planes of reflection (mirror-image) symmetry (one passes through the vertical axis, the other the horizontal axis). The stamens have often been considered as comprising two whorls, but as described elsewhere (15) and discussed in more detail below, they are treated here as a single whorl. Whorl 4 is occupied by a gynoecium with a bilocular ovary. For brevity, the identity of whorls in both species is indicated in sequence starting from whorl 1, so wild-type is sepal, petal, stamen, carpel.

GENETIC CONTROL OF FIRST STEPS IN FLORAL DEVELOPMENT

Phenotypic Classes of Mutants

Many physiological and genetical studies on the control of flowering have appeared (13). In most, a flowering plant is considered to be one that has both inflorescence and floral meristems, whereas a nonflowering plant has only vegetative meristems. A general conclusion from these studies is that induction of flowering depends on multiple complex interactions between the plant and its environment. Mutants affecting this process generally alter the time or environmental circumstances required for flowering.

Unfortunately, the transition from inflorescence to floral meristems has been relatively neglected in physiological studies on flowering because both meristems are generally induced together. However, this transition represents the first step specific to the floral developmental pathway and is therefore of central importance to understanding floral morphogenesis. Mutants unable to carry out the transition between inflorescence and floral meristems might be expected to produce proliferating inflorescence shoots in place of flowers. A well-characterized mutant showing this phenotype is *floricaula (flo)* in *Antirrhinum* (18). The *flo* mutant initiates vegetative growth and the transition to the inflorescence meristem in a manner similar to that of the wild type. Instead of flowers being produced in the axils of bracts, indeterminate shoots bearing further bracts are produced, each shoot having two opposite bracts at the base followed by a spiral of single bracts. Each of these shoots can in turn produce further shoots in the axils of its bracts, and this sequence can repeat itself indefinitely. The wild-type *flo* product is therefore necessary for the transition between inflorescence and floral meristems; in its absence, the inflorescence program is continually reiterated (Figure 3). The *flo* mutant can be considered homeotic because it results in the replacement of one structure (the flower) by a homologous structure (indeterminate shoot). The mutant can also be viewed as heterochronic because it results in continual reiteration of an early developmental program (inflorescence meristem).

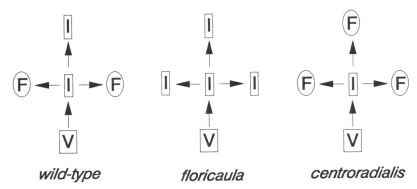

Figure 3 Types of meristems in wild-type, *floricaula* and *centroradialis* mutants of *Antirrhinum*. Abbreviations for meristems are: vegetative (V); inflorescence (I); floral (F).

Other mutants have been described in *Antirrhinum* that produce inflorescence-like shoots in the axils of bracts, but in many cases these shoots can eventually produce flowers (21, 57, 88, 99). It is not clear if the production of flowers reflects leakiness in the particular alleles of these genes or a specific property of the genes themselves. Mutations that produce proliferating inflorescence meristems have also been described in other species, such as *cauliflower-head* in alfalfa (12), *anantha* in tomato (41, 75), *leafy* in *Arabidopsis* (40), *ramosa-1* in maize (72, 81), and perhaps *veg* in pea (36, 82). Their phenotypes can appear to be quite different from that of *flo* because of differences in the wild-type inflorescence. For example, the *anantha* mutant of tomato produces highly branched indeterminate inflorescences with no flowers or bracts. This is presumably because in wild-type tomato, flowers arise at the apexes of a branching inflorescence and do not have subtending bracts. Interpretation of the *veg* mutant in pea is more complicated. The apical meristem of wild-type pea produces leaves initially and then undergoes a transition to produce leaves with inflorescences in their axils. In some lines of pea, this transition of the apical meristem is associated with other features, such as reduction in internode length. The apical meristem of the *veg* mutant appears to undergo a transition similar to that in the wild type (as judged by internode-length and by its interaction with certain genes affecting flowering time), but leafy side-branches with short internodes are produced in place of inflorescences (82). Whether this mutant is analogous to *flo* depends on which meristem in pea (apical or axillary) is considered to be homologous to the inflorescence meristem of *Antirrhinum*.

A second class of mutant has an effect almost the opposite of that described above because it promotes the conversion of inflorescence to floral meristems in positions where this would not normally occur. Plants with indeterminate inflorescences, such as *Antirrhinum* and *Arabidopsis,* do not produce terminal flowers, because the apical inflorescence meristem does not itself undergo the transition to a floral meristem. The *centroradialis* mutant in *Antirrhinum* produces terminal flowers (57), suggesting that the wild-type product of this gene prevents the conversion of the inflorescence apex to a floral meristem (Figure 3).

Molecular Analysis of Early Homeotic Genes

The *flo* gene of *Antirrhinum* has been isolated by transposon tagging and extensively characterized (18, 27). A plant homozygous for an allele called *flo*-613 was obtained from a general transposon-mutagenesis experiment. This plant had a typical *flo* phenotype and initially needed to be propagated vegetatively because of its lack of flowers. However, occasional flowers were observed on the propagated *flo*-613 mutant plants, and seed of these flowers sometimes gave wild-type progeny, showing that *flo*-613 was genetically unstable and could revert to a wild-type allele. It was shown that reversion to wild type correlated with excision of a particular copy of the transposon Tam3 in the genome, indicating that this copy was responsible for the mutant phenotype. It was possible to clone the sequences flanking this copy using Tam3 as a probe and thus isolate the *flo* locus.

The *flo* gene produces a transcript of about 1.6 kb that has the potential to encode a protein FLO of 396 amino acids. The protein contains a proline-rich N-terminus and an acidic region; both of these features have also been found in activation domains of transcription factors (65, 98). However, FLO shows no extensive homology to other proteins in current databases; thus although it is possible that FLO functions as a transcriptional activator, other roles cannot be excluded. In situ hybridization shows that *flo* is expressed from an early stage in wild-type inflorescences in a very specific temporal and spatial sequence. The earliest expression is in bract primordia and is followed by expression in sepal, petal, and carpel primordia; but no expression is detected in stamen primordia. Expression in each organ is transient and is not observed in later stages of development.

Taken together, these results suggest that *flo* not only acts as a switch between inflorescence and floral meristems but is also involved in directing specific patterns of gene expression in the early floral meristem. The expression of *flo* in certain primordia may be required to activate genes necessary for their normal development. Similarly, the absence of *flo* from whorl 3 may be necessary for normal stamen development. One important class of genes that may interact with *flo* in particular whorls is that of homeotic genes controlling whorl identity; a possible model for such interactions is discussed in detail below. However, it is important to emphasize that the expression of *flo* in a primordium need not imply a role for its development. For example, even though *flo* is expressed in bracts, it is not required for bract development: These organs apparently develop normally in *flo* mutants.

One of the earliest effects of *flo* is on phyllotaxis, because the first organs produced by a wild-type floral meristem comprise a whorl of five primordia, whereas in a *flo* mutant the axillary inflorescence meristem initiates two opposite primordia, followed by a spiral of single organs. Clearly the action of *flo* on phyllotaxis should precede the physical appearance of these primordia. One possibility is that *flo* expression in bract primordia gives a signal that affects the number of primordia generated by the axillary floral meristem. Another explanation is that expression in early floral meristems determines the number and arrangement of primordia produced.

Cell-Autonomy of Genes

As has already been mentioned, the *flo*-613 mutant occasionally produces wild-type flowers and sometimes whole inflorescence spikes on an otherwise mutant plant. In most cases, seed derived from these flowers give mainly mutant progeny, but sometimes the progeny segregate in the ratio 3 wild type:1 mutant. The cases of germinal transmission of the revertant phenotype suggest that excision of Tam3 from *flo*-613 occurred early in somatic development to give a heterozygous flower that was Flo^+/flo-613. However, the failure of germinal transmission in most cases shows that functional gametes can be produced from homozygous mutant tissue (*flo*-613/*flo*-613). One explanation for this is that FLO is not essential for floral development; other *flo*-independent pathways exist, allowing occasional flowers to be produced. This explanation does not seem to be correct because the *flo*-640 allele, which arose as an imprecise Tam3 excision from *flo*-613, has never been observed to produce flowers (27). A more probable explanation is that *flo* does not act in a cell-autonomous manner in the tissue giving rise to gametes. The lineage of cells giving rise to gametes is usually restricted to a particular cell layer in the flower (the LII) so that only events represented in this layer will be transmitted to progeny (48, 101). It is possible that restoration of FLO activity in the epidermal (LI) layer is sufficient to direct floral development, so that Tam3 somatic excisions that only affect the epidermis may allow development of a normal flower even though the genotype of its cells in the LII layer are still mutant.

Evolution of the Inflorescence

Some plant species have single flowers at the termini of shoots. In most species, however, flowers are clustered together in inflorescences that have been classified into two basic types: determinate and indeterminate (97, 106). In plants with determinate inflorescences, the apical inflorescence meristem is converted to a floral meristem and further floral meristems are produced in axillary positions or at the apexes of side branches (Figure 4). In contrast, the main inflorescence apex of indeterminate inflorescences (e.g. *Antirrhinum*, *Arabidopsis*) is not converted into a floral meristem.

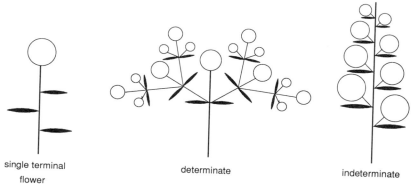

single terminal
flower

determinate

indeterminate

Figure 4 Examples of three different types of inflorescence. The particular determinate inflorescence shown is a dichasium and the indeterminate inflorescence is a raceme.

Several evolutionary schemes have been proposed for the derivation of the various inflorescence types from an ancestral condition (76, 84). One important feature of this evolutionary process is the change in timing and position at which inflorescence meristems switch to the floral pathway. The isolation of genes such as *flo*, which regulate this transition, now opens these problems to molecular analysis. For example, in plants with single terminal flowers there may be no distinct inflorescence meristem, and genes such as *flo* may be activated in the apex directly after vegetative growth. In determinate inflorescences, *flo* may be activated in the main apex and in the apexes of side branches. In indeterminate inflorescences, *flo* expression may be restricted to bracts and axillary meristems. The study of how *flo* is regulated in different plant species and also in mutants such as *centroradialis*, which convert indeterminate to determinate inflorescences, should provide considerable new insights into inflorescence evolution.

GENETIC CONTROL OF WHORL IDENTITY

Phenotypic Classes of Mutants

The most extensively documented cases of homeosis in flowers are those in which an organ in one whorl develops in the form typical of an organ from a different whorl. Such transformations in identity may affect a single organ, all the organs in a whorl, or several whorls of organs. This phenomenon was originally called metamorphy, and examples from diverse species have been compiled by several authors (63, 66, 70, 78, 107). Metamorphies are generally classified according to the particular fate assumed by organs: phyllody, sepalody, petalody, staminody, carpellody. Sepalody, for example, refers to sepal-like structures growing in place of one or more other organs in the flower. A general conclusion from these studies is that "all of the organs in the flower are capable in some plant or other, of developing in the form of any of the other organs present in the normal flower" (66). Although this is a valuable observation and confirms Goethe's proposition that all floral organs are homologous to each other, it is too general to be of use in constructing specific models for the control of organ identity. One problem with these extensive studies on metamorphy is their lack of selectivity. They cover many different plant species and include sporadic observations, environmentally induced abnormalities, and some heritable mutations. More recently, selective studies in particular species have been carried out, most notably in *Antirrhinum* and *Arabidopsis*, to determine systematically the types of metamorphic mutations that can occur (15, 18, 40, 54, 58, 88). These studies show that some phenotypes are repeatedly observed. In particular, a group of genes that affect the identity of whorls of organs, subsequently referred to as whorl identity genes, have been identified and characterized in detail. (Note that these genes do not affect the numbering of the whorls but only the identities of organs in the whorls.)

A major class of whorl identity genes contains those that affect the identities of organs in two adjacent whorls (Table 1). For ease of description the wild-type flower of *Antirrhinum* or *Arabidopsis* can be divided into three overlapping regions, each consisting of two adjacent whorls: A (whorls 1 and 2); B (whorls 2 and 3); C (whorls 3 and 4) (Figure 5). One class of mutants affects region A and gives carpels instead of sepals in whorl 1 and stamens in place of petals in whorl 2, giving the overall phenotype carpel, stamen, stamen, carpel. This is conferred by the semi-dominant *ovulata (ovu)* mutant in *Antirrhinum* and some *apetala2 (ap2)* mutants in *Arabidopsis*. A second class affects region B and gives sepals instead of petals in whorl 2 and carpels instead of stamens in whorl 3, giving the phenotype sepal, sepal, carpel,

Table 1 Phenotypes of some whorl identity mutants in *Antirrhinum* and *Arabidopsis*

genotype	phenotype				region affected
	whorl 1	whorl 2	whorl 3	whorl 4	
wild-type	sepal	petal	stamen	carpel	
ovu, ap2	carpel	stamen	stamen	carpel	A
def, glo, sep, pi, ap3	sepal	sepal	carpel	carpel[a]	B
pleni, ag	sepal	petal	petal	variable[b]	C

[a] Whorl 4 does not always develop in some of these mutants.
[b] In *Antirrhinum*, whorl 4 can be petaloid, sepaloid, carpeloid, or a mixture of these; in *Arabidopsis* this whorl contains sepals.
In both species extra petaloid or sepaloid whorls are produced within whorl 4.

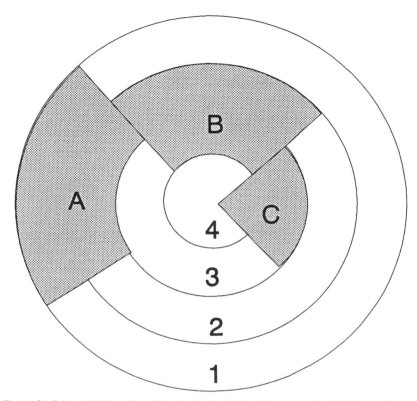

Figure 5 Schematic illustration of the 4 whorls (1–4) and 3 regions (A, B, C) of a flower.

carpel (whorl 4 does not always develop in these mutants). This class includes the *deficiens (def)*, *globosa (glo)*, and some *sepaloidea (sep)* mutants in *Antirrhinum* and *pistillata (pi)* and *apetala3 (ap3)* in *Arabidopsis*. A third class affects region C and gives petals instead of stamens in whorl 3 and sepals or variable structures in whorl 4. Usually these mutations also affect the number of whorls and give extra whorls of petals or sepals within whorl 4. Examples include the *pleniflora (pleni)* mutants of *Antirrhinum* and the *agamous (ag)* mutants of *Arabidopsis*. In *Antirrhinum* several other mutations with phenotypes similar to those mentioned above have also been described, but in many cases complementation tests have not yet been carried out or cannot be done because some of the lines are no longer available (18, 88). The observation of very similar classes of mutation in two taxonomically distant species, *Antirrhinum* and *Arabidopsis*, suggests that the mechanisms controlling whorl identity have been highly conserved in evolution. It remains to be demonstrated directly, however, that these similarities reflect the action of homologous genes in the two species.

Examples of some of these classes of mutation have also been described in other species, such as *Primula* (16), *Cheiranthus* (20), and *Mathiola* (86). Although many of these mutants have been known for a long time, the importance of their dual effects on two adjacent whorls has not generally been appreciated, partly because they have been buried among many other examples of metamorphies in general treatises and partly because the classification of abnormalities according to the fate of an organ (i.e. sepalody, petalody etc) has meant that each whorl was considered separately.

Models for the Genetic Control of Whorl Identity

The action of many gene functions in overlapping regions could give each whorl a unique combination of functions. For example, if genes acting in regions A, B, and C are required for three functions *a*, *b*, and *c*, respectively, then the combination of functions in the four whorls of wild type would be *a*, *ab*, *bc*, *c*. In principle, this might provide sufficient information to specify a distinct identity for organs in each whorl. An important constraint on such a combinatorial model is that it needs to account for the particular mutant phenotypes observed when *a*, *b*, or *c* is mutated. By this criterion, the simple model that the domains of *a*, *b*, and *c* expression are established independently of each other does not fit the data. For example, if *a* is required for sepal and petal development in region A, how can these organ types develop outside this region in certain mutants (e.g. mutants lacking *c* have petals in whorl 3; see above)?

Consistent models can be constructed assuming that only two functions (e.g. *a* and *b*) are established independently of each other (18). If only the interaction of *a* and *b* is considered, the combination of functions in the four wild-type whorls would be *a*, *ab*, *b*, *0*. Mutants lacking *a* would give *0*, *b*, *b*, *0*, consistent with the observed phenotype carpel, stamen, stamen, carpel. Mutants lacking *b* would give *a*, *a*, *0*, *0*, consistent with the observed phenotype sepal, sepal, carpel, carpel. A further prediction is that double mutants lacking *a* and *b* should give only carpels: *0*, *0*, *0*, *0*. Such double mutants have recently been described in *Arabidopsis* and have the expected phenotype although all four whorls do not usually develop (68). Thus, if only *a* and *b* are considered, the carpel appears to be a "ground state."

In order to incorporate the third function, *c*, in such a model, it is necessary to postulate that the *a* and *c* functions are not established independently but can influence each other's expression (18, 68). For example, if *c* inhibits *a* in region C, a mutant lacking *c* should be *a*, *ab*, *ab*, *a*, giving the predicted phenotype sepal, petal, petal, sepal. The first four whorls of *ag* mutants of

Arabidopsis have this phenotype. The *pleni* mutants in *Antirrhinum* are similar, although the fate of whorl 4 is variable in these mutants and can be sepaloid, petaloid, carpeloid, or a mixture of these. This result may reflect differences in the action of *ag* and *pleni* or differences in the particular alleles of these genes studied so far.

Further insights into the interactions between *a* and *c* have come from studies on double and triple mutants in *Arabidopsis* (68). Double mutants lacking *a* and *c* have the first four whorls with the phenotype cauline leaf, staminoid organ, staminoid organ, cauline leaf (the cauline leaves have some carpeloid features). The *a;c* double mutant therefore differs in all four whorls from either single mutant, indicating that in mutants lacking *c* the domain of *a* action extends to all whorls and vice versa. This outcome suggests that in a wild-type flower, *a* and *c* inhibit each other such that the action of each function is restricted to its respective region. Triple mutants lacking *a, b,* and *c* only produce cauline leaves with some carpeloid features, in contrast to the *a;b* double mutant, which only produces carpels. Thus considering *a* and *b* alone, the carpel is the "ground state"; but if *c* is also included, the cauline leaf is the "ground state." That is, one effect of the *c* function is to transform cauline leaf into carpel. Thus "ground state" is a relative term that depends on the genes being considered. In the absence of double and triple mutants in *Antirrhinum,* it is unclear whether the same interactions are true for this species; the possible interactions between the *a, b,* and *c* functions with *flo* (see below) suggest that the bract may be a "ground state" in *Antirrhinum,* which corresponds to the cauline leaf of *Arabidopsis.* This is because *flo* is probably required for activation of *a, b,* and *c* so that the identity of organs produced in a *flo* mutant (i.e. bracts) might be expected to be similar to those produced in an *a;b;c* triple mutant. However, it should be noted that a bract in *Antirrhinum* subtends a flower whereas in *Arabidopsis* the cauline leaf subtends an inflorescence; thus it is not clear if these two structures are truly homologous in the two species.

So far only one class of whorl identity genes has been considered in detail, but other types of homeotic transformations exist (66). The *apetala1* mutation in *Arabidopsis* affects region A and gives bract-like organs in whorl 1 and floral buds in whorl 2 (50). In other cases a single whorl may be affected. A common phenotype in horticultural varieties is growth of petals in place of stamens (66, 86), and this phenotype has also been observed in *Antirrhinum* (R. Carpenter, E. Coen, unpublished results). The *heptandra* mutant in *Digitalis purpurea* only affects whorl 2 and gives stamens in place of lower petals (42, 85, 91). The models presented above are therefore almost certainly oversimplified, and more sophisticated versions should be possible when the full range of molecular and genetic interactions among all of these genes is determined.

Molecular Analysis of Whorl Identity Genes

So far two different whorl identity genes have been isolated: *def* and *ag*. The *def* gene, required for the *b* function in *Antirrhinum*, was isolated using differential cDNA screening (94). Classical genetic studies have shown that a particular *def* allele (*deficiens*[globifera], also renamed *defA*-1) is unstable both in somatic and in germinal tissue (11, 44), indicating that this allele might be caused by a transposon insertion (24, 94). The strategy for isolating the *def* gene was based on the assumption that *def* expression was abolished or significantly reduced in the *defA*-1 mutant. A cDNA library was constructed from wild-type mRNA enriched for flower-specific transcripts. The library was screened with probes made from inflorescence mRNA of wild type or the *defA*-1 mutant. Several clones were isolated that hybridized to wild-type but not to mutant probes; these could be divided into 12 classes. In Southern blots, clones of one class revealed a band in the *defA*-1 mutant distinct from that of wild type. Somatic or germinal reversion of the *defA*-1 allele to wild type correlated with restoration of the wild-type band on Southern blots, proving that this class of clones hybridized to the *def* gene.

The *ag* gene, required for the *c* function in *Arabidopsis*, was isolated using T-DNA insertion mutagenesis (108); 136 independent transformants, each carrying one or more T-DNA insertions were obtained by cultivating *Agrobacterium tumefaciens* together with seed of *Arabidopsis* (32). One of these transformants carried a mutation in the *ag* locus, called *ag*-2, and genetic analysis showed that the *ag*-2 allele co-segregated with the kanamycin resistance marker encoded by the T-DNA. This suggested that the *ag*-2 mutation was caused by insertion of T-DNA into the *ag* locus. The sequences flanking the T-DNA of the *ag*-2 mutant were cloned and used as probes to isolate a cosmid carrying the *ag* locus from a wild-type plant. Complementation of the *ag*-2 mutant with this cosmid confirmed that it carried the entire wild-type *ag* gene.

The *def* and *ag* genes both produce transcripts of about 1.1 kb (94, 108). The sequence of corresponding cDNAs indicates that *def* and *ag* transcripts have the potential to encode proteins of 227 and about 290 amino acids, respectively, that share a homologous region of about 50 amino acids. The shared region also shows strong homology to a conserved region found in a family of transcription factors present in humans (SRF; 74) and yeast (MCM1; 77), suggesting that both *def* and *ag* encode related transcription factors. By using *def* and *ag* as low-stringency probes on flower cDNA libraries, additional genes have been isolated that share homology in the 50-amino-acid conserved domain (88, 108). In the case of *Antirrhinum* two of these genes have been shown to be good candidates for the homeotic genes, *globosa* and *squamosa* (88). Families of related transcription factors controlling a common developmental pathway have also been described for other organisms. For example, in *Drosophila* several different homeotic selector genes encode transcription factors with a homologous region (the homeobox) thought to be involved in binding to DNA (71). These homeotic selector genes are thought to interact in a combinatorial way to determine segment identity (33, 59). Thus the combinatorial models proposed for flower or insect development both appear to reflect interactions between related transcription factors.

Expression of both *def* and *ag* can be detected from early stages in wild-type flower development when primordia are becoming visible, mainly in the regions they affect (B and C, respectively). In the case of *def*, a low level of expression is also detected in whorls 1 and 4. Unlike *flo* gene expression, which is transient, *def* and *ag* expression is maintained until late stages of organ development. Expression of *ag* has also been studied in *ap2* mutants; as predicted from the above model, *ag* expression extends to all whorls of the flower in these mutants (68). This shows that in wild type, the *a* function acts at the level of gene expression to restrict *c* to region C.

The genetic and molecular analysis of the whorl identity genes has already provided some important insights into how floral development is controlled. These results, however, leave several important questions unanswered: How are the whorl identity genes regulated, at what time do they act, and what do the whorl identity genes themselves regulate?

Regulation of Whorl Identity Genes

Several models for the control of whorl identity have been proposed, but unfortunately most take little or no account of the extensive genetic data available (69). Many of the models propose that activities in the floral meristem change in a specific time sequence such that earlier-developing primordia have a fate different from that of later ones (28, 37, 45–47, 105). These models are therefore sequential because the consecutive growth or initiation of the primordia is essential for the determination of organ identity. Perhaps the best evidence for these sequential models comes from surgical alterations involving bisection or isolation of developing floral meristems (28, 43, 46, 51, 64, 93). These experiments show that it is difficult to perturb the temporal order in which the different organ types arise. Furthermore, if primordia are induced at abnormal positions, they adopt a fate typical of other primordia initiating at about the same time. Thus although organ identity can be uncoupled from position it cannot easily be uncoupled from time. These results were taken to suggest that the developing floral meristem passes through a succession of physiological states determining the formation of each type of organ (28, 46). However, because the experiments involve sequential regeneration, it is not clear whether the fixed direction of this succession of states reflects constraints on the process of regeneration or on the determination of organ identity. More sophisticated approaches involving transplantations rather than incisions have not so far proved possible.

Clearly the whorl identity mutations do uncouple both the time and location of primordium initiation from final developmental fate, because the types of organs in one or more whorls can be altered. There has therefore been much confusion in reconciling the sequential models with the whorl identity mutants (46, 69). For example, according to one type of sequential model, the activity of primordia in whorl 2 directs the fate of stamen primordia in whorl 3 (45), yet mutants lacking the a function have stamens in whorl 2 without any antecedent petals. This and many other examples described earlier show that the identity of an organ does not depend directly on the whorl it occupies or on the identity of the preceding or subsequent organs. The sequential models can only be reconciled with the genetic data if organ identity is an outcome rather than a cause of sequential processes. That is, activation or regulation of the whorl identity genes may depend on sequential processes, but the subsequent action of these genes in directing organ identity is not restricted by time or position in the floral meristem. The control of primordium fate therefore needs to be restated in genetic terms: How are the domains of the whorl identity genes established?

Two general types of model might be imagined for explaining the expression pattern of whorl identity genes. According to a purely spatial model, concentric fields could be set up in the early floral meristem, independently of the sequence of primordium initiation. The a, b, and c functions could be activated in the appropriate field domains and hence specify the fate of primordia growing out from the various regions of the meristem. This type of model is formally similar to models proposed for the control of segment identity in *Drosophila* (49). In the sequential type of model, which shares some features of the earlier models of flower development, the consecutive growth of the primordia would be essential for the establishment of the domains of activity. The a, b, and c functions would be activated in a manner that reflected the sequence of primordium initiation.

Some supporting evidence for a sequential model has come from studies on the *flo* gene of *Antirrhinum* (27). As described above, *flo* is expressed sequentially and appears to be activated in regions where primordia are being initiated (in some cases its activation seems to precede the physical appearance of primordia). A further feature of *flo* expression is that it occurs transiently in bract, sepal, petal, and carpel but not stamen primordia. According to one model (27), as *flo* expression progresses from initiating bract to sepal and then petal primordia it activates a in whorls 1 and 2, perhaps in combination with other gene-products specific to the floral meristem [stages (i)–(iii), Figure 6]. Before *flo* expression has been established in whorl 3, the b and c functions could be activated so that they become established in the presumptive whorl 3 cells. The combination of b and c (or b in the absence of a) could then inhibit *flo* expression and prevent *flo* from being established in the third whorl [stage (iii), Figure 6]. The factors activating b and c are

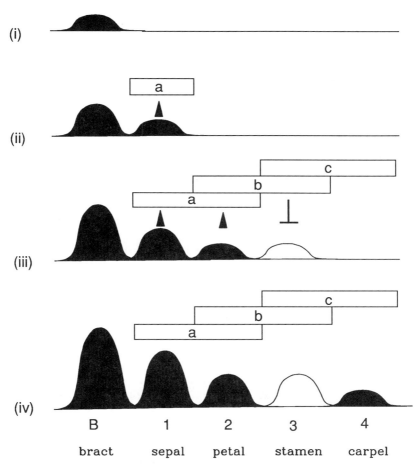

Figure 6   Model for the interaction of *flo* with the whorl identity functions *a, b, c;* (i)–(iv) show progressive stages of the development of bracts and whorls 1–4. The developmental stage of each primordium is indicated by a dome that increases in height with time. It is important to emphasize that each dome does not represent the physical appearance of the primordium but only its developmental stage. It is likely that some of the small domes represent early stages of development that occur before the physical appearance of the primordia. The petal and stamen primordia are shown to be at a similar stage because they appear almost simultaneously in development. Solid-filled domes indicate *flo* expression whereas white domes show no *flo* expression. The regions representing the *a, b, c* functions are illustrated by open boxes. In (ii) and (iii) arrows indicate possible activation of *a* by *flo* and (⊥) indicates inhibition of *flo*. See the text for a full explanation.

unknown but may depend on a separate cascade of events initiated by early *flo* activity. The absence of *c* from region A might be explained by the prior establishment of the *a* function in this region (*a* inhibits *c*; see above) by the time *c* is activated. In agreement with this view, the *a* function in *Arabidopsis* has been shown to be active before expression of *ag,* required for *c,* can be detected (68). Finally, *b* is inactivated in whorl 4, allowing expression of *flo;* but because *c* is already preestablished in this whorl, *flo* is unable to activate *a* [stage (iv)].

This model presents one plausible set of interactions, consistent with the data so far, although many other possibilities could be imagined. The model illustrates two important features of sequential systems. First, the fate of a whorl may depend on the temporal sequence in which gene expression is activated so that preestablishment of a gene function in one region can determine whether genes can be activated later on in that region. For example, in region A, *a* is activated early and prevents *c* from being established; but in region C, *c* is established early and prevents activation of *a.* Second, early and late genes need not interact in a simple hierarchical fashion. For example, even though *flo* is activated earliest in the floral meristem as a whole, in cells giving rise to whorl 3 the *b* and *c* functions could be established earlier than *flo* and therefore regulate its expression.

Timing of Gene Action

It is clear from the above discussion that the nature of genetic interactions depends on the time and duration of gene action. The timing of *def* gene action, required for the *b* function, has been studied in some detail. One of the effects of *def* mutations is that sepals grow in place of petals in whorl 2 (Table 1). The *def* gene is expressed at a nearly constant level throughout flower development (94), and some of the consequences of this expression for petal morphogenesis have been revealed by clonal analysis. From a transposon-mutagenesis experiment, an unstable allele, *def*-621, was obtained that shows distinct clonal patches of petal tissue on the sepals in whorl 2 (18). These patches are separated by sharp boundaries both from the surrounding sepal epidermal cells and from the underlying mesophyll tissue, and they can be explained by restoration of gene function (and hence petal morphology) as a result of somatic excision of a transposon from the *def* locus. Some of the patches comprise as few as four cells and presumably reflect restoration of *def* function within the last few cell divisions of organ development. These observations indicate that *def* can act at late stages to direct cells of whorl 2 towards a petal developmental program. However, the phenotypic effects of the *def* mutation can also be detected at early developmental stages, when petal and sepal primordia acquire distinct morphologies. Taken together, these results suggest that the *def* product is active in whorl 2 throughout its development, from the early stages when petal primordia become distinct to the final cell divisions of the petal. This is consistent with the observed expression of *def* throughout flower development (94).

Two further features concerning the timing of genetic interactions are revealed by the small clonal patches. First, because the clonal patches of *def*-621 occur only in sepals of whorl 2 and not whorl 1, there must be other genes that act specifically in whorl 2 throughout its development and that are necessary either for *def* expression or for the action of the *def* product. Two such genes may be *glo* and *sep,* because mutation of these can give a phenotype similar to that of the *def* mutant. Second, expression of *def* in whorl 2 early in development is not necessary for its later expression. This is because *def* can be correctly expressed late in development without any previous history of *def* activity in the organ. This phenomenon rules out models in which *def* is switched on by a transient early signal and then independently maintains its own expression.

It should be noted that the mode of *def* action in whorl 2 need not be the same as in whorl 3. The *def* mutant gives carpels in place of stamens in whorl 3, but no small patches of stamen tissue on these carpels have so far been observed in *def*-621 or *defA*-1 (88; R. Carpenter, E. Coen, unpublished). This lack may indicate that *def* does not act throughout development in whorl 3, or that it does not act cell-autonomously, or that its late action depends on early *def* function. The latter possibility is consistent with the proposed early inhibition of *flo* by the *b* function. Perhaps the action of *def* late in development is dependent on its early effect of inhibiting *flo*.

Additional evidence for the relatively late action of genes controlling organ development has come from studies on the effects of changing environmental conditions during flower morphogenesis. Certain *ap3* alleles in *Arabidopsis* confer a temperature-sensitive phenotype. By shifting the growth temperature during organogenesis, it has been shown that *ap3* can act in whorls 2 and 3 at relatively late stages of flower development (15). In *Impatiens balsamina,* the developmental fate of floral organs can be influenced at relatively late stages by certain light regimes (9, 10). If plants are transferred between noninductive (long daylength) and inductive (short daylength) conditions at certain times, primordia that would otherwise have developed into leaves can be directed towards petal development. If the primordia are transferred at a particular stage of their development, they produce leaves with petaloid areas. This result indicates that the developmental pathway of primordia can be redirected even after they have started along one course of development.

Although some genes act for extended periods of development, others appear to have a more transient action. The *flo* gene is transiently expressed in various primordia, precluding a very late role in development. Such expression is consistent with the view that *flo* is a primary regulator; it is also reminiscent of findings from studies of *Drosophila* segmentation that some of the early-acting regulatory genes are transiently expressed whereas the expression and activity of many homeotic selector genes are maintained for much longer periods of development (33, 71). Environmental-shift experiments with temperature-sensitive *ap2* alleles in *Arabidopsis* also suggest a transient period of early action (15). However, in this case the transience may reflect the temperature-sensitive period of the allele rather than the full duration of gene function.

Targets of Genes Controlling Whorl Identity

It is clear from the preceding discussions that the genes controlling whorl identity interact with each other, and therefore one such gene may be a target of another. For example, it has been proposed that *a* and *c* interact and that *flo* interacts with *a, b,* and *c*. In addition to these interactions, many other genes involved in organ morphogenesis must be directly or indirectly regulated by the genes controlling whorl identity. Because both direct and indirect effects are important and cannot be easily distinguished at present, these are collectively referred to here as target genes [in *Drosophila* they have been called "realizator genes" (33)].

Some clues about the nature of the interactions between homeotic and target genes come from studies of the *def* locus (18). Sepals and petals have distinct forms, cell-types, and colors. In *Antirrhinum*, a sepal grows as a relatively small separate organ terminating in a point; a petal is a larger organ united for part of its length with adjacent petals and terminating in a rounded lobe. The epidermal cells of a mature sepal are weakly pigmented with anthocyanin and have green underlying mesophyll cells. The petal typically has conical epidermal cells strongly pigmented with anthocyanin, and unpigmented underlying mesophyll cells. All these aspects of the organs are ultimately affected by the *def* locus because the *def* mutant has sepals in place of petals. The small clonal patches of petal tissue conferred by the *def*-621 allele are clearly identified by their distinct conical pigmented cells, but they do not affect the overall form of the organ because this has already been laid down by the growth of the primordium. Thus, whereas the effect of *def* on final cell types and color depends on late *def* activity, the form of the organ must depend on earlier *def* functions. This suggests that *def* affects different processes at different stages of development, so that at early stages it might affect cell division, shape, and size—and hence the form of the petal—whereas at later stages it might determine final cell types and pigmentation. This is consistent with the phenotypes of some weak *def* alleles (*def* [nicotianoides] and *def* [chlorantha]), which have second whorl organs that appear to be intermediate between sepals and petals in form, cell type, and color (44, 53).

As proposed above, the effect of *def* on target genes presumably changes with time during organogenesis so that different processes can be regulated at different stages. Analysis of these targets and their interactions presents a major challenge to understanding morphogenesis. There are several possible approaches to this problem. One is to use differential cDNA screening to analyze genes expressed in the wild type but not in *def* mutants. As described above, such a screen yielded 12 different classes of clones, one of which corresponded to the *def* transcript. The other 11 clones could represent target genes regulated by *def* (94).

One problem with this approach is that the expression of some important target genes may not be qualitatively dependent on *def* function but only quantitatively modulated by it. For example, anthocyanin pigmentation is most intense in petals of *Antirrhinum*, but it also occurs in all other floral organs at a lower level. Thus in "replacing" sepals with petals the *def* product ultimately modulates the expression pattern of anthocyanin genes, resulting in stronger pigmentation in whorl 2 organs. Even though modulated by *def*, the expression of anthocyanin genes is not completely dependent on *def* because in *def* mutants floral and nonfloral tissues (e.g. stems) are still pigmented. It is possible that many important target genes involved in other processes in development, such as those controlling cell division, shape, and size, are also expressed in all floral organs and are modulated to different degrees in each organ by homeotic genes. This possibility has been explored for the *string* locus of *Drosophila*, which encodes a gene involved in the regulation of the cell division cycle. It has been proposed that homeotic genes may regulate the quantity of *string* expression in different regions of the embryo and hence control early morphogenetic events (30). In this case, however, the role of cell division patterns in morphogenesis is unclear because even though ectopic expression of *string* can alter the normal timing of embryonic mitoses, this does not appear to interfere dramatically with pattern formation (31).

A possible approach to analyzing the genetic interactions of *def* with target genes is to study the genetic regulation of particular pathways affected by *def*. A good example is the anthocyanin pathway, because both the structural genes encoding enzymes required for anthocyanin biosynthesis and their interactions with regulatory genes have been studied (1, 25, 95, 96). By studying the interaction of these genes with those controlling morphogenesis a model regulatory network might be established. Other pathways such as those involved in the control of cell division might be studied by isolating and analyzing genes homologous to those already characterized from other well-defined systems (e.g. yeast or *Drosophila*).

Evolution of Floral Organs

In spite of the enormous diversity in the form, color, and structure of flowers in various plant species, most flowers have a conserved basic plan of construction: They consist of a series of concentric organs with sterile members on the outside and fertile ones towards the center. It is therefore possible that the overall expression patterns of homeotic genes controlling whorl identity have been highly conserved in evolution. This is confirmed by the striking similarities between the phenotypes conferred by genes described in *Arabidopsis* and *Antirrhinum*.

Most information on the evolutionary origin of the flower comes from comparative studies. The stamens and carpels of the flower are generally thought to have evolved by modification of leaves (sporophylls) bearing male or female sporangia (106). If male and female sporophylls were clustered together and protected by sterile leaves, the basic plan for the hermaphrodite flower would be constituted. The distinction between petals and sepals was presumably a later refinement reflecting the need to attract insect pollinators. According to this view, the basic expression patterns of many of the genes controlling whorl identity might have been established very early in floral evolution. The sequence homologies between some of the products of different whorl identity genes (e.g. DEF and AG) suggest that some of them may have arisen by ancient duplications of key regulatory genes. Some of the rather curious aspects of the genetic interactions described above may also have an evolutionary explanation. For example, the *flo* gene is expressed in all whorls except whorl 3. Perhaps *flo* was originally involved in the switch between the sex of sporophylls and was specifically switched off in male and on in female organs. The various roles of *flo* in initiation of floral development as a whole may have arisen subsequently to this primitive role in sex specification.

In contrast to the conserved basic plan of the flower, there is great diversity in the morphology, color, and number of floral organs. The central importance of flower morphology to the plant taxonomist is a reflection of this evolutionary variability. If the domains of the whorl identity genes are conserved, then much of the diversity in organ morphology presumably reflects changes in the way that common signals are interpreted by target genes. Flower pigmentation may again provide a useful model for understanding how such changes could arise.

Although the most intensely pigmented organs in *Antirrhinum* are the petals, in other species the bracts, sepals, or stamens are the most intensely colored organs. How have these differences in degrees of whorl pigmentation evolved? The biosynthesis of anthocyanin requires several specific enzymatic steps, so coordinate expression of a set of genes is required. Changes in the *cis*-acting regulatory regions of the biosynthetic genes would only change the activity of one gene at a time and would not alter expression in a coordinate manner. The different patterns of anthocyanin distribution would therefore be most easily explained by changes in the expression pattern of regulatory genes, known to control many of the genes encoding enzymes of the pathway. Such regulatory genes have been isolated and characterized from maize (62) and *Antirrhinum* (J. Goodrich, R. Carpenter, and E. S. Coen, unpublished results). In *Antirrhinum* the *delila* gene regulates the pattern of anthocyanin biosynthetic gene expression in all floral organs and in vegetative structures (1). The *delila* gene is highly homologous to the *R* locus gene family in maize, which also regulates anthocyanin genes. Both *delila* and *R* probably encode transcription factors, because their products have a domain with homology to the helix-loop-helix region found in a family of transcription factors. Changes in the direct or indirect interactions of genes such as *delila* and *R* with the whorl identity genes could result in more or less pigment gene expression in different whorls and could therefore account for the preferential pigmentation of distinct organs in diverse species. For example, a mutation in the *delila* promoter region that modified the interaction of *delila* with the *a, b* combination of whorl identity functions in whorl 2 could result in altered *delila* expression in this whorl, giving more or less intensely pigmented petals.

As noted above, pigmentation is only one type of target modulated by whorl identity genes, but its mode of evolution may be generally applicable to other targets. Two further examples of evolutionary variation are now considered briefly: congenital organ fusion and dicliny. The organs of primitive flowers are generally considered to have been separate from each other (97). In many species, however, the organs in one or more whorls are united at their margins for all or part of their length. Most commonly, the carpels are united (syncarpy), but the sepals or the petals can also be united (gamosepaly and sympetaly, respectively). For example, *Antirrhinum* has a syncarpous ovary comprising two united carpels and a sympetalous corolla consisting of five petals united for part of their length. *Arabidopsis* has a syncarpous ovary but separate petals. In most cases, united organs arise from continuous growth of regions between primordia (congenital fusion) rather than later fusion of originally separate organs (postgenital fusion). The whorl identity mutations indicate that the united growth of organs reflects their identity rather than the whorl they occupy. For example, extreme *ovu* mutants in *Antirrhinum* have united carpels in whorl 1 in place of separate sepals, and separate stamens in whorl 2 in place of united petals. This arrangement suggests that in the course of floral evolution, factors controlling growth of the regions between primordia in a whorl have been coupled in some way to the whorl identity functions. By analogy with the evolution of pigmentation, this coupling could have arisen through modulating the interactions between whorl identity genes and genes regulating cell division and growth in inter-primordial zones.

A further feature of primitive flowers is that they are hypothesized to have contained both male and female organs and were therefore hermaphrodite (monoclinous). Some plant species, however, have unisexual flowers (diclinous) containing either stamens or carpels but not both. Diclinous species may have male and female flowers on the same individual (monoecious) or on separate individuals (dioecious). Most unisexual flowers begin development with a meristem of hermaphrodite pattern; functional unisexuality then results from a failure of one organ type to develop fully (45). The genetic control of this process has been studied in maize, a monoecious species with male flowers produced in the tassel and female flowers in the ear. Several mutations have been described that produce inflorescences with some hermaphrodite (perfect) flowers (23, 72), indicating that the genes affected may be directly or indirectly involved in the repression of specific whorls in the flower. The evolution of dicliny could reflect changes in the way that such genes or their targets interact with genes controlling whorl identity so that the development of specific whorls could be repressed. Clearly this interaction would also depend on other factors because a particular whorl will only be affected in one type of inflorescence in monoecious species (i.e. tassel or ear) or in one type of individual in dioecious species.

The above examples concerning the evolution of pigmentation, organ fusion, and dicliny all illustrate the variety of ways in which the basic signals determined by genes controlling whorl identity might be variously interpreted in different species. Understanding the interactions between whorl identity genes and their targets should therefore provide valuable information about both the evolution and development of flowers.

GENETIC CONTROL OF WHORL NUMBER

Mutants Affecting Whorl Number

As described above, some of the whorl identity mutants also affect whorl number. For example, extreme mutants in some of the genes required for b can give fewer whorls than wild type; extreme def mutants give flowers with three whorls: sepal, sepal, carpel. Mutants in genes needed for the c function generally give an increase in whorl number and a more or less indeterminate growth pattern. In the case of the ag in Arabidopsis, the mutant phenotype is sepal, petal, petal, sepal, petal, petal, sepal, etc. This is sometimes described as a "flowers within flowers" arrangement because the sepals in whorls 4, 7, etc can also be considered as the first whorl of sepals of an enclosed flower (15). The pleni mutants in Antirrhinum are similar to ag but give variable petaloid, sepaloid, carpeloid organs in whorl 4.

These findings indicate that whorl identity genes can have two roles: the control of organ fate and the control of whorl number. The degree to which these separate roles are mediated by independent target genes is not fully understood. In all the mutants described above, determinate flowers have carpels in the final whorl whereas indeterminate flowers do not produce carpels. For example, although some def mutants in Antirrhinum have only three whorls, the third whorl contains carpels so the flower still has carpels in the final whorl. However, the simplistic view that organ identity (i.e. production of a carpel) itself controls whorl number does not appear to be true because *ovu* mutants have carpels in whorl 1 but still produce four whorls. It therefore seems likely that the control of whorl identity and that of whorl number reflect activations of functions that are at least partially separable.

The flower can be considered as homologous to a shoot with an imposed determinate growth pattern (see above; 4, 35). Support for this notion comes from the *flo* mutant phenotype, which has indeterminate shoots in place of flowers. According to this view, the control of whorl number could be restated as the regulation of a determinacy or "stop" function. Thus in wild-type flowers, determinacy is activated after four whorls are produced. In mutants with fewer whorls than wild type, determinacy could be activated prematurely. In flowers with more than four whorls there could be a delay or failure in the activation of determinacy. It is perhaps significant that the *ag* and *pleni* genes, required for determinacy, affect organ identity in whorl 4 whereas the other whorl identity genes do not. This phenomenon indicates that the determinacy function might be a localized activity which, once established in a whorl (i.e. in whorl 4 of wild type), prevents subsequent whorls from developing.

Another type of mutant affecting determinacy is the *flo* mutant of *Antirrhinum*, which produces proliferating inflorescence meristems in place of flowers. One explanation for this phenotype is that in initiating floral development, *flo* switches on a cascade of events ultimately leading to activation of the determinacy function. As *flo* is itself expressed in whorl 4 it is also possible that it plays a direct role in activating the determinacy function in this whorl. The interactions among *flo*, the whorl identity genes, and a determinacy function explain the "flowers within flowers" phenotype of *ag* mutants (27). According to the previously described model (Figure 6), the absence of the *c* function in *ag* mutants would be expected to result in expression of both *flo* and *a* in whorl 4 because *c* normally inhibits *a* in this whorl. Furthermore, the determinacy function would not be expressed because this requires *ag* expression. The combination of functions in whorl 4 would therefore be identical to that normally found in whorl 1 (Figure 6); hence *flo* could act in whorl 4 as it normally does in the first whorl to initiate a new "flower." This process would repeat itself in whorls 7, 10, etc. According to this view, *ag* mutants can be considered heterochronic because the combination of gene functions in late whorls recapitulates those found in earlier whorls.

Genes affecting both identity and number of homologous parts (i.e. segments) have also been described in *Drosophila*. However, unlike the floral mutants, the *Drosophila* mutants never produce an increase in the number of homologous parts but only a reduction in number (49). This may reflect a fundamental difference in the mechanism by which repetition of parts occurs in these two systems. In *Drosophila*, segmentation proceeds from an undivided structure of finite size, the egg, to a subdivided one, so that mutants failing to subdivide correctly give fewer segments. In flowers, whorl primordia are produced by sequential addition rather than by subdivision, providing the potential for indeterminate growth.

Evolution of Organ Number

The floral organs of *Antirrhinum* and *Arabidopsis* are in a whorled (cyclic) arrangement. In some species, however, floral organs can have a helical arrangement; sepals are produced at the base of the helix, and then petals, stamens, and carpels arise in sequence going up the helix. The number of each type of floral organ can be quite large in helical flowers and is often a representative of the Fibonacci series (3, 5, 8, 13, etc) such that an integral number of circuits or cycles of the floral meristem is occupied by one organ type (Figure 7).

In comparing the number of whorls and organs in different species, various nomenclature systems can be employed, none of which is completely satisfactory. It is perhaps simplest to use the term whorl to refer to all organs of a particular type for both helical and cyclic flowers. Thus the *Helleborus niger* flower, illustrated in Figure 7, would have 4 whorls: 5 sepals in whorl 1, 13 petals in whorl 2, 91 stamens in whorl 3, and 5 carpels in whorl 4. Species with only 3 types of organs (e.g. those without distinct sepals and petals) would have 3 whorls. This avoids the problem of discussing whether the 6 stamens of *Arabidopsis* flowers correspond to 1 or 2 whorls (15). However, if organ type is used to define whorl, how do we describe mutants that change organ identities or whorl number? A good operational system is to use the position of organs in the wild type to define the whorls so that if an organ appears in the position normally occupied by a stamen it would be in whorl 3 (assuming the flower has 4 whorls). Similarly, if additional organs are produced within the flower, compared to the wild type, they would be referred to as extra whorls, although it may not be easy in some cases to determine their number precisely. This operational definition of whorl is quite different from the more usual morphological definition of a whorl as "a group of organs originating from the same level on a stem," but it is more convenient for use in discussions of genetic control systems.

According to this nomenclature, it seems possible that the domains of expression of the various floral homeotic genes might correspond to similar whorls and regions in all flowering plants. This is because the domains should be a reflection of organ identity rather than the number or arrangement of organs. The similarity between mutants in *Arabidopsis* and *Antirrhinum*, which have different numbers and arrangements of organs in each whorl, confirms this view.

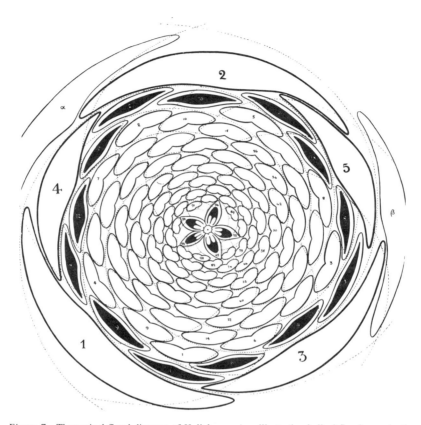

Figure 7   Theoretical floral diagram of *Helleborus niger* illustrating helical floral organization. The outside of the helix starts with the α and β prophylls, followed by 5 sepals, 13 petals *(solid-fill)*, 91 stamens, and 5 carpels. The number of stamens and carpels can vary between different specimens. Taken from Church (22).

By analogy with cyclic flowers, the total number of organs in helical flowers might also depend on the activation of a determinacy function. In those cases in which whorl 4 consists of a spiral of carpels, the function might be activated in the region where the final carpels of the helix develop. The development of helical flowers also raises some questions about how the distinctions between whorls are established. For example, in the illustration of *Helleborus niger* (Figure 7), the 13th primordium of whorl 2 develops into a petal but the next developing primordium develops into a stamen (i.e. 1st primordium of whorl 3). What mechanism regularly distinguishes the 13th from the next primordium? It seems likely that the distinction reflects spatial rather than temporal constraints, because 13 primordia complete a circuit of the meristem so that primordium 14 occupies a distinct spatial or contact zone. It should now be possible to test directly some of these speculations by using genes from *Antirrhinum* and *Arabidopsis* to study their counterparts in helical flowers.

GENETIC CONTROL OF DIFFERENCES BETWEEN ORGANS IN THE SAME WHORL

General Peloria

In actinomorphic flowers (e.g. *Arabidopsis*), all organs in the same whorl have very similar or identical morphologies. In zygomorphic flowers (e.g. *Antirrhinum*), one or more whorls contain organs with distinct morphologies. Examples of zygomorphic flowers include orchids, labiates (e.g. mint), and personates (e.g. *Antirrhinum, Digitalis*), the most striking features of which are the very distinct upper and lower petals. Many mutations have been described in zygomorphic species that can reduce or eliminate the differences between organs in a whorl and thus render the flower more or less actinomorphic. These are usually homeotic because all the organs in a mutant whorl resemble one particular organ (generally the upper or lower organ) of the corresponding wild-type whorl.

The first well-documented case of such variation in symmetry was in the common toadflax, *Linaria vulgaris*, a species of the same plant family as *Antirrhinum*. *Linaria* normally has zygomorphic flowers with two upper and three lower petals; the base of the lowest petal is extended to form a single hollow spur. In 1744, Linnaeus (60) described a peculiar plant, found on an island near Stockholm, that was identical in most respects with *Linaria* except that it had actinomorphic flowers with five spurs. Linnaeus based his system of plant taxonomy largely on floral structure, and accordingly the actinomorphic plant and *Linaria* should have belonged to entirely different taxonomic tribes. However, its close resemblance to *Linaria* in all other characters suggested to him that the actinomorphic form had arisen somehow from *Linaria*. This was a radical suggestion at a time when the fixity of species was generally accepted. Wrote Linnaeus, "This is certainly no less remarkable than if a cow were to give birth to a calf with a wolf's head" (38). These monstrous images led Linnaeus to name the plant *Peloria*, Greek for "monster."

In 1868, Charles Darwin described a similar phenomenon in *Antirrhinum* (29). He showed that when a plant with peloric (i.e. radially symmetrical) flowers was crossed to one with zygomorphic flowers, the hybrids were all zygomorphic. Self-pollination of the hybrids gave 90 zygomorphic and 37 peloric progeny, a near 3:1 ratio. Unfortunately, Darwin was not aware of Mendel's work and did not pursue the analysis further. Subsequent studies have shown that peloria in *Antirrhinum* can be caused by a single recessive allele, as might have been predicted from Darwin's data (11, 56).

Many other examples of peloria in species that normally have zygomorphic flowers have been described (8, 19, 29, 63, 70, 107). In most cases the petals of peloric flowers resemble one of the petals of the wild type, usually the uppermost or lowermost petals. For example, in the commonest form of peloria in *Linaria*, petals resembling the lowest lobe of the wild-type flower are represented five times, giving a flower with five spurs. In orchids, peloric flowers most frequently have labella in place of lateral petals of the same whorl (6). In early stages of wild-type orchid flower development the labellum is the uppermost organ of its whorl, although at later stages twisting of the gynoecium through 180° results in the labellum occupying the lowest position.

The most intensively studied example of peloria is in *Antirrhinum*. Several different *cycloidea (cyc)* mutations have been described that produce flowers with a more symmetrical appearance than those of the wild type (56). Extreme *cyc* mutations give radially symmetrical flowers with all petals resembling the lowest petal of the wild type, analogous to the peloric *Linaria*. Unlike the wild type, in which the uppermost stamen is aborted, all five stamen primordia develop fully in these mutants to give mature organs of lengths similar to those of the lower stamens of the wild type. This is similar to the peloric form of *Linaria* which also typically has five stamens (19, 60, 104). These *cyc* mutants also affect whorl 4 and give it a more symmetrical appearance than in the wild type. Thus, all organs in whorls 2, 3, and probably 4 resemble the lowest organs of the corresponding whorl in wild-type flowers. There are no major distinguishing features between upper and lower sepals in *Antirrhinum;* thus it is not possible to assess the effects of *cyc* alleles in whorl 1. However, whorl 1 is zygomorphic in some other species of the tribe Antirrhineae (100), suggesting that genes conferring zygomorphy (e.g. Cyc^+) can act in whorl 1. Furthermore, analysis of double mutants of *cyc* and *ovu* in *Antirrhinum* indicates that *cyc* can affect whorl 1 (R. Carpenter, E. Coen, unpublished results). Taken together, these observations indicate that *cyc* may be expressed in all whorls but is interpreted in distinct ways in different whorls (see the model below).

In addition to the phenotype described above, there are also many *cyc* mutations that confer intermediate phenotypes. The various *cyc* alleles can be arranged as a series going from those that give fully peloric flowers to those conferring an almost wild-type phenotype. These alleles can be divided into two classes based on genetic complementation tests (18, 56). Alleles from the same class fail to complement each other whereas alleles from different classes show partial complementation, giving almost wild-type F_1 hybrids with two small notches on the lower corolla lip. The F_2 from the notched hybrids gives 50% notched and 25% of each parental type, indicating linkage between the alleles from the two classes. This result suggests that *cyc* may be a complex locus composed of two interacting functional units. Alternatively, it is possible that there are two distinct *cyc* genes that are tightly linked on the same chromosome and whose products interact.

Polar-Coordinate Model for Zygomorphy

As noted above, many genes affect the identities of whorls of organs. It is thus possible to ask if the action of *cyc* on particular organs depends on the whorl they occupy. For example, abortion of the uppermost organ in whorl 3 of the wild type depends on the action of *cyc*, because in extreme *cyc* mutants all five stamen primordia develop fully. If stamens grow in whorl 2, as in *ovu* mutants, the upper two organs of this whorl are also vestigial or aborted. The two upper organs are therefore aborted whether stamens grow in whorls 2 or 3. This pattern suggests that *cyc* interacts with primordia in a similar way, irrespective of the whorl they occupy, and therefore that the fate of a primordium depends on a combinatorial interaction between functions determining whorl identity and the functions determining the differences between upper and lower organs. These observations have suggested a polar-coordinate model for the control of primordium fate (18; Figure 8). The action of the whorl identity functions varies along the radius (r) of the flower. The *cyc* function varies along the vertical (y) axis of the flower, its effect generally increasing from the lower to the upper parts of the axis. This results in a zygomorphic phenotype because for half of the flower, every organ has a unique specification, identical with its "mirror image" in the other half (Figure 8).

A further feature of the proposed model is that it can explain some unexpected effects of whorl identity mutations. For example, the uppermost carpel in the first whorl of *ovu* mutants is sterile and devoid of ovules, whereas the other carpels of this whorl contain ovules. This might be because the uppermost carpel is growing in a region of high *cyc* activity (i.e. at the top of the y-axis) not normally experienced by the wild-type carpels in whorl 4. Similarly, petals in the inner whorls of *pleni* mutants sometimes appear to be almost peloric, perhaps because in more central whorls there is less difference in *cyc* activity between upper and lower organs than in outer whorls.

Terminal Peloria

It is clear from classical botanical studies that zygomorphy is intimately connected with the position of flowers on the inflorescence. For example, the ray florets on the periphery of composite inflorescences (e.g. daisy) are strongly zygomorphic and have enlarged lower petals whereas the disc florets

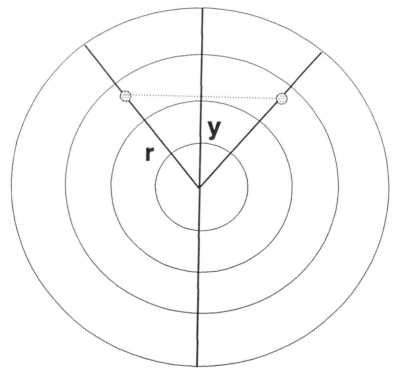

Figure 8 Polar coordinate model for zygomorphy. The four whorls are shown as concentric rings. Expression of whorl identity functions varies along the radius (r) and the *cyc* function varies along the vertical (y) axis. Two primordia with the same combination of functions, and hence the same developmental fate, are shown joined by a dotted horizontal line. Mutations eliminating the whorl identity functions result in some primordia in different whorls having similar specifications so that they adopt similar developmental fates. Mutations that abolish *cyc* function remove differential expression along the y-axis such that all primordia in a whorl adopt a fate similar to that of the lower primordia of the wild-type whorl.

are radially symmetrical. The most dramatic illustration of the effects of position is the phenomenon of terminal peloria.

Many species with zygomorphic flowers have indeterminate inflorescences (e.g. *Antirrhinum;* see above); however, exceptional plants that produce terminal flowers have been described in some of these species, and in most cases the terminal flowers exhibit radial symmetry (3, 19, 57, 78–80). This phenomenon is called terminal peloria because only the terminal flower is affected, in contrast to general peloria, which affects all flowers on the inflorescence. The inheritance of terminal peloria has been studied in *Antirrhinum* and shown to be caused by a single recessive allele at a locus called *centroradialis* (57). Similarly, a single recessive allele can confer terminal peloria in *Digitalis purpurea* (foxglove) (52), a member of the same taxonomic family as *Antirrhinum*. In both of these species, the terminal peloric flowers have petals resembling the lowest petal of the wild type, similar to the flowers seen in general peloria. However, the number of organs can be greater than 5. In the case of *D. purpurea* the large terminal flowers can have as many as 19 petals, 15 stamens, and a proliferation of central carpels (3).

The many instances of terminal peloria suggest that the establishment of zygomorphy requires flower meristems to be axillary to the main apex. An early biophysical model for production of asymmetry that might account for this correlation was proposed by A. P. de Candolle (17; cited in 34, 104). Lateral (axillary) flowers growing below a terminal flower could be subject to uneven pressures and grow outwards where the pressure is least, producing an asymmetrical flower. The terminal flower is symmetrical because "freedom from pressure or restriction on one side allows the flower to develop equally in all directions" (63). Although such a model might explain terminal peloria it is not easily reconciled with general peloria, in which symmetrical flowers are produced in axillary positions. Furthermore, there are many species with axillary actinomorphic flowers.

An alternative explanation can be proposed, based on the polar-coordinate model for zygomorphy described above. Flower meristems in axillary positions are in an asymmetrical environment with the main apex above them and a subtending bract below them. This arrangement could result in differences along the y-axis of the floral meristem and these may provide the basis for establishing the gradient of *cyc* activity and hence zygomorphy. Terminal flowers would be in a radially symmetrical environment and therefore unable to establish a y-axis and activate *cyc*. The greater number of organs in terminally peloric flowers might simply reflect the larger size of the main apex compared to axillary meristems.

Evolution of Zygomorphy

Taxonomic studies indicate that radial symmetry was likely the ancestral form of flowers and that zygomorphy has evolved several times in independent evolutionary lines (97). This idea is consistent with the observation that mutations to radial symmetry have been described and characterized in many zygomorphic species but that no well-defined mutations to zygomorphy have been described in actinomorphic species. Peloric mutations have therefore often been viewed as atavistic because they reveal a more ancestral condition (29, 107). In the case of terminal peloria there is a double atavism because both the flower shape (peloric) and the inflorescence type (determinate) are ancestral conditions.

It is clear from the previous section that the evolution of zygomorphy needs to be considered together with the evolution of the inflorescence. The most ancestral type of inflorescence is thought to be determinate, and indeterminate inflorescences with axillary flowers are thought to have evolved subsequently. As described above, axillary floral meristems are in an asymmetrical environment that could interact with genes controlling flower development. The *cyc* gene functions may therefore originally have acted in a symmetrical fashion in the flower, but the presence of an asymmetrical environment in axillary meristems allowed their activity and effects on the flower to become modified along the y-axis.

It is important to emphasize that zygomorphy has probably evolved many times independently in different plant taxa. Therefore the scheme proposed for *Antirrhinum* and its relatives may not be applicable to other plant groups. In some species it is clear that a completely different system is involved, because zygomorphy is imposed relatively late in floral development and depends on the effects of gravity (103). The latter phenomenon contrasts with most cases of zygomorphy, in which asymmetry is established early in flower development with an axis of symmetry oriented relative to the stem and not to external conditions. However, it is possible that in some cases factors originally controlled by a gravitropic system eventually became coupled to internal cues and that in this way a more typical type of zygomorphic system evolved.

CONCLUDING REMARKS

The sequential and additive nature of development, together with the direct relationship between overall form and the properties of cells, clearly provide a very flexible framework for the generation of great diversity in plant form. Rapid progress is now being made in the molecular and genetic analysis of homeotic genes, their genetic interactions, and their relationships with target genes. There can be little doubt that the results of this analysis will provide a new foundation for our understanding of floral development and evolution. Initially, this work is likely to concentrate on qualitative interactions, though perhaps a longer-term challenge will be to relate these findings in a quantitative way to the patterns of plant morphogenesis. Plants are amenable to such developmental studies; comparisons with animals, which have a distinct mode of development, should allow a general understanding of how the diversity of forms in nature is genetically controlled.

ACKNOWLEDGMENTS

I thank Keith Roberts for kindly providing the *Arabidopsis* drawings in Figure 2; Rosemary Carpenter, Robert Elliott, Jose Romero, Justin Goodrich, and

Desmond Bradley for helpful comments and discussions; and Sabine Hantke for translating some weighty German texts. I also thank Zsuzsanna Schwartz-Sommer and Elliot Meyerowitz for providing preprints of their work.

Literature Cited

1. Almeida, J., Carpenter, R., Robbins, T. P., Martin, C., Coen, E. S. 1989. Genetic interactions underlying flower color patterns in *Antirrhinum majus*. *Genes Dev.* 3:1758–67
2. Ambros, V., Horvitz, R. H. 1984. Heterochronic mutants of the nematode *Caenorhabditis elegans*. *Science* 226:409–16
3. Arber, A. 1932. Studies in flower structure. I. On a peloria of *Digitalis purpurea*, L. *Ann. Bot.* 46:929–39
4. Arber, A. 1950. *The Natural Philosophy of Plant Form*. Cambridge: Cambridge Univ. Press
5. Awasthi, D. K., Kumar, V., Murty, Y. S. 1984. Flower development in *Antirrhinum majus* L. (Scrophulariaceae) with a comment upon corolla tube formation. *Bot. Mag.* 97:13–22
6. Bateman, R. M. 1985. Peloria and pseudopeloria in British orchids. *Watsonia* 15:357–59
7. Bateson, W. 1894. *Materials for the Study of Variation*. London: Macmillan
8. Bateson, W., Bateson, A. 1891. On variations in the floral symmetry of certain plants having irregular corollas. *J. Linn. Soc. Bot.* 28:386–424
9. Battey, N. H., Lyndon, R. F. 1988. Determination and differentiation of leaf and petal primordia in *Impatiens balsamina*. *Ann. Bot.* 61:9–16
10. Battey, N. H., Lyndon, R. F. 1990. Reversion of flowering. *Bot. Rev.* 56:162–89
11. Baur, E. 1924. Untersuchungen über das Wesen, die Entstehung und die Vererbung von Rassenunterschieden bei *Antirrhinum majus*. *Bibl. Genet.* 4:1–70
12. Bayly, I. L., Craig, I. L. 1962. A morphological study of the X-ray induced cauliflower-head and single-leaf mutation in *Medicago sativa* L. *Can. J. Genet. Cytol.* 4:386–97
13. Bernier, G. 1988. The control of floral evocation and morphogenesis. *Annu. Rev. Plant. Physiol. Plant Mol. Biol.* 39:175–219
14. Blixt, S. 1972. Mutation genetics in *Pisum*. *Agric. Hort. Genet.* 30:1–293
15. Bowman, J. L., Smyth, D. R., Meyerowitz, E. M. 1989. Genes directing flower development in *Arabidopsis*. *Plant Cell* 1:37–52
16. Brieger, F. G. 1935. The developmental mechanics of normal and abnormal flowers in *Primula*. *Proc. Linn. Soc. London* 147:126–30
17. Candolle de, A. P. 1819. *Théorie Élémentaire de Botanique*. Paris. 2nd ed.
18. Carpenter, R., Coen, E. S. 1990. Floral homeotic mutations produced by transposon-mutagenesis in *Antirrhinum majus*. *Genes Dev.* 4:1483–93
19. Chavannes, E. 1833. *Monographie des Antirrhinées*. Paris: Treuttel & Wurtz
20. Chittenden, R. J. 1914. The rogue wallflower. *J. R. Hortic. Soc.* 40:83–87
21. Chittenden, R. J. 1928. Note on an abnormal *Antirrhinum*. *J. Genet.* 19:281–83
22. Church, A. H. 1908. *Types of Floral Mechanism, Part I*. Oxford: Clarendon Press
23. Coe, E. H., Neuffer, M. G. 1977. The genetics of corn. In *Corn and Corn Improvement*, ed. G. F. Sprague. Madison: Am. Soc. Agron.
24. Coen, E. S., Carpenter, R. 1986. Transposable elements in *Antirrhinum majus*: generators of genetic diversity. *Trends Genet.* 2:292–96
25. Coen, E. S., Carpenter, R., Martin, C. 1986. Transposable elements generate novel spatial patterns of gene expression in *Antirrhinum majus*. *Cell* 47:285–96
26. Coen, E. S., Robbins, T. P., Almeida, J., Hudson, A., Carpenter, R. 1989. Consequences and mechanisms of transposition in *Antirrhinum majus*. In *Mobile DNA*, ed. D. E. Berg, M. M. Howe, pp. 413–36. Washington, DC: Am. Soc. Microbiol.
27. Coen, E. S., Romero, J. M., Doyle, S., Elliott, R., Murphy, G., Carpenter, R. 1990. *floricaula*: A homeotic gene required for flower development in *Antirrhinum majus*. *Cell* 63:1311–22
28. Cusick, F. 1956. Studies of floral morphogenesis. I. Median bisections of

flower bud primordia in *Primula bulleyana* Forrest. *Trans. R. Soc. Edinburgh* 63:153–66

29. Darwin, C. 1868. *The Variation of Animals and Plants under Domestication.* London: J. Murray

30. Edgar, B. A., O'Farrell, P. H. 1989. Genetic control of cell division patterns in the *Drosophila* embryo. *Cell* 57:177–87

31. Edgar, B. A., O'Farrell, P. H. 1990. The three postblastoderm cell cycles of *Drosophila* embryogenesis are regulated in G2 by *string*. *Cell* 62:469–80

32. Feldmann, K. A., Marks, M. D., Christianson, M. L., Quatrano, R. S. 1989. A dwarf mutant of *Arabidopsis* generated by T-DNA insertion mutagenesis. *Science* 243:1351–54

33. Garcia-Bellido, A. 1977. Homeotic and atavic mutations in insects. *Am. Zool.* 17:613–29

34. Goebel, K. 1900. *Organography of Plants*, Vol. I, Transl. I. B. Balfour. Oxford: Clarendon Press

35. Goethe, J. W. 1790. *Versuch die Metamorphose der Pflanzen zu Erklären.* Gotha: C. W. Ettinger. Transl. A. Arber, 1946. Goethe's botany. *Chron. Bot.* 10:63–126

36. Gottschalk, W. 1979. A *Pisum* gene preventing transition from the vegetative to the reproductive stage. *Pisum Newsl.* 11:10

37. Green, P. B. 1988. A theory of inflorescence development and flower formation based on morphological and biophysical analysis in *Echeveria*. *Planta* 175:153–69

38. Gustafsson, Å. 1979. Linnaeus' peloria: the history of a monster. *Theor. Appl. Genet.* 54:241–48

39. Harte, C. 1974. *Antirrhinum majus* L. In *Handbook of Genetics*, ed. R. C. King, 2:315–31. New York: Plenum

40. Haughn, G. W., Somerville, C. R. 1988. Genetic control of morphogenesis in *Arabidopsis*. *Dev. Genet.* 9:73–89

41. Helm, J. 1951. Vergleichende Betrachtungen über die Entwicklung der Infloreszenz bei *Lycopersicon esculentum* Mill. und bei einer Röntgenmutante. *Züchter* 21:54–59

42. Henslow, G. 1882. Note on staminiferous corollas of *Digitalis purpurea* and *Solanum tuberosum*. *Linn. Soc. J. Bot.* 19:216–18

43. Hernandez, L. F., Palmer, J. H. 1988. Regeneration of the sunflower capitulum after cylindrical wounding of the receptacle. *Am. J. Bot.* 75:1253–61

44. Hertwig, P. 1926. Ein neuer Fall von

multiplem Allelomorphismus bei *Antirrhinum*. *Z. Indukt. Abstamm. Vererbungsl.* 41:42–47

45. Heslop-Harrison, J. 1964. Sex expression in flowering plants. *Brookhaven Symp. Biol.* 16:109–25

46. Hicks, G. S., Sussex, I. M. 1971. Organ regeneration in sterile culture after median bisection of the flower primordia of *Nicotiana tabacum*. *Bot. Gaz.* 132:350–63

47. Holder, N. 1979. Positional information and pattern formation in plant morphogenesis and a mechanism for the involvement of plant hormones. *J. Theor. Biol.* 77:195–212

48. Imai, Y. 1934. On the mutable genes of *Pharbitis*, with special reference to their bearing on the mechanism of budvariation. *J. Coll. Agric.* 12:479–523

49. Ingham, P. W. 1988. The molecular genetics of embryonic pattern formation in *Drosophila*. *Nature* 335:25–34

50. Irish, V. F., Sussex, I. M. 1990. Function of the *apetala-1* gene during *Arabidopsis* floral development. *Plant Cell* 2:741–53

51. Jensen, L. C. W. 1971. Experimental bisection of *Aquilegia* floral buds cultured *in vitro*. I. The effect of growth, primordia initiation, and apical regeneration. *Can. J. Bot.* 49:487–93

52. Keeble, F., Pellew, C., Jones, W. N. 1910. The inheritance of peloria and flower colour in foxgloves (*Digitalis purpurea*). *New Phytol.* 9:68–77

53. Klemm, M. 1927. Vergleichende morphologische und entwicklungsgeschichtliche Untersuchungen einer Reihe multipler Allelomorphe bei *Antirrhinum majus*. *Bot. Arch.* 20:423–74

54. Komaki, M. K., Okada, K., Nishino, E., Shimura, Y. 1988. Isolation and characterization of novel mutants of *Arabidopsis thaliana* defective in flower development. *Development* 104:195–203

55. Koornneef, M., van Eden, J., Hanhart, C. J., Stam, P., Braaksma, F. J., Feenstra, W. J. 1983. Linkage map of *Arabidopsis thaliana*. *J. Hered.* 74:265–72

56. Kuckuck, H. 1936. Über vier neue Serien multipler Allele bei *Antirrhinum majus*. *Z. Indukt. Abstamm. Vererbungsl.* 71:429–40

57. Kuckuck, H., Schick, R. 1930. Die Erbfaktoren bei *Antirrhinum majus* und ihre Bezeichnung. *Z. Indukt. Abstamm. Vererbungsl.* 56:51–83

58. Kunst, L., Klenz, J. E., Martinez-Zapater, J., Haughn, G. W. 1989. AP2 gene determines the identity of perianth

organs in flowers of *Arabidopsis thaliana*. *Plant Cell* 1:1195–1208

59. Lewis, E. B. 1978. A gene complex controlling segmentation in *Drosophila*. *Nature* 276:565–70

60. Linnaeus, C. 1744. De Peloria. Diss., Uppsala (Amoenitates Acad. 1749)

61. Lloyd, C. 1982. *The Cytoskeleton in Plant Growth and Development*. London: Academic

62. Ludwig, S. R., Wessler, S. R. 1990. Maize *R* gene family: tissue-specific helix-loop-helix proteins. *Cell* 62:849–51

63. Masters, M. T. 1869. *Vegetable Teratology: An Account of the Principle Deviations from the Usual Construction of Plants*. London: Ray Soc.

64. McHughen, A. 1980. The regulation of tobacco floral organ initiation. *Bot. Gaz.* 141:389–95

65. Mermod, N., O'Neill, E. A., Kelly, T. J., Tjian, R. 1989. The proline rich transcriptional activator of CTF/NF-1 is distinct from the replication and DNA binding domain. *Cell* 58:741–53

66. Meyer, V. 1966. Flower abnormalities. *Bot. Rev.* 32:165–95

67. Meyerowitz, E. M. 1989. *Arabidopsis*, a useful weed. *Cell* 56:263–69

68. Meyerowitz, E. M., Bowman, J. L., Brockman, L. L., Drews, G. N., Jack, T., et al. 1991. A genetic and molecular model for flower development in *Arabidopsis thaliana*. *Development*. In press

69. Meyerowitz, E. M., Smyth, D. R., Bowman, J. L. 1989. Abnormal flowers and pattern formation in floral development. *Development* 106:209–17

70. Moquin-Tandon, A. 1841. *Éléments de tératologie végétale, ou histoire abrégée des anomalies de l'organisation dans les végétaux*. Paris: J.-P. Loss

71. Morata, G., Macias, A., Urquia, N., Gonzalez-Reyes, A. 1990. Homeotic genes. *Sem. Cell Biol.* 1:219–27

72. Nickerson, N. H., Dale, E. E. 1955. Tassel modifications in maize. *Ann. Mo. Bot. Gard.* 42:195–211

73. Nilan, R. A. 1974. Barley. In *Handbook of Genetics*, ed. R. C. King, 2:93–110. New York: Plenum

74. Norman, C., Runswick, M., Pollock, R., Treisman, R. 1988. Isolation and properties of cDNA clones encoding SRF, a transcription factor that binds to the c-*fos* serum response element. *Cell* 55:989–1003

75. Paddock, E. F., Alexander, L. J. 1952. Cauliflower, a new recessive mutation in tomato. *Ohio J. Sci.* 52:327–34

76. Parkin, J. 1914. The evolution of the inflorescence. *Linn. J. Bot.* 42:511–604

77. Passmore, S., Maine, G. T., Elble, R., Christ, C., Tye, B. 1988. *Saccharomyces cerevisiae* protein involved in plasmid maintenance is necessary for mating of *MATa* cells. *J. Mol. Biol.* 204:593–606

78. Penzig, O. 1890–1894. *Pflanzen-Teratologie, Systematisch Geordnet*. Vols. I, II. Genua: A. Ciminago

79. Peyritsch, J. 1870. Über Pelorien bei Labiaten. *Sitzungsber. Akad. Wiss.* 62:1–27

80. Peyritsch, J. 1872. Über Pelorienbildungen. *Sitzungsber. Akad. Wiss.* 66:1–35

81. Postlethwaite, S. N., Nelson, O. E. 1964. Characterization of development in maize through the use of mutants. I. The *Polytypic (Pt)* and *ramosa 1 (ra1)* mutants. *Am. J. Bot.* 51:238–43

82. Reid, J. B., Murfet, I. C. 1984. Flowering in *Pisum*: a fifth locus, *Veg*. *Ann. Bot.* 53:369–82

83. Rick, C. M., Butler, L. 1956. Cytogenetics of the tomato. *Adv. Gen.* 8:267–381

84. Rickett, H. W. 1944. The classification of inflorescences. *Bot. Rev.* 10:187–231

85. Saunders, E. R. 1911. On inheritance of a mutation in the common foxglove (*Digitalis purpurea*). *New Phytol.* 10:47–63

86. Saunders, E. R. 1912. Double flowers. *J. R. Hort. Soc.* 38:469–82

87. Saunders, E. R. 1923. The bractless inflorescence of the Cruciferae. *New Phytol.* 22:150–56

88. Schwarz-Sommer, Z., Huijser, P., Nacken, W., Saedler, H., Sommer, H. 1990. Genetic control of flower development: homeotic genes in *Antirrhinum majus*. *Science*. 250:931–936

89. Sears, E. 1974. Wheat. In *Handbook of Genetics*, ed. R. C. King, 2:59–91. New York: Plenum

90. Sheridan, W. F. 1988. Maize development genetics: genes of morphogenesis. *Annu. Rev. Genet.* 22:353–85

91. Shull, G. H. 1912. Inheritance of the heptandra-form of *Digitalis purpurea* L. *Z. Indukt. Abstamm. Vererbungsl.* 6:257–67

92. Smyth, D. R., Bowman, J. L., Meyerowitz, E. M. 1990. Early flower development in *Arabidopsis*. *Plant Cell* 2:755–67

93. Soetiarto, S. R., Ball, E. 1969. Ontogenetical and experimental studies of *Portulaca grandiflora*. 2. Bisection of the meristem in successive stages. *Can. J. Bot.* 47:1067–76

94. Sommer, H., Beltrán, J., Huijser, P., Pape, H., Lönnig, W., et al. 1990. *De-*

ficiens, a homeotic gene involved in the control of flower morphogenesis in *Antirrhinum majus:* the protein shows homology to transcription factors. *EMBO J.* 9:605–13

95. Sommer, H., Bonas, U., Saedler, H. 1988. Transposon-induced alterations in the promoter region affect transcription of the chalcone synthase gene of *Antirrhinum majus. Mol. Gen. Genet.* 211:49–55

96. Sommer, H., Saedler, H. 1986. Structure of the chalcone synthase gene of *Antirrhinum majus. Mol. Gen. Genet.* 202:429–34

97. Stebbins, G. L. 1974. *Flowering Plants, Evolution above the Species Level.* Cambridge, MA: Harvard Univ. Press

98. Struhl, K. 1989. Helix-turn-helix, zincfinger, and leucine-zipper motifs for eukaryotic transcriptional regulatory proteins. *Trends Biochem. Sci.* 14:137–40

99. Stubbe, H. 1966. *Genetik und Zytologie von Antirrhinum L. sect. Antirrhinum. Veb.* Jena: Gustav Fischer

100. Sutton, D. A. 1988. *A revision of the tribe Antirrhineae.* Oxford: Br. Mus. Oxford Univ. Press

101. Tilney-Basset, R. A. E. 1986. *Plant Chimeras.* London: E. Arnold

102. Vilmorin, P., Bateson, W. 1912. A case of gametic coupling in *Pisum. Proc. R. Soc. London Ser. B* 84:9–11

103. Vöchting, H. 1886. Über Zygomorphie und deren Ursachen. *Pringsheim's Jahrb. Wiss. Bot.* 17:297–346

104. Vöchting, H. 1898. Über Blüthen-Anomalien. *Pringsheim's Jahrb. Wiss. Bot.* 31:1–120

105. Wardlaw, C. W. 1957. The floral meristem as a reaction system. *Proc. R. Soc. Edinburgh* 66:394–408

106. Weberling, F. 1989. *Morphology of Flowers and Inflorescences.* Cambridge: Cambridge Univ. Press

107. Wordsell, W. C. 1915–1916. *The Principles of Plant-Teratology.* London: Ray Soc. 2 Vols.

108. Yanofsky, M. F., Ma, H., Bowman, J. L., Drews, G. N., Feldmann, K. A., Meyerowitz, E. M. 1990. AGAMOUS: an *Arabidopsis* homeotic gene whose product resembles transcription factors. *Nature* 346:35–39

Annu. Rev. Plant Physiol. Plant Mol. Biol. 1991. 42:281–311
Copyright © 1991 by Annual Reviews Inc. All rights reserved

PROTEIN PHOSPHORYLATION IN GREEN PLANT CHLOROPLASTS*

John Bennett

International Centre for Genetic Engineering and Biotechnology, NII Campus, Shaheed Jeet Singh Marg, New Delhi-110067, India

KEY WORDS: protein kinase, phosphatase, thylakoid, envelope, light-harvesting chlorophyll protein

CONTENTS

*Abbreviations: Chl: chlorophyll; CP: chlorophyll-protein; DBMIB: 2,5-dibromo-3-methyl-6-isopropyl-*p*-benzoquinone; DCMU: 3-(3,4-dichlorophenyl)-1,1-dimethylurea; DNP-INT: the 2,4-dinitrophenyl ether of iodonitrothymol; HQNO: 2-(*n*-heptyl)-4-hydroxyquinoline N-oxide; LHC II: light-harvesting chlorophyll *a*/chlorophyll *b* pigment protein complex associated with photosystem II; PPDK: pyruvate,orthophosphate dikinase; PQ: plastoquinone; PS: photosystem; Rubisco: ribulose-1,5-*bis*phosphate carboxylase/oxygenase*

1040-2519/91/0601-0281$02.00

INTRODUCTION

Once considered of importance only in animal cells, protein phosphorylation is now recognized to be a ubiquitous regulatory mechanism, affecting many aspects of prokaryotic and eukaryotic metabolism, gene expression, and responses to environmental stimuli. Several aspects of plant growth and development have been shown to be regulated by protein phosphorylation (133), and many more examples can be expected in the near future. The highest concentration of plant phosphoproteins is found in the chloroplast (25). First detected in thylakoid membranes (15), phosphoproteins have also been located in other chloroplast compartments, namely the soluble phase or stroma (58), chloroplast ribosomes (65), the outer (146) and inner (144) envelope membranes, and the interenvelope space (144); none has thus far been located in the luminal space inside the thylakoid membranes. Of the more than 30 chloroplast phosphoproteins, only 14 have been identified, and in only two cases has a function for phosphorylation been established. Given the multiplicity of chloroplast phosphoproteins, it is likely that other aspects of the structure, function, and biogenesis of the chloroplast will prove to be controlled by this mechanism.

THYLAKOID PHOSPHOPROTEINS

The major function of the thylakoid membranes is to carry out photosynthetic electron transport and ATP synthesis. Four principal membrane-bound complexes contribute to this function: PS I, PS II, the cytochrome b/f complex, and ATP synthase (121). The major chloroplast phosphoproteins are all associated with PS II. They include the light-harvesting chlorophyll a/b proteins (LHC II, 25 and 27 kDa) (17) and four proteins of PS II core (8, 32, 34, and 43 kDa) (113, 115) (Figure 1).

LHC II

PROTEIN STRUCTURE   LHC II is a family of proteins encoded by nuclear *cab* genes (48). The proteins are synthesized in the cytoplasm as precursors (30–34 kDa) that are imported into the chloroplast and cleaved to their mature size (24–28 kDa) (93). It is not clear whether insertion into thylakoids occurs before or after cleavage (162). Each precursor contains an N-terminal extension known as a transit peptide, which is essential for import of the protein across the double envelope membrane (93). The most variable parts of pre-LHC II molecules are the transit peptide and the region between 6 and 15 residues after the cleavage site (48, 114). The cleavage site itself is highly conserved in higher plants, with cleavage occurring at a Met-Arg motif (112). LHC II contains three highly hydrophobic regions that form transmembrane

α-helixes in the mature pigment-protein complex (36). Chl a, Chl b, and carotenoids are noncovalently attached to each LHC II polypeptide. In higher plants the Chl a/Chl b is usually about $1:1$; in green algae it can range from $0.6:1$ to $4:1$ (151). The N-terminus of mature LHC II is on the stromal side of the thylakoid membrane (18), while the C-terminus is on the luminal side (31, 36). The phosphorylation site of each LHC II polypeptide is very close to the N-terminus, on a surface-exposed segment that is readily removed by trypsin (18, 118).

SEQUENCING OF PHOSPHORYLATION SITES   Phosphorylation-site sequences have been determined for five LHC IIs—the most abundant LHC II of pea (118), and four LHC IIs of spinach (112) (Table 1). The identification of such sites is not straightforward. First, it is difficult to separate completely the different forms of LHC II. Second, Edman degradation cannot be applied to the molecules because they are blocked by N-terminal acetylation (112). Mullet (118) solubilized LHC II from ^{32}P-labeled pea thylakoids, separated the different forms of LHC II by preparative isoelectric focusing, and sequenced radioactive partial typtic fragments of the major LHC II. Sufficient sequence data could be obtained to locate the phosphorylation site relative to the sequence predicted from a cloned pea cab gene (40) (Table 1). Mullet

Table 1 Currently identified chloroplast phosphoproteins

Protein	Species	Phosphorylation site	Reference
LHC II			
Type I 27 kDa	pea	?-[R,K]SAT(?P)T(?P)KK...	118
	spinach	Ac-RKT(P)AGKPKN...	*a
		Ac-RKT(P)AGKPKT...	*a
		Ac-RKS(P)AGKPKN...	*a
Type II 25 kDa	spinach	Ac-RRT(P)AKSVPQ...	*a
PS II			
psbH protein-8kDa	spinach	AT(P)QTVESSSR...	113
D1 32 kDa	spinach	Ac-T(P)AILERR...	115
D2 34 kDa	spinach	Ac-T(P)IAVGK...	115
CP43 43 kDa	spinach	Ac-T(P)LFNGTLTLAGR...	115
Stromal			
PPDK 94 kDa	maize	...TERGGMT(P)SH(P)AAVVAR..	135
Rubisco LS 54 kDa	spinach	—	58
Rubisco SS 14 kDa	spinach	—	92
Gal-3-PDH 38 kDa	spinach	—	66
Ribosomal			
L18	spinach	—	65
LS31	spinach	—	65

^{a}Reference 112 and unpublished data of H. P. Michel, P. Griffin, J. Shabanowitz, D. F. Hunt, and J. Bennett.

(118) concluded that phosphate was attached to Thr-5 and/or Thr-6. The chemical structure of the N-terminal blocking group was not determined.

Michel et al (112), working with spinach, took a different approach. They released the phosphorylation sites of LHC II from ^{32}P-labeled thylakoids by treatment with thermolysin or proteinase K. Phosphopeptides were then separated from other peptides by metal-ion affinity chromatography and from one another by reverse-phase high-performance liquid chromatography. This protocol yielded four distinct phosphopeptides, which were then sequenced by fast atom-bombardment mass spectrometry (83). The resulting sequences (Table 1) began with N-acetylarginine and carried phosphate on Thr or Ser at position 3.

More than 30 *cab* genes have been sequenced and divided into three groups. Type I and type II code for LHC IIs; type III encode a related set of Chl *a/b* proteins (LHC-I) which are associated with photosystem I (36, 130) and are not phosphorylated. LHC IIs of type I and type II are closely related but are differentiated on the basis of certain aspects of their overall sequence (48). Even in the first few amino acid residues after the cleavage site, two distinct patterns emerge which correspond to the same division. Type I LHC IIs usually begin with Arg-Lys-Thr/Ser-(2 residues)-Lys-, whereas type II LHC IIs begin with Arg-Arg-Thr-(1 residue)-Lys-. The sequenced spinach LHC IIs adhere to these consensus sequences for types I and II; the sequenced pea LHC II is one of the few exceptions. All sequenced *cab* genes appear to code for phosphorylatable proteins, except the tomato *cab* 1B gene, which lacks an hydroxy amino acid in this region. The use of synthetic peptides to explore the substrate specificity of LHC II kinase is discussed below.

A few *cab* sequences are now available for lower green plants. A *cab* gene from the moss *Physcomitrella patens* predicts a protein sequence similar to that of type II LHC IIs from higher plants, with the sequence M$^\mathrm{v}$RRTVSKSAG beginning 37 residues from the start of the open reading frame (107). Thus, the cleavage site (Met$^\mathrm{v}$Arg), the phosphorylation site, and the length of the transit peptide are all conserved. However, in green algae, *cab* genes predict sequences highly divergent from the higher-plant model. A gene from *Dunaliella salina* contains neither a conserved Met$^\mathrm{v}$Arg cleavage site nor an obvious phosphorylation site (107). These two features are also absent from a cloned *cab* gene from *D. tertiolecta*, and N-terminal sequencing of mature LHC IIs suggests that three other *cab* genes of this alga are similar (100). It should be noted that the four major LHC IIs of *D. tertiolecta* are not blocked at their N-termini. The sequence of a *cab* gene from *Chlamydomonas reinhardtii* predicts a nonconserved cleavage site but what seems recognizable as a phosphorylation site: G$^\mathrm{v}$KKTAAKAAA (84). Before these results can be interpreted fully, it will be necessary to sequence the phosphorylation sites of green algal LHC II; it is possible that the genes cloned so far do not encode phosphorylated proteins. Given the possibility of marked differences in N-

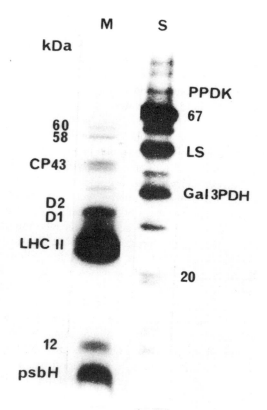

Figure 1 Chloroplast phosphoproteins. Isolated intact spinach chloroplasts were incubated with [^{32}P]orthophosphate in the light for 15 min. and then were washed, lysed by hypotonic shock, and separated into membrane (M) and soluble (S) fractions by centrifugation. Proteins were separated by sodium dodecylsulfate-polyacrylamide gel electrophoresis and phosphoproteins were detected by autoradiography (12 hr for track M, 120 hr for track S). PSII phosphoproteins are *psb* H protein (8 kDa), D1 (32 kDa), D2 (34 kDa), and CP43 (43 kDa).

terminal structure between higher-plant LHC IIs and green algal LHC IIs, it will be unwise to assume that LHC II phosphorylation is identical in higher plants and green algae.

PSII Core Proteins

PS II catalyzes the light-dependent oxidation of water and reduction of plastoquinone (PQ). It consists of a Chl *a*–containing reaction center of three proteins (D1, D2, and cytochrome b_{559}) surrounded by Chl *a*–containing proximal antennae (CP43, CP54) and various peripheral Chl *a/b* antennae, of which LHC II is by far the most abundant (14). Projecting into the lumen is a complex of four polypeptides (of 10, 18, 23, and 33 kDa) involved in water oxidation (121).

SEQUENCING OF PHOSPHORYLATION SITES   PS II particles stripped of LHC II contain four phosphoproteins (of 8, 32, 34, and 43 kDa). To identify rigorously these phosphoproteins and to locate their phosphorylation sites, Michel and coworkers (113, 115) purified and sequenced tryptic phosphopeptides derived from ^{32}P-labeled PS II particles (Table 1). The phosphopeptide derived from the 8-kDa phosphoprotein was sequenced by Edman degradation and proved to be the N-terminal peptide of the chloroplast *psb*H gene product (115), as judged by comparison with the published N-terminal sequence (51). The phosphoryl group was located on the second residue, a threonine. The other PS II phosphoproteins were blocked at the N-terminus and had to be sequenced by fast atom-bombardment mass spectrometry (113). In each case, the first residue was N-acetyl-O-phospho-threonine, establishing not only the identity of the blocking group but also locating the phosphoryl group on the first residue in each protein. When compared with sequences predicted from cloned chloroplast genes, each sequence permitted the identification of the phosphoproteins. The 32- and 34-kDa phosphoproteins were D1 and D2, respectively, but lacked the N-formylmethionyl residue predicted by the gene sequences. The 43-kDa protein was clearly CP43, but the sequence shown in Table 1 corresponds to an internal segment of the protein. How far it is from the N-terminus of the primary translation product depends on where translation begins. If translation begins at the preceding ATG, 14 residues would have to be removed from the protein to account for the sequence in Table 1. If translation begins at the preceding GTG, as seems to be the case for this gene in cyanobacteria (29), then only two residues (N-formyl-Met-Glu) would have to be removed. The location of the phosphorylation sites of *psb*H protein, D1, D2, and CP43 at the N-terminus has established the stromal location of this end of each molecule and has helped in establishing the orientation of the whole protein in the bilayer (113, 115).

USE OF MASS SPECTROMETRY IN ANALYSIS OF PHOSPHORYLATION SITES Although mass spectrometry is less commonly used for peptide sequencing than Edman degradation, it is uniquely powerful in certain situations. Several of its particular advantages are illustrated here: (*a*) It is not affected by the existence of N-terminal blocking groups and can indeed identify such groups; (*b*) it can unambiguously determine the number and identity of groups that modify amino acid side chains; and (*c*) in the case of phosphopeptides, it does not require radioactive phosphate, which can be either dispensed with or used merely as a low-specific-activity tag during peptide purification. The use of tandem mass spectrometry has disadvantages: (*a*) The peptide must be very clean and free of salts; (*b*) it is difficult to obtain sequences of peptides longer than about 12 residues; and (*c*) very hydrophilic peptides tend to shun the surface of the solution and are difficult to sample. (The latter difficulty can be overcome by chemical acetylation, but care needs to be exercised to determine the number of acetyl mass units added, in case the peptide already contains an acetyl group, as found for LHC II and PS II phosphopeptides.)

THYLAKOID PROTEIN KINASES

Phosphorylation of thylakoid proteins was first observed in isolated illuminated pea chloroplasts supplied with [^{32}P]orthophosphate (15). The light-dependence of this process was initially attributed solely to a requirement for photophosphorylation to convert [^{32}P]orthophosphate into [^{32}P]ATP. However, studies on isolated thylakoid membranes showed that addition of [γ-^{32}P]ATP did not abolish the requirement for light (2, 16). The fact that the electron transport inhibitor DCMU blocked protein phosphorylation (3, 16) indicated that light was activating a protein kinase through electron transport rather than through direct excitation. This notion received support from the finding that certain reducing agents such as dithionite (4), duroquinol (6), and reduced ferredoxin (16) could activate thylakoid protein phosphorylation in the dark. The fact that soluble, artificial substrates such as histone (37, 108) or peptide analogs of the LHC II phosphorylation site (22) are also phosphorylated in a redox-controlled manner establishes that it is the kinase and not the conformation of the protein substrate that is under redox control.

Redox Control

Several studies pointed to PQ as the key redox regulator of the thylakoid kinase. Dithionite, duroquinol, and reduced ferredoxin are known to reduce PQ in isolated thylakoids. Furthermore, DCMU blocks the reduction of PQ by preventing electron flow from PS II. However, these comments could equally well be made about several other redox components that follow PQ in the electron transport chain. The evidence implicating specifically PQ included the following points. First, the kinase is activated progressively by a long series of single-turnover flashes of light (4). This observation suggests that multiple molecules of the key redox regulator are involved. PQ is 10–25 times more abundant in thylakoids than other redox components, which are usually present as 1–2 molecules per electron transport chain (8, 121). Second, redox titrations of kinase activation in the dark give midpoint redox potentials close to that of PQ (75, 76, 117), with $n = 2$ (75) in the Nernst equation. PQ is one of the few two-electron carriers in the thylakoid. Third, the PQ antagonist DBMIB, which blocks electron flow from PQH$_2$ to the cytochrome *b/f* complex, does not inhibit phosphorylation at concentrations that block whole-chain electron transport (water to methyl viologen) (4, 42). This result suggests that kinase activation does not require reduction of electron transport components after the PQ pool.

For several years PQ was assumed to be the key redox activator of the thylakoid protein kinase, but this view has recently been modified as evidence emerged for two redox-controlled kinases that are under significantly different redox controls.

Evidence for More Than One Thylakoid Protein Kinase

INHIBITOR STUDIES   LHC II and the 8-kDa *psb*H gene product are the two most conspicuous thylakoid phosphoproteins (Figure 1). It has been noticed by several groups that the phosphorylation of these two proteins is not always similar. In particular, inhibitors such as sulfhydryl reagents (117) and the ATP analog fluorosulfonyl-benzoyladenosine (52) block LHC II phosphoryla-tion more effectively than phosphorylation of the 8-kDa protein. Subsequent studies have tended to confirm the differential inhibition of LHC II and the 8-kDa protein. Thus, PS II proteins continue to be phosphorylated under photoinhibitory conditions, whereas LHC II phosphorylation is abolished (140). In addition, bicarbonate stimulates phosphorylation of the 25-kDa LHC II and inhibits phosphorylation of the 8-kDa protein (153). However, phosphorylation of these and other thylakoid proteins is totally and equally blocked by DCMU, suggesting that if there are two thylakoid protein kinases, both are under redox control.

DIFFERENTIAL REDOX CONTROL   The two kinases differ also in redox control itself. DBMIB gives preferential inhibition of LHC II phosphorylation compared with phosphorylation of the PS II proteins (22, 52, 61). Under conditions that inhibit LHC II phosphorylation by 90%, only one other thylakoid protein (an unidentified 12-kDa protein) shows a comparable sensitivity, while phosphorylation of the four PS II phosphoproteins is in-hibited less than 10%. Phosphorylation of synthetic peptides resembling the phosphorylation site of LHC II is also inhibited by 90% (22).

DBMIB has been compared with two other inhibitors of the cytochrome complex (DNP-INT and HQNO) (23). Its effect is mimicked by DNP-INT, which also blocks the flow of electrons into the cytochrome *b/f* complex from the PQ pool (Figure 2). Its effect is not mimicked, however, by HQNO, which inhibits the flow of electrons out of the cytochrome complex into the PQ pool. These results implicate reduction of the cytochrome *b/f* complex by PQH$_2$ in activation of the LHC II kinase. Involvement of the cytochrome complex is suggested also by genetic evidence. Several mutants deficient in the cytochrome *b/f* complex have now been examined, and in each case

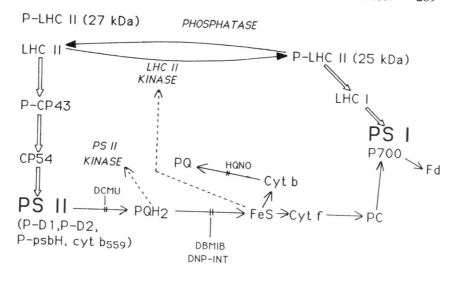

Figure 2 Activation of putative LHC II and PS II protein kinases by reduction of the cytochrome *b/f* complex and PQH$_2$, respectively, and regulation of location of 25-kDa LHC II. PS I is located principally in the stromal (unstacked) thylakoid membranes, while PS II is located principally in granal (stacked) membranes. P-protein: phosphorylated form of LHC II, CP43, D1, D2, or *psb*H gene product. Fd: ferredoxin; FeS: Rieske, iron-sulfur protein; PC: plastocyanin.

phosphorylation of LHC II is abolished or reduced while phosphorylation of the *psb*H protein and other PS II core proteins is largely unaffected. This finding was reported first for *Chlamydomonas* (105) and then for maize and *Lemna* (23, 42, 60, 61). Thus, both genetic analysis and studies with PQ antagonists suggest that phosphorylation of some or all LHC II molecules requires an active cytochrome *b/f* complex (or some factor associated with that complex), while phosphorylation of PS II core proteins does not.

How can this result be reconciled with the observation that DBMIB does not inhibit LHC II phosphorylation at concentrations that inhibit whole-chain electron transport (4)? It is interesting to note that DCMU gives 50% inhibition of LHC II phosphorylation at 1–3 μM, whereas 50% inhibition of electron transport can require as little as 50 nM DCMU. Thus, protein phosphorylation is less sensitive then electron transport to both DCMU and DBMIB, perhaps because electron transport measurements involve the flow of electrons through the PQ pool at a rate of approximately one per 10 ms. In contrast, activation of LHC II kinase can be accomplished by reduction of the PQ pool during the first 10–30 sec of a 1–5 min kinase assay. Thus, a quantity of inhibitor that gives 95% inhibition of rapid electron transport may still allow full reduction of the electron transport chain over the time scale of the kinase assay.

SUBSTRATE SPECIFICITY OF THYLAKOID PROTEIN KINASES   Additional evidence for two distinct redox-controlled thylakoid protein kinases comes from a comparison of the phosphorylation sites of LHC II and the four PS II core proteins (Table 1). The phosphorylation sites of D1, D2, CP43, and the 8-kDa phosphoprotein are located at or close to the N-terminal residue in a fairly neutral type of sequence (113, 115). In contrast, pea and spinach LHC IIs have been shown to be phosphorylated in an N-terminal region that is highly basic and also highly conserved among higher plants (112).

The sequence requirements for the phosphorylation of LHC II by pea and spinach thylakoids have been studied using synthetic peptide analogs (22, 112, 160). It is clear that the analogs are being phosphorylated by LHC II kinase, because their phosphorylation shows the same sensitivity to inhibitors such as DCMU and DBMIB as does LHC II phosphorylation itself. Incorporation of radioisotope from ATP is linear for at least 30 min in most cases, whether catalyzed by spinach thylakoids or pea thykakoids. Synthetic peptide analogs of the phosphorylation site of pea LHC II are phosphorylated preferentially on Thr-5 rather than Ser-3 or Thr-6 (22), a result consistent with in situ phosphorylation (118). To simplify the analysis, we have synthesized a large set of peptide analogs that are based on the spinach data but that contain only one hydroxyamino acid (112). A key point to emerge from this study is that the basic residues on either side of the phosphorylation site are important for the kinetics of the kinase reaction. When the lysyl residue on the C-terminal side of the phosphorylated threonine or the arginyl and lysyl residues on the N-terminal side are replaced by alanine, K_m increases 3–10 fold (decreased affinity between enzyme and substrate) and V_{max} decreases 2–8 fold. It is interesting that, of several classes of mammalian protein kinases, only protein kinase C also requires basic residues on either side of the phosphorylation site (94, 154).

The conserved N-acetylarginine is clearly important for the association of enzyme with substrate, but a comparison of acetylated and nonacetylated peptides indicates that the acetyl group itself does not significantly alter phosphorylation. We have conducted site-directed mutagenesis of several residues in and around the cleavage site (W. Buvinger, J. Bennett, unpublished results). While replacement of the conserved methionine by valine forces the protease to cleave at an unidentified site slightly downstream, replacement of the conserved arginine by either histidine or leucine has no discernible effect on processing. Thus, the arginyl residue is conserved more for its role in kinase recognition than for its role in protease recognition.

Purification

The first report of the purification of protein kinases from thylakoid membranes described two kinases with apparent molecular masses of 25 and 38 kDa (106, 109). The identification of the 38-kDa polypeptide as a kinase has since been questioned, and it has been suggested that what was actually purified was ferredoxin NADP reductase (44). The existence of the 25-kDa kinase has been confirmed (44), but it has a low affinity for histone, a commonly employed artificial substrate of the LHC II kinase. In contrast, a third candidate protein kinase (of 64 kDa) has a high affinity for histone; affinity chromatography on immobilized histone was a key step in its purification from solubilized thylakoids (43).

The 64-kDa kinase is to some extent associated with the cytochrome b/f complex during the early stages of purification (37, 43, 59). Restoration of PQH_2 to a purified aggregate of LHC II kinase and the cytochrome b/f complex stimulates LHC II phosphorylation in vitro (59). DBMIB inhibits activation of the aggregate. Antibody raised against the 64-kDa kinase inhibits phosphorylation of thylakoid proteins (41); this result would be consistent with the presence of a single protein kinase for LHC II and PS II or the existence of two immunologically related kinases. Purified 64-kDa kinase can also restore LHC II phosphorylation to a membrane in which LHC II phosphorylation has been inactivated by treatment with CHAPS detergent (41). The enzyme is reported to phosphorylate also the PS II proteins under these circumstances. This result could be taken to indicate that LHC II and PS II proteins are phosphorylated by the same kinase rather than two different kinases, but it is not an easy result to interpret because the rate of PS II phosphorylation was only 10% that of LHC II and the phosphorylated bands were not rigorously identified as PS II proteins. There appear to have been no studies to determine whether the 25-kDa kinase might be the PS II kinase and therefore capable of phosphorylating synthetic peptides resembling the phosphorylation site of the 8-kDa phosphoprotein.

THYLAKOID PROTEIN PHOSPHATASE

The half-time for dephosphorylation of LHC II has been determined in intact pea chloroplasts under various conditions (114). In each case, phosphorylation was allowed to proceed for 5 min in the light with [^{32}P]orthophosphate and was then arrested by addition of the uncoupler CCCP (carbonyl cyanide-m-chlorophenylhydrazone), the electron transport inhibitor DCMU, nonradioactive phosphate, or transfer to darkness. The kinetics of LHC II dephosphoryaltion were then ascertained. Irrespective of treatment, the half-time for dephosphorylation was 6–7 min. A comparable rate of dephosphorylation of LHC II is seen in isolated thylakoids suspended in buffered 0.01–10 mM $MgCl_2$. Inhibitors of electron transport do not inhibit dephosphorylation; but fluoride (18) and molybdate (126), classical phosphatase inhibitors, block dephosphorylation by more than 90% at moderate concentrations. For these reasons, it is assumed that dephosphorylation of LHC II is catalyzed by a phosphatase rather than by the protein kinase operating in reverse. Dephosphorylation of PS II phosphoproteins is 5–10 times slower than that of LHC II in isolated chloroplasts and in vivo and has a much higher Mg^{2+} ion requirement in vitro (114). Whether this finding reflects the operation of different phosphatases or the involvement of phosphorylation sites of very different sequence is not clear. Triton X-100 inhibits dephosphorylation in isolated thylakoids (18).

Some protein phosphatase activity may be removed from wheat thylakoids by washing with salt (152). Maximal release of phosphatase activity (50–60%) occurs at 0.3–0.5 M NaCl in 25 mM Tricine (pH 7.6), 1 mM KCl, 5 mM $MgSO_4$. The released activity is detectable using [^{32}P]-labeled histone III-S as substrate. Salt-washed thylakoids incorporated less radioactivity from [^{32}P]ATP into proteins than unwashed controls and dephosphorylated more slowly (half-time of 22 min compared with 14 min). Addition of solubilized phosphatase to unwashed thylakoids increased the rate of dephosphorylation of endogenous phosphoproteins (half-time of 9 min compared with 14 min in the unsupplemented, unwashed control).

SIGNIFICANCE OF THYLAKOID PROTEIN PHOSPHORYLATION

LHC II

In many green plants the photosynthetic membranes of the chloroplast are differentiated into stacked regions (grana) and unstacked regions (stroma lamellae) (8, 9, 121). The adhesive properties of thylakoids that lead to grana formation are known to reside at least in part in the N-terminus of LHC II (12, 118, 119, 136). Treatment of thylakoids with trypsin removes a number of residues from the N-terminus of LHC II and thereby abolishes the ability of thylakoids to stack in vitro in the presence of 100 mM NaCl or 5 mM $MgCl_2$. Trypsin treatment also fixes PS I and PS II in a state where absorbed light energy "spills over" with high probability from PS II to PS I; in the stacked state prior to trypsin treatment, spillover is much less probable (12, 150). These changes in the probability of spillover are readily monitored through changes in Chl fluorescence. As phosphorylation of LHC II occurs on the same segment that is removed by trypsin, it seemed possible that LHC II phosphorylation might affect the structure of grana and energy transfer between photosystems, which would be detectable through phosphorylation-dependent changes in Chl fluorescence.

PHOSPHORYLATION-DEPENDENT   CHANGES   IN   ENERGY   DISTRIBU-TION   Under conditions conducive to thylakoid protein phosphorylation, ATP causes a marked decline in Chl fluorescence at room temperature (24, 76, 78). The fluorescence decline is reversed under conditions conducive to dephosphorylation, and reversal is blocked by the phosphatase inhibitor NaF. As only PS II fluoresces strongly at room temperature, this result indicates that protein phosphorylation decreases energy transfer to PS II. When phosphorylated and dephosphorylated samples are examined for Chl fluorescence at 77 K (where the two photosystems fluoresce strongly and at different wavelengths), phosphorylation is seen to induce a decrease in emission from PS II and a roughly comparable increase in emission from PS I (24, 137, 155). Inclusion of internal standards for fluorescence yield confirms the absolute increase in PS I emission (137). The rates of ATP-dependent changes in chlorophyll fluorescence are correlated more closely with phosphorylation of 25-kDa LHC II than with that of the 27-kDa LHC II (85, 88, 89). Dephosphorylation of PS II proteins is too slow to explain the rapid reversal of PS II fluorescence that accompanies dephosphorylation of LHC II (149).

At low light intensities, the rates of PS I and PS II electron flow are proportional to intensity and to absorption cross-section. Under these conditions, measurements of electron flow through PS II consistently reveal a phosphorylation-dependent decrease in rate, with an average decrease of about 20% (49, 53, 103, 149). This decrease has been interpreted in terms of a reduction in absorption cross-section of PS II as a result of detachment of phosphorylated LHC II units.

A phosphorylation-dependent increase in the electron transport rate through PS I has also been seen, but less consistently. At low light intensities, P700, the reaction center of PS I, is oxidized at a rate proportional to its absorption cross-section. Some workers have detected a significant rise in the rate of P700 oxidation on phosphorylation (156), whereas others have observed no such rise, even though the same membranes showed a 15–25% reduction in the cross-section of PS II (49). This discrepancy remains unresolved. Horton & Black (77) failed to observe a faster oxidation of cytochrome f in phosphorylated thylakoids, although a faster rate would be expected if the antenna size of PS I had increased. Measurements of light-limited electron transport through PS I from a donor to a receptor usually reveal a phosphorylation-dependent increase in PS I activity (27, 53, 57, 78, 103). Increases range from 8% (103) to 40% (57), with 20% being an average value. The 40%

increase is reported for 650 nm light, the wavelength of maximal absorbance by Chl *b;* in contrast, 720-nm light (absorbed exclusively by red-shifted Chl *a* in the vicinity of P700) shows no phosphorylation-dependent increase in PS I rate (57). Photoacoustic measurements on photosynthetic systems permit dissipative processes of a thermal rather than fluorescent nature to be followed. Such measurements in vivo confirm that energy is transferred to PS I with greater probability in what is believed to be the phosphorylated state than in the dephosphorylated state (34).

The concept that protein phosphorylation is responsible for the ATP-dependent fluorescence quenching of PS II has recently been challenged. Islam (86) has shown that GTP can replace ATP as the phosphoryl group donor for thylakoid protein phosphorylation without producing the decline in PS II fluorescence seen with ATP. However, one of the problems in the interpretation of these experiments is the large difference between ATP and GTP with respect to K_m(NTP) and V_{max}; GTP is in fact a rather poor substrate compared with ATP. Islam (87) also used the ATP analog adenosine-5'-[γ-O-thio]triphosphate. This molecule induces partial PS II fluorescence quenching without apparently phosphorylating thylakoid proteins. However, the latter result may be suspect because ^{35}S is less efficiently detected by autoradiography and scintillation spectrometry than is ^{32}P, leading to underestimation of the incorporation of thiophosphate into protein. Nevertheless, the use of ATP analogs such as GTP and adenosine 5'-[γ-O-thio]triphosphate is a promising approach towards dissecting thylakoid protein phosphorylation into its elements.

LATERAL HETEROGENEITY OF THYLAKOIDS   How can LHC II phosphorylation lead to decreased energy transfer to PS II and increased energy transfer to PS I? Early hypotheses centered on spillover mechanisms; that is, phosphorylation of LHC II was considered to increase the probability that energy transferred from LHC II to a closed PS II unit would spill over into a neighboring PS I unit (4). However, thylakoid fractionation studies indicate that PS I and PS II units are not side-by-side in green-plant thylakoids: Stacked membranes are highly enriched for PS II, while unstacked membranes are highly enriched for PS I and ATP synthase (9). Cytochrome *b/f* complex is found in both stacked and unstacked membranes. The differential distribution of membrane components between the two thylakoid regions is referred to as lateral heterogeneity (9) and required new ideas concerning the mechanism of energy redistribution between the photosystems.

LHC II MOBILITY   Total LHC II is found preferentially (75–80%) in stacked membranes, where it functions as the peripheral antenna for PS II. However, a part of the LHC II pool is now recognized to be mobile and able to migrate laterally from the stacked to the unstacked regions (27, 98, 99, 101, 102, 104, 148). The mobile fraction is enriched in phosphorylated LHC II, especially the 25-kDa form. The available data suggest that (*a*) phosphorylation of 25- and 27-kDa LHC II units in stacked membranes leads to their detachment from PS II units and consequently a decrease in energy transfer to PS II, and (*b*) some of these detached LHC II units (mainly of 25 kDa) migrate from the stacked membranes to the unstacked membranes, where they can act as antennae for PS I. In one study of picosecond time-resolved Chl fluorescence in spinach, a fluorescence component was attributed to phosphorylated LHC II units that had separated from PS II but had not migrated and transferred energy to PS I (73). However, in a similar study on the green alga *Scenedesmus obliquus* (159), no such fluorescence component was observed and no evidence was obtained for an increase in the antenna size of PS I. Rather, the data indicated that mobile LHC II units transfer excitation energy to the so-called PS II β-centers located in the exposed membranes (β-centers are PS II units rendered slow by virtue of a small antenna size compared to the active PS II α-centers located in the stacked membranes). Since the activity of LHC II phosphorylation changes on a circadian basis in *S. obliquus* (68), it is possible that the pattern of energy distribution among PS I, PS II α-centers, and PS II β-centers is also variable throughout the day.

STATE TRANSITIONS   Under what physiological conditions is the regulation of energy distribution between the photosystems important? There are two related answers to this question. The first concerns the phenomenon of State I to State II transitions, the second the need to adjust the generation of ATP and reducing power to their consumption by carbon metabolism.

PS I and PS II operate in series to transfer electrons from water to ferredoxin with concomitant generation of ATP. Additional ATP is generated by cyclic electron flow around PS I. At subsaturating light intensities, it is advantageous to green plants to have a balanced distribution of excitation between PS I and PS II; such a distribution prevents overoxidation or over-reduction of the electron transport chain and ensures most efficient utilization of light. These ends cannot be achieved by a static arrangement of pigment-protein complexes because (*a*) the spectral quality of ambient light is variable, and (*b*) PS I and PS II exhibit different absorption spectra, chiefly as a result of the preponderance of Chl *b* molecules around PS II and the presence of red-shifted Chl *a* in PS I. Green algae and higher plants have the capacity to acclimate to changes in the spectral quality of light through a phenomenon known as State I to State II transitions (12, 28, 56, 122, 155).

The illumination conditions and kinetics of state transitions are comparable to those of reversible LHC II phosphorylation and associated fluorescence changes (155). When PS II is overexcited relative to PS I (e.g. under 645-nm light), chloroplasts are driven into State II. At the same time, the PQ pool and the cytochrome complex become reduced, the LHC II kinase is activated, and the imbalance in excitation is at least partially redressed, both as an immediate result of detachment of LHC II units from PS II α-centers and as a consequence of the migration of some of these LHC II units to PS I; thus, State II corresponds to phosphorylation of LHC II. Conversely, overexcitation of PS I relative to PS II (e.g. by 710-nm light) drives chloroplasts into State I. It leads also to oxidation of PQ and the cytochrome complex, inactivation of the kinase, and restoration of a larger antenna to PS II (partly at the expense of PS I) through dephosphorylation of LHC II; thus, State I corresponds to dephosphorylation of LHC II. The redox state of PQ/cytochrome b/f complex detects imbalances in excitation energy distribution between the photosystems, and reversible LHC II phosphorylation corrects those imbalances. Under natural conditions the most extreme difference in light quality is likely to be that between sunlight filtered through neighboring leaves (mainly PS I light) and direct, unfiltered sunlight (mainly PS II light). Reversible changes in LHC II phosphorylation in response to successive exposure to PS I light and PS II light have been observed in vivo and in vitro (19–21, 25, 47, 76, 126), but in *C. reinhardtii* the in vivo changes in LHC II phosphorylation are more dramatic if the redox state of the PQ pool is manipulated through changes in chlororespiration (47).

BALANCING CYCLIC AND LINEAR ELECTRON FLOW   Carbon metabolism is the major consumer of ATP and reducing power generated by photosynthetic electron transport. The demands for ATP and reducing power can vary independently with cellular metabolic requirements. This variability in demand is believed to be met by flexibility in the ratio of cyclic to linear electron flow, involving changes in the distribution of excitation energy between the photosystems.

Studies on intact maize protoplasts (54) and chloroplasts (82) indicate that when metabolism is altered experimentally to create a demand for additional ATP, the pH gradient across thylakoids is relaxed and LHC II phosphorylation is stimulated. Since this change in LHC II kinase activity appears to occur without a change in redox state of the PQ pool, it is suggested that the kinase responds to the change in pH gradient, possibly through the sensitivity of the cytochrome b/f complex to the gradient. Electron flow through the cytochrome complex is known to be inhibited by a high transmembrane pH gradient (142). Thus, the involvement of the cytochrome b/f complex in redox control of LHC II phosphorylation can be rationalized as allowing energy distribution to be modulated by changes in the pH gradient. However, the aggregation state of LHC II in thylakoids is also sensitive to the transmembrane pH gradient (62); this is another possible mechanism by which the pH gradient could influence LHC II phosphorylation in cells or isolated chloro-plasts.

HOW DOES LHC II MIGRATE?   It has been suggested that phosphorylated LHC II is repelled from grana by the negative charges on the phosphoryl groups of PS II core proteins (5). This hypothesis appears oversimplified. (*a*) It does not explain why half of phosphorylated LHC II (predominantly of 27 kDa) remains in grana. (*b*) It exaggerates the distance over which significant electrostatic interaction occurs within and between proteins (110). (*c*) It ignores the fact that the charge on the outside surface of PS II complexes is already negative in the absence of phosphorylation (12). (*d*) It ignores the existence of screening cations that must be present to stabilize the granal condition (12).

Brownian motion may explain quite adequately the mobility of LHC II. At the fastest, LHC II migrates from grana to exposed lamellae, a maximum straight-line distance of about 250 nm (116), in about 1 min (101, 104). The Einstein two-dimensional diffusion equation gives a lateral diffusion coefficient of 2.5×10^{-12} cm^2 sec^{-1}. Although very approximate, this figure is 70 times lower than the diffusion coefficient measured for the much larger cytochrome oxidase complex of the mitochondrial membrane, 1.5×10^{-10} cm^2 sec^{-1} (71). It is therefore more relevant to ask why mobile phosphorylated LHC IIs move so slowly between grana and exposed membranes, and, equally pertinently, why other phosphorylated LHC II units and all phosphorylated PS II units fail to migrate.

Grana stacks are stabilized by poorly understood interactions between LHC II units and PS II units. Some of these interactions occur within the lipid bilayer, while others occur across the gap between appressed thylakoid surfaces. Phosphorylation of LHC II units will disrupt interactions involving the N-terminus of these proteins, including presumably the local contribution to the adhesion between opposing thylakoid surfaces. If the phosphorylated LHC II unit is constrained by no other attractive force, it will be free to move under Brownian motion to partition itself between the stacked and unstacked membranes. Its Mg^{2+}-mediated affinity for PS I (161) could be an important factor in the partitioning process. On dephosphorylation, the protein partitions itself predominantly into the stacked membranes because of the affinity of its N-terminus for adhesion sites. Thus, the distribution of LHC II between the two membrane regions of the thylakoid would be determined by a combination of Brownian motion and purely local protein-protein interactions that were modulated by phosphorylation of the N-terminus of LHC II. Long-range interactions would be unnecessary. One factor that might tend to slow down LHC II migration is its tendency to form trimers (96, 136); a trimer might not migrate until all three of its N-termini have been phosphorylated.

PS II

The role of the phosphorylation of PS II core proteins is uncertain. In contrast to the N-terminus of LHC II, no special function has ever been ascribed to the N-termini of D1, D2, CP43, or the *psb*H gene product. One report indicated that phosphorylation altered the affinity of PS II for herbicides (141), but that observation could not be confirmed (74) and is made unlikely by the repeated observation that herbicides interact with the C-terminal halves of D1 and D2 (13). Several other effects of protein phosphorylation on PS II function have been reported: (*a*) increased stability (91) or reduced stability (72) of Q^-_B, the anionic semiquinone form of the secondary electron acceptor of PS II; (*b*) an increased negative surface-charge density near the primary acceptor, Q_A; (*c*) stimulation of a hydroxylamine-sensitive cyclic flow of electrons around PS II (79); (*d*) protection against photoinhibition (81); (*e*) decreased connectivity of PS II units (97); and (*f*) inhibition of light-saturated electron transfer in PS II (74, 80, 127). However, in none of these cases was the phosphorylation of PS II proteins themselves shown to be responsible for the observed effects.

Phosphorylated PS II units are almost entirely restricted to stacked membranes. Their phosphoryl groups line the 5-nm "partition" gap between the stacked membranes. This partition forms part of the pathway for the flow of protons (or the counterflow of OH^- ions) during photosynthesis (90, 131), with two protons entering PS II for each PQ reduced. Buffering in the partition is important because the reducing side of PS II operates most efficiently in the pH range 6.4–7.9 (45). Of all the ionizable groups in proteins, only the α-amino group at the N-terminus buffers in this range, but because its pK_a (about 8.2) is *above* the pH of the partition, it will be almost entirely protonated in the partition and therefore liable to retard hydroxyl ion diffusion (90, 131). By contrast, the phosphoryl group on threonine will buffer almost as well as the α-amino group, but, with its pK_a (about 6.5) *below* the pH of the partition, it will be largely unprotonated and thus much less effective in retarding hydroxyl ion diffusion. This argument offers a rationale not only for the phosphorylation of PS II core proteins but also for the presence of N-acetylation in three of these proteins and in LHC II itself (Table 1).

Analogs of Thylakoid Protein Phosphorylation

CYANOBACTERIA   Like green-plant chloroplasts, cyanobacteria employ two photosystems and undergo state transitions (56). However, the kinetics of the transitions in these prokaryotes are faster than those of chloroplasts, and LHC II is replaced by phycobilisomes. The latter consist of linear tetrapyrroles bound to soluble proteins, the three-dimensional structure of which has been solved by X-ray crystallography (63). They absorb in the visible range between the red and blue peaks of chlorophyll absorption and are attached to PS

II units on the protoplasmic side of the photosynthetically active regions of the cell membrane (124).

Protein phosphorylation has been observed in the photosynthetic membranes of cyanobacteria, both in vivo (138) and in vitro (7). Cells of *Synechococcus* 6301 were grown in the presence of [^{32}P]orthosphosphate under light absorbed preferentially by PS I or PS II (138). Under steady-state growth, a 15-kDa membrane protein and a 19-kDa phycobilisome protein were phosphorylated preferentially under PS II light, whereas a soluble 13-kDa protein was phosphorylated exclusively under PS II light. The identities of these proteins are unknown. Redox control of cyanobacterial protein kinase activity has not been clearly established. However, redox control has been observed for two related processes: redistribution of light energy during state transitions (120, 124), and randomization of PS II particles in the transition from State I to State II (124), when a fraction of phycobilisomes appear to move from PS II to PS I (120). The small distances involved in phycobilisome movement could explain the more rapid kinetics of state transitions in cyanobacteria compared with chloroplasts.

There is strong sequence homology between the *psb*H gene products of green algae, liverwort, and higher plants, on the one hand (46, 51, 70), and the corresponding genes from cyanobacteria, on the other. However, the cyanobacterial proteins lack the phosphorylation site (1, 95, 111). It will be instructive to determine which organisms contain and which lack this phosphorylation site. Is the phosphorylation site found only in green algae and higher plants, or is it found in all eukaryotic algae? It would also be instructive to know whether D1, D2, and CP43 are phosphorylated in cyanobacteria. The available gene sequences suggest that the phosphorylation site has been conserved, at least to the extent that threonines are present in the expected places and the neighboring sequences show some similarity with those of higher plants and green algae.

RED ALGAE   Like cyanobacteria, red algae contain phycobilisomes instead of LHC II and undergo state transitions. Some of the thylokoid proteins undergo phosphorylation, but there is as yet no evidence for redox control of phosphorylation of components involved in excitation energy capture and transfer (26).

PROCHLOROPHYCEAE   Chlorophyll *a/b* protein complexes related to LHC II have been found in the unusual prokaryotic group known as the Prochlorophyceae, which includes two genera, *Prochloron* and *Prochlorothrix*. Protein phosphorylation has been studied in *Prochlorothrix hollandica* (158). Externally added reductants such as NADPH enhance (and oxidants inhibit) the phosphorylation of 14 proteins (of 9–88 kDa) in isolated cell membranes.

The major phosphoprotein is a 29-kDa protein that does not co-electrophorese with the major 31-kDa apoprotein of LHC II. There is no correlation between state transitions and phosphorylation in vivo or in vitro.

STROMAL AND ENVELOPE PHOSPHOPROTEINS

Ninety percent of the radioactivity and the research that have gone into chloroplast phosphoproteins have gone into thylakoid proteins. However, there are more phosphoproteins in stroma and envelopes than in thylakoids. Even allowing that some of these proteins might be enzymes phosphorylated as part of their catalytic mechanism rather than for regulation, or are stromal proteins phosphorylated by the thylakoid protein kinases (25, 39), stromal and envelope phosphoproteins offer a considerable challenge for the future, not least because we understand these chloroplast compartments far less well than we understand thylakoids.

Stromal Phosphoproteins

PYRUVATE,ORTHOPHOSPHATE DIKINASE   In maize, CO_2 is concentrated into bundle sheaths by an intercellular malate-pyruvate shuttle. PEP and CO_2 in mesophyll cells are converted by cytosolic PEP carboxylase to malate, which diffuses to the chloroplasts of bundle sheath cells, where it is decarboxylated to yield CO_2 and pyruvate. The former is fixed into triose-phosphate, while pyruvate returns to the chloroplasts of mesophyll cells, to be regenerated into PEP by PPDK and re-initiate the shuttle. Several of the enzymes of the cycle are activated in response to light (50), but only PPDK (32) and the cytosolic PEP carboxylase (123) are regulated through reversible phosphorylation.

PPDK, a tetramer of 94-kDa subunits, catalyzes the conversion of pyruvate, ATP, and P_i to PEP, AMP, and PP_i (50). It is inactivated by phosphorylation of a threonyl residue and reactivated by dephosphorylation. The enzyme is also phosphorylated on an active-site histidine as part of its catalytic mechanism (32). The two phosphorylated residues are located on the same partial tryptic peptide (135) (Table 1). Thus, regulatory phosphorylation of PPDK occurs very close to the active site. Phosphohistidine is formed during partial catalysis with ATP plus P_i, or with PEP. Phosphorylation/ dephosphorylation of the threonine is catalyzed by a single bifunctional regulatory protein (10, 30, 33, 134). Phosphorylation is a unique reaction requiring ADP as phosphoryl group donor (11), and the protein substrate is the phosphohistidyl form of PPDK (32). Dephosphorylation is a phospho-transferase reaction requiring P_i and releasing PP_i (10); the substrate is the dephosphohistidyl form of PPDK (33). It is assumed that phosphorylation of the threonyl residue inhibits the overall reaction at the neighboring active site by altering the flexibility of the protein backbone. It is not clear how light activates the enzyme.

RIBULOSE-1,5-BISPHOSPHATE CARBOXYLASE/OXYGENASE Rubisco contains eight copies of each of two subunits: a 54-kDa large subunit encoded and synthesized in the chloroplast, and a 14-kDa small subunit encoded in the nucleus and synthesized as a 20-kDa precursor in the cytoplasm (93). Most studies of protein phosphorylation in leaves or stromal extracts have failed to reveal any more than a small amount of [^{32}P] incorporated into Rubisco and then mainly into the large subunit (25, 39, 58; Figure 1). However, a soluble extract of spinach chloroplasts was found to incorporate apparently high levels of ^{32}P from [γ-^{32}P]ATP into the large subunit in the presence of 40 μM spermidine (66). Since the data were not quantitative, it is impossible to calculate the percentage of large subunits phosphorylated in these experiments.

When [^{32}P]orthophosphate was supplied for 24 hr to spinach leaves or a moss, radioisotope was incorporated into the small subunit of Rubisco (92). Removal of the phosphoryl group with alkaline phosphatase resulted in a 70% decline in the biological activity of the enzyme and a dissociation of the holoenzyme into an octomer of large subunits and dimers of small subunits. The authors suggest that phosphorylation of small subunits is mandatory for reconstitution of holoenzyme and hence crucial for the activation of Rubisco. This potentially important observation should be confirmed. The small subunit of Rubisco is reported to be phosphorylated by protein kinases associated with chloroplast ribosomes (65) and the chloroplast envelope (146). Sequencing of the phosphorylation sites on Rubisco subunits would provide important additional information that could be interpreted in the light of the crystal structure of the protein, where each small subunit makes contact with three other small subunits and three large subunits (35).

Activation of Rubisco involves an enzyme, Rubisco activase, which has ATP-hydrolyzing activity (132). Although the role of ATP in Rubisco activation is unknown, attempts to link activation with phosphorylation of either Rubisco or the activase have not been successful.

RIBOSOMAL PROTEINS Two chloroplast ribosomal proteins are phosphorylated when isolated spinach chloroplasts are incubated in the light with [^{32}P]orthophosphate and a full complement of amino acids (65). The phosphoproteins are L18 of the large subunit and LS31, a protein apparently located on both subunits. The two proteins were phosphorylated on serine. A protein kinase associated with purified ribosomes phosphorylated principally the same two proteins and also L4 when supplied with [γ-^{32}P]ATP.

Envelope Phosphoproteins

The chloroplast is delimited by an envelope consisting of two membranes (67). The outer membrane is permeable to molecules of up to about 10 kDa, whereas the inner membrane is impermeable to even very small molecules. The intense metabolic flux between chloroplast and cytoplasm is mediated by specific transporters, such as the phosphate/triose phosphate translocator, ATP/ADP translocator, and a carboxylic acid transporter, all located on the inner envelope (67). During plastid development only a minority of plastid proteins are synthesized in the organelle; the majority enter the organelle as precursors by binding to the outer envelope and then being translocated across the two membranes (93).

OUTER ENVELOPE   Several unidentified intrinsic membrane proteins are phosphorylated when isolated outer envelopes are incubated with [γ-^{32}P]ATP. The two most conspicuous phosphoproteins (with molecular masses of 86 and 110 kDa) are also phosphorylated when intact chloroplasts are incubated with [^{32}P]orthophosphate (144, 146). A 70-kDa ATP-dependent protein kinase has been isolated from outer envelope membranes by a protocol that included solubilization with cholate and octylglucoside, and affinity chromatography on histone III-S Sepharose 4B (143). A GTP-dependent protein kinase has also been detected in the outer envelope (147).

Protein phosphorylation may be involved in the transport of proteins into the chloroplast. It is well established that ATP is required for import (55, 64, 128, 139); in addition, the phosphatase inhibitors fluoride and molybdate block import (55, 139). It is assumed that one or more receptor proteins recognize nuclear-encoded precursors of chloroplast proteins and facilitate their entry into the organelle (93, 129). A 51-kDa outer envelope protein is phosphorylated to an extent that is inversely correlated with import (69). Addition of increasing amounts of pyridoxal-5-phosphate to intact chloroplasts results in progressive inhibition of protein import and increased phosphorylation of the 51-kDa protein. Incubation of chloroplasts with a precursor protein leads to specific enhancement of phosphorylation of the 51-kDa protein; this effect is not seen in the presence of mature protein lacking transit sequence. Thermolysin treatment of intact chloroplasts abolishes import and also prevents phosphorylation of the 51-kDa protein. The K_m for ATP in the phosphorylation of the 51-kDa protein is about 5 μM. It is interesting that the ATP requirement for half-maximal binding of precursors to the envelope is also about 5 μM (125), whereas the ATP requirement for half-maximal import is about 300 μM (157). For both binding and import, a nonhydrolyzable ATP analog does not suffice, and the source of ATP must be internal to the organelle.

INNER ENVELOPE   Although the inner envelope is much more richly endowed with protein than the outer envelope (38), protein phosphorylation is less conspicuous in the former than in the latter (144). Two characteristic inner-envelope phosphoproteins (of 24 and 33 kDa) are labeled when either purified inner envelopes or "mixed" vesicles are incubated with micromolar levels of $[\gamma\text{-}^{32}P]$ATP. Mixed vesicles form when chloroplasts are lysed under hypotonic conditions. They consist of fusions between parts of the inner-envelope membrane and parts of the outer-envelope membrane. They are assumed also to contain some of the soluble material from the intermembrane space.

INTERENVELOPE SPACE   For many years the most conspicuous soluble phosphoprotein of the chloroplast [of 64 kDa in pea, 67 kDa in spinach (Figure 1)] was believed to be of stromal origin (58). It has been claimed that this protein originates from the space between the inner and outer envelope membranes, the first protein localized to that compartment (144). It is labeled in isolated mixed envelope vesicles but not in purified inner or outer envelopes. When mixed vesicles are sonicated, the phosphoprotein is released as a soluble protein. It is very rapidly labeled in both intact chloroplasts and in isolated mixed envelopes. Its phosphoryl group undergoes rapid turnover (half-time = 2 min) in the light or dark, and ^{32}P in the protein can be chased out in organello with orthophosphate or in soluble extracts with ATP (145). However, evidence has recently been presented that the 64-kDa phosphoprotein is actually the phosphoenzyme intermediate of stromal phosphoglucomutase (137a).

CONCLUDING REMARKS

Since their discovery 15 years ago, chloroplast phosphoproteins have received a great deal of attention, which has led to the identification of 14 out of an estimated minimum of 30 such proteins. Functions have been determined for phosphorylation of PPDK and the LHC II family, but not for phosphorylation of four PS II core proteins, the large and small subunits of Rubisco, glyceraldehyde-3-phosphate dehydrogenase, or two ribosomal proteins.

The most important deficiency in the study of chloroplast phosphoproteins is the lack of a genetic approach. This deficiency is slowly being corrected, chiefly through studies on mutants of the unicellular green alga *Chlamydomonas reinhardtii*. To the extent that protein phosphorylation in cyanobacteria is relevant to protein phosphorylation in green-plant chloroplasts, the application to cyanobacteria of molecular techniques such as transformation and replacement mutagenesis may provide insights into the structure, regulation, and physiological roles of photosynthetic protein kinases and their protein substrates.

ACKNOWLEDGMENTS

The author's research was supported by grants from the UK Science and Engineering Research Council and the US Departments of Agriculture and Energy.

Literature Cited

1. Abdel-Mawgood, L., Dilley, R. A. 1990. Cloning and nucleotide sequence of the *psb*H gene from cyanobacterium *Synechocystis* 6803. *Plant Mol. Biol.* 14:445–46
2. Alfonzo, R., Nelson, N., Racker, E. 1980. A light-dependent protein kinase activity of chloroplasts. *Plant Physiol.* 65:730–34
3. Allen, J. F., Bennett, J. 1981. Photosynthetic protein phosphorylation in intact chloroplasts: inhibition by DCMU and by the onset of CO$_2$ fixation. *FEBS Lett.* 123:67–70
4. Allen, J. F., Bennett, J., Steinback, K. E., Arntzen, C. J. 1981. Chloroplast protein phosphorylation couples plastoquinone redox state to distribution of excitation energy between photosystems. *Nature* 291:21–25
5. Allen, J. F., Holmes, N. G. 1986. A general model for regulation of photosynthetic unit function by protein phosphorylation. *FEBS Lett.* 202:175–81
6. Allen, J. F., Horton, P. 1981. Chloroplast protein phosphorylation and chlorophyll fluorescence quenching activation by tetramethyl-*p*-hydroquinone, an electron donor to plastoquinone. *Biochim. Biophys. Acta* 638:290–95
7. Allen, J. F., Sanders, C. E., Holmes, N. G. 1985. Correlation of membrane protein phosphorylation with excitation energy distribution in cyanobacterium *Synechococcus* 6301. *FEBS Lett.* 193:271–75
8. Anderson, J. M. 1986. Photoregulation of the composition, function and structure of thylakoid membranes. *Annu. Rev. Plant Physiol.* 37:93–136
9. Andersson, B., Anderson, J. M. 1980. Lateral heterogeneity in the distribution of chlorophyll-protein complexes of the thylakoid membranes of spinach chloroplasts. *Biochim. Biophys. Acta* 593:427–40
10. Ashton, A. R., Burnell, J. N., Hatch, M. D. 1984. Regulation of C$_4$ photosynthesis: inactivation of pyruvate, P$_i$ dikinase by ADP-dependent phosphorylation and activation by phosphorolysis. *Arch. Biochem. Biophys.* 230:492–503
11. Ashton, A. R., Hatch, M. D. 1983. Regulation of C$_4$ photosynthesis: physical and kinetic properties of active (dithiol) and inactive (disulfide) NADP-malate dehydrogenase from *Zea mays*. *Arch. Biochem. Biophys.* 227:406–15
12. Barber, J. 1982. Influence of surface charges on thylakoid structure and function. *Annu. Rev. Plant Physiol.* 33:261–95
13. Barber, J. 1987. Photosynthetic reaction centres: a common link. *Trends Biochem. Sci.* 12:321–26
14. Bassi, R., Rigoni, F., Barbato, R., Giacometti, M. 1988. Light-harvesting chlorophyll *a/b* protein (LHCII) populations in phosphorylated membranes. *Biochim. Biophys. Acta* 936:29–38
15. Bennett, J. 1977. Phosphorylation of chloroplast membrane polypeptides. *Nature* 269:344–46
16. Bennett, J. 1979. Chloroplast phosphoproteins. The protein kinase of thylakoid membranes is light-dependent. *FEBS Lett.* 103:342–44
17. Bennett, J. 1979. Chloroplast phosphoproteins. Phosphorylation of polypeptides of the light-harvesting chlorophyll protein complex. *Eur. J. Biochem.* 99:133–37
18. Bennett, J. 1980. Chloroplast phosphoproteins. Evidence for a thylakoid-bound phosphoprotein phosphatase. *Eur. J. Biochem.* 104:85–89
19. Bennett, J. 1983. Regulation of photosynthesis by reversible phosphorylation of the light-harvesting chlorophyll *a/b* protein. *Biochem. J.* 212:1–12
20. Bennett, J. 1984. Chloroplast protein phosphorylation and the regulation of

photosynthesis. *Physiol. Plant.* 60:583–90

21. Bennett, J. 1984. Thylakoid protein phosphorylation: in vitro and in vivo. *Biochem. Soc. Trans.* 12:771–74

22. Bennett, J., Shaw, E. K., Bakr, S. 1987. Phosphorylation of thylakoid proteins and synthetic peptide analogs: differential sensitivity to inhibition by a plastoquinone antagonist. *FEBS Lett.* 210:22–26

23. Bennett, J., Shaw, E. K., Michel, H. 1988. Cytochrome b_6f complex is required for phosphorylation of light-harvesting chlorophyll *a/b* complex II in chloroplast photosynthetic membranes. *Eur. J. Biochem.* 171:95–100

24. Bennett, J., Steinback, K. E., Arntzen, C. J. 1980. Chloroplast phosphoproteins: regulation of excitation energy transfer by phosphorylation of thylakoid membranes. *Proc. Natl. Acad. Sci. USA* 77:5253–57

25. Bhalla, P., Bennett, J. 1987. Chloroplast phosphoproteins: Phosphorylation of a 12-kDa stromal protein by the redox-controlled kinase of thylakoid membranes. *Arch. Biochem. Biophys.* 252:249–54

26. Biggins, J., Campbell, C. L., Bruce, D. 1984. Mechanism of the light state transition in photosynthesis. II. Analysis of phosphorylated polypeptides in the red alga *Porphyridium cruentum. Biochim. Biophys. Acta* 767:138–44

27. Black, M. T., Lee, P., Horton, P. 1986. Changes in topography and function of thylakoid membranes following membrane protein phosphorylation. *Planta* 168:330–36

28. Bonaventura, C., Myers, J. 1969. Fluorescence and oxygen evolution for *Chlorella pyrenoidosa. Biochim. Biophys. Acta* 189:366–83

29. Bricker, T. M. 1990. The structure and function of CPa-1 and CPa-2 in photosystem II. *Photosynth. Res.* 24:1–13

30. Budde, R. J. A., Ernst, S. M., Chollet, R. 1986. Substrate specificity and regulation of the maize (*Zea mays*) leaf ADP: protein phosphotransferase catalysing phosphorylation/inactivation of pyruvate, orthophosphate dikinase. *Biochem. J.* 236:579–84

31. Bürgi, R., Suter, F., Zuber, H. 1987. Arrangement of the light-harvesting chlorophyll *a/b* protein complex in the thylakoid membrane. *Biochim. Biophys. Acta* 890:346–51

32. Burnell, J. N., Hatch, M. D. 1984. Regulation of C$_4$ photosynthesis: identification of a catalytically important

histidine residue and its role in the regulation of pyruvate, P$_i$ dikinase. *Arch. Biochem. Biophys.* 231:175–82

33. Burnell, J. N., Hatch, M. D. 1985. Regulation of C$_4$ photosynthesis: purification and properties of the protein catalyzing ADP-mediated inactivation and P$_i$-mediated activation of pyruvate, P$_i$ dikinase. *Arch. Biochem. Biophys.* 237:490–503

34. Canaani, O., Barber, J., Malkin, S. 1984. Evidence that phosphorylation and dephosphorylation regulate the distribution of excitation energy between the two photosystems of photosynthesis in vivo: photoacoustic and fluorimetric study of an intact leaf. *Proc. Natl. Acad. Sci. USA* 81:1614–18

35. Chapman, M. S., Suh, S. W., Curmi, P. M. G., Cascio, D., Smith, W. W., et al. 1988. Tertiary structure of plant RuBisCO: domains and their contacts. *Science* 241:71–74

36. Chitnis, P. R., Thornber, J. P. 1988. The major light-harvesting complex of photosystem II: aspects of its molecular and cell biology. *Photosynth. Res.* 16:41–63

37. Clark, R. D., Hind, G., Bennett, J. 1985. Partial purification of a spinach thylakoid protein kinase that can phosphorylate light-harvesting chlorophyll *a/b* proteins. In *Molecular Biology of the Photosynthetic Apparatus,* ed. K. Steinback, S. Bonitz, C. J. Arntzen, L. Bogorad, pp. 259–67. Cold Spring Harbor, NY: Cold Spring Harbor Lab, 437 pp.

38. Cline, K., Keegstra, K., Staehelin, L. A. 1985. Freeze-fracture electron microscopic analysis of ultrarapidly frozen envelope membranes on intact chloroplasts and after purification. *Protoplasma* 125:111–13

39. Cortez, N., Lucero, H. A., Vallejos, R. H. 1987. Stromal serine protein kinase activity in spinach chloroplasts. *Arch. Biochem. Biophys.* 254:504–8

40. Coruzzi, A., Broglie, R., Cashmore, A., Chua, N.-H. 1983. Nucleotide sequences of two pea cDNA clones encoding the small subunit of ribulose 1,5-bisphosphate carboxylase and the major chlorophyll *a/b* binding thylakoid polypeptide. *J. Biol. Chem.* 258:1399–1402

41. Coughlan, S., Hind, G. 1987. Phosphorylation of thylakoid proteins by a purified kinase. *J. Biol. Chem.* 262:8402–8

42. Coughlan, S. J. 1988. Chloroplast thylakoid protein phosphorylation is in-

fluenced by mutations in the cytochrome *bf* complex. *Biochim. Biophys. Acta* 933:413–22

43. Coughlan, S. J., Hind, G. 1986. Purification and characterization of a membrane-bound protein kinase from spinach thylakoids. *J. Biol. Chem.* 261: 11378–85

44. Coughlan, S. J., Hind, G. 1986. Protein kinases of the thylakoid membrane. *J. Biol. Chem.* 261:14062–68

45. Crofts, A. R., Robinson, H. H., Snozzi, M. 1984. Reactions of quinones at catalytic sites: a diffusional role in H-transfer. In *Advances in Photosynthesis Research,* ed. C. Sybesma, 4:461–68. The Hague: Martinus Nijhoff/Dr. W. Junk. 777 pp.

46. Dedner, N., Meyer, H. E., Ashton, C., Wildner, G. F. 1988. N-terminal sequence analysis of the 8 kDa protein in *Chlamydomonas reinhardii. FEBS Lett.* 236:77–82

47. Delepelaire, P., Wollman, F.-A. 1985. Correlations between fluorescence and phosphorylation changes in thylakoid membranes of *Chlamydomonas reinhardtii* in vivo: a kinetic analysis. *Biochim. Biophys. Acta* 809:277–83

48. Demmin, D. S., Stockinger, E. J., Chang, Y. C., Walling, L. L. 1989. Phylogenetic relationships between the chlorophyll *a/b* binding protein (CAB) multigene family: an intra- and interspecific study. *J. Mol. Evol.* 29:266–79

49. Deng, X., Melis, A. 1986. Phosphorylation of the light-harvesting complex II in higher plant chloroplasts: effect on photosystem II and photosystem I absorption cross section. *Photobiochem. Photobiophys.* 13:41–52

50. Edwards, G. E., Nakamoto, H., Burnell, J. N., Hatch, M. D. 1985. Pyruvate, P_i dikinase and NADP-malate dehydrogenase in C_4 photosynthesis: properties and mechanism of light/dark regulation. *Annu. Rev. Plant Physiol.* 36:255–86

51. Farchaus, J., Dilley, R. A. 1986. Purification and partial sequence of the M_r 10,000 phosphoprotein from spinach thylakoids. *Arch. Biochem. Biophys.* 244:94–101

52. Farchaus, J., Dilley, R. A., Cramer, W. A. 1985. Selective inhibition of the spinach thylakoid LHC II protein kinase. *Biochim. Biophys. Acta* 809:17–26

53. Farchaus, J. W., Widger, W. R., Cramer, W. A., Dilley, R. A. 1982. Kinase-induced changes in electron transport rates of spinach chloroplasts. *Arch. Biochem. Biophys.* 217:362–67

54. Fernyhough, P., Foyer, C. H., Horton, P. 1984. Increase in the level of thylakoid protein phosphorylation in maize mesophyll chloroplasts by decrease in the transthylakoid pH gradient. *FEBS Lett.* 176:133–38

55. Flügge, U. I., Hinz, G. 1986. Energy dependence of protein translocation into chloroplasts. *Eur. J. Biochem.* 160:563–70

56. Fork, D. C., Satoh, K. 1986. The control by state transitions of the distribution of excitation energy in photosynthesis. *Annu. Rev. Plant Physiol.* 37:335–61

57. Forti, G., Vianelli, A. 1988. Influence of thylakoid protein phosphorylation on photosystem I photochemistry. *FEBS Lett.* 231:95–98

58. Foyer, C. H. 1985. Stromal protein phosphorylation in spinach (*Spinacia oleracea*) chloroplasts. *Biochem. J.* 231:97–103

59. Gal, A., Hauska, G., Hermann, R., Ohad, I. 1990. Interaction between LHCII kinase and cytochrome b_6/f: in vitro control of kinase activity. *J. Biol. Chem.* 265:19742–49

60. Gal, A., Shahak, Y., Schuster, G., Ohad, I. 1987. Specific loss of LHCII phosphorylation in the *Lemna* mutant 1073 lacking the cytochrome b_6/f complex. *FEBS Lett.* 221:205–10

61. Gal, A., Schuster, G., Frid, D., Canaani, O., Schwieger, H. G., et al. 1988. Role of cytochrome $b_6.f$ complex in the redox-controlled activity of *Acetabularia* thylakoid protein kinase. *J. Biol. Chem.* 263:7785–91

62. Garab, G., Leegood, R. C., Walker, D. A., Sutherland, J. C., Hind, G. 1988. Reversible changes in macroorganization of the light-harvesting chlorophyll *a/b* pigment-protein complex detected by circular dichroism. *Biochemistry* 27:2430–34

63. Glazer, A. N., Melis, A. 1987. Photochemical reaction centers: structure, organization, and function. *Annu. Rev. Plant Physiol.* 38:11–45

64. Grossman, R., Bartlett, S., Chua, N.-H. 1980. Energy-dependent uptake of cytoplasmically synthesized polypeptides by chloroplasts. *Nature* 285:625–28

65. Guitton, C., Dorne, A.-M., Mache, R. 1984. In organello and in vitro phosphorylation of chloroplast ribosomal proteins. *Biochem. Biophys. Res. Commun.* 121:297–303

66. Guitton, C., Mache, R. 1987. Phosphorylation in vitro of the large subunit of the ribulose-1,5-bisphosphate carboxylase and of the glyceraldehyde-

3-phosphate dehydrogenase. *Eur. J. Biochem.* 166:249–54

67. Heber, U., Heldt, H. W. 1981. The chloroplast envelope: structure, function, and role in leaf metabolism. *Annu. Rev. Plant Physiol.* 32:139–68

68. Heil, W. G., Senger, H. 1986. Thylakoid-protein phosphorylation during the life cycle of *Scenedesmus obliquus* in synchronous culture. *Planta* 167:233–39

69. Hinz, G., Flügge, U. I. 1988. Phosphorylation of a 51-kDa envelope membrane polypeptide involved in protein translocation into chloroplasts. *Eur. J. Biochem.* 175:649–59

70. Hird, S. M., Dyer, T. A., Gray, J. C. 1986. The gene for the 10 kDA phosphoprotein of photosystem II is located in chloroplast DNA. *FEBS Lett.* 209:181–86

71. Hochman, J., Ferguson-Miller, S., Schindler, M. 1985. Mobility in the mitochondrial electron transport chain. *Biochemistry* 24:2509–16

72. Hodges, M., Boussac, A., Briantais, J.-M. 1987. Thylakoid membrane protein phosphorylation modifies the equilibrium between photosystem II quinone electron acceptors. *Biochim. Biophys. Acta* 894:138–45

73. Hodges, M., Briantais, J.-M., Moya, I. 1987. The effect of thylakoid membrane reorganisation on chlorophyll fluorescence lifetime components: a comparison between state transitions, protein phosphorylation and the absence of Mg^{2+}. *Biochim. Biophys. Acta* 893:480–89

74. Hodges, M., Packham, N. K., Barber, J. 1985. Modification of photosystem II activity by protein phosphorylation. *FEBS Lett.* 181:83–87

75. Horton, P., Allen, J. F., Black, M. T., Bennett, J. 1981. Regulation of phosphorylation of chloroplast membrane polypeptides by the redox state of plastoquinone. *FEBS Lett.* 125:193–96

76. Horton, P., Black, M. 1980. Activation of adenosine 5' triphosphate-induced quenching of chlorophyll fluorescence by reduced plastoquinone. *FEBS Lett.* 119:141–44

77. Horton, P., Black, M. 1981. Light-induced redox changes in chloroplast cytochrome *f* after phosphorylation of membrane proteins. *FEBS Lett.* 132:75–77

78. Horton, P., Black, M. 1982. On the nature of the fluorescence decrease due to phosphorylation of chloroplast membrane proteins. *Biochim. Biophys. Acta* 680:22–27

79. Horton, P., Lee, P. 1983. Stimulation of a cyclic electron transfer pathway around photosystem 2 by phosphorylation of chloroplast thylakoid proteins. *FEBS Lett.* 162:81–84

80. Horton, P., Lee, P. 1984. Phosphorylation of chloroplast thylakoids decreases the maximum capacity of photosystem 2 electron transfer. *Biochim. Biophys. Acta* 767:563–67

81. Horton, P., Lee, P. 1985. Phosphorylation of chloroplast membrane proteins partially protects against photoinhibition. *Planta* 165:37–42

82. Horton, P., Lee, P., Fenryhough, P. 1990. Emerson enhancement, photosynthetic control and protein phosphorylation in isolated maize mesophyll chloroplasts; dependence upon carbon metabolism. *Biochim. Biophys. Acta* 1017:160–66

83. Hunt, D. F., Shabanowitz, J., Yates, J. R. III, Zhu, N.-Z., Russell, D. H., et al. 1987. Tandem quadrupole Fourier-transform mass spectrometry of oligopeptides and small proteins. *Proc. Natl. Acad. Sci. USA* 84:621–23

84. Imbault, P., Wittemer, C., Johanningmeier, U., Jacobs, J. D., Howell, S. H. 1988. Structure of the *Chlamydomonas reinhardtii cab* II-1 gene encoding a chlorophyll *a/b* protein. *Gene* 73:397–407

85. Islam, K. 1987. The rate and extent of phosphorylation of the two light-harvesting chlorophyll *a/b* binding protein complex (LHC-II) polypeptides in isolated spinach thylakoids. *Biochim. Biophys. Acta* 893:333–41

86. Islam, K. 1989. GTP-induced chloroplast membrane protein phosphorylation and photosystem II fluorescence changes: evidence for multiple protein kinase activities. *Biochim. Biophys. Acta* 974:261–66

87. Islam, K. 1989. Thylakoid protein phosphorylation and associated photosystem II fluorescence changes: a study with the ATP analogue adenosine-5-O-thiotriphosphate (ATP-γ-S). *Biochim. Biophys. Acta* 974:267–73

88. Islam, K., Jennings, R. C. 1985. Relative kinetics of quenching of photosystem II fluorescence and phosphorylation of the two light-harvesting chlorophyll *a/b* polypeptides in isolated spinach thylakoids. *Biochim. Biophys. Acta* 810:158–63

89. Jennings, R. C., Islam, K., Zucchelli, G. 1986. Spinach-thylakoid phosphorylation: studies on the kinetics of changes in photosystem antenna size, spill-over and phosphorylation of light-harvesting

chlorophyll *a/b* protein. *Biochim. Biophys. Acta* 850:483–89

90. Junge, W., McLaughlin, S. 1987. The role of fixed and mobile buffers in the kinetics of proton movement. *Biochim. Biophys. Acta* 890:1–5

91. Jursinic, P. A., Kyle, D. J. 1983. Changes in the redox state of the secondary acceptor of photosystem II associated with light-induced thylakoid protein phosphorylation. *Biochim. Biophys. Acta* 723:37–44

92. Kaul, R., Saluja, D., Sachar, R. C. 1986. Phosphorylation of small subunit plays a crucial role in the regulation of RuBPCase in moss and spinach. *FEBS Lett.* 209:63–70

93. Keegstra, K., Olsen, L. J. 1989. Chloroplastic precursors and their transport across the envelope membranes. *Annu. Rev. Plant Physiol. Plant Mol. Biol.* 40:471–501

94. Kikkawa, U., Kishimoto, A., Nishizuka, Y. 1989. The protein kinase C family: heterogeneity and its implications. *Annu. Rev. Biochem.* 58:31–44

95. Koike, H., Mamada, K., Ikeuchi, M., Inoue, Y. 1989. Low-molecular-mass proteins in cyanobacterial photosystem II: identification of *psb*H and *psb*K gene products by N-terminal sequencing. *FEBS Lett.* 244:391–96

96. Kühlbrandt, W. 1987. Three-dimensional crystals of the light-harvesting chlorophyll *a/b* protein complex from pea chloroplasts. *J. Mol. Biol.* 194:757–62

97. Kyle, D. J., Haworth, P., Arntzen, C. J. 1982. Thylakoid membrane protein phosphorylation leads to a decrease in connectivity between photosystem II reaction centers. *Biochim. Biophys. Acta* 680:336–42

98. Kyle, D. J., Kuang, T.-Y., Watson, J. L., Arntzen, C. J. 1984. Movement of a subpopulation of LHC-II from grana to stroma lamellae as a consequence of its phosphorylation. *Biochim. Biophys. Acta* 765:89–96

99. Kyle, D. J., Staehelin, L. A., Arntzen, C. J. 1983. Lateral mobility of the light-harvesting chlorophyll protein in chloroplast membranes controls excitation energy distribution in higher plants. *Arch. Biochem. Biophys.* 222:527–41

100. LaRoche, J., Bennett, J., Falkowski, P. G. 1990. Characterization of a cDNA encoding for the 28.5 kDa LHC II apoprotein from the unicellular marine chlorophyte. *Dunaliella tertiolecta. Gene.* 95:165–71

101. Larsson, U. K., Andersson, B. 1985.

Different degrees of phosphorylation and lateral mobility of two polypeptides belonging to the light-harvesting complex of photosystem II. *Biochim. Biophys. Acta* 809:396–402

102. Larsson, U. K., Jergil, B., Andersson, B. 1983. Changes in the lateral distribution of the light-harvesting chlorophyll-*a/b*—protein complex induced by its phosphorylation. *Eur. J. Biochem.* 136:25–29

103. Larsson, U. K., Ögren, E., Öquist, G., Andersson, B. 1986. Electron transport and fluorescence studies on the functional interaction between phospho-LHC II and photosystem I in isolated stroma lamellae vesicles. *Photobiochem. Photobiophys.* 13:29–39

104. Larsson, U. K., Sundby, C., Andersson, B. 1987. Characterization of two different subpopulations of LHCII: polypeptide composition, phosphorylation pattern and association with photosystem II. *Biochim. Biophys. Acta* 894:59–68

105. Lemaire, C., Girard-Bascou, J., Wollman, F.-A. 1987. Characterization of the B6/F complex subunits and studies on the LHC-kinase in *Chlamydomonas reinhardtii* using mutant strains altered in the B6/F complex. In *Progress in Photosynthesis Research,* ed. J. Biggins, 4:655–58. Dordrecht: Martinus Nijhoff. 858 pp.

106. Lin, Z.-F., Lucero, H. A., Racker, E. 1982. Protein kinases from spinach chloroplasts. I. Purification and identification of two distinct protein kinases. *J. Biol. Chem.* 257:12153–56

107. Long, Z., Wang, S.-Y., Nelson, N. 1989. Cloning and nucleotide sequence analysis of genes coding for the major chlorophyll-binding protein of the moss *Physcomitrella patens* and the halotolerant alga *Dunaliella salina. Gene* 76:299–312

108. Lucero, H. A., Cortez, N., Vallejos, R. H. 1987. Light modulation of serine and threonine phosphorylation in histone III by thylakoids. *Biochim. Biophys. Acta* 890:77–81

109. Lucero, H. A., Lin, Z.-F., Racker, E. 1982. Protein kinases from spinach chloroplasts. II. Protein substrate specificity and kinetic properties. *J. Biol. Chem.* 257:12157–60

110. Matthew, J. B., Weber, P. C., Salemme, F. R., Richards, F. M. 1983. Electrostatic orientation during electron transfer between flavodoxin and cytochrome *c. Nature* 301:169–72

111. Mayes, S. R., Barber, J. 1990. Nucleo-

tide sequence of the *psb*H gene of the cyanobacterium *Synechocystis* 6803. *Nucleic Acids Res.* 18:194

112. Michel, H., Buvinger, W. E., Hunt, D. F., Bennett, J. 1989. Redox control and substrate specificity of thylakoid protein phosphorylation. *Physiol. Plant.* 76:A21

113. Michel, H., Hunt, D. F., Shabanowitz, J., Bennett, J. 1988. Tandem mass spectrometry reveals that three photosystem II proteins of spinach chloroplasts contain N-acetyl-O-phosphothreonine at their NH₂ termini. *J. Biol. Chem.* 25:1123–30

114. Michel, H., Shaw, E. K., Bennett, J. 1987. Protein kinase and phosphatase activities of thylakoid membranes. In *Plant Membranes: Structure, Function, Biogenesis*, ed. C. J. Leaver, H. Sze, pp. 85–102. New York: Alan R. Liss. 461 pp.

115. Michel, H. P., Bennett, J. 1987. Identification of the phosphorylation site of an 8.3 kDa protein from photosystem II of spinach. *FEBS Lett.* 212:103–8

116. Millner, P. A., Barber, J. 1984. Plastoquinone as a mobile redox carrier in the photosynthetic membrane. *FEBS Lett.* 169:1–6

117. Millner, P. A., Widger, W. R., Abbott, M. S., Cramer, W. A., Dilley, R. A. 1982. The effect of adenine nucleotides on inhibition of the protein kinase by sulphydryl-directed reagents. *J. Biol. Chem.* 257:1736–42

118. Mullet, J. E. 1983. The amino acid sequence of the polypeptide segments which regulates membrane adhesion (grana stacking) in chloroplasts. *J. Biol. Chem.* 258:9941–48

119. Mullet, J. E., Arntzen, C. J. 1980. Simulation of grana stacking in a model membrane system mediated by a purified light-harvesting pigment-protein complex from chloroplasts. *Biochim. Biophys. Acta* 589:100–17

120. Mullineaux, C. W., Bittersmann, E., Allen, J. F. 1990. Picosecond time-resolved fluorescence emission spectra indicate decreased energy transfer from the phycobilisome to photosystem II in light-state 2 in the cyanobacterium *Synechococcus* 6301. *Biochim. Biophys. Act* 1015:231–42

121. Murphy, D. J. 1986. The molecular organization of the photosynthetic membrane in higher plants. *Biochim. Biophys. Acta* 864:33–94

122. Myers, J. 1971. Enhancement studies in photosynthesis. *Annu. Rev. Plant Physiol.* 22:289–312

123. Nimmo, G. A., Nimmo, H. G., Hamilton, I. D., Fewson, C. A., Wilkins, M. B. 1986. Purification of the phosphorylated night form and dephosphorylated day form of phosphoenolpyruvate carboxylase from *Bryophyllum fedtschenkoi*. *Biochem. J.* 239:213–20

124. Olive, J., M'Bina, I., Vernotte, C., Astier, C., Wollman, F.-A. 1986. Randomization of the EF particles in thylakoid membranes of *Synechocystis* 6714 upon transition from state I to state II. *FEBS Lett.* 208:308–12

125. Olsen, L. J., Theg, S. M., Selman, B. R., Keegstra, K. 1989. ATP is required for the binding of precursor proteins to chloroplasts. *J. Biol. Chem.* 264:6724–29

126. Owens, G. C., Ohad, I. 1982. Phosphorylation of *Chlamydomonas reinhardtii* chloroplast membrane proteins in vivo and in vitro. *J. Cell Biol.* 93:712–18

127. Packham, N. K. 1987. Phosphorylation of the 9 kDa photosystem II–associated protein and the inhibition of photosynthetic electron transport. *Biochim. Biophys. Acta* 893:259–66

128. Pain, D., Blobel, G. 1987. Protein import into chloroplasts requires a chloroplast ATPase. *Proc. Natl. Acad. Sci. USA* 84:3288–92

129. Pain, D., Kanwar, Y. S., Blobel, G. 1988. Identification of a receptor for protein import into chloroplasts and its localization to envelope contact zones. *Nature* 331:232–37

130. Pichersky, E., Brock, T. G., Nguyen, D., Hoffman, N. E., Peichulla, B., et al. 1989. A new member of the CAB gene family: structure, expression and chromosomal location of *Cab-8*, the tomato gene encoding the Type III chlorophyll *a/b* binding polypeptide of photosystem I. *Plant Mol. Biol.* 12:257–70

131. Polle, A., Junge, W. 1986. The slow rise of the flash-light-induced alkalization by photosystem II of the suspending medium of thylakoids is reversibly related to thylakoid stacking. *Biochim. Biophys. Acta* 848:257–64

132. Portis, A. R. Jr. 1990. Rubisco activase. *Biochim. Biophys. Acta* 1015:15–28

133. Ranjeva, R., Boudet, A. M. 1987. Phosphorylation of proteins in plants: regulatory effects and potential involvement in stimulus/response coupling. *Annu. Rev. Plant Physiol.* 38:73–93

134. Roeske, C. A., Chollet, R. 1987. Chemical modification of the bifunctional regulatory protein of maize leaf pyruvate,

orthophosphate dikinase. Evidence for two distinct active sites. *J. Biol. Chem.* 262:12575–82

135. Roeske, C. A., Kutny, R. M., Budde, R. J. A., Chollet, R. 1988. Sequence of the phosphothreonyl regulatory site peptide from inactive maize leaf pyruvate,orthophosphate dikinase. *J. Biol. Chem.* 263:6683–87

136. Ryrie, I. J., Anderson, J. M., Goodchild, D. J. 1980. The role of light harvesting chlorophyll *a/b* protein complex in chloroplast membrane stacking. Cation-induced aggregation of reconstituted proteoliposomes. *Eur. J. Biochem.* 107:345–54

137. Saito, K., Williams, W. P., Allen, J. F., Bennett, J. 1983. Comparison of ATP-induced and state 1/state 2-related changes in excitation energy distribution in *Chlorella vulgaris. Biochim. Biophys. Acta* 724:94–103

137a. Salvucci, M. E., Drake, R. R., Broadbent, K. P., Haley, B. E., Hanson, K. R., McHale, N. A. 1990. Identification of the 64 kilodalton chloroplast stromal phosphoprotein as phosphoglucomutase. *Plant Physiol.* 93:105–9

138. Sanders, C. E., Melis, A., Allen, J. F. 1989. In vivo phosphorylation of proteins in the cyanobacterium *Synechococcus* 6301 after chromatic acclimation to photosystem I or photosystem II light. *Biochim. Biophys. Acta* 976:168–72

139. Schindler, C., Hracky, R., Soll, J. 1987. Protein transport in chloroplasts: ATP is prerequisite *Z. Naturforsch. Teil C* 42:103–8

140. Schuster, G., Dewit, M., Staehelin, L. A., Ohad, I. 1986. Transient inactivation of the thylakoid photosystem II light-harvesting protein kinase and concomitant changes in intramembrane particle sizes during photoinhibition of *Chlamydomonas reinhardii J. Cell Biol.* 103:71–80

141. Shochat, S., Owens, G. C., Hubert, P., Ohad, I. 1982. The dichlorophenyldimethylurea-binding site in thylakoids of *Chlamydomonas reinhardii:* role of photosystem II reaction center and phosphorylation of the 32–35 kilodalton polypeptide in the formation of the high-affinity binding site. *Biochim. Biophys. Acta* 681:21–31

142. Slovacek, R. E., Hind, G. 1981. Correlation between photosynthesis and transthylakoid proton gradient. *Biochim. Biophys. Acta* 635:393–404

143. Soll, J. 1988. Purification and characterization of a chloroplast outer-envelope-bound, ATP-dependent protein kinase. *Plant Physiol.* 87:893–903

144. Soll, J., Bennett, J. 1988. Localization of a 64-kDa phosphoprotein in the lumen between the outer and inner envelopes of pea chloroplasts. *Eur. J. Biochem.* 175:301–7

145. Soll, J., Berger, V., Bennett, J. 1989. Adenylate effects on protein phosphorylation in the interenvelope lumen of pea chloroplasts. *Planta* 177:393–400

146. Soll, J., Buchanan, B. B. 1983. Phosphorylation of chloroplast ribulose bisphosphate carboxylase/oxygenase small subunit by an envelope-bound protein kinase in situ. *J. Biol. Chem.* 258:6686–89

147. Soll, J., Fischer, I., Keegstra, K. 1988. A GTP-dependent protein kinase is localized in the outer envelope membrane of pea chloroplasts. *Planta* 176:488–96

148. Staehelin, L. A., Arntzen, C. J. 1983. Regulation of chloroplast membrane function: protein phosphorylation changes the spatial organisation of membrane components. *J. Cell Biol.* 97:1327–37

149. Steinback, K. E., Bose, S., Kyle, D. J. 1982. Phosphorylation of the light-harvesting chlorophyll protein regulates excitation energy distribution between PS2 and PS1. *Arch. Biochem. Biophys.* 216:356–61

150. Steinback, K. E., Burke, J. J., Arntzen, C. J. 1979. Evidence for the role of surface-exposed segments of the light harvesting complex in cation-mediated control of chloroplast structure and function. *Arch. Biochem. Bioenerg.* 195:546–57

151. Sukenik, A., Wyman, K. D., Bennett, J., Falkowski, P. G. 1987. A novel mechanism for regulating the excitation of photosystem II in a green alga. *Nature* 327:704–7

152. Sun, G., Bailey, D., Jones, M. W., Markwell, J. 1989. Chloroplast thylakoid protein phosphatase is a membrane surface-associated activity. *Plant Physiol.* 89:238–43

153. Sundby, C., Larsson, U. K., Henrysson, T. 1989. Effects of bicarbonate on thylakoid protein phosphorylation. *Biochim. Biophys. Acta* 975:277–82

154. Taylor, S. S., Buechler, J. A., Yonemoto, W. 1990. cAMP-dependent protein kinase: framework for a diverse family of regulatory enzymes. *Annu. Rev. Biochem.* 59:971–1005

155. Telfer, A., Allen, J. F., Barber, J., Bennett, J. 1983. Thylakoid protein phos-

phorylation during state 1–state 2 transitions in osmotically shocked pea chloroplasts. *Biochim. Biophys. Acta* 722:176–81

156. Telfer, A., Bottin, H., Barber, J., Mathis, P. 1984. The effect of magnesium and phosphorylation of the light-harvesting chlorophyll *a/b* protein on the yield of P700 photooxidation in pea chloroplasts. *Biochim. Biophys. Acta* 764:324–30

157. Theg, S. M., Bauerle, C., Olsen, L. J., Selman, B. R., Keegstra, K. 1989. Internal ATP is the only energy requirement for the translocation of precursor proteins across chloroplastic membranes. *J. Biol. Chem.* 264:6730–36

158. van der Staay, G. W. M., Matthijs, H. C. P., Mur, L. R. 1989. Phosphorylation and dephosphorylation of membrane proteins from the prochlorophyte *Prochlorothrix hollandica* in fixed redox states. *Biochim. Biophys. Acta* 975:317–24

159. Wendler, J., Holzwarth, A. R. 1987. State transitions in the green alga *Scenedesmus obliquus* probed by time-resolved chlorophyll fluorescence spectroscopy and global data analysis. *Biophys. J.* 52:717–28

160. White, I. R., O'Donnell, P. J., Keen, J. N., Findlay, J. B. C., Miller, P. A. 1990. Investigation of the substrate specificity of thylakoid protein kinase using synthetic peptides. *FEBS Lett.* 269:49–52

161. Williams, R. S., Allen, J. F., Brain, A. P. R., Ellis, R. J. 1987. Effect of Mg^{2+} on excitation energy transfer between LHC II and LHC I in a chlorophyll-protein complex. *FEBS Lett.* 225:59–66

162. Yalovsky, S., Schuster, G., Nechushtai, R. 1990. The apoprotein precursor of the major light-harvesting complex of photosystem II (LHCIIb) is inserted primarily into stromal lamellae and subsequently migrates to the grana. *Plant Mol. Biol.* 14:753–64

Annu. Rev. Plant Physiol. Plant Mol. Biol. 1991. 42:313–49
Copyright © 1991 by Annual Reviews Inc. All rights reserved

CHLOROPHYLL FLUORESCENCE AND PHOTOSYNTHESIS: The Basics

G. H. Krause

Institute for Biochemistry of Plants, Heinrich Heine University Düsseldorf, D-4000 Düsseldorf 1, Germany

E. Weis

Institute of Botany, University of Münster, D-4400 Münster, Germany

KEY WORDS: lifetimes of chlorophyll fluorescence, photochemical energy conversion, photoinhibition of photosynthesis, photosystem II, quenching of chlorophyll fluorescence

CONTENTS

1040-2519/91/0601-0313$02.00

INTRODUCTION

Chlorophyll *a* fluorescence emitted by green plants reflects photosynthetic activities in a complex manner. Recent improvements of fluorescence measuring techniques have made the fluorescence method an important tool in basic and applied plant physiology research. Such methods include the introduction of amplitude modulated fluorescence recording systems and the refinement of lifetime analysis. In particular the use of fluorescence from intact plant leaves has increased as a unique nonintrusive method of monitoring photosynthetic events and judging the physiological state of the plant. The improved techniques have facilitated deeper insight into the mechanism of fluorescence emission and thus have resulted in a more precise interpretation of the emitted signals. However, basic explanations of fluorescence phenomena mostly rely on studies in vitro with isolated protoplasts, chloroplasts, thylakoid membranes, and particle preparations. The full interpretation of the complex signals emanating from intact photosynthetic organisms, particularly from leaves of higher plants, is still problematic. Certain measured parameters are of empirical value only.

In the last 15 years a number of reviews on chlorophyll fluorescence have appeared, stressing different aspects of the fluorescence phenomenon and covering most of the relevant literature. An early broad summary of the basic fluorescence characteristics was given by Papageorgiou (124). Fluorescence related to physical and primary photosynthetic events was reviewed by Lavorel & Etienne (94), Butler (22, 23), van Gorkom (159), Moya et al (111), and Holzwarth (67), the two latter dealing predominantly with fluorescence lifetimes. Biochemical and physiological implications of fluorescence are accentuated in the reviews by Krause & Weis (86), Briantais et al (20), Murata & Satoh (114), Renger & Schreiber (130), and Krause & Weis (87). The present article surveys current knowledge about the relationships between fluorescence emission and the physiology of photosynthesis, based on available biophysical and biochemical information. Our review is concerned only with oxygenic organisms, predominantly with higher plants; specific features of fluorescence in lower photosynthetic organisms are not considered. Also the rapidly growing application of fluorescence in detection and analysis of stress effects on plants (see 98, 147a), being highly important to ecophysiology, cannot be covered here.

BIOPHYSICAL BASIS OF FLUORESCENCE EMISSION FROM CHLOROPLASTS

Fluorescence as a Reaction Competing in the Deactivation of Excited Chlorophyll

In the photosynthetic apparatus, light is absorbed by the antenna pigments, and the excitation energy is transferred to the reaction centers of the two photosystems. There the energy drives the primary photochemical reactions that initiate the photosynthetic energy conversion. In low light under optimal conditions, the primary photochemistry occurs with high efficiency. From data of Björkman & Demmig (14) it can be calculated that more than 90% of absorbed light quanta are utilized by photosynthesis. The yield of primary photochemical energy conversion may be even higher. A minor competing process of deactivation of excited pigments is the emission of chlorophyll (Chl)[1]a fluorescence. At room temperature, most fluorescence is emitted by Chl a of photosystem (PS) II. In solution (e.g. in ether) Chl a exhibits a high fluorescence yield, ϕF, of about 30%. In contrast, maximum fluorescence yield in the photosynthetic apparatus, ϕF_M (observed when all reaction centers of PS II are "closed"), is only around 3% of the absorbed light (see 8). When all reaction centers are "open," the fluorescence yield, ϕF_0, is about five times lower (approximately 0.6%) owing to the competition with photochemistry.

Before further discussing this competition, we briefly describe the events of photosynthetic electron transport in the PS II reaction center (see 57). In the primary photochemical reaction, one electron is transferred from the pigment P_{680} in the first excited singlet state (P_{680}^*) to pheophytin a (equation 1). From there the electron is transferred to the primary quinone-type acceptor Q_A.

$$
\begin{array}{llll}
[Z & P_{680} & \text{Pheo} \quad Q_A \quad Q_B] & \textit{(open centers)} \\
 & \downarrow h\nu & \qquad\qquad 1 \\
[Z & P_{680}^* & \text{Pheo} \quad Q_A \quad Q_B] \\
 & \downarrow & \qquad\qquad 2 \ (3\text{ps}) \\
[Z & P_{680}^+ & \text{Pheo}^- Q_A \quad Q_B] \\
 & \downarrow & \qquad\qquad 3 \ (250\text{–}300 \ \text{ps}) \\
[Z & P_{680}^+ & \text{Pheo} \quad Q_A^- \quad Q_B] & \textit{(closed centers)} \\
 & \downarrow & \qquad\qquad 4 \ (20 \ \text{ns–}35 \ \mu s) \\
[Z^+ & P_{680} & \text{Pheo} \quad Q_A^- \quad Q_B] \\
 & \downarrow e^- & \qquad\qquad 5 \ (50 \ \mu s\text{–}1.3 \ \text{ms}) \\
[Z & P_{680} & \text{Pheo} \quad Q_A^- \quad Q_B] \\
 & \downarrow & \qquad\qquad 6 \ (100\text{–}200 \ \mu s) \\
[Z & P_{680} & \text{Pheo} \quad Q_A \quad Q_B^-] & \textit{(open centers)} \\
 & & \text{(time constants see reference 57)} & \qquad 1.
\end{array}
$$

[1]**Abbreviations:** Chl: chlorophyll; CP: chlorophyll protein complex; DCMU: 3-(3',4'-dichlorophenyl)-1,1-dimethylurea; LHC II (I): light harvesting chlorophyll a,b protein complex of photosystem II (I); PFD: photon flux density; Pheo: pheophytin; PQ: plastoquinone; PS: photosystem; Q_A, Q_B: quinone-type electron acceptors of photosystem II; Z: secondary electron donor of photosystem II

The charge separation creates a highly active oxidant, P_{680}^+. This receives an electron from the secondary donor Z, which has been identified as a tyrosine residue of the D1 protein. The oxidized donor, Z^+, is reduced by an electron from the water oxidation system. In a slower reaction, which may involve an iron atom, the electron is transmitted to the quinone Q_B. After receiving two electrons, Q_B binds two protons from the lumen side of the thylakoid membrane and merges into the plastoquinone/plastohydroquinone (PQ) pool.

The rate of fluorescence emission, F, is proportional to the absorbed light flux, I_a, and to the quotient of the rate constant of fluorescence, k_F, over the sum of rate constants, Σk_i, of all competing reactions that result in a return of the Chl molecule to the ground state (equation 2). The most important of those are the photochemical reaction (k_P), thermal deactivation (k_D), and excitation energy transfer to nonfluorescent pigments (k_T)—e.g. to antennae of PS I. From these considerations, the general equation for fluorescence yield is obtained:

$$F = I_a \cdot k_F/\Sigma k_i \qquad \qquad 2.$$

$$\Phi F = F/I_a = k_F/(k_F + k_D + k_T + k_P) \qquad \qquad 3.$$

As first proposed by Duysens & Sweers (41) and later elaborated by Butler and coworkers (see 22, 23), fluorescence yield is minimal (ΦF_0) when all reaction centers are in an active, "open" state (see equation 1). The experimentally observed high photon yield of photosynthesis demands that $k_P \gg k_F + k_D + k_T$. However, when Q_A is fully reduced (i.e. all reaction centers are in the state Z P_{680} Pheo Q_A^-), excitation of P_{680} cannot result in stable charge separation ($k_p = 0$). Then the maximum fluorescence yield, ΦF_M, is obtained.

By a similar approach, the potential yield of the photochemical reaction of PS II is obtained as

$$\Phi P_0 = k_P/(k_F + k_D + k_T + k_P) = (\Phi F_M - \Phi F_0)/\Phi F_M = F_V/F_M, \qquad 4.$$

where F_M is the maximum total fluorescence and F_V the maximum variable fluorescence emission ($F_V = F_M - F_0$). It should be noted that F_V does not represent an independent fluorescence component; the term only describes the change in fluorescence emission between two defined states. The F_V/F_M ratio has become an important and easily measurable parameter of the physiological state of the photosynthetic apparatus in intact plant leaves. The F_V/F_M ratio was found to stay in a remarkably narrow range (0.832 ± 0.004) among leaves of many different species and ecotypes (14). Under comparable experimental conditions, 0.106 ± 0.001 O_2 are evolved per quantum absorbed by the leaf (14). For equal distribution of excitation energy between PS II and I, this denotes transport in PS II of 0.85 e^- per quantum. Environmental stresses that affect PS II efficiency lead to a characteristic decrease in F_V/F_M.

From equations 3 and 4 an inverse relationship between rate of fluorescence emission and rate of photochemical reaction results. In the fluorescence induction curve, the rise of emission from F_0 to F_M reflects the reduction of Q_A (41). The complex kinetics of the fluorescence rise is discussed below. We also deal with the fact that fluorescence emission is influenced by various nonphotochemical quenching mechanisms.

By introducing rate constants of exciton transfer between antenna Chl and reaction centers and between PS II and PS I units, the bi- and tripartite models were developed (see 22, 23) and later extended by Strasser (152) to the general description of a polypartite model comprising all photosynthetic pigment systems. As discussed below, these models, in which a limitation of primary photochemistry by exciton migration is assumed, have been challenged recently on the basis of fluorescence lifetime analyses.

Lifetimes of Fluorescence

Analyses of fluorescence lifetimes have become increasingly important in the study of the organization and function of the photosynthetic apparatus. In particular, lifetime determinations appear useful for elucidating excitation energy transfer and the kinetics of primary photochemical processes. Application of lifetime measurements to physiological problems has just begun (see e.g. 12, 119) and will certainly increase in the near future. Because fluorescence lifetime analysis has been the topic of several recent reviews (66–68, 111), here we summarize the basic points only.

In principle, the lifetime of fluorescence is analyzed by measuring the fluorescence decline after brief exciting flashes of light, thus following the decline in exciton density of Chl. The decay can now be resolved to the time scale of a few picoseconds. It can be described by a sum of exponential (first-order) decay reactions:

$$F(t) = A_i \cdot e^{-t/\tau_i}, \qquad\qquad 5.$$

where F is the fluorescence intensity at the time t, A_i the amplitude (pre-exponential factor) and τ_i the lifetime. A_i depends on the wavelength of excitation and emission and is approximately proportional to the number of antenna molecules. The yield of each decay component of fluorescence is given by $\Phi F_i = A_i \cdot \tau_i$, and the total fluorescence yield by

$$\Phi F = \Sigma \Phi F_i = \Sigma A_i \cdot \tau_i = \bar{\tau} \cdot \Sigma A_i, \qquad\qquad 6.$$

where $\bar{\tau}$ is the average lifetime ($\bar{\tau} = \Sigma A_i \cdot \tau_i/\Sigma A_i$).

Usually, the decay kinetics can be resolved into three or more exponential components (cf 73). It appears that the number of components that can be distinguished increases with refinement of methods and increased signal-to-noise ratios (see 68). This fact points to the extremely complex nature of the fluorescence decay. Improvements have been achieved by "global" analysis—i.e. measurements of the decay at a set of different wavelengths (64, 66, 73). When the yields ΦF_i or amplitudes A_i of the individual components (characterized by a certain τ_i) are plotted versus the wavelength of emission or excitation, "decay-associated spectra" are obtained. Such methods provide considerably more information than single-decay analysis. The evaluation of data was further refined by so-called "target analysis" (96), in which the data are fitted to sets of rate constants of kinetic models describing the primary processes.

Recently, an ultrafast decay component ($\tau \approx 15$ ps) was resolved (68, 102) that has been attributed to fast reversible excitation equilibration between antennae and reaction centers. This contradicts the earlier "bipartite model" of Butler and coworkers (see 22, 23), which was based on the assumption that the primary photochemical reaction is limited by the rate of exciton migration from the antennae to the reaction center. In Butler's original model the rate constant of energy transfer to the "trap" was supposed to be much higher than the rate constant of back-transfer from reaction centers to the antennae, picturing the PS II unit as a "funnel." In recent years, evidence has accumulated that for oxygenic photosynthesis the photochemical reaction in both photosystems is "trap limited" rather than limited by exciton migration (135–137). Similar conclusions were drawn for the photosystem of purple bacteria (16). This supported the "equilibrium models" assuming quasi-equilibrium of excitation between antenna pigments and reaction center, as well as between singlet excited reaction center Chl and radical pair (P_{680}^+Pheo$^-$) in PS II. According to Holzwarth (68), the equilibration of excitons is about an order of magnitude faster than the process of charge separation. In such models, the reaction center appears as a "shallow trap" (135–137). The kinetic model suggested by Holzwarth and coworkers describes the primary photochemical events according to equation 7 (after 136), where Chl denotes antenna chlorophyll, k_A the rate constant of radiationless and radiative deactivation, and k_t and k_{-t} the rate constants for energy transfer between antennae and reaction center:

$$\begin{array}{llll} [\text{Chl } P_{680} & \text{Pheo} & Q_A] \\ h\nu \; \updownarrow \; k_A \\ \left[\text{Chl* } P_{680}\right. & \text{Pheo} & Q_A \\ k_t \; \updownarrow \; k_{-t} \\ \left.\text{Chl } P_{680}^* \right. & \text{Pheo} & Q_A \end{array}$$

7.

$$\begin{array}{cc} k_1 \;\; \Updownarrow \;\; k_{-1} & \text{step 1: charge separation} \\ [\text{Chl P}_{680}^{+} \quad \text{Pheo}^- \;\; Q_A] & \\ \downarrow \;\; k_2 & \text{step 2: charge stabilization} \\ [\text{Chl P}_{680}^{+} \quad \text{Pheo} \;\; Q_A^-] & \end{array}$$

In the exciton-equilibrated pigment system, the rate of primary charge separation is approximately proportional to the probability of P_{680} excitation. In turn, this probability for a given number of absorbed photons is inversely proportional to the antenna size. Thus, a component of the fluorescence decay can be attributed to the reversible charge separation (step 1, equation 7). The lifetime of this component was shown to be proportional to the number (N) of antenna pigments and is supposed to be about 250 ps for PS IIα units having about 180 Chl a and 70 Chl b molecules (102). According to the model of Schatz et al (136), a second component is related to the process of charge stabilization by electron transfer to Q_A (step 2 in equation 7, see also equation 1). The lifetimes of those components that are related to primary photochemistry increase with the closure of the traps, leading to the increased fluorescence yield in the F_M state (see equation 6). In the above model the increase in τ results predominantly from a decreased rate constant of primary charge separation in closed reaction centers (equation 8):

$$\begin{bmatrix} \text{Chl* P}_{680} \; \text{Pheo} \; Q_A^- \\ \Updownarrow \\ \text{Chl P}_{680}^{*} \; \text{Pheo} \; Q_A^- \end{bmatrix} \;\; \underset{k'_{-1}}{\overset{k'_1}{\rightleftharpoons}} \;\; [\text{Chl P}_{680}^{+} \; \text{Pheo}^- \; Q_A^-]$$

8.

The constant k'_1 (for closed reaction centers) was calculated to be lower than k_1 (for open reaction centers, see equation 7) by a factor of about six. This drastic effect is supposed to be caused by the electric field created by the negative charge on Q_A in the state $[P_{680} \text{ Pheo } Q_A^-]$ (cf 73). As a consequence, formation of the primary radical pair, $P_{680}^{+} \text{ Pheo}^-$ is restricted. Schatz et al (136) infer that the primary charge separation even becomes an endergonic reaction in the F_M state. This problem is discussed below with regard to the origin of fluorescence.

By "target" analysis, Holzwarth (68) resolved for cells of the green alga *Scenedesmus* six components for the decay of fluorescence both in the F_O and F_M state. Three of them show a strongly increased lifetime with closure of reaction centers (Table 1). Based on evidence from fluorescence amplitudes and their excitation spectra (68), four components were attributed to revers-

Table 1 Lifetimes of fluorescence from *Scenedesmus obliquus* (from Ref. 68, altered) (τ in ps)

	F_O	F_M	Assignment
τ_1	260–300	700–740	PS IIα
	100–130	120–150	PS IIβ
τ_2	440–550	1510–1520	PS IIα
	530–570	2250–2260	PS IIβ
τ_3	10–20	10–20	exciton equilibration, PS I, PS II
τ_4	94–96	114–118	PS I

ible charge separation (τ_1) and to charge stabilization (τ_2), respectively, of two PS II pools, namely PS IIα and PS IIβ units. The other two components were assigned to ultrafast exciton equilibration in both photosystems, and to PS I emission (Table 1).

It should be emphasized that these assignments must still be viewed as tentative. In recent reports (66, 168) the component with the longer lifetime at F_M ($\tau \approx 2.2$ ns) was attributed to PS IIα, whereas the component with $\tau \approx 1.2$ ns was assumed to represent PS IIβ emission, in contrast to the assignments in Table 1. However, even though results of lifetime measurements from different groups of workers are partly comparable, wide disagreement on interpretation exists. Moya and his coworkers (60, 62–64) concluded from their studies, particularly of *Chlamydomonas* mutants and of various protein/Chl complexes, that there is no need to explain the fluorescence decay on the basis of PS IIα/PS IIβ heterogeneity. Rather, they attribute the different lifetimes to specific domains of antenna pigments, which should exhibit different rates of exciton transfer to the reaction centers. For instance, the two long-lived components at F_M are both attributed to the light harvesting complex of PS II (LHC II). This view, however, presumes a limitation by exciton migration and cannot be reconciled with the trap-limited model discussed above.

In contrast to other workers who postulate at least four decay components of PS II, Keuper & Sauer (73) maintain that the PS II decay of thylakoids is sufficiently described by three exponential decays. The lifetimes of all three components continually increased with trap closure (Table 2). The authors attributed the fast component to the trapping kinetics of open PS II units, because its amplitude tended to approach zero towards F_M conditions. The slow component, which was absent at F_O is assumed to represent closed PS II units. Keuper & Sauer discuss that the changes in amplitude they observe are in contradiction to the general assumption of exciton transfer ("connectivity") between PS IIα units (72, 105). On the other hand, Holzwarth (67) argues

Table 2 Fluorescence lifetimes of spinach thylakoid membranes and their increase during closure of PS II reaction centers (after Ref. 73).

Component	Lifetime (ps)	Assignment
fast	170–300	open PS II centers
middle	400–1700	Q_B-nonreducing (PS IIβ ?) centers
slow	1700–3000	closed PS II centers

that in the present state, lifetime analysis cannot provide evidence for or against connectivity. The middle component of Keuper & Sauer (73) was suggested to present "Q_B-nonreducing" PS IIβ (see 103, 106), because the amplitude could not be influenced by the redox state of the PQ pool. The contradictions in the interpretations of fluorescence lifetime data await future clarification.

Origin of Fluorescence Emission

Whereas most agree that fluorescence at F_O is an emission by antenna Chl a molecules, the origin of variable fluorescence has been the subject of controversy. Butler and coworkers (see 22, 23) postulated that all fluorescence originates in the antennae. Variable fluorescence was assumed to arise owing to back-transfer of excitation energy from the closed reaction centers. Klimov et al (74, 75) established that Pheo is the primary electron acceptor of PS II and transmits electrons to the quinone Q_A. According to their hypothesis, variable fluorescence results from recombination of the radical pair P^+_{680} Pheo$^-$ in closed reaction centers (see equation 8) and either arises from emission by P^*_{680} after charge recombination, or from emission by antennae Chl a following exciton transfer from P^*_{680}. It was also suggested that Pheo emits fluorescence from closed reaction centers. Breton (18) attributed the emission band at 695 nm observed at 77 K to emanation from Pheo*. However, recent progress both in time-resolved fluorescence spectroscopy and in preparation of functional reaction centers of PS II provided strong evidence against the hypotheses of Klimov et al (74, 75) and Breton (18).

As discussed above, Schatz et al (136) deduced from the picosecond fluorescence decay in thylakoid membranes that in closed reaction centers the primary charge separation forming P^+_{680} Pheo$^-$ is strongly restrained by the electrostatic effect of Q_A^-. Similar conclusions were drawn by Schlodder & Brettel (138) from flash-induced absorbance changes related to the primary radical pair in PS II particles. Thus, the quantum yield of primary charge separation and therefore of recombination appears to be low in the presence of Q_A^-. It should be noted that the kinetic equilibrium model of Schatz et al

(136), in fact, includes the possibility that charge recombination contributes to fluorescence emission both in the F_0 and F_M state. However, the rate constants calculated for this model (96) would favor prompt fluorescence emission. Also Hodges & Moya (62, 63) concluded that their results from lifetime measurements cannot be explained by the charge-recombination hypothesis. They found in PS II–enriched preparations three fluorescence decay components with variable τ_i between F_0 and F_M [similar to Keuper & Sauer's finding (73) for thylakoids] in the range of 20 ps to 2 ns. The authors argue that the variable nature of τ_i, starting in the picosecond range, contradicts the assumption of a slow (nanosecond) recombination luminescence. Furthermore, a *Chlamydomonas* mutant lacking functional PS II reaction centers exhibited lifetime components very similar to those of the wild type in the F_M state (63). This also speaks against the origin of the long-lived components at F_M from charge recombination.

Further information on this topic has been obtained from fluorescence lifetime studies of isolated reaction centers. Such particles have recently been isolated by Nanba & Satoh (115) and Barber et al (7). The preparations are free of antenna chlorophyll protein complexes and of bound PQ and nonheme iron. The centers consist of five polypeptides: the D1/D2 heterodimer, cytochrome b-559 with one heme and two protein subunits, and a further chloroplast-encoded 4.8-kDa protein (see 6). The complex binds 4 Chl *a*, 2 Pheo, and 1 β-carotene. The reaction centers are able to perform the primary photochemical reaction but proved very unstable when illuminated in the presence of O_2. Considerable stabilization was achieved by applying anaerobic conditions (32). The centers emitted fluorescence (44% of total yield) of a very long lifetime (around 36.5 ns). This long-lived component was attributed to recombination of the radical pair P^+_{680} Pheo$^-$. Similarly, Govindjee et al (48) assigned a very slow decay component (5–20 ns) observed with isolated reaction centers to the charge recombination process (cf 108). They demonstrated that this component disappears when the centers are prereduced (P_{680}Pheo$^-$); in the state P^*_{680} Pheo$^-$ the exciton cannot initiate the charge separation/recombination process.

At first sight, these results are in agreement with Klimov's hypothesis (74, 75), showing that recombination of the radical pair P^+_{680} Pheo$^-$ can indeed give rise to light emission. However, the long lifetime (in the range of 5–35 ns) speaks against its role in emanation of variable fluorescence in the intact PS II. There is no indication of a significant contribution of such long-lived components in intact thylakoids (see 134 and Tables 1,2).

Obviously, charge recombination as a major deactivation pathway may take place only in the isolated reaction center complex which lacks bound quinones. In the intact system, the presence of Q_A favors electron transfer to this acceptor. When Q_A has been reduced, the rate of charge separation is supposedly strongly diminished, as outlined above. The 695-nm band of low-temperature fluorescence does not occur in the emission spectrum of the isolated reaction center complex (8, 158) and thus cannot result from Pheo*, as had been suggested by Breton (18) and van Dorssen et al (157). It has been shown (158) that this band originates from the CP47 core antenna complex. Thus, present evidence supports the assumption that most of the chlorophyll fluorescence in vivo (both at F_0 and F_M levels) emanates from the antenna system.

Fluorescence of PS II and PS I at Ambient and Low Temperatures

EMISSION AT ROOM TEMPERATURE   In fluorescence emission spectra of dilute suspensions of thylakoid membranes or intact isolated chloroplasts (see 86), a sharp peak around 685 nm with a broad shoulder at about 740 nm is observed. In leaf tissue with densely packed chloroplasts, light scattering and reabsorption effects strongly lower the 685-nm band relative to 740-nm emission (see 132). At room temperature, most of the fluorescence in both spectral regions is emanated from PS II. There is a minor contribution of emission by PS I, which is larger at the longer wavelengths than at 685 nm. According to data of Holzwarth (68), PS I (in *Scenedesmus* cells) contributes around 5% at 720 nm and 1–2% at 685 nm to the total fluorescence yield in the F_M state. From lifetime studies of fluorescence emitted at 5° C by PS I particles, Holzwarth et al (69) concluded that the primary photochemical reaction in PS I—like that in PS II—is trap limited.

Closure of PS I reaction centers does not contribute to variable fluorescence (see 20). This can be explained by the relative stability of P_{700} in the oxidized state, whereas P_{680}^+ is readily returned to the ground state by electron donation (cf equation 1). The reaction sequence in PS I probably is

$$P_{700} \, A \xrightarrow{h\nu} P_{700}^* \, A \longrightarrow P_{700}^+ \, A^- \underset{e^-}{\overset{}{\rightleftharpoons}} P_{700}^+ \, A \xrightarrow{e^-} P_{700} \, A, \qquad 9.$$

where A denotes the primary PS I electron acceptor. Depending on the rate of electron transfer, most of the "closed" PS I reaction centers are either in the state $P_{700}^+ \, A^-$ or $P_{700}^+ A$. The oxidized pigment P_{700}^+ (like P_{680}^+) still acts as a trap for excitation energy (which is dissipated as heat) and therefore represents a fluorescence quencher—i.e. upon closure of PS I reaction centers, the fluorescence yield remains low (118).

EMISSION SPECTRA AT LOW TEMPERATURES   Cooling of green leaf tissue, or of chloroplast or thylakoid suspensions drastically alters the emission spectra. A most conspicuous effect is the strong increase in PS I fluorescence

Table 3 Assignments of fluorescence emission bands at low temperatures (77–4 K).

Band	Assignment	References
F_{680}	LHC II	see 146
F_{685}	PS II core (CP43)	see 146
F_{695}	PS II core (CP47)	158
F_{720}	PS I core ("PS I-65")	see 146
F_{735}	PS I (LHC I?)	112, 113

in the far-red region (720–740 nm). When light scattering and reabsorption of fluorescence are minimized (see 20), three distinct bands are seen at 77 K, termed F_{685}, F_{695}, and F_{735} after their approximate emission maxima. The broad F_{735} band can be resolved into two emissions showing peaks at 714–725 nm (F_{720}) and 735–745 nm (F_{735}) (see 114). Further spectral changes, particularly in the relative peak sizes, occur upon cooling from 77 to 4 K. At about 4 K, a shoulder at 680 nm appears. Even though these temperature-induced changes are presently not well understood, the various low-temperature bands can be attributed to specific Chl/protein complexes (Table 3). Evidence for these assignments is now largely consistent thanks to studies by numerous workers using isolated particles and mutants devoid of certain complexes (see 20, 112–114, 146).

VARIABLE FLUORESCENCE AT LOW TEMPERATURES   PS II fluorescence at 77 K (F_{695}) exhibits a very similar F_V/F_M ratio (about 0.8) as at room temperature. The rise from F_0 to F_M is supposed to reflect reduction of Q_A in the photochemical reaction (see 20). However, in contrast to the case at ambient temperature, reoxidation of Q_A^- is extremely slow.

In the PS I band (F_{735}), the F_V/F_M ratio is much lower (below 0.3). According to Butler's hypothesis (22, 23), variable fluorescence (and part of F_0) emanating from PS I is a result of exciton transfer from PS II to PS I, termed "spillover." If the fluorescence yield of PS I is plotted vs that of PS II during the rise from F_0 to F_M, a straight line is obtained. The slope of this line is assumed to be proportional to the rate of spillover. From Butler's model it follows that there are no fluorescence changes related to the photochemical reaction in PS I (see above). The relationship between PS I and PS II fluorescence at 77 K has often been used to study the distribution of excitation energy between the two photosystems (for references see 20). However, interpretation of the spillover phenomenon appears problematic, as PS II is predominantly located in the appressed grana regions and PS I in the stroma lamellae of the thylakoids and border regions of grana (2). Efficient energy transfer from PS II to I appears possible only in the nonappressed membrane domains, where besides PS I supposedly PS IIβ units are present (see the following section). Thus, spillover might be dominated by energy transfer from PS IIβ to PS I and is expected to vary with changes in the ratio of PS IIα to PS IIβ.

FLUORESCENCE INDUCTION AND PS II HETEROGENEITY

Fluorescence Transient from F_0 to F_M

In the dark-adapted state of the photosynthetic apparatus, Q_A normally is fully oxidized. If continuous illumination is started, a fluorescence rise from F_0 is observed related to Q_A reduction (41). The induction signal represents a complex polyphasic process which in detail depends on experimental conditions. As discussed below, the fluorescence rise kinetics is influenced by: 1. PS II cooperativity, 2. PS II heterogeneity, 3. size of the PQ pool and rate of its reoxidation, 4. rate of electron transport beyond PS I including carbon metabolism, and 5. rate of electron donation to P_{680}^+.

When the actinic PFD is high enough to warrant a rate of Q_A reduction that is faster than reoxidation of Q_A (see equation 1), the F_M state is reached, representing full reduction of Q_A. However, at a given time during the rise from F_0 to F_M the variable fluorescence is not directly proportional to the redox state of Q_A. For a mixed population of open and closed centers, the variable fluorescence is lower than expected from the proportion of oxidized Q_A, presumably because of excitation energy transfer from closed to open centers (72). Moreover, heterogeneity of PS II influences the shape of the rise curve, as energy transfer seems possible between PS IIα but not between PS IIβ units (see the next section).

As electrons are passed on from Q_A^- via Q_B to PQ, the fluorescence rise is also related to the reduction of these electron carriers and has, in fact, been used to determine the size of the PQ pool (see 20). Oxidized PQ has been shown to act as a fluorescence quencher (160a). Thus, the true F_M can only be reached when the PQ pool becomes reduced. The rate of its reoxidation depends on electron transfer via PS I and on the final consumption of reducing equivalents in carbon metabolism and other metabolic reactions. In addition, PQH_2 reoxidation is controlled by the transthylakoid ΔpH (see below). The typical fluorescence induction signal exhibits a relatively fast rise from F_0 to F_I (I = "inflection") followed by a plateau or "dip" (F_D) and a slower rise to the "peak" (F_P) or fluorescence maximum (F_M). F_P is lower than F_M when full reduction of Q_A is not achieved.

REOXIDATION KINETICS OF Q_A^- The "dark decay" of variable fluorescence back to the F_0 state can be used to study the kinetics of electron transfer from Q_A^- to Q_B (and PQ) or more generally, the reoxidation kinetics of Q_A^- (24, 93, 133). In this method, by means of high-intensity single-turnover flashes approximately all Q_A (but not Q_B and PQ) is reduced. In the following dark period, the decline of fluorescence (induced by a weak modulated "measuring" light) is recorded and reflects the reoxidation of Q_A^-. The decay proceeds in three exponential phases with lifetimes in spinach thylakoids of about 500 μs, 10 ms, and 2 s (24). The fast phase (amplitude about 62%) is attributed to electron transport from Q_A^- to Q_B in centers that possess bound Q_B. The middle phase (18%) is thought to represent those centers that had no bound Q_B before the flash; thus this phase may indicate the kinetics of equilibration of PQ binding to the Q_B site of the D1 protein. The third, very slow component (20%) seems to represent the decay of the F_I level and has been attributed to PS II centers that are unable to transmit electrons to the PQ pool. The slow Q_A^- reoxidation apparent from this phase is thought to result from recombination between Q_A^- and the water oxidation system in the S_2 state.

The F_I Level and Inactive PS II

In recent years, evidence has accumulated for the interpretation first given by Lavergne (93) that the rise from F_0 to F_I (in low-incident light) reflects reduction of Q_A in centers that are not (or are only loosely) connected to Q_B and PQ (24, 26, 27, 103, 106). Convincing proof for the existence of inactive PS II centers in vivo comes from Chylla & Whitmarsh (27, see 121), who measured in intact leaves the dark "recovery" of the rapid electrochromic absorbance shift (ΔA_{518}). This signal is supposed to indicate charge separation in PS II and PS I reaction centers. It was found that about 20% of the amplitude of ΔA_{518} required a very long recovery time ($t_{1/2} = 1.7$ s). The kinetics of this slow phase was virtually identical with that of the F_I to F_0 fluorescence decay, indicating that the inactive centers belong to PS II. Their turnover rate was calculated to be only 0.3 e^- s^{-1}, as compared to the 200–300 e^- s^{-1} of active centers. The fraction of inactive PS II was estimated as about 30% in spinach leaves. Inactive PS II centers were detected in all of a series of higher plant species studied.

The concept of "Q_B-nonreducing" centers is further supported by the observation that artificial electron acceptors such as ferricyanide (106) or 2,5-dimethyl-p-benzoquinone (24), which cause rapid reoxidation of the PQ-pool, lower the variable fluorescence of thylakoids only to the F_I level. In contrast, 2,6-dichloro-p-benzoquinone can lower the variable fluorescence below F_I and virtually to F_0, thereby leading to enhanced photosynthetic O_2 evolution (24). It is assumed that the latter quinone accepts electrons directly from Q_A^-. This effect indicates that the inactive centers possess a functional water oxidation system but, owing to their inability to reduce PQ, do not significantly contribute to net electron transport.

Fluorescence Induction in High Light

It should be emphasized that the F_I level can serve as an indicator of inactive PS II only in relatively low actinic light. The slower rise from F_I to F_M then represents closure of active reaction centers concomitant with reduction of PQ. In higher light, the two rise phases overlap. Very high continuous light induces a fast rise (in about 800 μs) to a first high F_{I1} level, followed by a dip (F_D) and biphasic rise via F_{I2} to F_M in a total time of about 200 ms (116, 139). At F_{I1} (in saturating light) all Q_A is in the reduced state. The further fluorescence rise is supposed to be of nonphotochemical nature ("thermal phases"). At the F_{I1} level, an initial limitation of electron donation to P_{680}^+ is assumed. Accumulation of P_{680}^+ quenches fluorescence (see the next section). During the rise from F_D to F_{I2} this limitation then would be overcome. The slower phase from F_{I2} to F_M apparently reflects reduction of PQ (which quenches in the oxidized form; see 160a). This rise phase is prevented by electron acceptors that keep PQ oxidized or by 3-(3′,4′-dichlorophenyl)-1,1-dimethylurea (DCMU) which prevents electron transfer to PQ. These findings have practical implications, as they show that F_M cannot be reached in a shorter time than about 200 ms even in the highest actinic light. When fluorescence methods (e.g. the amplitude-modulated techniques) are applied, in which a high fluorescence level is reached by saturating light pulses (143), the peculiar rise kinetics in high light should be considered.

Rise in the Presence of DCMU and α/β Heterogeneity

DCMU blocks electron transfer from Q_A^- to PQ, probably by binding to the Q_B site of the D1 protein. Therefore, in the presence of DCMU the fluorescence rise from F_0 to F_M is much faster than in the absence of the inhibitor. The apparent F_0 level often is slightly increased by addition of DCMU. It appears that a small part of Q_A^- is not reoxidized even during dark periods of several minutes. A further effect of DCMU binding is a back-transfer of electrons in a fraction of the centers that are in the state $Q_A Q_B^-$ (see 159):

$$Q_A Q_B^- + DCMU \rightarrow Q_A^- DCMU + Q_B \qquad 10.$$

Also the true F_M level normally is not reached in the presence of DCMU. At least part of this quenching may be related to oxidized PQ (160a).

If these complications are neglected, the "complementary area" above the fluorescence induction curve (in other words, the integrated fluorescence deficit) at a given time is assumed to be proportional to the fraction of reduced

Q_A. The kinetics of "area growth" represents the closure of PS II reaction centers. The rate of Q_A reduction obtained from the area growth is proportional to PFD, antenna size, and quantum yield of the photochemical reaction. The maximum size of the area, A_{max} (obtained at a given PFD) is a measure of the quantity of centers photochemically active in Q_A reduction.

Melis & Homann (107; see also 106) found that the area growth can in semilogarithmic plots be resolved into two phases, α and β. Based on such analyses, the existence of two types of PS II units was postulated. The fast nonexponential α phase is attributed to connected PS IIα units, which are supposed to reside in the appressed grana membranes (2). The slow exponential β phase is assumed to represent PS IIβ units, located in the nonappressed membrane regions. The exponential kinetics of this phase indicates separate units without exciton migration between them. The β phase normally comprises about 20–35% of A_{max}. The biphasic kinetics of Q_A reduction was confirmed by absorbance measurements in the wavelength region of 250–350 nm (105, 155).

The two- to three-fold slower kinetics of primary photochemistry that is obvious from the β phase was first ascribed to less efficient quantum conversion. Later, this view was revised on grounds of the finding of a lower absorption cross section of PS IIβ (104, 105, 154). For spinach chloroplasts, the antenna size was estimated to be 250 ± 40 Chl $a+b$ for PS IIα and 120 ± 20 Chl $a+b$ for PS IIβ (104). Similar values were calculated for tobacco chloroplasts (154). Further studies (49, 55, 103, 109) indicated that PS IIα units possess besides their core antennae an inner and a peripheral Chl a/b LHC II. The peripheral LHC II antenna appears to be absent in PS IIβ. From recent lifetime analyses it was concluded that α and β centers also differ in their rate constants of primary charge separation (and recombination) and in their F_V/F_M ratio (96). Interconversion between PS IIβ and PS IIα is viewed as a dynamic property of the thylakoid membrane (53, 54, 101, 153). Reversible phosphorylation of part of the LHC II transforms α to β units, but transformation independent of phosphorylation (e.g. induced by heat stress) also seems possible (see the next section).

The concept of α/β heterogeneity is, however, still controversial (see also the section on fluorescence lifetimes, above). Based on the observation that high concentrations of DCMU diminish the β phase of fluorescence induction, it was argued that this phase merely reflects a type of PS II centers less sensitive to DCMU (59, 142). This conclusion was disputed (106) in view of chaotropic side effects of DCMU at high concentrations. More recently (147), the effect of high [DCMU] on fluorescence rise kinetics could not be reproduced.

A further problem presently under discussion is the relationship between PS II α/β and Q_B-nonreducing (inactive) units. The observation of an exponential fluorescence rise from F_0 to F_I (attributed to inactive centers) and indications of a smaller antenna system of inactive units (27) make it very likely that the Q_B-nonreducing units are of β type. This is supported by studies of Guenther et al (55), who found that PS IIβ in the green alga *Dunaliella salina* occurs in both an active and an inactive state. It was postulated that Q_B-nonreducing units represent a subpopulation of PS IIβ. In spinach thylakoids, PS IIβ and Q_B-nonreducing units appeared to be identical (106). Presumably, PS IIα and active and inactive PS IIβ are interconvertible (54, 56). Localization of active PS IIβ in the appressed thylakoid regions has been suggested (4). Recently the role of the inactive units has been proposed to be that of a reserve pool for assembly of active PS II (56), related either to development or to turnover of the D1 protein in the light and repair of "damaged" centers (53).

It should be noted that several further heterogeneities, deduced from fluorescence and absorbance studies and redox titrations, have been reported. Those phenomena might be related to the above inactive and active forms of PS II (for discussions see 4, 15).

FLUORESCENCE QUENCHING

The typical fluorescence induction signal of chloroplasts in vivo in continuous light is known as the Kautsky phenomenon (see 20): A decline in fluorescence yield follows the peak, F_P, or maximum, F_M, of emission. Normally, a polyphasic "quenching" is observed, sometimes interspersed by one or several secondary peaks, until a final steady-state level of fluorescence, "terminal" fluorescence, F_T, is reached in the span of minutes. In a more general sense, the term "quenching" denotes all processes that lower the fluorescence yield below its maximum. Commonly, the extent of quenching is expressed by quenching coefficients ($0 \leq q \leq 1$) indicating the quenched proportion of maximum fluorescence. If not defined otherwise, q refers specifically to quenching of maximum *variable* fluorescence, F_V (143): $q = (F_V - F_V')/F_V$, where F_V' denotes the variable fluorescence in the quenched state. But the quenching coefficients may also be used to describe lowering of fluorescence yield of F_0 or F_M. Alternatively, quenching can be expressed by the normalized "remaining" fluorescence level. For variable fluorescence emission, this is $F_V'/F_V = 1 - q$. Another definition of quenching (21, 37, 85), based on the Stern-Volmer equation, expresses the ratio of "quenched" to "remaining" fluorescence, $\Delta F/F' = (F - F')/F' = F/F' - 1$. Various mechanisms contribute to the quenching observed. The major routes to the lowering of fluorescence yield have been briefly reviewed recently (87). Resolution of quenching components provides important information on the functional state of the photosynthetic apparatus, and more specifically on the efficiency of PS II.

As already outlined (see equations 2, 3), the fluorescence yield is lowered owing to competition of the photochemical reaction with other pathways of de-excitation. Such *photochemical quenching* depends on the presence of Q_A in the oxidized state. The coefficient for photochemical quenching, q_P (or q_Q) denotes the proportion of excitons captured by open traps and being converted to chemical energy in the PS II reaction center. The reoxidation of Q_A^- thus causes quenching. It should be emphasized that q_P is supposedly larger than the proportion of oxidized Q_A because of energy transfer from closed to open units (except for the extreme cases $q = 0$ and $q = 1$).

However, the fluorescence yield can also be lowered by mechanisms not directly related to the redox state of Q_A. Such *nonphotochemical quenching*, q_N (or q_{NP}), may be caused in vivo under physiological conditions by three major mechanisms:

1. *"energy-dependent" quenching* (q_E) caused by the intrathylakoid acidification during light-driven proton translocation across the membrane;
2. *quenching related to "state 1–state 2" transition* (q_T) regulated by phosphorylation of LHC II;
3. *"photoinhibitory" quenching* (q_I) related to photoinhibition of photosynthesis.

Besides, several other nonphotochemical ways of quenching have been elucidated, which are probably not of great physiological significance.

Resolution of Quenching Components

In principle, the contribution of various components to total fluorescence quenching can be resolved by means of their relaxation kinetics. In isolated chloroplasts, photochemical and nonphotochemical quenching were first resolved by addition of DCMU in the light (85). This causes reversion of quenching in two phases ($t_{1/2} \approx 2$ s and 15 s). The fast phase represents relaxation of q_P due to reduction of the proportion of Q_A that was oxidized at the time of DCMU addition. The slower phase indicates reversion of q_N (most of it representing q_E) following the decay of the transthylakoid ΔpH. Alternatively, the application of saturating pulses (11, 17, 143; see 141) allows one to determine q_P and q_N at any point of the fluorescence induction signal, as well as during dark relaxation. By means of the pulses (duration about 1 s) all Q_A connected to active centers and the PQ pool become reduced.

More detailed studies using either technique revealed three distinct phases of relaxation of q_N (61, 71, 163; cf 36). The fastest phase ($t_{1/2} < 1$ min) is attributed to q_E, the second phase ($t_{1/2} \approx 8$ min) to q_T and the slow phase ($t_{1/2} \approx 40$ min) to q_I. It should be noted that the resolution of q_N can be problematic. Depending on the material, the relaxation of q_I is sometimes quite fast (97, 149) and may occur in several phases (71). Separation from q_T is not always possible (see 151). Also, the kinetics of q_E relaxation can vary. A slowly relaxing q_E component has been observed recently in *Dunaliella* (95).

Mechanism of Energy-Dependent Quenching

Energization of the thylakoids because of the build-up of a transmembrane Δ pH may lead to quenching of up to about 90% of F_V. Also some quenching of F_0 (up to 15–25%) has been reported. Such F_0 quenching in leaves, originally attributed to state transition (143), may rather be related to q_E in view of the fast dark relaxation observed (165). As demonstrated in Figure 1, the extent of the ΔpH- or energy-dependent quenching is linearly related to the intrathylakoid H^+ concentration (21, 91). From 77 K fluorescence analyses it was concluded (78) that the quenching is based on an increased rate constant of thermal deactivation in PS II (k_D in equation 3). The molecular mechanism of this quenching is still unknown. An ultrastructural change induced by intrathylakoid acidification and concomitant cation exchange has been postulated (for discussion of earlier literature see 20, 86). It should be

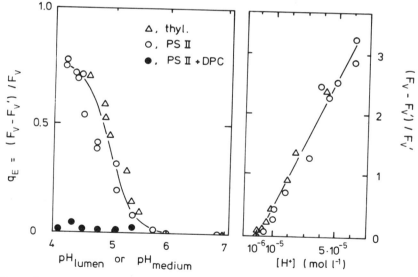

Figure 1 Relationship between energy-dependent fluorescence quenching and pH *(left)* or proton concentration *(right)*. Quenching was induced in thylakoids (Δ) by light-dependent H^+ uptake into the lumen or by suspending PS II particles in media of different pH (○). Inhibition of quenching by an artificial electron donor to PS II (diphenylcarbazide) is shown (●, *left*). Quenching is expressed as the fractional coefficient q_E *(left)* or according to the Stern-Volmer equation *(right)*. F_V = maximum variable fluorescence, $F_{V'}$ = variable fluorescence in the quenched state. Experiment by A. Krieger and E. Weis, unpublished.

noted that changes in high-energy quenching are usually slower than changes in the intrathylakoid $[H^+]$ (10).

However, even though the ΔpH is a strict requirement for q_E formation, other factors are involved. Antimycin A eliminates q_E without affecting the ΔpH (122). The inhibitor is known to block PS I–driven cyclic electron transport, possibly by affecting a ferredoxin-quinone reductase (110). This observation suggests that q_E is subjected to a redox control. A role of zeaxanthin in the thylakoids has been postulated for q_E. From studies of correlation between quenching capacity and zeaxanthin content (1, 38; see 37) it was concluded that q_E consists of two components. The major component in leaves, characterized by a decrease in F_O, is eliminated by dithiothreitol, an inhibitor of zeaxanthin formation. The second component (without F_O quenching) was suggested to be independent of zeaxanthin (cf 46). This component was found to be minor in intact leaves but a substantial effect in isolated chloroplasts. The mechanism of zeaxanthin-related quenching is seen in an interaction of the xanthophyll with antenna Chl causing an increase in k_D of equation 3 (see also discussion on q_I, below). According to Rees et al (129), the presence of zeaxanthin in isolated chloroplasts enhances q_E at intermediary ΔpH values, but the maximum q_E is not altered.

The relationship between $[H^+]$ and q_E, indeed, is not necessarily constant. It can be substantially influenced by stress factors such as heat (91) or freezing/thawing (151). In presence of tertiary amines such as dibucaine, the proton gradient and rate of photophosphorylation in thylakoids are lowered, indicating uncoupling, while photosynthetic control of electron transport and q_E are not affected (92).

Whether the quenching originates in the reaction center or in the antennae is still controversial. In a model of Weis & Berry (165), deduced from detailed data on q_P and q_E in leaves, and in a modified version of this model (166, 167), a photochemically active form (PS II_0, with open or closed centers) and an "energized", photochemically inactive and nonfluorescent form of PS II (PS II_e) are postulated. The populations of the two states are assumed to depend on the intrathylakoid H^+ concentration and the redox balance at the centers:

$$PS\ II_0(Q_A) \underset{}{\overset{e^-}{\rightleftharpoons}} PS\ II_0\ (Q_A^-) \underset{low\ [H^+]}{\overset{high\ [H^+]}{\rightleftharpoons}} PS\ II_e \qquad 11.$$

A strongly increased dissipative process in the reaction centers is suggested for PS II_e which would be responsible for decline both in the net photochemical yield and in fluorescence. A negative correlation between q_E and photon yield of open centers [calculated as ϕ_s/q_P, where ϕ_s is the measured apparent quantum yield of electron transport at steady state (165)] has been confirmed by several groups (71, 125, 127, 160, 167). Since most of the quenching occurs in the range between pH 5.5 and 4.5 in the thylakoid lumen, an apparent pK value of 4.5–5.0 has been assumed for the interconversion between PS II_o and PS II_e suggested in the model (88, 167; see also Figure 1).

However, this concept of pH-dependent control of PS II activity by conversion of centers to a "quenching state" was opposed by Horton and his coworkers. Mainly from the differential effect of antimycin A on fluorescence quenching and ΔpH they concluded that in addition to nonphotochemical decay processes, PS II efficiency is controlled by alternative cyclic photochemical processes, not necessarily related to fluorescence quenching (117, 123, 128).

An alternative model that again relates the quenching to decreased PS II efficiency was proposed by Genty et al (44, 45). This model presumes that fluorescence quenching occurs in the antennae. The ratio of the fluorescence parameters in the quenching state, $F_V'/F_M' = (F_M'-F_0')/F_M'$, is taken as a measure of quantum efficiency of open centers. Such assumption would not be valid for two populations of centers, as proposed by Weis & Berry (165). Though partially contradictory in their theoretical implications, both models allow full quantification on a purely empirical basis and can be applied to evaluation of complex photosynthetic behavior. Thus, both models can be used to calculate electron transport rate or efficiency of PS II from fluorescence signals (44, 45, 165, 167).

The concept of "antenna quenching," developed on the basis of Butler's model of energy transfer processes (see 22, 23), has often been used to explain fluorescence quenching under physiological conditions (see also 34, 128a). It is, however, in conflict with recent concepts based on time-resolved fluorescence spectroscopy, which assume that energy transfer within the antenna is very fast and fluorescence is mainly determined by photochemical events in the center (see the section above on lifetimes).

As a possible mechanism of ΔpH-dependent quenching at centers, a limitation of electron donation to P_{680} caused by intrathylakoid acidification has been suggested (139, 141). Evidence for a donor-side limitation of PS II at low pH comes from time-resolved spectroscopic studies of reduction of P_{680}^+ (138). In fact, quenching at low pH seems to depend on the redox balance on the donor side of PS II: Under moderately reducing condition, or in the presence of specific electron donors to the donor side of PS II, no quenching was seen at low pH in PS II particles (31, 88; see also Figure 1). When electron donation from the water splitting side is slowed down, excitation energy trapped by PS II may be dissipated by fast internal charge-recombination processes. An increased yield of recombination processes at low pH is indicated by μs-luminescence (140). From time-resolved fluorescence studies it has been concluded that in the high energy state, centers are indeed kept in an open state (with oxidized Q_A) even in high light (A. Krieger, I. Moya, E. Weis, unpublished).

Quenching Related to State Transition

Phosphorylation of part of the LHC II—leading to a transition from "state 1" to "state 2"—is supposed to cause detachment of this complex from the core antenna of PS II. Thereby the absorption cross section of PS II relative to PS I is lowered, resulting in decreased fluorescence emission expressed as q_T. F_V and F_O are quenched in the same proportion. The phosphorylation is controlled via the redox state of the intersystem electron chain. Detached phosphorylated LHC II was found to move from appressed to nonappressed membrane regions. Changes in the physical state of the thylakoid membranes (e.g. by mild heat treatment) may also cause detachment of the LHC II from the core antennae, and the fraction of PS IIβ increases. In this case, PS II units containing the core antennae and inner LHC II were found to move from appressed to nonappressed regions, leaving the peripheral LHC II in the grana region (153). Details about these well-studied phenomena have been reviewed elsewhere (20, 169).

Originally it was assumed that in "state 2," light energy absorbed by phosphorylated LHC II could be transferred to PS I, but this concept is still debated (169). Remarkable enhancement in PS I activity was found, however, upon phosphorylation, when PS II was converted to its β-form by exposure to moderately elevated temperature (156). Spillover may occur in a complex [PS IIβ·LHC-P·PS I] (65), but whether such a complex is formed under physiological conditions is not clear. Perhaps state regulation and spillover depend on a dynamic interaction between the physical state of the membrane (as affected, for example, by cation relations or temperature) and the metabolically controlled enzymatic phosphorylation of the LHC II.

The extent of q_T is usually much lower than that of q_E and q_I; maximum values of q_T for barley protoplasts seen in low light were around 0.2 (71). It has been reported that conditions of high light (and high ΔpH) that promote formation of q_E and q_I tend to suppress q_T (35, 42, 71). The physiological significance of state changes and their impact on regulation of photosynthetic electron transport are still not well understood.

Photoinhibitory Quenching

Photoinhibition of photosynthesis, observed within minutes to hours in response to excessive irradiation, is consistently related to quenching of F_V, whereas changes in F_O may vary. This effect has recently been reviewed (76). Low-temperature fluorescence spectroscopy indicates that q_I (like q_E) results from increased nonphotochemical de-excitation of pigments. Reversal of the quenching is correlated with recovery from photoinhibition. The quenching is often expressed as a decrease in the ratio F_V/F_M, recorded after a dark period of several minutes following high light exposure. (The dark interval is required to relax q_E and q_T, while q_I is not readily reversible in the dark.). The decrease in F_V/F_M in leaves was found in numerous studies to be linearly correlated with a decline in simultaneously measured optimal quantum yield of photosynthesis (e.g. 13, 34, 39, 148) or to loss of PS II photosynthetic activity of isolated thylakoids (84). A linear relation between quantum yield and F_V (not F_V/F_M) was, however, observed with kiwifruit leaves (52). One should keep in mind that light gradients within the leaf (162), causing gradients of photoinhibition, can influence such relationships.

HYPOTHETICAL MECHANISMS   Various hypotheses regarding the mechanism of photoinhibitory quenching are presently under discussion. One relates the major component of q_I (as well as of q_E; see above) to formation of zeaxanthin from violaxanthin in the xanthophyll cycle. High light is known to induce this reaction. Indeed, a close correlation between degree of fluorescence quenching (or calculated k_D values) and concentration of zeaxanthin in the thylakoids has been observed in various investigations reviewed lately (37). Zeaxanthin is suggested to facilitate increased thermal deactivation in the PS II antennae in a reversible fashion. On the other hand, it is unknown so far whether this specific xanthophyll species could cause quenching of singlet excited states. Several explanations have been discussed (see 37). There are also reports showing that a correlation between quenching and level of zeaxanthin is missing (43, 126).

It has been proposed (37) that the q_I and q_E mechanisms are in fact identical, as far as the zeaxanthin-related components are concerned. Their only difference would lie in the kinetics of formation and relaxation of quenching. However, this hypothesis neglects the fact that q_E relaxes readily in the dark or upon dissipation of the ΔpH by the action of uncouplers. In contrast, relaxation of q_I is controlled by light (see below). Thylakoids isolated from photoinhibited leaves show decreased PS II activity in the uncoupled state (30, 84, 120, 149). Even if the two mechanisms appear alike in terms of increase in energy dissipation, substantial differences in underlying mechanisms must be assumed.

A number of authors favor the hypothesis that the photoinhibitory quenching is initiated in the PS II reaction center. A transformation (of unknown type) of a fraction of centers to "quenchers" has been postulated (29). Such transformed centers would still act as energy traps, but they would be incapable of the normal photochemical reaction and would convert the excitation energy to heat. The fluorescence level of these centers would stay (approximately) at F_0, whereas the remaining centers are supposed to possess largely unaltered activity. There is, in fact, evidence against "antenna quenching": Increased k_D in the antennae should slow down the rate of Q_A reduction (indicated by the fluorescence rise in the presence of DCMU). Such an effect was not seen upon photoinhibition of thylakoids or intact chloroplasts (29, 84). Moreover, quenching in the antennae should result in inhibition of PS II–driven electron transport measured in low light, while little or no inhibition should be seen in high, saturating light. In contrast, thylakoids isolated from photoinhibited spinach leaves exhibited PS II activities lowered to about the same degree when measured in limiting or saturating light (84, 149).

Quenching in the antennae—expressed as a gradual increase in k_D (equation 3)—would be in agreement with the linear correlation between F_V/F_M and quantum efficiency of photosynthesis (13). However, a simple model (Ch. Giersch, G. H. Krause, unpublished), presuming two populations of centers—either photoinhibited ("quenched") or active—also results in a quasi-linear relationship between quantum yield and F_V/F_M, if one infers (a) α/β heterogeneity with predominant photoinhibition of α units, which probably is due to larger antennae (28, 84, 100), and (b) excitation energy transfer from active α units in the F_M state to photoinhibited α units. The variable response of F_0 to photoinhibition (e.g. 34, 148, 149) is obtained in this model from differences in k_D of the inhibited population.

In earlier studies, it has been inferred that the formation of q_I is directly related to damage and degradation of the D1 protein in the reaction center (see 90). Turnover of this protein in the light is obviously involved in photoinhibition. However, more recent data indicate that photoinhibition and concomitant quenching precede the dysfunction and degradation of D1 (25, 161; S. Schöner, G. H. Krause, unpublished). Moreover, it was demonstrated that degradation of D1 in *Chlamydomonas* (in low light in the presence of chloramphenicol, which suppresses resynthesis of D1) does *not* lead to the quenching of F_M that is characteristic for photoinhibition (19).

Molecular oxygen and/or reactive species derived from it may be involved in formation of q_I (e.g. 9, 131, 145). Also, presence of O_2 seems to be required for D1 degradation (e.g. 3, 89). However, anaerobiosis strongly enhances q_I formation. This effect is related to a conspicuous increase in F_0, indicating that photoinhibition under anaerobiosis is caused by a mechanism different from that in the presence of O_2 (80). The F_0 increase was ascribed by Setlik et al (145) to "permanently closed" reaction centers with a stable Q_A^-. In C3 plants, high O_2 levels indirectly diminish photoinhibition because they enhance photorespiration, which increases the drain of photosynthetic energy (see 79).

REVERSIBILITY OF PHOTOINHIBITION   Recovery from photoinhibition, as seen by relaxation of q_I, has been suggested to proceed *via* a "PS II repair

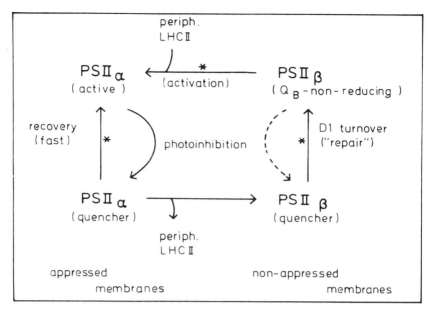

Figure 2 Scheme of hypothetical mechanism of photoinhibition and recovery (after 53, 54, altered). Stars denote possible points of light regulation.

cycle" involving transformation of "damaged" PS IIα to PS IIβ units, in which form the reaction center is restructured by D1 degradation and replacement by the freshly synthesized protein (53). In view of the fast initial phase of recovery ($t_{1/2} \approx 15$ min) that can be found, depending on material and conditions (97, 149), we speculate that photoinhibited ("quenching") centers may, in part, be reversed to the active ("fluorescent") state without D1 replacement. Another fraction (possessing damaged D1?) would follow the path of D1 turnover known to occur in the light (see scheme of Figure 2). It is interesting to note that the recovery is light-regulated in a complex manner. Recovery proceeds very slowly in darkness and is optimal in rather low light. There is evidence (51, 97, 150) that moderate light (much below the level causing detectable photoinhibition) restricts recovery. Thus the hypothesis (50) that "damage" and "repair" proceed simultaneously in high light, and that photoinhibition becomes apparent only when the rate of damage exceeds the capacity for repair, may not be valid under all circumstances.

In summary, the present evidence, though providing considerable insights, does not support a unifying hypothesis of photoinhibitory quenching. It cannot be excluded that both discussed mechanisms—quenching in the antennae and in the reaction center—operate in vivo, depending on conditions and the physiological state of the plant.

Further Quenching Mechanisms

THE MAGNESIUM EFFECT   The fluorescence emanating from thylakoid suspensions, particularly F_V, is strongly influenced by the cation concentration of the medium. This phenomenon is based on the lateral segregation of the two photosystems (at high Mg^{2+} levels) related to the stacking of membranes, which controls energy transfer from PS II to PS I. The effect has been reviewed previously (20). It seems that quenching due to lack of Mg^{2+}, a large and conspicuous effect *in vitro*, is not important *in vivo*.

QUENCHING BY PIGMENT RADICALS   Excitation energy captured by the radicalic or excited states of photosynthetic pigments is converted to heat, and fluorescence is quenched. Such molecular species are, for example, $Pheo^-$, P_{680}^+, Chl^*, and Car^+, which show a shift of their absorbance bands towards the far-red region of the spectrum. However, since these molecular species are usually very unstable, it seems unlikely that they significantly contribute to fluorescence quenching *in vivo*. Under anaerobic or strongly reducing conditions, in high actinic light a reversible fluorescence quenching is observed, which has been attributed to photoaccumulation of $Pheo^-$ (58, 74, 145). The explanation given for this effect is that fast electron donation to P_{680}^+ in the presence of prereduced Q_A leads to the state $P_{680}Pheo^-Q_A^-$, in which excitons will be converted to heat by $Pheo^-$.

It has been suggested that the radical P_{680}^+ is an efficient quencher of variable fluorescence (see 22). To explain the fluorescence induction kinetics in very high light, transient quenching by P_{680}^+ has been suggested (139). Because of its extremely high redox potential, P_{680}^+ is very short-lived, even when electron donation is inhibited, and normally it may never accumulate in the light. Photoaccumulation of oxidized carotene and the possible significance of this pigment species as a quencher in chloroplasts have been discussed by van Gorkom (159).

QUENCHING BY OXIDIZED PLASTIQUINONE   In the oxidized state, the PQ pool exerts a "static" quenching (160a). This is usually a rather minor effect, but quenching is enhanced by detergents [e.g. when used in particle preparations (see 159)].

Physiological Aspects of Quenching

Chlorophyll fluorescence quenching has to be seen in close context with regulation of photosynthesis and adjustment to external factors. Quenching phenomena are strongly influenced by environmental stresses. A complex interrelation between the various quenching components became apparent in numerous recent studies (e.g. 34, 36, 44, 45, 61, 71, 81, 83, 85, 165–167).

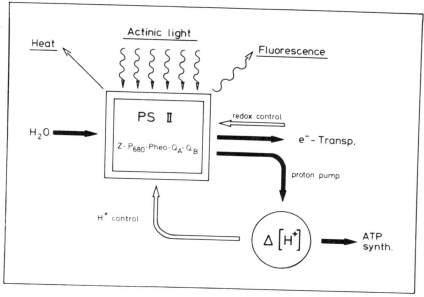

Figure 3 Scheme of PS II–dependent energy flux showing control by the proton gradient and redox state. Open arrows denote feedback control.

In leaves kept in low light, q_P is high (> 0.9), allowing optimal trap function (165). Maximum q_T may develop at low PFD, indicating transition to "state 2". The related change in absorption cross section is viewed as a regulatory mechanism that alters the distribution of excitation energy between PS II and PS I according to the requirement for optimal function of photosynthetic energy conversion. Such balance of excitation is important in low, limiting light (71). With increasing PFD, a growing proportion of absorbed light energy is not used in the photosynthetic process. An increased ΔpH may control q_T and promote q_E; a strong increase in q_N (mainly q_E) is observed. The response of q_E, indicating increased nonphotochemical energy dissipation, has been suggested to represent a dynamic property of the thylakoid membrane (81, 82, 83, 165). The q_E mechanism is supposed to open a pathway of controlled and harmless deactivation of excessively excited pigments, serving a protective function against adverse effects of high light (see scheme of Figure 3). It has been experimentally proven that in the presence of a high q_E photoinhibition is diminished (77, 82, 123).

Photoinhibition of PS II appears to be enhanced in the state of reduced Q_A. Thermal deactivation achieved at high q_E presumably increases the fraction of open centers present in the steady state (164, 165). Only when the PFD exceeds light saturation of photosynthetic CO_2 assimilation and q_E is also saturated does Q_A become largely reduced (q_P strongly declines). Under such

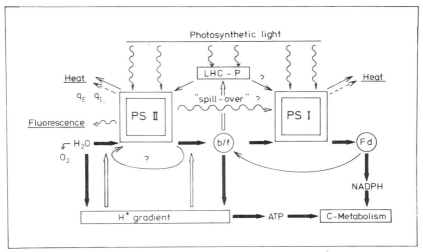

Figure 4 Scheme of the hypothetical relationships between photosynthetic energy conversion, PS II regulation, and fluorescence emission. Closed arrows denote photosynthetic energy conversion, open arrows regulation, broken arrows controlled increase in heat emission.

conditions, q_I is enhanced (71). This concept has been confirmed in a study with spinach leaves on effects of freezing/thawing and subsequent high light stress (151). Freezing stress was found to affect, besides CO_2 assimilation, the formation of q_E in response to the ΔpH. This caused under steady-state conditions in moderate light a decrease in q_P and corresponding increase in q_I; an inverse linear relation between q_P and q_I was seen. Formation of q_I is generally promoted by conditions that limit energy utilization in carbon metabolism or slacken "repair" processes. Then even moderate light leads to strong photoinhibitory quenching (52, 148; see 76).

Formation of q_I (like that of q_E) may be viewed as a protective mechanism of thermal energy dissipation, insofar as the effect is reversible (see scheme of Figure 2). It has been suggested (13, 34, 76, 87, 150) that the q_I mechanism, although it reduces photosynthetic activity, prevents gross destruction of the photosynthetic apparatus. Photoinhibitory quenching becomes effective within minutes in excessive light, when the capacity for q_E is exhausted. This would stabilize the photosynthetic system and allow long-term acclimatory processes to proceed. Notably, these regulatory phenomena occurring in PS II might also contribute to stabilization of PS I under extreme conditions (see 166, 167, 167a). The hypothetical role of the discussed mechanisms to optimize electron flow and energy balance in contrasting environmental conditions is schematically presented in Figure 4. In addition to the different quenching processes, this regulation may include direct control of linear electron transport *via* the ΔpH (167, 167a); it may also include dissipation of excess, potentially harmful energy by a futile cycle of electrons around PS II (70, 117, 128, 139, 140) and by accumulation of P^+_{700} in the reaction center of PS I (164, 167, 167a).

From the above consideration it follows that the quenching phenomena in intact leaves are strongly related to regulation and the steady-state condition of photosynthetic carbon metabolism. We refer to a recent review by Seaton & Walker (144) of this well-studied topic (see also 20).

CONCLUSIONS AND PERSPECTIVES

Our understanding of the physical and structural basis of fluorescence in relation to primary processes of photosynthesis has progressed in recent years together with our knowledge about the molecular architecture of the photosynthetic apparatus. Improved time-resolved spectroscopy has particularly helped this development. We have also made substantial progress in the characterization of various types of nonphotochemical fluorescence quenching and its relation to control of photosynthesis. Nevertheless, the molecular mechanisms of photoinhibitory and high-energy quenching, and of fluorescence quenching in various types of PS IIβ units, are still not well understood. Despite open questions regarding underlying mechanisms, progress in techniques to separate nonphotochemical and photochemical quenching *in vivo* has opened new applications of fluorescence spectroscopy in basic and applied plant biology. In combination with other techniques such as photosynthetic gas exchange and far-red spectroscopy that detects the redox state of PS I, fluorescence spectroscopy can be used to study regulation and control parameters of photosynthesis in leaves. The analysis of complex fluorescence transients (e.g. fluorescence oscillations) may help us to understand the network of various feedback interactions between primary processes of energy conversion and the biochemical reactions of carbon metabolism in intact plants (see 70, 144).

More work is required to improve the reliability of these approaches, and to simplify their application in the field. The various expressions that have now been developed to predict the quantum efficiency of PS II, photosynthetic electron transport, and gas exchange from fluorescence quenching in intact plants are still empirical. Obviously, more general expressions are required on a biophysical and/or biochemical basis. There is still disagreement with respect to the value and physical meaning of empirical parameters, yet these expressions may be regarded as first steps towards using fluorescence spectroscopy for small- and large-scale scanning of biochemical activity and photosynthetic fluxes in intact plants or canopies. Digitized video techniques have recently been introduced for detecting fluorescence. Images of fluorescence quenching have been obtained from small leaf areas (33). Small-scale fluorescence scanning may be used in the future to investigate photosynthesis in relation to various physiological processes at the whole-plant level, to address the source-sink problem in plants, and to study canopy photosynthesis. For large-scale scanning and for vegetation analyses first attempts have been made to use sunlight- or laser-induced fluorescence detected from the ground, as well as from aircraft (see 40, 99, 130). Extension of fluorescence measurements to large-scale spectroscopy may be useful in basic and applied environmental research, such as mapping of the photosynthetic activity of terrestrial and marine vegetation. In future developments, chlorophyll fluorescence will certainly remain an ever-refined instrument for basic photosynthesis research, but we can also expect a wide diversification of practical applications in the plant sciences.

ACKNOWLEDGMENT

The authors thank the Deutsche Forschungsgemeinschaft for support.

Literature Cited

1. Adams, W. W. III, Demmig-Adams, B., Winter, K. 1990. Relative contributions of zeaxanthin-unrelated types of 'high-energy-state' quenching of chlorophyll fluorescence in spinach leaves exposed to various environmental conditions. Plant Physiol. 92:302–9
2. Anderson, J. M., Melis, A. 1983. Localization of different photosystems in separate regions of chloroplast membranes. Proc. Natl. Acad. Sci. USA 80:745–49
3. Arntz, B., Trebst, A. 1986. On the role of the Q_B protein of PS II in photoinhibition. FEBS Lett. 194:43–49
4. Baker, N. R., Webber, A. N. 1987. Interactions between photosystems. Adv. Bot. Res. 13:2–66
5. Baltscheffsky, M., ed. 1990. Current Research in Photosynthesis, Vols. 1, 2, 4. Dordrecht: Kluwer
6. Barber, J. 1989. Isolation of the photosystem two reaction centre: a step forward in understanding how photosynthetic organisms split water. In Photoconversion Processes for Energy and Chemicals, ed. D. O. Hall, G. Grassi, pp. 172–84. Amsterdam: Elsevier
7. Barber, J., Chapman, D. J., Telfer, A. 1987. Characterisation of a PS II reaction centre isolated from the chloroplasts of Pisum sativum. FEBS Lett. 220:667–73
8. Barber, J., Malkin, S., Telfer, A. 1989.

The origin of chlorophyll fluorescence in vivo and its quenching by the photosystem II reaction centre. Philos. Trans. R. Soc. London Ser. B 323:227–39
9. Barényi, B., Krause, G. H. 1985. Inhibition of photosynthetic reactions by light. A study with isolated spinach chloroplasts. Planta 163:218–26
10. Bilger, W., Heber, U., Schreiber, U. 1988. Kinetic relationship between energy dependent fluorescence quenching, light scattering, chlorophyll luminescence and proton pumping in intact leaves. Z. Naturforsch. Teil C 43:877–87
11. Bilger, W., Schreiber, U. 1986. Energy-dependent quenching of dark-level chlorophyll fluorescence in intact leaves. Photosynth. Res. 10:303–8
12. Bittersmann, E., Holzwarth, A. R., Agel, G., Nultsch, W. 1988. Picosecond time-resolved emission spectra of photoinhibited and photobleached Anabaena variabilis. Photochem. Photobiol. 47:101–5
13. Björkman, O. 1987. Low-temperature chlorophyll fluorescence in leaves and its relationship to photon yield of photosynthesis in photoinhibition. In Photoinhibition, ed. D. J. Kyle, C. B. Osmond, C. J. Arntzen, pp. 123–44. Amsterdam: Elsevier
14. Björkman, O., Demmig, B. 1987. Photon yield of O_2 evolution and chloro-

phyll fluorescence characteristics at 77K among vascular plants of diverse origins. *Planta* 170:489–504

15. Black, M. T., Brearley, T. H., Horton, P. 1986. Heterogeneity in chloroplast photosystem II. *Photosynth. Res.* 8:193–207

16. Borisov, A. Y. 1990. Energy migration in purple bacteria. The criterion for discrimination between migration- and trapping-limited photosynthetic units. *Photosynth. Res.* 23:283–89

17. Bradbury, M., Baker, N. R. 1984. A quantitative determination of photochemical and non-photochemical quenching during the slow phase of the chlorophyll fluorescence induction curve of bean leaves. *Biochim. Biophys. Acta* 765:275–81

18. Breton, J. 1982. Hypothesis—The F695 fluorescence of chloroplasts at low temperature is emitted from the primary acceptor of photosystem II. *FEBS Lett.* 147:16–20

19. Briantais, J.-M., Cornic, G., Hodges, M. 1988. The modification of chlorophyll fluorescence of *Chlamydomonas reinhardtii* by photoinhibition and chloramphenicol addition suggests a form of PS II less susceptible to degradation. *FEBS Lett.* 236:226–30

20. Briantais, J.-M., Vernotte, C., Krause, G. H., Weis, E. 1986. Chlorophyll *a* fluorescence of higher plants: chloroplasts and leaves. See Ref. 47, pp. 539–83

21. Briantais, J.-M., Vernotte, C., Picaud, M., Krause, G. H. 1979. A quantitative study of the slow decline of chlorophyll *a* fluorescence in isolated chloroplasts. *Biochim. Biophys. Acta* 548:128–38

22. Butler, W. L. 1977. Chlorophyll fluorescence: a probe for electron transfer and energy transfer. In *Encyclopedia of Plant Physiology*, ed. A. Trebst, M. Avron, 5:149–67. Berlin: Springer-Verlag

23. Butler, W. L. 1978. Energy distribution in the photochemical apparatus of Photosynthesis. *Annu. Rev. Plant Physiol.* 29:345–78

24. Cao, J., Govindjee. 1990. Chlorophyll *a* fluorescence transient as an indicator of active and inactive Photosystem II in thylakoid membranes. *Biochim. Biophys. Acta* 1015:180–88

25. Chow, W. S., Osmond, C. B., Huang, L. K. 1989. Photosystem II function and herbicide binding sites during photoinhibition of spinach chloroplasts *in vivo* and *in vitro*. *Photosynth. Res.* 21:17–26

26. Chylla, R. A., Garab, G., Whitmarsh, J. 1987. Evidence for slow turnover in a

fraction of photosystem II complexes in thylakoid membranes. *Biochim. Biophys. Acta* 894:562–71

27. Chylla, R. A., Whitmarsh, J. 1989. Inactive photosystem II complexes in leaves. *Plant Physiol.* 90:765–72

28. Cleland, R. E., Melis, A. 1987. Probing the events of photoinhibition by altering electron-transport activity and light-harvesting capacity in chloroplast thylakoids. *Plant Cell Environ.* 10:747–52

29. Cleland, R. E., Melis, A., Neale, P. J. 1986. Mechanism of photoinhibition: photochemical reaction center inactivation in system II of chloroplasts. *Photosynth. Res.* 9:79–88

30. Critchley, C. 1981. Studies on the mechanism of photoinhibition in higher plants. I. Effects of high light intensity on chloroplast activities in cucumber adapted to low light. *Plant Physiol.* 67:1161–65

31. Crofts, J., Horton, P. 1990. The effect of pH on photosynthesis in photosystem 2 particles. See Ref. 5, pp. 391–94

32. Crystall, B., Booth, P. J., Klug, D. R., Barber, J., Porter, G. 1989. Resolution of a long lived fluorescence component from D1/D2/cytochrome b-559 reaction centres. *FEBS Lett.* 249:75–78

33. Daley, P. F., Raschke, K., Ball, J. T., Berry, J. A. 1989. Topography of photosynthetic activity of leaves obtained from video images of chlorophyll fluorescence. *Plant Physiol.* 90:1233–38

34. Demmig, B., Björkman, O. 1987. Comparison of the effect of excessive light on chlorophyll fluorescence (77K) and photon yield of O_2 evolution in leaves of higher plants. *Planta* 171:171–84

35. Demmig, B., Cleland, R. E., Björkman, O. 1987. Photoinhibition, 77K chlorophyll fluorescence quenching and phosphorylation of the light-harvesting chlorophyll-protein complex of photosystem II in soybean leaves. *Planta* 172:378–85

36. Demmig, B., Winter, K. 1988. Characterisation of three components of non-photochemical quenching and their response to photoinhibition. *Aust. J. Plant Physiol.* 15:163–77

37. Demmig-Adams, B. 1990. Carotenoids and photoprotection in plants: a role for the xanthophyll zeaxanthin. *Biochim. Biophys. Acta.* 1020:1–24

38. Demmig-Adams, B., Adams, W. W. III, Heber, U., Neimanis, S., Winter, K., et al. 1990. Inhibition of zeaxanthin formation and of rapid changes in radiationless energy dissipation by di-

thiothreitol in spinach leaves and chloroplasts. *Plant Physiol.* 92:293–301

39. Demmig-Adams, B., Adams, W. W. III, Winter, K., Meyer, A., Schreiber, U., et al. 1989. Photochemical efficiency of photosystem II, photon yield of O_2 evolution photosynthetic capacity, and carotenoid composition during the midday depression of net CO_2 uptake in *Arbutus unedo* growing in Portugal. *Planta* 177:377–87

40. Doerffer, R. 1988. Remote sensing of sunlight induced phytoplankton fluorescence. See Ref. 98, pp. 269–74

41. Duysens, L. N. M., Sweers, H. E. 1963. Mechanism of the two photochemical reactions in algae as studied by means of fluorescence. In *Studies on Microalgae and Photosynthetic Bacteria,* Jpn. Soc. Plant Physiol., pp. 353–72. Tokyo: Univ. Tokyo Press

42. Fernyhough, P., Foyer, C. H., Horton, P. 1984. Increase in the level of thylakoid protein phosphorylation in maize mesophyll chloroplasts by decrease in the transthylakoid pH gradient. *FEBS Lett.* 176:133–38

43. Foyer, C. H., Dujardyn, M., Lemoine, Y. 1990. Turnover of the xanthophyll cycle during photoinhibition and recovery. See Ref. 5, 2:491–94

44. Genty, B., Briantais, J.-M., Baker, N. R. 1989. The relationship between the quantum yield of photosynthetic electron transport and quenching of chlorophyll fluorescence. *Biochim. Biophys. Acta* 990:87–92

45. Genty, B., Harbinson, J., Briantais, J.-M., Baker, N. R. 1990. The relationship between non-photochemical quenching of chlorophyll fluorescence and the rate of photosystem 2 photochemistry in leaves. *Photosynth. Res.* 25:249–57

46. Gilmore, A. M., Yamamoto, H. Y. 1990. Zeaxanthin formation in q_E-inhibited chloroplasts. See Ref. 5, 2:495–98

47. Govindjee, Amesz, J., Fork, D. J., eds. 1986. *Light Emission by Plant and Bacteria.* New York: Academic. 638 pp.

48. Govindjee, van de Ven, M., Preston, C., Seibert, M., Gratton, E. 1990. Chlorophyll a fluorescence lifetime distributions in open and closed photosystem II reaction center preparations. *Biochim. Biophys. Acta* 1015:173–79

49. Greene, B. A., Staehelin, L. A., Melis, A. 1988. Compensatory alterations in the photochemical apparatus of a photoregulatory, chlorophyll *b*-deficient mutant of maize. *Plant Physiol.* 87:365–70

50. Greer, D. H., Berry, J. A., Björkman, O. 1986. Photoinhibition of photosynthesis in intact bean leaves: role of light and temperature, and requirement for chloroplast-protein synthesis during recovery. *Planta* 168:253–60

51. Greer, D. H., Laing, W. A. 1988. Photoinhibition of photosynthesis in intact kiwifruit *(Actinidia deliciosa)* leaves: effect of light during growth on photoinhibition and recovery. *Planta* 175:355–63

52. Greer, D. H., Laing, W. A., Kipnis, T. 1988. Photoinhibition of photosynthesis in intact kiwifruit *(Actinidia deliciosa)* leaves: effect of temperature. *Planta* 174:152–58

53. Guenther, J. E., Melis, A. 1990. The physiological significance of photosystem II heterogeneity in chloroplasts. *Photosynth. Res.* 23:105–9

54. Guenther, J. E., Melis, A. 1990. Dynamics of photosystem II heterogeneity in *Dunaliella salina (green algae).* *Photosynth. Res.* 23:195–203

55. Guenther, J. E., Nemson, J. A., Melis, A. 1988. Photosystem stoichiometry and chlorophyll antenna size in *Dunaliella salina (green algae).* *Biochim. Biophys. Acta* 934:108–17

56. Guenther, J. E., Nemson, J. A., Melis, A. 1990. Development of photosystem II in dark grown *Chlamydomonas reinhardtii.* A light-dependent conversion of PS II_β, Q_B-nonreducing centers to PS II_α, Q_B-reducing form. *Photosynth. Res.* 24:35–46

57. Hansson, Ö., Wydrzynski, T. 1990. Current perceptions of photosystem II. *Photosynth. Res.* 23:131–62

58. Heber, U., Kobayashi, Y., Leegood, R. C., Walker, D. A. 1985. Low fluorescence yield in anaerobic chloroplasts and stimulation of chlorophyll *a* fluorescence by oxygen and inhibitors that block electron flow between photosystems II and I. *Proc. R. Soc. London Ser. B* 225:41–53

59. Hodges, M., Barber, J. 1986. Analysis of chlorophyll fluorescence induction kinetics exhibited by DCMU-inhibited thylakoids and the origin of α and β centers. *Biochim. Biophys. Acta* 848:239–46

60. Hodges, M., Briantais, J.-M., Moya, I. 1987. The effect of thylakoid membrane reorganisation on chlorophyll fluorescence lifetime components: a comparison between state transitions, protein phosphorylation and the absence of Mg^{2+}. *Biochim. Biophys. Acta* 893:480–89

61. Hodges, M., Cornic, G., Briantais, J.-M. 1989. Chlorophyll fluorescence from

spinach leaves: resolution of non-photochemical quenching. *Biochim. Biophys. Acta* 974:289–93

62. Hodges, M., Moya, I. 1987. Modification of room-temperature picosecond chlorophyll fluorescence kinetics in photosystem-II-enriched particles by photochemistry. *Biochim. Biophys. Acta* 892:42–47

63. Hodges, M., Moya, I. 1987. Time-resolved chlorophyll fluorescence studies on photosynthetic mutants of *Chlamydomonas reinhardtii:* origin of the kinetic decay components. *Photosynth. Res.* 13:125–41

64. Hodges, M., Moya, I. 1988. Time-resolved chlorophyll fluorescence studies on pigment-protein complexes from photosynthetic membranes. *Biochim. Biophys. Acta* 935:41–52

65. Holzwarth, A. R. 1987. A model for the functional antenna organization and energy distribution in the photosynthetic apparatus of higher plants and green algae. In *Progress in Photosynthesis Research,* ed. J. Biggins, 1:53–60. Dordrecht: Martinus Nijhoff

66. Holzwarth, A. R. 1988. Time resolved chlorophyll fluorescence. See Ref. 98, pp. 21–31

67. Holzwarth, A. R. 1991. Excited state kinetics in chlorophyll systems and its relationship to the functional organization of the photosystem. In *The Chlorophylls CRC Handbook,* ed. H. Scheer. Boca Raton: CRC Press. In press

68. Holzwarth, A. R. 1990. The functional organization of the antenna systems in higher plants and green algae as studied by time-resolved fluorescence techniques. See Ref. 5, 2:223–30

69. Holzwarth, A. R., Haehnel, W., Ratajczak, R., Bittersmann, E., Schatz, G. H. 1990. Energy transfer kinetics in photosystem I particles isolated from *Synechococcus sp.* and from higher plants. See Ref. 5, 2:611–14

70. Horton, P. 1989. Interactions between electron transport and carbon assimilation: regulation of light-harvesting and photochemistry. In *Photosynthesis,* ed. W. R. Briggs, pp. 393–406. New York: Alan R. Liss

71. Horton, P., Hague, A. 1988. Studies on the induction of chlorophyll fluorescence in isolated barley protoplasts. IV. Resolution of non-photochemical quenching. *Biochim. Biophys. Acta* 932:107–15

72. Joliot, A., Joliot, P. 1964. Etude cinétique de la réaction photochimique libérant l'oxygene au cours de la photosynthése. *C.R. Acad. Sci. Paris* 258:4622–25

73. Keuper, H. J. K., Sauer, K. 1989. Effect of photosystem II reaction center closure on nanosecond fluorescence relaxation kinetics. *Photosynth. Res.* 20:85–103

74. Klimov, V. V., Klevanik, A. V., Shuvalov, V. A., Krasnovsky, A. A. 1977. Reduction of pheophytin in the primary light reaction of photosystem II. *FEBS Lett.* 82:183–86

75. Klimov, V. V., Krasnovskii, A. A. 1981. Pheophytin as a primary electron acceptor in photosystem II reaction centres. *Photosynthetica* 15:592–609

76. Krause, G. H. 1988. Photoinhibition of photosynthesis. An evaluation of damaging and protective mechanisms. *Physiol. Plant.* 74:566–74

77. Krause, G. H., Behrend, U. 1986. ΔpH-Dependent chlorophyll fluorescence quenching indicating a mechanism of protection against photoinhibition of chloroplasts. *FEBS Lett.* 200:298–302

78. Krause, G. H., Briantais, J.-M., Vernotte, C. 1983. Characterization of chlorophyll fluorescence quenching in chloroplasts by fluorescence spectroscopy at 77 K. I. ΔpH-Dependent quenching. *Biochim. Biophys. Acta* 723:169–75

79. Krause, G. H., Cornic, G. 1987. CO_2 and O_2 interactions in photoinhibition. In *Photoinhibition,* ed. D. J. Kyle, C. B. Osmond, C. J. Arntzen, pp. 169–96. Amsterdam: Elsevier

80. Krause, G. H., Köster, S., Wong, S. C. 1985. Photoinhibition of photosynthesis under anaerobic conditions studied with leaves and chloroplasts of *Spinacia oleracea* L. *Planta* 165:430–38

81. Krause, G. H., Laasch, H. 1987. Energy-dependent chlorophyll fluorescence quenching in chloroplasts correlated with quantum yield of photosynthesis. *Z. Naturforsch. Teil C* 42:581–84

82. Krause, G. H., Laasch, H. 1987. Photoinhibition of photosynthesis. Studies on mechanisms of damage and protection in chloroplasts. In *Progress in Photosynthesis Research,* ed. J. Biggins, 4:19–26. Dordrecht: Martinus Nijhoff

83. Krause, G. H., Laasch, H., Weis, E. 1988. Regulation of thermal dissipation of absorbed light energy in chloroplasts indicated by energy-dependent fluorescence quenching. *Plant Physiol. Biochem.* 26:445–52

84. Krause, G. H., Somersalo, S., Zumbusch, E., Weyers, B., Laasch, H. 1990. On the mechanism of photoinhibition in chloroplasts. Relationship between changes in fluorescence and activ-

ity of Photosystem II. *J. Plant Physiol.* 136:472–79

85. Krause, G. H., Vernotte, C., Briantais, J.-M. 1982. Photoinduced quenching of chlorophyll fluorescence in intact chloroplasts and algae. Resolution into two components. *Biochim. Biophys. Acta* 679:116–24

86. Krause, G. H., Weis, E. 1984. Chlorophyll fluorescence as a tool in plant physiology. II. Interpretation of fluorescence signals. *Photosynth. Res.* 5:139–57

87. Krause, G. H., Weis, E. 1988. The photosynthetic apparatus and chlorophyll fluorescence. An introduction. See Ref. 98, pp. 3–11

88. Krieger, A., Weis, E. 1990. pH-Dependent quenching of chlorophyll fluorescence in isolated PS II particles: dependence on the redox potential. See Ref. 5, 1:563–69

89. Kuhn, M., Böger, P. 1990. Studies on the light-induced loss of the D1 protein in photosystem-II membrane fragments. *Photosynth. Res.* 23:291–96

90. Kyle, D. J. 1987. The biochemical basis for photoinhibition of photosystem II. In *Photoinhibition*, ed. D. J. Kyle, C. B. Osmond, C. J. Arntzen, pp. 197–226. Amsterdam: Elsevier

91. Laasch, H. 1987. Non-photochemical quenching of chlorophyll *a* fluorescence in isolated chloroplasts under conditions of stressed photosynthesis. *Planta* 171:220–26

92. Laasch, H., Weis, E. 1989. Photosynthetic control, "energy-dependent" quenching of chlorophyll fluorescence and photophosphorylation under influence of tertiary amines. *Photosynth. Res.* 22:137–46

93. Lavergne, J. 1974. Fluorescence induction in algae and chloroplasts. *Photochem. Photobiol.* 20:377–86

94. Lavorel, J., Etienne, A. L. 1977. In vivo chlorophyll fluorescence. In *Primary Processes of Photosynthesis. Topics in Photosynthesis*, ed. J. Barber, 2:203–68. Amsterdam: Elsevier

95. Lee, C. B., Rees, D., Horton, P. 1990. Non-photochemical quenching of chlorophyll fluorescence in the green alga *Dunaliella*. *Photosynth. Res.* 24:167–73

96. Lee, C.-H., Roelofs, T. A., Holzwarth, A. R. 1990. Target analysis of picosecond fluorescence kinetics in green algae: characterization of primary processes in photosystem IIα and β. See Ref. 5, 1:387–90

97. Le Gouallec, J. L., Cornic, G., Briantais, J. M. 1991. Chlorophyll fluorescence and photoinhibition in a tropical

rainforest understory plant. *Photosynth. Res.* In press

98. Lichtenthaler, H. K., ed. 1988. *Application of Chlorophyll Fluorescence.* Dordrecht: Kluwer. 366 pp.

99. Lichtenthaler, H. K. 1988. Remote sensing of chlorophyll fluorescence in oceanography and in terrestrial vegetation: an introduction. See Ref. 98, pp. 287–97

100. Mäenpää, P., Andersson, B., Sundby, C. 1987. Difference in sensitivity to photoinhibition between photosystem II in the appressed and non-appressed thylakoid regions. *FEBS Lett.* 215:31–36

101. Mäenpää, P., Aro, E., Somersalo, S., Tyystjärvi, E. 1988. Rearrangement of the chloroplast thylakoid at chilling temperature in the light. *Plant Physiol.* 87:762–66

102. McCauley, S. W., Bittersmann, E., Mueller, M., Holzwarth, A. R. 1990. Picosecond chlorophyll fluorescence from higher plants. See Ref. 5, 2:297–300

103. Melis, A. 1985. Functional properties of photosystem IIβ in spinach chloroplasts. *Biochim. Biophys. Acta* 808:334–42

104. Melis, A., Anderson, J. M. 1983. Structural and functional organization of the photosystems in spinach chloroplasts. Antenna size, relative electron transport capacity and chlorophyll composition. *Biochim. Biophys. Acta* 724:473–84

105. Melis, A., Duysens, L. N. M. 1979. Biphasic energy conversion kinetics and absorbance difference spectra of photosystem II of chloroplasts. Evidence for different photosystem II reaction centers. *Photochem. Photobiol.* 29:373–82

106. Melis, A., Guenther, J. E., Morrissey, P. J., Ghirardi, M. L. 1988. Photosystem II heterogeneity in chloroplasts. See Ref. 98, pp. 33–43

107. Melis, A., Homann, P. H. 1976. Heterogeneity of the photochemical centers in system II of chloroplasts. *Photochem. Photobiol.* 23:343–50

108. Mimuro, M., Yamazaki, I., Itoh, S., Tamai, N., Satoh, K. 1988. Dynamic fluorescence properties of D1-D2-cytochrome *b*-559 complex isolated from spinach chloroplasts: analysis by means of the time-resolved fluorescence spectra in picosecond time range. *Biochim. Biophys. Acta* 933:478–86

109. Morrissey, P. J., Glick, R. E., Melis, A. 1989. Supramolecular assembly and function of subunits associated with the chlorophyll *ab* light-harvesting complex II (LHC II) in soybean chloroplasts. *Plant Cell Physiol.* 30:335–44

110. Moss, D. A., Bendall, D. S. 1984. Cyclic electron transport in chloroplasts. The Q-cycle and the site of action of antimycin. *Biochim. Biophys. Acta* 767:389–95

111. Moya, I., Sebban, P., Haehnel, W. 1986. Lifetime of excited states and quantum yield of chlorophyll *a* fluorescence *in vivo*. See Ref. 47, pp. 161–90

112. Mukerji, I., Sauer, K. 1989. Temperature-dependent steady-state and picosecond kinetic fluorescence measurements of a photosystem I preparation from spinach. In *Photosynthesis*, ed. W. H. Briggs, pp. 105–22. New York: Alan R. Liss

113. Mukerji, I., Sauer, K. 1990. A spectroscopic study of a photosystem I antenna complex. See Ref. 5, 2:321–24

114. Murata, N., Satoh, K. 1986. Absorption and fluorescence emission by intact cells, chloroplasts, and chlorophyll-protein complexes. See Ref. 47, pp. 137–59

115. Nanba, O., Satoh, K. 1987. Isolation of a photosystem II reaction center containing D1 and D2 polypeptides and cytochrome *b*-559. *Proc. Natl. Acad. Sci. USA* 84:109–12

116. Neubauer, C., Schreiber, U. 1987. The polyphasic rise of chlorophyll fluorescence upon onset of strong continuous illumination. I. Saturation characteristics and partial control by the photosystem II acceptor side. *Z. Naturforsch. Teil C* 42:1246–54

117. Noctor, G., Horton, P. 1990. Uncoupler titration of energy-dependent chlorophyll fluorescence quenching and photosystem II photochemical yield in intact pea chloroplasts. *Biochim. Biophys. Acta* 1016:228–34

118. Nuijs, A. M., Shuvalov, V. A., van Gorkom, H. J., Plijter, J. J., Duysens, L. N. M. 1986. Picosecond absorbance-difference spectroscopy on the primary reactions and the antenna-excited states in photosystem I particles. *Biochim. Biophys. Acta* 850:310–18

119. Nultsch, W., Bittersmann, E., Holzwarth, A. R., Agel, G. 1990. Effects of strong light irradiation on antennae and reaction centres of the cyanobacterium *Anabaena variabilis*. A time-resolved fluorescence study. *J. Photochem. Photobiol. B* 5:481–94

120. Ögren, E., Öquist, G. 1984. Photoinhibition of photosynthesis in *Lemna gibba* as induced by the interaction between light and temperature. II. Photosynthetic electron transport. *Physiol. Plant* 62:187–92

121. Ort, D. R., Whitmarsh, J. 1990. Inactive photosystem II centers: a resolution of discrepancies in photosystem II quantitation? *Photosynth. Res.* 23:101–4

122. Oxborough, K., Horton, P. 1987. Characterisation of the effects of antimycin A upon high energy state quenching of chlorophyll fluorescence (qE) in spinach and pea chloroplasts. *Photosynth. Res.* 12:119–28

123. Oxborough, K., Horton, P. 1988. A study of the regulation and function of energy-dependent quenching in pea chloroplasts. *Biochim. Biophys. Acta* 934:135–43

124. Papageorgiou, G. 1975. Chlorophyll fluorescence: an intrinsic probe of photosynthesis. In *Bioenergetics of Photosynthesis*, ed. Govindjee, pp. 320–66. New York: Academic

125. Peterson, R. B., Sivak, M. N., Walker, D. A. 1988. Relationship between steady-state fluorescence yield and photosynthetic efficiency in spinach leaf tissue. *Plant Physiol.* 88:158–63

126. Pfündel, E., Strasser, R. J. 1990. Chlorophyll *a* fluorescence (77 K) and zeaxanthin formation in leaf discs *(Nicotiana tabacum)* and isolated thylakoids *(Lactuca sativa)*. See Ref. 5, 2:503–6

127. Quick, P., Scheibe, R., Stitt, M. 1989. Use of tentoxin and nigericin to investigate the possible contribution of ΔpH to energy dissipation and the control of electron transport in spinach leaves. *Biochim. Biophys. Acta* 974:282–88

128. Rees, D., Horton, P. 1990. The mechanisms of changes in photosystem II efficiency in spinach thylakoids. *Biochim. Biophys. Acta* 1016:219–27

128a. Rees, D., Noctor, G. D., Horton, P. 1990. The effect of high-energy-state excitation quenching on maximum and dark level chlorophyll fluorescence yield. *Photosynth. Res.* 25:199–211

129. Rees, D., Young, A., Noctor, G., Britton, G., Horton, P. 1989. Enhancement of the ΔpH-dependent dissipation of excitation energy in spinach chloroplasts by light-activation: correlation with synthesis of zeaxanthin. *FEBS Lett.* 256:85–90

130. Renger, G., Schreiber, U. 1986. Practical applications of fluorometric methods to algae and higher plant research. See Ref. 47, pp. 587–619

131. Richter, M., Rühle, W., Wild, A. 1990. Studies on the mechanism of photosystem II photoinhibition. II. The involvement of toxic oxygen species. *Photosynth. Res.* 24:237–43

132. Rinderle, U., Lichtenthaler, H. K. 1988. The chlorophyll fluorescence ratio

F690/F735 as a possible stress indicator. See Ref. 98, pp. 189–96

133. Robinson, H. H., Crofts, A. R. 1983. Kinetics of the oxidation-reduction reaction of the photosystem II quinone acceptor complex, and the pathway for deactivation. *FEBS Lett.* 153:221–26

134. Roelofs, T. A., Holzwarth, A. R. 1990. On a presumed long-lived relaxed radical pair state in closed photosystem II. See Ref. 5, 1:443–46

135. Schatz, G. H., Brock, H., Holzwarth, A. R. 1987. Picosecond kinetics of fluorescence and absorbance changes in photosystem II particles excited at low photon density. *Proc. Natl. Acad. Sci. USA* 84:8414–18

136. Schatz, G. H., Brock, H., Holzwarth, A. R. 1988. Kinetic and energetic model for the primary processes in photosystem II. *Biophys. J.* 54:397–405

137. Schatz, G. H., Holzwarth, A. R. 1986. Mechanism of chlorophyll fluorescence revisited: prompt or delayed emission from photosystem II with closed reaction centers? *Photosynth. Res.* 10:309–18

138. Schlodder, E., Brettel, K. 1988. Primary charge separation in closed photosystem II with a lifetime of 11 ns. Flash absorption spectroscopy with oxygen evolving photosystem II complexes from *Synechococcus*. *Biochim. Biophys. Acta* 933:22–34

139. Schreiber, U., Neubauer, C. 1987. The polyphasic rise of chlorophyll fluorescence upon onset of strong continuous illumination. II. Partial control by the photosystem II donor side and possible ways of interpretation. *Z. Naturforsch. Teil C* 42:1255–64

140. Schreiber, U., Neubauer, C. 1989. Correlation between dissipative fluorescence quenching at photosystem II and 50 μs recombination luminescence. *FEBS Lett.* 258:339–42

141. Schreiber, U., Neubauer, C. 1990. O_2-dependent electron flow, membrane organization and the mechanism of nonphotochemical quenching of chlorophyll fluorescence. *Photosynth. Res.* 25:279–93

142. Schreiber, U., Pfister, K. 1982. Kinetic analysis of the light-induced chlorophyll fluorescence rise curve in the presence of dichlorophenyldimethylurea. Dependence of the slow-rise component on the degree of chloroplast intactness. *Biochim. Biophys. Acta* 680:60–68

143. Schreiber, U., Schliwa, U., Bilger, W. 1986. Continuous recording of photochemical and non-photochemical chlorophyll fluorescence quenching with a new

type of modulation fluorometer. *Photosynth. Res.* 10:51–62

144. Seaton, G. G. R., Walker, D. A. 1990. Chlorophyll fluorescence as a measure of photosynthetic carbon metabolism. *Proc. R. Soc. London Ser. B.* 242:29–35

145. Setlik, I., Allakhverdiev, S. I., Nedbal, L., Setlikova, E., Klimov, V. V. 1990. Three types of photosystem II photoinactivation. 1. Damaging processes on the acceptor side. *Photosynth. Res.* 23:39–48

146. Siefermann-Harms, D. 1988. Fluorescence properties of isolated chlorophyll-protein complexes. See Ref. 98, pp. 45–54

147. Sinclair, J., Spence, S. M. 1990. Heterogeneous photosystem 2 activity in isolated spinach chloroplasts. *Photosynth. Res.* 24:209–20

147a. Snel, J. F. H., van Kooten, O., eds. 1990. The use of chlorophyll fluorescence and other non-invasive spectroscopic techniques in plant stress physiology. *Photosynth. Res.* (Spec. Iss.) 25:146–332

148. Somersalo, S., Krause, G. H. 1989. Photoinhibition at chilling temperature. Fluorescence characteristics of unhardened and cold-acclimated spinach leaves. *Planta* 177:409–16

149. Somersalo, S., Krause, G. H. 1990. Reversible photoinhibition of unhardened and cold-acclimated spinach leaves at chilling temperatures. *Planta* 180:181–87

150. Somersalo, S., Krause, G. H. 1990. Photoinhibition at chilling temperatures and effects of freezing stress on cold-acclimated spinach leaves in the field. A fluorescence study. *Physiol. Plant.* 79:617–22

151. Somersalo, S., Krause, G. H. 1990. Effects of freezing and subsequent light stress on photosynthesis of spinach leaves. *Plant Physiol. Biochem.* 28:467–75

152. Strasser, R. J. 1986. Mono-, bi-, and polypartite models in photosynthesis. *Photosynth. Res.* 10:255–76

153. Sundby, C. A., Melis, A., Mäenpää, P., Andersson, B. 1986. Temperature-dependent changes in the antenna size of photosystem II. Reversible conversion of photosystem II_{α} to photosystem II_{β}. *Biochim. Biophys. Acta* 851:475–83

154. Thielen, A. P. G. M., van Gorkom, H. J. 1981. Quantum efficiency and antenna size of photosystems $II\alpha$, $II\beta$ and I in tobacco chloroplasts. *Biochim. Biophys. Acta* 635:111–20

155. Thielen, A. P. G. M., van Gorkom, H.

J. 1981. Energy transfer and quantum yield in photosystem II. *Biochim. Biophys. Acta* 637:439–46

156. Timmerhaus, M., Weis, E. 1990. Regulation of photosynthesis: α- to β-conversion of photosystem II and thylakoid protein phosphorylation. See Ref. 5, 2:771–74

157. van Dorssen, R. J., Breton, J., Plijter, J. J., Satoh, K., van Gorkom, H. J., Amesz, J. 1987. Spectroscopic properties of the reaction center and of the 47 kDa chlorophyll protein of photosystem II. *Biochim. Biophys. Acta* 893:267–74

158. van Dorssen, R. J., Plijter, J. J., Dekker, J. P., den Ouden, A., Amesz, J., van Gorkom, H. J. 1987. Spectroscopic properties of chloroplast grana membranes and of the core of photosystem II. *Biochim. Biophys. Acta* 890:134–43

159. van Gorkom, H. 1986. Fluorescence measurements in the study of photosystem II electron transport. See Ref. 47, pp. 267–89

160. van Wijk, K. J., van Hasselt, P. R. 1990. The quantum efficiency of photosystem II and its relation to nonphotochemical quenching of chlorophyll fluorescence: the effect of measuring- and growth-temperature. *Photosynth. Res.* 25:233–40

160a. Vernotte, C., Etienne, A. L., Briantais, J.-M. 1979. Quenching of the system II chlorophyll fluorescence by the plastoquinone pool. *Biochim. Biophys. Acta* 545:519–27

161. Virgin, I., Styring, S., Andersson, B. 1988. Photosystem II disorganization and manganese release after photoinhibition of isolated spinach thylakoid membranes. *FEBS Lett.* 233:408–12

162. Vogelmann, T. C., Bornman, J. F., Josserand, S. 1989. Photosynthetic light gradients and spectral régime within

leaves of *Medicago sativa*. *Philos. Trans. R. Soc. London Ser. B* 323:411–21

163. Walters, R. G., Horton, P. 1990. The use of light pulses to investigate the relaxation in the dark of chlorophyll fluorescence quenching in barley leaves. See Ref. 5, 1:631–34

164. Weis, E., Ball, J. T., Berry, J. A. 1987. Photosynthetic control of electron transport in leaves of *Phaseolus vulgaris*: evidence for regulation of photosystem 2 by the proton gradient. In *Progress in Photosynthesis Research*, ed. J. Biggins, 2:553–56. Dordrecht: Martinus Nijhoff

165. Weis, E., Berry, J. A. 1987. Quantum efficiency of photosystem II in relation to 'energy'-dependent quenching of chlorophyll fluorescence. *Biochim. Biophys. Acta* 894:198–208

166. Weis, E., Lechtenberg, D. 1988. Steady state photosynthesis in intact plants as analyzed by chlorophyll fluorescence and far-red spectroscopy. See Ref. 98, pp. 71–76

167. Weis, E., Lechtenberg, D. 1989. Fluorescence analysis during steady-state photosynthesis. *Philos. Trans. R. Soc. London Ser. B* 323:253–68

167a. Weis, E., Lechtenberg, D., Krieger, A. 1990. Physiological control of primary photochemical energy conversion in higher plants. See Ref. 5, 4:307–12

168. Wendler, J., Holzwarth, A. R. 1987. State transitions in the green alga *Scenedesmus obliquus* probed by time-resolved chlorophyll fluorescence spectroscopy and global data analysis. *Biophys. J.* 52:717–28

169. Williams, W. P., Allen, J. F. 1987. State 1 / State 2 changes in higher plants and algae. *Photosynth. Res.* 13:19–45

Annu. Rev. Plant Physiol. Plant Mol. Biol. 1991. 42:351–71

CIRCADIAN RHYTHMS AND PHYTOCHROME

Peter John Lumsden

Department of Applied Biology, Lancashire Polytechnic, Preston, Lancashire PR1 2TQ, United Kingdom

KEY WORDS: endogenous oscillator, entrainment, light, photoperiodism

CONTENTS

INTRODUCTION

The natural environment is characterized by periodicities—seasonal, lunar, tidal, and diurnal (day/night)—and organisms display a range of biological rhythms in physiology, development, and behavior that match these environmental rhythms. Biological functions become entrained to environmental variables. Nocturnal animals, for example, restrict their activity to periods of

351

darkness (13, 97). Endogenous rhythmic functions are usually synchronized by environmental stimuli but persist when the stimuli are absent. In this sense the organism anticipates environmental changes; it behaves without having to wait for a specific periodic signal—i.e. it shows predictive control (82).

Rhythmicity appears to be a general property of biological processes in eukaryotes, ranging from long behavioral rhythms to such high-frequency biochemical oscillations as glycolysis. The most extensively studied biological rhythms are those entrained by the diurnal light-dark (LD) cycle to a period of 24 hr. These are called circadian rhythms (*circa* = about, *diem* = day). All eukaryotic organisms appear to have evolved circadian rhythms— for example in enzyme activity, body temperature, activity, mitotic index, and (in plants) CO_2 exchange, ion uptake, and leaf movement (2, 12, 22, 107, 122). Circadian rhythmicity in prokaryotes has not yet been demonstrated (40). Under constant conditions the rhythms continue to oscillate with a period of about, but not exactly, 24 hr, the free-running period (FRP). Other features of circadian periodicity have been comprehensively covered elsewhere (2, 7, 13, 30, 40, 97), but the underlying mechanisms are still unknown.

In plants, circadian rhythms may facilitate the coordination of metabolism with the light period, during which photosynthetically active radiation is available. They also provide a mechanism for measuring the length of day/ night central to season detection in photoperiodism. Phytochrome is probably the photoreceptor involved in entrainment of such rhythms by LD cycles (78), but the mechanism of phytochrome action is unknown.

In many rhythmic phenomena light does more than entrain. Here I discuss these responses to light in terms of the interaction of light with the circadian system and the action of phytochrome. I first review the formal properties of circadian rhythms and the data on the action of phytochrome.

FEATURES OF CIRCADIAN RHYTHMS

Circadian rhythms are the observable result of a complex system. Like the hands of a clock, they are coupled to an underlying oscillatory process, the circadian oscillator. Entrainment and phase shifting of rhythms by light result from the action of light on the oscillator (87), which, being coupled to the rhythmic function, alters its phase. I refer to a single oscillator for convenience; several models suggest that the rhythmic output (the circadian rhythm) results from the coupling of several oscillators.

Investigations of circadian systems must consider the components of the oscillatory system; coupling of the oscillator to the observed rhythm; the biochemical components of the rhythmic function; and, since entrainment occurs after light acts on the oscillator, the photoreceptor and its mode of action and coupling to the oscillator (Figure 1). Three main experimental

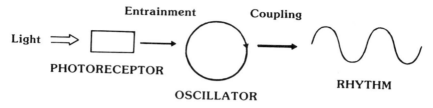

Figure 1   Model for generation of circadian rhythms by coupling to an underlying oscillator, and entrainment by light acting on the oscillator.

approaches can therefore be adopted. One can (*a*) perturb the oscillator mechanism and identify the site of action of the perturbing agent; (*b*) identify the components of the driven rhythm and work back along the transduction chain to the oscillatory mechanism; or (*c*) identify the photoreceptor and work forward along the transduction chain to the oscillatory mechanism (48).

The Endogenous Oscillator

The oscillator has avoided detection for many years. Its operation can only be deduced from studies of the behavior of rhythmic phenomena assumed to be coupled to it—for example, in the marine unicellular alga *Gonyaulax polyhedra,* which displays circadian rhythms in cell division, photosynthetic capacity, bioluminescence, and stimulated light emission. Although these rhythms all peak at different times of the day, they show a constant phase relationship with each other, and phase-shifting light treatments shift all four rhythms similarly, indicating a common underlying oscillator. Blocking the rhythm of photosynthesis with dichloro-methyl-urea (DCMU) did not affect the rhythm of bioluminescence, and blocking the rhythm of luminescence with puromycin did not affect the rhythm of photosynthesis. Further, when inhibition was removed, the respective rhythm resumed in phase with the untreated controls (68).

MATHEMATICAL MODELS   Since the biochemical basis of endogenous oscillators is still unknown, we use mathematical models based on the formal characteristics of the observed rhythms to provide a conceptual view of the oscillator. Models include simple harmonic oscillators, relaxation oscillators, and limit cycles. Limit cycles comprise state variables (which smoothly affect each other's rates of change and characterize the state of the oscillation) and parameters (which are not part of the oscillating system but influence the way the variables change and hence control the dynamics of the oscillation). Oscillations do not depend on initial conditions, and when perturbed the system returns to its original amplitude and frequency. The trajectory of this return depends on the degree of displacement. In short, the system displays stability. Limit cycles also possess unique states to which they can be driven, and from which normal rhythmicity does not develop: the so-called singularity (82, 127).

Negative feedback of linear systems can also result in oscillations (117). A theoretical feedback model, with no specific biophysical-biochemical processes but with a built-in delay block to produce 24-hr periodicity, was proposed (52) that accurately predicted results for the *Kalanchoe* petal-movement rhythm, including phase shifting. A feedback model was also proposed for the rhythm of leaf movement in *Trifolium* (99). Such systems can also display limit-cycle behavior.

BIOCHEMICAL MODELS   It seems likely that all aspects of cell chemistry are susceptible to circadian rhythmicity without necessarily being part of the oscillator mechanism. Thus although rhythmicity might be abolished altogether by an effect on the coupling between oscillator and rhythm elements, or on the elements of the rhythm per se (Figure 1), to demonstrate that a process is part of the basic oscillatory mechanism, alterations in it must affect the period or phase of an observed rhythm.

At the biochemical level, several models have been advanced (30, 31, 40). These fall into several, by no means mutually exclusive, classes. They include simple molecular models, involving changes in the properties of molecules themselves (e.g. phosphorylation); biochemical feedback, based on a variety of elements (e.g. the glycolytic oscillator, cyclic AMP, mitochondrial Ca^{2+} cycle); the chronon model, involving sequential transcription of polycistronic DNA; and membrane models. The membrane models all incorporate some kind of feedback (31); some involve active transport of ions across a membrane until a threshold is reached, after which leakage occurs until the concentration of the ion is low enough for active transport to resume. Others incorporate protein synthesis; membrane proteins are synthesized that allow transport of monovalent ions, specifically K^+ in the model of Burgoyne (16); when the concentration reaches a threshold, synthesis or transport of the proteins stops, and only begins again when the ion concentration falls.

Application of inhibitors such as cycloheximide indicates that protein synthesis on 80S ribosomes is important for the function of some circadian oscillators. In *Gonyaulax*, pulsed applications of cycloheximide caused phase advances and delays in the bioluminescence rhythm very similar to those caused by light (29). A mathematical model that involved three components—an enzyme, its product, and a translational inhibitor, the activity of which depended on the product and which in turn controlled the rate of enzyme synthesis—showed limit-cycle behavior with the appropriate state variables (28). Further, simulated phase-response curves gave results similar to experimentally observed results, suggesting that agents such as light, temperature, and inhibitors of protein synthesis might be acting on a common target, possibly protein synthesis. 80S ribosomal synthesis may be a state variable; it is also possible that protein synthesis affects a component of the oscillator but is not itself a component.

Whether the mechanism of the oscillator involves membrane transport or protein synthesis or both, some variable must be affected by light, giving rise to effects on phase.

Single or Multiple Oscillators?

Circadian rhythms, then, are driven by an underlying oscillator, which may comprise a number of mechanisms. A single common molecular basis has sometimes been assumed, but this assumption is only based on the formal and functional similarities among observable behaviors. In organisms with neural connections there is clear evidence for a number of centers, a hierarchy of oscillators; some—e.g. the suprachiasmatic nuclei in mammals and the pineal gland in birds—provide especially strong synchronization. In plants, which lack neural connections, synchronization between organs seems unlikely. In several systems studied, rhythms persist in vitro in small pieces of tissue (35, 94, 125), indicating that all the necessary components for rhythmicity are present in individual organs and (in some cases) individual cells. Within individual cells there might also be more than one light-sensitive oscillator. In some unicellular systems (e.g *Gonyaulax*), several rhythms are driven by one oscillator. However, in *Chlamydomonas, Euglena,* and *Chlorella* it was concluded that the rhythms of extracellular pH, autokinesis, and CO_2 turnover were controlled by different oscillators (45). In higher plants, too, there may be common or separate oscillators. In *Pharbitis* it was concluded that separate oscillators control leaf movement and photoperiodic response (6), but leaf movement in soybean (8) and *Coleus* (37) showed very similar phase relations to that of the rhythmic flowering response to night breaks, arguing for a common oscillator. Evidence for a circadian oscillator at one locale (e.g. cellular membrane) does not preclude the existence of a different system, located elsewhere within the cell or in different cells or tissues, that controls other rhythmic phenomena (99).

Coupling to Rhythms

Investigating the coupling of oscillator to rhythmic phenomenon involves following a temporal chain of events to its origin. In animals the synthesis of melatonin, involved in synchronizing the circadian system, is rhythmic and is a result of rhythmic N-acetyltransferase (NAT) activity, particularly in the pineal. The rhythm of NAT activity seems to be regulated by a rhythm in intracellular cAMP (109). The next step, the source of this oscillation, is now being sought.

In *Gonyaulax* the activity of luciferase, responsible for the circadian rhythm of bioluminescence, is regulated in a circadian way. Changes in the amount of luciferin binding protein (LBP) also show a circadian rhythm. Further, it was shown that the level of LBP mRNA did not undergo fluctuation, and that the rhythm in protein level was the consequence of changes in translation (70). This rhythm might be caused by the periodic production of translation factors. Translational control might in fact be a type of basic oscillatory mechanism comprising many different regulatory proteins, each controlling the synthesis of the next, but each also able to direct the translation of nonclock proteins such as LBP. Such a mechanism would of course be sensitive to inhibitors of protein synthesis.

In *Euglena* the rhythms of photosynthesis and cell shape can become uncoupled from the circadian oscillator, so that they no longer display rhythmicity. For example, application of agents that chelated Ca^{2+}, blocked Ca^{2+} channels, or interfered with the inositol trisphosphate and diacylglycerol pathways appeared to uncouple the rhythms from the oscillator (61). This observation would be consistent with an oscillatory mechanism that produced a rhythmic change in cellular Ca^{2+} (36).

Responses to Light

Following the pathway from light to the circadian system—i.e. investigating the interaction between light and oscillator in mechanistic terms—requires the identification of the photoreceptor(s) and the subsequent signal transduction pathways. First, though, the actions of light must be understood.

ENTRAINMENT   Under constant conditions, circadian rhythms free run with a periodicity slightly different from 24 hr. Under daily LD cycles the rhythm becomes entrained, adopting the 24-hr period of the diurnal cycle, so that the rhythmic process has a constant phase relationship with the LD cycle—i.e. a particular phase of the rhythm coincides with a particular phase of the LD cycle.

Entrainment occurs when light phase-shifts the oscillator, causing a change in the phase of the rhythm (87). The extent and direction of the shift depend on the phase at which the light is given. Plotting shift against phase at which light was given during darkness (DD) produces a phase response curve (PRC): Light early in the subjective night causes a phase delay, while light in the late subjective night or early subjective day causes a phase advance (97). During the subjective day there is often a period of no response. Entrainment represents an equilibrium, where the amount of phase shift induced each cycle is equal to the difference between the natural period of 24 hr (T) and the FRP (t). This conclusion is supported by experiments with skeleton photoperiods, where a complete photoperiod is simulated with pulses of light corresponding to the times of dawn and dusk. The phase relationship of the rhythm with the light signals is the same for intact and skeleton photoperiods, entrainment being the net result of the shifts caused by the two pulses (87). Skeletons have successfully simulated intact photoperiods in a number of circadian rhythms—e.g. CO_2 output (39)—but the interpretation of skeleton entrainment in photoperiodic time measurement is doubtful (43), because pulses of light have effects in addition to control of phase (79, 114).

Photoperiods longer than about 12 hr, however, cannot be simulated by skeletons; the rhythm assumes a different phase relationship to that with the complete photoperiod, the shorter interval between the pulses being "read" as the photoperiod. Thus entrainment cannot always be explained by the actions of light at the beginning and end of the photoperiod; an effect of continuous light on the oscillator must be assumed. In some rhythms the transition to darkness also affects phase, and so is presumably involved in entrainment (58, 67, 125). The rhythm is sensitive to darkness during the subjective day, when it is insensitive to light. Rhythms involved in photoperiodic timing are often phased or set by the onset of darkness (123). In other rhythms the onset of darkness seems not to affect phase (19, 33). Thus the precise way in which entrainment is achieved varies between rhythms. A full analysis requires detailed PRCs and dose responses to show how much light is needed to generate a light-on, phase-shift, or light-off response. Generally, the phase of rhythms in plants is shifted only after quite long exposures, indicating a low sensitivity of the oscillator to light. This might be expected because plants are exposed to daylight, and we may assume they do not require a highly sensitive entraining system (49).

CONTINUOUS LIGHT   When organisms are exposed to light for longer than 12 hr, rhythmicity is often abolished; when they are returned to darkness, their rhythms resume at the phase normally occurring at hour 12 of a 12:12 LD cycle, circadian time (CT) 12 (107).

However, many rhythms persist under continuous light (LL) (13, 22), although period length is usually altered. An understanding of the PRC can be useful here; if the delay part of the PRC is greater than the advance part the period will be lengthened, and vice versa (49, 86). Effects of LL of different quality can be understood in the same way: The effect on period length depends on the PRC associated with the particular wavelength of light. In *Albizzia* the PRC with blue was of a shape different from that obtained with red light (96). In *Coleus* the period of the leaf-movement rhythm was shortened in continuous red and lengthened in continuous blue (38) light, whereas in *Gonyaulax* the bioluminescence period was shorter in blue than in red (90) light. Care must therefore be taken when analyzing PRCs obtained in white light, since in such a regime more than one photoreceptor may provide an input to the oscillator.

In cases where rhythmicity appears to have been abolished, three possibilities can be suggested. First, the oscillator may be unaffected, but continuous light causes loss of expression by an effect on the rhythmic process (108). In other cases light changes the dynamics of the oscillator such that it continues to oscillate only within a small sector of the circadian cycle, dramatically reducing the oscillations of the associated rhythm. In several insects (e.g. *Drosophila* and *Sarcophaga*), eclosion occurs at a constant time from L-off when the light period is longer than 12 hr, but residual rhythmicity is detectable. One hypothetical explanation for this observation is that the oscillator continues in LL but migrates to a distinct light-limit cycle, which effectively oscillates around one phase position. Following transfer to darkness the oscillator would move to the dark-limit cycle, taking practically the same time to reach it regardless of its position on the light-limit cycle; periodicity would be barely, if at all, discernable, and the rhythm would appear to resume from a fixed phase point regardless of light-period length (83). A similar suggestion has made for the apparent suspension in LL of the circadian rhythm involved in photoperiodic time measurement (114). Finally, the oscillator may really be stopped in LL. This is conceivable if the oscillator involves ion channels kept open in LL (77, 126).

DIRECT EFFECTS   As well as acting on the oscillator to control phase, light can interact with the observed rhythmic process. Thus light can bring about closure of leaflets of nyctinastic plants (32, 88, 94, 95), inhibition of floral induction in short-day plants (81, 122, 123), enhancement of gene expression (104, 115), and opening of stomata (128). Many of these responses have been shown to be phytochrome mediated, but may only occur at certain phases of the rhythm.

PHYTOCHROME

Responses

Two types of response to light involving phytochrome can be identified. The first is the inductive response to red light (R), given during darkness, which is reversible by far-red light (FR), with respective action maxima of 660 and 730 nm corresponding to the absorption maxima of Pr and Pfr. Pfr is the active element. In some responses the Pfr involved is unstable and is rapidly lost; in others, Pfr is quite stable since even after several hours of darkness FR, which would photoconvert Pfr to Pr, causes a change in response—e.g. stem elongation increase (27) and floral induction inhibition (93). Floral induction can actually be inhibited at certain times by both R (producing Pfr) and FR (removing Pfr) (55, 93), indicating the coexistence of two physiologically active pools of phytochrome, termed labile and stable. Spectrophotometric evidence supports the existence of two pools with different kinetics following exposure to R—one showing rapid loss of Pfr (labile), the other having stable Pfr (91).

The second response type is usually referred to as the high irradiance response (HIR). It requires the presence of Pfr over a prolonged period of light. The most-studied response is the control of stem elongation, which in etiolated tissue shows action maxima in the blue, red, and far-red regions (98). As the tissue de-etiolates, the peak of responsiveness to FR decreases, probably owing to loss of labile phytochrome, while the responsiveness to R remains (3).

Despite a wealth of physiological data and numerous kinetic models, the primary action of phytochrome in these responses is still unknown. Since the successful purification of the native molecule (121), many studies have sought information on the mechanism of action in the molecule itself. These efforts have been thoroughly reviewed elsewhere (20, 34, 51) and are only briefly summarized here.

Molecular Species

Immunochemical techniques have shown that green tissue contains phytochrome immunologically distinct from that in etiolated tissue (1, 103, 113, 116). By convention these two species are referred to as type I and type II. Type I predominates in etiolated tissue; it is synthesized in the dark and undergoes rapid degradation in the light. Its synthesis is also reduced in the light by down-regulation of its mRNA (20), although the extent of down-regulation varies between species. Type II predominates in light-grown tissue; it has a longer half-life than type I, and because it is not subject to down-regulation, it is quite stable in light. Phytochrome cDNAs for phytochrome from a number of species have been produced. In *Avena* at least four phytochrome genes are present (42). In *Arabidopsis* three distinct cDNA sequences have been identified, termed *phy* A, *phy* B and *phy* C (101); *phy* A appears to code for type I phytochrome.

Do the different molecular species correspond to the physiologically labile and stable pools, and do they have the same mechanism of action? Data from phytochrome mutants are consistent with the attractive hypothesis that type I is functionally equivalent to labile, and type II to stable phytochrome (34). Other data, however, indicate that type I phytochrome is not necessarily labile; the cellular environment of the phytochrome may affect its behavior. Membrane-associated phytochrome was found to be less susceptible to spectral degradation in vitro than soluble phytochrome, when incubated in either the Pr (75) or the Pfr form (102). In both experiments the level of type II phytochrome would have been too low to detect spectrally, indicating differential behavior of type I phytochrome. Binding of phytochrome through different domains could also affect its kinetic behavior; binding of phytochrome to monoclonal antibodies specific for the N-terminal domain increased the rate of dark reversion from Pfr of pea phytochrome in vitro (64). Thus while physiological responses may be related to labile and stable phytochrome, care must be taken in equating physiological pools with biochemical species.

What of reactions in continuous light? Conventionally the HIR in etiolated seedlings, with a predominant action maximum in the far red, has been assigned to the action of type I. However, in green tissue type II phytochrome will usually be most abundant. A recent action spectrum for the long-day promotion of flowering in wheat (17) showed action maxima in the red (660 nm) and far red (716 nm), typical of an HIR (3). Further, type I phytochrome was detected immunochemically in plants treated with far red during the day extension; the "action spectrum" for type I phytochrome level corresponded closely with that for flowering in the far red (18). This indicates, first, that type I phytochrome can occur in continuous light in green tissue, and second, that if the action maximum at 660 nm corresponds to the action of type II Pfr, then different molecular species can be active in the same response.

Phytochrome Action

To understand the chain of events leading from photon capture and phototransformation to the biological response, attempts have been made to study what occurs soon after formation of Pfr. Several events have been demonstrated (5, 47, 88) that support a membrane site for primary action of phytochrome (66). However, the primary sequence of the type I monomer indicates it to be a soluble protein. Nevertheless, for a long time there has been evidence that even in dark-grown material about 5% of the phytochrome is recovered in a membrane fraction (92). Alternatively, if phytochrome is not a membrane-bound protein, irradiation with R could lead to association of Pfr with a membrane-bound receptor. Although it was found that in vivo irradiation increased the proportion of phytochrome recovered in a membrane fraction (the pelletable phytochrome), this phenomenon was shown to be an electrostatic attraction; it appears to be part of the process leading to destruction. Another type of in vivo R-induced association has been reported (75, 124) that exhibits R/FR reversibility and reciprocity, the criteria for an induction response. Further, the binding data reported indicated that cytosolic phytochrome could bind to a membrane fraction (specifically to a postulated membrane protein), as a PfrPr heterodimer. This result is important, because purified phytochrome exists in solution as a dimer (50), and a reexamination of experimental work indicates that it also exists as a dimer in vivo (10). Phytochrome dimers must thus be incorporated into models of phytochrome action. Only one model currently does this: Van Der Woude's (119, 120) model incorporates PfrPfr homodimers, PrPr homodimers, PfrPr heterodimers, and a specific receptor X. The model closely simulates fluence-response curves for a number of low and very low fluence responses and predicts action spectra for the HIR with peaks in the red and far-red regions.

The responses to light mediated by phytochrome are thus still consistent with a primary site of action at a membrane (9). Subsequent steps could then lead to the changes in gene expression and development that have been the subject of much recent study (20). Efforts are needed, however, to identify the primary event(s) following phytochrome phototransformation. A putative binding partner, for example, may be present at very low concentrations (121). Possible differences in action of different molecular species of phytochrome may be more significant in inductive responses than in responses to continuous light.

INTERACTION OF PHYTOCHROME WITH CIRCADIAN RHYTHMS

Three types of response involving circadian rhythms and phytochrome can be identified (Figure 2). First, control of phase is achieved by light acting on the oscillator. Second, light can have a direct effect on the rhythmic function, but only at particular phases of the cycle. Third, removing Pfr may have a more general effect that is not phase dependent. Here phytochrome is not acting to transduce a light signal, and strictly speaking is not interacting with the circadian rhythm.

Rhythms Involving Movement of Ions

The most often studied rhythms are those of leaf and leaflet movement, which occur in a number of species. These movements persist in DD, and (in *Phaseolus,* for example) in LL also. The basis for movement is a periodic change in membrane properties, with K^+ transport leading to changes in turgor of cells on the opposite side of the pulvinus (95, 100). Leaf-movement rhythms can be entrained to various cycle lengths (13), and for some, PRCs have been obtained (14, 105). The leaflet movement of *Samanea* was shifted by 5 min R, and was reversible by FR, indicating phytochrome action. *Phaseolus,* however, requires at least 2 hr of light to phase-shift the leaf-movement rhythm. Here a similar PRC was obtained by treatment with a high concentration of K^+ (15). Stomatal opening is another process affected by changes in turgor, involving movement of ions including K^+ (128). The circadian rhythm of stomatal opening in *Xanthium pennsylvanicum* also re-

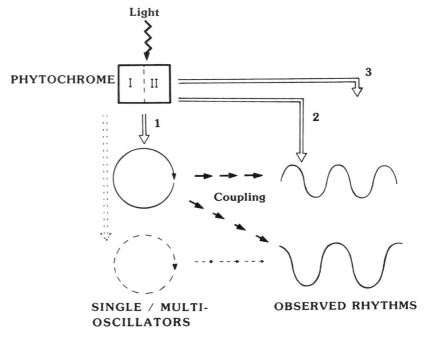

Figure 2   Model for the interactions of phytochrome with circadian rhythms: 1. response to light acting on one or more oscillators to control phase; 2. light acting directly on a component of the driven rhythm; 3. a requirement for Pfr.

quired prolonged exposures to FR or repeated exposure to R to elicit a phase shift (65). Young leaves were much more sensitive than old ones.

In *Lemna gibba* G3 a rhythm of K^+ uptake and release from solution can be measured directly (59). A complete PRC was obtained for the *Lemna* uptake rhythm with 15 min of light (58), and entrainment was demonstrated with pulses of light as brief as 1 min every 24 hr. R was the most effective wavelength to bring about entrainment, but the effect could not be reversed with FR as rhythmicity and uptake were abolished after FR. This result indicates that Pfr must be present to maintain the uptake and hence the expression of the rhythm. A similar loss of rhythmicity was seen in the CO_2 exchange rhythm of *Bryophyllum* (39), while in *Phaseolus* plants moved to FR from DD no rhythm of leaf movement was observed (62).

Direct effects of light on these rhythms can be observed. Leaflets can be induced to open by exposure to blue and to close by exposure to red (32) light, depending on the phase of the cycle at which the light is given (95). Agents that interfere with calcium transport through calcium channels have been shown to block red-induced closure in *Albizzia* (72). In *Samanea*, brief exposures to white light during darkness produced an increase in the concentration of inositol phosphates and a decrease in phosphatidyl inositol bisphosphate (71), consistent with stimulation of the phosphatidyl inositol (PI) cycle, known to act as a second messenger in animals (4). However, light was given at a time when blue promotes leaflet opening and red phase-shifts the rhythm, so the results did not indicate whether the response was specific to the actions of blue or red light. Nevertheless, generation of such messengers could, for example, affect the H^+ ATPase involved in ion transport (95), resulting in ion pumping and hence leaflet movement, but only at permitted phases of the circadian cycle. Similar second messengers could act on appropriate components of the oscillator to effect changes in phase.

Metabolic Rhythms

Oscillations in components of CAM metabolism are well documented; rhythmic CO_2 exchange is the overt manifestation. In *Bryophyllum* kept in DD there is periodic transport of malate into the vacuole (126), but there is also a persistent rhythm in the activity of phosphoenol pyruvate carboxylase due to phosphorylation and dephosphorylation (76). Rhythms in nitrate reductase activity, phased from light-on, have been recorded in a number of species (19, 60, 118), but whether they were due to activation (e.g. by uptake of nitrate) or to de novo synthesis was not determined. Recent studies have demonstrated oscillations in nitrate reductase mRNA, protein, and enzyme activity (26).

A number of studies have shown oscillations in the small subunit of ribulose-1,5-bisphosphate carboxylase (SSU) and the structural chlorophyll *a*/*b*-binding protein (Cab). An initial report by Kloppstech (56) showed that in peas the amount of translatable mRNA for these proteins increased at the end of the dark period under LD cycles. Under LL the same pattern persisted, showing the rhythm to be circadian and not simply imposed by the diurnal LD cycle. Similar results, using probe hybridization, have been reported for *Cab* in a number of other species (73, 85, 112). In tomato, of ten types of gene analyzed, only *Cab* showed oscillations (69). There is evidence that these rhythms are rhythms in transcription: Chimaeric constructs with the upstream coding region of *Cab* fused to the chloramphenicol acetyltransferase (CAT) coding region still showed circadian expression in transformed tobacco, while constructs with a −124 5'-deletion mutant of *Cab-1* fused to the CaMV 35S enhancer resulted in a constant level of transcript (53).

The *Cab* rhythm in tomato was entrained to various LD cycles, even to 6:6 (84). When plants grown in LL from sowing were transferred to different durations of darkness and then back to LL, a 24-hr rhythm was observed, the phase of which was constant from the time of light-on (89), suggesting that the light-on signal is the important one. There is direct evidence for the involvement of phytochrome in the control of phase of the *Cab* rhythm in etiolated *Phaseolus vulgaris* (Figure 2). The rhythm was initiated by 2 min R; a second R pulse 24 hr later had no effect, but a second R pulse 36 hr after the first completely reset the rhythm (112). This response was reversed when the R was followed immediately by 5 min FR.

The expression of *Cab* is also promoted directly by phytochrome at the level of transcription, but the way phytochrome acts is more complex than simple induction. A R/FR induction response can be seen in etiolated tissue (104), but the presence of Pfr seems to be required for sustained expression (53).

It may seem logical that *Cab* should be expressed in anticipation of dawn, but since the concentration of Cab protein remained constant throughout the diurnal cycle (85), the rhythm may not be of vital importance under normal conditions. However, it may prove to be a useful model for investigating underlying mechanisms. If the *cis*- and *trans*-acting factors regulating the circadian expression can be identified, the transduction chain might be followed back to the primary output from the oscillator, while the sensitivity to Pfr of phase control might enable us to follow the transduction chain from photoreceptor to oscillator (Figure 1).

Photoperiodism

The photoperiodic induction of flowering, being the most studied type of response, is discussed here. To measure the duration of light or dark requires a time-measuring system and a photoreceptor. In the case of short-day plants (SDP), Bunning's hypothesis (11) that time measurement involves a circadian rhythm, a certain portion of whose cycle must pass in darkness for induction to occur, is now generally accepted (44). The rhythm (the biochemical nature of this rhythm has not been elucidated) is revealed by changes in the flowering response (*a*) to increasing durations of darkness (21, 74), or (*b*) to light interruptions of an otherwise inductive dark period (21, 111). To time the duration of darkness the rhythm must be phased from the end of the light period. Although a number of models have been proposed (41), the simplest explanation is that after a certain duration of LL the rhythm becomes "suspended," only being released at the onset of darkness. Thus the rhythm is effectively reset each cycle; phase is constant from the light-on signal and beyond a certain duration of LL is constant from the light-off signal (63, 80, 123). Some evidence suggests that the oscillator continues during LL (106); a light-limit cycle has been proposed to explain how this continuation could still lead to timing from a light-off signal (114).

Evidence that phytochrome is the photoreceptor in the control of phase (Figure 2) is still not definitive. In terms of LL, R was most effective in preventing the onset of dark timing (110). In *Pharbitis* seedlings, phase shifts have been obtained with brief exposures to R (63, 114), the sensitivity to phase shifting being different at different phases, but the response has been reversed by FR (P. J. Lumsden, unpublished observations). However, sensitivity to light clearly differs between species. In *Chenopodium* seedlings 2 hr of R had no effect on the flowering rhythm, 6 hr caused phase shifts, and more than 12 hr was needed for phase to be constant from L-off (54). In this case the shifts caused by 6-hr photoperiods were shown to entrain the rhythm such that the inductive phase would occur during darkness.

The observed circadian rhythm of flowering response to night-break light involves a direct action of phytochrome (Figure 2, Response 2). At about 8–10 hr after the onset of darkness the rhythm passes through a phase where light causes maximum inhibition of flowering. This response is clearly mediated by phytochrome, being R/FR reversible in a number of species (81, 122), although in others longer exposures are needed to inhibit flowering. This dual action of light constitutes an external coincidence mechanism for photoperiodism.

In addition, the third action of phytochrome can be identified, in this case a requirement for Pfr during darkness following short or low-light photoperiods. Removal of Pfr by FR prevents induction (93). Although this action of phytochrome is also located in the leaf (57), nothing is known of the processes involved. This phytochrome is stable and is likely to be of type II.

Since the rhythm is timed from L-off, there must be a mechanism for detecting the L-to-D transition. One possibility is that rapid reversion of Pfr from unstable PfrPr heterodimers would occur, leading to release of the rhythm. The stable Pfr would be in the form of PfrPfr homodimers (B. Thomas, submitted). This would require the cell to recognize heterodimers and homodimers.

The interaction of phytochrome and rhythms in long-day plants (LDP) is less well understood. To obtain flowering, long daily exposures to light are required, and light with added FR is most effective. Clearly a circadian rhythm is involved, since there is a rhythmic change in flowering response to the time of addition of FR during the cycle (24). The phase of the rhythm is also shifted by FR (25), but precisely how entrainment occurs under normal LD cycles is unclear.

SUMMARY

Light acting through phytochrome can interact with some component of the oscillator, of which there are probably a number of different types, to bring about entrainment. In some cases phase can be changed through phytochrome acting in an inductive manner, while in others prolonged exposure to light is required. Entrainment is thus a complex phenomenon, requiring details of

PRCs and dose responses for a full formal analysis. Although rapid phase responses in some tissues may not reflect entrainment under normal conditions, such systems offer a possible approach to the mechanisms of both the oscillator and photoreceptor; but care is needed to distinguish between components of the oscillatory mechanism, the variables, and parameters that affect the variables. Light acting through phytochrome can also have a direct effect on certain rhythms, but only at phases specified by the nature of the rhythm.

Changes in sensitivity of the oscillator to phase shifting and of rhythms to direct effects may be due to differences in their accessibility to light at different phases. They may also be affected by changes in sensitivity of the photoreceptor (23), due, for example, to availability of its receptor. This could be the result of another output from the oscillator; in etiolated tissues, rhythms in phytochrome pelletability (46) and photoconversion cross-section (55) have been reported, which could reflect rhythmic changes in a membrane locale, but do not imply that phytochrome itself is part of the oscillator (41).

There is now much information on the biochemical and molecular properties of the phytochrome molecule. Identification, using techniques such as anti-idiotype antibodies, of the site(s) of action of phytochrome, and of subsequent steps in the transduction pathway(s) may yield information about the oscillator(s), coupling, etc, and provide the approach best able to uncover the secrets of circadian rhythms.

Literature Cited

1. Abe, H., Yamamoto, K. T., Nagatani, A., Furuya, M. 1985. Characterisation of green tissue-specific phytochrome isolated immunochemically from pea seedlings. *Plant Cell Physiol.* 26:1387–99

2. Aschoff, J., ed. 1965. *Circadian Clocks*. Amsterdam: North-Holland. 479 pp.

3. Beggs, C. J., Holmes, M. G., Jabben, M., Schafer, E. 1980. Action spectra for the inhibition of hypocotyl growth by continuous irradiation in light and dark grown *Sinapis alba* L. seedlings. *Plant Physiol.* 66:615–18

4. Berridge, M. J. 1987. Inositol trisphosphate and diacylglycerol: two interacting second messengers. *Annu. Rev. Biochem.* 56:159–93

5. Blakeley, S. D., Thomas, B., Hall, J. L., Vince-Prue, D. 1983. Regulation of swelling of etiolated-wheat-leaf-protoplasts by phytochrome and gibberellic acid. *Planta* 158:416–21

6. Bollig, I. 1977. Different circadian rhythms regulate photoperiodic flowering response and leaf movement in *Pharbitis nil* (L.) Choisy. *Planta* 135:137–42

7. Brady, E., ed. 1982. *Biological Timekeeping*. Cambridge: Cambridge Univ. Press. 197 pp.

8. Brest, D. E., Hoshizaki, T., Hamner, K. C. 1971. Rhythmic leaf movements in Biloxi soybeans and their relation to flowering. *Plant Physiol.* 47:676–81

9. Briggs, W. R., Rice, H. V. 1972. Phytochrome: chemical and physical properties and mechanism of action. *Annu. Rev. Plant Physiol.* 23:293–334

10. Brockmann, J., Rieble, S., Kazarinova-Fukshansky, N., Seyfried, M., Schäfer, E. 1987. Phytochrome behaves as a dimer in vivo. *Plant Cell Environ.* 10:105–11

11. Bunning, E. 1936. Die endogene Tagesrhythmik als Grundlage der photo-

periodischen Reaktion. *Ber. Dtsch. Bot. Ges.* 54:590–607

12. Bunning, E. 1956. Endogenous rhythms in plants. *Annu. Rev. Plant Physiol.* 7:71–90

13. Bunning, E. 1973. *The Physiological Clock.* London; English Univ. Press. 258 pp. 3rd ed.

14. Bunning, E., Moser, I. 1966. Response Curven bei der circadianen Rhythmik von *Phaseolus. Planta* 69:101–10

15. Bunning, E., Moser, I. 1973. Light-induced phase shifts of circadian leaf movements of *Phaseolus:* comparison with the effects of potassium and ethyl alcohol. *Proc. Natl. Acad. Sci. USA* 70:3387–89

16. Burgoyne, R. D. 1978. A model for the molecular basis of circadian rhythms involving monovalent ion-mediated translational control. *FEBS Lett.* 94:17–19

17. Carr-Smith, H. D., Thomas, B., Johnson, C. B. 1989. An action spectrum for the effects of continuous light on flowering in wheat. *Planta* 179:428–32

18. Carr-Smith, H. D., Thomas, B., Johnson, C. B., Plumpton, C., Butcher, G. 1991. The kinetics of type I phytochrome in green light grown wheat (*Triticum aestivum* L.). *Planta.* In press

19. Cohen, A. S., Cumming, B. G. 1974. Endogenous rhythmic activity of nitrate reductase in a selection of *Chenopodium rubrum. Can. J. Bot.* 52:2351–60

20. Colbert, J. T. 1988. Molecular biology of phytochrome. *Plant Cell Environ.* 11:305–18

21. Cumming, B. G., Hendricks, S. B., Borthwick, B. A. 1965. Rhythmic flowering responses and phytochrome changes in a selection of *Chenopodium rubrum. Can. J. Bot.* 43:825–53

22. Cumming, B. G., Wagner, E. 1968. Rhythmic processes in plants. *Annu. Rev. Plant Physiol.* 19:381–416

23. Daan, S. 1982. Circadian rhythms in animals and plants. See Ref. 7, pp. 11–32

24. Deitzer, G. F., Hayes, R., Jabben, M. 1979. Kinetics and time dependence of the effect of far red light on the photoperiodic induction of flowering in Wintex barley. *Plant Physiol.* 64:1015–21

25. Deitzer, G. F., Hayes, R., Jabben, M. 1982. Phase shift in circadian rhythm of floral promotion by far-red energy in *Hordem vulgare* L. *Plant Physiol.* 69:597–601

26. Deng, M.-D., Moreaux, T., Leydecker, M.-T., Caboche, M. 1990. Nitrate-reductase expression is under the control of a circadian rhythm and is light inducible in *Nicotiana tabacum* leaves. *Planta* 180:257–61

27. Downs, R. J., Hendricks, S. B., Borthwick, H. A. 1957. Photoreversible control of elongation of pinto beans and other plants under normal conditions of growth. *Bot. Gaz.* 118:199–208

28. Drescher, K., Cornelius, G., Rensing, L. 1982. Phase response curves obtained by perturbing different variables of a 24 hr model oscillator based on translational control. *J. Theor. Biol.* 94:345–53

29. Dunlap, J. C., Taylor, W., Hastings, J. W. 1980. The effects of protein synthesis inhibitors on the *Gonyaulax* clock. I. Phase-shifting effects of cycloheximide. *J. Comp. Physiol.* 138:1–8

30. Edmunds, L. N. Jr. 1988. *Cellular and Molecular Bases of Biological Clocks.* New York: Springer-Verlag. 497 pp.

31. Engelmann, W., Schrempf, M. 1981. Membrane models for circadian rhythms. *Photochem. Photobiol. Rev.* 5:49–86

32. Fondeville, J. C., Schneider, M. J., Borthwick, H. A., Hendricks, S. B. 1967. Photocontrol of *Mimosa pudica* L. leaf movement. *Planta* 75:228–38

33. Frosch, S., Wagner, E. 1973. Endogenous rhythmicity and energy transduction. II. Phytochrome action and the conditioning of rhythmicity of adenylate kinase, NAD- and NADP-linked glyceraldehyde-3-phosphate dehydrogenase in *Chenopodium rubrum* by temperature and light intensity cycles during germination. *Can. J. Bot.* 51:1521–28

34. Furuya, M. 1989. Molecular properties and biogenesis of phytochrome I and II. *Adv. Biophys.* 25:133–67

35. Gorton, H. L., Williams, W. E., Binns, M. E., Gemmell, C. N., Leheny, E. A., et al. 1989. Circadian stomatal rhythms in epidermal peels from *Vicia faba. Plant Physiol.* 90:1329–34

36. Goto, K., Laval-Martin, D., Edmunds, L. N. Jr. 1985. Biochemical modeling of an autonomously oscillatory circadian clock in *Euglena. Science* 228:1284–88

37. Halaban, R. 1968. The flowering response of *Coleus* in relation to photoperiod and the circadian rhythm of leaf movement. *Plant Physiol.* 43:1894–98

38. Halaban, R. 1969. Effects of light quality on the circadian rhythm of leaf movement of a short-day plant. *Plant Physiol.* 44:973–77

39. Harris, P. J. C., Wilkins, M. B. 1978. Evidence of phytochrome involvement in the entrainment of the circadian

rhythm of carbon dioxide metabolism in *Bryophyllum*. *Planta* 138:271–78

40. Hastings, J. W., Schweiger, H.-G., eds. 1976. *The Molecular Basis of Circadian Rhythms*. Berlin: Dahlem Konferenzen

41. Heide, O. M. 1977. Photoperiodism in higher plants: an interaction of phytochrome and circadian rhythms. *Physiol. Plant.* 39:25–32

42. Hershey, H. P., Barker, R. F., Idler, K. B., Lissemore, J. L., Quail, P. H. 1985. Analysis of cloned cDNA and genomic sequences for phytochrome: Complete amino acid sequences for two gene products expressed in etiolated *Avena*. *Nucleic Acids Res.* 13:8543–59

43. Hillman, W. S. 1964. Endogenous circadian rhythms and the response of *Lemna perpusilla* to skeleton photoperiods. *Am. Nat.* 98:323–28

44. Hillman, W. S. 1976. Biological rhythms and physiological timing. *Annu. Rev. Plant Physiol.* 27:159–79

45. Hoffmans-Hohn, M., Martin, W., Brinkmann, K. 1984. Multiple periodicities in the circadian system of unicellular algae. *Z. Naturforsch. Teil C* 39:791–800

46. Jabben, M., Schafer, E. 1976. Rhythmic oscillations of phytochrome and its pelletability in *Cucurbita pepo* L. *Nature* 259:114–15

47. Jaffe, M. J. 1968. Phytochrome-mediated bioelectric potentials in mung bean seedlings. *Science* 162:1016–17

48. Johnson, C. H., Hastings, J. W. 1986. The elusive mechanism of the circadian clock. *Am. Sci.* 74:29–36

49. Johnson, C. H., Hastings, J. W. 1989. Circadian phototransduction: Phase resetting and frequency of *Gonyaulax* cells in red light. *J. Biol. Rhythms* 4:417–37

50. Jones, A. M., Quail, P. H. 1986. Quaternary structure of 124-kilodalton phytochrome from *Avena sativa* L. *Biochemistry* 25:2987–95

51. Jordan, B. R., Partis, M. D., Thomas, B. 1986. The biology and molecular biology of phytochrome. *Oxf. Surv. Plant Mol. Biol.* 3:315–62

52. Karlsson, H. G., Johnsson, A. 1972. A feedback model for biological rhythms. II. Comparisons with experimental results, especially on the petal rhythm of *Kalanchoe*. *J. Theor. Biol.* 36:175–94

53. Kay, S. A., Nagatani, A., Keith, B., Furuya, M., Chua, N-H. 1989. Rice phytochrome is biologically active in transgenic tobacco. *The Plant Cell* 1:775–82

54. King, R. W., Cumming, B. G. 1972. Rhythms as photoperiodic timers in the control of flowering in *Chenopodium rubrum* L. *Planta* 103:281–301

55. King, R. W., Schafer, E., Thomas, B., Vince-Prue, D. 1982. Photoperiodism and rhythmic response to light. *Plant Cell Environ.* 5:395–404

56. Kloppstech, K. 1985. Diurnal and circadian rhythmicity in the expression of light-induced plant nuclear messenger RNAs. *Planta* 165:502–6

57. Knapp, P. H., Sawhney, S., Grimmet, M., Vince-Prue, D. 1986. Site of perception of the far-red inhibition of flowering in *Pharbitis nil* Choisy. *Plant Cell Physiol.* 27:1147–52

58. Kondo, T. 1982. Phase control of the potassium uptake rhythm by the light signals in *Lemna gibba* G3. *Z. Pflanzenphysiol.* 107:395–407

59. Kondo, T., Tsudzuki, T. 1980. Participation of a membrane system in the potassium uptake rhythm in a duckweed, *Lemna gibba* G3. *Plant Cell Physiol.* 21:627–35

60. Lillo, C. 1984. Circadian rhythmicity of nitrate reductase activity in barley leave. *Physiol. Plant.* 61:219–23

61. Lonergan, T. A. 1990. Steps linking the photosynthetic light reactions to the biological clock require calcium. *Plant Physiol.* 93:110–15

62. Lorcher, L. 1958. Die Wirkung verschiedener Lichtqualitäten auf die endogene Tagesrhythmik von *Phaseolus*. *Z. Bot.* 46:209–41

63. Lumsden, P. J., Furuya, M. 1986. Evidence for two actions of light in the photoperiodic induction of flowering in *Pharbitis nil. Plant Cell Physiol.* 27:1541–51

64. Lumsden, P. J., Nagatani, A., Yamamoto, K. T., Furuya, M. 1985. Effect of monoclonal antibodies on in vitro P_{FR} dark reversion of pea phytochrome. *Plant Cell Physiol.* 26:1313–22

65. Mansfield, T. A. 1965. The low intensity reactions of stomata: effects of red light on rhythmic stomatal behaviour in *Xanthium pennsylvanicum*. *Proc. R. Soc. London Ser. B* 162:567–74

66. Marme, D. 1977. Phytochrome: membranes as possible sites of primary action. *Annu. Rev. Plant Physiol.* 28:173–98

67. Martin, E. S., Meidner, H. 1972. The phase-response of the dark stomatal rhythm in *Tradescantia virginia* to light and dark treatments. *New Phytol.* 71:1045–54

68. McMurray, L., Hastings, J. W. 1972.

No desynchronisation among four circadian rhythms in the unicellular alga *Goyaulax polyhedra. Science* 175:1137–39

69. Meyer, H., Thienel, U., Piechulla, B. 1989. Molecular characterisation of the diurnal/circadian expression of the chlorophyll *a/b* binding proteins in leaves of tomato and other dicotyledenous and monocotyledenous plant species. *Planta* 180:5–15

70. Morse, D., Milos, P. M., Roux, E., Hastings, J. W. 1989. Circadian regulation of luminescence in *Gonyaulax* involves translational control. *Proc. Natl. Acad. Sci. USA* 86:172–76

71. Morse, M. J., Crain, R. C., Satter, R. L. 1988. Light stimulated inositol phospholipid turnover in *Samanea saman. Proc. Natl. Acad. Sci. USA* 84:7075–78

72. Moysset, L., Simon, E. 1989. Role of calcium in phytochrome-controlled nyctinastic movements of *Albizzia lophantha* leaflets. *Plant Physiol.* 90:1108–14

73. Nagy, F., Kay, S., A., Chua, N-H. 1988. A circadian clock regulates transcription of the wheat *Cab-1* gene. *Genes Dev.* 2:376–82

74. Nanda, K. K., Hamner, K. C. 1962. Investigations on the effect of 'light break' on the nature of the endogenous rhythm in the flowering response of Biloxi soybean (*Glycine max* L. Merr.). *Planta* 58:164–74

75. Napier, R. M., Smith, H. 1987. Photoreversible association of phytochrome with membranes. II. Reciprocity tests and a model for the binding reaction. *Plant Cell Environ.* 10:391–96

76. Nimmo, G. A., Nimmo, H. G., Fewson, C. A., Wilkins, M. B. 1984. Diurnal changes in the properties of phosphoenol pyruvate carboxylase in *Bryophyllum* leaves: a possible covalent modification. *FEBS Lett.* 178:199–203

77. Ninnemann, H. 1979. Photoreceptors for circadian rhythms. *Photochem. Photobiol. Rev.* 4:207–66

78. Njus, D., Sulzman, F. M., Hastings, J. W. 1974. Membrane model for the circadian clock. *Nature* 248:116–20

79. Oota, Y. 1984. Physiological function of night interruption of *Lemna paucicostata* 6746: action of light as a phaser on the photoperiodic clock. *Plant Cell Physiol.* 25:323–31

80. Papenfuss, H. D., Salisbury, F. B. 1967. Aspects of clock resetting of flowering of *Xanthium. Plant Physiol.* 42:1562–68

81. Parker, M. W., Hendricks, S. B., Borthwick, H. A., Scully, N. J. 1946. Action spectrum for the photoperiodic control of floral initiation of short-day plants. *Bot. Gaz.* 108:1–26

82. Pavlidis, T. 1973. *Biological Oscillators: Their Mathematical Analysis.* New York: Academic. 207 pp.

83. Peterson, E. L., Saunders, D. S. 1980. The circadian eclosion rhythm in *Sarcophaga argyrostoma*: a limit cycle representation of the pacemaker. *J. Theor. Biol.* 86:265–77

84. Piechulla, B. 1989. Changes of the diurnal and circadian (endogenous) mRNA oscillations of the chlorophyll *a/b* binding protein in tomato leaves during altered day/night (light/dark) regimes. *Plant Mol. Biol.* 12:317–27

85. Piechulla, B., Gruissem, W. 1987. Diurnal mRNA fluctuations of nuclear and plastid genes in developing tomato fruits. *EMBO J.* 6:3593–99

86. Pittendrigh, C. S. 1979. Some functional aspects of circadian pacemakers. In *Biological Rhythms and Their Central Mechanism,* ed. M. Suda, O. Hayaishi, H. Nakagawa, pp. 3–12. Amsterdam: Elsevier/North-Holland

87. Pittendrigh, C. S., Minis, D. 1964. The entrainment of circadian oscillations by light and their role as photoperiodic clocks. *Am. Nat.* 98:261–94

88. Racusen, R. H., Satter, R. L. 1975. Rhythmic and phytochrome-regulated changes in transmembrane potential in *Samanea* pulvini. *Nature* 255:408–10

89. Riesselmann, S., Piechulla, B. 1990. Effect of dark phases and temperature on the chlorophyll *a/b* binding protein oscillations in tomato seedlings. *Plant Mol. Biol.* 14:605–16

90. Roenneberg, T., Hastings, J. W. 1988. Two photoreceptors control the circadian clock of a unicellular alga. *Naturwissenschaften* 75:206–7

91. Rombach, J., Bensink, J., Katsura, N. 1982. Phytochrome in *Pharbitis nil* during and after de-etiolation. *Physiol. Plant.* 56:251–58

92. Rubinstein, B., Drury, K. S., Park, R. B. 1969. Evidence for bound phytochrome in oat seedlings. *Plant Physiol.* 44:105–9

93. Saji, H., Takimoto, A., Furuya, M. 1982. Spectral dependence of night-break effect on photoperiodic floral induction in *Lemna paucicostata* 441. *Plant Cell Physiol.* 26:623–29

94. Satter, R. L., Galston, A. W. 1971. Phytochrome-controlled nyctinasty in

Albizzia julibrissin. III. Interaction between an endogenous rhythm and phytochrome in control of potassium flux and leaflet movement. *Plant Physiol.* 48:740–46

95. Satter, R. L., Galson, A. W. 1981. Mechanisms of control of leaf movements. *Annu. Rev. Plant Physiol.* 32: 83–110

96. Satter, R. L., Guggino, S. E., Lonergan, T. A., Galston, A. W. 1981. The effects of blue and far red light on rhythmic leaflet movements in *Samanea* and *Albizzia.* *Plant Physiol.* 67:965–68

97. Saunders, D. S. 1977. *An Introduction to Biological Rhythms.* Glasgow: Blackie. 170 pp.

98. Schafer, E. 1976. The high irradiance reaction. In *Light and Plant Development,* ed. H. Smith, pp. 44–59. London: Butterworths

99. Scott, B. I. H., Gulline, H. F. 1972. Natural and forced oscillations in the leaf of *Trifolium repens.* *Aust. J. Biol. Sci.* 25:61–76

100. Scott, B. I. H., Gulline, H. F. 1975. Membrane changes in a circadian system. *Nature* 254:69–70

101. Sharrock, R. A., Quail, P. H. 1989. Novel phytochrome sequences in *Arabidopsis thaliana:* structure, evolution and differential expression of a plant regulatory photoreceptor family. *Gene Dev.* 3:1745–57

102. Shimazaki, Y., Furuya, M. 1980. Spectral properties of soluble and pelletable phytochrome from epicotyls of etiolated pea seedlings. *Planta* 149:313–17

103. Shimazaki, Y., Pratt, L. H. 1985. Immunochemical detection with rabbit polyclonal and mouse monoclonal antibodies of different pools of phytochrome from etiolated and green *Avena* shoots. *Planta* 164:333–44

104. Silverthorne, J., Tobin, E. M. 1987. Phytochrome regulation of nuclear gene expression. *BioEssays* 7:18–23

105. Simon, E., Satter, R. L., Galston, A. W. 1976. Circadian rhythmicity in excised *Samanea* pulvini. II. Resetting the clock by phytochrome conversion. *Plant Physiol.* 58:421–25

106. Spector, C., Paraska, J. R. 1973. Rhythmicity of flowering in *Pharbitis nil.* *Physiol. Plant.* 29:402–5

107. Sweeney, B. M. 1969. *Rhythmic Phenomena in Plants.* London: Academic. 147 pp.

108. Sweeney, B. M. 1979. Bright light does not immediately stop the circadian clock of *Gonyaulax.* *Plant Physiol.* 64:341–44

109. Takahashi, J. S., Menaker, M. 1984. Multiple redundant circadian oscillators within the isolated avian pineal gland. *J. Comp. Physiol.* 154:435–40

110. Takimoto, A. 1967. Spectral dependence of different light reactions associated with photoperiodic response in *Pharbitis nil.* *Bot. Mag.* 80:213–22

111. Takimoto, A., Hamner, K. C. 1964. Effect of temperature and preconditioning on photoperiodic response of *Pharbitis nil.* *Plant Physiol.* 40:852–54

112. Tavladoraki, P., Kloppstech, K., Argyroudi, A.-A. 1989. Circadian rhythm in the expression of the mRNA coding for the apoprotein of the light-harvesting complex of photosystem II. *Plant Physiol.* 90:665–72

113. Thomas, B., Crook, N. E., Penn, S. E. 1984. An enzyme-linked immunosorbent assay for phytochrome. *Physiol. Plant.* 60:409–15

114. Thomas, B., Lumsden, P. J. 1984. Photoreceptor action and photoperiodic induction in *Pharbitis nil.* In *Light and the Flowering Process,* ed. D. Vince-Prue, B. Thomas, K. E. Cockshull, pp. 107–21. London: Academic. 301 pp.

115. Tobin, E., Silverthorne, J. 1985. Light regulation of gene expression in higher plants. *Annu. Rev. Plant Physiol.* 36: 569–93

116. Tokuhisa, J. G., Quail, P. H. 1985. Phytochrome in green tissue: spectral and immunochemical evidence for two distinct molecular species of phytochrome in light-grown *Avena sative* L. *Planta* 164:521–28

117. Tyson, J. J., Alivisatos, S. G. A., Grun, F., Pavlidis, T., Richter, O., et al. 1976. See Ref. 40, pp. 85–108

118. Upcroft, J. A., Done, J. 1976. Circadian rhythm of nitrate reductase (NADH) activity in wheat seedlings grown in continuous light. *Aust. J. Plant Physiol.* 3:421–28

119. Van Der Woude, W. J. 1985. A dimeric mechanism for the action of phytochrome: evidence from photothermal interactions in lettuce seed germination. *Photochem. Photobiol.* 42:655–61

120. Van Der Woude, W. J. 1987. Application of the dimeric model of phytochrome action to the high irradiance responses. In *Phytochrome and Photoregulation in Plants,* ed. M. Furuya, pp. 249–58. Tokyo: Academic. 354 pp.

121. Vierstra, R. D., Quail, P. H. 1982. Purification and initial characterisation of 124-kilodalton phytochrome from *Avena.* *Biochemistry* 22:2498–505

122. Vince-Prue, D. 1975. *Photoperiodism in Plants*. London: McGraw-Hill. 444 pp.

123. Vince-Prue, D. 1983. Photomorphogenesis and flowering. In *Encyclopedia of Plant Physiology*, ed. W. Shropshire Jr., H. Mohr, 16B:457–90. Berlin: Springer-Verlag

124. Watson, P. J., Smith, H. 1982. Integral association of phytochrome with a membranous fraction from *Avena* shoots: red/far-red photoreversibility and in vitro characterisation. *Planta* 154:121–27

125. Wilkins, M. B. 1960. The effect of light upon plant rhythms. *Coldspring Harbor Symp. Quant. Biol.* 25:115–29

126. Wilkins, M. B. 1983. The circadian rhythm of carbon dioxide metabolism in *Bryophyllum:* the mechanism of phase shifts induced by thermal stimuli. *Planta* 157:471–80

127. Winfree, A. T. 1970. Integrated view of resetting a circadian clock. *J. Theor. Biol.* 35:159–89

128. Zeiger, E., Farquhar, G. D., Cowan, I. R. 1987. *Stomatal Function*. Stanford: Stanford Univ. Press. 503 pp.

Annu. Rev. Plant Physiol. Plant Mol. Biol. 1991. 42:373–92

CARBON IN N_2 FIXATION: Limitation or Exquisite Adaptation

C.P. Vance

USDA-ARS Plant Science Research, and the Department of Agronomy and Plant Genetics, University of Minnesota, St. Paul, Minnesota 55108-1013

G.H. Heichel

Department of Agronomy, University of Illinois, Urbana, Illinois 61801-4798

KEY WORDS: plant-microbe interaction, oxygen limitation, organic acids, amino acids, root nodules

CONTENTS

INTRODUCTION

Nodule carbon metabolism and the energy photosynthate requirements for symbiotic N_2 fixation have been fertile research topics for many years because of the continuing debate over whether N_2 or NO_3^- reduction is more bene-

1040-2519/91/0601-0373$02.00

ficial to crop productivity (38, 50, 58, 85, 98, 99). Interest in the topic continues, driven by several questions of scientific and technological importance. Do the carbon requirements for symbiotic N_2 fixation limit productivity of legumes? How does the host plant nodule, functioning for all practical purposes in an O_2-limited environment, provide sufficient carbon to the microsymbiont to fuel nitrogenase (N_2ase)? Will transforming non-N_2-fixing species to fix N_2 impair their productivity? Recent observations that leghemoglobin (Lb) may be present in all plant species (4, 9), that rhizobia can elicit nodule-like structures on rice and other monocots (1, 13), and that single gene changes in the legume host can result in supernodulating plants (12, 47) provide incentives for understanding the role of carbon pathways in N_2 fixation beyond those traditionally advanced for legumes.

The objectives of this review are to examine whether N_2ase activity is carbon limited, to describe plausible mechanisms whereby plant metabolism adapted to an O_2-limited environment to provide carbon for N_2 fixation, and to delineate how O_2 diffusion into the nodule may be regulated. It is beyond the scope of this article to detail each aspect of all events impinging upon nodule carbon and nitrogen metabolism. For a more complete bibliography on nodule ureide and amide metabolism (8, 27, 61, 100, 112, 126), energy and maintenance of O_2 concentration (3, 25, 58, 62, 63, 103, 136, 138), and plant gene expression during nodule development (33, 82, 126, 128), major reviews are available.

PHOTOSYNTHETIC CARBON LIMITATIONS AND N_2 FIXATION

Carbon Costs of N_2 Fixation

In 1804 de Saussure, by means of CO_2 fertilization, obtained evidence that carbon assimilation might limit dry matter accumulation of legumes. This result found practical applications in many crop species after 1916, when Reidel patented and commercialized a burner to CO_2-enrich (fertilize) fields of barley, lupin, potatoes, and beans (see 135). Beans fertilized with CO_2 had 50% more dry weight and five times as many nodules as controls. By 1919, the hypothesis that carbon from photosynthesis was linked to symbiotic dinitrogen (N_2) fixation of legumes was established. In 1933, Wilson et al (135) provided the first definitive results on the effects of CO_2 fertilization on N accumulation by red clover and alfalfa. In general, CO_2 fertilization for 40–50 days promoted dry matter accumulation more than accumulation of total reduced N. Nodule numbers were increased by CO_2 fertilization, but plant N concentration declined.

Four decades later, Hardy and colleagues (36, 38) observed "multifold increases in nitrogen fixation" by means of short- (6 hr) and long-term (67 days) CO$_2$ fertilization of field-grown soybean. The results indicated a stimulation of nodule mass and N accumulation consistent with the results of Wilson. In addition, the results indicated that N$_2$ase-specific activity in the nodule was responsive to short- and long-term CO$_2$ fertilization. The response of N$_2$ase-specific activity to CO$_2$ fertilization indicated that photosynthesis may limit N$_2$ fixation by (a) supply and/or availability of carbon substrates and reductant for N$_2$ase, and/or (b) quantity of N$_2$ase in the nodules. Attempts to verify these results and to extend them to other species have met only limited success. In 11 of 12 experiments, long-term CO$_2$ fertilization (3–15 weeks) increased nodule mass of soybean (30), pea (67, 80, 86), white and red clovers (67, 80), alfalfa (80), black locust (83), black alder (83), and Russian olive (83). The stimulation of N$_2$ase-specific activity by short-term CO$_2$ fertilization (4-6 hr) was verified once in pea, but neither in soybean nor in another pea experiment. The stimulation of N$_2$ase-specific activity by long-term CO$_2$ fertilization was observed in only 1 of 12 experiments.

Additional evidence indicates that N$_2$ase activity of intact legume plants is seldom affected by treatments that perturb the photosynthetic capacity of the shoot. Although abnormally short photoperiods at low photon flux density can reduce nodule activity (80), normal and extended photoperiods do not (130). Within photoperiods normally used for plant growth, diurnal patterns of nodule activity are independent of day-night-day changes in shoot photosynthetic CO$_2$ assimilation (71, 102, 134). Interruption of the photosynthate supply by decapitation or phloem-girdling often dramatically and quickly reduces nodule activity. These treatments seriously perturb metabolite transport, pool sizes, and source-sink relations of the plant and cannot be unambiguously interpreted in terms of physiological mechanism(s) (discussed below).

Overall, the results indicated that legume nodules seldom have excess N$_2$ase capacity that can be demonstrated or activated by increasing the capacity of the plant to conduct photosynthesis. The observation that N$_2$ase-specific activity is rarely enhanced by short-term CO$_2$ fertilization indicates that the supply (rate of production) of photosynthetic products is sufficient to provide energy and reductant for N$_2$ase, and sufficient for the synthesis of N$_2$ase. This position is strengthened by evidence that selecting pea for increased rate of leaf CO$_2$ assimilation did not increase N$_2$ fixation in the field (65). The published evidence convincingly indicates that legumes and other N$_2$-fixing species respond to an increased supply of photosynthetic products over the long term by a coordinated increase in the mass of all plant organs, including nodules. Moreover, selecting alfalfa for greater nodule mass resulted in a proportional increase in the mass of all plant organs (18). Although sinks and sources for C and N respond in concert to increased supply of photosynthetic products, there is evidence that N accumulation lags C accumulation as photosynthesis increases (135). Whether this lag in N accumulation reflects the true requirements of the plant for N assimilates or signals an inability of the nodule to metabolize sufficient C to satisfy unmet needs for reduced N is unknown.

Like all synthetic processes, N_2 fixation imposes an energy burden on the plant (78, 85, 131). When values for over 35 determinations (35, 58, 66, 97, 99, 131) are averaged, this cost approximates 6 milligrams of C per milligram of N reduced. Scientific concern arises regarding this cost because of the perception that N_2 fixation requires photosynthate from the plant that might otherwise be partitioned to yield. Implicit in this perception is the assumption that transforming nonlegumes to fix N_2 will be detrimental to yield. To date, no rigorous experimental or circumstantial evidence supports the hypothesis that N_2 fixation reduces yield potential. Indeed, field experiments on legume communities rarely show an advantage to final yield from supplemental fertilization with various forms of combined N. Spaced plants grown in controlled or field environments may respond differently for reasons that cannot be interpreted unambiguously. Despite the substantial research on the C costs of N_2 fixation (66, 85, 99, 103), relatively few attempts have been made to determine whether the C costs of N_2 fixation are actually related to legume productivity.

In fact, in at least four studies where dry matter production and C costs for N_2 fixation were evaluated, lower C costs of N_2 fixation (more "efficient" C usage) were found for those plants accumulating less dry matter. For 35-day-old peas nodulated by rhizobial strains containing various symbiotic plasmids, Skot et al (107) found lower C costs of N_2 fixation for plants that accumulated 48% less shoot dry matter than for plants with 38% higher C costs. Comparisons of two contrasting alfalfa-*Rhizobium* symbioses grown through two harvest-regrowth cycles showed 17% less total plant dry matter was accumulated with nodules that had 37% lower C costs of N_2 fixation than with higher-yielding plants having "less efficient" nodule C costs (123). Furthermore, supernodulating mutants of pea and soybean that accumulate substantially less dry matter than the wild type have C costs for N_2 fixation that are equal to or less than those of the wild type (94, 101). These results prompt us to question whether N_2 fixation penalizes legume productivity. The limited evidence currently available suggests that C costs of N_2 fixation are not directly related to legume productivity when both characters are measured over the lifetime of the plant.

Carbon Transport, Partitioning, and Availability

Mechanisms regulating photosynthate transport to nodules, the conversion and movement of catabolites of photosynthate from the plant nodule cytosol to bacteroids, and the partitioning of the photosynthate stream to competing organs are not well understood. It seems clear though, that on an organismal basis photosynthate is allocated to the strongest sinks (sink strength = size × activity). Although root nodules are very active sinks, they are a relatively small proportion of the mass of a crop community; for this reason photosynthate distribution to vegetative apices and new leaves has priority over distribution to nodules in most legume species (10, 17, 54, 108).

Current photosynthate is transported to the nodule in the form of sucrose (32, 55, 90, 110, 111) and is stored in the plant fraction of nodules as starch (42, 54, 55, 110, 130). Starch storage in tissues is generally interpreted as excess carbohydrate. Actively N$_2$-fixing root nodules of all legume species contain starch in both infected and uninfected cells (17, 32, 42, 54, 78). Starch accumulates in leaves and roots during the day, although N$_2$ase activity occurs unabated throughout both day and night. Furthermore, shoot carbohydrate pools have been identified as the primary C source involved in maintaining nodule activity during darkness (32, 54, 130). Root starch reserves appear to be unavailable for N$_2$ fixation. Lastly, nodule starch has a slow turnover rate. Nodule carbohydrate reserves are metabolized only when external, particularly shoot, reserves are depleted (42, 54, 130). These observations indicate not only that photosynthate in excess of the need of nodules is being produced and transported to the nodule, but also that photosynthate production is excess to the needs of other growing organs. An alternative, as yet untested hypothesis is that phloem resistance within the roots and root nodules limits transport of photosynthate, resulting in temporary storage elsewhere (103, 119). Why are shoot reserves preferentially mobilized in comparison to nodule reserves? These observations are inconsistent with the popular concept that reserves are mobilized from sites close to, rather than distant from, utilization sites. The partitioning of photosynthate into storage pools and its subsequent mobilization are important to N$_2$ fixation as well as to other growth and maintenance activities. Further experiments on nodulated legumes at the organismal level on pool size, distribution, and residence time of storage products are needed before we will understand these processes in relation to N$_2$ fixation.

Not only do reserve carbohydrate pools reflect excess photosynthate for N$_2$ fixation, but soluble sugar pools in nodules do as well (24, 35, 110, 118). While concentrations of soluble sugars decline in nodules when plants are given treatments that reduce N$_2$ase activity, the kinetics of N$_2$ase inhibition and sugar depletion show that N$_2$ase activity declines more rapidly and to a greater extent than soluble sugar concentrations. Furthermore, there are inconsistencies in root and nodule sugar concentrations and N$_2$ase activity of soybean (92, 110) and pigeon pea (24). A preponderance of available evidence indicates that nodules lack excess N$_2$ fixation capacity that is responsive to photosynthate supply.

If import and storage of C exceed demand, then what limits nodule function? Several authors (21, 25, 58, 63, 89) have suggested that there are constraints upon plant C utilization within the nodule. Because mechanistic interpretations of the availability concept are difficult, formulating and testing hypotheses are hard. There may be limitations on the conversion of sucrose to metabolites (organic acids) used by bacteroids to reduce N_2. Perhaps exchange of appropriate metabolites across cell and organelle (e.g. peribacteroid) membranes limits C and N availability to host plant and bacteroid? Export of reduced N from nodules may be limited, thus allowing end products to accumulate, resulting in reduced efficiency. Whatever mechanisms are involved, it has become apparent over the last few years that O_2 availability plays an important role in nodule C assimilation and N_2 fixation.

NODULE OXYGEN DIFFUSION

While the bacteroid enzyme N_2ase requires substantial energy in the form of ATP for N_2 reduction (16 ATP per N_2 reduced), the enzyme is irreversibly denatured by O_2 and is therefore functional only in low-O_2 environments (3, 25, 50, 58, 62, 131). This paradox is resolved by several exquisite adaptive features during the development of root nodule symbiosis (3, 103, 138): (a) elaboration by the plant of an O_2 diffusion barrier in the nodule cortex which limits influx of O_2 to infected cells; (b) synthesis of the O_2-binding protein leghemoglobin (Lb) within nodules, which facilitates O_2 diffusion to bacteroids and perhaps plant mitochondria within the infected zone; (c) plant redirection of glycolysis to malate with subsequent reductive formation of succinate under microaerobic conditions; (d) bacteroid utilization of C4-dicarboxylic acids rather than mono- and disaccharides to fuel N_2ase; and (e) bacteroid ATP formation coupled to a high-O_2-affinity terminal oxidase.

At external O_2 pressures of 21 kPa (normal atmospheric) the O_2 concentration around nodules is 260 μM while the internal nodule O_2 concentration is quite low (10–40 nM) (2, 44, 63, 79, 121, 138). Modeling of gas and carbohydrate fluxes by Tjepkema & Yocum (120) and Sinclair & Goudriaan (106) led to the proposition that a continuous shell of H_2O 45–80 μm thick would be a sufficient barrier to restrict O_2 diffusion. Dakora & Atkins (19) proposed that, depending upon external pO_2, the diffusion barrier varies from 7.5–72 μm. A 1–3 cell layer zone with few intercellular spaces could comprise such a barrier. Oxygen microelectrode measurements showed the remarkable drop in nodule O_2 concentration occurred over a 1–5 cell layer in the inner cortex within 100–150 μm of the exterior surface of the nodule (121, 138). The layers of cells restricting O_2 diffusion to the nodule interior appear tightly packed with few intercellular spaces through which free O_2 can diffuse

(121, 138). Several excellent reviews have eloquently addressed the metabolic and physiological implications of the O$_2$ diffusion barrier to legume nodule function (25, 62, 63, 136, 138), and thus we mention only the salient features here.

A substantial body of evidence from Witty, Minchin, Sheehy, and colleagues (76, 77, 104, 105, 137), and Layzell and colleagues (44–46, 62–64), demonstrated that nodules display a variable O$_2$ diffusion resistance barrier associated with electron allocation and N$_2$ase function. The group of Witty, Minchin, and Sheehy (76, 104, 105, 137) first clearly demonstrated the implications of a variable O$_2$ diffusion resistance to nodule physiology by showing that a lack of damage to N$_2$ase, following decreases in nodule respiration by exposure to acetylene (C$_2$H$_2$) or Ar:O$_2$, occurred because of increased resistance to O$_2$ diffusion. The magnitude of decline in N$_2$ase activity and respiration resulting from C$_2$H$_2$ could be alleviated by increasing the O$_2$ concentration surrounding nodules. The fact that intact nodule N$_2$ase of pea and clover was tolerant of 80% O$_2$ concentrations clearly implicated a variable diffusion resistance to O$_2$ in prevention of damage to N$_2$ase. Furthermore, these investigators demonstrated that shoot removal and disruption of the root environment led to decreased N$_2$ase activity and increased O$_2$ diffusion resistance. Hartwig et al (37) confirmed the change in nodule O$_2$ diffusion upon defoliation of clover and postulated that increased O$_2$ diffusion resistance was the main factor limiting N$_2$ase after defoliation. Decreases in N$_2$ase activity due to shoot removal and root disturbance could be ameliorated by increasing O$_2$ around nodules. Minchin and colleagues (77, 137) further showed that the standard assay for N$_2$ase activity, C$_2$H$_2$ reduction, resulted in a significant error due to increased O$_2$ diffusion resistance — a finding that led them to question the technique. It is now generally accepted that C$_2$H$_2$ reduction assays on disturbed plants in closed systems do not reflect true N$_2$ase activity and must be interpreted with caution.

Although initial studies could not establish a variable O$_2$ diffusion barrier in soybean nodules, Layzell and colleagues (44–46, 62–64) conclusively demonstrated the phenomenon. They extended the concept by showing a rapid readjustment (within minutes) of nodule internal O$_2$ concentration to increased external pO$_2$. Furthermore, sequential ramping of pO$_2$ in 10% increments resulted in a stimulation of N$_2$ase activity, suggesting nodules are O$_2$ limited under ambient conditions. More recently they showed that NO$_3^-$ treatments, stem-girdling, shading, and disturbance result in both decreased internal O$_2$ concentrations in nodules and reduced N$_2$ase activity. Since decreased N$_2$ase activity could be relieved by increasing external pO$_2$, they concluded that N$_2$ase was O$_2$ limited by the above treatments. Weisz & Sinclair (133) also documented and confirmed the presence of an O$_2$ diffusion barrier in soybeans.

What biochemical and/or morphological factors might regulate a rapid-response, variable-O_2-diffusion-resistance barrier? Currently two working hypotheses are being tested (25, 48, 63, 136). The first postulates that changes in an O_2-diffusion-resistance barrier could be a function of altering the length of the water-filled diffusion path by changes in osmotic potential. Water movement into and out of intercellular spaces in response to osmoticum would increase and decrease the water-filled path length. A corollary to this would be that nodule O_2 diffusion may be regulated similar to the way gas diffusion through stomata is. Cells involved in the diffusion barrier could swell and shrink owing to fluxes of K^+, H^+, Cl^-, and malate into and out of the vacuole (87). Since the O_2 barrier is thought to occur over a few cell layers in the cortex, methods will have to be devised to analyze this zone biochemically and structurally. Such techniques have been developed to study C metabolism of guard cells (84). Likewise, recent attempts to evaluate the biochemistry of cortical cells have met with some success (16, 53). Simple analysis of total nodule fluxes in amino acids, ureides, sugars, and organic acids as the O_2 diffusion barrier changes will be difficult to interpret and may only confuse the issue.

The second and most recent hypothesis is that some form of glycoprotein occludes the intercellular spaces of the diffusion-barrier zone, restricting O_2 diffusion (48, 136). Although it is difficult to visualize how a mechanism involving proteins might have rapid and variable responses, one could speculate on posttranslational modifications of preexisting molecules. Interesting with respect to physical constraints upon nodule O_2 diffusion is the observation that nodules contain large quantities of bound forms of γ-aminobutyric and other amino acids (11, 60). Could these be involved in such phenomena?

Any discussion of O_2 concentrations in root nodules would be incomplete without some mention of Lb and bacteroid terminal oxidase. How can bacteroids generate ATP to fuel N_2 reduction by N_2ase in such a low-O_2 environment? Oxygen transport to bacteroids is facilitated through the O_2 binding protein Lb (2-4, 6). This process occurs at O_2 concentrations sufficient for the bacteroid terminal oxidase to function, but below that which would inactivate N_2ase. The critical aspects required of Lb for this function are its acute O_2 affinity, fast O_2 turnover rates, and high concentration (it is the most abundant soluble protein in nodules). Additionally, the free O_2 concentration in nodules is near 20 nM while Lb bound O_2 may be 10^4 to 10^5 greater, thus prompting expeditious transfer of O_2 to bacteroids by facilitated diffusion (3). A related but unique adaptation of the bacteroid is the induction of a high-O_2-affinity cytochrome system during symbiosis. The terminal oxidase of the N_2-fixing bacteroid, unlike the free living bacterium, functions at O_2 concentrations of 5–10 nM (6). This allows bacteroid ATP production to occur in a low-O_2

environment. Facilitated diffusion via Lb may also play a role in generation of ATP by mitochondria under low-O$_2$ concentrations in the infected cells (115, 117).

O$_2$-LIMITED CARBON METABOLISM

Malate as a Reductive End Product

The standard dogma regarding glycolysis, derived from animal and microbial studies, states that glucose is degraded to pyruvate, which is then imported into mitochondria and converted to acetyl-CoA to be condensed with oxaloacetate to form citrate and initiate the TCA cycle. While this certainly occurs, an alternative branch (Figure 1) in the glycolytic pathway leading to

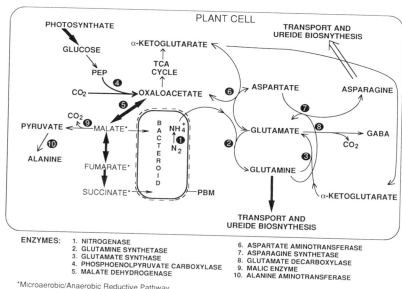

ENZYMES:
1. NITROGENASE
2. GLUTAMINE SYNTHETASE
3. GLUTAMATE SYNTHASE
4. PHOSPHOENOLPYRUVATE CARBOXYLASE
5. MALATE DEHYDROGENASE
6. ASPARTATE AMINOTRANSFERASE
7. ASPARAGINE SYNTHETASE
8. GLUTAMATE DECARBOXYLASE
9. MALIC ENZYME
10. ALANINE AMINOTRANSFERASE

*Microaerobic/Anaerobic Reductive Pathway

Figure 1 Carbon flow and enzymes involved in synthesis of the primary organic and amino acids in legume root nodules. Reductive synthesis of malate, fumarate, and succinate may be a crucial feature of O$_2$-limited C metabolism in nodules by being important in both utilization of photosynthate and provision of C necessary to fuel N$_2$ase. Cytosolic pH could be maintained through generation of CO$_2$ by malic enzyme (ME) and glutamate decarboxylase (GDC), which would provide a sink for H$^+$ released during hydration of the CO$_2$ required for phosphoenolpyruvate carboxylase (PEPC). This pH stat would result in accumulation of alanine and γ-aminobutyric acid (GABA). Carbon flow between organic and amino acids would be modulated by malate dehydrogenase (MDH) and aspartate aminotransferase (AAT).

malate may be of equal or more importance in plants (20, 59). Lance & Rustin (59) presented four compelling arguments to support the thesis that malate is a primary product of glycolysis in plants: "(a) plants have large vacuoles where high concentrations of malate are stored; (b) the enzyme phosphoenolpyruvate carboxylase (PEPC) shows universal distribution in plant cytosol; (c) malate can be oxidized by two routes in plant mitochondria, malate dehydrogenase (MDH) and malic enzyme (ME); and (d) pyruvate is not readily oxidized by plant mitochondria." Not only is malate production important in normal plant metabolism, but production of C_4-dicarboxylic acids may also represent a potential plant adaptation to microaerobic and anaerobic conditions (20, 26, 34, 61). Reductive microaerobic/anaerobic C flow to malate and succinate is well recognized in bivalve molluscs (40, 41) and has been implicated in resistance to anoxia in anaerobic metabolism of *Selenastrum minutum* (129), rice (74, 75), *Echinochloa* (74, 96), *Sesbania* (122) and *Spartina alterniflora* (73). Of particular interest, Vanlerberghe et al (129) noted that the anaerobic shift to reductive pathways for C_4-dicarboxylic acids in *Selenastrum* increased in response to NH_4^+. They proposed that NH_4^+ availability may modulate the reductive pathway during O_2 limitation. LaRue et al (61) and Witty et al (138) suggested that fermentative pathways to succinate may contribute to nodule N and C metabolism. Apparently this not only occurs, but also could be another key adaptation during symbiosis.

Nodule Organic Acids

Analysis of nodule organic acids (Table 1) in both ureide- and amide-forming legume species supports the possible involvement of reductive synthesis of C_4-dicarboxylic acids. Malate, succinate, and fumarate accumulate to μmol

Table 1 Concentration ranges of the most predominant amino acids and organic acids in legume root nodules.

Amino acid	Concentration range		Organic acid	Concentration range	
	nmol·GFW[a]	μg·GFW[b]		μmol·GFW	μg·GFW
aspartate	200–1500	(518)	malate	4.1–5.4	(540)
asparagine	900–29000	(14)	succinate	0.9–3.8	(60)
glutamine	300–850	(49)	fumarate	1.3–7.0	(90)
glutamate	300–4300	(239)	malonate	2.5–13.0	(1500)
alanine	300–3000	(47)	citrate	0.4–0.6	(142)
γ-aminobutyric	400–8000	(71)	α-ketoglutarate	0.5–0.8	—
serine	300–700	(94)			

[a] Data presented as nmol·GFW^{-1} were compiled from 27, 51, 55, 56, 95, 114, and 119.
[b] Data presented as μg·GFW^{-1} are from 111.

quantities in nodules; concentrations of other TCA cycle acids are much less (26, 51, 55, 56, 95, 111, 114). Organic acid labeling profiles when nodules are either incubated with $^{14}CO_2$ or receive [^{14}C]-sugars and [^{14}C]-ASP strengthen the hypothesis. Malate, fumarate, succinate, and the amino acid ASP are rapidly labeled when nodules are exposed to $^{14}CO_2$ (14, 52, 68, 95, 109, 127); this is conclusive evidence for their synthesis through PEPC, MDH, and aspartate aminotransferase (AAT). Furthermore, label derived from CO_2 fixation is rapidly incorporated into bacteroids in the form of malate and succinate (95). When a $^{14}CO_2$ pulse to nodules is followed by a cold chase, up to 70% of the acid stable ^{14}C is respired within 20 min (14, 52). These data indicate that C fixed via PEPC is incorporated into organic acids and is readily available for bacteroid respiration. Ineffective nodules that lack N$_2$ase activity and have low NH$_4^+$ concentrations have much lower rates of CO_2 fixation and lower concentrations of organic acids than do effective symbioses (68, 95). Likewise, treatments that reduce N$_2$ase activity (and NH$_4^+$) usually reduce nodule CO_2 fixation and organic acid synthesis (52, 57, 68, 109). The fixation of CO_2 via PEPC and subsequent incorporation of label into organic acids are intimately related to the production of NH$_4^+$ by effective nodules. When nodules are exposed to [^{14}C]-glucose, organic acids are labeled, but much less rapidly than when exposed to $^{14}CO_2$ (57). Label from photosynthate also appears in the organic acid pool, but less rapidly than label from nodule dark CO_2 fixation (54, 55, 90). Lastly, nodules can quickly metabolize [^{14}C]-ASP to succinate, fumarate, and malate (109), showing the critical roles of PEPC, MDH, and AAT in maintaining the flux of C between the organic and amino acid pools in nodules.

Enzyme analyses show that the plant cytosol of nodules has copious amounts of both activity and enzyme protein for PEPC, MDH, and AAT (16, 27, 28, 53, 116, 132), which are required to funnel C towards malate (34, 72, 126). Activities of these enzymes increase coincident with N$_2$ase and Lb during effective nodule development, and their activities are 4–10 times higher in nodules than in roots (16, 28, 53, 132). Like organic acid concentrations, PEPC, MDH, and AAT activities are reduced in ineffective nodules and nodules where treatments (NO$_3^-$, shoot removal, metabolic inhibitors) result in inhibition of N$_2$ase (28, 100, 127). As mentioned previously, such treatments also affect the O$_2$ diffusion barrier. Currently lacking in the enzymatic scheme is evidence for the reduction of fumarate to succinate. Rawsthorne & LaRue (89) were unsuccessful in the only attempt (to the authors' knowledge) to detect an ATP-generating fumarate reductase. However, negative results are difficult to interpret. Perhaps the high concentrations of malonate in nodules necessitate the involvement of some unusual succinate dehydrogenase in this step.

Several pieces of evidence make it apparent that effective bacteroids utilize plant-produced C_4-dicarboxylic acids to provide energy for N_2ase activity: (a) bacteroids are generally limited in carbohydrate metabolism and lack a full glycolytic pathway (16, 53, 90), and mutants incapable of utilizing glucose and fructose still form effective nodules (31); (b) *Rhizobium* mutants unable to take up malate, succinate, and fumarate form ineffective nodules (5, 29, 43, 93), radiolabeled succinate and malate synthesized in the plant cytosol via PEPC are directly incorporated into bacteroids and respired (52, 95), ME activity in bacteroids allows direct use of malate (7, 15), and bacteroids have a high-affinity uptake system for C_4-dicarboxylic acids but not sugars (5, 22, 29, 31, 70); and (c) peribacteroid membranes (PBM, plant-synthesized membrane surrounding bacteroids) have C_4-dicarboxylic acid transporters capable of mediating a high flux of malate, succinate, and fumarate, but lack an uptake system for sugars and amino acids except ASP (21, 23, 39, 70, 124, 125). The uptake and use by bacteroids of the most prevalent plant organic acids produced under low O_2 conditions reflect the ultimate in adaptation for symbiosis.

A Role for GABA

A novel feature of root nodule N metabolism is the synthesis and accumulation of γ-aminobutyric acid (GABA). Nodules of all legumes evaluated have relatively high concentrations of free GABA (Table 1). Even more interesting is that nodules of many species (11, 60) accumulate bound forms of GABA amounting to as much as 20% of the total N content of the organ. Since GABA is not usually incorporated into proteins, what explains the abundance of this compound in nodules? Kahn et al (50) and McDermott et al (69) suggested that synthesis of GABA provides an alternative route for succinate biosynthesis when and if α-ketoglutarate dehydrogenase is inhibited. A second possible explanation involves GABA in cytoplasmic pH regulation during microaerobic and anaerobic conditions (20, 41, 113). Operation of a reductive pathway to malate requires high rates of dark CO_2 fixation by PEPC. Assuming HCO_3^- serves as a substrate for nodule PEPC, acidosis resulting from excess H^+ produced during hydration of CO_2 could be toxic to host cells (20, 40, 41, 59). Malic enzyme (ME), known to occur in bacteroids and thought to be involved in initial bacterial metabolism of malate (7, 15), could contribute to alleviating this problem through generation of CO_2 (20, 40, 59). However, the detection of ME in the plant fraction of nodules has been inconsistent. Since GABA is derived by α-decarboxylation of glutamate, its accumulation may be a significant sink for H^+ (20, 81, 91). Accumulation of GABA has been implicated in pH balance of *Rosa* suspension-culture cells exposed to UV radiation (81), maize root tip cells exposed to sodium azide (91), and in resistance to anoxia in maize (91), radish (113), rice (75), *Chlorella* (20), and *Echinochloa* (74, 75). An accompanying feature of the GABA pathway is concomitant production of alanine through alanine aminotransferase (20, 40, 41, 113). The accumulation of high concentrations of free alanine in nodules (Table 1) lends added credence to the hypothesis that GABA production may be an adaptation to regulate pH during anaerobiosis.

Unresolved Issues

Pertinent questions, ranging from whole plant to molecular, remain in assembling a comprehensive understanding of C metabolism in N$_2$ fixation. 1. Regulation and functioning of nodule mitochondria remain obscure. Are nodule mitochondria O$_2$ limited? When assayed in 20% O$_2$, nodule mitochondria are fully competent (22, 88, 117). However, when assayed in low O$_2$, oxidation of succinate and malate is limited (89, 115, 117). Inclusion of Lb in O$_2$-limited assays stimulates oxidation of organic acids (115, 117), suggesting Lb may facilitate O$_2$ diffusion to mitochondria as well as to bacteroids. 2. Malonate accumulates to extremely high concentrations in the plant cytosol of nodules (Table 1), yet its functions are not obvious. Bacteroids do not appear to take up malonate (95, 111). Does malonate affect nodule mitochondria? 3. Kahn et al (50) suggested that N nutrition in symbiotic N$_2$ fixation involves the plant providing amino acids to the bacteroids in exchange for NH$_4^+$. This model remains to be verified. The limited evidence to date shows that the PBM does not transport most amino acids (21, 23, 70, 125). Does this evidence negate the need for a shuttle mechanism (like malate-aspartate) in nutrient exchange? Are MDH and AAT linked in the PBM to channel nutrients between the plant and bacteroids? 4. The enzyme PEPC plays a crucial role in nodule metabolism (14, 52, 68), yet we do not have an accurate estimate of its contribution to legume carbon budgets. Calculations based on radiolabeling with ^{14}CO$_2$ show large disparities in contribution (52, 68). Since isotope fractionation occurs with various carboxylation reactions (34, 72), measurements of ^{13}C:^{12}C ratios in various organs and constituents of nodulated and non-nodulated legumes may give an accurate picture of the contribution of dark CO$_2$ fixation to legume productivity. Metabolic inhibitors have been extremely useful in dissecting the importance of various enzymes to N metabolism. Jenkins (49) recently developed a specific inhibitor of PEPC. Use of this compound in studies of nodule N and C metabolism may prove especially insightful. 5. Kouchi et al (54, 55) and Snapp & Vance (109) postulated that two pools of C occur in nodules; one provides C for N assimilation while the other is funneled to bacteroids for N$_2$ase. Confirmation of these pools and assessment of their relative importance remain. 6. In ureide-forming legume species the uninfected nodule cells are integrally involved in N assimilation (8, 27, 100, 112). No comparable functions have been demonstrated in amide-forming species. In fact, cellular fractionation studies of nodules in amide species have been neglected. 7. Only four genes (glutamine and asparagine synthetase, uricase, sucrose synthase) involved in nodule C and N metabolism have been isolated and characterized (33, 82, 128). Alfalfa nodule AAT, PEPC, and aldolase genes have been isolated but not yet characterized (C. P. Vance and J. S. Gantt, unpublished). Other genes involved in metabolism must be characterized. Furthermore, it is not clear whether these genes are regulated by organogenesis, O_2 and NH_4^+ concentration, or a combination of factors. 8. Finally, we have presented arguments supporting glycolysis to malate with further reduction to succinate; however, further labeling and enzyme studies are required in order to substantiate this concept. Although this is by no means an exhaustive list, it is evident from this brief register that several years of fruitful research remain.

CONCLUSIONS

Inclusively the data are consistent with the hypothesis that N_2 fixation is not limited by photosynthate, but rather by utilization of C within the nodule. Features elaborated to protect N_2ase from O_2 during symbiosis have necessitated modifications in host-plant and bacteroid C and N metabolism. These modifications may have resulted in an energy burden for symbiotic N_2 fixation greater than that required in free-living systems. The large C requirement imposed by symbiosis may be a consequence of limited ATP production by the plant resulting from low O_2 induced fermentative pathways leading to C_4-dicarboxylic acid synthesis. While the perception that symbiotic N_2 fixation is energetically inefficient may be true in solely thermodynamic terms, the evolution of exquisite adaptations that facilitate the process is extremely efficient in biological terms. The putative yield loss due to acquisition of N through symbiosis remains to be established.

ACKNOWLEDGMENTS

The authors thank their many colleagues both inside and outside the United States for preprints of their work and for valuable discussions leading to this manuscript. We are also grateful to Marilyn Hole for her assistance in preparing the article. Research in C.P.V.'s laboratory directed toward the topic of this review has been supported by USDA-CRGO Grant 87-CRCR-1-2588.

Literature Cited

1. Al-Mallah, M. K., Davey, M. R., Cocking, E. C.1989. Formation of nodular structures on rice seedlings by rhizobia. *J. Exp. Bot.* 40:473-78
2. Appleby, C. A. 1969. Properties of leghemoglobin in vivo and its isolation as ferrous oxyleghemoglobin. *Biochim. Biophys. Acta* 188:222-29
3. Appleby, C. A. 1984. Leghemoglobin and *Rhizobium* respiration. *Annu. Rev. Plant Physiol.* 35:443-78
4. Appleby, C. A., Bogusz, D., Dennis, E. S., Peacock, W. J. 1988. A role for hemoglobin in all plant roots? *Plant Cell Environ.* 11:359-67
5. Arwas, R., McKay, I. A., Rowney, F. R. P., Dilworth, M. J., Glenn, A. R. 1985. Properties of organic acid mutants of *Rhizobium leguminosarum* strain 300. *J. Gen. Microbiol.* 131:2059-66
6. Bergersen, F. J., Turner, G. L. 1975. Leghemoglobin and the supply of O2 to nitrogen fixing root nodule bacteroids: presence of two oxidase systems and ATP production at low free O$_2$ concentrations. *J. Gen. Microbiol.* 91:345-54
7. Bergersen, F. J., Turner, G. L. 1990. Bacteroid from soybean nodules: accumulation of poly-β-hydroxybutyrate during supply of malate and succinate in relation to N$_2$ fixation in flow-chamber reactions. *Proc. R. Soc. London Ser. B* 240:39-59
8. Blevins, D. G. 1989. An overview of nitrogen metabolism in higher plants. In *Plant Nitrogen Metabolism*, ed. J. E. Poulton, J. T. Romero, E. E. Conn, pp. 1-41. New York: Plenum (ISBN 0-306-433-22-2)
9. Bogusz, D., Appleby, C. A., Landsmann, J., Dennis, E. S., Trinick, M. J., Peacock, W. J. 1988. Functioning hemoglobin genes in a nonnodulating plant. *Nature* 331:178-80
10. Boller, B. C., Heichel, G. H. 1983. Photosynthate partitioning in relation to nitrogen fixation capability of alfalfa. *Crop Sci.* 23:655-59
11. Butler, G. W., Bathurst, N. O. 1958. Free and bound amino acids in legume root nodules: bound γ-aminobutyric acid in the genus *Trifolium*. *Aust. J. Biol. Sci.* 11:529-37
12. Carroll, B. J., McNeil, D. L., Gresshoff, P. M. 1985. Isolation and properties of soybean (*Glycine max* L.) mutants that nodulate in the presence of high nitrate concentrations. *Proc. Natl. Acad. Sci. USA* 82:4162-66
13. Cocking, E. C., Al-Mallah, M. K., Benson, E., Davey, M. R. 1990. Nodulation of non-legumes by rhizobia. In *Nitrogen Fixation: Achievements and Objectives*, ed. P. Gresshoff, E. Roth, G. Stacy, W. Newton, pp. 813-23. New York-London: Chapman Hall
14. Coker, G. T., Schubert, K. R. 1981. Carbon dioxide fixation in soybean roots and nodules: Characterization and comparison with N$_2$ fixation and composition of xylem exudate during early nodule development. *Plant Physiol.* 67:691-96
15. Copeland, L., Quinnel, R. G., Day, D. A. 1989. Malic enzyme in bacteroids from soybean nodules. *J. Gen. Microbiol.* 135:2005-11
16. Copeland, L., Vella, J., Hong, A. 1989. Enzymes of carbohydrate metabolism in soybean nodules. *Phytochemistry* 28:57-61
17. Cralle, H. T., Heichel, G. H. 1985. Interorgan photosynthate partitioning in alfalfa. *Plant Physiol.* 79:381-85
18. Cralle, H. T., Heichel, G. H., Barnes, D. K. 1987. Photosynthate partitioning in plants of alfalfa populations selected for high and low nodule mass. *Crop Sci.* 27:96-100
19. Dakora, F. D., Atkins, C. A. 1990. Effect of pO$_2$ during growth on the gaseous diffusional properties of cowpea (*Vigna unguiculata* L. Walp.). *Plant Physiol.* 93:956-61
20. Davies, D. D. 1980. Anaerobic metabolism and the production of organic acids. In *The Biochemistry of Plants: An Advanced Treatise*, ed. P. K. Stumpf, E. E. Conn, 2:581-611. New York: Academic (ISBN 0-12-675402-0)
21. Day, D. A. 1990. Nutrient exchange across the peribacteroid membrane. See Ref 13, pp. 219-26
22. Day, D. A., Price, G. D., Gresshoff, P. M. 1986. Isolation and oxidative properties of mitochondria and bacteroids from soybean root nodules. *Protoplasma* 134:121-29
23. Day, D. A., Price, G. D., Udvardi, M. K. 1989. The membrane interface of the *Bradyrhizobium japonicum*-*Glycine max* symbiosis: peribacteroid units from soybean nodules. *Aust. J. Plant Physiol.* 16:69-84
24. Dabas, S., Swaraj, K., Sheoran, I. S. 1988. Effect of source removal on functioning of the nitrogen fixation of pigeon pea (*Cajanus cajan* L.). *J. Plant Physiol.* 132:690-94

25. Denison, R. F., Weisz, P. R., Sinclair, T. R. 1988. Oxygen supply to nodules as a limiting factor in symbiotic N_2 fixation. In *World Crops: Cool Season Food Legumes*, ed. R. J. Summerfield, pp. 767-73. London: Kluwer (ISBN 9-02-47364-12)

26. De Vries, G. E., In'T Veld, P., Kijne, J. W. 1980. Production of organic acids in *Pisum* sativum root nodules as a result of oxygen stress. *Plant Sci. Lett.* 20:115-23

27. Duke, S. H., Henson, C. A. 1985. Legume nodule carbon utilization in the synthesis of organic acids for production of transport amides and amino acids. In *Nitrogen Fixation and CO_2 Metabolism*, ed. P. W. Ludden, J. E. Burris, pp. 293-302. New York: Elsevier (ISBN 0-444-00953-1)

28. Egli, M. A., Griffith, S. M., Miller, S. S., Anderson, M. P., Vance, C. P. 1989. Nitrogen assimilating enzyme activity and enzyme protein during development and senescence of effective and plant controlled ineffective alfalfa nodules. *Plant Physiol.* 91:898-904

29. Finan, T. M., Wood, J. M., Jordon, D. C. 1983. Symbiotic properties of C4-dicarboxylic acid transport mutants of *Rhizobium leguminosarum*. *J. Bacteriol.* 154:1403-13

30. Finn, G. A., Brun, W. A. 1982. Effect of atmospheric CO_2 enrichment on growth, nonstructural carbohydrate content, and root nodule activity in soybean. *Plant Physiol.* 69:327-31

31. Glenn, A. R., McKay, I. A., Arwas, R., Dilworth, M. J. 1984. Sugar metabolism and the symbiotic properties of carbohydrate mutants of *Rhizobium leguminosarum*. *J. Gen. Bacteriol.* 130: 239-45

32. Gordon, A. J., Ryle, G. J. A., Mitchell, D. F., Powell, C. E. 1985. The flux of ^{14}C labelled photosynthate through soybean root nodules during N_2 fixation. *J. Exp. Bot.* 36:756-59

33. Govers, F., Nap, J. P., Van Kammen, A., Bisseling, T. 1987. Nodulins in the developing root nodule. *Plant Physiol. Biochem.* 25:309-22

34. Guy, R. D., Vanlerberghe, G. C., Turpin, D. H. 1989. Significance of phosphoenolpyruvate carboxylase during ammonium assimilation. *Plant Physiol.* 89:1150-57

35. Hansen, A. P., Gresshoff, P. M., Pate, J. S., Day, D. D. 1990. Interactions between irradiance levels, nodulation and nitrogenase activity of soybean cv. Bragg and a supernodulating mutant. *J. Plant Physiol.* 136:172-79

36. Hardy, R. W. F., Havelka, U. D. 1975. Photosynthate as a major factor limiting nitrogen fixation by field grown legumes with emphasis on soybeans. In *Symbiotic Nitrogen Fixation in Plants*, ed. P. S. Nutman, pp. 421-39. London: Cambridge Univ. Press (ISBN 0-521-20645-6)

37. Hartwig, U., Boller, B., Nosberger, J. 1987. Oxygen supply limits nitrogenase activity of clover nodules after defoliation. *Ann. Bot.* 59:285-91

38. Havelka, U. S., Hardy, R. W. F. 1976 Legume N_2 fixation as a problem in carbon nutrition. In *Proceedings 1st International Symposium Nitrogen Fixation*, ed. W. E. Newton, C. J. Nyman, 2:456-75. Pullman: Washington State Univ. Press

39. Herrada, G., Puppo, A., Rigaud, J. 1989. Uptake of metabolites by bacteroid containing vesicles and by free bacteroids from french bean nodules. *J. Gen. Microbiol.* 135:3165-71

40. Hochachka, P. W., Mommsen, T. P. 1983. Protons and anaerobiosis. *Science* 219:1391-97

41. Hochachka, P. W., Somero, G. N. 1984. *Biochemical Adaptations*, pp. 145-81. Princeton: Princeton Univ. Press (ISBN 0-691-08343-6)

42. Hostak, M. S., Henson, C. A., Duke, S. H., Vandenbosch, K. A. 1987. Starch granule distribution between cell types of alfalfa nodules as affected by symbiotic development. *Can. J. Bot.* 65:1108-15

43. Humbeck, C., Werner, D. 1989. Delayed nodule development in succinate transport mutants of *Bradyrhizobium japonicum*. *J. Plant Physiol.* 134:276-83

44. Hunt, S., Gaito, S. T., Layzell, D. B. 1988. Model of gas exchange and diffusion in legume nodules II. Characterization of the diffusion barrier and estimations of the concentrations of CO_2, H2 and N_2 in the infected cell. *Planta* 173:128-41

45. Hunt, S., King, B. J., Canvin, D. T., Layzell, D. B. 1987. Steady and nonsteady state gas exchange characteristics of soybean nodules in relation to the oxygen diffusion barrier. *Plant Physiol.* 84:164-72

46. Hunt, S., King, B. J., Layzell, D. B. 1989. Effects of gradual increases in O_2 concentration on nodule activity in soybeans. *Plant Physiol.* 91:315-21

47. Jacobsen, E., Feenstra, W. J. 1984. A new pea mutant with efficient nodulation in the presence of nitrate. *Plant Sci. Lett.* 33:337-44

48. James, E., Brewin, N. K., Chudek, J. A., Minchin, F. R., Sprent, J. I. 1990. Glycoprotein in intercellular spaces may control oxygen diffusion in soybean nodules. See Ref, p. 743

49. Jenkins, C. L. D. 1989. Effects of the phosphoenolpyruvate carboxylase inhibitor 3,3-dichloro-2-(dihydroxyphosphinoylmethy) propenoate on photosynthesis. Plant Physiol. 89:1231-37

50. Kahn, M. L., Kraus, J., Somerville, J. E. 1985. A model of nutrient exchange in the Rhizobium-legume symbiosis. In Nitrogen Fixation Research Progress, ed. H. J. Evans, P. J. Bottomley,W. J. Newton, pp. 193-99. Dordrecht: Martinus Nijhoff (ISBN 90-247-3255-7)

51. Kawai, S., Mori, S. 1984. Detection of organic acids in soybean by high performance liquid chromatography. Soil Sci. Plant Nutr. 30:261-66

52. King, B. J., Layzell, D. B., Canvin, D. T. 1986. The role of dark carbon dioxide fixation in root nodules of soybean. Plant Physiol. 81:200-5

53. Kouchi, H., Fukai, K., Katagiri, H., Minamisawa, K., Tajima, S. 1988. Isolation and enzymological characterization of infected and uninfected cell protoplasts from root nodules of Glycine max. Physiol. Plant. 73:327-34

54. Kouchi, H., Nakaji, K., Yoneyama, T., Ishizuka, J. 1985. Dynamics of carbon photosynthetically assimilated in nodulated soybean plants under steady state conditions. 3. Time course study on ^{14}C incorporation into soluble metabolites and respiratory evolution of CO$_2$ from roots and nodules. Ann. Bot. 56:333-46

55. Kouchi, H., Yoneyama, T. 1986. Metabolism of [^{13}C]-labelled photosynthate in plant cytosol and bacteroids of root nodules of Glycine max. Physiol. Plant. 68:238-44

56. Lafontaine, P. J., Lafreniere, C., Chalifour, F. P., Dion, P., Antoun, H. 1989. Carbohydrate and organic acid composition of effective and ineffective root nodules of Phaseolus. Physiol. Plant. 76:507-13

57. Laing, W. A., Christeller, J. T., Sutton, W. D. 1979. Carbon dioxide fixation by lupine root nodules. II. Studies with ^{14}C-labelled glucose, the pathway of glucose catabolism and the effects of some treatments that inhibit nitrogen fixation. Plant Physiol. 63:450-54

58. Lambers, H., Visser, R. D. 1984. Energy metabolism in nodulated roots. In Advances in Nitrogen Fixation Research, C. Veeger, W. E. Newton, pp. 453-60. Dordrecht: Martin Nijhoff/Dr. D. W. Junk (ISBN 90- 247-2906-8)

59. Lance, C., Rustin, P. 1984. The central role of malate in plant metabolism. Physiol. Veg. 22:625-41

60. Larher, F., Goas, G., LeRudulier, D., Gerard, J., Hamelin, J. 1983. Bound 4-aminobutyricacid in root nodules of Medicago sativa and other nitrogen fixing plants. Plant Sci. Lett. 29:315-26

61. LaRue, T. A., Peterson, J. B., Tajima, S. 1984. Carbon metabolism in legume nodules. See Ref. 58, pp. 437-43

62. Layzell, D. 1990. Oxygen and regulation of N$_2$ fixation in legume nodules. Physiol. Plant. 80:322-27

63. Layzell, D. B., Hunt, S., Moloney, A. H. M., Fernando, S. M., Diaz del Castillo, L. 1990. Physiological, metabolic and environmental implications of O$_2$ regulation in legume nodules. See Ref. 13, pp. 21-33

64. Layzell, D. B., Hunt, S., Palmer, G. R. 1990. Mechanism of nitrogenase inhibition in soybean nodules. Pulse modulated spectroscopy indicates nitrogenase activity is limited by O$_2$. Plant Physiol. 92:1102-7.

65. Mahon, J. D. 1982. Field evaluation of growth and nitrogen fixation in peas selected for high and low photosynthetic CO$_2$ exchange. Can. J. Plant Sci. 62:5-17.

66. Mahon, J. D. 1983. Energy relationships. In Nitrogen Fixation. Vol 3. Legumes W. J. Broughton, pp. 299-327. Oxford: Clarendon Press (ISBN 0-19-854555-X)

67. Masterson, C. L., Sherwood, M. T. 1978. Some effects of increased atmospheric carbon dioxide on white clover (Trifolium repens) and pea (Pisum sativum). Plant Soil 49:421-26

68. Maxwell, C. A., Vance, C. P., Heichel, G. H., Stade, S. 1984. CO$_2$ fixation in alfalfa and birdsfoot trefoil root nodules and partitioning of ^{14}C to the plant. Crop Sci. 24:257-64

69. McDermott, T. R., Griffith, S. M., Vance, C. P., Graham, P. H. 1989. Carbon metabolism in Bradyrhizobium bacteriods. FEMS Microbiol. Rev. 63:327-40

70. McRae, D. G., Miller, R. W., Berndt, W. B., Joy, K. 1989. Transport of C$_4$-dicarboxylates and amino acids by Rhizobium meliloti bacteroids. Mol. Plant-Microbe Interact. 2:273-78

71. Mederski, H. J., Streeter, J. G. 1977. Continuous, automated acetylene reduction assays using intact plants. Plant Physiol. 59:1076-81

72. Melzer, E., O'Leary, M. H. 1987. Anapleurotic CO_2 fixation by phosphoenolpyruvate carboxylase in C3 plants. *Plant Physiol.* 84:58-60

73. Mendelssohn, I. A., McKee, K. L., Patrick, W. H. 1981. Oxygen deficiency in *Spartina alterniflora* roots: metabolic adaptation to anoxia. *Science* 214:439-41

74. Menegus, F., Cattaruzza, L., Chersi, A., Fronza, G. 1989. Differences in the anaerobic lactate-succinate production and in the changes of cell sap pH for plants with high and low resistance to anoxia. *Plant Physiol.* 90:29-32

75. Menegus, F., Cattaruzza, L., Chersi, A., Selva, A., Fronza, G. 1988. Production and organ distribution of succinate in rice seedling during anoxia. *Physiol. Plant.* 74:444-49

76. Minchin, F. R., Sheehy, J. E., Ines-Minguez, M., Witty, J. F. 1985. Characterization of the resistance to oxygen diffusion in legume nodules. *Ann. Bot.* 55:53-60

77. Minchin, F. R., Sheehy, J. E., Witty, J. F. 1986. Further errors in the acetylene reduction assay: effects of plant disturbance. *J. Exp. Bot.* 37:1581- 91

78. Minchin, F. R., Summerfield, R. J., Hadley, P., Roberts, E. H., Rawsthorne, S. 1981. Carbon and nitrogen nutrition of nodulated roots of grain legumes. *Plant Cell Environ.* 4:5-26

79. Monroe, J. D., Owens, T. G., LaRue, T. A. 1989. Measurement of the fractional oxygenation of leghemoglobin in intact detached pea nodules by reflectance spectroscopy. *Plant Physiol.* 91:598-602

80. Murphy, P. M. 1986. The effect of light and atmospheric carbon dioxide concentration on nitrogen fixation by herbage legumes. *Plant Soil* 95:399-409

81. Murphy, T. M., Matson, G. B., Morrison, S. L. 1983.Ultraviolet-stimulated $KHCO_3$ efflux from rose cells: regulation of cytoplasmic pH. *Plant Physiol.* 73:20-24

82. Nap, J. P., Bisseling, T. 1990. The roots of nodulins. *Physiol. Plant.* 79:407-14

83. Norby, R. J. 1987. Nodulation and nitrogenase activity in nitrogen-fixing woody plants stimulated by CO_2 enrichment of the atmosphere. *Physiol. Plant.* 71:77-82

84. Outlaw, W. R. Jr. 1982. Carbon metabolism in guard cells. In *Cellular and Subcellular Localization in Metabolism*, ed. L. L. Creasy, G. Hrazdina, pp. 185-97. New York: Plenum (ISBN 0-306-41023-0)

85. Phillips, D. A. 1980. Efficiency of symbiotic nitrogen fixation in legumes. *Annu. Rev. Plant. Physiol.* 31:29-49

86. Phillips, D. A., Newell, K. D., Hassell, S. A., Felling, C. E. 1976. The effect of CO_2 enrichment on root nodule development and symbiotic N_2 reduction in *Pisum sativum* L. *Am. J. Bot.* 63:356-62

87. Raschke, K., Hedrich, R. 1987. Stomatal movement: ion transport and carbon metabolism. In *Models in Plant Physiology and Biochemistry*, D. W. Newman, K. G. Wilson, pp. 33-37. Boca Raton: CRC (ISBN 0-849-34342-9)

88. Rawsthorne, S., LaRue, T. A. 1986. Preparation and properties of mitochondria from cowpea. *Plant Physiol.* 81:1092-96

89. Rawsthorne, S., LaRue, T. A. 1986. Metabolism under microaerobic conditions of mitochondria from cowpea nodules. *Plant Physiol.* 81:1097-102

90. Reibach, P. H., Streeter, J. G. 1983. Metabolism of ^{14}C-labelled photosynthate and distribution of enzymes of glucose metabolism in soybean nodules. *Plant Physiol.* 72:634-40

91. Reid, R. J., Loughmann, B. C., Ratcliffe, R. G. 1985. ^{31}P NMR measurements of cytoplasmic pH changes in maize root tips. *J. Exp. Bot.* 36:889-97

92. Riggle, B. D., Wiebold, W. J., Kenworthy, W. J. 1984. Effect of photosynthate source-sink manipulation on dinitrogen fixation of male-fertile and male-sterile soybean lines. *Crop Sci.* 24:5-8

93. Ronson, C. W., Lyttleton, P., Robertson, J. G. 1981. C_4-dicarboxylate transport mutants of *Rhizobium trifolii* form ineffective nodules. *Proc. Natl. Acad. Sci. USA.* 78:4284-88

94. Rosendahl, L., Vance, C. P., Miller, S. S., Jacobsen, E. 1989. Nodule physiology of a supernodulating pea mutant. *Physiol. Plant.* 77:606-12

95. Rosendahl, L., Vance, C. P., Pederson, W. B. 1990. Products of dark CO_2 fixation in pea root nodules support bacteroid metabolism. *Plant Physiol.* 93:12-19

96. Rumpho, M. E., Kennedy, R. A. 1983. Activity of the pentose phosphate pathway and glycolytic pathways during anaerobic germination of *Echinochloa crus-galli* (barnyard grass) seed. *J. Exp. Bot.* 34:893-902

97. Ryle, G. J. A., Powell, C. E., Gordon, A. J. 1986. Defoliation in white clover: nodule metabolism, nodule growth and maintenance, and nitrogenase functioning during growth and regrowth. *Ann. Bot.* 57:263-71

98. Salsac, L., Drevon, J. J., Zengbe, M., Cleyet-Marcel, J. C., Obaton, M. 1984. Energy requirement of symbiotic nitrogen fixation. *Physiol. Veget.* 22:509-21

99. Schubert, K. R. 1982. The energetics of biological nitrogen fixation. In *Workshop Summaries — I*, p. 31. Rockville, MD: Am. Soc. Plant Physiol.

100. Schubert, K. R. 1986. Products of biological nitrogen fixation in higher plants: synthesis, transport, and metabolism. *Annu. Rev. Plant Physiol.* 37:539-74

101. Schuller, K. A., Minchin, F. R. Gresshoff, P. M. 1988. Nitrogenase activity and oxygen diffusion in nodules of soybean cv. Bragg and a supernodulating mutant: effects of nitrate. *J. Exp. Bot.* 39:865-77

102. Schweitzer, L. E., Harper, J. E. 1980. Effects of light, dark, and temperature on root nodule activity (acetylene reduction) of soybeans. *Plant Physiol.* 65:51-56

103. Sheehy, J. E. 1987. Photosynthesis and nitrogen fixation in legume plants. *CRC Crit. Rev. Plant Sci.* 5:121-59

104. Sheehy, J. E., Minchin, F. R., Witty, J. F. 1983. Biological control of the resistance to oxygen flux in nodules. *Ann. Bot.* 52:565-71

105. Sheehy, J. E., Minchin, F. R., Witty, J. F. 1985. Control of nitrogen fixation in a legume nodule: an analysis of the role of oxygen diffusion in relation to nodule structure. *Ann. Bot.* 55:549-62

106. Sinclair, T. R., Goudriaan, J. 1981. Physical and morphological contraints on transport in nodules. *Plant Physiol.* 67:143-45

107. Skot, L., Hirsch, P. R., Witty, J. F. 1986. Genetic factors in *Rhizobium* affecting the symbiotic carbon costs of N₂ fixation and host plant biomass production. *J. Appl. Bacteriol.* 61:239-46

108. Small, J. G. C., Leonard, O. A. 1969. Translocation of ¹⁴C-labeled photosynthate in nodulated legumes as influenced by nitrate nitrogen. *Am. J. Bot.* 56:187-94

109. Snapp, S., Vance, C. P. 1986. Asparagine biosynthesis in alfalfa *Medicago sativa* L. root nodules. *Plant Physiol.* 82:390-95

110. Streeter, J. G. 1981. Seasonal distribution of carbohydrates in nodules and stem exudates from field grown soya bean plants. *Ann. Bot.* 48:441-50

111. Streeter, J. G. 1987. Carbohydrate, organic acid, and amino acid composition of bacteroids and cytosol from soybean nodules. *Plant Physiol.* 85:768-73

112. Streeter, J. G., Salminen, S. O. 1985. Carbon metabolism in legume nodules. See Ref. 50, pp. 277-83

113. Streeter, J. G., Thompson, J. F. 1972. In vivo and in vitro studies on γ-aminobutyric acid metabolism with the radish plant (*Raphanus sativus* L.) *Plant Physiol.* 49-579-84

114. Stumpf, D. K., Burris, R. H. 1981. Organic acid content of soybean: age and source of nitrogen. *Plant Physiol.* 68:989-91

115. Suganuma, N., Kitou, M., Yamamoto, Y. 1987. Carbon metabolism in relation to cellular organization of soybean root nodules and respiration of mitochondria aided by leghemoglobin. *Plant Cell Physiol.* 28:113-22

116. Suganuma, N., Yamamoto, Y. 1987. Carbon metabolism related to nitrogen fixation in soybean root nodules. *Soil Sci. Plant Nutr.* 33:79-91

117. Suganuma, N., Yamamoto, Y. 1987. Respiratory metabolism of mitochondria in soybean root nodules. *Soil Sci. Plant Nutr.* 33:93-101

118. Swaraj, K., Kuhad, M. S., Garg, O. P. 1986. Dark treatment effects on symbiotic nitrogen fixation and related processes in *Cicer arietinum* L. chickpea. *Environ. Exp. Bot.* 26:31-38

119. Ta, T. C., Macdowall, F. D. H., Faris, M. A. 1987. Utilization of carbon from shoot photosynthesis and nodule CO₂ fixation in the fixation and assimilation of nitrogen by alfalfa root nodules. *Can. J. Bot.* 65:2537-41

120. Tjepkema, J. D., Yocum, C. S. 1973. Respiration and oxygen transport in soybean nodules. *Planta* 115:59-72

121. Tjepkema, J. D., Yocum, C. S. 1974. Measurement of oxygen partial pressure within soybean nodules by oxygen microelectrodes. *Planta* 119:351-60

122. Trinchant, J. C., Rigaud, J. 1989. Alternative energy yielding substrates for bacteroids isolated from stem and root nodules of *Sesbania rostrata* submitted to O₂ restricted conditions. *Plant Sci.* 59:141-49

123. Twary, S. N., Heichel, G. H. 1991. Carbon costs of dinitrogen fixation associated with dry matter accumulation in alfalfa. *Crop Sci.* 31: In press

124. Udvardi, M. K., Price, G. D., Gresshoff, P. M., Day, D. A. 1988. A dicarboxylate transporter on the peribacteroid membrane of soybean nodules. *FEBS Lett.* 231:36-40

125. Udvardi, M. K., Salom, C. L., Day, D. A. 1988. Transport of L-glutamate across the bacteroid membrane but not the peribacteroid membrane from soy-

bean root nodules. *Mol. Plant-Microbe Interact.* 1:250-54
126. Vance, C. P. 1990. Symbiotic nitrogen fixation: recent genetic advances. In *The Biochemistry of Plants: An Advanced Treatise*, ed. P. K. Stumpf, E. E. Conn, 16:43-88. New York:Academic
127. Vance, C. P., Boylan, K. L. M., Maxwell, C. A., Heichel, G. H., Hardman, L. L. 1985. Transport and partitioning of CO_2 fixed by root nodules of ureide and amide producing legumes. *Plant Physiol.* 78:774-78
128. Vance, C. P., Egli, M. A., Griffith, S. M., Miller, S. S. 1988. Plant regulated aspects of nodulation and N_2 fixation. *Plant, Cell, Environ.* 11:413-27
129. Vanlerberghe, G. C., Horsey, A. K., Weger, H. G., Turpin, D. H. 1989. Anaerobic carbon metabolism by the tricarboxylic acid cycle. *Plant Physiol.* 91:1551-57
130. Walsh, K. B., Vessey, J. K., Layzell, D. B. 1987. Carbohydrate supply and N_2 fixation in soybean. *Plant Physiol.* 85:137-44
131. Warembourg, F. R., Roumet, C. 1989. Why and how to estimate the cost of symbiotic N_2 fixation. A progressive approach based on the use of ^{14}C and ^{15}N isotopes. *Plant Soil* 115:167-77
132. Waters, J. K., Karr, D. B., Emerich, D.

W. 1985. Malate dehydrogenase from *Rhizobium japonicum* 311b-143 bacteroids and *Glycine max* root-nodule mitochondria. *Biochemistry* 24:6479-86
133. Weisz, P. R., Sinclair, T. R. 1987. Regulation of soybean nitrogen fixation in response to rhizosphere oxygen. *Plant Physiol.* 87:906-10
134. Williams, L. E., DeJong, T. M., Phillips, D. A. 1982. Effect of changes in shoot carbon-exchange rate on soybean root nodule activity. *Plant Physiol.* 69:432-36
135. Wilson, P. W., Fred, E. B., Salmon, M. R. 1933. Relation between carbon dioxide and elemental nitrogen assimilation in leguminous plants. *Soil Sci.* 35:145-65
136. Witty, J. F., Minchin, F. R. 1990. Oxygen diffusion in legume root nodules. See Ref. 13, pp. 285-92
137. Witty, J. F., Minchin, F. R., Sheehy, J. E., Ines-Minquez, M. 1984. Acetylene-induced changes in the oxygen diffusion resistance and nitrogenase activity of legume root nodules. *Ann. Bot.* 53:13-20
138. Witty, J. F., Minchin, F. R., Skot, L., Sheehy, J. E. 1986. Nitrogen fixation and O_2 in legume root nodules. *Oxford Surv. Plant Mol. Cell Biol.* 3:275-314

Annu. Rev. Plant Physiol. Plant Mol. Biol. 1991. 42:393-422

THE SELF-INCOMPATIBILITY GENES OF BRASSICA:
Expression and Use in Genetic Ablation of Floral Tissues

June B. Nasrallah, Takeshi Nishio, and Mikhail E. Nasrallah

Section of Plant Biology, Division of Biological Sciences, Cornell University, Ithaca, New York 14853

KEY WORDS: pollen tube, self-incompatibility, *slg* gene, *Brassica*

CONTENTS

1040-2519/91/0601-0393$02.00

INTRODUCTION

Flower development is an area where the study of the processes of development and of cell–cell communication in plants can be nicely combined. The development of the flower involves the differentiation of complex organ systems that act to promote and support gamete development in ovary and anther. The flower is also the site of crucial interactions between the haploid gametophyte (pollen and pollen tube) and the diploid tissues of the pistil. Innumerable morphological and cytological studies describe aspects of the ontogeny of the female and male reproductive structures. The germination of pollen and growth of pollen tubes in compatible pollinations, and the inhibition of these processes in incompatible pollinations have also been the subject of extensive cytological and genetic studies (reviewed in 38, 57, 89). Recently the molecular basis of the flowering process has attracted a flurry of research activity. These studies have led to the isolation of (*a*) genes involved in the establishment of patterns in the flower (115, 140) and (*b*) genes expressed in particular structures of the flower, specifically in pistils, anthers, and pollen. Among the latter group are genes, isolated from crucifer and solanaceous plants, that are associated with recognition of pollen in self-incompatible species and are related to the controlling self-incompatibility locus, the S-locus (3, 80).

Aspects of the molecular biology of self-incompatibility systems of ornamental tobacco (10, 18, 19, 26, 71) and *Brassica* (83) have been reviewed. This article focuses on recent advances in the study of the S-locus genes and related members of the S-multigene family of *Brassica*. After providing a brief overview of the cellular interactions that occur between pollen and stigma, and a description of the development of the tissues in which these genes operate, we consider the emerging features of pollen recognition in *Brassica*. The concept of single-gene control of self-incompatibility is reexamined in light of recent molecular genetic data demonstrating that the S-locus has a complex organization with at least two transcriptional units. Molecular data bearing on two S-locus related (SLR) loci are also considered. Of the four expressed genes known to date, one S-locus derived gene, the *SLG* gene, serves as a paradigm for the analysis of the tissue-specific expression of members of the S-multigene family. The tight regulation of the pattern of expression has allowed use of the *SLG* promoter to target toxin genes in order to effect the ablation of specific tissues of the flower. The successful demonstration of genetic ablation schemes in plants suggests that this approach could be utilized widely not only in the analysis of morphogenetic events in complex multicellular plant organs, but also in the engineering of male and female sterility.

POLLINATION AND POLLEN TUBE GROWTH

The sequence of events precipitated once pollen grains are delivered to the surface of the stigma has been described at the morphological and cytological levels (40, 57, 131). Pollination involves pollen capture, hydration, and pollen tube growth (22, 104). The pollen grain, in a process akin to the germination of a fungal spore, produces a tube that emerges from one of the pores. In crucifers, the pollen tube invades the stigmatic papillae through the action of digestive enzymes and grows within the secondary papillar cell wall (22, 27, 53). At the base of the papillar cell, the pollen tube grows intercellularly in the transmitting tissues of the stigma, style, and ovary. In the ovary, the tube grows over the surface of the septum, penetrates the funiculus, and enters the unfertilized ovule through the micropyle to effect fertilization (see Figure 1).

Pollen tubes grow by unidirectional extension of the tip (97). The growing region at the tip is very active and contains many vesicles that can be seen to fuse with the membrane, releasing wall components and enzymes (57). The transmitting tissue is thought to provide the mechanical and nutritive environments that promote normal pollen tube growth. Since pollen tubes are known to grow for several hours in the gynoecium, ample opportunities exist for interactions between these tubes and tissues along the length of the pistil. Signals of a chemical (102), electrical (112), and mechanical (124) nature have been suggested to induce the directionality of pollen tube growth, but the evidence for tropisms is conflicting. More recently, based on monitoring of movement of latex beads through the stylar transmitting tissue of three species, Sanders & Lord (103) have suggested that the intercellular matrix may also play an active role in the directed growth of pollen tubes towards the ovary.

Pollination precipitates many changes in the tissues of the pistil. Some of the documented changes include increases in respiratory rate, auxin level, and enzyme content (101), and cytological changes in the ovary (66, 73) and in the style of some species (21, 131). It is thought that in response to pollination the ovary emits a long-distance signal that stimulates the growth of pollen tubes while still in the style (75, 76).

Interactions Between the Male Gametophyte and the Pistil

Specific cell–cell interactive events are important components of pollination and fertilization processes. The stigma efficiently screens pollen that belongs to other species. Plants have evolved physiological mechanisms that allow them to arrest interspecific pollination, mechanisms that preserve gametes and maintain reproductive fitness. One such mechanism is based on incompatibil-

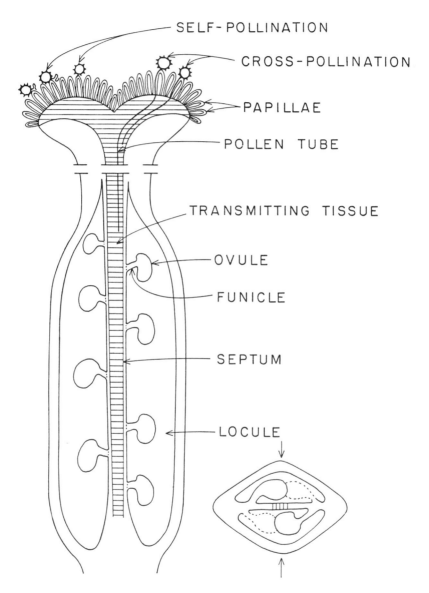

Figure 1   Schematic diagram of a *Brassica* pistil. The transverse view is at the level of the ovary.

ity reactions between pollen and pistil. Lewis & Crowe (63) observed that certain interspecific matings are blocked in one direction only, a phenomenon known as unilateral incompatibility. Many flowering plants have also evolved intraspecific genetic mechanisms that allow the discrimination between "self" and "nonself" pollen. This genetic barrier to self-fertilization, termed self-incompatibility, specifically interrupts the pathway of self-pollen development in the pistil (reviewed in 39, 89). In many families, including crucifers and Solanaceae, self-incompatibility is controlled by one locus, the S-locus, at which multiple alleles, numbering as many as 60 alleles in *Brassica* (92), have been described. The self-incompatibility response is regulated during the development of the flower and is typically acquired at 1–2 days before anthesis (100). In several species, the self-compatibility of young flower buds can be exploited effectively for the production of S-allele homozygotes (110).

Self-Incompatibility Systems: Gametophytic and Sporophytic Determination of Pollen Phenotype

The control of pollen incompatibility phenotype has been used as a basis for the classification of self-incompatible species into two types, gametophytic and sporophytic (38, 89). In gametophytic systems, the incompatibility phenotype of a pollen grain is determined by the haploid gene complement that is passed along to each microspore after meiosis. East & Mangelsdorf (25) advanced the hypothesis that self-incompatibility is brought about by genetic identity at the S-locus. Pollen tubes fail to develop normally where the S alleles are matched in pollen and pistil. In general, one-locus gametophytic systems, such as those operative in solanaceous plants (wild tomato, diploid potato, petunia, and ornamental tobacco) are characterized by pollen tube inhibition during growth in the transmitting tissue of the style. In an incompatible pollination, pollen germination is normal and growth of the pollen tubes, at least in the initial phase, is also normal but is later arrested at some distance into the style. Species with sporophytic control of pollen phenotype are exemplified by the mustard family, the Brassicaceae. In this family, the self-incompatibility phenotype of pollen is determined by the genetic constitution of the sporophyte or the diploid plant on which the pollen is borne. In self-incompatible mustards and radishes, inhibition occurs at the surface of the stigma. Within minutes after the initial contact between pollen and the papillar cells (the outer cells of the stigma that capture pollen), rapid cell–cell communications signal the arrest of pollen germination (93). In contrast to a compatible pollination in which several hundred pollen tubes germinate and invade the stigmatic tissue, an incompatible pollination results in the complete inhibition of pollen development or the drastic reduction in the number of developed pollen tubes. Pollen germination may take place in certain genotypes, but the pollen tubes often have abnormal morphology; more important, they fail to invade the wall of the papillae (22, 27). A rejection response is also manifested in the papillar cells by the deposition of the β-1,3-glucan callose at the site of contact with incompatible pollen (38).

DEVELOPMENTAL STUDIES OF REPRODUCTIVE STRUCTURES

The above description of pollen germination and pollen tube growth has focused attention on the stigma and the transmitting tissue as two tissues of the pistil where critical cell–cell recognition events take place, and on the importance of sporophytic and gametophytic contributions to the pollen's response in its interaction with the pistil. Progress in deciphering the basis of these cell–cell interactions, and of the genes involved, requires an understanding of the development of stigma and transmitting tissue, on the one hand, and of pollen on the other. While cytological and morphological studies have been used in the past to describe the development of the male and female reproductive structures, a molecular genetic approach has been applied more recently to relate the differentiation state of the pistil and anther/pollen to tissue-specific gene expression. It is one expectation of this approach that the identification of genes expressed exclusively in these tissues would pinpoint (a) functions specific to the pistil as a conduit for pollen capture and development and (b) functions specifically associated with the pollen's ability to respond to the stigmatic environment, its unique mode of growth, and its role in fertilization. Here we present a brief overview of pistil and pollen development, with special emphasis on crucifers and solanaceous plants.

Tissues of the Pistil with Special Reference to Pollination Functions

The female reproductive organ of the flower, the pistil, is diagrammed in Figure 1. The stigma, located at the distal end of the pistil, is specialized to serve as a receptacle for pollen grains. The papillar cells of the stigma surface form a dense layer in Brassicaceae, and are sparsely distributed in Solanaceae. Below the papillae lie a subepidermal secretory zone and a parenchymatous ground tissue. The secretory zone converges into the solid style as a central core of transmitting tissue surrounded by a parenchymatous cortex and by the epidermis. The style connects the stigma to the ovary, which contains ovules borne on placental tissue within two locules. The pistil of crucifers and Solanaceae is comprised of two fused carpels (11, 108). Fusion occurs immediately after the initiation of the two carpel primordia very early in bud development. At a slightly later stage of development, placental fusion within the ovary results in the formation of two locules(106–108).

Both of these fusion events depend on interactions between the epidermal cell layers of the fusing organs (132, 134), and involve redifferentiation of epidermal cells in the young carpels, apparently through the action of diffusible factors (111). Cells along the line of carpel fusion differentiate into the transmitting tissue. The analysis of graft chimeras of *Datura,* a member of the Solanaceae, has shown that the cells of the stigma surface and of the transmitting tissue, as well as cells of the ovarian placental epidermis, are all derived from the same epidermal (L1) layer (106).

The cells of the transmitting tissue are metabolically active cells that secrete various components into the extracellular matrix of the style (9, 43, 58, 105). The chemical composition of the stylar transmitting matrix has been histochemically analyzed in a number of plant species. It was found in general to be rich in pectic compounds and other polysaccharides, and in complex proteins including glycoproteins and lipoproteins (57, 59, 98). The apparent absence of proteins in the matrix of crucifers was however noted by Hill & Lord (43). Not all of these proteins are involved in specific functions relating to pollen tube growth. It has been shown recently, for example, that the transmitting tissue is a site of expression, during the course of normal development, of genes associated with stress responses, such as genes encoding pathogenesis-related proteins (65) and heat-shock proteins (23).

The method of differential screening of recombinant cDNA libraries has been successfully applied to the isolation of genes that exhibit specific or preferential expression in pistils. Genes associated with the operation of self-incompatibility were isolated from *Brassica oleracea* (80) and *Nicotiana alata* (3) as stigma-specific and pistil-specific sequences respectively. The *Brassica* genes are discussed at length in the remainder of this review. Other genes preferentially expressed in pistils have been isolated from tomato (31) and tobacco (32). However, with the exception of one gene that was exclusively expressed in the ovules, these genes were also expressed to variable degrees in other plant organs, suggesting that they are not involved in a pistil-specific function (31).

The Male Gametophyte

As a result of its developmental history, pollen exhibits both gametophytically and sporophytically determined characteristics (reviewed recently in 67, 68). Gametophytic determination is a consequence of the action of the haploid genome of the pollen grain itself. Sporophytic determination, obeying the rules of diploid genetics, is thought to result if the relevant genes are expressed pre-meiotically in the pollen mother cells (95) or post-meiotically in the cells of the sporophytic tissues surrounding the developing pollen. Post-meiotic sporophytic expression is believed, based strictly on cytological observations, to be the result of gene activity in the tapetum, the sporophytic cell layer of the anther that lines the locules in which the microspores develop (38).

The contributions of the gametophyte and sporophyte during pollen development are best illustrated by the formation of the pollen wall (36–38, 41, 42). The inner and outer cell walls of the pollen are synthesized by the pollen itself, and proteins in the inner layer of the pollen wall are gametophytic in origin. Some of these proteins, identified as hydrolytic enzymes, accumulate around the pores and are implicated in the emergence of the pollen tube (38, 40). The tapetum is thought to be responsible for completing the formation of the pollen outer wall, the exine, by filling its cavities with a variety of compounds, including carbohydrates and lipids as well as proteins. The properties of the tapetum have been reviewed recently (2) and are only briefly summarized here. Evidence for the importance of the tapetum in pollen development comes from the analysis of nuclear and cytoplasmic mutations leading to male sterility (54, 62, 64).

One function assigned to the tapetum is the production of the nutrients required for pollen development. During meiosis and the microspore tetrad stage, an amplification in the DNA content of tapetal cells occurs and leads to a multinuclear and/or polyploid condition (30). At the same time, and apparently in preparation for its subsequent intense secretory activity, an increase in the metabolic activity of these cells takes place, as suggested both by an increase in the number and size of endomembranes and organelles, and by an increase in ^{3}H-uridine incorporation into RNA (136). The tapetum apparently provides nutrients for the developing microspores by diffusion of simple metabolites through the callosic layer that surrounds the tetrads. Enzymes, specifically glucanases, responsible for the dissolution of the wall surrounding the tetrads are also synthesized by the tapetal cells (116).

The tapetum then undergoes a period of intense synthetic and secretory activity, ending with the disintegration of the tapetal cells, usually soon after the first mitotic division in pollen. Electron microscopic observations have suggested that the contents of the degenerated tapetal cells move onto the surface of the pollen grains (38, 41, 42). The sporopollenins [carotenoid derivatives that impregnate the pollen exine (37)], lipids, and proteins are the major synthetic products of the tapetum (41, 42, 133). Examples of proteins of known function that are synthesized and secreted by the tapetal cells are enzymes of phenylpropanoid synthesis for the pollen exine (7, 35). On the basis of these cytological observations it has also been suggested that the S-locus products of sporophytic self-incompatibility systems are produced by the tapetum and then incorporated into the pollen exine (38).

Genes preferentially or exclusively expressed in the gametophyte or in sporophytic tissues of the anther at earlier stages of pollen development have been isolated (reviewed in 68) and include meiosis-specific sequences (5), tapetal-specific sequences (32, 113, 114), and sequences expressed in mature pollen either preferentially (128, 130) or specifically (15, 33, 117). While the function of these genes is not known, a subset of those expressed in mature pollen appear to be involved in pollen germination and/or pollen tube growth. Such a function was suggested by the persistence of their transcripts in pollen tubes after germination, and by the nucleotide sequence similarity shared by some of these genes with genes encoding enzymes that hydrolyze cell wall components. In particular, pollen-expressed genes isolated from tomato and *Oenothera* exhibited sequence similarity to bacterial pectate lyase (137) and polygalacturonase (15) genes, respectively. Should the protein products of these genes have the expected enzymatic activity, it will be of interest to determine if their substrates are specific to floral tissues, and whether they function in loosening the cell wall during pollen tube elongation, or in facilitating pollen tube growth by digesting matrix components in the transmitting tissue.

MOLECULAR ANALYSIS OF SELF-INCOMPATIBILITY

To date, the molecular analysis of pollen-pistil interactions has focused on genes encoding glycoproteins thought to be involved in the recognition of self-pollen in two plant families, the Brassicaceae and the Solanaceae. In *Brassica*, the 60 known alleles at the S-locus have been associated with a class of stigmatic glycoproteins, the S-locus-specific glycoproteins (SLSG) (81, 87, 88, 91, 100). These glycoproteins are in general basic molecules, are produced as multiple molecular weight forms in the range of 55–65 kilodaltons (81), and exhibit extensive molecular weight and charge polymorphisms in different S-genotypes of *Brassica*. SLSG polymorphisms cosegregate perfectly with the corresponding S-alleles in F_2 populations of *B. oleracea*, *B. campestris*, and *Raphanus* species (44, 80, 82, 88, 90). The concentration of SLSG increases as the pistil matures, and maximal synthesis rates are attained at the onset of self-incompatibility in the developing stigma (78). SLSG are primarily localized in the papillar cells of the stigma surface, and are secreted into the papillar cell walls, where they accumulate (52). As secreted products, SLSG must coat the surfaces of papillae and from that vantage point could easily diffuse onto the pollen grains soon after pollination.

In Solanaceae, pistil glycoproteins, designated the S-associated glycoproteins, have been identified in *Nicotiana alata* (12), *Petunia hybrida* (14, 50), *Lycopersicon peruvianum* (70), *Solanum tuberosum*, (56) and *S. chacoense* (138). In *N. alata*, the S-associated glycoproteins are localized primarily in the intercellular matrix of the stylar transmitting tissue and at lower levels in the papillar cells of the stigma and in the placental epidermis of the ovary (20). This localization is consistent with inhibition of pollen within the transmitting tissue of the style in the Solanaceae. Somewhat limited genetic data are available on the segregation of these S-associated molecules with SI phenotype (26). The association of these molecules with self-incompatibility is based largely on in vitro pollen growth assays in which S-associated glycoproteins isolated from one S-genotype appear to have some general inhibitory effect on pollen tube growth (34, 49), an effect related perhaps to their ribonuclease activity (72).

DNA sequences encoding the *Brassica* SLSG have been isolated from several S-allele homozygous strains (16, 61, 79, 80, 127). In Solanaceae, cDNA clones encoding the S-associated proteins have been isolated from *N. alata* (3, 4, 55). DNA sequences have also been isolated from *Petunia inflata* (1) and *P. hybrida* (17) based on sequence homology with the *N. alata* genes. In the remainder of this review, we focus on the S-locus genes and related genes of *Brassica*.

The S-Multigene Family of Brassica

The *Brassica* genome contains multiple genes related to the *SLSG* structural gene (24, 84). The occurrence of approximately twelve genomic regions with homology to the *SLSG*-cDNA was demonstrated by the isolation of the corresponding genomic clones. Together, these genes constitute the S-multigene family, so designated to indicate the sequence relatedness of its members to S-locus linked genes. As discussed below, one subset of genes in the family encodes secreted glycoproteins that are localized in the cell walls of the stigma papillae, while others are predicted to encode putative receptor protein kinases.

The analysis of restriction fragment length polymorphisms of F_2 populations segregating for different S-allele pairs has demonstrated that among the multiple S-related genes, some are linked to the S-locus while others are independently inherited. To date, the expression of four members of the S-gene family has been demonstrated. Two of these expressed genes reside at the S-locus. These are the *SLSG* structural gene, which we refer to as the S-Locus Glycoprotein (*SLG*) gene, and a putative receptor protein kinase gene. Two other expressed genes are unlinked to the S-locus but are loosely linked to each other and separated by a distance of approximately 18.5 cM (11a). These S-locus related (*SLR*) genes have been designated *SLR1* and *SLR2*.

Allelism at the S-Locus and the SLG Gene

Many of the S-alleles of *B. oleracea* have been arranged in a complex dominance series based on their genetic behavior relative to other alleles in stigma and pollen (121). Briefly, two major classes of S-alleles can be recognized. High-activity (class I) alleles exhibit strong incompatibility phenotype in which an average of 0–10 pollen tubes develop per self-pollinated stigma, and are placed relatively high on the dominance scale of S-alleles. Low-activity (class II) alleles have weak incompatibility phenotype in which 10–30 pollen tubes develop per self-pollinated stigma, and exhibit recessive and competitive interactions in pollen. Class I is exemplified by the S_6, S_{13}, S_{14}, S_{22}, S_{29} and S_{63} alleles, and class II includes the S_2, S_5 and S_{15} alleles. The two classes can also be distinguished on the basis of the reactivity of their SLSG with MAbH8, a monoclonal antibody raised against SLSG from the S_6 genotype: Class I alleles react with MAbH8 while class II alleles do not (52). It is interesting that, based on this epitope polymorphism, S-alleles of *B. campestris* and of the related genus *Raphanus* can be similarly grouped into MAbH8 positive and negative subsets of alleles. In *B. campestris* for example, the S_8 allele and the S_{12} allele have the characteristics of class I and class II alleles, respectively, and in a wild population of *R. raphanistrum* the class I S-alleles outnumbered class II alleles by a ratio of 2:1 (M. E. Nasrallah, unpublished data). On this basis, S-alleles can have diverged from one another more within a species than between species. Despite this divergence, class I and class II variants are considered allelic based on transmission genetic criteria.

SLG genes, defined by their linkage to the S-locus, are highly variable in different S-allele homozygotes. The extent of DNA and amino acid variability between *SLG* alleles depends on the alleles being compared (Table 1). Among class I alleles, the *SLG* genes share over 90% DNA sequence similarity and 80% similarity overall at the amino acid level (79, 127). More substantial sequence variability is observed when class I and class II alleles are compared, underscoring the distinct nature of the two classes. The *SLG* gene

Table 1 Similarity of derived protein sequences at the highly polymorphic *SLG* locus (S_8 is from *B. campestris;* all other alleles are from *B. oleracea*). The highly divergent S_2 allele is representative of a subset of alleles known to be pollen recessive now designated class II.

	S_2	S_6	S_{13}	S_{14}	S_8	Classification
S_2		67	68	62	66	class II
S_6			84	81	83	class I
S_{13}				79	90	class I
S_{14}					78	class I
S_8						class I

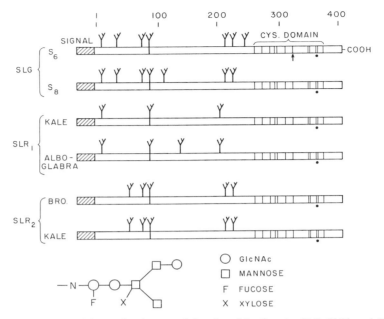

Figure 2   Diagram of the predicted structural domains of the *Brassica SLG, SLR1,* and *SLR2* genes. The signal peptide, the cysteine-rich domain and the N-glycosylation sites (trees) are shown. The arrow marks a polymorphic cysteine residue, and the dot marks a conserved N-glycosylation site within the cysteine-rich region. The structure of the glycan chain was determined by Takayama et al (119).

derived from the S_2 allele shares less than 70% overall DNA sequence similarity with class I alleles and is 32% diverged at the amino acid level (16) (Table 1). It should be noted that allelism of the class I *SLG* genes is defined in molecular terms by hybridization to a small, unique 3' terminus in the transcript and associated genomic DNA (84). Interestingly, the 3' sequences of class I and class II *SLG* genes do not share any homology, and at this point we cannot exclude the possibility that class I and class II alleles are pseudo-alleles.

The protein encoded by the *SLG* gene is a polypeptide produced from a single exon and composed of 435 amino acids on average. The N-terminal 31 amino acids form a signal peptide associated with the transport of this protein across the membrane of the endoplasmic reticulum. The functional protein thus comprises roughly 405 amino acids and contains several potential N-glycosylation sites for the attachment of glycan chains (Asn-X-Thr/Ser, where X is any amino acid). The *SLG* alleles differ by a large number of sub-stitutions, and by small deletions and insertions of a few amino acids. In addition, each allele analyzed to date exhibits a distinctive array of potential N-glycosylation sites (Figure 2), and stigma SLSG isolated from different

S-allele homozygotes have been shown by chemical deglycosylation to differ in the number of glycan chains (129).

Based on the analysis of total stigma glycoproteins from *B. campestris*, Takayama et al (119) concluded that the specificity of pollen recognition is not a function of the structure of the glycan moieties, since no differences in carbohydrate composition between different S-genotypes could be discerned. Based on the direct amino acid sequence analysis of purified stigma SLSG (120), these researchers have further concluded that the glycosylated amino acid residues are located in the amino terminal half of the SLSG molecule. In Figure 2, we therefore show only the potential N-glycosylation sites that lie outside the carboxy-terminal cysteine-rich domain. The structure of the glycan side chains of SLSG was determined (119) to be of the complex type commonly found in plant glycoproteins (Figure 2). Such a conclusion is also suggested by transformation experiments, in which *N. tabacum* plants transformed with *SLG* genes produced correctly modified SLSG with the molecular weight and charge characteristic of the introduced *SLG* allele (51, 74, 129).

The S-Locus Related (SLR) Genes

The *SLR1* and *SLR2* genes share several features with the *SLG* gene. All three genes have coding regions uninterrupted by introns (61, 84, 126). The primary translational products of allelic variants of *SLG* and *SLR1* exhibit on average approximately 65% sequence similarity in pairwise comparisons. A more complicated relationship is observed between the *SLR2* and *SLG* genes. The *SLG* genes of class I alleles are less than 70% similar to *SLR2* (11a). However, the *SLG* gene derived from the class II allele S_2 (16) is >90% and 85% similar to *SLR2* at the DNA and amino acid levels, respectively. This sequence similarity between different members of the S multigene family can be misleading. A case in point is that of a DNA sequence recently reported as being the *SLG* sequence encoded by the S_5 allele (109), but which is in fact an *SLR2* sequence (11a). Thus, the identification of newly isolated sequences as allelic variants of specific members of the *S*-gene family cannot rely solely on sequence comparisons, but should be accompanied by genetic data.

The sequence similarity between the *SLG* and *SLR* genes is reflected in the similar structures of their protein products. Figure 2 shows in diagrammatical form the structure of the polypeptides encoded by two alleles of the *SLG* gene (a *B. oleracea* S_6-derived allele, and a *B. campestris* S_8-derived allele), two alleles of the *SLR1* gene derived from the *B. oleracea* kale and alboglabra cultivars, and two alleles of the *SLR2* gene derived from the *B. oleracea* broccoli and kale cultivars. The presence of a hydrophobic signal peptide at the amino terminus of the polypeptide suggests that like SLSG, the SLR1 and SLR2 proteins are secreted; in fact, SLR1 protein was shown by immunogold labeling at the ultrastructural level to be localized, like SLSG, in the papillar cell wall (129). At their carboxy terminus, the polypeptide products of each gene contain 11 cysteine residues arranged in the same precise order. A 12th cysteine residue (marked by an arrow in Figure 2) is found in most but not all sequences analyzed.

The *SLR* genes, like *SLG*, contain a number of potential N-glycosylation sites, with *SLG* typically exhibiting the highest, and *SLR1* the lowest number of such sites (Figure 2). Two of these sites are conserved in all genes analyzed: One is located in the amino terminal third of the molecule, and another is embedded within the cysteine domain. That the products of these genes are in fact glycosylated in the stigma has been demonstrated for *SLR1* protein (129) and for SLR2 protein (T. Tantikanjana and J. B. Nasrallah, unpublished data). Differential glycosylation is thought to be at least partially responsible for the molecular weight and charge heterogeneity characteristic of SLSG and SLR glycoproteins (81, 129). The glycan side chains of the three glycoproteins are presumably similar, since a glycan structure similar to that described above for SLSG was also described for a stigmatic glycoprotein believed to be SLR1 (48).

A significant difference between the *SLG* gene and the two *SLR* genes, in addition to the differences in their linkage relationships to the S-locus, is in the extent of sequence variability exhibited by allelic variants of each gene. In contrast to the extensive polymorphism at the S-locus, the associated variability observed among alleles of the *SLG* gene and the extensive molecular-weight and charge polymorphisms of their SLSG products, *SLR1* and *SLR2* genes isolated from different *Brassica* strains and cultivars bred for homozygosity at the S-locus are highly conserved. In several kale strains that differ in their S alleles, the *SLR1* sequences predict identical proteins (61), and the stigmatic SLR1 glycoproteins showed no molecular-weight polymorphisms by immunoblot analysis (129). Within the alboglabra cultivar of *B. oleracea*, a comparison of two *SLR1* sequences derived from strains homozygous for the S_{29} and S_{63} self-incompatibility alleles (126) showed 99% conservation at the DNA level within the protein coding sequence and the 3' untranslated region. Ten of thirteen nucleotide substitutions resulted in amino acid substitutions, one of which involved the gain/loss of a potential N-glycosylation site. Only a silent substitution was allowed at one of the cysteines, underscoring the structural and/or functional importance of these residues. A similar degree of sequence conservation was observed at the *SLR2* locus. *SLR2* genes isolated from three different *B. oleracea* strains were 99% identical (11a).

The ubiquity of the *SLR* genes is further demonstrated by inter-specific comparisons that show conservation across species of *Brassica*, at least for the *SLR1* gene: An *SLR1* sequence isolated from a self-compatible strain of *B. campestris* shows 97% sequence similarity to the *B. oleracea SLR1* genes in

602 residues sequenced (61). SLR1 proteins were detected, by immunoblot analysis with antibodies raised against *B. oleracea SLR1*, in self-incompatible and self-compatible strains belonging to several *Brassica* species and to the related genus *Raphanus* (129). No molecular-weight polymorphisms were evident within species, but some variation in molecular weight was observed between species.

Based on the conservation of their sequences in different strains of *Brassica*, and the fact that they are unlinked to the S-locus, the *SLR1* and *SLR2* genes are not directly implicated in the determination of S-allelic specificity. The contribution of the SLR1 and SLR2 glycoprotein products to pollen recognition is yet to be determined. The obvious differences in selective pressure operating on S-locus genes and *SLR* genes, to generate diversity in the first case and to maintain near identity in the second case, must be related to the very different biological functions of the corresponding gene products. The extreme conservation of *SLR1* and *SLR2* sequences, and their expression to high levels even in self-compatible strains in which the *SLG* gene is either not functional or not expressed (61), suggests for the SLR protein products a fundamental role in pollination events in *Brassica*.

The involvement of at least some members of the *S*-gene family in general pollination processes is also suggested by the occurrence of sequences related to the *Brassica S*-gene family in other genera of the crucifer family that have no reported incidence of self-incompatibility. *Arabidopsis thaliana*, for example, contains approximately six such genes (84). Four of these genes were shown to contain open reading frames, but no expression of these genes was detected (99). It is tempting to speculate that some of these sequences may be the homologs of the *Brassica SLR* genes and that they fulfill a basic pollination-related function. In view of the ubiquity of the *SLR* genes in crucifers, we suggest that an *SLR*-like gene may represent the ancestral gene from which self-incompatibility genes have evolved by gene duplication.

Molecular Complexity of the S-Locus

The analysis of the *Brassica* S-multigene family has demonstrated that the S-locus, although behaving as a single Mendelian locus, has a complex organization. This conclusion stems from our observation, in F_2 populations segregating for different S-alleles, that self-incompatibility phenotype cosegregated with at least two polymorphic restriction fragments (16, 61, 84). In our laboratory, a total of over 300 F_2 plants have been analyzed to date in crosses between different S-allele homozygotes (class I × class I and class I × class II) and showed perfect cosegregation of these RFLPs and incompatibility phenotype. We recently described a structure consisting of a duplicated gene pair for the S-locus in an S_2 homozygote of *B. oleracea* (16).

This organization is a general feature of other S-alleles. The actual physical size of the S-locus remains to be determined, and will depend on the number of genes located at that locus. At a minimum, the size of the locus would correspond to the physical distance between the pair of S-locus genes. This distance is estimated to be greater than 40 kb, since no overlap of the flanking regions was observed in any of the genomic clones containing these genes.

The structure of one of the S-locus genes, the *SLG* gene, has been described in detail above. The second S-locus gene has been recently shown to be a functional gene and to contain, in addition to a region of S-homology, a region with sequence similarity to protein kinases (J. C. Stein, B. Howlett, M. E. Nasrallah, and J. B. Nasrallah, in preparation). This gene is similar to a putative receptor protein kinase gene of unknown function that has recently been isolated from maize roots (135). The maize gene has a structure consisting of a cytoplasmic domain containing sequence homology to the catalytic domains of protein kinases, a transmembrane domain, and an extracellular domain with sequence homology to *SLG*. While neither the maize nor the *Brassica* genes have been shown as yet to have kinase activity, the structures of these genes point to the existence of signal transduction pathways based on phosphorylation of protein substrates, and suggest that cell–cell communication in different plant tissues may be based on the action of a common class of related receptor protein kinases. We designate the second S-locus gene as the putative S-locus Receptor protein Kinase (*SRK*) gene.

The *SLG* and *SRK* genes derived from one S-allele homozygous genotype share approximately 94% DNA sequence identity and 90% amino acid sequence identity in their S-regions. It is likely that these genes act in concert to determine S-allelic specificity, and this sequence similarity may be essential for their function. A mechanism might thus exist that preserves the linkage relationship of the two S-locus genes. Additionally, the sequence homogeneity of these genes must somehow be maintained, possibly by frequent gene conversion, as has been suggested to explain the co-evolution of linked family members of other multigene families in higher eukaryotes (47). The requirement for at least two S-locus genes in the determination of self-incompatibility phenotype provides a molecular explanation for the general failure of strategies aimed at generating new allelic specificities by mutagenesis (see 89 for review). Several years ago, Fisher explained this failure by postulating that a newly generated S-allele remains incompatible with each of its two parental alleles (29). He also proposed an S-locus structure of even greater complexity than suggested here. To explain the extensive naturally occurring diversity of S-alleles, he visualized the S-locus as consisting of a chromosomal region with 10 or 20 antigenically active points which, upon recombination, could generate more than a thousand combinations, each of which represented an S-allele.

EXPRESSION OF *SLG* IN PISTILS AND ANTHERS

An important question is whether the isolated S-locus genes have the spatial and temporal pattern of expression expected of genes involved in the determination of SI phenotype in stigma and pollen. Genetic models of the pollen-stigma interaction of self-incompatibility predict expression of the S-locus genes in the two cell types of the flower that participate in the incompatibility response (for review see 38). In crucifers, as discussed above, the inhibition of incompatible pollen at the stigma surface predicts that the S-locus genes are expressed in the papillar cells of the stigma surface. Additionally, sporophytic control of pollen SI phenotype in this family predicts that the S-locus is expressed in sporophytic cells of the anther (6).

The analysis of *Brassica* S-locus gene expression during pollen development has lagged behind studies of gene expression in pistil tissue, and has been complicated by the expression of multiple related genes and the low level of expression of these genes in anther tissue. Recently however, the application of sensitive methods for the detection of gene expression has demonstrated that the two *Brassica* S-locus linked genes, the *SLG* gene (123, 125; T. Sato, M. K. Kandasamy, J. B. Nasrallah, M. E. Nasrallah, in preparation) and the *SRK* gene (J. C. Stein, B. Howlett, M. E. Nasrallah, and J. B. Nasrallah, in preparation) are expressed in specific cells of the pistil and anther. The expression of the *SLG* gene has been studied extensively and is reviewed here.

A detailed picture of *SLG* promoter activity has been obtained through the analysis of endogenous gene expression in *Brassica*, and through the analysis of plants transformed with *SLG* genes or with a chimeric gene consisting of the *SLG* promoter fused to the β-glucuronidase (GUS) reporter gene. A compilation of results obtained in *Brassica* and in transgenic plants belonging to three genera has demonstrated 1. the exclusive activity of the *SLG* promoter in two tissues of the flower, the pistil and the anther; and 2. significant species-specific differences in the temporal and spatial patterns of transgene expression.

Expression in Crucifers

In the *Brassica* stigma, the *SLG* gene was shown by in situ hybridization to be expressed specifically in the papillar cells of the stigma in keeping with the inhibition of incompatible pollen at the stigma surface in this genus (84). Figure 3 shows a series of sections taken from buds at different stages of development along a *Brassica* inflorescence (Figure 3A) and hybridized in situ to ^{35}S-labeled single-stranded "antisense" RNA probes transcribed in vitro from *SLG*-cDNA. No transcripts are evident in very young buds in which stigmatic papillar cells are in their early stages of differentiation (buds

17 and 15 in Figure 3). In later stages of bud development, *SLG* transcripts are detected in the stigmatic papillar cells. As the stigma matures, and while the differentiation and elongation of the papillar cells is occurring, a hybridization signal of increasing intensity is obtained (buds 13, 11, 9, 7, 5, and 3). No hybridization signal above background is evident in the underlying cells of the stigma, style and ovary, or in anthers.

In transgenic *B. oleracea* (T. Sato, M. K. Kandasamy, J. B. Nasrallah, M. E. Nasrallah, in preparation) and *Arabidopsis* (125), the *SLG* genes and the *SLG-GUS* fusion were similarly expressed at high levels in the papillar cells of the stigma. In plants expressing the *SLG-GUS* fusion, the sensitivity of assays for GUS activity allowed the demonstration of *SLG*-promoter activity in anthers. GUS staining was detected in the tapetum at much lower levels than in the stigma, and during a narrow developmental window after meiosis. Tapetal expression was first detected in correlation with the beginning of exine deposition; it disappeared coincident with tapetal cell degeneration. This sporophytic pattern of *SLG*-promoter activity in the tapetum implies that sporophytic control of SI phenotype in crucifers is based on the expression of the S-locus postmeiotically in the tapetum rather than on premeiotic expression as has also been suggested (95). In addition, the low level of activity exhibited by the *SLG* promoter in anthers may account for the fact that detection of the *SLG*-gene protein product in anthers of self-incompatible *Brassica* strains has met with little success to date.

Expression in Transgenic Tobacco and the Relationship of Gametophytic and Sporophytic Incompatibilities

A very different pattern of *SLG*-promoter activity was observed in transgenic *N. tabacum*. The highest levels of transgene expression were observed in the transmitting tissue of the style (51, 74, 123). A lower level of expression was also observed over the secretory cells in the basal transition region of the stigma and over the placental epidermal cells adjoining the ovules. It is possible that the activity of the *SLG* promoter represents a cell lineage–specific pattern of expression in the tobacco pistil, because, as described above, the various tissues involved are all derived from the same epidermal L1 layer of the young carpels. In the anthers of tobacco plants transformed with the *SLG*-GUS fusion, expression was detected only in mature pollen grains after the initiation of tapetal cell breakdown, and not in any sporophytic tissue of the anther.

The pattern of *SLG*-promoter activity observed in the pistil and pollen of transgenic tobacco reflects the expectations for the expression of endogenous self-incompatibility genes in Solanaceae. On the one hand, maximal expres-

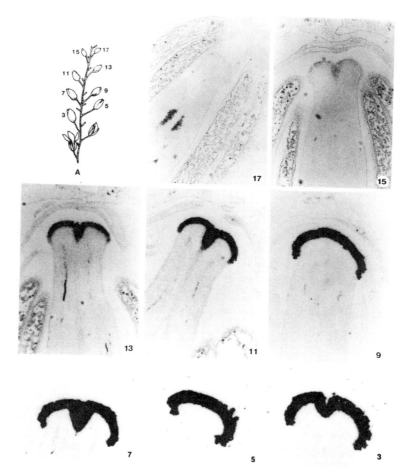

Figure 3   *SLG* gene expression in the developing *Brassica* flower.

The diagram of an inflorescence with buds at various stages of development is shown in (A). The numbers on the micrographs designate the bud stage in (A) from which the tissue section was obtained.

Paraffin-embedded sections were hybridized in situ with an ^{35}S-labeled single-stranded RNA probe synthesized in vitro from *SLG* cDNA. After a 2-day exposure, the sections were developed, stained with toluidene blue and examined by bright-field microscopy. Specific hybridization to transcripts in the papillar cells of the stigma is detected first in bud 13, which corresponds to the late uninuclear microspore stage.

sion in the transmitting tissue of the style is consistent with stylar inhibition of incompatible pollen in self-incompatible solanaceous plants. On the other hand, gametophytic expression of the *SLG* promoter in transgenic tobacco pollen is consistent with the timing of expression expected for the S-locus in its self-incompatible relative *N. alata*. This remarkable species-specific regulation of the *SLG* gene implies a degree of evolutionary relatedness between gametophytic and sporophytic self-incompatibility systems, at least at the level of gene regulation.

To some degree, also, the pattern of expression of the *Brassica SLG* genes in the transgenic tobacco pistil is similar to the expression of the S-associated genes cloned from self-incompatible *N. alata* (4, 20). It should be noted, however, that while this observation might suggest the conservation of regulatory signals, the similarity between the two sets of genes is limited to expression in pistil tissue. In contrast to the activity of the *Brassica SLG* promoter during pollen development, the S-associated genes cloned from *N. alata* are not expressed in male reproductive tissues (26). This lack of expression in anthers and pollen also characterizes genes cloned from other Solanaceae based on sequence homology with the *N. alata* S-associated genes. In *Petunia hybrida*, for example, no expression of such genes was observed in anthers and pollen even after the application of highly sensitive methods such as RNA blots capable of detecting sequences present at $10^{-4}\%$ of total mRNA and transient assays of β-glucuronidase reporter-gene activity (17).

GENETIC INTERACTIONS IN SELF-INCOMPATIBILITY RESPONSES

The Analysis of Self-Compatible Lines

An understanding of the function of S-locus genes and their involvement in self-incompatibility requires a knowledge of the genetic factors that affect S-locus function. In systems exhibiting sporophytic self-incompatibility, it is known that the activity of S-alleles is dependent on allelic interactions at the S-locus and on the action of genetic factors unlinked to this locus. In S-allele heterozygotes exhibiting dominant interactions, the recessive allele produces SLSG that is somehow inactivated. While nothing is known about the mechanism of dominance in heterozygous S-genotypes, an activation by posttranslational modification may be necessary for the gene product to be rendered functional.

The analysis of naturally occurring self-compatible variant strains has shown that the genetic breakdown of self-incompatibility can result from mutations in genes unlinked to the S-locus (Table 2). In *Brassica*, these

Table 2 Summary of genetic interactions: suppressor and modifier genes of *SLG*

Gene symbol	Description	Phenotypic effect	Reference
SLG	S-locus/*SLSG* structural gene	self-incompatibility	84
SUP1	incompletely dominant suppressor of incompatibility in *B. oleracea*	reduction in SLSG self-compatibility	85
sup2	recessive suppressor of incompatibility in *B. campestris*	reduction in SLSG self-compatibility	86
mod1	recessive modifier of incompatibility in *B. campestris*	processing of SLSG? self-compatibility	46

mutations are accompanied by the loss of the incompatibility response in the stigma but not in the pollen (45, 46, 85, 86, 122). In some self-compatible strains, self-compatibility resulted from the action of unlinked suppressor (*sup*) genes showing recessive (86), incompletely dominant (85), or dominant inheritance (122). In these cases, the self-compatible phenotype was associated with a reduction of SLSG in the stigma, perhaps as a result of decreased protein or transcript stability, or of down-regulation of the *SLSG* structural gene itself. In other self-compatible strains such as yellow sarson, a natural cultivar of *B. campestris*, the breakdown of S-locus function was attributed to the action of an unlinked recessive modifier gene designated *mod* (45). In self-compatible plants homozygous for the *mod* locus, SLSG was detected at normal levels in the stigma (M. E. Nasrallah, unpublished data).

The conclusion stemming from the analysis of the *mod* locus and the dominance interactions of S-alleles is that the presence of SLSG in the stigma is not sufficient in itself to impart a self-incompatibility phenotype. But is SLSG necessary for the development of specific incompatibility responses? The most compelling evidence for the involvement of SLSG in pollen-stigma recognition in *Brassica* is the large body of genetic evidence demonstrating the cosegregation of SLSG polymorphisms with the inheritance of incompatibility reactions in F_2 populations (44, 80, 82, 87, 88, 90). Additionally, a reduction in the level of SLSG is usually associated with the loss of incompatibility phenotype. This association has been observed not only in some of the previously described self-compatible strains (Table 2), but also, as discussed below, in transgenic *Brassica* plants that exhibit altered pollination phenotype.

The Analysis of Transgenic Plants

Brassica transformation experiments have been performed with *SLG* genes derived from class I alleles of *B. oleracea* and *B. campestris*. The breeding work required to assess self-incompatibility phenotype fully is still in its early stages, but it is already clear that the outcome of these experiments is dependent on the genetic background of the recipient plants and on the donor. The introduction of these class I *SLG* genes into self-compatible *B. napus* var. "Westar" plants did not generate self-incompatible plants (T. Nishio, M. K. Kandasamy, J. B. Nasrallah, and M. E. Nasrallah, in preparation), and the transformation of class II–containing *B. oleracea* plants with class I *SLG* genes produced no change in S-allelic specificity (T. Sato, M. K. Kandasamy, D. J. Paolillo, J. B. Nasrallah, and M. E. Nasrallah, in preparation). These results confirm the inference from the genetic analysis of self-compatible strains described above that the expression of the *SLG* gene is not sufficient in itself to confer self-incompatibility phenotype. However, transformation of self-incompatible *B. oleracea* with an *SLG* gene from a *B. campestris* class I S-allele resulted in the perturbation of self-incompatibility phenotype, and the generation of fully self-compatible transgenic plants (125a). The self-compatible transgenic phenotype was associated with a drastic reduction in the level of endogenous SLSG. The suppression of endogenous *SLG* gene expression is similar to the recently reported and poorly understood phenomenon of sense inhibition of gene expression in transgenic plants (60, 69, 77). It is interesting to note that the production of self-compatible transgenic phenotypes mimics the generation of the self-compatible amphidiploid genus *B. napus* following naturally occurring fusions between the genomes of the self-incompatible species *B. oleracea* and *B. campestris*. The basis of self-compatibility in *B. napus* is not understood, but may be a result of the co-suppression of the S-locus genes from each genome.

An implication of the discovery of at least two transcriptional units at the S-locus is that the determination of self-incompatibility phenotype requires at a minimum the expression of the two S-locus linked genes, *SLG* and *SRK*. A determination of the interactions of these genes, and of their contributions to self-incompatibility phenotype and allelic specificity, will therefore involve the transformation of *Brassica* not only with *SLG*, but also with the *SRK* gene either alone or in combination with *SLG*. By a combination of these transformation experiments and biochemical analyses, it should also be possible to decipher the underlying mechanisms of cell–cell signaling between pollen and papillar cells. In particular, these experiments will test the suggestion that pollen recognition is based on a cascade of events involving auto-phosphorylation of "S-receptor kinases" and/or phosphorylation of protein substrates by the SRK gene product.

GENETIC ABLATION OF FLORAL TISSUES

As a Tool for the Study of Plant Development

The tight regulation of the *SLG* gene during flower development was exploited to test the feasibility of using genetic ablation as an experimental tool in dissecting plant developmental processes. Genetic ablation, a method by which cell death is genetically targeted to specific tissues, has been a powerful tool in the analysis of cell lineages in mammals (reviewed in 28).

In designing *SLG* promoter–toxic gene fusions, the diphtheria toxin A (DT-A) gene was used. This gene fusion has been used to direct cell death genetically in a variety of mammalian tissues (8, 13, 94). DT-A blocks protein synthesis through ADP-ribosylation of elongation factor 2 in many organisms, including plants (96). A single molecule of DT-A can cause cell death (139), so DT-A can serve as a very sensitive indicator of gene expression. Additionally, the effects of DT-A expression are cell autonomous, since DT-A is not internalized in the absence of the DT-B chain.

Cells transformed with the *SLG* promoter fused to a toxic gene should be able to survive, proliferate, organize shoot and root meristems, and give rise to normal plants. In fact, transformation of *N. tabacum* with the SLG/DT-A gene fusion resulted in the production at high frequency of transgenic plants that underwent normal differentiation and produced flowers, a demonstration that, even at this level of sensitivity, the *SLG* promoter is not active in vegetative tissues, petals, and sepals. In transgenic flowers, only pistil tissues and pollen were ablated. The expression of the toxic gene fusion had profound effects on pistil development and function. Varying levels of transgene expression were obtained in different transgenic plants and resulted in a series of distinct phenotypes and structural abnormalities that depended on the timing of cell death along the temporal progression of morphogenesis. Carpel and placental fusion were disrupted by toxic gene expression in the primordial epidermis, resulting at the extreme in bicarpellate flowers and in multiple locules in the ovary. Even when carpel fusion was complete, an additional effect of toxic gene expression was the inhibition of stylar elongation and the production of stunted pistils. The observation that the ablated pistils of mature flowers did not support pollen germination, while those of young buds allowed germination but not pollen tube elongation, provides experimental evidence for the suggestion that the secretions of the stigma and style promote pollen hydration, germination, and pollen tube growth, and that the transmitting tissue of the style provides functions indispensable to the development of the pollen tube (57).

Toxic gene expression also affected pollen development. In contrast to the uniformly spherical pollen grains observed in untransformed control plants, pollen from the open flowers of toxic-gene transformants was a mix of spheres and small collapsed grains. The strictly gametophytic activity of the *SLG* promoter in tobacco was verified by the observation that only a subset, typically 50–75%, of pollen grains had the ablated phenotype and were nonfunctional in germination medium or in situ after pollination. Pollen grains that did not inherit the toxic-gene fusion were morphologically and functionally normal. In analogous experiments, a genetic ablation strategy that used a tapetal-specific promoter fused to a ribonuclease gene was recently applied to generate male-sterile plants (65a).

Thus, genetically directed cell killing using promoters specifically expressed in different cell types (32), combined with the ease with which transgenic plants can be generated, should allow the study of a broad range of temporal events in development. Together with laser microbeam ablation, and the more limited approach of generating chimeric plants (118), genetic ablation provides an important tool for studying the function, origins, and interrelations of plant cell lineages.

As a Tool for Crop Improvement

Genetic ablation methods have practical applications in plant improvement. The fact that the self-incompatibility genes are expressed in anthers and pistils has allowed the utilization of these genes to generate sterile plants by ablation of tissues essential for normal fertilization. In these plants, the corolla of the flower is perfectly normal in appearance. In ornamental crops such as *Petunia* and ornamental tobacco, flowers represent the organ of greatest economic value, and the generation of sterile varieties that may then be propagated on a large scale by meristem culture is of great advantage. In cultivars that rely on fruit development, the engineering of male-sterile lines by genetic ablation of pollen is potentially useful in the breeding of hybrid varieties. In this case, the restoration of fertility can be envisioned through the use of inducible factors.

ACKNOWLEDGMENTS

We thank Drs. M. K. Kandasamy for Figure 1, S. M. Yu for the in situ hybridization photographs, and T. Tantikanjana for Figure 2. We thank Dr. K. Dwyer for her numerous contributions to the work described in this review.

Literature Cited

1. Ai, Y., Singh, A., Coleman, C. E., Ioerger, T. R., Kheyr-Pour, A., Kao, T. H. 1990. Self-incompatibility in *Petunia inflata*: isolation and characterization of cDNAs encoding three S-allele associated proteins. *Sex. Plant Reprod.* 3:130-38

2. Albertini, L., Souvre, A., Audran, J. C. 1987. Le tapis de l'anthere et ses relations avec les microsporocytes et les grains de pollen. *Rev. Cytol. Biol. Veget.- Bot.* 10:211-42

3. Anderson, M. A., Cornish, E. C., Mau, S. L., Williams, E. G., Hoggart, R., et al. 1986. Cloning of cDNA for a stylar glycoprotein associated with expression of self-incompatibility in *Nicotiana alata*. *Nature* 321:38-44

4. Anderson, M. A., McFadden, G. I., Bernatzky, R., Atkinson, A., Orpin, T., et al. 1989. Sequence variability of three alleles of the self-incompatibility gene of *Nicotiana alata*. *Plant Cell* 1:483-91

5. Appels, R., Bouchard, R. A., Stern, H. 1982. cDNA clones from meiotic-specific poly(A)+ RNA in *Lilium*: homology with sequences in wheat, rye, and maize. *Chromosoma* 85:591-602

6. Bateman, A. J. 1955. Self-incompatibility systems in angiosperms. III. Cruciferae. *Heredity* 9:52-68

7. Beerhues, L., Forkmann, G., Schopker, H., Stotz, G., Wiermann, R. 1989. Flavanone 3-hydrolase and dihydroflavonol oxygenase activites in anthers of *Tulipa* . The significance of the tapetum fraction in flavonoid metabolism. *J. Plant Physiol.* 133:743-46

8. Behringer, R. R., Mathews, L. S., Palmiter, R. D., Brinster, R. L. 1988. Dwarf mice produced by genetic ablation of growth hormone-expression ex-pressing cells. *Genes Dev.* 2:453-61

9. Bell, J., Hicks, G. 1976. Transmitting tissue in the pistil of tobacco: light and electron microscopic observations. *Planta* 131:187-200

10. Bernatzky, R., Anderson, M. A., Clarke, A. E. 1988. Molecular genetics of self-incompatibility in flowering plants. *Dev. Genet.* 9:1-12

11. Boeke, J. H. 1971. Location of the post-genital fusion in the gynoecium of *Capsella bursa-pastoris* (L.) med. *Acta Bot. Neerl.* 20:570-76

11a. Boyes, D. C., Chen, C. H., Tantikanjana, T., Esch, J. J., Nasrallah, J. B. 1991. Isolation of a second S-locus-related cDNA from *Brassica oleracea*: genetic relationships between the S locus and two related loci. *Genetics* 127:221-28

12. Bredemeijer, G. M. M., Blaas, J. 1981. S-specific proteins in styles of self-incompatible *Nicotiana alata*. *Theor. Appl. Genet.* 59:185-90

13. Breitman, M. L., Clapoff, S., Rossant, J., Tsui, L. C., Glode, L. M., et al. 1987. Genetic ablation: targeted expression of a toxin gene causes microphthalmia in transgenic mice. *Science* 238:1563-65

14. Broothaerts, W. J., van Laere, A., Witters, R., Preaux, G., Decock, B., et al. 1989. Purification and N-terminal sequencing of style glycoproteins associated with self-incompatibililty in *Petunia hybrida*. *Plant Mol. Biol.* 14:93-102

15. Brown, S. M., Crouch, M. L. 1990. Characterization of a gene family abundantly expressed in *Oenothera organensis* pollen that shows sequence similarity to polygalacturonase. *Plant Cell* 2:263-74

16. Chen, C. H., Nasrallah, J. B. 1990. A new class of S sequences defined by a pollen recessive self-incompatibility allele of *Brassica oleracea*. *Mol. Gen. Genet.* 222:241-48

17. Clark, K. R., Okuley, J. J., Collins, P. D., Sims, T. L. 1990. Sequence variability and developmental expression of S-alleles in self-incompatible and pseudo-self-compatible *Petunia*. *Plant Cell* 2:815-26

18. Clarke, A. E., Anderson, M. A., Atkinson, A., Bacic, A., Ebert, P. R., et al. 1989. Recent developments in the molecular genetics and biology of self-incompatibility. *Plant Mol. Biol.* 13:267-71

19. Cornish, E. C., Anderson, M. A., Clarke, A. E. 1988. Molecular aspects of fertilization in flowering plants. *Annu. Rev. Cell Biol.* 4:209-28

20. Cornish, E. C., Pettitt, J. M., Bonig, I., Clarke, A. E. 1987. Developmentally controlled expression of a gene associated with self-incompatibility in *Nicotiana alata*. *Nature* 326:99-102

21. Dashek, W. V., Thomas, H. R., Rosen, W. G. 1971. Secretory cells of lily pistils: electron microscope cytochemistry of canal cells. *Am. J. Bot.* 58:909-20

22. Dickinson, H. G., Lewis, D. 1973. Cytochemical and ultrastructural differences between intraspecific compatible and incompatible pollinations in *Raphanus*. *Proc. Roy. Soc.* 184:21-28

23. Duck, N., McCormick, S., Winter, J. 1989. Heat shock protein hsp70 cognate gene expression in vegetative and reproductive organs of *Lycopersicon esculentum. Proc. Natl. Acad. Sci. USA* 86:3674-78

24. Dwyer, K. G., Chao, A., Cheng, B., Chen, C. H., Nasrallah, J. B. 1989. The *Brassica* self-incompatibility multigene family. *Genome* 31:969-72

25. East, E. M., Mangelsdorf, A. J. 1925. A new interpretation of the hereditary behavior of self-sterile plants. *Proc. Natl. Acad. Sci. USA* 11:166-71

26. Ebert, P. R., Anderson, M. A., Bernatky, R., Altschuler, M., Clarke, A. E. 1989. Genetic polymorphism of self-incompatibilty in flowering plants. *Cell* 56:255-62

27. Elleman, C. J., Willson, C. E., Sarker, R. H., Dickinson, H. G. 1988. Interaction between the pollen tube and stigmatic cell wall following pollination in *Brassica oleracea. New Phytol.* 109:111-17

28. Evans, G. A. 1989. Dissecting mouse development with toxigenics. *Genes Dev.* 3:259-63

29. Fisher, R. A. 1961. A model for the generation of self-sterility alleles. *J. Theor. Biol.* 1:411-14

30. Franceschi, V. R., Horner, H. T. 1979. Nuclear condition of the anther tapetum of *Ornithogalum caudatum* during microsporogenesis. *Cytobiologie* 18:413-21

31. Gasser, C. S., Budelier, K. A., Smith, A. G., Shah, D. M., Fraley, R. T. 1989. Isolation of tissue-specific cDNAs from tomato pistils. *Plant Cell* 1:15-24

32. Goldberg, R. B. 1988. Plants: novel developmental processes. *Science* 240:1460-67

33. Hanson, D. D., Hamilton, D. A., Travis, J. L., Bashe, D. M., Mascarenhas, J. P. 1989. Characterization of a pollen-specific cDNA clone from *Zea mays* and its expression. *Plant Cell* 1:173-79

34. Harris, P. J., Weinhandl, J. A., Clarke, A. E. 1989. Effect on in vitro pollen growth of an isolated style glycoprotein associated with self-incompatibility in *Nicotiana alata. Plant Physiol.* 89:360-67

35. Herdt, E., Sutfield, R., Wiermann, R. 1978. The occurrence of enzymes involved in phenylpropanoid metabolism in the tapetum fraction of anthers. *Cytobiologie* 17:433-41

36. Heslop-Harrison, J. 1968. Pollen wall development. *Science* 161:230-37

37. Heslop-Harrison, J. 1971. The pollen wall: structure and development. In *Pollen: Development and Physiology,* ed. J. Heslop-Harrison, pp. 75-98. London: Butterworths

38. Heslop-Harrison, J. 1975. Incompatibility and the pollen stigma interaction. *Annu. Rev. Plant Physiol.* 26:403-25

39. Heslop-Harrison, J. 1975. The physiology of the pollen grain surface. *Proc. R. Soc. London Ser. B* 190:275-99

40. Heslop-Harrison, J. 1987. Pollen germination and pollen tube growth. *Int. Rev. Cytol.* 107:1-77

41. Heslop-Harrison, J., Heslop-Harrison, Y., Knox, R. B., Howlett, B. J. 1973. Pollen-wall proteins: "gametophytic" and "sporophytic" fraction in the pollen wall of the Malvaceae. *Ann. Bot.* 37:403-12

42. Heslop-Harrison, J., Knox, R. B., Heslop-Harrison, Y. 1974. Pollen-wall proteins: exine-held fractions associated with the incompatibility response in Cruciferae. *Theor. Appl. Genet.* 44:133-37

43. Hill, J. P., Lord, E. M. 1987. Dynamics of pollen tube growth in the wild radish *Raphanus raphanistrum* (Brassicaceae). II. Morphology, cytochemistry and ultrastructure of transmitting tissues, and path of pollen tube growth. *Am. J. Bot.* 74:988-97

44. Hinata, K., Nishio, T. 1978. Stigma proteins in self-incompatible *Brassica campestris* L. and self-incompatible relatives, with special reference to S-allele specificity. *Jpn. J. Genet.* 53:27-33

45. Hinata, K., Okasaki, K. 1986. Role of the stigma in the expression of self-incompatibility in crucifers in view of genetic analysis. In *Biotechnology and Ecology of Pollen,* ed. G. B. Mulcahy, D. L. Mulcahy, E. Ottaviano, pp. 185-90. Berlin/New York/London: Springer-Verlag

46. Hinata, K., Okasaki, K., Nishio, T. 1983. Gene analysis of self-compatibility in *Brassica campestris* var yellow sarson (a case of recessive epistatic modifier). In Proc. 6th Int. Rapeseed Conf., Paris, I:354-59

47. Hood, L., Campbell, J. H., Elgin, S. C. R. 1975. The organization, expression and evolution of antibody genes and other multigene families. *Annu. Rev. Genet.* 9:305-53

48. Isogai, A., Takayama, S., Shiozawa, H., Tsukamoto, C., Kanbara, T., et al. 1988. Existence of a common glycoprotein homologous to S-glycoproteins in two self-incompatible homozygotes of

Brassica campestris. Plant Cell Physiol. 29:1331-36

49. Jahnen, W., Lush, W. M., Clarke, A. E. 1989. Inhibition of in vitro pollen tube growth by isolated S-glycoproteins of *Nicotiana alata. Plant Cell* 1:501-10

50. Kamboj, R. K., Jackson, J. F. 1986. Self-incompatibility alleles control a low molecular weight, basic protein in pistils of *Petunia hybrida. Theor. Appl. Genet.* 71:815-19

51. Kandasamy, M. K., Dwyer, K. D., Paolillo, D. J., Doney, R. C., Nasrallah, J. B., Nasrallah, M. E. 1990. *Brassica* S-proteins accumulate in the intercellular matrix along the path of pollen tubes in transgenic tobacco pistils. *Plant Cell* 2:39-49

52. Kandasamy, M. K., Paolillo, D. J., Faraday, C. D., Nasrallah, J. B., Nasrallah, M. E. 1989. The S locus specific glycoproteins of *Brassica* accumulate in the cell wall of developing stigma papillae. *Develop. Biol.* 134:462-72

53. Kanno, T., Hinata, K. 1969. An electron microscopic study of the barrier against pollen-tube growth in self-incompatible Cruciferae. *Plant Cell Physiol.* 10:213-16

54. Kaul, M. L. H. 1988. *Male Sterility in Higher Plants.* New York: Springer-Verlag

55. Kheyr-Pour, A., Bintrim, S. B., Ioerger, T. R., Remy, R., Hammond, S. A., Kao, T. H. 1990. Sequence diversity of pistil S-proteins associated with gametophytic self-incompatibility in *Nicotiana alata. Sex. Plant Reprod.* 3:88-97

56. Kirch, H. H., Uhrig, H., Lottspeich, F., Salamini, F., Thompson, R. D. 1989. Characterization of proteins associated with self-incompatibility in *Solanum tuberosum. Theor. Appl. Genet.* 78:581-88

57. Knox, R. B. 1984. Pollen-pistil interactions. In *Encyclopaedia of Plant Physiology*, ed. H. F. Linskens, J. Heslop-Harrison, 17:508-92. Berlin/Heidelberg/New York: Springer

58. Kroh, M. 1973. Nature of the intercellular substance of stylar transmitting tissue. In *Biogenesis of Plant Cell Wall Polysaccharides*, ed. F. Loewus, pp. 195-205. New York: Academic

59. Kroh, M., Munting, A. J. 1967. Pollen germination and pollen tube growth in *Diplotaxis tenuifolia* after cross pollination. *Acta Bot. Neerl.* 16:182-87

60. Krol, A. R. van der, Mur, L. A., Beld, M., Mol, J. N. M., Stuitje, A. R. 1990. Flavonoid genes in *Petunia*: addition of a limited number of gene copies may lead to a suppression of gene expression. *Plant Cell* 2:291-99

61. Lalonde, B., Nasrallah, M. E., Dwyer, K. D., Chen, C. H., Barlow, B., Nasrallah, J. B. 1989. A highly conserved *Brassica* gene with homology to the S-locus specific glycoprotein structural gene. *Plant Cell* 1:249-58

62. Laughnan, J. R., Gabay-Laughnan, S. 1983. Cytoplasmic male sterility in maize. *Annu. Rev. Genet.* 117:27-48

63. Lewis, D., Crowe, L. K. 1958. Unilateral incompatibility in flowering plants. *Heredity* 12:233-56

64. Lonsdale, D. M. 1987. Cytoplasmic male sterility: a molecular perspective. *Plant Physiol. Biochem.* 25:265-71

65. Lotan, T., Ori, N., Fluhr, R. 1989. Pathogenesis-related proteins are developmentally regulated in tobacco flowers. *Plant Cell* 1:881-87

65a. Mariani, C., de Beuckeleer, M., Truettner, J., Leemans, J., Goldberg, R. B., 1990. Induction of male sterility in plants by a chimaeric ribonuclease gene. *Nature* 347:737-41

66. Mascarenhas, J. P. 1975. The biochemistry of angiosperm pollen development. *Bot. Rev.* 41:259-314

67. Mascarenhas, J. P. 1989. The male gametophyte of flowering plants. *Plant Cell* 1:657-64

68. Mascarenhas, J. P. 1990. Gene activity during pollen development. *Annu. Rev. Plant Physiol. Plant Mol. Biol.* 41:317-38

69. Matzke, M. A., Primig, M., Trnovsky, J., Matzke, A. J. M. 1989. Reversible methylation and inactivation of marker genes in sequentially transformed tobacco plants. *EMBO J.* 8:643-49

70. Mau, S. L., Williams, E. G., Atkinson, A., Anderson, M. A., Cornish, E. C., et al. 1986. Style proteins of a wild tomato (*Lycopersicon peruvianum*) associated with expression of self-incompatibility. *Planta* 169:184-91

71. McClure, B. A., Haring, V., Ebert, P. R., Anderson, M. A., Bacic, A., Clarke, A. E. 1990. Molecular genetics and biology of self-incompatibility in *Nicotiana alata* an ornamental tobacco. *Aust. J. Plant Physiol.* 17:345-53

72. McClure, B. A., Haring, V., Ebert, P. R., Anderson, M. A., Simpson, R. J., et al. 1989. Style self-incompatibility gene products of *Nicotiana alata* are ribonucleases. *Nature* 342:955-57

73. Mogensen, H. L., Suthar, H. K. 1979. Ultrastructure of the egg apparatus of *Nicotiana tabacum* before and after fertilization. *Bot. Gaz.* 140:168-79

74. Moore, H. M., Nasrallah, J. B. 1990. A *Brassica* self-incompatibility gene is expressed in the stylar transmitting tissue of transgenic tobacco. *Plant Cell* 2:29-38

75. Mulcahy, D. L., Curtis, P. S., Snow, A. A. 1983. Pollen competition in a natural population. In *Handbook of Pollination Biology*, ed. C. E. Jones, R. E. Little, pp. 330-37. New York: Van Nostrand Reinhold

76. Mulcahy, G. B., Mulcahy, D. L. 1985. Ovarian influence on pollen tube growth, as indicated by the semivivo technique. *Am. J. Bot.* 72:1078-80

77. Napoli, C., Lemieux, C., Jorgensen, R. 1990. Introduction of a chimeric chalcone synthase gene into *Petunia* results in reversible co-suppression of homologous genes in trans. *Plant Cell* 2:279-89

78. Nasrallah, J. B., Doney, R. C., Nasrallah, M. E. 1985. Biosynthesis of glycoproteins involved in the pollen-stigma interaction of incompatibility in developing flowers of *Brassica oleracea* L. *Planta* 165:100-7

79. Nasrallah, J. B., Kao, T. H., Chen, C. H., Goldberg, M. L., Nasrallah, M. E. 1987. Amino-acid sequence of glycoproteins encoded by three alleles of the S locus of *Brassica oleracea*. *Nature* 326:617-19

80. Nasrallah, J. B., Kao, T. H., Goldberg, M. L., Nasrallah, M. E. 1985. A cDNA clone encoding an S-locus specific glycoprotein from *Brassica oleracea*. *Nature* 318:263-67

81. Nasrallah, J. B., Nasrallah, M. E. 1984. Electrophoretic heterogeneity exhibited by the S-allele specific glycoproteins of *Brassica*. *Experientia* 40:279-81

82. Nasrallah, J. B., Nasrallah, M. E. 1986. Molecular markers of self-incompatibility in *Brassica*. In *Biotechnology and Crop Improvement and Protection*, Monogr. No. 34, ed. P. R. Day, pp.83-89. Cambridge: British Crop Protection Council

83. Nasrallah, J. B., Nasrallah, M. E. 1989. The molecular genetics of self-incompatibility in *Brassica*. *Annu. Rev. Genet.* 23:121-39

84. Nasrallah, J. B., Yu, S. M., Nasrallah, M. E. 1988. Self-incompatibility genes of *Brassica oleracea*: expression, isolation and structure. *Proc. Natl. Acad. Sci. USA* 85:5551-55

85. Nasrallah, M. E. 1974. Genetic control of quantitative variation in self-incompatibility proteins detected by immunodiffusion. *Genetics* 76:45-50

86. Nasrallah, M. E. 1989. The genetics of self-incompatibility in *Brassica* and the effects of suppressor genes. In *Plant Reproduction: From Floral Induction to pollination*, ed. E. M. Lord, G. Bernier, 1:146-55. Rockville, MD: Am. Soc. Plant Physiol.

87. Nasrallah, M. E., Wallace, D. H. 1967. Immunogenetics of self-incompatibility in *Brassica oleracea* L. *Heredity* 22:519-27

88. Nasrallah, M. E., Wallace, D. H., Savo, R. M. 1972. Genotype, protein, phenotype relationships in self-incompatibility of *Brassica*. *Genet. Res.* 20:151-60

89. Nettancourt, D. d. 1977. *Incompatibility in Angiosperms*. Berlin: Springer-Verlag. 230 pp

90. Nishio, T., Hinata, K. 1977. Analysis of S-specific proteins in stigmas of *Brassica oleracea* L. by isoelectric focusing. *Heredity* 38:391-96

91. Nishio, T., Hinata, K. 1982. Comparative studies on S-glycoproteins purified from different S-genotypes in self-incompatible *Brassica* species. I. Purification and chemical properties. *Genetics* 100:641-47

92. Ockendon, D. J. 1974. Distribution of self-incompatibility alleles and breeding structure of open pollinated cultivars of Brussels sprouts. *Heredity* 33:159-71

93. Ockendon, D. J., Gates, P. J. 1975. Growth of cross- and self-pollen tubes in the styles of *Brassica oleracea*. *New Phytol.* 75:155-60

94. Palmiter, R. D., Behringer, R. R., Quaife, C. J., Maxwell, F., Maxwell, I. H., Brinster, R. L. 1987. Cell lineage ablation in transgenic mice by cell-specific expression of a toxin gene. *Cell* 50:435-43

95. Pandey, K. K. 1970. Time and site of the S-gene action, breeding systems and relationships in incompatibility. *Euphytica* 19:364-72

96. Pappenheimer, A. M., Gill, D. M. 1972. Inhibition of protein synthesis by activated diphtheria toxin. In *Molecular Mechanisms of Antibiotic Action on Protein Biosynthesis*, ed. E. Munoz, F. Garcia-Fernandez, D. Vasquez, pp. 134-39. Amsterdam: Elsevier

97. Picton, J. M., Steer, M. W. 1982. A model for the mechanism of tip extension in pollen tubes. *J. Theor. Biol.* 98:15-20

98. Pluijm, J. E. van der, Linskens, H. F. 1966. Feinstruktur der Pollen-Schläuche im Griffel von *Petunia*. *Züchter* 36:220-24

99. Pruitt, R. E. 1990. Abstr. 4th Int. Conf.

Arabidopsis Res., ed. D. Schweizer, K. Peuker, J. Loidl, p. 74. Vienna: Univ. Vienna

100. Roberts, I. N., Stead, A. D., Ockendon, D. J., Dickinson, H. G. 1979. A glycoprotein associated with the acquisition of the self-incompatibility system by maturing stigmas of *Brassica oleracea*. *Planta* 146:179-83

101. Roggen, H. P. J. R. 1967. Changes in enzymeactivities during the pregamete phase in *Petunia hybrida*. *Acta Bot. Neerl.* 16:1-31

102. Rosen, W. G. 1968. Ultrastructure and physiology of pollen. *Annu. Rev. Plant Physiol.* 19:435-62

103. Sanders, L. C., Lord, E. M. 1989. Directed movement of latex particles in the gynoecia of three species of flowering plants. *Science* 243:1606-8

104. Sarker, R. H., Elleman, C. J., Dickinson, H. G. 1988. Control of pollen hydration in *Brassica* requires continued protein synthesis, and glycosylation is necessary for intraspecific incompatibility. *Proc. Natl. Acad. Sci. USA* 85: 4340-44

105. Sassen, M. M. A. 1974. The stylar transmitting tissue. *Acta Bot. Neerl.* 23:99-108

106. Satina, S. 1944. Periclinal chimeras in *Datura* in relation to development and structure (A) of the style and stigma (B) of calyx and corolla. *Am. J. Bot.* 31:493-502

107. Satina, S., Blakeslee, A. F. 1943. Periclinal chimeras in *Datura* in relation to the development of the carpel. *Am. J. Bot.* 30:453-62

108. Sattler, R. 1973. *Organogenesis of Flowers. A Photographic Atlas*. Toronto: Univ. Toronto Press

109. Scutt, C. P., Gates, P. J., Gatehouse, J. A., Boulter, D., Croy, R. R. D. 1990. A cDNA encoding an S-locus specific glycoprotein from *Brassica oleracea* plants containing the S5 self-incompatibility allele. *Mol. Gen. Genet.* 220:409-13

110. Shivanna, K. R., Heslop-Harrison, Y., Heslop-Harrison, J. 1978. The pollen-stigma interaction: bud pollination in the Cruciferae. *Acta Bot. Neerl.* 27:107-19

111. Siegel, B. A., Verbeke, J. A. 1989. Diffusible factors essential for epidermal cell redifferentiation in *Catharanthus roseus*. *Science* 244:580-82

112. Sinyukin, A. M., Britikov, E. A. 1967. Action potentials in the reproductive system of plants. *Nature* 215:1278-80

113. Smith, A. G., Gasser, C. S., Budelier, K. A., Fraley, R. T. 1990. Identification and characterization of stamen- and tapetum-specific genes from tomato. *Mol. Gen. Genet.* 222:9-16

114. Smith, A. G., Hinchee, M. A., Horsch, R. 1987. Cell and tissue specific expression localized by in situ hybridization in floral tissues. *Plant Mol. Biol. Rep.* 5:237-41

115. Sommer, H., Beltrán, J. P., Huijser, P., Pape, H., Lönnig, W. E., Schwarz-Sommer, Z. 1990. *Deficiens*, a homeotic gene involved in the control of flower morphogenesis in *Antirrhinum majus*: the protein shows homology to transcription factors. *EMBO J.* 9:605-13

116. Stieglitz, H. 1977. Role of β-1,3-glucanase in postmeiotic microspore release. *Dev. Biol.* 57:87-97

117. Stinson, J. R., Eisenberg, A. J., Willing, R. P., Pe, M. E., Hanson, D. D., Mascarenhas, J. P. 1987. Genes expressed in the male gametophyte of flowering plants and their isolation. *Plant Physiol.* 83:442-47

118. Sussex, I. M. 1989. Developmental programming of the shoot meristem. *Cell* 56:225-29

119. Takayama, S., Isogai, A., Tsukamoto, C., Ueda, Y., Hinata, K., et al. 1986. Structure of carbohydrate chains of S-glycoproteins in *Brassica campestris* associated with self-incompatibility. *Agric. Biol. Chem.* 50:1673-76

120. Takayama, S., Isogai, A., Tsukamoto, C., Ueda, Y., Hinata, K., et al. 1987. Sequences of S-glycoproteins, products of the *Brassica campestris* self-incompatibility locus. *Nature* 326:102-4

121. Thompson, K. F., Taylor, J. P. 1966. Non-linear dominance relationships between S alleles. *Heredity* 21:345-62

122. Thompson, K. F., Taylor, J. P. 1971. Self-compatibility in kale. *Heredity* 27: 459-71

123. Thorsness, M. K., Kandasamy, M. K., Nasrallah, M. E., Nasrallah, J. B. 1991. A *Brassica* S-locus gene promoter targets toxic gene expression and cell death to the pistil and pollen of transgenic *Nicotiana*. *Dev. Biol.* 143:173-84

124. Tilton, V. R., Horner, H. T. 1980. Stigma, style, and obturator of *Ornithogalum caudatum* (Liliaceae) and their function in the reproductive process. *Am. J. Bot.* 67:1113-31

125. Toriyama, K., Thorsness, M. K., Nasrallah, J. B., Nasrallah, M. E. 1991. A *Brassica* S-locus gene promoter directs sporophytic expression in the anther tapetum of transgenic *Arabidopsis*. *Dev. Biol.* 143:427-31

125a. Toriyama, K., Stein, J. C., Nasrallah,

M. E., Nasrallah, J. B. 1991. Transformation of *Brassica oleracea* with an S-locus gene from *B. campestris* changes the self-incompatibility phenotype. *Theor. Appl. Genet.* In press

126. Trick, M. 1990. Genomic sequence of a *Brassica* S locus related gene. *Plant Mol. Biol.* 15:203-5

127. Trick, M., Flavell, R. B. 1989. A homozygous S genotype of *Brassica oleracea* expresses two S-like genes. *Mol. Gen. Genet.* 218:112-17

128. Twell, D., Wing, R., Yamaguchi, J., McCormick, S. 1989. Isolation and expression of an anther-specific gene from tomato. *Mol. Gen. Genet.* 218:240-45

129. Umbach, A. L., Lalonde, B. A., Kandasamy, M. K., Nasrallah, J. B., Nasrallah, M. E. 1990. Immunodetection of protein glycoforms encoded by two independent genes of the self-incompatibility multigene family of *Brassica*. *Plant Physiol.* 93:739-47

130. Ursin, V. M., Yamaguchi, J., McCormick, S. 1989. Gametophytic and sporophytic expression of anther-specific genes in developing tomato anthers. *Plant Cell* 1:727-36

131. Vasil, I. K., Johri, M. M. 1964. The style, stigma, and pollen tube. I. *Phytomorphology* 14:352-69

132. Verbeke, J. A., Walker, D. B. 1985. Rate of induced cellular dedifferentiation in *Catharanthus roseus*. *Am. J. Bot.* 72:1314-17

133. Vithanage, H. I., Knox, R. B. 1976. Pollen-wall proteins: quantitative cytochemistry of the origins of intine and exine enzymes in *Brassica oleracea*. *J. Cell Sci.* 21:423-35

134. Walker, D. B. 1975. Postgenital carpel fusion in *Catharanthus roseus* (Apocynaceae). I. Light and scanning electron microscopic study of gynoecial ontogeny. *Am. J. Bot.* 62:457-67

135. Walker, J., Zhang, R. 1990. Relationship of a putative receptor kinase from maize to the S-locus glycoproteins of *Brassica*. *Nature* 345:743-46

136. Williams, E. G., Heslop-Harrison, J. 1979. A comparison of RNA synthetic activity in the plasmodial and secretory types of tapetum during the meiotic interval. *Phytomorphology* 29:370-81

137. Wing, R. A., Yamaguchi, J., Larabell, S. K., Ursin, V. M., McCormick, S. 1989. Molecular and genetic characterization of two pollen-expressed genes that have sequence similarity to pectate lyases of the plant pathogen *Erwinia*. *Plant Mol. Biol.* 14:17-28

138. Xu, B., Grun, P., Kheyr-Pour, A., Kao, T. H. 1990. Identification of pistil-specific proteins associated with three self-incompatibility alleles in *Solanum chacoense*. *Sex. Plant Reprod.* 3:54-60

139. Yamaizumi, M., Mekada, E., Uchida, T., Okada, Y. 1978. One molecule of diphtheria toxin fragment A introduced into a cell can kill the cell. *Cell* 15:245-50

140. Yanofsky, M. F., Ma, H., Bowman, J. L., Drews, G. N., Feldmann, K. A., Meyerowitz, E. M. 1990. The protein encoded by the *Arabidopsis* homeotic gene *agamous* resembles transcription factors. *Nature* 346:35-39

Annu. Rev. Plant Physiol. Plant Mol. Biol. 1991. 42:423-66

PHYSIOLOGICAL AND MOLECULAR STUDIES OF LIGHT-REGULATED NUCLEAR GENES IN HIGHER PLANTS

William F. Thompson

Departments of Botany and Genetics, North Carolina State University, Raleigh, North Carolina 27695-7612

Michael J. White

Department of Botany, North Carolina State University, Raleigh, North Carolina 27695-7612

KEY WORDS: gene expression, photobiology, transcription, posttranscriptional control, translational control, posttranslational control

CONTENTS

1040-2519/91/0601-0423$02.00

INTRODUCTION

Light is essential to normal plant growth, both because it provides energy for photosynthesis and because it provides many of the environmental signals that regulate plant development. The transition from an etiolated seedling to a fully green plant is as dramatic a developmental event as any other in the higher-plant life cycle. During this process, the expression of many genes is affected in many different ways. Light interacts with endogenous developmental programs to modulate these gene responses, often by acting through two or more different photoreceptors.

This review is concerned with how gene responses to light are controlled physiologically and biochemically. It is by no means exhaustive. We deal almost exclusively with nuclear genes, and offer only a general overview of the burgeoning literature on transcriptional control elements. We emphasize the diversity and complexity of light responses at several levels of gene expression, and the utility of the physiological approach in molecular and genetic studies.

Photoreceptors

PHYTOCHROME   Higher plants contain several major light receptors that respond to light intensity, spectral quality, and state of polarization. Collectively, these receptors can absorb photons over a wide range of wavelengths, ranging from the far-red to the ultraviolet. The major light receptor absorbing in the red and far-red portions of the electromagnetic spectrum is phytochrome. This chromoprotein was first isolated on the basis of its red and far-red reversible absorption spectrum (42). Phytochrome exists in two spectrophotometric forms, the red-absorbing (P_r) form, and the far-red-absorbing (P_{fr}) form. These two forms may be reversibly interconverted by irradiation, which leads to isomerization of the phytochrome chromophore, the tetrapyrrole biliverdin (reviewed in 240). Dark-grown plants contain the P_r form, often referred to as the inactive form.

Absorption of red light will convert P_r into P_{fr}, leading to many morphological, physiological, and biochemical changes. The latter include changes in transcription, and in mRNA and protein abundances (Table 1). Far-red

Table 1 Light-regulated nuclear genes in higher plants

Gene[a]	Protein	Receptor[b]	Response[c]	Reference[d]
Genes encoding Plastid Proteins				
CabI	LHCI	R + B + UV	+, r	210, 220–222, 275, 318
CabII	LHCII	R + B + UV	+, r	178, 210, 264, 315, 318
—	ELIP[e]	?	+, r	215
GapA	gly.-3-phos.[h] DH[h] A	?	+	257
GapB	gly.-3-phos. DH B	?	+	257
—	gln synthase 2	R	+	88, 296, 316
—	nitrite reductase	?	+, r	30
—	nitrite reductase 2	R	+	251
—	PEPC	?	+	208
—	Phosphoribulokinase	?	+	231
Pcr	PChlide reductase	R	−[f]	7, 70, 199, 200
PetE	plastocyanin	R + B	+	170, 210
PetF(Fed-1)	ferredoxin	?	+	82, 312
PsaD	subunit II, PSI	R + B	+, r	107, 210
PsbO (Oee1)	33-kD OEC[g]	?	+, r	107, 154, 256
PsbP (Oee2)	23-kD OEC	R	+	256, 319
PsbQ (Oee3)	16-kD OEC	?	+	256
PsbR (ST-LS1)	10-kD OEC	?	+	87, 278
RbcS	SSU	R + B + UV	+, r	78, 84, 96, 215, 318
Rca	Rubisco activase	?	+	328
Genes for Enzymes of Flavonoid Biosynthesis				
Chs	chalcone-synthase	R + B + UV	+	47, 155, 214
Pal	Phe amm.-lyase	UV	+	47, 162
—	4-coum.:CoA lig.	UV	+	47, 162
Other Genes				
—	Asn synthase 1	R	−	303
—	β-sub.mit. ATPase	?	−, r	225
—	biotin-binding protein	?	−, r	107
Cam	calmodulin	?	−	327
Din	?	?	−	316
GapC	gly.-3-phos. DH C	?	+	257
—	heat shock proteins (4)	?	r	215
NR11	?	R	−	297
NR18	?	R	−	297
NR300	?	R	−	297
—	nitrate reductase		+, r	30, 100
—	nitrate reductase 1–4	R	+	251
—	thionins	R + B	−	232
pEA214	?	R	+	138, 139
pEA238	?	R	+	138, 139

Table 1   (*continued*)

Gene[a]	Protein	Receptor[b]	Response[c]	Reference[d]
pEA303	?	R	+	138
pEA315	?	R	+	138, 139
pEA215	?	R + B	+	178, 315
pEA25	?	R + B	−	178, 315
pEA207 (Blec1)	bud lectin	B	−	83, 178, 315
Phy	phytochrome	R	−	66, 300
rRNA		?	+	12, 103, 199, 262
TubA	α-tubulin	?	r	223
TubB	β-tubulin	R*	−, r	41, 64, 225

[a] Nomenclature for nuclear-encoded plastid proteins follows reference 121 where possible. (—) indicates no standardized gene nomenclature available.

[b] Photoreceptor designations are as follows: R = affected by red light, mediated by phytochrome; R* = affected by red light, not far-red reversible; B = affected by blue light and/or UV-A light, likely mediated by cryptochrome; UV = affected by UV or UV-B light, likely mediated by UV-B receptor; ? = photoreceptor not determined.

[c] (+) = increases in light; (−) = decreases in light or increases in darkness; r = fluctuates following a diurnal or circadian rhythm.

[d] References cited for CabII, RbcS, and Phy include reviews; all others are original sources.

[e] ELIP = Early Light-Induced Protein.

[f] Both PChlide reductase protein and Pcr mRNA levels decrease after light treatment of barley seedlings (see the introduction in text), but only the protein is affected in cress (147).

[g] OEC = Oxygen Evolving Complex of photosystem II.

[h] Nonstandard abbreviations: Gly.-3-phos. DH = glyceraldehyde-3-phosphate dehydrogenase; Phe amm. lyase = phenylalanine ammonia lyase; 4-coum. : CoA lig. = 4-coumarate : CoA ligase; β-sub. mit. ATPase = β subunit of the mitochondrial ATPase

wavelengths (e.g. 720–760 nm) will convert most of the P_{fr} form back into the P_r form. Thus far-red will often cancel the effects of preceding red light. Although phytochrome is primarily a red and far-red receptor, both P_r and P_{fr} forms also absorb in the blue region of the spectrum, and to a limited extent at other wavelengths as well (e.g. 191). Thus even green "safelights" are capable of producing some photoconversion and eliciting sensitive responses.

Various types of phytochrome molecule may also be distinguished on a physiological, biochemical, or immunological basis. For example, the most abundant phytochrome in etiolated tissues (Type I) can be distinguished from the most abundant phytochrome in green tissues (Type II) immunologically, by peptide mapping, and on the basis of amino acid sequence (reviewed in 99). Type I phytochrome undergoes proteolytic degradation when illumination converts it to the P_{fr} form, while Type II phytochrome contents in green tissue are relatively stable in the light (99). A 124-kDa phytochrome molecule predominates in etiolated oat seedlings, but the species in green tissue has a molecular mass of 118 kDa (299). This 124-kDa Type I molecule is sensitive to light-induced proteolysis, and corresponds to the *PhyA* gene in oat (253, 255, 299).

Multiple phytochrome genes have recently been identified in several species (reviewed in 300), and multiple transcripts for a single phytochrome I gene have been documented in pea (243, 301). For example, the *Arabidopsis* genome contains four to five phytochrome-related gene sequences (255), and genomic clones for three phytochrome loci have been isolated from maize (57). In cases where a single phytochrome gene has been reported, such as rice (141) and zucchini (171), it is likely that other genes exist that encode Type II phytochrome.

OTHER PHOTORECEPTORS   In addition to phytochrome, the major red-light absorbing pigments in higher plants are chlorophylls *a* and *b*, and their precursor, protochlorophyllide (PChlide). Like the phytochrome chromophore biliverdin, they are also tetrapyrroles and possess absorption maxima in the blue region of the spectrum as well as the red. The photoconversion of PChlide into chlorophyllide (Chlide) is mediated by the enzyme NADPH-PChlide oxidoreductase (henceforth PChlide reductase) and requires NADPH and light (114, 294). Thus dark-grown plants do not contain chlorophyll but do contain a small pool of PChlide. There are two forms of photoconvertible PChlide, with absorption maxima of approximately 638 and 650 nm (85, 114, 294). Both absorption spectra overlap the 667-nm absorption maximum of the P_r form of phytochrome (240). Thus the major criterion for distinguishing the actions of PChlide from those of phytochrome is the far-red reversibility of phytochrome responses. The photoconversion of PChlide leads to chlorophyll synthesis, the formation of green chloroplasts containing photosystems I and II, and the initiation of photosynthesis. Thus PChlide may be considered a light receptor, not only because its photoconversion dramatically alters plant metabolism via the process of photosynthesis, but also because its photoconversion is required for the assembly of the photosynthetic apparatus.

Other light receptors in higher plants include a blue/ultraviolet-A (UV-A) light receptor, and an ultraviolet-B (UV-B) light receptor (84, 191, 210, 214, 318). The blue and UV-A light receptor is frequently referred to as "cryptochrome." These receptors are postulated primarily on the basis of numerous responses that possess action spectra with maxima in the blue or UV (reviewed in 228, 252). Neither blue nor UV light receptors have yet been isolated, although both flavoproteins and carotenoids have been postulated as candidates for blue-light receptors (228, 252).

INTERACTIONS   The various photoreceptors interact with each other in complex fashions. Photoconversion of two or more light receptors is often required to obtain maximum levels of gene expression, as with *Cab*, *RbcS*, and certain enzymes of flavonoid synthesis (Table 1). In other cases, one light receptor may act to potentiate its own action (e.g. 14, 194) or the actions of a second light receptor (e.g. 84, 194, 214). Thus brief light pretreatments may decrease the fluence needed to elicit a response to a subsequent light treatment, increase the magnitude of the response, or alter the lag time between receptor photoconversion and realization of the response.

In at least some instances, blue or UV light treatments will enhance sensitivity to phytochrome but not elicit a response on their own. One well-studied example is phytochrome-induced anthocyanin synthesis in seedlings (reviewed in 194, 272). Evidence for the blue-light/UV-A receptor acting through phytochrome is further supported by studies on light-regulation of nucleus-encoded transcripts in the *aurea* mutant of tomato (210), which has greatly reduced levels of phytochrome in etiolated seedlings. Levels for several of these transcripts in wild-type tomato show a blue-light-induced increase beyond the levels produced by red light alone. However, neither red nor blue light leads to accumulation of these transcripts in the *aurea* mutant, showing that blue light is ineffective in the absence of phytochrome in this system (210). Nevertheless, blue light (or the blue/UV-A receptor) appears to be effective by itself in some systems. For example, blue light strongly increases the response of chalcone synthase transcription to UV light (around 310 nm), but blue light alone induces chalcone synthase mRNA to a low but significant level (214).

Light Responses

DE-ETIOLATION AND GREENING   Seedlings grown in the dark, or plant tissues that develop for a prolonged period in darkness, accumulate large amounts of phytochrome and PChlide reductase relative to green plants. Upon illumination of an etiolated plant, P_r is converted to P_{fr}, PChlide is converted to Chlide, signal transduction is initiated, and the abundance of phytochrome and PChlide reductase mRNAs is greatly decreased (7, 16, 65, 67, 68, 171). This decrease in mRNA levels is due at least in part to rapid and dramatic decreases in the transcription of both phytochrome and PChlide reductase genes (172, 199, 200). The mRNA decrease, together with proteolysis of the proteins, leads to substantially reduced amounts of phytochrome and PChlide reductase in light-grown plants (59, 115, 177, 242), and the residual phytochrome consists primarily of the light-stable or so-called green-tissue phytochrome (59, 99, 299). A light-stable pool of a PChlide reductase(-like) protein located in the plasmalemma has also been reported (79, 80, 123).

With further illumination, chlorophyll and carotenoid levels increase, altering the internal light regime of the plant. These photosynthetic pigments partially block absorption of light by the P_r form of phytochrome and blue-light receptors. Thus the accumulation of photosynthetic pigments alters the quality and intensity of the light absorbed by the photoreceptors. The effects of photosynthetic pigment accumulation, light quality, or light intensity on internal light regimes, photoreceptor function, gene expression, and photosynthetic acclimation have become increasingly important topics in recent investigations (e.g. 6, 55, 56, 75, 109, 134, 135, 311).

DIURNAL AND CIRCADIAN RHYTHMS   Green plants or plants grown under a diurnal light regime also exhibit biological rhythms. PChlide reductase undergoes a diurnal variation in level (115), and many genes exhibit diurnal or circadian rhythms in expression. Circadian rhythms in transcript abundance have been demonstatrated for *Cab* genes (92, 107, 152, 205, 275), nitrate reductase (100), early light-inducible protein (152, 215), four nucleus-encoded heat shock genes (215), and to a lesser extent for *RbcS, rbcL, psbA, TubA, PsbO (Oee1), PsaD* and a biotin-binding protein (107, 152, 223, 225). These rhythms interact with photoreceptor systems to control gene expression. For example, the phase and amplitude of *Cab* transcript rhythms are determined by light signals (107, 146, 166, 186, 223, 224, 283). Experiments with red or far-red light given at the end of the photoperiod indicate that phytochrome interacts with the circadian clock to control the amplitude of *Cab* transcript accumulation (205). It appears that the time of the dark-to-light transition (224), or the time of the light-to-dark transition (166), are involved in determining the times of minimum and maximum *Cab* transcript accumulation.

DIVERSITY OF RESPONSES

Etiolated Seedlings

The most extensive photobiological studies have been carried out with etiolated seedlings. Advantages of the etiolated system include the absence of complications arising from the presence of chlorophyll and a functional photosynthetic apparatus. In addition, phytochrome responses, which offer a well-characterized photoreceptor system and the experimentally convenient property of photoreversibility, are often more prominent in etiolated material.

Even such an apparently simple system can exhibit a great diversity of responses, however. Various responses, including the induction of different genes, can differ with respect to light requirement, time course, and induction kinetics. The capacity for a particular response can also vary with development, reflecting the interaction of light signals with an underlying developmental program.

FLUENCE REQUIREMENTS   Most plastid protein genes exhibit a normal low-fluence phytochrome response, in which induction by red light is fully reversible by subsequent far-red light. However, accumulation of *Cab* mRNA is mediated partly by a very-low-fluence (VLF) response, as well as by a typical low-fluence response (127, 140). The VLF response is orders of magnitude more sensitive than normal responses; it can be induced by irradiations converting less than 1% of the phytochrome to P_{fr}, and is saturated by far-red light pulses producing only about 3% P_{fr} (140). Very few genes other than *Cab* exhibit such a sensitive response. Those that do include *phyA* (67), *Pcr* (198, 199, 200), and probably β-tubulin (66). It is interesting that whereas *Cab* mRNA increases, the latter mRNAs all decrease in abundance.

Cab mRNA accumulation shows a VLF response in several species other than pea. The evidence is strongest for barley (198, 199, 200), but both wheat (204), and *Arabidopsis* (136) show partial induction by FR pulses. However, there appears to be no significant VLF effect on *Cab* transcription in oat seedlings (172); and although there is a strong VLF response for total *Cab* mRNA in pea buds (127, 140), some members of this gene family also lack a VLF response, responding to red light pulses only in the low fluence range (M. J. White, B. Fristensky, D. Falconet, and W. F. Thompson, unpublished). In contrast, a single wheat *Cab* gene, *Cab*-1, exhibits both VLF (induction by a FR pulse) and LF (photoreversible) responses. Similar results were obtained when this gene was transferred to tobacco (204), where it was shown that regulatory sequences sufficient for both responses are contained in a relatively short region 5' to the start of transcription (203). Further dissection of the *Cab*-1 promoter may determine whether or not different *cis*-acting elements are required for the two components of its phytochrome response.

Why do *Cab* genes so frequently show VLF responses when most other genes do not? Selection has probably favored mechanisms ensuring that *Cab* proteins accumulate rapidly whenever the plant begins to accumulate chlorophyll, since light energy absorbed by free chlorophyll cannot be dissipated readily and can cause severe damage to plastids through photooxidation (see 209). Production of small amounts of *Cab* mRNA in response to very low fluences of light, such as might be encountered a few centimeters below the soil surface (175, 286), would be one way of preparing for rapid synthesis of *Cab* proteins prior to seedling emergence from the soil and the beginning of rapid chlorophyll accumulation. This hypothesis fails to explain why *Cab* mRNA is not synthesized in the dark, as are the mRNAs for certain other chlorophyll apoproteins encoded by the plastid genome (202; see the section below on Translational and Posttranslational Controls). However, one can speculate that constitutive synthesis of a superabundant mRNA like *Cab* might unduly deplete the limited energy reserves of an etiolated seedling. In addition, if *Cab* proteins were to be produced too rapidly after illumination they might compete for chlorophyll with the reaction center and other antenna proteins. Such competition might interfere with photosystem assembly and lead to photodamage early in the greening process (cf 202). It should be possible to test the latter idea by using chimaeric *Cab* gene constructs to modify the level and regulation of *Cab* mRNA in transgenic plants.

INDUCTION AND ESCAPE KINETICS   The rates at which different mRNAs accumulate can differ dramatically. For example, it is possible to distinguish several different accumulation patterns in etiolated pea buds responding to a single pulse of red light. Some transcripts show delayed accumulation, while *Cab*, *RbcS*, and several other mRNAs begin to increase immediately and accumulate steadily over the 24-hr period following the pulse (139). Transcripts encoding GS2, a plastidic glutamine synthase, follow a similar time course (88). In contrast, *Fed*-1 mRNA (referred to as "pEA46" in 139) increases rapidly (within 1–2 hr) and then remains relatively constant.

Perhaps more significantly, the rate at which the need for P_{fr} was lost also differed dramatically in this group of genes. For most mRNAs, including *Cab* and *RbcS*, increasing the time between the inductive red-light pulse and a subsequent far-red-light treatment resulted in a gradual decrease in the effectiveness of the far-red light (127, 139). However in other cases, including *Fed*-1, far-red remained effective for at least several hours following the red-light pulse (82, 139). The *Fed*-1 response is of special interest, since it can be reversed by far-red light given after mRNA accumulation is complete. Thus the continued presence of *Fed*-1 mRNA requires the continued presence of P_{fr}, in contrast to the inductive nature of most other phytochrome responses.

COMPETENCE   Cells that respond to light (and not all cells do) respond in different ways at different times. These differences are manifested as spatial and temporal patterns of responsiveness that are independent of light and thought to reflect the operation of an underlying developmental program. A classic example comes from early work by Wagner & Mohr (313), who showed that phytochrome-induced synthesis of anthocyanins occurs in subepidermal cells of the mustard seedling hypocotyl. Only a single layer of cells is competent to respond in this way; no other cells accumulate anthocyanins. Competence can also vary with time or developmental stage within a single cell type. Thus, for example, anthocyanin synthesis in mustard cotyledons can be induced by P_{fr} only after a certain point in germination (276). Light treatments prior to this time have no effect as long as any P_{fr} formed is converted back to P_r before competence is attained. A few hours later, however, the same light treatments produce a very strong phytochrome response. For more discussion of these and other examples, the reader is referred to reviews by Mohr (190) and Schopfer (249).

Such issues are of fundamental importance for studies of gene regulation, since programmed changes in competence may alter the molecular apparatus regulating a given gene at different stages of plant development or in different cell types. To take an extreme example, the pea *Fed*-1 gene can be expressed in both roots and cotyledons of transgenic tobacco seedlings under the control of the CaMV 35S promoter. This gene has been shown to contain internal *cis* elements capable of conferring light responsiveness in the absence of a light-responsive promoter (90). However, only leaf or cotyledon cells are competent to recognize these elements (289). Similarly, Schmid et al (247) showed that the activity of a bean *Chs* promoter increased upon illumination in cotyledons and decreased near the root-hypocotyl junction where expression was high in the dark. Averaging these effects might well obscure both light responses. Clearly, analogous situations involving different cell types within a single organ would be more difficult to detect, and thus more confusing to analyze by conventional techniques.

PLASTID FACTOR   One of the major determinants of competence is the so-called "plastid factor" that is thought to be produced by plastids and required for expression of certain nuclear genes whose products are plastid-related (209, 282). The molecular nature of this factor(s) remains elusive, although it appears not to be a protein (211). Its role is inferred from experiments with plants lacking the carotenoid pigments that normally protect against photooxidative destruction of plastids. In carotenoid-deficient mutants (33, 181), or in plants treated with inhibitors of carotenoid biosynthesis such as Norflurazon (212, 233), plastid development proceeds normally as long as the plants are grown in dim light. Under these conditions a normal complement of cytoplasmic mRNAs accumulate. However, when such plants are exposed to strong light, photooxidation quickly destroys chlorophyll and most other chloroplast components. Under these conditions the cytoplasmic mRNAs encoding *Cab* and several other plastid proteins decrease to low levels, even though it can be shown that photooxidative damage is confined to the organelle (181, 212, 233). Both in vitro experiments and work with chimaeric genes in transgenic plants suggest that the block is at the level of transcription, suggesting that the plastid factor is a positive regulator of transcription whose production ceases rapidly when plastids are destroyed by photooxidation [(122, 277); see also reviews by Oelmüller (209) and Taylor (282)].

A certain degree of plastid development appears necessary for formation of the plastid factor, since chloramphenicol can prevent its appearance when applied early enough to block plastid development but is ineffective when applied later (211). Schmidt et al (248) showed that the time at which the *RbcS* and *Cab* genes begin to acquire competence to respond to phytochrome corresponds roughly to the time at which the plastid factor first becomes detectable. However, although the plastid factor is necessary for these responses, it is apparently not sufficient. A role for other, gene-specific, factors can be inferred from the fact that the time course for appearance and disappearance of competence differs for *RbcS* and *Cab*.

Green Plants

Many light-induced mRNAs decline when green plants are placed in the dark for extended periods, and increase again upon re-illumination. Such assays are particularly useful in work with transgenic plants, since primary transformants must be regenerated and grown in the light. However, the light responses observed in dark-adaptation assays are quite different from those of etiolated seedlings. Dark-grown and green leaves contain very different populations of phytochromes, with much lower levels of Type I phytochrome and thus proportionately greater amounts of other phytochrome species (99, 300). These changes would be expected to affect the nature of phytochrome responses in green material. At present we have relatively little detailed physiological information on gene responses to phytochrome in green plants, in part because the blue/UV-A photoreceptor assumes much greater importance in "dark adaptation" experiments with green tissue. Jenkins & Smith (134) found that changes in the phytochrome photoequilibrium had no effect on the abundance of *RbcS* and *Cab* mRNAs in green pea seedlings, in contrast to the strong phytochrome responses of these genes in etiolated material. Fluhr & Chua (96) showed that red-light treatments that would have induced *RbcS* gene expression in dark-grown peas had little effect on dark-adapted green leaves, but that blue light could induce large increases in transcript abundance. Since the blue-light response was reduced by subsequent far-red light, they concluded that a co-action of phytochrome and the blue-light receptor was involved (see 193, 210 for more on the co-action concept). A blue-light effect was also obtained by Clugston et al (63) using a shorter dark-adaptation period and more defined irradiation conditions. These investigators could not clearly establish a role for phytochrome in their system, however. Low-fluence-rate red light was ineffective, and higher fluence rates might also exert their effect via photosynthetic pigments. If, as these results seem to indicate, phytochrome plays only a minor role in the gene responses of green tissue, it is not surprising that Chory et al (53) found normal levels of light-responsive gene expression in light-grown leaves of several *Arabidopsis* mutants with low or undetectable levels of spectrally active phytochrome. This observation must be interpreted with caution, however, since spectral assays are limited to type I phytochrome and the phytochrome status of the light-grown seedlings remains unknown.

TRANSCRIPTIONAL CONTROLS

In vitro experiments

Light effects on transcription were first demonstrated with "run-on" assays in which nuclei isolated from light-treated and control plants are allowed to synthesize RNA in vitro and the radioactive transcripts from specific genes subsequently quantitated by hybridization to cloned DNAs. Following the original report by Gallagher & Ellis (102) of a white-light effect on *Cab* and *RbcS* transcription in pea, many papers have described responses of these and other genes to phytochrome, the blue/UV-A receptor, or UV-B irradiation (e.g. 12, 28, 47, 102, 103, 172, 198, 200, 262; reviewed in 65, 261, 290, 292, 298).

It is clear that the run-on assay reveals transcription-related effects in a wide variety of systems, and the basic conclusion from these experiments — that light affects transcription of certain plant genes — has been elegantly corroborated by work with transgenic plants (see below). However, it is still not at all clear exactly what the run-on assay measures. In the absence of chain initiation (which does not seem to occur in isolated nuclei), RNA synthesis must result from elongation of RNA chains initiated in vivo prior to nuclear isolation. Thus it is often assumed that the activity of the in vitro system is proportional to the number of polymerases engaged on a given template. However, experience with animal systems has shown that a better indication of the number of engaged polymerases can be obtained by using agents such as heparin, sarkosyl, or high salt. These treatments dissociate most chromosomal proteins but do not disrupt active transcription complexes (e.g. 234). These complexes can then elongate RNA chains with minimal interference from chromosomal proteins. In plants, we are aware of only two examples in which this approach was taken. Chappell & Hahlbrock (47) showed that UV irradiation of parsley suspension culture cells increased transcription of *Pal* and *Chs* in both intact and salt-lysed nuclei, indicating that the light treatment may have increased the initiation rate in vivo, and thus the number of polymerases transcribing these genes. However, when barley nuclei were treated with heparin, in vitro RNA synthesis increased five fold and *Cab* or *Pcr* transcriptional activity was no longer affected by prior in vivo red-light treatments (200). The latter result supports the hypothesis that the phytochrome effect detected in this assay is a change in chromatin structure that enhances the rate of RNA chain elongation.

If the run-on assay is sensitive to factors affecting elongation, as the barley results suggest, quantitative interpretations are open to question because elongation rates in vitro cannot be assumed to accurately reflect those in vivo. On the other hand, if the assay measures only the frequency of in vivo initiation, it remains unclear to what extent this parameter controls the overall process of RNA synthesis in living cells. Elongation rates in vivo could well vary from gene to gene or as a function of development. We conclude that run-on transcription can be used as a qualitative indicator of changes in certain (unknown) aspects of transcriptional activity, but that its use as a quantitative measure of RNA synthetic activity is highly questionable.

Transgenic Plants

Control of transcription can be demonstrated by showing that nontranscribed regulatory sequences confer light responsiveness on a normally nonresponsive reporter gene. This approach was first used to demonstrate light responses mediated by *RbcS* and *Cab* promoters (126, 195, 265, 295). These early experiments used transformed callus cultures that had to be grown for prolonged periods to demonstrate light effects. Since different media were required to support growth in the dark and permit plastid development in the light, it was difficult to separate light effects from the nutritional and hormonal variables. However, these experiments did clearly demonstrate the ability of upstream sequences to regulate gene expression. It is now routine to work with transgenic plants rather than with callus cultures, and a number of other genes have been analyzed. This literature has been extensively reviewed (e.g. 19, 97, 106, 133, 161, 206, 261, 264, 290, 317), and space considerations preclude detailed discussion here. In general, however, transcriptional regulatory sequences are located in promoter- and enhancer-like regions 5' to the start of transcription. Elements at or near the 3' ends of certain genes have been reported to influence transcriptional activity (e.g. 77) but have not yet been shown to regulate light responses per se.

DEFINING *cis*- ACTING ELEMENTS   Showing that light responsiveness could be conferred by large regions 5' to the start of transcription was at least conceptually simple, but defining individual elements within these regions has proven a more difficult task. Difficulties arise largely because promoters turn out to be complex assemblages of interacting elements.[1] The complexity makes it hard to separate (for example) elements that act as general enhancers from others that act in response to light or developmental signals.

Many of the known promoter or enhancer elements were first defined as short (10–20 bp) sequences that seemed to have been conserved in evolution. For example, sequences now known to be binding sites for the transcription factor GT-1 were first identified as conserved sequences in several *RbcS* genes (97, 111, 160). A similar approach has been used to identify a variety of other potential regulatory sequences (e.g. 46, 108, 118, 176). Elements can also be defined in terms of their ability to interact with sequence-specific DNA-binding proteins, using nuclear extracts in combination with gel mobility shift and DNase-I footprinting assays (e.g. 108, 111), or by means of "genomic footprinting" experiments that measure the effect of bound proteins

---

[1]It has been realized for many years that coordinating gene expression in eukaryotic organisms would require arrays of regulatory elements at each locus. For example, soon after their discovery of repetitive DNA, Britten & Davidson (39, 74) proposed that interspersed repetitive sequences might serve as *cis* elements for a multivalent system with formal properties very similar to those now being characterized.

on the ability of dimethylsulfate to methylate guanosine residues in the DNA of living cells (e.g. 250).

Once a conserved sequence or a protein binding site has been identified, its function must be established in vivo. To date, most such information has been derived from loss-of-function deletion and mutagenesis experiments (e.g. 46, 112). One complication associated with this approach is that promoters often contain multiple elements that are functionally redundant (76, 159, 305). In addition, mutational analysis in the context of a single promoter may not resolve elements mediating light responses or cell specificity from those that simply alter the overall level of expression (106). To approach such questions it is necessary to carry out gain-of-function experiments with individual elements. Thus, mutational analysis showed that binding sites for the nuclear factors ASF-2 and GT-1 were both essential components of light-responsive promoters, but could not further distinguish their functions (46, 94, 105, 159). However, clear differences were revealed by gain-of-function experiments in which tetramers of these binding sites were separately fused to a truncated viral promoter. In this way the ASF-2 site was shown to confer leaf-specific expression that was independent of light, while the GT-1 site could mediate both light responses and tissue specificity (165, 167). The availability of such simple model promoters has tremendous implications for future studies of *trans*-acting factors and the events that bring about changes in their activity.

TRANSCRIPTION FACTORS   Proteins capable of binding to specific DNA sequences are under study in a number of laboratories. Among the best known are GT-1, which was originally discovered in studies with pea *RbcS* genes and nuclear extracts and binds to sequences containing GGTTAA (111, 112), and GBF, which was discovered in studies using other *RbcS* genes in *Arabidopsis* and tomato nuclear extracts and binds to sequences including CACGTG (108). Neither of these factors has been purified, so it is as yet unknown whether they represent a single protein species or a family of related proteins.

There is evidence that the association of a GBF-like protein with the parsley *Chs* promoter is altered in vivo during induction of *Chs* transcription by UV light (250), and experiments discussed above (165, 167) show that synthetic binding sites for GT-1 confer light-responsive and organ-specific expression. There is thus little doubt that these factors are somehow involved in light-induced transcription. However, there is no consistent change in GBF or GT-1 activity in extracts of plants exposed to light or darkness (108, 111, 274). Although this observation may at first appear paradoxical, there are several ways in which a protein might participate in light responses without altering its in vitro binding ability. Certain readily reversible posttranslational modifications, such as phosphorylation, might not be preserved in vitro but might strongly influence DNA binding and/or the ability to activate transcription in vivo (cf 71). Alternatively, or in addition, light responses might be mediated by other factors that modify transcription by interacting with DNA binding proteins but do not themselves bind DNA and are therefore not detectable in binding assays.

In addition to GT-1 and GBF, several other factors binding to light-responsive promoters are under study in several laboratories (see 106 for a review). Some, such as ASF-2 (165), appear to determine organ specificity rather than light responsiveness, while there are preliminary reports that others may change in abundance or activity in response to light (see 43 and unpublished data cited in 106). The significance of the latter observation is unclear, however, since the binding sequences can be eliminated from the pea *RbcS*-3A promoter without affecting its light responsiveness (106). We have much yet to learn.

PROMOTER COMPLEXITY   Promoter elements in natural genes do not function in isolation. Natural promoters are complex multicomponent structures in which the arrangement of elements and interactions between the factors binding to them must be considered. There is great potential for combinatorial diversity in such systems, which may contribute to the diversity of light responses described in the preceding section (e.g. 19a, 246). Even more complex interactions may arise from the presence of enhancer-like elements upstream of the promoter proper. Although most of the functional data presently available are for sequences within a few hundred bases of the transcription start site, in many cases it is known that additional regulatory sequences are present further upstream, and that the relative importance of these elements may differ at different times in development (e.g. 159, 161, 206, 246, 305).

These examples of complex interactions provide some indication of what is likely to be much greater biological complexity. To date, most attention has been focused on defining a few elements in relation to a simple light/dark response. Much more must be learned before we will be able to assess the extent to which the many subtle differences in the light responses of different genes can be explained in terms of promoter architecture. However, we now have a clear conceptual basis for thinking about such problems, and progress should continue to be rapid. Among the more interesting questions that can now be approached is the extent to which different light responses are mediated by different *trans*-acting factors.

POSTTRANSCRIPTIONAL CONTROLS

Although modulation of transcriptional activity is the most commonly documented way of controlling mRNA abundance, posttranscriptional processes can also be regulated. In animal systems there are many examples of developmental modulations of RNA processing or RNA stability that play important roles in determining the steady-state mRNA levels (see 13, 34, 60, 148, 287 for recent reviews). However, work with plants has lagged well behind work with animals in this area, and our knowledge is still fragmentary. Much of what we know concerns genes in the plastid genome, which are regulated very differently from nuclear genes and are not reviewed here. The reader is encouraged to consult several recent reviews (117, 120, 201).

Comparison of Transcription and mRNA Abundance

EXAMPLES OF DISCREPANCIES   For nuclear genes, most of the evidence for posttranscriptional mechanisms consists of discrepancies between in vitro transcription data and in vivo mRNA levels. The first such data were reported for phytochrome (*phyA*) mRNA, which declines very rapidly when etiolated oat seedlings are given a pulse of red light (67). Initial attempts to measure transcription rates suggested an approximately 3-fold decrease following red light, too little to account for the 10-fold reduction in *phyA* message (229). Similar logic suggests that the phytochrome-mediated decrease in the level of *Pcr* mRNA in barley seedlings may also be mediated at least partly at the posttranscriptional level (38, 198). In the latter case, changing the mRNA level requires higher fluences than are needed to affect transcription, another indication that the two processes are not tightly coupled.

Meagher and his colleagues have reported discrepancies between *RbcS* transcription and mRNA data in soybeans at two different stages of development. When young seedlings were transferred from light to darkness, transcription declined more rapidly than mRNA abundance (258), which is consistent with the notion that the mRNA might have a longer half-life in the dark. However, the opposite effect was observed in mature plants, where a relatively small decrease in transcription accompanied a sharp drop in mRNA level (288). In this case, the authors suggested that *RbcS* mRNA in mature leaves is preferentially degraded in the dark.

Marrs & Kaufman (179) have suggested that blue light can act both to stimulate *Cab* mRNA transcription and to increase its turnover in pea seedlings. Using pea seedlings grown in low-intensity red light to saturate phytochrome responses, they showed that run-on transcription of *Cab* genes is stimulated above the red-light control in nuclei isolated following either a low- or high-fluence pulse of blue light. *Cab* mRNA levels also increase in response to low-fluence blue-light irradiation, as would be expected, but do not change in response to a high-fluence pulse (315). One interpretation of these data is that the high-fluence blue-light reaction stimulates *Cab* mRNA turnover as well as transcription, and that the two effects cancel each other out.

TECHNICAL PROBLEMS   As already noted, much uncertainty is associated with the use of in vitro transcription to make quantitative estimates of in vivo RNA synthesis rates. This consideration alone introduces great uncertainty in any attempt to interpret "discrepancies" such as those just described. Furthermore, since absolute synthesis rates cannot usually be calculated for specific mRNAs it is usually necessary to resort to comparisons of the relative increases or decreases in transcriptional activity and mRNA abundance. Such comparisons are extremely sensitive to background problems. These are not always predictable, but will usually have a disproportionate effect when signal strength is low, and when dot- or slot-blot hybridization is used rather than methods that detect bands of discrete molecular weight.

Some such artifact may have affected the early data on *phyA* gene transcription discussed previously in this section. More recent work from the same group has indicated a much larger relative decrease in *phyA* transcription following a red-light pulse, and the new data are in excellent agreement with the changes in mRNA abundance (172). The difference between the two sets of data could not be explained fully, but much of it seems to have been associated with the use of double-stranded plasmid clones as hybridization targets for run-on transcripts in the earlier work, as compared with single-stranded M-13 phage clones in the subsequent investigation. It seems reasonable that the background hybridization might have been higher in the former case, preventing observation of the full transcriptional decrease.

The Ferredoxin System

Messenger RNA encoding ferredoxin I in peas shows strong light responses, increasing by at least 5–10 fold after red-light treatment or when transferred to white light (138, 139, 289). Recent experiments from our laboratory (90, 91, 291) raise the possibility that this response is controlled at a posttranscriptional level. Experiments with chimaeric genes in transgenic tobacco show that *cis*-acting elements sufficient to confer light-responsive mRNA accumulation are located within the transcribed portion of the gene rather than in the promoter. Since *Fed*-1 includes no introns, the active elements must be located in the message sequence. Internal *cis* elements could conceivably act as enhancers of transcriptional initiation, although most internal enhancers in animal genes are located in introns (e.g. 98). Internal elements could also act on transcriptional elongation and/or termination (e.g. 23, 326). However, transgenic tobacco plants containing these *Fed*-1 constructs show large changes in mRNA abundance in the absence of detectable changes in run-on transcription activity (289). While not conclusive, these data favor the idea that *Fed*-1 mRNA abundance may be regulated largely at a posttranscriptional level. If so, this system will provide a convenient model for detailed studies of posttranscriptional light responses.

TRANSLATIONAL AND POSTTRANSLATIONAL CONTROLS

Light acts at many levels of protein biosynthesis, function, and degradation. Although some light-induced transcriptional and posttranscriptional responses are relatively rapid, they are less direct and thus usually less rapid in their effect on protein function as compared to photoregulated steps occurring in the later stages of protein biosynthesis. Thus translational and posttranslational photoregulation allow more immediate response to the environment, and probably provide additional adaptational flexibility as well.

Light may affect translation, protein folding and transport, posttranslational modification, association with cofactors, and protein degradation. Direct and indirect effects on enzyme activity also occur, as in the regulation of Calvin cycle enzymes by the ferredoxin-thioredoxin system (reviewed in 40). In the case of nucleus-encoded organelle proteins, light may also affect steps such as protein import, post-import processing, and insertion into organelle membranes or internal compartments. Examples of the latter include the insertion of *Cab* proteins into the thylakoid membranes (see the section below on *Cab* proteins as a model system), and targeting of plastocyanin into the thylakoid lumen (268, 269). Many of the best-known examples of translational and posttranslational photoregulation involve proteins of the photosynthetic apparatus, and it is these proteins we emphasize in the following discussion.

Translational Control

It is often difficult to distinguish translational controls from certain posttranscriptional or posttranslational controls. Translational control by light may be postulated if an mRNA is present in dark-grown plants but the corresponding protein is absent and protein turnover is not occurring, or if mRNA levels remain stable upon a change in illumination, but protein levels are rapidly increased or decreased in the absence of changes in the rate of protein turnover. Further insight may be gained by determining if the mRNA is associated with polysomes, and by testing the translational competence of the polysomes. Berry et al (26, 27) used such an approach to study the rapid depression of Rubisco large subunit (LSU) and small subunit (SSU) synthesis that occurs upon transfer of light-grown amaranth (or pea, 36) seedlings into darkness. RNAs for both subunits remained bound to polysomes in darkness, but were not translated in vivo, although they were translatable in vitro (25, 26, 27). Thus light-regulation of translation can occur both in the chloroplast and in the cytosol. The mechanism by which light regulates chloroplast translation is unknown in higher plants, but appears to be ATP dependent in *Chlamydomonas* (187).

Unlike the situation in light-grown amaranth, *rbcL* and *RbcS* RNAs are not associated with polysomes in dark-grown amaranth seedlings, reflecting an absence of translational initiation (25). Other data indicate that illumination causes a general stimulation of plastid translational initiation and plastid translation activity (104, 150). Illumination of 4.5-day-old dark-grown barley seedlings induces recruitment of *rbcL* and *psbA* transcripts into chloroplast polysomes (150). Illumination of 8-day-old dark-grown barley causes *rbcL* and *psbD-psbC* transcripts to be shifted into larger polysomes, and increases plastid polysome content (104, 150). Gamble et al (104) showed that barley plastid protein synthesis was equally stimulated by red or white light, and that the effect of a red pulse is partially attenuated by far-red light given immediately after the red light. These experiments indicate that phytochrome is one receptor involved in the activation of barley plastid protein synthesis (104).

Light regulation may occur at steps in translation other than initiation. In dark-grown barley seedlings, for example, the synthesis of CPI polypeptides encoded by *psaA* and *psaB* is arrested at the level of translational elongation (149, 158). This translational block is removed by the synthesis of Chl *a* in vivo and in isolated plastids, allowing CPI accumulation (149, 151, 158, 306). Light-induced accumulation of CPI apoprotein is not far-red reversible and does not occur in Chl-deficient mutants despite the presence of the relevant mRNA (149, 163). All these data indicate that PChlide is the light receptor controlling translation of CPI (149, 163).

It seems likely that PChlide photoconversion regulates CPI and the other Chl *a* proteins by virtue of its effect on chlorophyll synthesis, with protein accumulation depending on the number of chlorophyll molecules bound to a given apoprotein. However, other modes of action for PChlide photoconversion cannot yet be completely excluded. For example, the products of photoconversion could act allosterically to influence other aspects of plastid development and photomorphogenesis. Such possibilities should not be dismissed without experimentation, especially given the existence of an extraplastidic PChlide-reductase(-like) molecule in the plasmalemma (79, 80).

Using an in vitro translation system from lysed plastids, Eichacker et al (89) have shown that the conversion of PChlide to Chlide *a* is a necessary but not a sufficient condition for the accumulation of CPI apoprotein. In addition, phytol pyrophosphate must be added to allow Chl *a* synthesis. Similar Chl synthesis requirements were obtained for the Chl *a* apoproteins CP47 and CP43, and a protein that comigrates with the D1 protein (89). However, pulse-chase labeling experiments suggest that failure of CP43 and D1 to accumulate in the dark may result from rapid turnover (202). The above data constitute an important milestone in photobiology, by providing direct molecular evidence of a light receptor function for PChlide. Taken together with those of Gamble et al (104), these data indicate that different light receptors can mediate translational control of different proteins. It will be interesting to determine the extent to which photoreceptor interactions are involved in this process as well.

Posttranslational Control

Following translation, the nascent polypeptide chain is folded and transported to its cellular destination. Other proteins such as chaperonins or heat-shock(-like) proteins may assist in this process, and a number of plant heat-shock proteins have been found in chloroplasts (48, 124, 153, 180, 307, 308). One example of a chaperonin involved in light-regulated protein biosynthesis and assembly is the Rubisco subunit-binding protein (15, 110, 124, 238, 239). This chaperonin forms stable associations not only with Rubisco subunits but also with a number of other polypeptides imported into chloroplasts (174). Transcripts of several nucleus-encoded heat-shock proteins are known to follow circadian rhythms similar to those of light-induced proteins (215). However, the extent to which light regulates the synthesis of these proteins remains to be determined.

The process of protein biosynthesis or degradation may be accompanied by posttranslational modifications such as proteolytic processing, methylation, acetylation of amino groups, glycosylation, ubiquitination, etc, some of which may be light-dependent. For example, phytochrome is aggregated and ubiquitinated following illumination, and this process appears to be part of a degradatory pathway (131, 132, 253). Posttranslational modifications such as attachment of cofactors are critical for CAB protein accumulation and function, and are discussed in further detail below. In addition, phosphorylation of certain CAB polypeptides found in LHCII is associated with the distribution of excitation energy between photosystems I and II (21, 22, 35, 81, 86, 273, 293).

CAB proteins as a Model System

The most-studied nucleus-encoded chloroplast proteins are the Chl *a/b*-binding proteins (CAB proteins) and SSU, and their biosynthetic pathways and functional regulation have been discussed in a number of recent reviews (35, 78, 143, 144, 176, 282, 293). Therefore we concentrate here on light-regulated steps in posttranslational CAB protein accumulation. The major posttranslational steps in CAB protein biosynthesis are chloroplast import, post-import processing, insertion into the thylakoid membranes, complexation with photosynthetic pigments, proteolytic processing, and assembly into the correct photosystem. A number of these steps, as well as CAB protein accumulation and degradation, are directly or indirectly regulated by light acting through phytochrome, a blue-light receptor, PChlide, and the process of photosynthesis.

The first indirect light requirement for chloroplast import of nucleus-encoded proteins is ATP (119; for recent reviews see 143, 144, 188). Flügge & Hinz (95) showed that transport of the precursor protein (SSU in this case) was strictly light dependent, or required externally added or indirectly imported (Mg)ATP in the dark (95). Such chloroplast import experiments are typically done with isolated chloroplasts, and nonphotosynthetic sources of ATP might circumvent this apparent light requirement under some circumstances in vivo. Use of cytosolic ATP, or molecules such as dihydroxyacetone phosphate (95) that give rise to ATP upon being metabolized, could explain the uptake of polypeptides into nonphotosynthetic plastids (32, 188).

After entering the chloroplast, CAB polypeptides must be transported to the thylakoid membranes. The mechanism by which transport occurs is unknown, but recent results indicate that a soluble stromal intermediate is involved (62). The insertion of a 31-kDa LHCII precursor polypeptide into the thylakoid is accompanied by a stromal processing activity (1, 52, 169) that cleaves at one or two sites of the amino terminus to give the mature 25- and 26-kDa polypeptides (58). This stromal activity is induced during light-dependent plastid development, along with an increase in the receptivity of the plastid membranes (52). These data show that light-induced development is required for both stromal and thylakoid components involved in LHCII assembly and processing. Insertion of LHCII precursor polypeptides into the thylakoid can precede proteolytic processing (50, 51, 52), but insertion does not necessarily precede processing under conditions where the maturation protease is not limiting (62) or when mature chloroplasts are used (51). Integration of LHCII precursor polypeptides into the thylakoids requires light and/or ATP (61). This process is inhibited by ionophores or uncouplers that collapse the *trans*-thylakoid proton electrochemical gradient (62), indicating that the light requirement reflects a requirement for photosynthesis.

Insertion of CAB polypeptides into thylakoids is accompanied by attachment of photosynthetic pigments, whose synthesis in turn requires a number of light-regulated steps. Regulation of Chl protein accumulation by pigment synthesis is not unique to the CAB proteins, as the Chl *a* proteins CPI, CP47, and CP43 require Chl *a* to accumulate as well (89, 149, 202, 280). The accumulation of the Chl *a* proteins may in turn affect CAB protein assembly and integration into the photosystems.

The cofactors required for LHCII assembly are Chl *a*, Chl *b*, and at least two of the three xanthophylls, neoxanthin, violaxanthin, and lutein (218, 226). Other CAB polypeptides likely have similar cofactor requirements. A major light requirement for Chl synthesis is at the level of PChlide conversion (as discussed in the introduction). However, other important light-regulated steps in Chl synthesis or degradation are controlled by phytochrome and, to a lesser extent, the blue-light receptor (reviewed in 29, 137, 189, 192). Light-regulated steps include the light-induced increase in the rate of 5-aminolevulinic acid synthesis, and the conversion of 5-aminolevulinic acid into porphobilinogen (reviewed in 17, 128, 129, 137). Light determines whether monovinyl or divinyl PChlide biosynthesis predominates in etiolated seedlings (45, 302) and in green plants grown under a diurnal light cycle (44). The nature of this light-regulated accumulation of monovinyl or divinyl PChlide is plant species dependent (e.g. 44), and suggests a species-dependent light-regulation of divinyl porphyrin reductase(s).

The accumulation of Chl b is strongly light dependent, more so than is Chl a synthesis, and much more so than are most phytochrome or other light receptor responses. For example, high Chl a/b ratios are found at the onset of greening and eventually decrease to a value near 3 as greening is completed. The effect of P_{fr} on Chl b synthesis is stronger than, and independent of, the P_{fr} effect on Chl a synthesis (137, 213, 309). It also appears that substantial Chl b synthesis takes place only if the accumulation of Chlide a (or a Chlide a ester in equilibrium with Chlide a) exceeds a threshold level (e.g. 213). Thus P_r to P_{fr} and PChlide to Chlide photoconversions have differing effects on Chl a and Chl b synthesis, with greater degrees of receptor photoconversion stimulating Chl b synthesis preferentially. The differing effects of these photoconversions on Chl b synthesis are not yet fully understood. The view we summarize here is that the various light receptor effects on Chl b synthesis are closely associated with posttranslational regulation of Cab biosynthesis, and the nature of this posttranslational regulation depends on the light regime. The light regimes discussed below all result in some degree of greening. In this respect they approximate a plant's natural environment and photoreceptor status more closely than the single-light-pulse experiments used in many photoreceptor studies.

The regulatory effects of Chl b accumulation are most dramatically seen in systems in which Chl a accumulates but Chl b does not. For instance, plants grown under very low intensity or short intermittent light regimes (e.g. 2 min every 2hr) will accumulate considerable amounts of Chl a but little or no Chl b (11, 69, 304). Certain CAB polypeptides, especially those of LHCII, will not accumulate in the absence of Chl b despite the presence of their RNAs. The absence of these CAB polypeptides is most dramatically seen in plants grown under intermittent light (304, 310, 322) or in Chl b–less mutants (8, 9, 18, 113, 188a, 241, 263, 284, 285, 320, 321). The failure of these CAB polypeptides to accumulate under these conditions results primarily from posttranslational degradation (9, 18, 20), but translational control by light may also be involved (267). However, intermittent-light plants and Chl b–less mutants do differ in some aspects of posttranslational CAB polypeptide accumulation. For example, some LHCI polypeptides will accumulate in Chl b–less mutants (285, 320, 321), but not in plants grown under intermittent light regimes that strongly limit Chl b accumulation (304, 322). These data suggest a light requirement for LHCI accumulation that is not a Chl b requirement (322).

If newly greened plant tissues that have accumulated both Chl a and Chl b are transferred into darkness, Chl b and LHCII are degraded (5, 10, 69, 281). Under certain light regimes this dark-destruction of LHCII and Chl b is accompanied by resynthesis of Chl a and the Chl a reaction center proteins (5, 10, 281). This has been reported for both photosystem II and LHCII (10), and for photosystem I and LHCI (5), in etiolated bean leaves exposed to continuous light and then transferred back into darkness. In summary, experiments have shown that the (rate of) accumulation of Chl b is dependent on the light regime and plays a critical role in determining which CAB polypeptides will accumulate and which will be degraded. However, in cases such as the dark-transfer experiments, the levels of Chl b may be a consequence of translational or posttranslational regulation of CAB protein levels, rather than a factor determining CAB protein abundance.

SIGNAL TRANSDUCTION

One or more signal response chains are likely required between a photoreceptor and its final effect on gene expression. Although it is conceivable that a chromoprotein like phytochrome might interact with DNA, there is little evidence for phytochrome in the nucleus (182, 227, 270, 271), and attempts to measure interactions with DNA have been negative (172, 230). Furthermore, some sort of amplification system seems necessary to account for responses produced by as few as 20–30 molecules per cell of P_{fr} (172, 245). In addition, the diversity and subtlety of photoresponses are more easily reconciled with a network of signal transduction chains than with any type of simple photoreceptor:gene interaction. Unfortunately, rather little is yet known concerning the signal transduction mechanisms that alter gene expression. The purpose of this section is therefore less to review what is known than to highlight approaches that may lead to future progress in this area.

Multiple Transduction Chains

The most direct evidence for multiple (or branched) signal response chains comes from studies in which different photoreceptors have been shown to influence expression of the same gene in the same tissue (see the introduction and Table 1). As yet none of these systems have been studied in enough detail to determine at what level the chains merge, and whether separate *cis*-acting DNA elements are required for responses mediated by different photoreceptors. However, genetic and molecular techniques are becoming available to address these questions. The operation of different transduction chains can also be deduced from the behavior of different genes responding to the same stimulus. For example, different phytochrome-responsive genes may lose their requirement for P_{fr} ("escape from photoreversibility") at different rates. Since P_{fr} is by definition at the head of any phytochrome signal-response chain, responses that escape at different rates are thought to display fundamental differences in signal transduction mechanisms (190, 249). On this basis, we (139, 290) distinguished at least three categories of response — rapid escape, delayed escape, and little or no escape — among 9 genes in etiolated peas. These results indicate differences in the rate of the early, P_{fr}-dependent steps in the pathways leading to different gene responses.

More recently, Sun & Tobin (279) have shown that the phytochrome system regulating *Cab* gene expression operates normally in the *Arabidopsis* long hypocotyl mutants *hy*-3 and *hy*-5. However, these and other *hy* mutants show reduced sensitivity to phytochrome in their hypocotyl growth responses, so the phytochrome effects on growth must result from a chain of events that is at least partially different from that affecting *Cab* gene expression. This example illustrates the utility of genetic analysis in defining signal transduction pathways (3). The combination of genetic and molecular approaches now being undertaken in various laboratories will provide much exciting information in the near future.

Biochemical events

ROLE OF PROTEIN SYNTHESIS   Direct responses occur when the signal transduction chain from the photoreceptor involves only pre-existing cellular components, while indirect responses require activation of other genes whose products are required to affect the gene of interest. One way to distinguish between direct and indirect effects is to measure the sensitivity of the final response to inhibitors of protein synthesis. Where the product being measured is mRNA, direct responses should be initiated in the absence of protein synthesis, while indirect responses should be blocked.

So far, this test has been applied to relatively few gene systems. Results from more than one laboratory are available only for *Cab* genes. Merkle & Schäfer (185) reported experiments in which a combination of cycloheximide (CHI) and chloramphenicol (CAP) inhibiting over 95% of ^{35}S-methionine incorporation was used to block protein synthesis in etiolated barley leaves. Under these conditions CHI inhibited continued increases in *Cab* mRNA beyond 4–6 hours after the light pulse, but the initial increase was unaffected. Thus while protein synthesis may be required to maintain a response for long periods (or perhaps to accumulate mRNA above a certain level), transduction of the red-light signal did not depend on protein synthesis.

In contrast, Lam et al (168) concluded that *Cab* and *RbcS* induction is sensitive to CHI in etiolated wheat seedlings. In these experiments *Cab* mRNAs were not detectable until 5 hr after a red-light pulse, and CHI prevented their accumulation. In transgenic tobacco, CHI inhibited expression from the Cab-1 promoter (measured 10 or 20 hr after etiolated seedlings were transferred to continuous white light) but somewhat increased expression from the 35S promoter. Generally similar results were obtained for RbcS transcript accumulation in pea seedlings and transgenic tobacco.

The results with the 35S promoter indicate that even relatively extended CHI treatments do not simply block RNA synthesis. Thus the simplest interpretation is that a labile protein factor is required for Cab transcription but not for activity of the 35S promoter. However, such a factor might be required only for high-level transcription and would not necessarily be involved in transducing the light signal. Thus the results of Lam et al do not conflict with those of Merkle & Schäfer, since Merkle & Schäfer concluded that only the initiation of the Cab response to red light was independent of protein synthesis. Results similar to those of Merkle & Schäfer (185) have also been reported by Marrs & Kaufman for blue-light induction of Cab transcription (179).

Several mRNAs that decline in response to light also do so in the presence of CHI. Notable among these is phyA mRNA encoding a type I dark-abundant phytochrome. Lissemore & Quail (172) showed that red-light-induced decline of phyA transcription occurs normally in oat leaf sections incubated with a combination of CHI+CAP, indicating that this response does not require protein synthesis. Together with the speed of this response (lag period of less than 5 min), this result supports the notion that the transduction chain is relatively simple. Lam et al (168) reported similar results for the red-light induced decline in Pcr mRNA in wheat seedlings, while Marrs & Kaufman (179) showed a CHI-independent decrease in run-on transcription activity for the Blec-1 gene (formerly "pEA207"; see 83) in response to blue light.

In none of the above experiments was protein synthesis inhibited by 100%. Formally, therefore, signal transduction might still depend on synthesis of some particularly resistant protein, or on one normally synthesized in vast excess. Apart from these somewhat unlikely possibilities, however, the results summarized above indicate that at least the initial stages of a light response can be obtained in the absence of protein synthesis for all genes thus far examined. The number of genes is still small, and this conclusion must therefore still be considered preliminary. However at present it appears that signal transduction mechanisms leading to gene activation use machinery already available in dark-grown cells.

MEMBRANES AND SECOND MESSENGERS   The earliest phases of signal transduction are more conveniently studied in conjunction with rapid, immediately visualized responses. For light signals, such responses include changes in membrane potential and modulation of ion fluxes across the plasmalemma and/or organelle membranes (236). Another rapid response is the so-called "sequestering" of phytochrome, in which red light causes immunocytochemically detectable phytochrome to become associated with discrete, possibly membrane-associated, structures. It is now believed that this reaction is connected with phytochrome destruction (182, 227, 244, 270, 271).

Prominent among cellular events that can be rapidly induced by light are changes in Ca^{++} flux. Ca^{++} ionophores can substitute for P_{fr} in some phytochrome-mediated growth responses, and calmodulin antagonists can block several such responses (125). Roux and colleagues (73, 235, 237) have suggested that altered Ca^{++} concentrations might lead to changes in the phosphorylation status of nuclear proteins, either by activating nuclear protein kinases or by affecting Ca^{++}-dependent regulatory proteins like calmodulin. Evidence for this idea includes the observation that "calmodulin antagonists" (which act on most proteins that undergo Ca^{++} induced conformational changes) blocked light-stimulated phosphorylation of several proteins in pea nuclei (72), and that light treatment also stimulated the activity of a calmodulin-activated nucleoside triphosphatase (49). In this connection it is interesting that Datta & Cashmore (71) have reported that phosphorylation reversibly inhibits the binding of a nuclear factor to an AT-rich DNA sequence found in RbcS and a number of other gene promoters.

Lam et al (164) reported that calmodulin antagonists could block light-induced expression of Cab genes in dark-adapted soybean suspension culture cells. Thus some calmodulin-dependent process is probably required for Cab gene expression in this system. Whether or not calmodulin activation is sufficient to induce expression is less clear. Treatment with a Ca^{++} ionophore failed to induce the mRNA in the dark, but there was no positive control for the effectiveness of the ionophore, and it is possible that effective concentrations were not achieved at the appropriate site within the cell. Alternatively, more than one transducer may regulate Cab gene expression.

Signal transduction mechanisms involving inositol phospholipid turnover are ubiquitous in animal cells (24) and have been the subject of much interest by plant physiologists. In animals, signals such as light or hormones activate G-proteins (207), which cause an increase in the activity of phospholipase C, which in turn leads to breakdown of inositol phospholipids and production of inositol trisphosphate (IP_3) and diacylglycerol (DAG). Cascades of secondary biochemical responses occur when IP_3 triggers increases in free Ca^{++} levels and when protein kinase C is activated by DAG. It is not yet clear whether such a system operates in the same way in plant cells, particularly in relation to light stimuli (130). Morse et al (196, 197) reported rapid light-induced changes in the concentrations of inositol phosphates and phospholipids in the pulvini of *Samanea* leaflets. On this basis they proposed that a signaling system similar to that in animal cells mediates light effects on ion flux and leaflet movement in this system. However, in plasma membrane preparations from sunflower hypocotyls Memon & Boss (184) observed a decrease in phospholipid kinase activity following brief in vivo illuminations. Since they also measured a decrease in the activity of a plasma membrane ATPase known to be sensitive to inositol phospholipids, these authors proposed that inositol phospholipids may themselves be active regulatory intermediates in plant systems, and that signaling may not always require the production of IP_3 and DAG. Further work is clearly required to assess the significance of such systems in light responses.

Recent evidence indicates that a GTP-binding protein resembling the G-proteins of animal transduction systems (207) may participate in blue-light signal transduction in pea buds. Warpeha et al (314; and L. S. Kaufman, personal communication) found that irradiating purified plasma membrane fractions with blue light resulted in a rapid increase in GTPase activity that was associated with enhanced binding of GTPγS. The activity was not responsive to red light, and the system showed no ATPase activity. When an azido derivative of GTP was added to the membranes during irradiation it labeled a 40-kDa protein. Antibodies to a visual system G-protein, transducin, also recognized a 40-kDa protein from these membranes. Thus some type of a blue-light-activated G-protein system seems to be present in pea buds. Although its function remains to be defined, it is reasonable to suppose that it is involved in signal transduction for at least some blue-light responses.

PROTEIN PHOSPHORYLATION AND PHOTOPERCEPTION   Protein phosphorylation reactions other than those regulated by Ca^{++}/calmodulin or inositol phospholipids may also be important in light responses. One such example involves phytochrome itself. Lagarias and his colleagues have reported a polycation-stimulated protein kinase activity that copurifies with phytochrome and phosphorylates a highly conserved serine-rich region at the N-terminus of the molecule (183, 323–325). These observations have raised the possibility that phytochrome itself may have kinase activity. Although this point is still controversial (e.g. 116), the possibility that phytochrome is at least closely associated with, and regulates, a protein kinase deserves further investigation (266).

Briggs and coworkers (101, 259) have recently discovered that blue-light-stimulated changes in phosphorylation of at least two different proteins can be detected both in vivo in pea stem sections and in vitro in isolated membrane preparations. One whose behavior has been studied intensively has a molecular mass of approximately 120 kDa and nearly comigrates with phytochrome, although the two proteins can be separated both on SDS gels and by their responses to blue and red light. Similar proteins have been detected by Western blot and phosphorylation analysis of preparations from a variety of other plant species, including *Arabidopsis* and maize (37). Curiously, however, membranes from the apical buds of pea seedlings, in which Warpeha et al (314) observed the GTP binding activity mentioned above, showed no phosphorylation in the 120-kDa size range.

Fluence response curves for phosphorylation changes following in vivo irradiations of pea stem sections are very similar to those for phototropism. The phosphorylation effect is extremely rapid, occurring within a few seconds of irradiation at most, and so could represent an early step in a blue-light transduction chain. Also consistent with this hypothesis is the observation that Triton treatment of the membrane preparations does not affect the quantum efficiency of the response, suggesting that both the photoreceptor and the phosphorylation machinery must be present in the same micelle (37, 260). Further work on these last two systems should prove extremely interesting, since it should soon be possible to isolate the relevant proteins and clone the genes encoding them.

Genetic Approaches

It seems likely that the study of mutants will become more and more important as increasing attention is focused on molecular events in the signal transduction process. Several laboratories are currently working on selection schemes based on chimaeric genes in which light-responsive promoters are fused to the coding sequences of selectable reporter genes. It is hoped that such systems will facilitate isolation of signal transduction mutants and thus aid in obtaining a complete genetic analysis.

A variety of photomorphogenetic mutants in tomato, *Arabidopsis*, cucumber, and pea have already proven useful in such studies. Perhaps the best characterized of these is *aurea* in tomato (156). Etiolated *aurea* seedlings have greatly reduced levels of phytochrome as measured either by spectral activity or by immunological detection of the apoprotein (216). However, normal levels of phytochrome mRNA are present (254), and the defect in phytochrome production is therefore thought to reflect instability of the protein.

Consistent with the reduction in detectable phytochrome, light induction of *Cab* and several other mRNAs is severely reduced in etiolated *aurea* seedlings (210, 254). However, when *aurea* and wild-type plants are grown in the light they both show similar photoreversible growth responses to end-of-day far-red-light treatments, indicating that a phytochrome system is functional in light-grown material. The most likely interpretation of these data is that the mutation results in the elimination of a dark-abundant, photolabile phytochrome pool responsible for initiating synthesis of the photosynthetic apparatus, while a separate, light-stable phytochrome pool — unaffected by the *aurea* mutation — controls the end-of-day response (2, 173). Other photo-morphogenetic mutants can be interpreted as affecting the physiologically stable phytochrome pool. Examples include *lv* in pea and *lh* in cucumber, both of which lack responses to end-of-day far-red irradiations (reviewed in 300).

A series of long-hypocotyl mutants have been reported in *Arabidopsis* (53, 157) and partially characterized in terms of their physiological and gene responses. The mutants *hy-1*, *hy-2*, and *hy-6* have reduced levels of photoreversible phytochrome (but normal levels of phytochrome apoprotein) in etiolated seedlings and show retarded greening and reduced expression of photosynthetic genes in etiolated seedlings (53, 217). These mutants behave in a manner generally similar to *aurea* in tomato, although it is not yet known if they exhibit the end-of-day responses characteristic of physiologically stable phytochrome. In contrast, *hy-3*, *hy-4*, and *hy-5* have normal levels of spectrally active phytochrome in etiolated seedlings and might be considered similar to the pea and cucumber mutants mentioned above. If so, they would be expected to lack end-of-day responses.

A particularly interesting tomato mutant, *hp* (high pigment), has recently been described (4, 219). This mutant has a normal level of phytochrome in etiolated tissues, but when plants are grown in the light they are shorter in stature, with darker green leaves and higher levels of anthocyanins than the wild type. Based on these data and on physiological studies of *aurea/hp* double mutants, the authors suggest that *hp* is a signal transduction mutant that increases sensitivity to a labile form of phytochrome. It is interesting to note the similarity between this mutant phenotype and transgenic plants overexpressing a phytochrome gene (31, 142, 145), since on a priori grounds increasing the amount of phytochrome might be expected to have effects similar to those of increasing sensitivity. Additional studies of the *hp* mutation will clearly be of great interest, perhaps in conjunction with antisense and overexpression experiments.

Another class of potential signal transduction mutants deserving of exten-sive investigation are the *det* mutants first reported by Chory et al (54). These mutants develop many of the morphological and biochemical features of light-grown plants — including partially developed plastids, the presence of several transcripts encoding photosynthetic components, and high levels of anthocyanin — even when grown in constant darkness. Since *det* mutations are recessive, it is difficult to argue that the mutation inactivates a positive regulator of light-induced development. Chory et al consequently suggested that a developmental program characteristic of light grown plants is normally repressed in etiolated seedlings. According to this hypothesis, *det* mutants would contain a defective gene for a *trans*-acting repressor.

It is a novel concept to suppose that "light-induced" development is repressed in the dark, rather than activated by light. However, in other respects the Chory hypothesis fits extremely well with previous concepts of developmental competence, and with the many differences between light responses of etiolated and green plants. We may suppose that as long as the "light" developmental program is repressed the plant uses a particular set of receptors and signal transduction chains to sense the presence of light. However, once the initial light signal has been received and the "light" program initiated a different set of receptors (such as the stable phytochrome pool and one or more blue/UVA receptors) becomes more important. In this context the *det* gene product can be seen as part of a switch mechanism controlling a developmental transition as dramatic as any other in the plant life cycle. Determining how such genes are affected by light, and how they in turn affect development, should be particularly rewarding.

ACKNOWLEDGMENTS

We begin by thanking Pamela and Adrian for their patience and support, without which this paper and many others would never have been written. We also thank the many colleagues around the world who provided papers and preprints, and the several colleagues in our laboratory who provided constructive comments on various versions of the manuscript. Preparation of this review was supported by NSF grant DCB-8817276 and NIH grant GM43108, as well as by funds from North Carolina State University.

Literature Cited

1. Abad, M. S., Clark, S. E., Lamppa, G. K. 1989. Properties of a chloroplast enzyme that cleaves the chlorophyll a/b binding protein precursor. *Plant Physiol.* 90:117-24

2. Adamse, P., Jaspers, P. A. P., Bakker, J. A., Wesselius. J. C., Heeringa, G. H., et al. 1988. Photophysiology of a tomato mutant deficient in labile phytochrome. *J. Plant Physiol.* 133:436-40

3. Adamse, P., Kendrick, R. E., Koornneef, M. 1988. Photomorphogenetic mutants of higher plants. *Photochem. Photobiol.* 48:833-41

4. Adamse, P., Peters, J. L., Jaspers, P. A. P., van Tuinen, A., Koornneef, M., et al. 1989. Photocontrol of anthocyanin synthesis in tomato seedlings: a genetic approach. *Photochem. Photobiol.* 50: 107-11

5. Akoyunoglou, A., Akoyunoglou, G. 1985. Reorganization of thylakoid components during chloroplast development in higher plants after transfer to darkness. Changes in PSI unit components and in cytochromes. *Plant Physiol.* 79:425-31

6. Anderson, J. M. 1988. Thylakoid membrane organization in sun/shade acclimation. *Aust. J. Plant Physiol.* 15:11-26

7. Apel, K. 1981. The protochlorophyllide holochrome of barley (*Hordeum vulgare* L.): Phytochrome-induced decrease of translatable mRNA coding for the NADPH: protochlorophyllide oxidoreductase. *Eur. J. Biochem.* 120:89-93

8. Apel, K., Kloppstech, K. 1978. The plastid membranes of barley (*Hordeum vulgare*). Light-induced appearance of mRNA coding for the apoprotein of the light-harvesting chlorophyll a/b protein. *Eur. J. Biochem.* 85:581-88

9. Apel, K., Kloppstech, K. 1980. The effect of light on the biosynthesis of the light-harvesting chlorophyll a/b protein. *Planta* 150:426-30

10. Argyroudi-Akoyunoglou, J. H., Akoyu-

noglou, A., Kalosakas, K., Akoyunoglou, G. 1982. Reorganization of photosystem II unit in developing thylakoids of higher plants after transfer to darkness. Changes in chlorophyll b, light-harvesting protein content and grana stacking. *Plant Physiol.* 70:1242-48

11. Argyroudi-Akoyunoglou, J. H., Akoyunoglou, G. 1970. Photoinduced changes in the chlorophyll a to chlorophyll b ratio in young bean plants. *Plant Physiol.* 46:247-49

12. Baerson, S., Kaufman, L. S. 1990. Increased rRNA gene activity during a specific window of early pea leaf development. *Mol. Cell. Biol.* 10:842-45

13. Bandziulis, R. J., Swanson, M. S., Dreyfuss, G. 1989. RNA-binding proteins as developmental regulators. *Genes Dev.* 3:431-37

14. Barnett, L., Clugston, C., Jenkins, G. 1987. Two phytochrome-mediated effects of light on transcription of genes encoding the small subunit of ribulose-1,5-bisphosphate carboxylase-oxygenase in dark grown pea (*Pisum sativum*) plants. *FEBS Lett.* 224:287-90

15. Barraclough, R., Ellis, E. 1980. Protein synthesis in chloroplasts. IX. Assembly of newly synthesized large subunits into ribulose bisphosphate carboxylase in isolated intact pea chloroplasts. *Biochim. Biophys. Acta* 608:19-31

16. Batschauer, A., Apel, K. 1984. An inverse control by phytochrome of the expression of two nuclear genes in barley (*Hordeum vulgare* L.). *Eur. J. Biochem.* 143:593-97

17. Beale, S. I. 1990. Biosynthesis of the tetrapyrrole pigment precursor, γ-aminolevulinic acid, from glutamate. *Plant Physiol.* 93:1273-1279

18. Bellemare, G., Bartlett, S. G., Chua, N. H. 1982. Biosynthesis of chlorophyll a/b-binding polypeptides in wild type and chlorina f2 mutant of barley. *J. Biol. Chem.* 257:7762-67

19. Benfey, P. N., Chua, N.-H. 1989. Regulated gene expression in transgenic plants. *Science* 244:174-81

19a. Benfey, P. N., Chua, N.-H. 1990. The cauliflower mosaic virus 35S promoter: combinatorial regulation of transcription in plants. *Science* 250:959-66

20. Bennett, J. 1981. Biosynthesis of the light-harvesting chlorophyll a/b protein. Polypeptide turnover in darkness. *Eur. J. Biochem.* 118:61-70

21. Bennett, J. 1983. Regulation of photosynthesis by reversible phosphorylation of the light-harvesting chlorophyll a/b protein. *Biochem. J.* 212:1-13

22. Bennett, J., Shaw, E. K., Michel, H. 1988. Cytochrome b6f complex is required for phosphorylation of light-harvesting chlorophyll a/b complex II in chloroplast photosynthetic membranes. *Eur. J. Biochem.* 171:95-100

23. Bentley, D. L., Groudine, M. 1988. Sequence requirements for premature termination of transcription in the human *c-myc* gene. *Cell* 53:245-56

24. Berridge, M. 1987. Inositol trisphosphate and diacylglycerol: two interacting second messengers. *Annu. Rev. Biochem.* 56:159-93

25. Berry, J. O., Breiding, D. E., Klessig, D. F. 1990. Light-mediated control of translational initiation of ribulose 1,5-bisphosphate carboxylase in amaranth cotyledons. *Plant Cell* 2:795-803

26. Berry, J. O., Carr, J. P., Klessig, D. F. 1988. mRNAs encoding ribulose-1,5-bisphosphate carboxylase remain bound to polysomes but are not translated in amaranth seedlings transferred to darkness. *Proc. Natl. Acad. Sci. USA* 85:4190-94

27. Berry, J. O., Nikolau, B. J., Carr, J. P., Klessig, D. F. 1986. Translational regulation of light-induced ribulose 1,5-bisphosphate carboxylase gene expression in amaranth. *Mol. Cell. Biol.* 6:2347-53

28. Berry-Lowe, S. L., Meagher, R. B. 1985. Transcriptional regulation of a gene encoding the small subunit of ribulose-1,5-bisphosphate carboxylase in soybean tissue is linked to the phytochrome response. *Mol. Cell. Biol.* 5:1910-17

29. Biswal, B., Choudhury, N. K. 1986. Photocontrol of chlorophyll loss in papaya leaf discs. *Plant Cell Physiol.* 27:1439-44

30. Bowsher, C. G., Long, D. M., Oaks, A., Rothstein, S. J. 1991. The effect of light/dark cycles on expression of nitrate assimilatory genes in maize roots and shoots. *Plant Physiol.* In press

31. Boylan, M. T., Quail, P. H. 1989. Oat phytochrome is biologically active in transgenic tomatoes. *Plant Cell* 1:765-73

32. Boyle, S. A., Hemmingsen, S. M., Dennis, D. T. 1990. Energy requirement for the import of protein into plastids from developing endosperm of *Ricinus communis*. *Plant Physiol.* 92:151-54

33. Bradbeer, J., Atkinson, Y., Borner, T., Hagemann, R. 1979. Cytoplasmic synthesis of plastid polypeptides may be controlled by plastid-synthesized RNA. *Nature* 279:816-17

34. Brawerman, G. 1989. mRNA Decay: finding the right targets. *Cell* 57:9-10

35. Brecht, E. 1986. The light-harvesting chlorophyll a/b-protein complex II of higher plants: results from a twenty-year research period. *Photochem. Photobiol.* 12:37-50

36. Breiding, D. E., Klessig, D. F. 1991. Light-regulated translation of ribulose 1,5-bisphosphate carboxylase in pea seedlings. *Plant Physiol.* In press

37. Briggs, W., Short, T. 1990. Possible early events in the transduction of light signals. In *Phytochrome Properties and Biological Action*, ed. B. Thomas. Berlin: Springer-Verlag

38. Briggs, W. R., Mösinger, E., Schäfer, E. 1988. Phytochrome regulation of greening in barley — effects on chlorophyll accumulation. *Plant Physiol.* 86:435-40

39. Britten, R., Davidson, E. 1969. Gene regulation for higher cells: a theory. *Science* 165:349-57

40. Buchanan, B. 1984. The ferredoxin/thioredoxin system: a key element in the regulatory function of light in photosynthesis. *BioScience* 34:378-83

41. Bustos, M. M., Guiltinan, M. J., Cyr, R. J., Ahdoot, D., Fosket, D. E. 1989. Light regulation of β-tubulin gene expression during internode devlopment in soybean (*Glycine max* L. Merr.). *Plant Physiol.* 91:1157-61

42. Butler, W. L., Norris, K. H., Siegelman, H. W., Hendricks, S. B. 1959. Detection, assay, and preliminary purification of the pigment controlling photoresponsive development of plants. *Proc. Natl. Acad. Sci. USA* 45:1703-8

43. Buzby, J., Yamada, T., Tobin, E. 1990. A light-regulated DNA-binding activity interacts with a conserved region of a *Lemna gibba* rbcS promoter. *Plant Cell* 2:805-14

44. Carey, E. E., Rebeiz, C. A. 1985. Chloroplast biogenesis 49. Differences among angiosperms in the biosynthesis and accumulation of monovinyl and divinyl protochlorophyllide during photoperiodic greening. *Plant Physiol.* 79:1-6

45. Carey, E. E., Tripathy, B. C., Rebeiz, C. A. 1985. Chloroplast biogenesis 51. Modulation of monovinyl and divinyl protochlorophyllide biosynthesis in light and darkness in vitro. *Plant Physiol.* 79:1059-63

46. Castresana, C., Garcia-Luque, I., Alonso, E., Malik, V. L., Cashmore, A. R. 1988. Both positive and negative regulatory elements mediate expression of a photoregulated CAB gene from *Nicotiana plumbaginifolia*. *EMBO J.* 7:1929-1936

47. Chappell, J., Hahlbrock, K. 1984. Transcription of plant defense genes in response to UV light or fungal elicitor. *Nature* 311:76-78

48. Chen, Q., Lauzon, L. M., DeRocher, A. E., Vierling, E. 1990. Accumulation, stability, and localization of a major chloroplast heat-shock protein. *J. Cell Biol.* 110:1873-83

49. Chen, Y.-R., Roux, S. 1986. Characterization of nucleoside triphosphatase activity in isolated pea nuclei and its photoreversible regulation by light. *Plant Physiol.* 81:609-13

50. Chitnis, P. R., Harel, E., Kohorn, B. D., Tobin, E. M., Kohorn, J. P. 1986. Assembly of the precursor and processed light-harvesting chlorophyll a/b protein of *Lemna* into the light-harvesting complex II of barley etiochloroplasts. *J. Cell Biol.* 102:982-88

51. Chitnis, P. R., Morishige, D. T., Nechushtai, R., Thornber, J. P. 1988. Assembly of the barley light-harvesting chlorophyll a/b proteins in barley etiochloroplasts involves processing of the precursor on thylakoids. *Plant. Mol. Biol.* 11:95-107

52. Chitnis, P. R., Nechushtai, R., Thornber, J. P. 1987. Insertion of the precursor of the light-harvesting chlorophyll a/b-protein into the thylakoids requires the presence of a developmentally regulated stromal factor. *Plant Mol. Biol.* 10:3-11

53. Chory, J., Peto, C. A., Ashbaugh, M., Saganich, R., Pratt, L., et al. 1989. Different roles for phytochrome in etiolated and green plants deduced from characterization of *Arabidopsis thaliana* mutants. *Plant Cell* 1:867-80

54. Chory, J., Peto, C. A., Feinbaum, R., Pratt, L., Ausubel, F. 1989. *Arabidopsis thaliana* mutant that develops as a light-grown plant in the absence of light. *Cell* 58:991-99

55. Chow, W. S., Goodchild, D. J., Miller, C., Anderson, J. M. 1990. The influence of high levels of brief or prolonged supplementary far-red illumination during growth on the photosynthetic characteristics, composition and morphology of *Pisum sativum* chloroplasts. *Plant Cell Env.* 13:135-45

56. Chow, W. S., Qian, L., Goodchild, D. J., Anderson, J. M. 1988. Photosynthetic acclimation of *Alocasia macrorrhiza* (L.) G. Don to growth irradiance: structure, function and composition of chloroplasts. *Aust. J. Plant Physiol.* 15:107-22

57. Christensen, A. H., Quail, P. H. 1989.

Structure and expression of a maize phytochrome-encoding gene. *Gene* 85: 381-90

58. Clark, S. E., Abad, M. S., Lamppa, G. K. 1989. Mutations at the transit peptide-mature protein junction separate two cleavage events during chloroplast import of the chlorophyll a/b-binding protein. *J. Biol. Chem.* 264:17544-50

59. Clarkson, D. T., Hillman, W. S. 1967. Apparent phytochrome synthesis in *Pisum* tissue. *Nature* 4:468-70

60. Cleveland, D. W., Yen, T. J. 1989. Multiple determinants of eukaryotic mRNA stability. *New Biol.* 1:121-26

61. Cline, K. 1988. Light-harvesting chlorophyll a/b protein. Membrane insertion, proteolytic processing, assembly into LHCII, and localization to appressed membranes occurs in chloroplast lysates. *Plant Physiol.* 86:1120-26

62. Cline, K., Fulsom, D. R., Viitanen, P. V. 1989. An imported thylakoid protein accumulates in the stroma when insertion into thylakoids is inhibited. *J. Biol. Chem.* 264:14225-232

63. Clugston, C., Barnett, L., Urwin, N., Jenkins, G. 1990. Photoreceptors controlling transcription of rbcS genes in green leaf tissue of *Pisum sativum.* *Photochem. Photobiol.* 52:23-28

64. Colbert, J. T. 1988. Molecular biology of phytochrome. *Plant Cell Environ.* 11:305-18

65. Colbert, J. T., Costigan, S. A., Avissar, P., Zhao, Z. 1991. Regulation of phytochrome gene expression. *J. Iowa Acad. Sci.* In press

66. Colbert, J. T., Costigan, S. A., Zhao, Z. F. 1990. Photoregulation of β-tubulin mRNA abundance in etiolated oat and barley seedlings. *Plant Physiol.* 93: 1196-1202

67. Colbert, J. T., Hershey, H. P., Quail, P. H. 1983. Autoregulatory control of translatable phytochrome mRNA levels. *Proc. Natl. Acad. Sci. USA* 80:2248-52

68. Cotton, J. L. S., Ross, C. W., Byrne, D. H., Colbert, J. T. 1990. Downregulation of phytochrome mRNA abundance by red light and benzyladenine in etiolated cucumber cotyledons. *Plant Mol. Biol.* 14:707-14

69. Cuming, A. C., Bennett, J. 1981. Biosynthesis of the light-harvesting chlorophyll a/b protein. Control of messenger RNA activity by light. *Eur. J. Biochem.* 118:71-80

70. Darrah, P. M., Kay, S. A., Teakle, G. R., Griffiths, W. T. 1990. Cloning and sequencing of protochlorophyllide reductase. *Biochem. J.* 265:789-98

71. Datta, N., Cashmore, A. R. 1989. Binding of a pea nuclear protein to promoters of certain photoregulated genes is modulated by phosphorylation. *Plant Cell* 1:1069-77

72. Datta, N., Chen, Y.-R., Roux, S. 1985. Phytochrome and calcium stimulation of protein phosphorylation in isolated pea nuclei. *Biochem. Biophys. Res. Commun.* 128:1403-8

73. Datta, N., Roux, S. 1986. Regulation of enzymes in isolated plant nuclei. *Bioessays* 5:120-23

74. Davidson, E., Britten, R. 1973. Organization, transcription, and regulation in the animal genome. *Q. Rev. Biol.* 48:565-613

75. Davies, E. C., Jordan, B. R., Partis, M. D., Chow, W. S. 1987. Immunochemical investigation of thylakoid coupling factor protein during photosynthetic acclimation to irradiance. *J. Exp. Bot.* 38:1517-27

76. Davis, M. C., Yong, M.-H., Gilmartin, P., Goyvaerts, E., Kuhlemeier, C., et al. 1990. Minimal sequence requirements for the regulated expression of rbcS-3A from *Pisum sativum* in transgenic tobacco plants. *Photochem. Photobiol.* 52:43-50

77. Dean, C., Favreau, M., Bond-Nutter, D., Bedbrook, J., Dunsmuir, P. 1989. Sequences downstream of translation start regulate quantitative expression of two petunia rbcS genes. *Plant Cell* 1:201-08

78. Dean, C., Pichersky, E., Dunsmuir, P. 1989. Structure, evolution, and regulation of rbcS genes in higher plants. *Annu. Rev. Plant Physiol. Plant Mol. Biol.* 40:415-39

79. Dehesh, K., Klaas, M., Häuser, I., Apel, K. 1986. Light-induced changes in the distribution of the 36000-M_r polypeptide of NADPH-protochlorophyllide oxidoreductase within different cellular compartments of barley (*Hordeum vulgare* L.) I. Localization by immunoblotting in isolated plastids and total leaf extracts. *Planta* 169:162-71

80. Dehesh, K., van Cleve, B., Ryberg, M., Apel, K. 1986. Light-induced changes in the distribution of the 36000-M_r polypeptide of NADPH-protochlorophyllide oxidoreductase within different cellular compartments of barley (*Hordeum vulgare* L.) II. Localization by immunogold labelling in ultrathin sections. *Planta* 169:172-83

81. Deng, X., Melis, A. 1986. Phosphorylation of the light-harvesting complex II in

higher plant chloroplasts: effect on photosystem II and photosystem I absorption cross section. *Photobiochem. Photobiophys.* 13:41-52

82. Dobres, M. S., Elliott, R. C., Watson, J. C., Thompson, W. F. 1987. A phytochrome-regulated pea transcript encodes ferredoxin I. *Plant Mol. Biol.* 8:53-59

83. Dobres, M. S., Thompson, W. F. 1989. A developmentally regulated bud specific transcript in pea has sequence similarity to seed lectins. *Plant Physiol.* 89:833-38

84. Drumm-Herrell, H., Mohr, H. 1988. Mode of coaction between UV-A and light absorbed by phytochrome in control of appearance of ribulose-1,5-bisphosphate carboxylase in the shoot of milo (*Sorghum vulgare* Pers.). *Photochem. Photobiol.* 4:599-604

85. Dujardin, E. 1984. Protochlorophyllide photoreduction in plants: some comments on recent data. *Isr. J. Bot.* 33:83-92

86. Dunahay, T. G., Schuster, G., Staehelin, L. A. 1987. Phosphorylation of spinach chlorophyll-protein complexes. CPII*, but not CP29, CP27, or CP24, is phosphorylated in vitro. *FEBS Lett.* 215 (1):25-30

87. Eckes, P., Rosahl, S., Schell, J., Willmitzer, L. 1986. Isolation and characterization of a light-inducible, organ-specific gene from potato and analysis of its expression after tagging and transfer into tobacco and potato shoots. *Mol. Gen. Genet.* 205:14-22

88. Edwards, J. W., Coruzzi, G. M. 1989. Photorespiration and light act in concert to regulate the expression of the nuclear gene for chloroplast glutamine synthetase. *Plant Cell* 1:241-48

89. Eichacker, L. A., Soll, J., Lauterbach, P., Rüdiger, W., Klein, R. R., et al. 1990. In vitro synthesis of chlorophyll a in the dark triggers accumulation of chlorophyll a-apoproteins in barley etioplasts. *J. Biol. Chem.* 265:13566-600

90. Elliott, R. C., Dickey, L. F., White, M. J., Thompson, W. F. 1989. *cis*-Acting elements for light regulation of pea ferredoxin I gene expression are located within transcribed sequences. *The Plant Cell* 1:691-98

91. Elliott, R. C., Pedersen, T. J., Fristensky, B., White, M. J., Dickey, L. F., et al. 1989. Characterization of a single copy gene encoding ferredoxin I from pea. *The Plant Cell* 1:681-90

92. Ernst, D., Apfelbock, A., Bergmann, A., Weyrauch, C. 1990. Rhythmic regulation of the light-harvesting chlorophyll a/b protein and the small subunit of ribulose-1,5-bisphosphate carboxylase mRNA in rye seedlings. *Photochem. Photobiol.* 52:29-33

93. Deleted in proof

94. Fang, R.-X., Nagy, F., Sivasubramaniam, S., Chua, N.-H. 1989. Multiple *cis* regulatory elements for maximal expression of the cauliflower mosaic virus promoter in transgenic plants. *Plant Cell* 1:141-50

95. Flügge, U. I., Hinz, G. 1986. Energy dependence of protein translocation into chlororoplasts. *Eur. J. Biochem.* 160:563-70

96. Fluhr, R., Chua, N.-H. 1986. Developmental regulation of two genes encoding ribulose bisphosphate carboxylase small subunit in pea and transgenic petunia plants: phytochrome response and blue light induction. *Proc. Natl. Acad. Sci. USA* 83:2358-62

97. Fluhr, R., Kuhlemeier, C., Nagy, F., Chua, N.-H. 1986. Organ-specific and light-induced expression of plant genes. *Science* 232:1106-12

98. Ford, A. M., Watt, S. M., Furley, A. J. W., Molgaard, H. V., Gleaves, M. F. 1988. Cell lineage specificity of chromatin conformation around the immunoglobulin heavy chain enhancer. *EMBO J.* 7:2393-99

99. Furuya, M. 1989. Molecular properties and biogenesis of phytochrome I and II. *Adv. Biophys.* 25:133-67

100. Galangau, F., Daniel-Vedele, F., Moureaux, T., Dorbe, M.-F., Leydecker, M.-T., et al. 1988. Expression of leaf nitrate reductase genes from tomato and tobacco in relation to light-dark regimes and nitrate supply. *Plant Physiol.* 88:383-88

101. Gallagher, S., Short, T. W., Ray, P., Pratt, L., Briggs, W. 1988. Light-mediated changes in two proteins found associated with plasma membrane fractions from pea stem sections. *Proc. Natl. Acad. Sci. USA* 85:8003-7

102. Gallagher, T. F., Ellis, R. J. 1982. Light-stimulated transcription of genes for two chloroplast polypeptides in isolated leaf nuclei. *EMBO J.* 1:1493-98

103. Gallagher, T. F., Jenkins, G. I., Ellis, R. J. 1985. Rapid modulation of transcription of nuclear genes encoding chloroplast proteins by light. *FEBS Lett.* 186:241-45

104. Gamble, P. E., Klein, R. R., Mullet, J. E. 1989. Illumination of eight-day-old dark-grown barley seedlings activates chloroplast protein synthesis; evidence

for regulation of translation initiation. In *Photosynthesis*, ed. W. Briggs, pp. 285-98. New York: Alan R. Liss

105. Gidoni, D., Brosio, P., Bond-Nutter, D., Bedbrook, J., Dunsmuir, P. 1989. Novel *cis*-acting elements in petunia *Cab* gene promoters. *Mol. Gen. Genet.* 215:337-44

106. Gilmartin, P. M., Sarokin, L., Memelink, J., Chua, N.-H. 1990. Molecular light switches for plant genes. *Plant Cell* 2:369-78

107. Giuliano, G., Hoffman, N. E., Ko, K., Scolnik, P. A., Cashmore, A. R. 1988. A light-entrained circadian clock controls transcription of several plant genes. *EMBO J.* 7:3635-42

108. Giuliano, G., Pichersky, E., Malik, V. S., Timko, M. P., Scolnik, P. A., et al. 1988. An evolutionarily conserved protein binding sequence upstream of a plant light-regulated gene. *Proc. Natl. Acad. Sci. USA* 85:7089-93

109. Glazer, A. N., Melis, A. 1987. Photochemical reaction centers: structure, organization and function. *Annu. Rev. Plant Physiol.* 38:11-45

110. Goloubinoff, P., Gatenby, A. A., Lorimer, G. H. 1989. GroE heat-shock proteins promote assembly of foreign prokaryotic ribulose bisphosphate carboxylase oligomers in *Escherichia coli.* *Nature* 337:44-47

111. Green, P. J., Kay, S. A., Chua, N.-H. 1987. Sequence-specific interactions of a pea nuclear factor with light-responsive elements upstream of the rbcS-3A gene. *EMBO J.* 6:2543-2549

112. Green, P. J., Yong, M.-H., Cuozzo, M., Kano-Murakami, Y., Silverstein, P., et al. 1988. Binding site requirements for pea nuclear protein factor GT-1 correlate with sequences required for light-dependent transcriptional activation of the *rbcS-3A* gene. *EMBO J.* 7:4035-44

113. Greene, B. A., Allred, D. R., Morishige, D. T., Staehelin, L. A. 1988. Hierarchical response of light harvesting chlorophyll-proteins in a light-sensitive chlorophyll b-deficient mutant of maize. *Plant Physiol.* 87:357-64

114. Griffiths, W. T. 1978. Reconstitution of protochlorophyllide formation by isolated etioplast membranes. *Biochem. J.* 174:681-92

115. Griffiths, W. T., Kay, A. S., Oliver, R. P. 1985. The presence of photoregulation of protochlorophyllide reductase in green tissue. *Plant Mol. Biol.* 4:13-22

116. Grimm, R., Gast, D., Rüdiger, W. 1989. Characterization of a protein-kinase activity associated with phytochrome from etiolated oat (*Avena sativa* L.) seedlings. *Planta* 178:199-206

117. Grinsven, M. Q. J. M. v., Kool, A. J. 1988. Plastid gene regulation during development: an intriguing complexity of mechanisms. *Plant Mol. Biol. Report.* 6:213-39

118. Grob, U., Stuber, K. 1987. Discrimination of phytochrome dependent light-inducible from non-light-inducible plant genes. Prediction of a common light-responsive element (LRE) in phytochrome dependent light-inducible genes. *Nucleic Acids Res.* 15:9957-9972

119. Grossman, A., Bartlett, S., Chua, N.-H. 1980. Energy-dependent uptake of cytoplasmically-synthesized polypeptides by chloroplasts. *Nature* 285:625-28

120. Gruissem, W. 1989. Chloroplast gene expression: how plants turn their plastids on. *Cell* 56:161-70

121. Hallick, R. B. 1989. Proposals for the naming of chloroplast genes. II. Update to the nomenclature of genes for thylakoid membrane polypeptides. *Plant Mol. Biol. Report.* 7:266-75

122. Harkins, K., Jefferson, R., Kavanagh, T., Bevan, M., Galbraith, D. 1990. Expression of photosynthesis-related gene fusions is restricted by cell type in transgenic plants and in transfected protoplasts. *Proc. Natl. Acad. Sci. USA* 87:816-20

123. Hauser, I., Dehesh, K., Apel, K. 1987. Light-induced changes in the amounts of the 36000-M_r polypeptide of NADPH-protochlorophyllide oxidoreductase and its mRNA in barley plants grown under a diurnal light/dark cycle. *Planta* 170:453-60

124. Hemmingsen, S. M., Woolford, C., van der Vies, S. M., Tilly, K., Dennis, D. T., et al. 1988. Homologous plant and bacterial proteins chaperone oligomeric protein assembly. *Nature* 333:330-34

125. Hepler, P. K., Wayne, R. O. 1985. Calcium and plant development. *Annu. Rev. Plant Physiol.* 36:397-439

126. Herrera-Estrella, L., Van den Broeck, G., Maenhaut, R., Van Montagu, M., Schell, J., et al. 1984. Light-inducible and chloroplast-associated expression of a chimaeric gene introduced into *Nicotiana tabacum* using a Ti plasmid vector. *Nature* 310:115-20

127. Horwitz, B., Thompson, W., Briggs, W. 1988. Phytochrome regulation of greening in *Pisum. Plant Physiol.* 86:299-305

128. Huang, L., Bonner, B. A., Castelfranco, P. A. 1989. Regulation of 5-aminolevulinic acid synthesis in developing chloroplasts. II. Regulation of ALA-synthesizing capacity by phytochrome. *Plant Physiol.* 90:1003-8

129. Huang, L., Castelfranco, P. A. 1989. Regulation of 5-aminolevulinic acid synthesis in developing chloroplasts. I. Effect of light/dark treatments in vivo and in organello. *Plant Physiol.* 90:996-1002

130. Irvine, R. F. 1990. Messenger gets the green light. *Nature* 346:700-1

131. Jabben, M., Shanklin, J., Vierstra, R. D. 1989. Red light–induced accumulation of ubiquitin-phytochrome conjugates in both monocots and dicots. *Plant Physiol.* 90:380-84

132. Jabben, M., Shanklin, J., Vierstra, R. D. 1989. Ubiquitin-phytochrome conjugates. Pool dynamics during in vivo phytochrome degradation. *J. Biol. Chem.* 264:4998-5005

133. Jenkins, G. 1988. Photoregulation of gene expression in plants. *Photochem. Photobiol.* 48:821-32

134. Jenkins, G., Smith, H. 1985. Red : far-red ratio does not modulate the abundance of transcripts for two major chloroplast polypeptides in light-grown *Pisum sativum* terminal shoots. *Photochem. Photobiol.* 42:679-84

135. Jordan, B. R., Dharmasiri, S., Hopley, J., Le Fay, J. 1989. Changes in *psaA* and *psaB* transcript levels during wheat leaf development under different irradiances. *Plant Physiol. Biochem.* 27:769-76

136. Karlin-Neuman, G. A., Sun, L., Tobin, E. M. 1988. Expression of light-harvesting chlorophyll a/b-protein genes is phytochrome-regulated in etiolated *Arabidopsis thaliana* seedlings. *Plant Physiol.* 88:1323-1331

137. Kasemir, H. 1983. Light control of chlorophyll accumulation in higher plants. In *Encyclopedia of Plant Physiology*, ed. A. Pirson, M. H. Zimmerman, 16:662-86. Berlin: Springer-Verlag

138. Kaufman, L. S., Briggs, W. R., Thompson, W. F. 1985. Phytochrome control of specific mRNA levels in developing pea buds: the presence of both very low fluence and low fluence responses. *Plant Physiol.* 78:388-93

139. Kaufman, L. S., Roberts, L. R., Briggs, W. R., Thompson, W. F. 1986. Phytochrome control of specific mRNA levels in developing pea buds. Kinetics of accumulation, reciprocity, and escape kinetics of the low fluence response. *Plant Physiol.* 81:1033-38

140. Kaufman, L. S., Thompson, W. F., Briggs, W. R. 1984. Different red light requirements for phytochrome induced accumulation of *Cab* RNA and rbcS RNA. *Science* 226:1447-49

141. Kay, S. A., Keith, B., Shinozaki, K., Chye, M., Chua, N.-H. 1989. The rice phytochrome gene: structure, auto-regulated expression and binding of GT-1 to a conserved site in the 5' upstream region. *Plant Cell* 1:351-60

142. Kay, S. A., Nagatani, A., Keith, B., Deak, M., Furuya, M., et al. 1989. Rice phytochrome is biologically active in transgenic tobacco. *Plant Cell* 1:775-82

143. Keegstra, K. 1989. Transport and routing of proteins into chloroplasts. *Cell* 56:247-53

144. Keegstra, K., Olsen, L. J. 1989. Chloroplastic precursors and their transport across the envelope membranes. *Annu. Rev. Plant Physiol.* 40:471-501

145. Keller, J. M., Shaulklin, J., Vierstra, R. D., Hershey, H. P. 1989. Expression of a functional monocotyledonous phytochrome in transgenic tobacco. *EMBO J.* 8:1005-12

146. Kellmann, J. W., Pichersky, E., Piechulla, B. 1990. Analysis of the diurnal expression patterns of the tomato chlorophyll a/b binding protein genes. Influence of light and characterization of the gene family. *Photochem. Photobiol.* 52:35-41

146a. Kendrick, R. E., Kronenberg, G. H. M., eds. 1986. *Photomorphogenesis in Plants*. Dordrecht: Martinus Nijhoff/Dr. W. Junk

147. Kittsteiner, U., Paulsen, H., Schendel, R., Rüdiger, W. 1990. Lack of light regulation of NADPH: protochlorophyllide oxidoreductase mRNA in cress seedlings (*Lepidium sativum* L.). *Z. Naturforsch.* 45:1077-79

148. Klausner, R. D., Harford, J. B. 1989. *cis-trans* Models for post-transcriptional regulation. *Science* 246:870-72

149. Klein, R. R., Gamble, P. E., Mullet, J. E. 1988. Light-dependent accumulation of radiolabeled plastid-encoded chlorophyll a-apoproteins requires chlorophyll a. *Plant Physiol.* 88:1246-56

150. Klein, R. R., Mason, H. S., Mullet, J. E. 1988. Light-regulated translation of chloroplast proteins. I. Transcripts of *psaA-psaB*, *psbA*, and *rbcL* are associated with polysomes in dark-grown and illuminated barley seedlings. *J. Cell Biol.* 106:289-301

151. Klein, R. R., Mullet, J. E. 1986. Regulation of chloroplast-encoded chlorophyll-binding protein translation during higher plant chloroplast biogenesis. *J. Biol. Chem.* 261:11138-45

152. Kloppstech, K. 1985. Diurnal and circadian rhythmicity in the expression of light-induced plant nuclear messenger RNAs. *Planta* 165:502-6

153. Kloppstech, K., Meyer, G., Schuster, G., Ohad, I. 1985. Synthesis, transport and localization of a nuclear coded 22-kd heat shock protein in the chloroplast membranes of pea and *Chlamydomonas reinhardtii*. *EMBO J.* 4:1901-1909

154. Ko, K., Granell, A., Bennett, J., Cashmore, A. R. 1990. Isolation and characterization of cDNAs from *Lycopersicon esculentum* and *Arabidopsis thaliana* encoding the 33 kDa protein of the photosystem II-associated oxygen-evolving complex. *Plant Mol. Biol.* 14:217-27

155. Koes, R. E., Spelt, C. E., Mol, J. N. M. 1989. The chalcone synthase multigene family of *Petunia hybrida* (V30): differential, light-regulated expression during flower development and UV light induction. *Plant Mol. Biol.* 12:213-25

156. Koornneef, M., Cone, J. W., Dekens, R. G., O'Hearne-Robers, E. G., Spruit, C. J. P., et al. 1985. Photomorphogenetic responses of long hypocotyl mutants of tomato. *J. Plant Physiol.* 120:153-65

157. Koornneef, M., Rolff, E., Spruit, C. J. P. 1980. Genetic control of light-inhibited hypocotyl elongation in *Arabidopsis thaliana* (L.) Heynh. *Z. Planzenphysiol.* 100S:147-60

158. Kreuz, K., Dehesh, K., Apel, K. 1987. The light-dependent accumulation of the P700 chlorophyll a protein of photosytem I reaction center in barley. Evidence for translational control. *Eur. J. Biochem.* 159:459-67

159. Kuhlemeier, C., Cuozzo, M., Green, P. J., Goyvaerts, E., Ward, K., et al. 1988. Localization and conditional redundancy of regulatory elements in *rbcS-3A*, a pea gene encoding the small subunit of ribulose bisphosphate carboxylase. *Proc. Natl. Acad. Sci. USA* 85:4662-66

160. Kuhlemeier, C., Fluhr, R., Green, P. J., Chua, N.-H. 1987. Sequences in the pea *rbcS-3A* gene have homology to constitutive mammalian enhancers but function as negative regulatory elements. *Genes Dev.* 1:247-55

161. Kuhlemeier, C., Green, P. J., Chua, N.-H. 1987. Regulation of gene expression in higher plants. *Annu. Rev. Plant Physiol.* 38:221-57

162. Kuhn, D. N., Chappell, J., Boudet, A., Hahlbrock, K. 1984. Induction of phenylalanine ammonia-lyase and 4-coumarate:CoA ligase in cultured plant cells by UV light or fungal elicitor. *Proc. Natl. Acad. Sci. USA* 81:1102-6

163. Laing, W., Kreuz, K., Apel, K. 1988. Light-dependent, but phytochrome-independent, translational control of the accumulation of the P700 chlorophyll-a protein of photosystem I in barley (*Hordeum vulgare* L.). *Planta* 176:269-76

164. Lam, E., Benedyk, M., Chua, N.-H. 1989. Characterization of phytochrome-regulated gene expression in a photoautotrophic cell suspension: possible role for calmodulin. *Mol. Cell. Biol.* 9:4819-23

165. Lam, E., Chua, N.-H. 1989. ASF-2: A factor that binds to the cauliflower mosaic virus 35S promoter and a conserved GATA motif in *Cab* promoters. *Plant Cell* 1:1147-56

166. Lam, E., Chua, N.-H. 1989. Light to dark transition modulates the phase of antenna chlorophyll protein gene expression. *J. Biol. Chem.* 264:20175-76

167. Lam, E., Chua, N.-H. 1990. GT-1 binding sites confer light-responsive expression in transgenic tobacco. *Science* 248:471-74

168. Lam, E., Green, P. J., Wong, M., Chua, N.-H. 1989. Phytochrome activation of two nuclear genes requires cytoplasmic protein synthesis. *EMBO J.* 8:2777-83

169. Lamppa, G. K., Abad, M. S. 1987. Processing of a wheat light-harvesting chlorophyll a/b protein precursor by a soluble enzyme from higher plant chloroplasts. *J. Cell. Biol* 105:2641-48

170. Last, D., Gray, J. 1990. Synthesis and accumulation of pea plastocyanin in transgenic tobacco. *Plant Mol. Biol.* 14:229-38

171. Lissemore, J. L., Colbert, J. T., Quail, P. H. 1987. Cloning of cDNA for phytochrome from etiolated *Cucurbita* and coordinate photoregulation of the abundance of two distinct phytochrome transcripts. *Plant Mol. Biol.* 8:485-96

172. Lissemore, J. L., Quail, P. H. 1988. Rapid transcriptional regulation by phytochrome of the genes for phytochrome and chlorophyll a/b binding protein in *Avena sativa*. *Mol. Cell. Biol.* 8:4840-50

173. Lopez-Juez, E., Nagatani, A., Buurmeijer, W. F., Peters, J. L., Furuya,

M., et al. 1990. Response of light-grown wild-type and *aurea* mutant tomato plants to end-of-day far red light. *J. Photochem. Photobiol., B.* 4:391-405

174. Lubben, T. H., Donaldson, G. K., Viitanen, P. V., Gatenby, A. A. 1989. Several proteins imported into chloroplasts form stable complexes with the GroEL-related chloroplast molecular chaperone. *Plant Cell* 1:1223-230

175. Mandoli, D., Ford, G., Waldron, L., Nemson, J., Briggs, W. 1990. Some spectral properties of several soil types: implications for photomorphogenesis. *Plant, Cell Environ.* 13:287-94

176. Manzara, T., Gruissem, W. 1988. Organization and expression of the genes encoding ribulose-1,5-bisphosphate carboxylase in higher plants. *Photoynthesis Res.* 16:117-39

177. Mapleston, R. E., Griffiths, W. T. 1980. Light-modulation of the activity of protochlorophyllide reductase. *Biochem. J.* 189:125-33

178. Marrs, K. A., Kaufman, L. S. 1989. Blue light regulation of transcription for nuclear genes in pea. *Proc. Natl. Acad. Sci. USA* 86:4492-95

179. Marrs, K. A., Kaufman, L. S. 1991. Rapid transcriptional regulation of the *Cab* and *pEA207* gene families in peas by blue light in the absence of cytoplasmic protein synthesis. *Planta.* In press

180. Marshall, J. S., DeRocher, A. E., Keegstra, K., Vierling, E. 1990. Identification of heat shock protein *hsp70* homologues in chloroplasts. *Proc. Natl. Acad. Sci. USA* 87:374-78

181. Mayfield, S. P., Taylor, W. C. 1984. Carotenoid deficient maize seedlings fail to accumulate light-harvesting chlorophyll a/b binding protein (LHCP) mRNA. *Eur. J. Biochem.* 144:79-84

182. McCurdy, D. W., Pratt, L. H. 1986. Immunogold electron microscopy of phytochrome in *Avena*: Identification of intracellular sites responsible for sequestering and enhanced pelletability. *J. Cell Biol.* 103:2541-50

183. McMichael, R., Lagarias, J. 1990. Phosphopeptide mapping of *Avena* phytochrome phosphorylated by protein kinases in vitro. *Biochemistry* 29:3872-878

184. Memon, A., Boss, W. 1990. Rapid light-induced changes in phosphoinositide kinases and H⁺-ATPase in plasma membranes of sunflower hypocotyls. *J. Biol. Chem.* 265:14817-21

185. Merkle, T., Schäfer, E. 1988. Effect of cycloheximide on the inverse control by

phytochrome of the expression of two nuclear genes in barley. *Plant Cell Physiol.* 29:1251-254

186. Meyer, H., Thienel, U., Piechulla, B. 1989. Molecular characterization of the diurnal/circadian expression of the chlorophyll a/b-binding proteins in leaves of tomato and other dicotyledonous and monocotoledonous plant species. *Planta* 180:5-15

187. Michaels, A., Herrin, D. L. 1991. Translational regulation of chloroplast gene expression during the light-dark cell cycle of *Chlamydomonas*: evidence for control by ATP/energy supply. *Biochem. Biophys. Res. Commun.* 170:1082-88

188. Mishkind, M. L., Scioli, S. E. 1988. Recent developments in chloroplast protein transport. *Photosynth. Res.* 19:153-84

188a. Mogen, K., Eide, J., Duysen, M., Eskins, K. 1990. Chloramphenicol stimulates the accumulation of light-harvesting chlorophyll a/b protein II by affecting posttranscriptional events in the chlorina CD3 mutant wheat. *Plant Physiol.* 92:1233-40

189. Mohr, H. 1980. Interaction between blue light and phytochrome in photomorphogenesis. In *The Blue Light Syndrome*, ed. H. Senger, pp. 97-109. Berlin: Springer-Verlag

190. Mohr, H. 1983. Pattern specification and realization in photomorphogenesis. In *Encylopedia of Plant Physiology*, N. S., ed. W. Shropshire, H. Mohr, 16:336-57. Berlin: Springer-Verlag

191. Mohr, H. 1984. Criteria for photoreceptor involvement. In *Techniques in Photomorphogenesis*, ed. H. Smith, M. Holmes, pp. 13-42. London: Academic

192. Mohr, H. 1984. Phytochrome and chloroplast development. In *Chloroplast Biogenesis*, ed. N. R. Baker, J. Barber. Amsterdam: Elsevier

193. Mohr, H. 1986. Coaction between pigment systems. See Ref. 146a, pp. 547-64

194. Mohr, H., Drumm-Herrel, H. 1983. Coaction between phytochrome and blue/UV light in anthocyanin synthesis in seedlings. *Physiol. Plant.* 58:408-14

195. Morelli, G., Nagy, F., Fraley, R. T., Rogers, S. G., Chua, N.-H. 1985. A short conserved sequence is involved in the light-inducibility of a gene encoding ribulose 1,5-bisphosphate carboxylase small subunit of pea. *Nature* 315:200-4

196. Morse, M., Crain, R., Coté, G., Satter, R. 1990. Light-stimulated inositol phos-

pholipid turnover in *Samanea saman* pulvini. Increased levels of diacylglycerol. *Plant Physiol.* 89:724-27

197. Morse, M., Crain, R., Satter, R. 1987. Light-stimulated inositol phospholipid turnover in *Samanea saman* pulvini. *Proc. Natl. Acad. Sci. USA* 84:7075-78

198. Mösinger, E., Batschauer, A., Apel, K., Schäfer, E., Briggs, W. R. 1988. Phytochrome regulation of greening in barley: effects on mRNA abundance and transcriptional activity of isolated nuclei. *Plant Physiol.* 86:706-10

199. Mösinger, E., Batschauer, A., Schäfer, E., Apel, K. 1985. Phytochrome control of in vitro transcription of specific genes in isolated nuclei from barley (*Hordeum vulgare*). *Eur. J. Biochem.* 147:137-42

200. Mösinger, E., Batschauer, A., Vierstra, R., Apel, K., Schäfer, E. 1987. Comparison of the effects of exogenous native phytochrome and in vivo irradiation on in vitro transcription in isolated nuclei from barley (*Hordeum vulgare*). *Planta* 170:505-14

201. Mullet, J. E. 1988. Chloroplast development and gene expression. *Annu. Rev. Plant Physiol. Plant Mol. Biol.* 39:475-502

202. Mullet, J. E., Klein, P. G., Klein, R. R. 1990. Chlorophyll regulates accumulation of the plastid-encoded chlorophyll apoproteins CP43 and D1 by increasing apoprotein stability. *Proc. Natl. Acad. Sci. USA* 87:4038-42

203. Nagy, F., Boutry, M., Hsu, M.-Y., Wong, M., Chua, N.-H. 1987. The 5' proximal region of the wheat *Cab*-1 gene contains a 268 bp enhancer-like sequence for phytochrome response. *EMBO J.* 6:2537-42

204. Nagy, F., Kay, S. A., Boutry, M., Hsu, M.-Y., Chua, N.-H. 1986. Phytochrome-controlled expression of a wheat *Cab* gene in transgenic tobacco seedlings. *EMBO J.* 5:1119-24

205. Nagy, F., Kay, S. A., Chua, N.-H. 1988. A circadian clock regulates transcription of the wheat *Cab*-1 gene. *Genes Dev.* 2:376-82

206. Nagy, F., Kay, S. A., Chua, N.-H. 1988. Gene regulation by phytochrome. *Trends Genet.* 4:37-42

207. Neer, E. J., Clapham, D. 1988. Roles of G protein subunits in transmembrane signalling. *Nature* 333:129-34

208. Nelson, T., Harpster, M. H., Mayfield, S. P., Taylor, W. C. 1984. Light-regulated gene expression during maize leaf development. *J. Cell Biol.* 98:558-64

209. Oelmüller, R. 1989. Photooxidative de-

struction of chloroplasts and its effect on nuclear gene expression and extraplastidic enzyme levels. *Photochem. Photobiol.* 49:229-39

210. Oelmüller, R., Kendrick, R. E., Briggs, W. R. 1989. Blue-light mediated accumulation of nuclear-encoded transcripts coding for proteins of the thylakoid membrane is absent in the phytochrome-deficient aurea mutant of tomato. *Plant Mol. Biol.* 13:223-32

211. Oelmüller, R., Levitan, I., Bergfeld, R., Rajasekhar, V. K., Mohr, H. 1986. Expression of nuclear genes as affected by treatments acting on plastids. *Planta* 168:482-92

212. Oelmüller, R., Mohr, H. 1986. Photooxidative destruction of chloroplasts and its consequences for expresssion of nuclear genes. *Planta* 167:106-13

213. Oelze-Karow, H., Kasemir, H., Mohr, H. 1978. Control of chlorophyll b formation by phytochrome and a threshold level of chlorophyllide a. In *Chloroplast Development*, ed. G. Akoyunglou, J. H. Argyroudi-Akoyunglou, 2:787-92. Elsevier/North-Holland Biomedical

214. Ohl, S., Hahlbrock, K., Schäfer, E. 1989. A stable blue light-derived signal modulates ultraviolet-light-induced activation of the chalcone synthase gene in cultured parsley cells. *Planta* 177:228-36

215. Otto, B., Grimm, B., Ottersbach, P., Kloppstech, K. 1988. Circadian control of the accumulation of mRNAs for light- and heat-inducible chloroplast proteins in pea (*Pisum sativum* L.). *Plant Physiol.* 88:21-25

216. Parks, B. M., Jones, A. M., Adamse, P., Koornneef, M., Kendrick, R. E., et al. 1987. The *aurea* mutant of tomato is deficient in spectro-photometrically and immunochemically detectable phytochrome. *Plant Mol. Biol.* 9:97-107

217. Parks, B. M., Shanklin, J., Koornneef, M., Kendrick, R. E., Quail, P. H. 1989. Immunochemically detectable phytochrome is present at normal levels but is photochemically nonfunctional in the *hy 1* and *hy 2* long hypocotyl mutants of *Arabidopsis*. *Plant Mol. Biol.* 12:425-37

218. Paulsen, H., Rumler, U., Rüdiger, W. 1990. Reconstitution of pigment-containing complexes from light-harvesting chlorophyll a/b-binding protein overexpressed in *Escherichia coli*. *Planta* 181:204-11

219. Peters, J., van Tuinen, A., Adamse, P., Kendrick, R. E., Koornneef, M. 1989. High pigment mutants of tomato exhibit

high sensitivity for phytochrome action. *J. Plant Physiol.* 134:661-66

220. Pichersky, E., Brock, T. G., Nguyen, D., Hoffman, N. E., Piechulla, B., et al. 1989. A new member of the *Cab* gene family: structure, expression, and chromosomal location of *Cab-8*, the tomato gene encoding the Type III chlorophyll a/b binding polypetptide of photosystem I. *Plant Mol. Biol.* 12:257-70

221. Pichersky, E., Hoffman, N. E., Bernatzky, R., Piechulla, B., Tanksley, S. D., et al. 1987. Molecular characterization and genetic mapping of DNA sequences encoding the Type I chlorophyll a/b-binding polypeptide of photosystem I in *Lycopersicon esculentum* (tomato). *Plant Mol. Biol.* 9:205-16

222. Pichersky, E., Tanksley, S. D., Piechulla, B., Stayton, M. M., Dunsmuir, P. 1988. Nucleotide sequence and chromosomal location of *Cab-7*, the tomato gene encoding the type II chlorophyll a/b-binding polypeptide of photosystem I. *Plant Mol. Biol.* 11:69-71

223. Piechulla, B. 1988. Plastid and nuclear mRNA fluctuations in tomato leaves-diurnal and circadian rhythms during extended dark and light periods. *Plant Mol. Biol.* 11:345-53

224. Piechulla, B. 1989. Changes of the diurnal and circadian (endogenous) mRNA oscillations of the chlorophyll a/b binding protein in tomato leaves during altered day/night (light/dark) regimes. *Plant Mol. Biol.* 12:317-27

225. Piechulla, B., Gruissem, W. 1987. Diurnal mRNA fluctuations of nuclear and plastid genes in developing tomato fruits. *EMBO J.* 6:3593-99

226. Plumley, F. G., Schmidt, G. W. 1987. Reconstitution of chlorophyll a/b light-harvesting complexes: xanthophyll-dependent assembly and energy transfer. *Proc. Natl. Acad. Sci. USA* 84:146-50

227. Pratt, L. 1986. Localization within the plant. See Ref. 146a, pp. 61-82

228. Presti, D. E. 1983. The photobiology of carotenes and flavins. In *The Biology of Photoreception*, ed. D. J. Cosens, D. Vince-Prue, pp. 133-80. Cambridge: Cambridge Univ. Press

229. Quail, P. H., Gatz, C., Hershey, H., Jones, A., Lissemore, J. L., et al. 1987. Molecular biology of phytochrome. In *Phytochrome and Photoregulation in Plants*, ed. M. Furuya, pp. 23-37. Orlando, Florida: Academic

230. Quail, P. H., Colbert, J. P., Peters, N. K., Christensen, A. H., Lissemore, J. L. 1987. Phytochrome and the regula-tion of expression of its genes. *Philos. Trans. R. Soc. London Ser. B* 314:469-80

231. Raines, C. A., Longstaff, M., Lloyd, J. C., Dyer, T. A. 1989. Complete coding sequence of wheat phosphoribulokinase: Developmental and light-dependent expression of the mRNA. *Mol. Gen. Genet.* 220:43-48

232. Reimann-Philipp, U., Behnke, S., Batschauer, A., Schäfer, E., Apel, K. 1989. The effect of light on the biosynthesis of leaf-specific thionins in barley, *Hordeum vulgare. Eur. J. Biochem.* 182:283-89

233. Reiss, T., Bergfeld, R., Link, G., Thien, W., Mohr, H. 1983. Photooxidative destruction of chloroplasts and its consequences for cytosolic enzyme levels and plant development. *Planta* 159:518-28

234. Rougvie, A. E., Lis, J. T. 1988. The RNA polymerase II molecule at the 5' end of the uninduced *hsp70* gene of *D. melanogaster* is transcriptionally engaged. *Cell* 54:795-804

235. Roux, S. 1984. Ca^{+2} and phytochrome action in plants. *BioScience* 34:25-29

236. Roux, S. 1986. Phytochrome and membranes. See Ref. 146a, pp. 115-34

237. Roux, S., Guo, Y., Li, H. 1990. Characterization of two calcium-dependent protein kinases implicated in stimulus-response coupling in plants. *Curr. Topics Plant Biochem. Physiol.* 9:129-40

238. Roy, H. 1989. Rubisco assembly: a model system for studying the mechanism of chaperonin action. *Plant Cell.* 1:1035-42

239. Roy, H., Bloom, M., Milos, P., Monroe, M. 1982. Studies on the assembly of large subunits of ribulose bisphosphate carboxylase in isolated pea chloroplasts. *J. Cell Biol.* 94:20-27

240. Rüdiger, W. 1987. Phytochrome: the chromophore and photoconversion. *Photobiochem. Photobiophys.* 49(Suppl.):217-27

241. Ryrie, I. J. 1983. Immunological evidence for apoproteins of the light-harvesting chlorophyll-protein complex in a mutant of barley lacking chlorophyll b. *Eur. J. Biochem.* 131:149-55

242. Santel, H.-J., Apel, K. 1981. The protochlorophyllide holochrome of barley (*Hordeum vulgare* L.) The effect of light on the NADPH:protochlorophyllide oxidoreducatase. *Eur. J. Biochem.* 120:95-103

243. Sato, N. 1988. Nucleotide sequence and expression of the phytochrome gene in

Pisum sativum: differential regulation by light of multiple transcripts. *Plant Mol. Biol.* 11:697-710

244. Schäfer, E. 1987. Primary action of phytochrome. In *Phytochrome and Photoregulation in Plants*, ed. M. Furuya, pp. 279-89. London: Academic

245. Schäfer, E., Briggs, W. R. 1986. Photomorphogenesis from signal perception to gene expression. *Photobiochem. Photobiophys.* 12:305-20

246. Schindler, U., Cashmore, A. 1990. Photoregulated gene expression may involve ubiquitous DNA binding proteins. *EMBO J.* 9:3415-27

247. Schmid, J., Doerner, P., Clouse, S., Dixon, R., Lamb, C. 1990. Developmental and environmental regulation of a bean chalcone synthase promoter in transgenic tobacco. *Plant Cell* 2:619-31

248. Schmidt, S., Drumm-Herrel, H., Oelmüller, R., Mohr, H. 1987. Time course of competence in phytochrome-controlled appearance of nuclear-encoded plastidic proteins. *Planta* 170:400-7

249. Schopfer, P. 1984. Photomorphogenesis. In *Advanced Plant Physiology*, ed. M. Wilkins, pp. 380-407. London: Pitman

250. Schulze-Lefert, P., Dangl, J. L., Becker-André, M., Hahlbrock, K., Schulz, W. 1989. Inducible in vivo DNA footprints define sequences necessary for UV light activation of the parsley chalcone synthase gene. *EMBO J.* 8:651-56

251. Schuster, C., Mohr, H. 1990. Appearance of nitrite-reductase mRNA in mustard seedling cotyledons is regulated by phytochrome. *Planta* 181:327-34

252. Senger, H., Schmidt, W. 1986. Cryptochrome and UV receptors. See Ref. 146a, pp. 137-58

253. Shanklin, J., Jabben, M., Vierstra, R. D. 1987. Red light-induced formation of ubiquitin-phytochrome conjugates: Identification of possible intermediates of phytochrome degradation. *Proc. Natl. Acad. Sci. USA* 84:359-63

254. Sharrock, R. A., Parks, B. M., Koornneef, M., Quail, P. H. 1988. Molecular analysis of the phytochrome deficiency in an aurea mutant of tomato. *Mol. Gen. Genet.* 213:9-14

255. Sharrock, R. A., Quail, P. H. 1989. Novel phytochrome sequences in *Arabidopsis thaliana*: structure, evolution, and differential expression of a plant regulatory photoreceptor family. *Genes Dev.* 3:1745-57

256. Sheen, J.-Y., Sayre, R. T., Bogorad, L.

1987. Differential expression of oxygen-evolving polypeptide genes in maize leaf cell types. *Plant Mol. Biol.* 9:217-26

257. Shih, M.-C., Goodman, H. M. 1988. Differential expression of nuclear genes encoding chloroplast and cytosolic glyceraldehyde-3-phosphate dehydrogenase in *Nicotiana tabacum*. *EMBO J.* 7:893-98

258. Shirley, B. W., Meagher, R. B. 1990. A potential role for RNA turnover in the light regulation of plant gene expression: ribulose-1,5-bisphosphate carboxylase small subunit in soybean. *Nucleic Acids Res.* 18:3377-85

259. Short, T., Briggs, W. R. 1990. Characterization of a rapid, blue light–mediated change in detectable phosphorylation of a plasma membrane protein from etiolated pea (*Pisum sativum* L.) seedlings. *Plant Physiol.* 92:179-85

260. Short, T. W., Gallagher, S., Briggs, W. R. 1991. Protein phosphorylation as a possible signal transduction step for blue light–mediated phototropism in pea (*Pisum sativum* L.) epicotyls. *Curr. Topics Plant Biochem. Physiol.* 9:In press

261. Silverthorne, J., Tobin, E. M. 1987. Phytochrome regulation of nuclear gene expression. *Bioessays* 7:18-23

262. Silverthorne, J., Tobin, E. M. 1984. Demonstration of transcriptional regulation of specific genes by phytochrome action. *Proc. Natl. Acad. Sci. USA* 81:1112-16

263. Simpson, D. J., Machold, O., Hoyer-Hansen, G., von Wettstein, D. 1985. *Chlorina* mutants of barley (*Hordeum vulgare* L.). *Carlsberg Res. Commun.* 50:223-28

264. Simpson, J., Herrera-Estrella, L. 1990. Light-regulated gene expression. *Crit. Rev. Plant Sci.* 9:95-109

265. Simpson, J., Timko, M. P., Cashmore, A. R., Schell, J., Van Montagu, M., et al. 1985. Light-inducible and tissue-specific expression of a chimaeric gene under control of the 5' flanking sequence of a pea chlorophyll a/b-binding protein gene. *EMBO J.* 4:2723-29

266. Singh, B. R., Song, P.-S. 1990. Phytochrome and protein phosphorylation. *Photochem. Photobiol.* 52:249-54

267. Slovin, J. P., Tobin, E. M. 1982. Synthesis and turnover of the light-harvesting chlorophyll a/b-protein in *Lemna gibba* grown with intermittent red light: possible translational control. *Planta* 154:465-72

268. Smeekens, S., Weisbeek, P., Robinson,

C. 1990. Protein transport into and within chloroplasts. *Trends Biochem. Sci.* 15:73-76

269. Smeekens, S., Bauerle, C., Hageman, J., Keegstra, K., Weisbeek, P. 1986. The role of the transit peptide in the routing of precursors toward different chloroplast compartments.*Cell* 46:365-75

270. Speth, V., Otto, V., Schäfer, E. 1987. Intracellular localization of phytochrome and ubiquitin in red light-irradiated oat coleoptiles by electron microscopy. *Planta* 171:332-38

271. Speth, V., Otto, V., Schäfer, E. 1987. Intracellular localization of phytochrome in oat coleoptiles by electron microscopy. *Planta* 168:299-304

272. Sponga, F., Deitzer, G. F., Mancinelli, A. L. 1986. Cryptochrome, phytochrome, and the photoregulation of anthocyanin production under blue light. *Plant Physiol.* 82:952-55

273. Staehelin, L. A., Arntzen, C. J. 1983. Regulation of chloroplast membrane function: protein phosphorylation changes the spatial organization of membrane components. *J. Cell Biol.* 97:1327-337

274. Staiger, D., Kaulen, H., Schell, J. 1989. A CACGTG motif of the *Antirrhinum majus* chalcone synthase promoter is recognized by an evolutionarily conserved nuclear protein. *Proc. Natl. Acad. Sci. USA* 86:6930-934

275. Stayton, M., Brosio, P., Dunsmuir, P. 1989. Photosynthetic genes of *Petunia* (Mitchell) are differentially expressed during the diurnal cycle. *Plant Physiol.* 89:776-82

276. Steinitz, B., Schäfer, E., Drumm, H., Mohr, H. 1979. Correlation between far-red absorbing phytochrome and response in phytochrome-mediated anthocyanin synthesis. *Plant Cell Environ.* 2:159-63

277. Stockhaus, J., Schell, J., Willmitzer, L. 1989. Correlation of the expression of the nuclear photosynthetic gene *ST-LS1* with the presence of chloroplasts. *EMBO J.* 8:2445-51

278. Stockhaus, J., Schell, J., Willmitzer, L. 1989. Identification of enhancer and silencer sequences in the upstream region of the light-regulated nuclear photosynthetic gene *ST-LS1*. *Plant Cell* 1:805-13

279. Sun, L., Tobin, E. M. 1990. Phytochrome-regulated expression of genes encoding light-harvesting chlorophyll A/B protein in two long hypocotyl mutants and wild type plants of *Arabidopsis thaliana*. *Photochem. Photobiol.* 52:51-56

280. Sutton, A., Sieburth, L. E., Bennett, J. 1987. Light-dependent accumulation and localization of photosystem II proteins in maize. *Eur. J. Biochem.* 164:571-78

281. Tanaka, A., Tsuji, H. 1983. Formation of chlorophyll-protein complexes in greening cucumber cotyledons in light and then in darkness. *Plant Cell Physiol.* 24:101-8

282. Taylor, W. C. 1989. Regulatory interactions between nuclear and plastid genomes. *Annu. Rev. Plant Physiol. Plant Mol. Biol.* 40:211-33

283. Taylor, W. C. 1989. Transcriptional regulation by a circadian rhythm. *Plant Cell* 1:259-64

284. Terao, T., Katoh, S. 1989. Synthesis and breakdown of the apoproteins of light-harvesting chlorophyll a/b protein in chlorophyll b–deficient mutants of rice. *Plant Cell Physiol.* 30:571-80

285. Terao, T., Matsuoka, M., Katoh, S. 1988. Immunological quantitation of proteins related to light-harvesting chlorophyll a/b protein complexes of the two photosystems in rice mutants totally and partially deficient in chlorophyll b. *Plant Cell Physiol.* 29:825-34

286. Tester, M., Morris, C. 1987. The penetration of light through soil. *Plant Cell Environ.* 10:281-86

287. Theil, E. 1990. Regulation of ferretin and transferrin receptor mRNAs. *J. Biol. Chem.* 265:4771-74

288. Thompson, D. M., Meagher, R. B. 1990. Transcriptional and post-transcriptional processes regulate expression of RNA encoding the small subunit of ribulose bisphosphate carboxylase differently in petunia and in soybean. *Nucleic Acids Res.* 18:3621-29

289. Thompson, W., Elliott, R., Dickey, L., Gallo, M., Pedersen, T. 1990. Unusual features of the light respose system regulating ferredoxin gene expression. In *Phytochrome Properties and Biological Action*, ed. B. Thomas. Berlin: Springer-Verlag

290. Thompson, W. F. 1988. Photoregulation: diverse gene responses in greening seedlings. *Plant Cell Environ.* 11:319-28

291. Thompson, W. F., Elliott, R. C., Dickey, L. F., Fristensky, B. R., White, M. J. 1990. Light regulated expression of the *Fed-1* gene in pea involves *cis*-acting elements within the transcription unit. In *Mechanisms of Plant Perception and Response to Environmental Stimuli*, ed. T. Thomas, A. Smith, pp. 3-17. Bristol: Br. Soc. Plant Growth Regul.

292. Thompson, W. F., Kaufman, L. S.,

Watson, J. C. 1985. Induction of plant gene expression by light. *BioEssays* 3:153-59

293. Thornber, J. P., Peter, G. F., Chitnis, P. R., Nechushtai, R., Vainstein, A. 1988. The light-harvesting complex of photosystem II of higher plants. In *Light Energy Transduction in Photosynthesis: Higher Plant and Bacterial Models*, ed. S. E. Stevens, D. A. Bryant, pp. 137-54. Rockville, MD: Am. Soc. Plant Physiol.

294. Thorne, S. W. 1971. The greening of etiolated bean leaves 1. The initial photoconversion process. *Biochim. Biophys. Acta* 226:113-27

295. Timko, M. P., Kausch, A. P., Castresana, C., Fassler, J., Herrera-Estrella, L., et al. 1985. Light regulation of plant gene expression by an upstream enhancer-like element. *Nature* 318:579-82

296. Tingey, S. V., Tsai, F.-Y., Edwards, J. W., Walker, E. L., Coruzzi, G. M. 1988. Chloroplast and cytosolic glutamine synthase are encoded by homologous nuclear genes which are differentially expressed in vivo. *J. Biol. Chem.* 263:9651-57

297. Tobin, E., Brusslan, J., Buzby, J., Karlin-Neuman, G., Kehoe, D., et al. 1990. Approaches to understanding phytochrome regulation of transcription in *Lemna gibba* and *Arabidopsis thaliana*. In *Plant Molecular Biology*, ed. R. Hermann. New York: Plenum

298. Tobin, E. M., Silverthorne, J. 1985. Light regulation of gene expression in higher plants. *Annu. Rev. Plant Physiol.* 36:569-93

299. Tokuhisa, J. G., Quail, P. H. 1989. Phytochrome in green tissue: partial purification and characterization of the 118-kilodalton phytochrome species from light-grown *Avena sativa* L. *Photochem. Photobiol.* 50:143-52

300. Tomizawa, K.-I., Nagatani, A., Furuya, M. 1990. Phytochrome genes: studies using the tools of molecular biology and photomorphogenetic mutants. *Photochem. Photobiol.* 51:265-76

301. Tomizawa, K.-I., Sato, N., Furuya, M. 1989. Phytochrome control of multiple transcripts of the phytochrome gene in *Pisum sativum*. *Plant Mol. Biol.* 12:295-99

302. Tripathy, B. C., Rebeiz, C. A. 1988. Chloroplast biogenesis 60. Conversion of divinyl protochlorophyllide to monovinyl protochlorophyllide in green(ing) barley, a dark monovinyl/light divinyl plant species. *Plant Physiol.* 87:89-94

303. Tsai, F.-Y., Coruzzi, G. 1990. Dark-induced and organ-specific expression of two asparagine synthetase genes in *Pisum sativum*. *EMBO J.* 9:323-32

304. Tzinas, G., Argyroudi-Akoyunoglou, J. H., Akoyunoglou, G. 1987. The effect of the dark interval in intermittent light on thylakoid development: photosynthetic unit formation and light harvesting protein accumulation. *Photosyn. Res.* 14:241-58

305. Ueda, T., Pichersky, E., Malik, V. S., Cashmore, A. R. 1989. Level of expression of the tomato *rbcS-3A* gene is modulated by a far upstream promoter element in a developmentally regulated manner. *Plant Cell* 1:217-27

306. Vierling, E., Alberte, R. S. 1983. Regulation of synthesis of the photosystem I reaction center. *J. Cell Biol.* 97:1806-14

307. Vierling, E., Harris, L. M., Chen, Q. 1989. The major low-molecular-weight heat shock protein in chloroplasts shows antigenic conservation among diverse higher plant species. *Mol. Cell. Biol.* 9:461-68

308. Vierling, E., Mishkind, M. L., Schmidt, G. W., Key, J. L. 1986. Specific heat shock proteins are transported into chloroplasts. *Proc. Natl. Acad. Sci. USA* 83:361-65

309. Virgin, H. I. 1986. Action spectra for chlorophyll formation during greening of wheat leaves in continuous light. *Physiol. Plant.* 66:277-82

310. Viro, M., Kloppstech, K. 1982. Expression of genes for plastid membrane proteins in barley under intermittent light conditions. *Planta* 154:18-23

311. Vogelmann, T. C., Bornman, J. F., Josserand, S. 1989. Photosynthetic light gradients and spectral regime within leaves of *Medicago sativa*. *Philos. Trans. R. Soc. London Ser. B.* 323:411-22

312. Vorst, O., van Dam, F., Oosterhoff-Teertstra, R., Smeekens, S., Weisbeek, P. 1990. Tissue-specific expression directed by an *Arabidopsis thaliana* pre-ferredoxin promoter in transgenic tobacco plants. *Plant Mol. Biol.* 14:491-99

313. Wagner, E., Mohr, H. 1966. 'Primäre' und 'sekundäre' Differenzierung im Zusammenhang mit der Photomorphogenese von Keimpflanzen (*Sinapis alba* L.). *Planta* 71:204-21

314. Warpeha, K. M. F., Hamm, H., Kaufman, L. S. 1990. A blue-light-induced plasma-membrane-associated GTP binding protein in pea. *Plant Physiol. (Suppl.)* 93:30, Abstr. 164

315. Warpeha, K. M. F., Marrs, K. A., Kaufman, L. S. 1989. Blue light regula-

tion of specific transcript levels in *Pisum sativum*. *Plant Physiol.* 91:1030-35

316. Watanabe, A., Kawakami, N., Azumi, Y. 1989. Gene expression in senescing leaves. In *Cell Separation in Plants*, ed. D. J. Osborne, M. B. Jackson. Berlin: Springer-Verlag

317. Watson, J. C. 1989. Photoregulation of gene expression in plants. In *Plant Biotechnology*, ed. S. Kung, C. J. Arntzen, pp. 161-205. Boston: Butterworths

318. Wehmeyer, B., Cashmore, A. R., Schäfer, E. 1990. Photocontrol of the expression of genes encoding chlorophyll a/b binding proteins and small subunit of ribulose-1,5-bisphosphate carboxylase in etiolated seedlings of *Lycopersicon esculentum* (L.) and *Nicotiana tabacum* (L.). *Plant. Physiol.* 93:990-97

319. Wenng, A., Ehmann, B., Schäfer, E. 1989. The 23 kDa polypeptide of the photosynthetic oxygen-evolving complex from mustard seedlings (*Sinapis alba* L.). *FEBS Lett.* 246:140-44

320. White, M. J., Green, B. R. 1987. Polypeptides belonging to each of the three major chlorophyll a+b protein complexes are present in a chlorophyll b–less barley mutant. *Eur. J. Biochem.* 165:531-35

321. White, M. J., Green, B. R. 1987b. Immunological studies on the chlorophyll a+b antenna complexes of photosystem I and photosystem II. In *Progress in Photosynthesis Research*, ed. J. Biggins, pp. 193-96. Boston: Martinus-Nijhoff Publishers

322. White, M. J., Green, B. R. 1988. Intermittent-light chloroplasts are not developmentally equivalent to chlorina f2 chloroplasts in barley. *Photosynth. Res.* 15:195-203

323. Wong, Y.-S., Cheng, H.-C., Walsh, D., Lagarias, J. 1986. Phosphorylation of *Avena* phytochrome in vitro as a probe of light-induced conformational changes. *J. Biol. Chem.* 261:12089-97

324. Wong, Y.-S., Lagarias, J. 1989. Affinity labeling of *Avena* phytochrome with ATP analogs. *Proc. Natl. Acad. Sci. USA* 89:3469-73

325. Wong, Y.-S., McMichael, R., Lagarias, J. 1989. Properties of a polycation-stimulated protein kinase associated with purified *Avena* phytochrome. *Plant Physiol.* 91:709-18

326. Wright, S., Bishop, J. M. 1989. DNA sequences that mediate attenuation of transcription from the mouse proto-oncogene *myc*. *Proc. Natl. Acad. Sci. USA* 86:505-9

327. Zielinski, R. E. 1987. Calmodulin mRNA in barley (*Hordeum vulgare* L.). *Plant Physiol.* 84:937-43

328. Zielinski, R. E., Werneke, J. M., Jenkins, M. E. 1989. Coordinate expression of rubisco activase and rubisco during barley leaf cell development. *Plant Physiol.* 90:516-21

Annu. Rev. Plant Physiol. Plant Mol. Biol. 1991. 42:467–506

GLYCEROLIPID SYNTHESIS:
Biochemistry and Regulation[1]

John Browse

Institute of Biological Chemistry, Washington State University, Pullman, Washington 99164-6340

Chris Somerville

MSU-DOE Plant Research Laboratory, Michigan State University, East Lansing, Michigan 48824

KEY WORDS: *Arabidopsis*, desaturation, fatty acid synthesis, membrane lipids, phospholipid synthesis, oilseeds

CONTENTS

[1]**Abbreviations used**: ACP, Acyl Carrier Protein; DAG, diacylglycerol; DGD, digalactosyldiacylglycerol; FAS, fatty acid synthase; MGD, monogalactosyldiacylglycerol; PA, phosphatidic acid; PC, phosphatidylcholine; PE, phosphatidylethanolamine; PG, phosphatidylglycerol; PI, phosphatidylinositol; PS, phosphatidylserine; SL, sulfoquinovosyldiacylglycerol (sulfolipid); TAG, triacylglycerol; X:Y, a fatty acyl group containing X carbon atoms and Y double bonds (*cis* unless specified). Double bond positions are indicated relative to the methyl carbon of the fatty acid chain (e.g. $\omega 6$) or the carboxyl end of the chain (e.g. $\Delta 5$)

INTRODUCTION

This is an exciting time to be working on plant lipid metabolism. There is an increasingly sophisticated appreciation of how the chemical properties and molecular shapes of lipid molecules affect the biogenesis and function of the various membranes of the cell and, thus, of how they contribute to the normal growth and development of plants and other organisms. Recent descriptions of the involvement of phosphoinositides (12, 50, 59) and methyl jasmonate (51) in cellular signal transduction, the demonstration that lipid composition can determine chilling and freezing tolerance (119, 177, 194) and our better understanding of how thylakoid fatty acid composition may affect the thermal tolerance of photosynthetic light reactions (139) exemplify the advances being made. Likewise there is widespread interest in genetically modifying the pathways of seed triacylglycerol synthesis to provide new plant oils both for food and for industrial uses (127). These developments have underscored the need for detailed and specific knowledge of the pathways, enzymes, and mechanisms that regulate plant glycerolipid metabolism. We review these topics here.

At present, there are only a few indications of how the pathways of glycerolipid synthesis are regulated to provide the complement of molecules required for the correct assembly and functioning of the various membranes of the cell. In large measure, the meagerness of our knowledge results from our inability to solubilize and purify the membrane-bound enzymes involved—a prerequisite to a biochemical characterization of their kinetic and regulatory properties. However, through the use of both biochemical and genetic approaches, a cogent model of glycerolipid metabolism has been developed that forms a useful basis for interpreting evidence about the regulation of both membrane lipid synthesis and seed triacylglycerol synthesis in plants.

We have tried to make this review intelligible to a broad audience of plant scientists. Some knowledge of lipid molecular structures and the distribution of lipids among different membranes of the cell is necessary, but this information can be found in many texts or in several books devoted to these and related topics (68, 180, 181).

HISTORICAL PERSPECTIVE

As in many other aspects of metabolism, radiotracer experiments in vivo have provided the basic framework for the pathways of lipid synthesis. Early studies on algae and higher plants provided the first evidence for desaturation of 18:1 to 18:2 on both PC and MGD (2, 66, 81). Subsequently, the detailed kinetics of [^{14}C]acetate, $^{14}CO_2$ and [^{3}H]glycerol labeling in leaves of higher plants demonstrated the flow of label through PC to chloroplast lipids (141, 170). This information provided the starting point for the elucidation of the two-pathway model (145) that forms the contemporary view of leaf lipid metabolism.

The second major set of results that led to the development of the two-pathway hypothesis came from labeling experiments with isolated spinach chloroplasts. The critical observations were that the products of chloroplast lipid synthesis—PA, DAG and acylCoA—are the postulated substrates for the prokaryotic and eukaryotic pathways and that the relative amounts of these products could be altered by additions to the medium, such as glycerol-3-P (144, 145, 150).

More recently, mutant analysis has provided a complementary method to circumvent the problems of studying the membrane-bound enzymes of lipid metabolism by traditional biochemical techniques. The isolation and characterization of a collection of mutants of *Arabidopsis* with alterations in the fatty acid composition of their leaf lipids (22) has provided considerable, though indirect, information about the desaturation steps leading to the synthesis of polyunsaturated membrane lipids (21, 23, 101). In addition, the mutants provide a novel approach toward understanding the role of lipids in membrane function (77, 102, 107) and may soon allow the isolation of key genes of lipid metabolism by gene tagging or chromosome walking (62).

Of necessity much of the enzymological work carried out before about 1980 used crude subcellular fractions or partially purified enzymes. The usual difficulties of working with such undefined systems, combined with the difficulties of presenting lipid soluble substrates, resulted in a mixture of useful information and artefacts. In the last decade, however, the availability of high-performance chromatography systems and other advances in protein purification techniques has permitted the purification to apparent homogeneity of a number of the soluble enzymes of fatty acid and glycerolipid synthesis (Table 1). In several cases, cDNA or genomic clones corresponding to the genes encoding these enzymes have been obtained. However, in contrast to the considerable success in characterizing the soluble enzymes, only a few

Table 1   Enzymes of glycerolipid synthesis purified from higher plants

Enzyme	Source	Cloned	References
acetyl-CoA carboxylase	parsley, soybean, rapeseed	no	33, 48, 72
enoyl-ACP reductase	rapeseed	no	165
3-ketoacyl-ACP synthase I	rapeseed, soybean, barley	yes	91, 97, 105
3-ketoacyl-ACP synthase II	soybean	no	97
3-ketoacyl-ACP reductase	avocado	no	156
stearoyl-ACP desaturase	avocado, safflower, soybean	yes	36, 97, 152, 155
glycerol-3-P acyltransferase	squash	yes	124
acyl-ACP thioesterase	soybean, rapeseed	no	73, 97
UDP-galactose : DAG galactosyltransferase	spinach	no	190
CTP : choline-P cytidyltransferase	castor bean	no	197

membrane-bound enzymes of lipid metabolism have been purified from plant sources. These include CTP:choline-P cytidyltransferase purified from castor bean (197) and UDP-galactose:DAG galactosyltransferase purified from spinach chloroplast envelopes (190). The development of methods for purifying other membrane-bound enzymes remains a major challenge for the forthcoming decade.

THE TWO PATHWAYS OF LIPID SYNTHESIS IN LEAVES

Overview

The recognition that the leaves of higher plants utilize two distinct pathways for glycerolipid synthesis is one of the most important discoveries in plant lipid metabolism during the past decade (145, 149). The essence of the two-pathway model (Figure 1) is that 16:0 and 18:1 fatty acids synthesized de novo in the chloroplast (128, 144) may either be used directly in the chloroplast envelope for production of chloroplast lipids via the prokaryotic pathway (14, 40, 46, 69, 145) or be exported from the chloroplast as CoA esters (13) to enter the eukaryotic pathway at extrachloroplast sites, particularly in the endoplasmic reticulum (110). In all higher plants, a proportion of the diacylglycerol moiety of PC synthesized by the eukaryotic pathway is returned to the chloroplast where it contributes to the production of thylakoid lipids (70, 149, 170, 201).

In many families of angiosperms (designated "18:3" plants) PG is the only

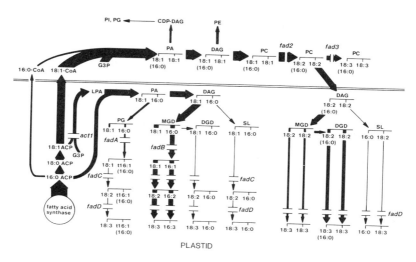

Figure 1   An abbreviated diagram of the two-pathway scheme of glycerolipid biosynthesis in the 16:3 plant *Arabidopsis*. See text for details. Widths of the lines show the relative fluxes through different reactions.

major product of the prokaryotic pathway, and the remaining chloroplast lipids are synthesized entirely by the eukaryotic pathway (149). However, in the members of some primitive angiosperm families (designated "16:3" plants), both pathways contribute to the synthesis of MGD and, to a lesser extent, SL and DGD (28, 162). Because MGD formed via the prokaryotic pathway contains a high level of 16:3 at the sn-2 position (147, 162), these species are distinguishable from other angiosperms (18:3 plants) whose MGD contains predominantly α-linolenate (83).

The evidence leading to the elaboration of the two-pathway hypothesis has been reviewed in this series (149). Here we use the framework to consider recent advances in our understanding of cellular lipid metabolism.

Fatty Acid Synthesis

In all plants studied, de novo fatty acid synthesis has been found localized almost exclusively in the plastids of cells (127). The recent discovery of low levels of acyl-ACP in plant mitochondria (38) raises the possibility that ACP-dependent fatty acid synthesis or modification may also take place in this organelle. Although it will be interesting to discover the function of this ACP pool, any fatty acid biosynthetic activity in the mitochondrion is quantitatively insignificant by comparison with that of the plastid pathway.

Two sources of acetyl-CoA for fatty acid synthesis have been identified. A plastid pyruvate dehydrogenase complex provides acetyl-CoA from glycolysis products in nongreen plastids (44) and also in the leaf chloroplasts of many species (31). It has also been suggested that in spinach leaves, free acetate generated by mitochondrial pyruvate dehydrogenase and acetyl-CoA hydrolase diffuses to the chloroplast where it is readily converted to acetyl-CoA by acetyl-CoA synthetase (104). However, even in spinach leaves, it is likely that a substantial proportion of the acetyl-CoA used in fatty acid synthesis is derived via the plastid pyruvate dehydrogenase (104). Further evidence against a major role for the mitochondrial pyruvate dehydrogenase lies in the fact that the enzyme is less active in illuminated leaves than in the dark (29) whereas chloroplast fatty acid synthesis is light dependent (146,147).

The synthesis of malonyl-CoA by acetyl-CoA carboxylase is the first committed step in fatty acid synthesis, and in animals it is well established that acetyl-CoA carboxylase is the rate-limiting step for fatty acid synthesis (93). The animal enzyme is subject to positive regulation by citrate and phosphorylation, and to inhibition by long-chain acyl-CoA (93). Like the animal carboxylase, the plant enzyme is a single polypeptide containing the three functional domains—biotin carboxylase, carboxyl carrier protein (which contains a biotin prosthetic group), and carboxyl transferase—found as separate proteins in E. coli. When care is taken to limit degradation by endogenous proteinases, a plant enzyme is purified as a high-molecular-weight multimer of identical subunits of about 220–240 kd (33, 48, 72). However, it is not yet clear whether this high-molecular-weight form is the only isozyme present in plants. Relatively little is known about the regulation of acetyl-CoA carboxylase in plants, but the recent discovery that the enzyme is the target for aryloxyphenoxypropionate and cyclohexanedione herbicides (30, 132) should provide useful new tools for analysis.

Each cycle in fatty acid synthesis is initiated by the condensation of a fatty acyl group (which is linked by a thioester bond to the active site of 3-ketoacyl-ACP synthase) with malonyl-ACP to produce a 3-ketoacyl-ACP and CO_2. The cycle continues with the sequential action of 3-ketoacyl-ACP reductase, 3-hydroxyacyl-ACP dehydratase, and enoyl-ACP reductase to form a new acyl chain that is two carbons longer than at the start of the cycle. The individual activities have been separated from extracts of several plants (75, 157, 158, 166), indicating that the plant fatty acid synthase is comparable to the dissociable prokaryotic Type II rather than the multienzyme complexes (Type I) found in yeast and animals. The degree of association of the Type II proteins in vivo is not known for either plants or bacteria. In view of the existence of Type I enzymes in other organisms, it is tempting to speculate that the individual polypeptides of plant FAS may exist in vivo as a supramolecular structure similar to that proposed for the citric acid cycle (176) or the Calvin cycle (60). However, the balance of evidence suggests that the degree of association is not very great. In the yeast and animal Type I synthases, each of the activities is represented by only a single protein domain. In plants and *E. coli,* however, there are three forms of 3-ketoacyl-ACP synthase and two forms of enoyl-ACP reductase, with different iso-zymes acting at different stages in the sequence from acetate to 16:0 and 18:0 fatty acids (85, 157, 158). There are also different isoforms of ACP (129) and malonyl-CoA:ACP transacylase (64). Further information on these aspects can be found in recent reviews (67, 127). It seems unlikely that these diverse proteins could be brought together into a single complex that was at all similar to the yeast or animal fatty acid synthases. Furthermore, the fatty acid synthase enzymes do not appear to be present in plants in stoichiometric amounts. For example, the enoyl-ACP reductase, which has been purified from several species, was shown to be a homotetramer of 35 kDa and was calculated to represent 0.79% of the total protein in developing *Brassica napus* seeds (165). In contrast, the 3-ketoacyl-ACP synthase I purified from the same tissue was shown to be a homodimer of 43 kDa and to make up only 0.01% of total protein (105).

The major difference between current views of the plant fatty acid synthase and the traditional description of just a few years ago (67) is the discovery of a third isoform of 3-ketoacyl-ACP synthase (KAS III). In both *E. coli* (80) and plants (85), KAS III appears to be responsible for the first condensation of acetate with malonyl-ACP. Furthermore, the enzyme appears to use acetyl-CoA rather than acetyl-ACP as the C_2 substrate, thus obviating for acetyl-CoA:ACP transacylase. In addition to the multiple isozymes of KAS, there are also different isoforms of ACP (63, 76, 129); malonyl-CoA:ACP trans-acylase exists in two forms in soybean leaf but in only one form in developing seeds (63). Although the ACP isoforms may be involved in regulating flux of fatty acids between the prokaryotic and eukaryotic pathways (63), the func-tional significance of the isoforms is not yet known. The observation that ACP is present in multiple isoforms in all vascular and nonvascular plants (6) but present in only one form in unicellular algae and cyanobacteria suggests that the isoforms may have a tissue-specific function.

Regulation of Fatty Acid Synthesis

Based on in vitro measurements of the activity of the various partial reactions of fatty acid synthesis, acetyl-CoA:ACP transacylase was proposed to be rate limiting to overall fatty acid synthesis (159). However, because it is now thought that the transacylase is bypassed by 3-ketoacyl-ACP synthase III (85), there no longer seems a compelling reason to consider this step as rate limiting in plant fatty acid synthesis.

Recent results suggest that carboxylation of acetyl-CoA is rate limiting for plant fatty acid synthesis in vivo, making it likely that acetyl-CoA carboxylase will prove to be the enzyme subject to regulation. The analysis is based on the concept that when the overall flux through a pathway is reduced, the concentrations of the intermediate immediately prior to the regulated step will increase, while the intermediates of subsequent steps will decrease in concentration. Post-Beittenmiller et al (137) exploited the observation that fatty acid synthesis is 5–8-fold higher in light than in darkness (25). When fatty acid synthesis by spinach leaves was reduced by shifting plants from the light to darkness, the concentration of acetyl-ACP (but not of other acyl-ACPs) increased; acetyl-CoA and malonyl-CoA concentrations were not measured directly; but if it is assumed that these compounds are in equilibrium with acetyl-ACP and malonyl-ACP, respectively (137), then the result implicates acetyl-CoA carboxylase (rather than KAS III or another enzyme) as the most strongly regulated step during the shift from light to dark.

In addition to this direct regulation of fatty acid synthesis by the control of malonyl-CoA synthesis, it seems likely (as noted below) that the components of the fatty acid synthase are tightly regulated to permit coordination of the production of fatty acids with their utilization for glycerolipid synthesis. Although the mechanism of this regulation is not known, it appears capable of compensating for potentially disruptive changes to the fatty acid synthesis machinery. For example, increasing ACP-I levels four-fold in tobacco by overexpression of a spinach ACP-I gene had no detectable effect on lipid composition, overall fatty acid composition, or the lipid to dry weight ratio of the transgenic plants (138).

Desaturation of Stearoyl-ACP

Stearoyl-ACP is efficiently desaturated by a stromal stearoyl-ACP desaturase. Palmitoyl-ACP is not a substrate for desaturation, so 16:0-ACP and 18:1-ACP are the major products of plastid fatty acid synthesis. With only a few exceptions (74), plant lipids contain very low levels of 18:0, indicating that the activity of this desaturase is generally not limiting. The stearoyl-ACP desaturase from safflower was found to be a soluble dimer of 38-kDa subunits that utilizes ferredoxin as an electron donor (108). Recently, the 18:0-ACP desaturase has been isolated from several sources (36, 97, 152, 155), and cDNAs corresponding to the genes from castor bean, cucumber, soybean, and safflower have been cloned (152, 155, 202). Active desaturase protein was produced upon expression of the safflower cDNA in *E. coli* (152) and the castor bean cDNA in yeast (155). The polypeptides encoded by the cDNA clones exhibited atypically high sequence identity between the various plant sequences, suggesting that there is strong conservation of sequence. However, there was no apparent amino acid sequence identity to the stearoyl-CoA desaturase from vertebrates or fungi (155) or to the $\Delta 12$ desaturase from *Synechocystis* (194), suggesting that they are independently evolved.

The Prokaryotic Pathway

In most higher plants, the first enzyme of the prokaryotic pathway, the stromal acyl-ACP:glycerol-3-P acyltransferase, is highly specific for 18:1-ACP (54), while the second acylation, catalyzed by a membrane-bound acyl-ACP-1:lysoPA acyltransferase, specifically utilizes 16:0-ACP to yield 1-18:1,2-16:0-PA (Figure 1). Chloroplast PG is synthesized from prokaryotic PA in all higher plants (149). In 16:3 plants, a prokaryotic DAG pool is formed from PA by the action of PA phosphatase (46, 87). In 18:3 plants, such a DAG pool is not formed, so PG is the only prokaryotic lipid in these plants; a DAG pool derived exclusively from the eukaryotic pathway is used for the synthesis of other chloroplast lipids (57, 70).

The most abundant chloroplast lipid, MGD, is formed from DAG by UDP-galactose:DAG galactosyltransferase. This enzyme has been reported to be located in the inner envelope membrane of the 16:3 plant spinach (14) but in the outer envelope of the 18:3 plant pea (39). However, the pea enzyme is not destroyed by thermolysin, a protease that inactivates other outer envelope enzymes, suggesting that the galactosyl transferase may be located in the inner envelope in both 16:3 and 18:3 plants (69). The enzyme has now been purified to apparent homogeneity (190) and, in a separate study, has been shown to require glycerolipids to reconstitute activity (40). There now seems to be general agreement that the synthesis of DGD is achieved by dismutation of two MGD molecules (69). In a recent study, no evidence could be found for the participation of a postulated UDP-galactose:MGD galactosyltransferase (69). The MGD:MGD galactosyltransferase is sensitive to thermolysin treatment of intact chloroplasts from both 16:3 and 18:3 plants, which suggests that it is in the outer envelope in both cases (69).

One of the long-standing problems in plant lipid metabolism is the mechanism of biosynthesis of sulfolipid, 1',2'-di-O-acyl-3'-O-(6-deoxy-6-sulfo-α-D-glucopyranosyl)-sn-glycerol. Since the elucidation of the structure by Benson and colleagues more than 30 years ago, no enzymatic step had been identified between inorganic sulfate and SL until the recent demonstration by Heinz et al (71) that the final step in SL biosynthesis is catalyzed by a UDP-sulfoquinovose:1,2-diacylglycerol 3-O-α-D-sulfoquinovosyltransferase. In contrast to the galactosyltransferase involved in MGD synthesis, the sulfoquinovosyltransferase cannot utilize exogenously supplied substrate, so it must have access to the stromal side of the inner chloroplast envelope. However, both enzymes apparently use the same pool of DAG since exogenous UDP-galactose was found to inhibit SL biosynthesis in intact organelles, presumably by competition for DAG (88). The pathway by which plants synthesize UDP-sulfoquinovose remains a matter of speculation (98, 99, 117).

The Chloroplast Desaturases

All the lipids derived from the prokaryotic pathway contain only 18:1 and 16:0 fatty acids when they are first synthesized in the chloroplast envelope (Figure 1). In vitro evidence for desaturation of these complex lipids in higher plants first came from labeling experiments using isolated spinach chloroplasts (147). More recently, substantial information concerning the substrate specificity and the regulation of the desaturases has been obtained by the analysis of four classes of *Arabidopsis* mutants—each one deficient in a specific desaturation step.

The 16:0 present at the sn-2 position of lipid molecules derived from the prokaryotic pathway is desaturated in two different ways. In most green plants 30–70% of the 16:0 at the sn-2 position of PG is converted to Δ3-*trans*-16:1. This fatty acid structure is unusual because the double bond is very close to the carboxyl group and because it is the *trans* rather that the *cis* isomer. The *fadA* mutants of *Arabidopsis* (22) lack Δ3-*trans*-16:1. Since there are no other changes in the fatty acid composition of the mutants, apart from a corresponding increase in the amount of 16:0 in PG, it was concluded that the *fadA* locus encodes a desaturase that acts specifically on 16:0 at the sn-2 position of PG. The only other green plants known to lack Δ3-*trans*-16:1 are members of the orchid family (143).

The presence of 16:3 in MGD (and to a much lesser extent DGD) but not in PG or SL, implies that the conversion of 16:0 to ω9-*cis*-16:1 occurs only on galactolipids and is most probably specific for the sn-2 position of MGD. A mutant of *Arabidopsis* (*fadB*) that apparently lacks the 16:0 desaturase contains no 16:3 in MGD or DGD but instead shows increased levels of 16:0 in MGD and several other lipids (101). Evidently, the insertion of this first double bond at the ω9 position is required before other desaturases can act on the fatty acid chain. The proportions of the various polar lipids in leaves showed no major change between the *fadB* mutant and wild type, but in MGD of the mutant the proportion of 16:0 (14%) was very much lower than the proportion of 16:3 in MGD of wild-type *Arabidopsis* (31%). Labeling experiments showed that this difference was the result of both a reduced synthesis of prokaryotic MGD and an accelerated turnover of 16:0-MGD in the mutant plants. The loss of prokaryotic MGD was largely compensated by increased synthesis of MGD via the eukaryotic pathway, and 16:0 lost from MGD appeared to be used for the synthesis of DGD and PC (20, 101). As discussed below, the fact that inactivation of a single desaturase can result in such extensive changes in the pathways of glycerolipid synthesis provides evidence for considerable regulation of these processes.

In contrast to the two very specific desaturases controlled by the *fadA* and *fadB* genes, the desaturases controlled by the *fadC* and *fadD* loci desaturate fatty acyl chains on many different lipids. When grown at temperatures above 26°C, mutants with lesions at the *fadD* locus exhibited an 80% decrease in the amount of 16:3 and a 60% decrease in the amount of 18:3 fatty acids in their leaf lipids and a corresponding increase in the amount of 16:2 and 18:2 (23). In the *fadC* mutant, 16:3 and 18:3 are also both reduced but, in this case, it is the 16:1 and 18:1 precursors that are increased (21). In the absence of purified desaturases, it has nevertheless been possible to draw important conclusions about the enzymes from straightforward comparisons of the fatty acid compositions of lipids from the *fadC*, *fadD*, and wild-type plants. Since the desaturases controlled by the *fadC* and *fadD* genes act on both, 16-carbon and 18-carbon fatty acid chains, they must determine the site of desaturation relative either to the existing double bond(s) or to the methyl end of the acyl chain. Both enzymes act on acyl groups at either the sn-1 or sn-2 positions of the glycerol backbone, and both enzymes catalyze desaturation of all the major glycerolipids of the chloroplast membranes (21, 23, 125). These results are consistent with a model for the ω6 and ω3 desaturases as enzymes that reside mainly in the hydrophobic region of the bilayer (such that the active sites have little or no interaction with the hydrophilic headgroups of the membrane lipids) and that desaturate any fatty acids that meet the requirement of having a specific distance between an existing double bond and the methyl end of the acyl chain. Analysis of the mutants has also demonstrated the degree of substrate specificity exhibited by the desaturases. As noted above, neither the ω6 nor the ω3 desaturase acts on 16:0-MGD that accumulates in the *fadB* mutant. However, in the *fadC* mutant, in addition to the accumulation of 16:1 and 18:1, trace amounts of what appear to be ω9, ω3 isomers of 16:2 and 18:2 are also observed (21). The presence of these isomers presumably reflect the fact that the ω3 desaturase can act on 16:1 and 18:1 at greatly reduced efficiency.

The effect of the *fadD* mutation was only fully expressed when plants were grown at temperatures above 26°C. At 18°C, the fatty acid composition of mutant and wild type were similar (23). The subsequent isolation of additional mutant alleles at the *fadD* locus with the same properties (S. Hugly and C. R. Somerville, unpublished) renders it unlikely that this effect is due to a temperature-sensitive mutation. Instead, it appears that suppression of the mutant phenotype at low temperature is due to induction or activation of a second plastid ω3 desaturase by low temperature (17).

A further unexpected and intriguing property of the *fadD* mutant is that the amount of 18:3 is reduced to a similar extent (about 45% reduction at temperatures above about 26°C) in all the major leaf lipids (i.e. MGD, DGD, SL, PG, PC, PE, PI) (23). Since 16:3 is thought to be synthesized from 16:2 only in the plastid, the desaturase controlled by the *fadD* locus must be located within the plastid. However, there is no PE in the chloroplast, and levels of PI and PC there are very low (45, 149). Thus, it is apparent that either the same desaturase is located in both compartments or the 18:3 on PE and the other lipids of the extra chloroplast membranes is synthesized in the chloroplast and transported to the cytoplasm. In principle, the mechanism that transfers lipids from the endoplasmic reticulum to the chloroplast could also transfer in the opposite direction. However, as noted below, interpretation of the effects of *fadD* on unsaturation of extrachloroplast membranes is some-what complicated by the fact that there is now clear evidence that a separate endoplasmic reticulum ω3 desaturase (controlled by the *fad3* gene) plays an important role in controlling fatty acid unsaturation in leaves of higher plants (24, 126).

Very recently, in vitro assays have been developed for the chloroplast lipid–linked desaturases. Studies utilizing a CHAPS solubilized chloroplast preparation demonstrated the requirement for ferredoxin and (in the dark) NAD(P)H as cofactors for desaturation of 18:1 to 18:2 and 18:3 (153). More recently, purified envelope membranes have been shown to be the major site of the desaturases that accept several different glycerolipids as substrates (154). It is not yet clear whether the thylakoid membranes contain any glycerolipid desaturase activity.

Although none of the genes encoding higher-plant glycerolipid desaturases has been cloned, the elegant use of genetic techniques has permitted the cloning of the *desA* gene of *Synechocystis* PCC 6803 which encodes the desaturase responsible for inserting a double bond at the $\Delta 12$ position of membrane acyl groups in this cyanobacterium. Murata and colleagues ex-ploited the fact that mutations at *desA* prevent growth of the cyanobacteria at low temperature (24°C) (195). This permitted them to clone the desaturase gene from a genomic library of *Synechocystis* by screening for clones capable of complementing the cold-sensitive phenotype (194). The cloned *desA* gene has already been useful in studies of cyanobacteria (194), and it may prove to be a successful heterologous probe for higher-plant desaturases.

Because of the development of methods for isolating genes in *Arabidopsis* by chromosome walking and gene tagging techniques (62), the *fad* mutants may eventually be useful in permitting the isolation of the corresponding genes by genetic methods. Alternatively, since many mutations lead to changes in amount or electrophoretic mobility of proteins, it may be possible to identify the gene products in the mutants by high-resolution two-dimensional electrophoretic techniques. An analysis of the *fadD* mutant by these techniques indicated a specific decrease in the amount of a 90-kDa chloroplast protein (18). However, the amount of this polypeptide was also reduced following treatment of wild-type plants with the herbicide SAN 9785 (17), raising the possibility that the change in amount of polypeptide is an effect rather than a cause of the reduced amount of polyunsaturation in the *fadD* mutants. Alternatively, Brockman et al (17) speculated that the effects of SAN 9785 on both fatty acid composition and the amount of the polypeptide may be due to a specific effect of the herbicide on stability of the desaturase and related proteins.

The Eukaryotic Pathway

The first committed step of the eukaryotic pathway is the hydrolysis of 16:0-ACP and 18:1-ACP to free fatty acids. A single acyl-ACP thioesterase activity has been purified to homogeneity from soybean seeds (97) and oilseed rape (73), and available evidence also suggests the presence of a single enzyme activity in leaf chloroplasts (130). As discussed below, the implication that both 16:0 and 18:1 free fatty acids are produced by the same enzyme is not readily reconciled with changes observed in the ratio of 16-carbon to 18-carbon fatty acids in several *Arabidopsis* mutants; however, at present, there is no firm evidence for a second thioesterase for long-chain acyl-ACPs. The free fatty acids move through the two membranes of the chloroplast envelope and are converted to CoA thioesters by an acyl-CoA synthetase in the outer envelope. The soluble acyl-CoA thioesters are used for synthesis of PA at extrachloroplast sites including, most importantly, the endoplasmic reticulum (163, 170). Lipid synthesis also occurs to some extent in the mitochondria (55) and possibly at other cellular sites (110). In contrast to the plastid isozymes, the acyl-CoA-specific acyltransferases of the endoplasmic reticulum only produce PA with an 18-carbon fatty acid at the sn-2 position; 16:0, when present, is confined to the sn-1 position (53). This PA gives rise to the phospholipids such as PC, PE, PI, PG, and PS that are characteristic of the various extrachloroplast membranes (110).

PHOSPHOLIPID SYNTHESIS Until recently, it has been thought that PC was synthesized by the same pathways used in animals and yeast (7, 32). These three pathways are: transfer of choline from CDP-choline to DAG, sequential methylation of PE, and an exchange reaction involving exchange of free choline for another headgroup. In animal cells the main route is thought to be via the nucleotide pathway, involving the sequential action of choline kinase, cytidyltransferase, and cholinephosphotransferase (Figure 2). The cytidyltransferase reaction has been proposed to be regulatory step in animals because of the unusual properties of the enzyme (134). The enzyme may exist as a soluble form or bound to the endoplasmic reticulum (84), and it undergoes a phosphorylation-dephosphorylation cycle.

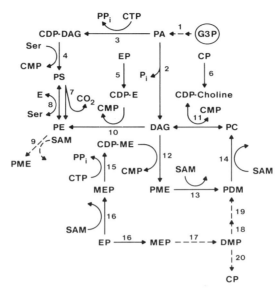

Figure 2   The pathways of phosphatidylcholine and phosphatidylethanolamine synthesis. The reactions that are of uncertain significance or appear not to occur in some plants are indicated with dashed lines. The enzymes or reactions are: [1] G3P and LPA acyltransferase; [2] PA phosphatase; [3] CTP:phosphatidate cytidyltransferase; [4] CDP-diacylglycerol:serine-O-phosphatidyltransferase; [5] CTP:ethanolaminephosphate cytidyltransferase; [6] CTP:cholinephosphate cytidyltransferase; [7] phosphatidylserine decarboxylase; [8] phosphatidylethanolamine:serine phosphatidyltransferase; [9] SAM:PE N-methyltransferase; [10] CDP-ethanolamine:DAG ethanolaminephosphotransferase; [11] CDP-choline:DAG cholinephosphotransferase; [12] CDP-N-methylethanolamine:DAG N-methylethanolaminephosphotransferase; [13] SAM:N-methyl-PE N-methyltransferase; [14] SAM:N,N-dimethyl-PE N-methyltransferase; [15] N-methylethanolaminephosphate cytidyltransferase; [16] SAM:ethanolaminephosphate methyltransferase; [17] methylation of N-methylethanolaminephosphate; [18, 19] formation of CDP:N,N-dimethyethanolamine and condensation with DAG to form PDM and CMP; [20] methylation of N,N-dimethylethanolaminephosphate to choline phosphate.

In animal cells, PC can also be synthesized by the sequential methylation of PE by a single 18.3-kDa enzyme that catalyzes the transfer of all three methyl groups from S-adenosylmethionine (192). Recent studies indicate either that plant cells do not have an enzyme that catalyzes this reaction or that it is of minor quantitative significance (42). Rather, Datko & Mudd have proposed the following two reactions:

$$\text{ethanolamine} + \text{ATP} \rightarrow \text{ethanolamine-P}$$
$$\text{ethanolamine-P} + \text{SAM} \rightarrow \text{N-methylethanolamine-P}$$

The subsequent fate of the N-methylethanolamine-P (MEP) appears to vary in different species. In *Lemna*, the MEP appeared to be methylated to choline-P (41, 114). In soybean suspension cells the MEP was first transferred to DAG to form phosphatidylmethylethanolamine, presumably via the action of CTP:choline-P cytidyltransferase (196, 197) and CDP-choline:DAG cholinephosphotransferase (192), then methylated to produce PC (43, 115, 116). Carrot cell suspensions exhibited all of the reactions found in both soybean and *Lemna*. Since each of the three plant species appeared to use a different combination of reactions, further work will be required to determine the degree of polymorphism for this pathway in higher plants. It should also be noted that different cell cultures may represent tissues in different developmental states. Thus, the differences among the three sources of tissue used by Mudd & Datko could be due to the use of different pathways in the various tissues within plants.

In light of the evidence for these modified pathways for PC synthesis, it is now necessary to judge their importance relative to the cholinephosphotransferase pathway. In seed tissues, and perhaps also in leaves, the high turnover of PC is mediated by the reversible cholinephosphotransferase reaction (169, 172). Labeling studies (115, 116) suggest that in the presence of choline, the cholinephosphotransferase pathway dominates. This is consistent with the observation that in extracts from *Lemna*, soybean, and carrot cells grown on choline, activity of the enzyme that methylates ethanolamine-P is suppressed (115, 116). However, the significance of choline feeding studies is somewhat uncertain because the in vivo source of choline is not known. Recent experiments concerning the source of choline (94, 114) have provided equivocal evidence concerning the quantitative significance of serine as a precursor of the choline moiety of phosphatidylcholine. Mudd & Datko (115, 116) have suggested that the major source of choline is the synthesis of PC by the methylation of ethanolamine-P, the ethanolamine arising by decarboxylation of serine. However, the observation that choline-P is normally transported in the xylem from the roots to other tissues at rates sufficient to support PC synthesis in expanding leaves (106) raises the possibility that in many tissues of the plant exogenous choline-P is normally the substrate for PC synthesis by the nucleotide pathway. Considerable progress has been made with the purification and characterization of choline kinase (95, 96). The properties of cholinephosphotransferase have been studied in microsomal preparations (172). Further investigation of the activities and tissue distributions of these enzymes should help to assess the importance of these reactions in PC synthesis.

Investigators have obtained relatively little new information about the synthesis of phospholipids other than PC since the last review of phospholipid synthesis in this series (110). Phosphatidylethanolamine can be synthesized in plants by transfer of ethanolamine from CDP-ethanolamine to DAG, by decarboxylation of PS, or by an exchange reaction in which the exchange of free ethanolamine with the headgroup of another phospholipid is catalyzed by a membrane-bound enzyme (110, 160). The nucleotide pathway is considered the quantitatively most important pathway (110, 111), and no function has yet been established for the exchange reaction (161). Phosphatidylserine is a ubiquitous but minor component (less than 1%) of plant lipids. It is thought to be synthesized only by the exchange reaction in animals (7). On the other hand, yeast cells may only utilize the synthesis from CDP-DAG. In higher plants, PS appears to be synthesized by either an exchange reaction or a reaction of L-serine with CDP-DAG (110). A uniquely interesting property of PS is that in all of eighteen species of higher plants studied, PS contained very-long-chain saturated fatty acids that were not found in any other class of lipid (120). It was suggested that the fatty acids in PS may be precursors of epidermal lipids and that PS serves as a carrier. Studies of the acyl-CoA elongase from leek have not yet provided any evidence for such a role (11). Phosphatidylinositol is synthesized in the endoplasmic reticulum from CDP-diacylglycerol and myo-inositol by phosphatidylinositol synthase (110), and is converted by sequential phosphorylation to phosphatidylinositol-4-phosphate, and phosphatidylinositol-4,5-bisphosphate. The PI synthetase from plants has not been characterized. However, a recent study of the molecular species of PI in potato tuber, pea leaf, and germinating soybeans indicated that the enzyme had an apparent substrate preference for 16:0/18:2 PI (90), as was previously found for cholinephosphotransferase and ethanol-aminephosphotransferase (89).

In mammalian cells, the hydrolysis of polyphosphoinositides produces potent second messengers in response to a variety of stimuli. Recent evidence concerning the potential involvement of phosphoinositides in a similar role in plant cells has been the subject of several recent reviews (15, 50, 113). The relatively limited biochemical information concerning the regulation of biosynthesis or the properties of the enzymes involved may also be found in these reviews.

SYNTHESIS OF CHLOROPLAST LIPIDS BY THE EUKARYOTIC PATH-WAY   Phosphatidylcholine is the major structural lipid of almost all the extrachloroplast membranes of the cell, but it is also a substrate for desaturation, and it is rapidly turned over to provide DAG for the synthesis of chloroplast lipids (Figure 1). In vivo labeling results (170) and the description of a microsomal 18:1-PC desaturase from pea leaves (173) provided evidence that PC is the predominant substrate for 18:1 desaturation on the eukaryotic pathway. In microsomal preparations from seed tissue the 18:1-PC desaturase has been shown to use electrons transferred from NAD(P)H via cytochrome

b_5 (174). Recently, we and others have isolated mutants of *Arabidopsis* deficient in 18:1 (*fad2*) or 18:2 (*fad3*) desaturation in leaves, roots, and seeds (82, 103). The difference in fatty acid composition between the *fad2*, *fad3*, and wild-type lines, as well as in vitro assays of desaturase activity, are consistent with the suggestion that the *fad2* and *fad3* mutations are in the structural genes encoding the major eukaryotic 18:1 and 18:2 desaturases, respectively (103). Analysis of the fatty acid composition of individual leaf and root lipids indicated that the fatty acid compositions of all the major extrachloroplast phospholipids were affected to a similar extent in the mutants (M. Miquel and J. Browse, unpublished). Thus, these desaturases are responsible for the synthesis of polyunsaturated fatty acids for most of the phospholipids of the cell.

The biochemical and genetic analysis of these desaturases does not give a complete picture of how lipid synthesis and desaturation relate to the biogenesis of the different membrane systems of the cell. For instance, there is no direct information on whether desaturation only occurs in the endoplasmic reticulum or whether the *fad2* and *fad3* gene products are also targeted to other membranes. The possibility remains that additional desaturase isozymes are located at sites other than the endoplasmic reticulum, although these would necessarily be of relatively minor quantitative significance. Data on the fatty acid composition of root tissues of wild-type and mutant *Arabidopsis* indicate that the *fad2* mutants are 90% deficient in 18:1 desaturation while the *fad3* mutants are 80% deficient in 18:2 desaturation (M. M. McConn and J. Browse, unpublished). Thus, either the currently available mutant alleles are leaky (e.g. the mutant protein retains a low amount of activity) or there are, indeed, additional desaturase isozymes that make a minor contribution to 18:2 and 18:3 synthesis. Analysis of the fatty acid composition of subcellular fractions from *fad2*, *fad3*, and wild-type plants should help determine whether different membrane systems of the cell, such as the mitochondria or the plasma membrane, contain distinct 18:1 and/or 18:2 desaturases.

In leaf mesophyll cells, a major proportion of the PC synthesized in the endoplasmic reticulum is further metabolized to provide DAG for the synthesis of galactolipids and SL for chloroplast membrane biogenesis. The mechanism of lipid transfer between the endoplasmic reticulum and the chloroplast is not known. In plants and in other organisms, lipid transfer proteins (LTPs) have been postulated to transfer lipids between different cellular membranes (3), although in only one case (5) has a role in vivo been clearly demonstrated. Lipid transfer proteins have been isolated from plants (3) and shown to mediate the in vitro transfer of PC from liposomes to chloroplasts (109), although it has not been possible to demonstrate the cholinephosphotransferase (or phospholipase C) activity required to convert PC to DAG in the chloroplast envelope (149). However, the involvement of LTPs has recently become somewhat uncertain owing to information deduced from the structure of cDNA clones for maize (189), spinach (10) LTPs, and a barley LTP (9, 16) for which the gene had previously been cloned (for reasons unrelated to its possible function) (118). The polypeptides encoded by these cDNA clones have signal sequences that do not appear on the mature protein. In vitro, these signal sequences direct the translated polypeptides into the endoplasmic reticulum (10). According to current models of synthesis and processing of soluble proteins, the plant LTPs that have been studied would be expected either to be located in the vacuole or to be secreted from the cell into the cell wall by the secretory pathway. In either case, such a location would be inconsistent with a role in intracellular lipid transport. A final conclusion on this point must await the results of immunolocalization experiments.

Whatever the mechanism by which lipid moves from the endoplasmic reticulum to the chloroplast, a property of the *fadD* mutant of *Arabidopsis* is interpreted as evidence that the transfer must be reversible to some extent. As noted earlier, although the *fadD* mutation specifically affects the activity of a plastid desaturase (17, 23, 125, 126), the amount of 18:3 is reduced to a similar extent (about 45% reduction at temperatures above 28°C) in both the chloroplast and extrachloroplast lipids (23). These results are most simply explained by invoking some transfer of 18:3 from the chloroplast to the endoplasmic reticulum in the wild type. However, because one does not find significant quantities of lipids containing unsaturated 16-carbon fatty acids in extraplastid membranes, it is necessary to propose that only lipid molecules synthesized by the eukaryotic pathway are transferred from the chloroplast to the endoplasmic reticulum.

LIPID METABOLISM IS HIGHLY REGULATED

Acyltransferases Regulate Lipid Acyl Composition

The acyl composition of lipids is determined by three factors: the substrate specificity of the acyltransferases, the pool of acyl donors available, and modifications, such as desaturation, made to the acyl group after esterification to the glycerol moiety. Evidence has accumulated that because of species differences in substrate specificity, the acyltransferases can significantly affect acyl composition. A well-developed example involves the first enzyme of the prokaryotic pathway, the stromal acyl-ACP:glycerol-3-P acyltransferase. In many plant species, the enzyme has a distinct substrate preference for 18:1-ACP (54), whereas in other species it will utilize 16:0-ACP (or 18:0-ACP) to varying extents (56). In either case, the resulting lysophophatidic acid is acylated by a membrane-bound acyl-ACP:1-lysoPA acyltransferase that is specific for 16:0-ACP (Figure 1).

The significance of the substrate specificity of the G3P-acyltransferase arises from the observation that many chilling-sensitive species accumulate PG in which both the sn-1 and sn-2 positions contain saturated or *trans*-unsaturated fatty acids (119, 121). Murata has proposed that the presence of such PG molecular species in the chloroplast (or plastid) envelope results in the formation of gel phase domains at chilling temperatures, and that lateral phase separation within the envelope membranes is a direct cause of chilling sensitivity. This hypothesis is not universally applicable since some chilling-sensitive plants have PG profiles similar to those of chilling-resistant plants (122). However, to date all plants containing greater than 60% 16:0 + 18:0 + 16:1-*trans* fatty acids in PG have been shown to be chilling sensitive.

Murata and colleagues have purified three plastid isozymes of the G3P-acyltransferase (AT1, AT2, AT3) from squash, a chilling-sensitive plant (124). Analysis of the substrate specificity of these isozymes indicated that AT1 utilized 18:1 in preference to 16:0 or 18:0 under certain conditions, but that AT2 and AT3 showed essentially no discrimination among these three substrates. Antibodies raised against the squash acyltransferases were used to isolate cDNA clones from squash and *Arabidopsis* (a chilling-resistant species) (79, 124). Expression of the squash cDNAs in chilling-resistant plants is expected to increase the proportion of disaturated PG and thereby provide the biological materials to test the effect of quantitative differences in the amount of disaturated PG on chilling sensitivity.

As discussed in a later section, the regulation of the fatty acid composition of triacylglycerols in oil seeds is a second area where the substrate specificities of acyltransferases have an important role. Also, recent measurements of the acyl-ACP content of isolated chloroplasts have indicated that as much as 30–40% of the ACP is esterified to long-chain fatty acids (148). Thus it may be the activity of the glycerol-3-phosphate acyltransferase or the availability of glycerol-3-phosphate (58) that limits chloroplast lipid synthesis rather than the availability of fatty acids.

Regulation of Lipid Desaturation

The fact that plant membranes are not fully unsaturated suggests that regulation of the desaturase enzymes controls membrane properties. The implication is that the level of unsaturation is limited by the amount of the various desaturases, or that the enzymes are subject to some form of feedback regulation. Evidence pertaining to this question was obtained by exploiting the fact that the heterozygous plants resulting from crosses between desaturase mutants and wild type have only one active copy of a particular *fad* gene. The results of such analysis for the *Arabidopsis* desaturase mutants indicated that the various desaturases are regulated by different mechanisms. For the *fadA*, *fadD*, and *fad3* mutations, the heterozygotes have compositions almost exactly intermediate between the wild type and the homozygous mutants (22, 23; J. Browse, unpublished). This surprising finding suggests that the level of gene expression is the principle factor regulating these reactions, which are the final steps in each series of desaturation. By contrast, heterozygotes of *fadB*, *fadC*, and *fad2* have fatty acid compositions much more similar to that of the wild type than to that of the homozygous mutant parent (21, 101, M. Miquel and J. Browse, unpublished). Thus, either these enzymes are normally produced at levels in excess of the amount needed or the activity of these enzymes has been up-regulated in the heterozygotes to compensate for the enzymatic defect. The observation that some desaturation steps are controlled by gene dosage has important consequences for attempts to modify plant lipid composition by genetic engineering or related methods. It implies that overall level of membrane (or storage lipid) unsaturation might be increased by overexpressing desaturase genes in transgenic plants.

A related question pertains to the changes in fatty acid composition that occur in response to growth temperature. These changes, which occur in leaf and seed tissues, involve an increase in unsaturation with decreasing temperature (140). Mechanisms relying on changes in the amount or activity of desaturase enzymes have been established in certain microbes, but in the case of higher plants, the situation remains unclear. For example, it is difficult to evoke an adaptive response involving desaturases whose level of activity is primarily determined by gene dosage. One hypothesis that is not dependent on adaptive regulation of desaturase activity relates the rate of desaturation to the overall rate of fatty acid and lipid synthesis. When the temperature response of desaturation is less than that of fatty acid synthesis the extent of lipid unsaturation will change with temperature as expected (26). However, not all data on the response of desaturase activity to changes in temperature support such a simple explanation (34, 35). The development and improvement of techniques for in vitro assay of the various desaturases (153) should facilitate the resolution of this issue.

Regulation of Acyl Flux Between the Two Pathways

The original elaboration of the two-pathway scheme (149) raised the question of how much each pathway contributes in 16:3 plants to the synthesis of chloroplast membrane lipids. The question can be answered at least partially by labeling experiments (28). Because lipids derived by the prokaryotic pathway contain a 16-carbon fatty acid at position sn-2 of the glycerol backbone, while lipids from the eukaryotic pathway have an 18-carbon fatty acid at this position, it is possible to use positional analysis of leaf lipids to calculate the relative fluxes and product distributions through the two pathways (28). In *Arabidopsis* leaves, approximately 38% of newly synthesized fatty acids enter the prokaryotic pathway. Of the 62% exported as acyl-CoA species to enter the eukaryotic pathway, 56% (34% of the total) is ultimately reimported into the chloroplast. Thus, in *Arabidopsis*, almost equal amounts of the chloroplast lipids are produced by each pathway.

Several suggestions have been made about how the flux into the two pathways might be regulated. Guerra et al (63) proposed that ACP isoforms could be involved. Oleoyl-ACP II appeared to be a preferred substrate for acylation of glycerol-3-phosphate, whereas oleoyl-ACP I was preferred in the thioesterase reaction. Thus, it was suggested that flux between the two pathways could be partially regulated by differential expression of the two isoforms. However, this concept seems inconsistent with the observation that ACP I is the main isoform in spinach seeds, where the ratio of eukaryotic to prokaryotic lipid synthesis is relatively high.

A somewhat surprising feature of 16:3 plants is that the proportions of the various chloroplast lipids synthesized by the prokaryotic and eukaryotic pathways are quite different. All the chloroplast PG is synthesized by the prokaryotic pathway in spinach and *Arabidopsis*, and this pathway predominates in the synthesis of MGD and SL (28, 149). On the other hand, DGD is derived largely from the eukaryotic pathway even though MGD is the immediate precursor for DGD synthesis (69). Furthermore, the MGD, DGD, and SL synthesized via the eukaryotic pathway each shows a fatty acid composition distinct from that of PC, their common precursor. This suggests that the enzymes involved possess quite different selectivities. Synthesis of MGD and DGD discriminates against molecular species containing 16:0, but eukaryotic DGD contains more 16:0 than eukaryotic MGD. Eukaryotic SL is enriched in 16:0 relative to PC (28). All these observations point to regulation of the fatty acid composition (and thus of the molecular structure) of membranes.

These observations can be explained most simply by invoking particular molecular species specificities for each of the enzymes involved in lipid synthesis and perhaps by assuming separate pools of chloroplast DAG in the two pathways. However, such a simple model cannot explain the extensive alterations in lipid metabolism that occur in some of the *Arabidopsis* mutants. For example, the *act1* mutants contain less than 5% of the wild-type activity of the plastid 18:1-ACP:glycerol-3-P acyltransferase. This deficiency does not result in the accumulation of precursors (16:0-ACP and 18:1-ACP) upstream of the enzyme lesion but instead causes redirection of lipid metabolism so that the eukaryotic pathway predominates in these mutants, as it does in 18:3 plants (100). However, this redirection has little effect on the amount of each lipid that accumulates or the total glycerolipid content of mutant leaves relative to wild type (100, 19). This is because the enhanced flux through the eukaryotic pathway in the mutant is accompanied by a change in the proportion of individual lipids synthesized by this pathway (19). Thus, the amounts of lipids (such as PE and PI) in the extrachloroplast membranes of the mutant are similar to those in the wild type, but to compensate for the loss of the prokaryotic pathway, the amount of PC synthesized must be increased about two-fold to provide the DAG moieties required for normal levels of glycerolipid synthesis by the chloroplast. The net amount of PC that accumulates is apparently only that required for the synthesis of the extrachloroplast membranes, because the level of this lipid in the mutant is only slightly higher than in the wild type. The partitioning of eukaryotic DAG to MGD, DGD, and SL is also greatly altered in the mutant and, finally, the proportion of 16-carbon fatty acids synthesized is reduced from 30% in the wild type to less than 18% in the mutant (100). Clearly, the synthetic reactions of the eukaryotic pathway, including synthesis of PC and export of lipid from the endoplasmic reticulum membranes, are regulated in concert to provide the complement of lipids required for correct biogenesis of the thylakoid membranes (and by implication of the other membranes of the cell also).

The synthesis of almost normal levels of chloroplast PG in the *act1* mutants reveals correspondingly powerful regulation of the prokaryotic pathway. In all higher plants, the prokaryotic pathway is responsible for chloroplast PG synthesis, and in 16:3 plants PG synthesis involves the same pool of PA used for the synthesis of prokaryotic galactolipids and SL (1, 142). Nevertheless, the *act1* mutants contain at least 70% as much PG as the wild type, even though they exhibit less than 5% of the wild-type level of chloroplast glycerol-3-P acyltransferase activity. It appears, therefore, that in the mutant, the small amount of PA formed is used preferentially for PG synthesis at the expense of DAG formation by the PA phosphatase (100).

Arabidopsis mutants with defects in plastid desaturases also have a reduced flux into the prokaryotic pathway. In the *fadC* mutant (deficient in the plastid 16:1/18:1 desaturase) and the *fadB* mutant (deficient in the plastid 16:0-MGD desaturase), the flux of fatty acids into prokaryotic MGD is reduced by 30% and 20%, respectively (21, 101). In addition, and particularly in the *fadB* mutant, turnover (relative to wild-type) of the prokaryotic species of MGD is accelerated. In both these mutants, the loss of prokaryotic MGD is accurately compensated for by increased synthesis of MGD by the eukaryotic pathway (21, 101). We have suggested that these differences in lipid metabolism represent a regulatory mechanism that ameliorates the consequences of reduced desaturation of prokaryotic lipids. For example, without the changes in lipid metabolism described here, MGD of the *fadC* mutant would contain an average of 2.8 double bonds per molecule, compared with 5.8 for the wild type. The altered regulation of MGD synthesis results in the *fadC* mutant's containing an average of 3.7 double bonds per MGD molecule (21).

Examples of the modulation of the prokaryotic and eukaryotic pathways are not limited to the *Arabidopsis* mutants. In the 16:3 plants *Brassica napus* (86) and *Atriplex lentiformis* (133), synthesis of prokaryotic MGD is decreased by increasing temperature. Growth of *A. lentiformis* at elevated temperatures led to the complete loss of the 16:3 which was largely replaced, in MGD, by 18-carbon fatty acids (133). By contrast, recent studies of lipid synthesis in *Arabidopsis* tissue culture have indicated that the flux through the eukaryotic pathway is stimulated at low temperature (17).

Regulation of Fatty Acid Chain Length

The primary product of fatty acid synthesis, 16:0-ACP, can undergo one of three competing reactions: hydrolysis by acyl-ACP thioesterase, elongation and desaturation to 18:1-ACP, or transfer to lyso-PA (Figure 1). Some mechanisms that regulate the differential hydrolysis of 16:0 and 18:1-ACP would seem to be essential if the synthesis of glycerolipids containing 16-carbon and 18-carbon acyl groups is to be modulated. For example, quantitative analysis of the *Arabidopsis act1* mutants showed that essentially all the 16:0 diverted from the prokaryotic pathway was elongated and desaturated to 18:1-ACP such that the apparent ratio of 16:0-ACP to 18:1-ACP hydrolyzed by the thioesterase was reduced from 0.24 to 0.15 (100).

Regulation of chain length is also inferred from recent studies of the *fad2* mutants deficient in activity of the endoplasmic reticulum 18:1-PC desaturase (103). These mutants show a striking 50% reduction in the level of saturated fatty acids (mainly 16:0) of extrachloroplast membranes compared with wild-type plants (M. Miquel and J. Browse, unpublished). This appears to reflect a regulated response of the mutant to unfavorably elevated levels of 16:0/18:1 molecular species in the membranes. This suggestion is based on the analogous changes in lipid metabolism of the *fadB*, *fadC*, and *act1* mutants discussed above, and on preliminary evidence implicating 16:0/18:1 molecular species in membrane damage at low temperatures (177). Indeed, even though the proportion of 16:0/18:1 in membranes of the *fad2* mutants is substantially lower than would be expected on the basis of wild-type lipid composition, the mutants still show a chilling-sensitive phenotype (M. Miquel and J. Browse, unpublished).

What Regulates the Amount of a Particular Membrane?

All the evidence discussed above, and particularly the analysis of the *act1* mutants, suggests that the entire system of fatty acid and glycerolipid synthesis is regulated to meet the demand for particular lipid molecular structures for optimal membrane function. This observation raises the fundamental questions of precisely what signals this demand, and what transmits the signal from the many different membranes of the cell to the major sites of lipid and fatty acid synthesis in the endoplasmic reticulum and the chloroplast envelope and stroma. A possible clue with respect to the first of these questions comes from the finding that overexpression of several bacterial membrane proteins in *E. coli* leads to increased lipid synthesis and the formation of membrane tubules or cisterna that accommodate the excess membrane protein (193, 199). Similar results have been obtained in animal cells (37). By analogy, we propose a model in which the physiological and biochemical needs of the cell (for light harvesting membranes, for cell expansion, for respiration, etc) lead first of all to the activation of membrane protein synthesis. Accumulation of these proteins then regulates lipid metabolism through an unknown, but clearly very powerful, mechanism.

SEED LIPID SYNTHESIS

Lipids in the form of triacylglycerols are widely found as a major energy reserve in seeds and fruits. Vegetable oils from seed crops constitute one of the world's most important plant commodities. The major use of these oils is for human consumption, but a significant proportion find use in manufacturing industries, particularly in the production of detergents, coatings, plastics, and specialty lubricants. For both food and industrial applications, it is the fatty acid composition of the oil that determines its usefulness and, therefore, its commercial value.

Because of their economic importance, extensive information has been accumulated about the fatty acid compositions of triacylglycerols from the seeds of many species (74, 200). A large variety of different fatty acid structures is found in nature (65, 74), but just five account for 90% of the commercial vegetable oil produced. These fatty acids (16:0, 18:0, 18:1, 18:2, 18:3) are the ones also found most commonly in the membrane lipids. In addition, certain plant species have the capacity to produce specialized fatty acids that are used for triacylglycerol synthesis but that are largely excluded from membrane lipids. For example, triacylglycerols of oilseed rape (*Brassica napus*) and crambe (*Crambe abyssinica*) contain long-chain, monounsaturated fatty acids (20:1, 22:1). Palm kernel oil, coconut oil, and the oils of several *Cuphea* species contain high levels of medium-chain saturated fatty acids (10:0, 12:0, 14:0). Fatty acids containing hydroxyl groups, triple bonds, or conjugated (rather than methylene-interrupted) double bonds are each found in the oils of a few plant species (74). Investigators are currently interested in modifying oil composition by mutation breeding and altering the products of seed lipid metabolism by the use of cloned genes. The eventual application of DNA technology to the problem of producing specialized seed oils for the many different uses that exist is critically dependent on a complete understanding of the biochemistry and regulation of triacylglycerol synthesis.

Cellular Organization of Triacylglycerol Synthesis

The major reactions in seed triacylglycerol synthesis are shown in Figure 3. The scheme was drawn to describe lipid metabolism in developing *Arabidopsis* seeds, but it is applicable to many oilseed species. *Arabidopsis* seed lipids contain substantial proportions of both unsaturated 18-carbon fatty acids (30% 18:2, 20% 18:3) and very-long-chain fatty acids (22% 20:1) derived from 18:1. In this respect, *Arabidopsis* is a good model for the biochemistry of both 18:2/18:3-rich oilseeds and those species containing longer fatty acids.

In seeds, as in leaves, 16:0-ACP and 18:1-ACP are the major products of plastid fatty acid synthesis (Figure 3, reaction 1) and 18:0-ACP desaturase activity (Figure 3, reaction 2). 18:0-ACP is an intermediate, but is normally found only as a minor component in triacylglycerols and other plant lipids, although some fats such as cocoa butter contain a high proportion of 18:0 (200). Presumably in most plants the thioesterase (or thioesterases; see above) involved in acyl-ACP hydrolysis is specific for 16:0 and 18:1 (73, 97). By analogy with leaf chloroplasts, the free fatty acid products move through the plastid envelope and are converted to acyl-CoA thioesters on the outer envelope. Certainly, acyl-CoAs are the primary substrates for subsequent reactions in other cellular compartments (187). They are used for the synthe-

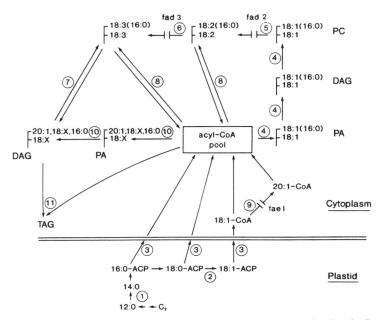

Figure 3   A simplified scheme for the reactions of triacylglycerol synthesis in oilseeds. See text for details.

sis of PC (Figure 3, reaction 4) which is the major substrate for 18:1 and presumably 18:2 desaturation (Figure 3, reactions 5 and 6) by microsomal enzymes (27, 183, and see previous discussion). The synthesis of PC from diacylglycerol (DAG) by choline phosphotransferase (Figure 3, reaction 7) is reversible (172), so that in many oilseeds PC is a direct precursor of the highly unsaturated species of DAG used for triacylglycerol (TAG) synthesis (169). However, the acyl-CoA pool in the seed is a much more complex collection of fatty acyl groups than that in leaf mesophyll cells. Exchange of 18:1 from CoA with the fatty acids at position sn-2 of PC (Figure 3, reaction 8) provides inputs of 18:2 and 18:3 (184, 185). Elongation (Figure 3, reaction 9) of 18:1-CoA to 20:1- and 22:1-CoA also occurs in *Arabidopsis*, as it does in rapeseed and other cruciferous oilseeds (182). Obviously, synthesis of DAG (Figure 3, reaction 10) via phosphatidic acid (PA) may also involve these other components of the acyl-CoA pool (182), as does the final acylation (Figure 3, reaction 11) of DAG to form TAG (188). Thus, the pool of DAG used for TAG synthesis is fed by both phosphatidic acid phosphatase and choline phosphotransferase.

In oilseeds that accumulate 18-carbon fatty acids, the glycerol-3-P acyl-transferase resembles the endoplasmic reticulum enzyme from leaves in transferring both 16-carbon and 18-carbon fatty acids to the sn-1 position. Incubation of safflower microsomes with an equimolar mixture of different acyl-CoA species and glycerol-3-P indicated that selectivity of acylation was in the order 16:0 > 18:1 = 18:2 > 18:0 (78). The second acylation catalyzed by acyl-CoA:lysophosphatidic acid acyltransferase is selective for 18:2 over 18:1 and almost completely excludes 16:0 and 18:0 (61). These in vitro results are consistent with the compositions observed for membrane phospholipids, for DAGs, and for the sn-1 and sn-2 positions of triacylglycerols.

The Formation of Oilbodies

Triacylglycerols accumulate in discrete subcellular organelles, the oilbodies. The structure and ontogeny of oilbodies have been contentious issues (169, 187). The questions are relevant to the biochemistry of oil synthesis because in different oilseeds both the endoplasmic reticulum and the oilbody have been suggested as the sites of reactions leading to TAG synthesis (169, 179). One hypothesis for the development of oilbodies (198) suggests that they arise by deposition of TAG between the two leaflets of the endoplasmic reticulum and are then budded off to form separate organelles. This model is consistent with results which indicate that oilbodies are bounded by a half-unit membrane (167, 203). Such a half-unit structure is to be expected because the hydrophobic TAG molecules will interact with the fatty acid tails rather than the polar head groups of phospholipid molecules.

A direct demonstration that oilbodies arise from the endoplasmic reticulum was later made by Stobart & Stymne (179). These authors demonstrated that microscopically pure microsomal membrane preparations from safflower seeds, when supplied with glycerol-3-P and acyl-CoA, are capable of triacyl-glycerol synthesis at rates in excess of 40 nmol/100 nmol membrane PC/hr. Such rates are comparable with the maximum rate of TAG deposition (65 nmol/100 nmol PC/hr) observed for developing safflower seeds in vivo (172). In the in vitro system used by Stobart & Stymne (179), TAGs were released from the membranes and accumulated as small droplets in the medium. It is not clear whether such free oil droplets are formed in vivo. One explanation for their formation in vitro is that the microsomal membranes do not synthe-size the phospholipid components (necessary substrates were not components of the assay mixture) required for membrane expansion and oilbody forma-tion. Evidence for the existence of a half-unit membrane in vivo include (*a*) the fact that oilbodies from developing cotyledons are not observed to coalesce either in vivo or during centrifugation after tissue homogenization (167, 203) and (*b*) the results of direct analysis (167, 203).

Suggestions that TAG synthesis in vivo gives rise to naked oil droplets in the cytoplasm of cells (8) may be largely explained by the fact that the major proteins of the oilbody membranes are synthesized on cytoplasmic ribosomes and inserted in the membranes after the main period of oilbody production and close to seed maturity (123, 191). Oilbody fractions from crambe and rapeseed have been reported to have activity for 18:1 elongation and TAG synthesis (168). However, the evidence outlined above and the description of some of the enzyme activities (131, 188) and intermediates of TAG synthesis (52) in microsomal fractions from several species make it likely that the endoplasmic reticulum is the principal site of all the reactions involved. The observation of activity in both fractions may merely reflect the formation of oilbodies from the endoplasmic reticulum membranes.

The Role of Phosphatidylcholine

A key problem in understanding storage lipid metabolism is deciding the relative contributions of choline phosphotransferase and PA phosphatase (reactions 7 and 10 in Figure 3) to the pool of DAG used for TAG synthesis. The activity of the reversible acyl-CoA:lysoPC acyltransferase (reaction 8 in Figure 3) is extremely high in microsomal preparations from developing seeds of safflower, sunflower and linseed (178, 185–187). Furthermore, detached cotyledons of safflower incorporate [^{14}C]oleate into the sn-2 position of PC via acyltransferase far more rapidly than they incorporate the label into the sn-1 position by de novo synthesis of PC (61). These observations led Stymne & Stobart to suggest that 18:2 (and 18:3) produced on PC was rapidly returned to the acyl-CoA pool and entered triacylglycerol via PA and DAG—the traditional Kennedy pathway (92). The arguments for this sequence of reactions are reviewed in reference 187. On the other hand, Slack and coworkers have suggested that in developing linseed, safflower, and soybean cotyledons, the DAG and PC pools are essentially kept in equilibrium by the action of choline phosphotransferase (168, 171) and that the major flux of glycerol and fatty acids proceeds via PC. At present, the available evidence indicates that in most oilseeds, all the reactions of PC and TAG synthesis occur in a single compartment—the endoplasmic reticulum (see above). Thus, determining the relative importance of the two alternative routes is obviously difficult.

Perhaps the data most pertinent to this issue come from experiments using radioactive glycerol (rather than [^{14}C]acetate or [^{14}C]fatty acids) to follow the kinetics of labeling in vivo or in microsomal preparations. When linseed cotyledons were incubated with [^{3}H]glycerol (171), label entered DAG linearly from zero time and accumulated in PC after only a short lag (less than 5 min). If the major route of triacylglycerol were directly from PA and DAG, then glycerol label should appear in TAG at least as rapidly as in PC. In fact, there was a considerable lag (35–40 min) associated with the appearance of [^{3}H]glycerol label in the TAG fraction (Figure 4). These results suggest that the DAG species first synthesized from radioactive glycerol are used preferentially for PC synthesis while TAG synthesis via acyl-CoA:DAG acyltransferase is specific for [^{3}H]-DAG species that accumulate later in the time course. In support of this concept, at early times, [^{3}H]glycerol label was found in molecular species of DAG containing 18:1, while species containing 18:3 became predominant only after 50–90 min of labeling (Figure 4). Thus, in intact linseed cotyledons, synthesis of polyunsaturated DAG and TAG proceeds predominantly via PC rather than depending on the direct use of 18:2- and 18:3-CoA (derived from PC by acyl exchange) for PA and DAG synthesis via the Kennedy pathway. Comparable in vivo labeling data are not available for other species, but safflower microsomes incubated with [^{14}C]glycerol-3-P and 18:2-CoA accumulated label in PC more rapidly than in TAG (178), again indicating that PC is an important intermediate in the flux of glycerol in TAG synthesis as well as in the flux of acyl groups. In principle, the diversion of DAG into the PC pool could be controlled by assuring that the acyl-CoA:DAG acyltransferase in these oilseeds exhibits selectivity for highly unsaturated DAG molecules over species containing 18:1. In contrast to the results for oilseeds, microsomes prepared from avocado mesocarp showed essentially no labeling of PC during TAG synthesis from [^{14}C] glycerol (178). This is consistent with the fact that the mesocarp tissue accumulates 18:1 rather than 18:2 or 18:3 in TAG. Similarly, as discussed below, plants that accumulate unusual fatty acids in their oil do so via PA and DAG alone. It is clear, therefore, that the relative contributions to

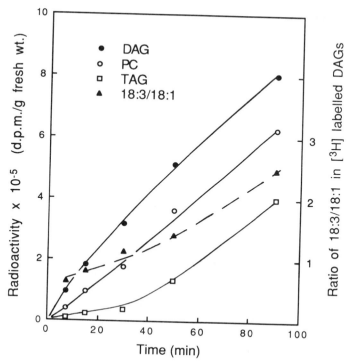

Figure 4 The appearance of radioactivity in lipids of developing linseed cotyledons incubated with [2-³H]glycerol together with the ratio of 18:3 to 18:1 in ³H-labeled molecular species of DAG (data are from 171).

TAG synthesis by the alternative pathways shown in Figure 3 vary among different plant species.

The Synthesis of Uncommon Fatty Acids

The seed oils of many plants are characterized by the presence of one or more fatty acids that are largely or completely excluded from the membrane lipids of the seed tissue. In terms of metabolism these present two broad questions: How is the fatty acid synthesized and how is it specifically targeted to triacylgylcerols? We first review recent information about the synthesis of some uncommon fatty acids and then describe current models for restricting these structures to the triacylglycerols.

Many plant species, including older varieties of oilseed rape, accumulate 20:1 and 22:1 fatty acids. Both genetic (47) and biochemical evidence (182) indicates that these fatty acids are formed by chain elongation of 18:1-CoA by an enzyme system associated with either the endoplasmic reticulum or the oil body membrane (see above). The elongation presumably requires the same series of reactions as fatty acid synthesis (condensation with malonyl-CoA and the subsequent reduction, dehydration, and reduction steps), but none of the activities has been purified. In oilseed rape, the levels of 20:1 and 22:1 can be independently varied by selection (47) and it is therefore possible that distinct isozymes (for one or more of the reactions) are used for elongation of 18:1 to 20:1 and of 20:1 to 22:1. Seeds of meadowfoam elongate 18:1 to 20:1 and 22:1 producing the same ω9 isomers found in oilseed rape and other species. However, Meadowfoam oil also contains isomers of 20:1 and 22:1 in which the double bond is at the Δ5 position. These fatty acids are formed from 16:0 by two or three cycles of elongation, and the resulting 20:0 and 22:0 are then desaturated by a specific acyl-CoA desaturase (136).

Oils from coconut and oil palm kernels are the main sources of medium-chain fatty acids (mainly 10:0 and 12:0) that are used to produce surfactants and detergents. Many species of *Cuphea* also accumulate seed oils rich in medium-chain fatty acids such that the predominant fatty acids in seeds of this genus range from 8:0 in *C. hookeriana* to 18:2 in *C. racemosa*. In specialized animal tissues, such as the mammary gland, that produce fatty acids with less than 16 carbons, a specific short-chain thioesterase is responsible for cleaving the fatty acid from the ACP arm of the Type I fatty acid synthase (151). Until very recently, attempts to demonstrate such short-chain thioesterases in plants failed. For example, the thioesterase activity present in coconut is quite specific for long-chain fatty acids (130). One alternative hypothesis for chain termination was based on experiments in which the addition of high concentrations of ACP to in vitro fatty acid synthase reactions increased the proportion of short-chain fatty acids produced (164). However, the over-production of ACP in transgenic plants did not lead to any change in fatty acid composition (138), suggesting that the in vitro results were artefacts. Similarly, it is not clear whether the experimental manipulation of chain length brought about by varying acetyl-CoA and malonyl-CoA levels in vitro (164, 166) have any relevance in vivo.

It has now been shown that extracts from developing cotyledons of California bay (whose oil contains 58% 12:0) hydrolyzed 12:0-ACP and 18:1-ACP at high rates while showing only low activity with other acyl-ACPs (164). The description of this activity suggests that chain length–specific thioesterases may indeed be the mechanism used for chain termination of fatty acid synthesis in the plants that produce medium-chain fatty acids.

Targeting Uncommon Fatty Acids to Triacylglycerols

Although it is not conclusive, the available evidence suggests that all the unusual fatty acids described above are components of the acyl-CoA pool and are used for TAG synthesis by reactions of the Kennedy pathway. This is true even for 18:1(12-OH), which appears to be synthesized on PC (112). It now seems likely that the reactions of TAG synthesis by the Kennedy pathway occur alongside those of phospholipid synthesis in the endoplasmic reticulum rather than in a separate compartment such as the differentiated oilbody itself (see above). If this is so, then the unusual fatty acids must be confined to TAGs by the specificities of the acyltransferases and of the enzymes (such as the cholinephosphotransferase) that are involved in phospholipid synthesis from DAG. For example, if the cholinephosphotransferase specifically excluded DAG molecular species containing unusual fatty acids, then these would accumulate in the DAG pool and could be used for TAG synthesis by a nonspecific DAG acyltransferase.

The observation that many oilseeds such as castor bean and *Cuphea lanceolata* accumulate TAGs containing almost exclusively 18:1(12-OH) or 10:0, respectively, at each position of the gylcerol poses the complementary question of how typical membrane fatty acids are excluded from the triacylglycerols. Bafor et al (4) have investigated these two questions in *C. lanceolata* by determining the substrate preferences of acyltransferase activities in microsomal preparations and relating these to the lipid compositions found in vivo. These results are consistent with the formation of both 10:0-lysoPA (LPA) and 18:1-LPA by one or more microsomal glycerol-3-P acyltransferases. The microsomes were also able to transfer 10:0 from 10:0-CoA to either of these LPA species, but 18:1-CoA could only be used for acylation of 18:1 LPA. The mixed-chain 18:1/10:0 species are only minor components in *C. lanceolata* lipids, and Bafor et al (4) argue that the ratio of different LPA and acyl-CoA precursors in vivo may limit the synthesis of 18:1/10:0PA. In any event, the DAG pool of developing *C. lanceolata* cotyledons was found to contain a bimodal distribution of species—10:0/10:0DAG, and species containing two long-chain fatty acids. This bimodal distribution makes it possible to exclude medium-chain fatty acids from PC by having the cholinephosphotransferase only accept long-chain DAG species as substrates. (It is, of course, also necessary to assume that the acyl-CoA:lysoPC acyltransferase also discriminates against 10:0CoA.) Conversely, the DAG acyltransferase uses almost exclusively 10:0/10:0DAG but appears to use both 10:0 and long-chain acyl-CoAs for the synthesis of TAG (4).

Complete analysis of the specificities and selectivities of the enzymes involved in seed oil synthesis is difficult. The concentrations of substrates in vivo are not known, and it is generally not possible to present lipid substrates to the enzymes in the same form as they are found in vivo—i.e. as components of the membranes. Nevertheless, more data from a range of oilseed species are needed in order to further test these hypotheses of specific lipid exclusion.

CONCLUDING REMARKS

Although much remains to be learned about plant lipid metabolism, significant progress has been made on many fronts during the past few years. Application of techniques for cloning genes for enzymes involved in lipid metabolism should permit continued progress toward resolving issues pertaining to the subcellular localization of the various reactions and the mechanisms that regulate the activity of key enzymes. The availability of the cloned genes will also permit the direct application of a great deal of the specific knowledge to the creation of useful genetic modifications of lipid metabolism in crop species. Indeed, considering the chemical diversity of lipids and the many industrial uses to which plant lipids could be applied, we consider the genetic engineering of plant lipid metabolism to be one of the most promising areas for future application of this powerful technology (175).

ACKNOWLEDGMENTS

We are grateful to many of our colleagues who have provided ideas and perspectives on the work discussed in this review, including preprints of their publications. We also thank the Department of Energy, the National Science Foundation, the United States Department of Agriculture, and Pioneer Hi-Bred International for support of research in our laboratories.

Literature Cited

1. Andrews, J., Mudd, J. B. 1985. Phosphatidylglycerol synthesis in pea chloroplasts. *Plant Physiol.* 79:259-65
2. Appleby, R. S., Safford, R., Nichols, B. W. 1971. The involvement of lecithin and monogalactosyl diglyceride in linoleate synthesis by green and blue-green algae. *Biochim. Biophys. Acta* 248:205-11
3. Arondel, V., Kader, J. C. 1990. Lipid transfer in plants. *Experientia* 46:579-85
4. Bafor, M., Jonsson, L., Stobart, A. K., Stymne, S. 1990. Regulation of triacylglycerol biosynthesis in embryos and microsomal preparations from the developing seeds of *Cuphea lanceolata*. *Biochem. J.* 272:31-88
5. Bankaitis, V. A., Aitken, J. R., Cleves, A. E., Dowhan, W. 1990. An essential role for a phospholipid transfer protein in yeast Golgi function. *Nature* 347:561-62
6. Battey, J. F., Ohlrogge, J. B. 1989. A comparison of the metabolic fate of fatty acids of different chain lengths in developing oilseeds. *Plant. Physiol.* 90:835-40
7. Bell, R. M., Coleman, R. A. 1980. Enzymes of glycerolipid synthesis in eukaryotes. *Annu. Rev. Biochem.* 49:459-87
8. Bergfeld, R., Hong, Y.-N., Kühnl, T., Schopfer, P. 1978. Formation of oleosomes (storage lipid bodies) during embryo-genesis and their breakdown during development in cotyledons of *Sinapsis alba* L. *Planta* 143:297-307
9. Bernhard, W., Somerville, C. R. 1989. Coidentity of putative amylase inhibitors from barley and millet with phospholipid transfer proteins inferred from amino acid sequence homology. *Arch. Biochem. Biophys.* 269:695-97
10. Bernhard, W. R., Thoma, S., Botella, J., Somerville, C. R. 1991. Isolation of a cDNA clone for spinach lipid transfer protein and evidence that the protein is synthesized by the secretory pathway. *Plant Physiol.* 95:164-70
11. Bessoule, J. J., Lessire, R., Cassagne, C. 1989. Partial purification of the acyl-CoA elongase of *Allium porrum* leaves. *Arch.Biochem. Biophys.* 268:475-84
12. Blatt, M. R., Thiel, G., Trentham, D. R. 1990. Reversible inactivation of K^+

channels of *Vicia* stomatal guard cells following the photolysis of caged inositol 1,4,5-trisphosphate. *Nature* 346:766-69

13. Block, M. A., Dorne, A. J., Joyard, J., Douce, R. 1983. The acyl CoA synthetase and acyl-CoA thioesterase are located on the outer and inner membrane of the chloroplast envelope, respectively. *FEBS Lett.* 153:377-80

14. Block, M. A., Dorne, A. J., Joyard, J., Douce, R. 1983. Preparation and characterization of membrane fractions enriched in outer and inner envelope membranes from spinach chloroplasts. *J. Biol. Chem.* 258:13273-86

15. Boss, W. F. 1989. Phosphoinositide metabolism: its relation to signal transduction in plants. In *Second Messengers in Plant Growth and Development*, ed. W. F. Boss, D. J. Morre, pp. 29-56. New York: Alan R. Liss, Inc.

16. Breu, V., Guerbette, F., Kader, J. C., Kannangara, C., Svensson, B., Von Wettstein-Knowles, P. 1989. A 10 kd barley basic protein transfers phosphatidylcholine from liposomes to mitochondria. *Carlsberg Res. Commun.* 54:81-84

17. Brockman, J. A., Norman, H. A., Hildebrand, D. F. 1990. Effects of temperature, light and a chemical modulator on linoleate biosynthesis in mutant and wild type *Arabidopsis* calli. *Phytochemistry* 29:1447-53

18. Brockman, J. A., Hildebrand, D. F. 1990. A polypeptide alteration associated with a low linolenate mutant of *Arabidopsis*. *Plant Physiol. Biochem.* 28: 11-16

19. Browse, J., Hugly, S., Miquel, M., Somerville, C. 1990. Elucidating lipid metabolism using mutants of *Arabidopsis*. In *Plant Molecular Biology*, ed. R. Herrmann, B. Larkins. NATO ASI Series. London: Portland Press. In press

20. Browse, J., Kunst, L., Hugly, S., Somerville, C. 1989. Modifications to the two pathway scheme of lipid metabolism based on studies of *Arabidopsis* mutants. In *Biological Role of Plant Lipids*, ed. P. A. Biacs, K. Gruiz, T. Kremmer, pp. 335-39. London: Plenum

21. Browse, J., Kunst, L., Anderson, S., Hugly, S., Somerville, C. R. 1989. A mutant of *Arabidopsis* deficient in the chloroplast 16:1/18:1 desaturase. *Plant Physiol.* 90:522-29

22. Browse, J., McCourt, P., Somerville, C. R. 1985. A mutant of *Arabidopsis* lacking a chloroplast-specific lipid. *Science* 227:763-65

23. Browse, J., McCourt, P., Somerville, C. R. 1986. A mutant of *Arabidopsis* deficient in C18:3 and C16:3 leaf lipids. *Plant Physiol.* 81:859-64

24. Browse, J., Miquel, M., Somerville, C. 1991. Genetic approaches to understanding plant lipid metabolism. In *Plant Lipid Biochemistry, Structure and Utilization*, ed. P. Quinn, J. L. Harwood. London: Portland Press. In press

25. Browse, J., Roughan, P. G., Slack, C. R. 1981. Light control of fatty acid synthesis and diurnal fluctuations of fatty acid composition in leaves. *Biochem. J.* 196:347-54

26. Browse, J. A., Slack, C. R. 1983. The effect of temperature and oxygen on the rates of fatty acid synthesis and oleate desaturation in safflower (*Carthamus tinctorius*) seed. *Biochim. Biophys. Acta* 753:145-52

27. Browse, J. A., Slack, C. R. 1981. Catalase stimulates linoleate desaturation in microsome preparations from developing linseed cotyledons. *FEBS Lett.* 131:111-14

28. Browse, J., Warwick, N., Somerville, C. R., Slack, C. R. 1986. Fluxes through the prokaryotic and eukaryotic pathways of lipid synthesis in the 16:3 plant *Arabidopsis thaliana*. *Biochem. J.* 235:25-31

29. Budde, R. J., Randall, D. D. 1990. Pea leaf mitochondrial pyruvate dehydrogenase complex is inactivated in vivo in a light-dependent manner. *Proc. Natl. Acad. Sci. USA* 87:673-76

30. Burton, J. D., Gronwald, J. W., Somers, D. A., Connelly, J. A., Gegenbach, B. G., Wyse, D. L. 1987. Inhibition of plant acetyl-CoA carboxylase by the herbicides sethoxydim and haloxyfop. *Biochem. Biophys. Res. Commun.* 148: 1039-44

31. Camp, P. J., Randall, D. D. 1985. Purification and characterization of the pea chloroplast pyruvate dehydrogenase complex. *Plant Physiol.* 77:571-77

32. Carman, G. M., Henry, S. H. 1989. Phospholipid biosynthesis in yeast. *Annu. Rev. Biochem.* 58:635-69

33. Charles, D. J., Cherry, J. H. 1986. Purification and characterization of acetyl-CoA carboxylase from developing soybean seeds. *Phytochemistry* 25: 1067-71

34. Cheesbrough, T. M. 1989. Changes in the enzymes for fatty acid synthesis and desaturation during acclimation of developing soybean seeds to altered growth temperature. *Plant Physiol.* 90: 760-64

35. Cheesbrough, T. M. 1990. Decreased

growth temperature increases soybean stearoyl-ACP desaturase activity. *Plant Physiol.* 93:555-59

36. Cheesbrough, T. M., Cho, S. H. 1990. Purification and characterization of soybean stearoyl-ACP desaturase. See Ref. 24

37. Chin, D. J., Luskey, K. L., Anderson, R. G., Faust, J. R., Goldstein, J. L., Brown, M. S. 1982. Appearance of crystalloid endoplasmic reticulum in compactin resistant Chinese hamster cells with a 500-fold increase in 3-hydroxy-3-methylglutaryl-coenzyme A reductase. *Proc. Natl. Acad. Sci. USA.* 79:1185-89

38. Chuman, L., Brody, S. 1989. Acyl carrier protein is present in the mitochondrion of plants and eukaryotic microorganisms. *Eur. J. Biochem.* 184:643-49

39. Cline, K., Keegstra, K. 1983. Galactosyltransferases involved in galactolipid biosynthesis are located in the outer membrane of pea chloroplast envelopes. *Plant Physiol.* 71:366-72

40. Covès, J., Joyard, J., Douce, R. 1988. Lipid requirement and kinetic studies of solubilized UDP-galactose:diacylglycerol galactosyltransferase activity from spinach chloroplast envelope membranes. *Proc. Natl. Acad. Sci. USA* 85:4966-70

41. Datko, A. H., Mudd, S. H. 1986. Uptake of choline and ethanolamine by *Lemna paucicostata* Hegelm. 6746. *Plant Physiol.* 81:285-88

42. Datko, A. H., Mudd, S. H. 1988. Enzymes of phosphatidylcholine synthesis in *Lemna*, soybean and carrot. *Plant Physiol.* 88:1338-48

43. Datko, A. H., Mudd, S. H. 1988. Phosphatidylcholine synthesis: differing patterns in soybean and carrot. *Plant Physiol.* 88:854-61

44. Dennis, D. T., Miernyk, J. A. 1982. Compartmentation of nonphotosynthetic carbohydrate metabolism. *Annu. Rev. Plant Physiol.* 33:27-50

45. Dorne, A. J., Joyard, J., Douce, R. 1990. Do thylakoids really contain phosphatidylcholine? *Proc. Natl. Acad. Sci. USA.* 87:71-74

46. Douce, R., Joyard, J. 1979. Structure and function of the plastid envelope. *Adv. Bot. Res.* 7:1-116

47. Downey, R. K., Craig, R. M. 1964. Genetic controls of fatty acid biosynthesis in rape seed (*Brassica napus* L.). *J. Am. Oil Chem. Soc.* 41:475-78

48. Egin-Buhler, B., Ebel, J. 1983. Improved purification and further characterization of acetyl-CoA carboxylase

from cultured cells of parsely (*Petroselinum hortense*). *Eur. J. Biochem.* 133:335-39

49. Deleted in proof

50. Einspahr, K. J. Thompson, G. A. Jr. 1990. Transmembrane signaling via phosphatidylinositol 4,5-bisphosphate hydrolysis in plants. *Plant Physiol.* 93:361-66

51. Farmer, E. E., Ryan, C. A. 1990. Interplant communication: Airborne methyl jasmonate induces synthesis of proteinase inhibitors in plant leaves. *Proc. Natl. Acad. Sci. USA* 87:7713-16

52. Fehling, E., Murphy, D. J., Mukherjee, K. D. 1990. Biosynthesis of triacylglycerols containing very long chain monounsaturated acyl moieties in developing seeds. *Plant Physiol.* 94:492-98

53. Frentzen, M. 1990. Comparison of certain properties of membrane bound and solubilized acyltransferase activities of plant microsomes. *Plant Sci.* 69:39-48

54. Frentzen, M., Heinz, E., McKeon, T. A., Stumpf, P. K. 1983. Specificities and selectivities of glycerol-3-phosphate acyltransferase from pea and spinach chloroplasts. *Eur. J. Biochem.* 129:629-36

55. Frentzen, M., Neuburger, M., Joyard, J., Douce, R. 1990. Intraorganelle localization and substrate specificities of the mitochondrial acyl-CoA:sn-glycerol-3-phosphate O-acyltransferase and acyl-CoA:1-acyl-sn-glycerol-3-phosphate O-acyltransferase from potato tubers and pea leaves. *Eur. J. Biochem.* 187:395-402

56. Frentzen, M., Nishida, I., Murata, N. 1987. Properties of the plastidial acyl-ACP:glycerol-3-phosphate acyltransferase from the chilling sensitive plant squash (*Cucurbita moschata*). *Plant Cell Physiol.* 28:1195-201

57. Gardiner, S. E., Roughan, P. G. 1983. Relationship between fatty-acyl composition of diacylgalactosylglycerol and turnover of chloroplast phosphatidate. *Biochem. J.* 210:949-52

58. Gee, R. W., Byerrum, R. U., Gerber, D. W., Tolbert, N. E. 1988. Dihydroxyacetone phosphate reductase in plants. *Plant Physiol.* 86:98-103

59. Gilroy, S., Read, N. D., Trewavas, A. J. 1990. Elevation of cytoplasmic calcium by caged calcium or caged inositol trisphosphate initiates stomatal closure. *Nature* 3346:769-71

60. Gontero, B., Subramani, S., Ricard, J. 1988. A functional five-enzyme complex of chloroplasts involved in the Calvin cycle. *Eur. J. Biochem.* 173:437-43

61. Griffiths, G., Stobart, A. K., Stymne, S. 1985. The acylation of sn-glycerol-3-phosphate and the metabolism of phosphatidate in microsomal preparations from the developing cotyledons of safflower (*Carthamus tinctorius* L.) seed. *Biochem. J.* 230:379-88

62. Grill, E., Somerville, C. R. 1991. Construction and characterization of a yeast artificial chromosome library of *Arabidopsis* which is suitable for chromosome walking. *Molec. Gen. Genet.* In press

63. Guerra, D. J., Ohlrogge, J. B., Frentzen, M. 1986. Activity of acyl carrier protein isoforms in reactions of plant fatty acid metabolism. *Plant Physiol.* 82:448-53

64. Guerra, D. J., Ohlrogge, J. B. 1986. Partial purification and characterization of two forms of malonyl-CoA:acyl carrier protein transacylase from soybean leaf tissue. *Arch. Biochem. Biophys.* 246:274-85

65. Gunstone, F. D. 1986. Fatty acid structure. In *The Lipid Handbook*, ed. F. D. Gunstone, J. L. Harwood, F. B. Padley, pp. 1-24. New York: Chapman and Hall

66. Gurr, M. I., Robinson, M. P., James, A. T. 1969. The mechanism of formation of polyunsaturated fatty acids by photosynthetic tissue. *Eur. J. Biochem.* 9:70-78

67. Harwood, J. L. 1988. Fatty acid metabolism. *Annu. Rev Plant Physiol.* 39:101-38

68. Harwood, J. L., Russell, N. J. 1984. *Lipids in Plants and Microbes.* London: George Allen and Unwin. 162 pp.

69. Heemskerk, J. W. M., Storz, T., Schmidt, R. R., Heinz, E. 1990. Biosynthesis of digalactosyldiacylglycerol in plastids from 16:3 and 18:3 plants. *Plant Physiol.* 93:1286-94

70. Heinz, E., Roughan, P. G. 1983. Similarities and differences in lipid metabolism of chloroplasts isolated from 18:3 and 16:3 plants. *Plant Physiol.* 72:273-79

71. Heinz, E., Schmidt, H., Hoch, M., Jung, K. H., Binder, H., Schmidt, R. 1989. Synthesis of different nucleoside 5'-diphospho-sulfoquinovoses and their use for studies on sulfolipid biosynthesis in chloroplasts. *Eur. J. Biochem.* 184:445-53

72. Hellyer, A., Bambridge, H. E., Slabas, A. R. 1986. Plant acetyl CoA carboxylase. *Biochem Soc. Trans.* 14:565-68

73. Hellyer, A., Slabas, A. R. 1991. Acyl-ACP thioesterase from oil seed rape: purification and characterization. See Ref. 24. In press

74. Hilditch, T. P., Williams, P. N. 1964. *The Chemical Constitution of Natural Fats.* London: Chapman & Hall. 745 pp.

75. Hoj, P. B., Mikkelsen, J. D. 1982. Partial separation of individual enzyme activities of an ACP-dependent fatty acid synthetase from barley chloroplasts. *Carlsberg Res. Commun.* 47:119-41

76. Hoj, P. B., Svendsen, I. 1984. Barley chloroplasts contain 2 acyl carrier proteins coded for by different genes. *Carls. Res. Commun.* 49:483-92

77. Hugly, S., Kunst, L., Browse, J., Somerville, C. R. 1989. Enhanced thermal tolerance and altered chloroplast ultrastructure in a mutant of *Arabidopsis* deficient in lipid desaturation. *Plant Physiol.* 90:1134-42

78. Ichihara, K. 1984. Sn-glycerol-3-phosphate acyltransferase in a particulate fraction from maturing safflower seeds. *Arch. Biochem. Biophys.* 232:685-98

79. Ishizaki, O., Nishida, I., Agata, K., Eguchi, G., Murata, N. 1988. Cloning and nucleotide sequence of cDNA for the plastid glycerol-3-phosphate acyltransferase from squash. *FEBS Lett.* 238:424-30

80. Jackowski, S., Rock, C. O. 1989. Acetoacetyl-acyl carrier protein synthase, a potential regulator of fatty acid biosynthesis in bacteria. *J. Biol. Chem.* 262:7927-31

81. James, A. T. 1963. The biosynthesis of long-chain saturated and unsaturated fatty acids in isolated plant leaves. *Biochim. Biophys. Acta* 70:9-19

82. James, D. W., Dooner, H. K. 1990. Isolation of EMS-induced mutants in *Arabidopsis* altered in seed fatty acid composition. *Theor. Appl. Genet.* 80:241-45

83. Jamieson, G. R., Reid, E. H. 1971. The occurrence of hexadeca-7,19,13-trienoic acid in the leaf lipids of angiosperms. *Phytochemistry* 10:1837-43

84. Jamil, H., Yao, Z., Vance, D. E. 1990. Feedback regulation of CTP:choline cytidyltransferase translocation between cytosol and endoplasmic reticulum by phosphatidylcholine. *J. Biol. Chem.* 265:4332-39

85. Jaworski, J. G., Clough, R. C., Barnum, S. R. 1989. A cerulenin insensitive short chain 3-ketoacyl-acyl carrier protein synthase in *Spinacia oleracea* leaves. *Plant Physiol.* 90:41-44

86. Johnson, G., Williams, J. P. 1989. Effect of growth temperature on the biosynthesis of chloroplastic galactosyldiacylglycerol molecular species in *Brassica napus* leaves. *Plant Physiol.* 91:924-29

87. Joyard, J., Douce, R. 1977. Site of synthesis of phosphatidic acid and diacylglycerol in spinach chloroplasts. *Biochim. Biophys. Acta* 486:273–85
88. Joyard, J., Blée, E., Douce, R. 1986. Sulfolipid synthesis from $^{35}SO_4$ [1-^{14}C]acetate in isolated spinach chloroplasts. *Biochim. Biophys. Acta* 879: 78-87
89. Justin, A. M., Demandre, C., Mazliak, P. 1987. Choline and ethanolaminephosphotransferase from pea leaf and soybean discriminate 1-palmitoyl-2-linoleoyldiacylglycerol as a preferred substrate. *Biochim. Biophys. Acta* 922:364-71
90. Justin, A. M., Demandre, C., Mazliak, P. 1989. Molecular species synthesized by phosphatidylinositol synthases from potato tuber, pea leaf and soya bean. *Biochim. Biophys. Acta* 1005:51-55
91. Kauppinen, S. 1991. Molecular cloning of the gene(s) coding for barley b-ketoacyl-ACP synthetase 1. See Ref. 24. In press
92. Kennedy, E. P. 1961. Biosynthesis of complex lipids. *Fed. Proc. Am. Soc. Exp. Biol.* 20:934-40
93. Kim, K. H., Lopez-Casillas, F., Bai, D. H., Luo, X., Pape, M. E. 1989. Role of reversible phosphorylation of acetyl-CoA carboxylase in long-chain fatty acid synthesis. *Fed. Proc. Fed. Am. Soc. Exp. Biol.* 3:2250-54
94. Kinney, A. J., Moore, T. S. 1986. Phosphatidylcholine synthesis in castor bean endosperm. I. metabolism of L-serine. *Plant Physiol.* 84:78-81
95. Kinney, A. J., Moore, T. S. 1987. Phosphatidylcholine synthesis in castor bean endosperm. The localization and control of CTP:choline-phosphate cytidyltransferase activity. *Arch. Biochem. Biophys.* 259:15-21
96. Kinney, A. J., Moore, T. S. 1988. Phosphatidylcholine synthesis in castor bean endosperm: characteristics and reversibility of the choline kinase reaction. *Arch. Biochem. Biophys.* 260:102-8
97. Kinney, A. J., Yadav, N., Hitz, W. D. 1991. Purification and characterization of fatty acid biosynthetic enzymes from developing soybeans. See Ref. 24. In press
98. Kleppinger-Sparace, K. F., Mudd, J. B. 1990. Bosynthesis of sulfoquinovosyldiacylglycerol in higher plants. *Plant Physiol.* 93:256-63
99. Kleppinger-Sparace, K. F., Mudd, J. B. 1987. Biosynthesis of sulfoquinovosyldiacylglycerol in higher plants. The incorporation of $^{35}SO_4$ by intact chloroplasts in darkness. *Plant Physiol.* 84:682-87
100. Kunst, L., Browse, J., Somerville, C. R. 1988. Altered regulation of lipid biosynthesis in a mutant of *Arabidopsis* deficient in chloroplast glycerol phosphate acyltranferase activity. *Proc. Natl. Acad. Sci. USA* 85:4143-47
101. Kunst, L., Browse, J., Somerville, C. R. 1989. A mutant of *Arabidopsis* deficient in desaturation of palmitic acid in leaf lipids. *Plant Physiol.* 90:943-47
102. Kunst, L., Browse, J., Somerville, C. R. 1989. Enhanced thermal tolerance in a mutant of *Arabidopsis* deficient in palmitic acid unsaturation. *Plant Physiol.* 91:401-8
103. Lemieux, B., Miquel, M., Somerville, C., Browse, J. 1990. Mutants of *Arabidopsis* with alterations in seed lipid fatty acid composition. *Theor. Appl. Genet.* 80:234-40
104. Liedvogel, B. 1985. Acetate concentration and chloroplast pyruvate dehydrogenase complex in *Spinacea oleracea* leaf cells. *Z. Naturforsch. Teil C* 40:182-85
105. MacKintosh, R. W., Hardie, D. G., Slabas, A. R. 1989. A new assay procedure to study the induction of b-ketoacyl-ACP synthase I and II, and the complete purification of β-ketoacyl-ACP synthase I from developing seeds of oilseed rape (*Brassica napus*). *Biochim. Biophys. Acta* 1002:114-24
106. Martin, B., Tolbert, N. E. 1983. Factors which affect the amount of inorganic phosphate, phosphorylcholine and phosphorylethanolamine in xylem exudate of tomato plants. *Plant Physiol.* 73:464-70
107. McCourt, P., Kunst, L., Browse, J., Somerville, C. R. 1987. The effects of reduced amounts of lipid unsaturation on chloroplast ultrastructure and photosynthesis in a mutant of *Arabidopsis*. *Plant Physiol.* 84:353-61
108. McKeon, T. A., Stumpf, P. K. 982. Purification and characterization of the stearoyl-acyl carrier protein desaturase and the acyl-acyl carrier protein thioesterase from maturing seeds of safflower. *J. Biol. Chem.* 257:12141-47
109. Miquel, M., Block, M. A., Joyard, J., Dorne, A. J., Dubac, J. P., et al. 1987 Protein-mediated transfer of phosphatidylcholine from liposomes to spinach chloroplast envelope membranes. *Biochim Biophys Acta* 937:219-28
110. Moore, T. S. 1982. Phospholipid biosynthesis. *Annu. Rev. Plant Physiol.* 33:235-39
111. Moore, T. S. 1987. Phosphoglyceride synthesis in endoplasmic reticulum. *Methods Enzymol.* 148:585-96
112. Moreau, R., Stumpf, P. K. 1982. Sol-

ubilization and characterization of an acyl-coenzyme A O-lysophospholipid acyltransferase from the microsomes of developing safflower seeds. *Plant Physiol.* 69:1293-97

113. Morse, M. J., Satter, R. L., Crain, R. C., Cote, G. G. 1989. Signal transduction and phosphatidylinositol turnover in plants. *Physiol. Plant.* 76:118-21

114. Mudd, S. H., Datko, A. H. 1986. Phosphoethanolamine bases as intermediates in phosphatidylcholine synthesis by *Lemna. Plant Physiol.* 82:126-35

115. Mudd, S. H., Datko, A. H. 1989a. Synthesis of methylated ethanolamine moieties, regulation by choline in *Lemna. Plant Physiol.* 90:296-305

116. Mudd, S. H., Datko, A. H. 1989b. Synthesis of methylated ethanolamine moieties, regulation by choline in soybean and carrot. *Plant Physiol.* 90: 306-10

117. Mudd, J. B., Kleppinger-Sparace, K. 1987. Sulfolipid. In *The Biochemistry of Plants*, Vol. 9, pp. 275-89. New York: Academic

118. Mundy, J., Rogers, J. C. 1986. Selective expression of a probable amylase/protease inhibitor in barley aleurone cells: comparison to the barley amylase/subtilisin inhibitor. *Planta* 196:51-63

119. Murata, N., Nishida, I. 1990. Lipids in relation to chilling sensitivity of plants. In *Chilling Injury of Horticultural Crops*, ed. C. Y. Wang. Boca Raton, FL: CRC Press. In press

120. Murata, N., Sato, N., Takahashi, N. 1984. Very-long chain saturated fatty acids in phosphatidylserine from higher plant tissues. *Biochim. Biophys. Acta* 795:1475-80

121. Murata, N., Sato, N., Takahashi, N., Hamazaki, Y. 1982. Compositions and positional distributions of fatty acids in phospholipids from leaves of chilling sensitive and chilling-resistant plants. *Plant Cell Physiol.* 23:1071-79

122. Murata, N., Yamaya, J. 1984. Temperature dependent phase behaviour of phosphatidylglycerols from chilling sensitive and chilling resistant plants. *Plant Physiol.* 74:1016-24

123. Murphy, D. J., et al. 1990. Structure, function and possible biotechnological uses of oleosins (oil body proteins) from oilseeds. See Ref. 24. In press

124. Nishida, I., Tasaka, Y., Shiraishi, H., Okada, K., Shimura, Y., et al. 1990. The genomic gene and cDNA for the plastidial   glycerol-3-phosphate acyltransferase of *Arabidopsis.* See Ref. 24. In press

125. Norman, H. A., St. John, J. B. 1986.

126. Norman, H. A., St. John, J. B. 1987. Differential effects of a substituted pyridazinone, BASF 13-338, on pathways of monogalactosyldiacylglycerol synthesis in *Arabidopsis. Plant Physiol.* 85:684-88

127. Ohlrogge, J., Browse, J., Somerville, C. R.The genetics of plant lipids. *Biochim. Biophys. Acta.* In press

128. Ohlrogge, J. B., Kuhn, D. N., Stumpf, P. K. 1979. Subcellular localization of acyl carrier protein in leaf protoplasts of *Spinacia oleracea. Proc. Natl. Acad. Sci. USA* 76:1194-98

129. Ohlrogge, J. B., Kuo, T. M. 1985. Plants have isoforms for acyl carrier protein that are expressed differently in different tissues. *J. Biol. Chem.* 260:8032-37

130. Ohlrogge, J. B., Shine, W. E., Stumpf, P. K. 1978. Characterization of plant acyl-ACP and acyl-CoA hydrolases. *Arch. Biochem. Biophys.* 189:382-91

131. Oo, K. C., Huang, A. H. C. 1989. Lysophosphatidate acyltransferase activities in the microsomes from palm endosperm, maize scutellum and rapeseed cotyledons of maturing seeds. *Plant Physiol.* 91:1288-95

132. Parker, W. B., Marshall, L. C., Burton, J. D., Sommers, D. A., Wyse, D. L., et al. 1990. Dominant mutations causing alterations in acetyl-coenzyme A carboxylase confer tolerance to cyclohexanedione and aryloxyphenoxypropionate herbicides in maize. *Proc. Natl. Acad. Sci. USA* 87:7175-79

133. Pearcy, R. W. 1978. Effect of growth temperature on the fatty acid composition of the leaf lipids in *Atriplex lentiformis* (Torr.) Wats. *Plant Physiol.* 61: 484-86

134. Pelech, S. L., Vance, D. E. 1984. Regulation of phosphatidylcholine biosynthesis. *Biochim. Biophys. Acta* 779:217-51

135. Pollard, M. R., Anderson, L., Fan, C., Hawkins, D. J., Davies, H. M. 1990. A specific acyl-ACP thioesterase implicated in laurate production of *Umbellularia californica.* See Ref. 24. In press

136. Pollard, M. R., Stumpf, P. K. 1980. Biosynthesis of C_{20} and C_{22} fatty acids by developing seeds of *Limnanthes alba. Plant Physiol.* 66:649-55

137. Post-Beitenmiller, M. A., Jaworski, J., Ohlrogge, J. B. 1991. In vivo pools of

free and acylated acyl carrier protein in spinach: evidence for sites of regulation of fatty acid biosynthesis. *J. Biol. Chem.* 266:1858-65

138. Post-Beittenmiller, M. A., Schmid, K. M., Ohlrogge, J. B. 1989. Expression of holo and apo forms of spinach acyl carrier protein-I in leaves of transgenic tobacco. *Plant Cell* 1:889-99

139. Quinn, P. J., Joo, F., Vigh, L. 1989. The role of unsaturated lipids in membrane structure and stability. *Prog. Biophys. Mol. Biol.* 53(2):71-103

140. Raison, J. K. 1986. Alterations in the physical properties and thermal responses of membrane lipids: correlations with acclimation to chilling and high temperature. In *Frontiers in Membrane Research in Agriculture*, ed. J. B. St. John, E. Berlin, P. C. Jackson, pp. 383-401. Ottowa: Rowman and Allanheld

141. Roughan, P. G. 1970. Turnover of the glycerolipids of pumpkin leaves. The importance of phosphatidylcholine. *Biochem. J.* 117:1-8

142. Roughan, P. G. 1985. Cytidine triphosphate–dependent, acyl-CoA–independent synthesis of phosphatidylglycerol by chloroplasts isolated from spinach and pea. *Biochim. Biophys. Acta* 835:527-32

143. Roughan, P. G. 1986. A simplified isolation of phosphtidylglycerol. *Plant Sci.* 43:57-62

144. Roughan, P. G., Holland, R., Slack, C. R. 1979. On the control of long-chain fatty acid synthesis in isolated intact spinach (*Spinacia oleracea*) chloroplasts. *Biochem. J.* 184:193-202

145. Roughan, P. G., Holland, R., Slack, C. R. 1980. The role of chloroplasts and microsomal fractions in polar lipid synthesis from [1-^{14}C]acetate by cell-free preparations from spinach (*Spinacia oleracea*) leaves. *Biochem. J.* 188:17-24

146. Roughan, P. G., Slack, T., Beevers, H. 1980. On the light dependency of fatty acid synthesis by isolated spinach chloroplasts. *Plant Sci. Lett.* 18:221-28

147. Roughan, P. G., Mudd, J. B., McManus, T. T., Slack, C. R. 1979. Linoleic and α-linolenate synthesis by isolated spinach (*Spinacia oleracea*) chloroplasts. *Biochem. J.* 184:571-74

148. Roughan, P. G., Nishida, I. 1990. Concentrations of long chain acyl-acyl carrier proteins during fatty acid synthesis by chloroplasts isolated from pea (*Pisum sativum*), safflower (*Carthamus tinctorius*) and amaranthus (*Amaranthus lividus*) leaves. *Arch. Biochem. Biophys.* 276:38-46

149. Roughan, P. G., Slack, C. R. 1982. Cellular organization of glycerolipid metabolism. *Annu. Rev. Plant Physiol.* 33:97-123

150. Roughan, P. G., Slack, C. R., Holland, R. 1976. High rates of [1-^{14}C]acetate incorporation into the lipid of isolated spinach chloroplasts. *Biochem. J.* 158:593-601

151. Sasaki, G. C., Cheesbrough, V., Kolattukudy, P. E. 1988. Nucleotide sequence of the S-acyl fatty acid synthase thioesterase gene and its tissue specific expression. *DNA* 7:449-57

152. Scherer, D. E., Kridl, J. C., Young, H. L., Aken, F. V., Kenny, J. W., et al. 1991. Cloning, expression in *E. coli* and characterization of a stearoyl-ACP desaturase from safflower seeds. See Ref. 24. In press

153. Schmidt, H., Heinz, E. 1990. Involvement of ferredoxin in desaturation of lipid-bound oleate in chloroplasts. *Plant Physiol.* 94:214-20

154. Schmidt, H., Heinz, E. 1990. Desaturation of oleoyl groups in envelope membranes from spinach chloroplasts. *Proc. Natl. Acad. Sci. USA.* 87:9477-80

155. Shanklin, J., Somerville, C. R. 1990. The cDNA clones for stearoyl-ACP desaturase from higher plants are not homologous to yeast or mammalian genes encoding stearoyl-CoA desaturase. *Proc. Natl. Acad. Sci. USA.* In press

156. Sheldon, P. S., Keckwick, R. G. O., Didebottom, C. M., Smith, C. G., Slabas, A. R. 1990. 3-Oxoacyl-(acyl carrier protein) reductase from avocado (*Persea americana*) fruit mesocarp. *Biochem. J.* 271:713-20

157. Shimakata, T., Stumpf, P. K. 1982. The prokaryotic nature of the fatty acid synthetase of developing *Carthamus tinctorius* L. (safflower) seeds. *Arch. Biochem. Biophys.* 217:144-54

158. Shimakata, T., Stumpf, P. K. 1982. Isolation and function of spinach leaf β-ketoacyl[acyl carrier protein] synthetases. *Proc. Natl. Acad. Sci. USA.* 79:5808-12

159. Shimakata, T., Stumpf, P. K. 1983. The purification and function of acetyl coenzyme A:acyl carrier protein transacylase. *J. Biol. Chem.* 258:3592-98

160. Shin, S., Moore, T. S. 1990. Phosphatidylethanolamine synthesis by castor bean endosperm. A base exchange reaction. *Plant Physiol.* 93:148-53

161. Shin, S., Moore, T. S. 1990. Phosphatidylethanolamine synthesis by castor bean endosperm. Membrane bilayer dis-

tribution of phosphatidylethanolamine synthesized by the ethanolaminephosphotransferase and ethanolamine exchange reactions. *Plant Physiol.* 93:154-59

162. Siebertz, H. P., Heinz, E. 1977. Labeling experiments on the origin of hexa- and octa-decatrienoic acids in galactolipids of leaves. *Z. Naturforsch. Teil C* 32:193-205

163. Simpson, E. E., Williams, J. P. 1979. Galactolipid synthesis in *Vicia faba* leaves. Sites of fatty acid incorporation into the major glycerolipids. *Plant Physiol.* 63:674-76

164. Slabas, A., Roberts, P., Ormesher, J. 1982. Characterization of fatty acid synthesis in a cell free system from developing oilseed rape. In *Biochemistry and Metabolism of Plant Lipids*, ed. J. F. G. M. Wintermans, P. J. C. Kuipers, pp. 251-56. Amsterdam: Elsevier. 606 pp.

165. Slabas, A. R., Cottingham, I. R., Austin, A., Hellyer, A., Safford, R., Smith, C. G. 1990. Immunological detection of NADH-specific enoyl-ACP reductase from rape seed (*Brassica napus*)— induction, relationship of α and β polypeptides, mRNA translation and interaction with ACP. *Biochim. Biophys. Acta* 1039:181-88

166. Slabas, A. R., Harding, J., Hellyer, A., Sidebottom, C., Gwynne, H., et al. 1984. Enzymology of plant fatty acid synthesis. In *Structure Function and Metabolism of Plant Lipids*, ed. P. A. Siegenthaler, W. Eichenberger, pp. 3-10. Amsterdam: Elsevier. 634 pp.

167. Slack, C. R., Bertaud, W. S., Shaw, B. D., Holland, R., Browse, J. A., Wright, H. 1980. Some studies on the composition and surface properties of oil bodies from oil seed cotyledons. *Biochem. J.* 190:551-61

168. Slack, C. R., Browse, J. A. 1984. Lipid synthesis during seed development. In *Seed Physiology*, ed. D. Murray, 1:209-44. Sydney: Academic

169. Slack, C. R., Campbell, L. C., Browse, J. A., Roughan, P. G. 1983. Some evidence for the reversibility of choline phosphotransferase-catalysed reaction in developing linseed cotyledons in vivo. *Biochim. Biophys. Acta* 754:10-20

170. Slack, C. R., Roughan, P. G. 1975. The kinetics of incorporation in vivo of [^{14}C]acetate and ^{14}CO$_2$ into the fatty acids of glycerolipids in developing leaves. *Biochem. J.* 152:217-28

171. Slack, C. R., Roughan, P. G., Balasingham, N. 1978. Labeling of glycerolipids in the cotyledons of developing oilseeds by [1-^{14}C]acetate and [2-^{3}H]glycerol. *Biochem. J.* 179:421-33

172. Slack, C. R., Roughan, P. G., Browse, J. A., Gardiner, S. E. 1985. Some properties of cholinephosphotransferase from developing safflower cotyledons. *Biochim. Biophys. Acta* 833:438-48

173. Slack, C. R., Roughan, P. G., Terpstra, J. 1976. Some properties of a microsomal oleate desaturase from leaves. *Biochem. J.* 155:71-80

174. Smith, M. A., Cross, A. R., Jones, O. T. G., Grifiths, W. T. Stymne, S., Stobart, K. 1990. Electron-transport components of the 1-acyl-2-oleoyl-sn-glycerol-3-phosphocholine Δ^{12}-desaturase (Δ^{12}-desaturase) in microsomal preparations from developing safflower (*Carthamus tinctorius* L.) cotyledons. *Biochem. J.* 272:23-29

175. Somerville, C. R., Browse, J. A. 1991. Plant lipids: metabolism, mutants and membranes. *Science*. In press

176. Srere, P. A., Sumegi, B., Sherry, A. D. 1987. Organizational aspects of the citric acid cycle. *Biochem. Soc. Symp.* 54:173-82

177. Steponkus, P. L., Lynch, D. V., Uemura, M. 1990. The influence of cold acclimation on the lipid composition and cryobehavior of the plasma membrane of isolated rye protoplasts. *Philos. Trans. R. Soc. Lond. Ser. B* 326:571-83

178. Stobart, A. K., Stymne, S. 1985. The regulation of the fatty-acid composition of the triacylglycerols in microsomal preparations from avocado mesocarp and the developing cotyledons of safflower. *Planta* 163:119-25

179. Stobart, A. K., Stymne, S., Hoglund, S. 1986. Safflower microsomes catalyse oil accumulation in vitro: a model system. *Planta* 169:33-37

180. Stumpf, P. K., Conn, E. E. 1980. The *Biochemistry of Plants*, Vol 4. New York: Academic 693 pp.

181. Stumpf, P. K., Conn, E. E. 1987. See Ref. 180, Vol. 9. 363 pp.

182. Stumpf, P. K., Pollard, M. R. 1983. Pathways of fatty acid biosynthesis in higher plants with particular reference to developing rapeseed. In *High and Low Erucic Acid Rapeseed Oils*, ed. J. K. G. Kramer, F. D. Sauer, W. J. Pigden. Toronto: Academic

183. Stymne, S., Appelqvist, L.-A. 1978. The biosynthesis of linoleate from oleoyl-CoA via oleoyl-phosphatidylcholine in microsomes of developing safflower seeds. *Eur. J. Biochem.* 90:223-29

184. Stymne, S., Stobart, A. K. 1985. Oil

synthesis in-vitro in microsomal membranes from developing cotyledons of *Linum usitatissimum* L. *Planta* 164:101-4

185. Stymne, S., Stobart, A. K. 1984. Evidence for the reversibility of the acyl-CoA:lysophosphatidylcholine acyltransferase in the microsomes of developing safflower cotyledons and rat liver. *Biochem. J.* 223:305-14

186. Stymne, S., Stobart, A. K., Glad, G. 1983. The role of the Acyl-CoA pool in the synthesis of polyunsaturated 18-carbon fatty acids and triacylglycerol production in the microsomes of developing safflower seeds. *Biochim. Biophys. Acta* 752:198-208

187. Stymne S., Stobart, A. K. 1987. Triacylglycerol biosynthesis. See Ref. 180, pp. 175-211

188. Sun, C., Cao, Y.-Z., Huang, A. H. C. 1988. Acyl coenzyme-A preference of the glycerol phosphate pathway in the microsomes from the maturing seeds of palm, maize and rapeseed. *Plant Physiol.* 88:56-60

189. Tchang, F., This, P., Stiefel, V., Arondel, V., Morch, M. D., et al. 1988. Phospholipid transfer protein: full-length cDNA and amino acid sequence in maize. *J. Biol. Chem.* 263:16849-55

190. Teucher, T., Heinz, E. 1990. Purification of UDP-galactose:diacylglycerol galactosyltransferase from chloroplast envelopes. See Ref. 24. In press

191. Vance, V. B., Huang, A. H. C. 1987. The major protein from lipid bodies of maize. Characterization and structure based on cDNA cloning. *J. Biol. Chem.* 262:11275-79

192. Vance, D. E., Ridgway, N. D. 1988. The methylation of phosphatidylethanolamine. *Prog. Lipid Res.* 27:61-79

193. von Meyenburg, K., Jorgensen, B. B., van Deurs, B. 1984. Physiological and morphological effects of overproduction of membrane-bound ATP synthase in *Escherichia coli* K-12. *EMBO J.* 3:1791-97

194. Wada, H., Gombos, Z., Murata, N. 1990. Enhancement of chilling tolerance of a cyanobacterium by genetic manipulation of fatty acid desaturation. *Nature* 347:200-3

195. Wada, H., Murata, N. 1989. Synechocystis PCC6803 mutants defective in desaturation of fatty acids. *Plant Cell Physiol.* 30:971-78

196. Wang, X., Moore, T. S. 1990. Phosphatidylcholine biosynthesis in castor bean endosperm: purification and properties of CTP:choline-phosphate cytidyltransferase from castor bean endosperm. *Plant Physiol.* 93:250-55

197. Wang, X., Moore, T. S. 1989. Partial purification and characterization of CTP:choline-phosphate cytidyltransferase from castor bean endosperm. *Arch. Biochem. Biophys.* 274:338-47

198. Wanner, G., Tormanek, H., Theimer, R. R. 1981. The ontogeny of lipid bodies (spherosomes) in plant cells. *Planta* 151:109-23

199. Weiner, J. H., Lemire, B. D., Elmes, M. L., Bradley, R. D., Scraba, D. G. 1984. Overproduction of fumarate reductase in *Escherichia coli* induces a novel intracellular lipid-protein organelle. *J. Bacteriol.* 158:590-96

200. Weiss, T. J. 1983. *Food Oils and Their Uses.* Westport, CT: AVI Publishing. 310 pp. 2nd ed.

201. Williams, J. P., Watson, G. R., Leung, S. 1976. Galactolipid synthesis in leaves. II. Formation and desaturation of long chain fatty acids in PC, PG and the galactolipids. *Plant Physiol.* 57:179-84

202. Yadev, N. S., Hitz, W. 1990. Cloning and bacterial expression of soybean cDNA for steraroyl-ACP desaturase. See Ref. 24. In press

203. Yatsu, L. Y., Jacks, T. J. 1972. Spherosome membranes. *Plant Physiol.* 49:937-43

Annu. Rev. Plant Physiol. Plant Mol. Biol. 1991. 42:507–28
Copyright © 1991 by Annual Reviews Inc. All rights reserved

CONTROL OF NODULIN GENES IN ROOT-NODULE DEVELOPMENT AND METABOLISM

Federico Sánchez, Jaime E. Padilla, Héctor Pérez, and Miguel Lara

Unidad de Biología Molecular y Biotecnología Vegetal, Universidad Nacional Autónoma de México, Cuernavaca, Morelos, México

KEY WORDS: Regulation of plant genes during legume-*Rhizobium* symbiosis

CONTENTS

507

INTRODUCTION

In the *Rhizobium*-legume interaction, an exchange of molecular signals regulates the expression of genes essential for infection, nodule development, and function (10, 79, 82, 84, 102, 108, 158; see Figure 1). Legume nodule formation has been divided into three major stages: "preinfection," "infection and nodule formation," and "nodule function" (Figure 1; 114, 148). Nodulins are plant proteins that accumulate specifically in nodules. Early nodulin genes are detected during the "infection and nodule formation" stage. The expression of late nodulin genes starts at about the onset of nitrogen fixation (96). Recently, "true" nodulins have been found to be expressed in parts of the plant distinct from the nodule (2, 12, 113). Nevertheless, the concept of nodulins still provides a framework, for following the *Rhizobium*-legume interaction. Indeed, the study of nodulins has helped to define the various stages of the nodulation process (114). Many nodulins have been isolated (30), and we have begun to understand their role in symbiosis. Several reviews on nodulins and nodulin genes have appeared in recent years (50, 53, 96, 142, 147). The present chapter is an overview of relevant data concerning (*a*) plant morphogenesis and nodulin gene expression elicited by *Rhizobium* symbiotic signals and (*b*) diverse mechanisms for nodulin regulation during nitrogen fixation.

MOLECULAR SIGNALING AND RECOGNITION

Plant Symbiotic Signaling

Plant-bacteria interactions begin with distant chemical signaling in the rhizosphere (35, 57, 84, 102). Substances interchanged at this stage modulate the early responses and set conditions for the initiation of symbiosis (102, 108, 109, 158). In legumes, pterocarpanes and isoflavonoids can function as phytoalexins, as signal molecules (for a review see 102, 108), and probably as endogenous regulators of polar auxin transport (63). In exudates and extracts from legume roots, flavones, isoflavones, and flavanones were identified as the inducing molecules for rhizobial chemotaxis and for the expression of *(Brady) Rhizobium* nodulation *(nod)* genes (36, 82, 102; Figure 1, *top*). That these substances are active at very low concentrations (10^{-7}–10^{-8} M) and stimulate bacterial *nod* gene expression within minutes of exposure suits them to the situation in the rhizosphere. The inducer substances are exuded specifically between the meristem and the zone of emerging root hairs, the area most responsive to nodulation in soybean (10, 17, 36).

Rhizobial Signals

Bacterial signals are necessary for eliciting early plant responses and normal nodule development. The initial stages of the interaction between rhizobia and their leguminous host plants include deformation and curling of plant root

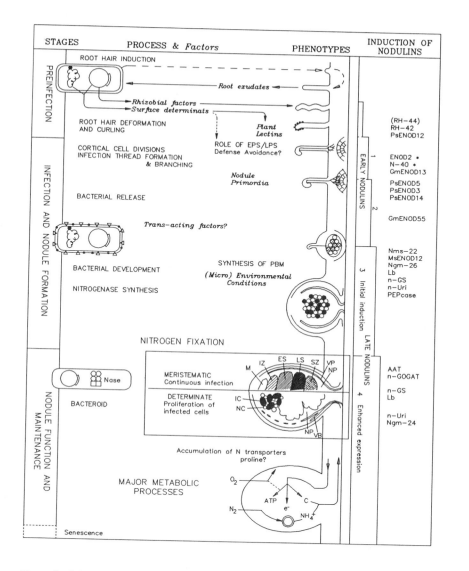

Figure 1 Schematic representation of developmental and metabolic events associated to nodulin expression. Distinct stages, processes, and factors *(left* and *center)* are shown in relation to nodule developmental phenotypes. The figure also depicts the sequence of expression of nodulins and nodule-enhanced activities *(right)* in an arbitrary time scale. At least four steps in nodulin induction are indicated (1–4), with representative examples of several species (* = ubiquitous nodulin type). Nodule types are represented in the middle (MERISTEMATIC: M = meristem; IZ = infection, ES = early symbiotic, LS = late symbiotic and SZ = senescing zones; VP = vascular bundle; NP = nodule parenchyma. DETERMINATE: IC = infected and NC = uninfected cell; VB = vascular bundle; NP = nodule parenchyma/inner cortex.) Major metabolic processes *(bottom)* are described as linked pathways of carbon, nitrogen, and oxygen metabolism, in which several nodulins participate.

hairs (10, 138), the initiation of cell division in the root cortex, and induction of a root-nodule meristem (38). Specific bacterial genes participate in these processes and fall into four general groups: *nodD*, common nodulation or *nodABC* genes, host range *(hsn)* genes, and genes involved in polysaccharide production. Recent reviews describe the structure, organization, and regulation of the symbiotic genes of *Rhizobium, Bradyrhizobium* and *Azorhizobium* (81, 86).

(*Brady*)*Rhizobium* produces heat-stable low-molecular-weight factors that cause three changes in the plant (*a*) root hair deformation and branching (Had), (*b*) thick and short roots (Tsr), and (*c*) root hair induction (Hai) (61, 148, 158). The *nodABC* genes are involved in producing small heat-stable compounds that cause root hair deformation that could be homologous to these factors. Histochemical data on *R. meliloti nod* genes-*uidA* (GUS activity) fusions assayed in alfalfa indicate that *nodABC* genes are expressed in different stages of nodule development (124). Furthermore, the NodA and NodB proteins produce compounds that stimulate mitosis in plant protoplasts (117). In addition, these genes transformed into tobacco produced abnormal growth patterns (116).

Signals that result in morphogenetic responses in the specific legume-host have been identified (61, 79). One of them, NodRm-1, was purified following a root hair deformation (Had) phenotype bioassay, positive in alfalfa (homologous host) and negative in vetch (heterologous host). The chemical nature of this factor has been determined. NodRm-1 is a sulfated beta-1,4-tetrasaccharide of D-glucosamine (M_r 1,102) in which three amino groups were acetylated and one was acylated with a C_{16} bis-unsaturated fatty acid. This factor is active at nanomolar concentrations (79). The fact that NodRm-1 is a tetrasaccharide is relevant because oligosaccharines act as plant morphogens (134); plant and microbial cell wall degradation fragments (oligosaccharides) are active in the nanomolar range as elicitors of phytoalexin production (35); and finally, saccharide-binding proteins are implicated in *Rhizobium* root attachment and host-specific recognition (27, 32, 57). A suggestion that plant elements may be modified by *Rhizobium* signals is considered in the next section.

Rhizobium Elicitors, Calcium, and Plant Cytoskeleton

Rhizobium produces soluble factors that may trigger both deformation and tip growing branches in root hairs of leguminous (61, 79, 117) and some nonleguminous plants (24). Also, the presence of spontaneous (136) as well as chemically induced nodules (60) suggests that the bacterium modulates fundamental capacities already present in plants. Recent reports suggest that *Rhizobium* could mediate redirection of apical growth in root hairs (70); a computer model predicts that hair curling would be produced by dominance of

Rhizobium-induced growth over normal tip growth (138). Pollen tubes, fungal hyphae, and root hairs grow by apical extension (119). In fungal hyphae, Ca^{2+}-calmodulin gradients play an important role in normal tip growth (33, 149), and modifications of calcium levels are related to abnormal tip-branching patterns in some fungi (33). Root hairs also show increased calcium and calmodulin concentrations (105) in the growing tip at the beginning of apical extension (58). Ca^{2+}-calmodulin levels are key regulators of growth and morphogenesis in animals and plants (85, 88). Alone or combined they modulate cytoplasmic streaming (71) and cytoskeleton organization (126). The cytoskeleton is an essential organelle of the plant cytoplasm and has a key role in cell division (80, 118), cytoplasmic streaming, and normal tip growth and elongation (68, 157). Root hairs infected by *Rhizobium* show increased or redistributed calcium levels, especially around the area of origin of the infection thread (123), and many dictyosomes and microtubules are observed near the forming thread (137). The inoculated root hairs also show many cytoplasmic strands and significantly increased cytoplasmic streaming (155). Thus cytoskeleton modifications might explain many of the morphological effects found in early stages of plant-*Rhizobium* interaction (like Hac and Had). The appealing possibility that Nod signals could trigger calcium-level modulations, which would in turn significantly alter plant cytoskeletal organization, deserves further evaluation.

INFECTION AND NODULE ORGANOGENESIS

Early nodulin genes are expressed during infection and nodule morphogenesis (50; Figure 1). Plant glycoproteins such as lectins (27, 70), although present in the preinfection stage, cannot be considered nodulins because they are not specifically expressed in the nodule.

Infection and Root Hair (RH) Proteins

The biochemical and genetic particularities of root hairs are studied because bacterial infections tend to start in these cells (10, 125). During the invasion process, through the action of localized changes in root-hair wall synthesis, a new cell wall develops around the penetration site, ahead of the invading bacteria (10, 70). Five to ten specific RH-proteins were identified as in vitro translation products (IVTP) by comparing whole-root and root hair RNA preparations (49, 107). By means of a fluorescent labeling method, a group of RH-specific proteins was detected in the outer surface of soybean root hair cells (107). In pea, two RH-proteins (RH-42 and RH-44) and a nodulin transcript were characterized. RH-42 is an acidic polypeptide detected 20 hr after inoculation with *Rhizobium* containing (at least) functional *nodABC* genes. By contrast levels of RH-44, already present in these cells, showed a marked increase. RH-44 seems to be related to deformation of root hairs elicited by soluble rhizobial factors, and RH-42 may be associated with root hair curling produced by bacterial inoculation (49), but the functions of these proteins have not been established. The other pea gene induced upon *Rhizobium* inoculation was identified by hybridizing root hair mRNA against the PsENOD12 cDNA clone. This gene is also expressed in young nodules of pea, but only if they contain infection threads (113; see Figure 1 and the section on early nodulin genes induced in nodule infection).

Nodulins Involved in Nodule Organogenesis

From data derived from empty nodules (lacking rhizobia) it is inferred that some early nodulin genes are active only in association with nodule organogenesis. Two-dimensional polyacrylamide gel analysis (2D-PAGE) of IVTP from nodule mRNA of pea (53), soybean (51), vetch (93), and alfalfa (34, 98) showed two to five early nodulins in each case. A correlation has been found between nodule meristem induction and the presence of the early soybean nodulins Ngm-38, Ngm-41, and probably Ngm-44 (51). The first early nodulin gene to be characterized was isolated from soybean. This gene, called *ENOD2* encodes for a protein with a putative N-terminal signal peptide followed by a sequence containing repeating proline-rich pentapeptide motifs (46); it is expressed in the inner cortex (nodule parenchyma) of determined and meristematic nodules (140, 141). Homologous *ENOD2* genes or transcripts have been reported in such other legume species as pea (141), alfalfa (34), vetch (93), *Sesbania* (128), common bean (111), and lupine (77). This finding suggests that this protein has a conserved role in legume nodules. In soybean, alfalfa, and common bean this early nodulin is expressed in empty nodules that contain neither infection threads nor intracellular bacteria (34, 46, 101). Such nodules are elicited on legume roots by different *Rhizobium* and *Bradyrhizobium* mutants affected in surface polysaccharides (81), and by *Agrobacterium* strains carrying *nod* genes (52). This gene is also expressed in spontaneous nodules (136) and in "pseudonodules" induced by inhibitors of polar auxin transport such as 2,3,5-triiodobenzoic acid (TIBA) or N-(1-naphthyl)-phthalamic acid (NPA) (60). Expression of the *ENOD2* gene in these empty nodules indicates a role in nodule morphogenesis rather than a function in the infection process. Evidence that the nodule parenchyma of soybean nodules is a physical barrier limiting oxygen diffusion to central nodule tissue (25) and the putative cell wall location of ENOD2 led to the proposal that this protein might help to create particular properties of the cortical cell layer (141). Other nodulins associated with nodule organogenesis are soybean *GmENOD13* (47) and a group of related products: Nps-40' from pea (52, 53), Nvs-40 from vetch (53, 93), and Nms-30 from alfalfa (34, 39).

Early Nodulin Genes Induced in Nodule Infection

Early nodulins have been characterized in soybean and pea (47, 113, 114). Pea nodules (meristematic type; see Figure 1) contain cells at different stages of development: the youngest cells adjacent to the apical meristem and the oldest cells at the basal root attachment point. This organization makes it possible to compare the expression patterns of nodulin genes in different cell types during nodule development. By using in situ hybridization, Scheres et al (114) localized *PsENOD12, PsENOD5, PsENOD3,* and *PsENOD14* transcripts in sections of nodules from plants of different ages. Leghemoglobin (Lb) transcript detection was used as a marker for the expression of late nodulin genes, and the *nifH* gene as a marker for expression of bacterial nitrogenase genes. *PsENOD12* mRNA is present adjacent to the meristem in the invasion zone. *PsENOD5* transcript reaches a maximum in the early symbiotic zone, where the first plant cells that contain bacteria are present. *PsENOD3* and *PsENOD14* messengers reach their maximum concentration in the youngest layers of the late symbiotic zone. In older cells of this zone when *PsENOD3* and *PsENOD14* mRNA are decreasing, Lb transcript is at its maximum and the nitrogenase mRNA can first be detected in the bacteroid. In conclusion *PsENOD5, PsENOD3,* and *PsENOD14* mRNAs are only present in the cells containing rhizobia, while *PsENOD12* mRNA is present in all the cells of the invasion zone. Particularly *PsENOD3* and *PsENOD14* mRNA concentrations mark the stage at which the infected cell is fully differentiated into a functional nodule cell.

Two *PsENOD12* transcripts, encoded by different genes (14), have been detected both during nodule infection and in stems and flowers (113). During the infection process they are observed in the root hairs, the nodule invasion zone, and in the root cortex. In addition, analysis of the deduced protein sequence suggests that this nodulin is a cell wall component. These observations have led to the proposal that *PsENOD12* may have a role both in the "preparation" of the cells for infection thread growth and in thread formation itself.

Although the amino acid sequence of the *PsENOD12* protein resembles those of other hydroxyproline-rich glycoproteins that accumulate in plant tissue after wounding or with pathogenic interactions, *PsENOD12* mRNA is not accumulated in fungus-inoculated plants (113). Both common and host-specific *nod* genes of *Rhizobium* are essential for eliciting *PsENOD12* mRNA expression in root hairs of peas, either inoculated or treated with a supernatant of a flavonoid-induced *R. leguminosarium* bv. *viciae* culture (113). Furthermore, addition of the purified NodRm-2-factor (79) to roots produces root hair deformation, cortical cell division, and *PsENOD12* and *PsENOD5* message accumulation (14). An exciting open question is whether the bacterial factors produce these effects directly or are mediated by a second signal(s) elicited in the plant.

EXPRESSION AND REGULATION OF LATE NODULINS

Late nodulins comprise the best-known nodule-specific proteins, which are induced shortly before nitrogen fixation (96, 147). They include enzymes involved in nitrogen assimilation, in carbon metabolism, and in amide and ureide biogenesis (40, 120, 144). Proteins present in the peribacteroid membrane (PBM) (45, 147) and the leghemoglobins (Lbs) (discussed below) (3, 154) also belong to this group. Several enzymes involved in these metabolic processes have been identified as nodulins; others are common to several plants organs but show enhanced activity in nodules.

Enzymes of Ammonia Assimilation

Levels of plant glutamine synthetase (GS) and glutamate synthase (GOGAT) activity increase along with nitrogenase activity during nodule formation. Both enzymes are responsible for assimilating the ammonia derived from the N_2 reduction in the bacteroid (6, 92).

GLUTAMINE SYNTHETASE   In legume plants, multiple isoenzymic GS forms result from the presence of different polypeptides. This GS subunit heterogeneity is associated with the expression of a multigene family (39, 42, 59, 133). Different GS polypeptides have been characterized (56, 74, 121, 133). In root nodules two types can be distinguished: those that are expressed as nodule-specific proteins (GS-n) and those whose expression is significantly increased in symbiosis (42). In both cases, GS genes are specifically expressed or enhanced before nitrogenase is active (39, 59, 100, 121). Results derived from *Lotus corniculatus* plants transformed with a common bean GS-n gene *(gln-γ)* fused to a *uidA* reporter gene showed that the β-glucuronidase activity is detected after nodule emergence and only in infected cells (43). Previously, it had been demonstrated that nodule-specific GS genes are expressed in Fix$^-$ nodules containing intracellular bacteria (100, 121). However, in nodules where cells were not infected or were unable to maintain peribacteroid membrane (PBM), GS expression was not detected (39, 122). In addition, experiments with nodules from plants growing in an atmosphere where N_2 ($N_2:O_2$) was substituted by argon (Ar:O_2) showed lower activity and an early decline in the expression of GS (22). In conclusion, the internalization of *Rhizobium* might generate developmental and/or metabolic conditions for the specific or enhanced expression of plant GS.

On the other hand, further increase in GS levels coincides with the onset of nitrogenase activity. In soybean, ammonia provided externally to roots induces the expression of GS (59). Furthermore, a soybean GS promoter fused to *uidA* was also activated by ammonia in transformed *L. corniculatus* roots (G. H. Miao et al, submitted for publication). In contrast in common bean, ammonia externally added failed to induce GS genes in roots and nodules (23). Recent studies showed that in nodulated root systems of common bean grown in an $AR:O_2$ atmosphere, the activity of the nodule GS-γ isoenzyme was reduced (85%) along with that of nitrogenase, although the GS-β isoenzyme was unaffected (22). In common bean plants, grown in an enriched CO_2 atmosphere (1000 ppm), nitrogenase activity increased two- to three-fold; a significant reduction of the GS-γ, but not of the GS-β polypeptide, was also observed (99). Taking these results together, it seems that both high and low nitrogenase activities can be correlated with a reduction of GS-γ. These data suggest that the carbon/nitrogen balance within the nodule, and not ammonia availability per se, may be the main modulating factor of GS expression.

GLUTAMATE SYNTHASE (GOGAT)   In legume nodules, as in other plant tissues, two types of GOGAT have been purified and characterized. One of these enzymes is ferredoxin dependent (Fd-GOGAT), whereas the other utilizes NADH as reductant (1, 21, 130). Root nodules of common bean express two *NADH-GOGAT* isoenzymes, in contrast with alfalfa nodules where only one NADH-GOGAT is found. The alfalfa enzyme and the NADH-GOGAT II of common bean are not detected in roots or leaves from either legume (1, 21). These two enzymes are initially expressed independently of nitrogenase induction. However, in nonfixing nodules of alfalfa the activity of NADH-GOGAT is less than 15% of the activity in effective nodules. This result shows that symbiotically fixed nitrogen is essential for the maximal induction of nodule GOGAT activity (21, 40).

Enzymes of Carbon Metabolism

Carbon flux to the nodule is necessary to sustain nodule and bacterial respiration, to enable synthesis of ATP and reductants for nitrogenase activity, and to provide a carbon skeleton source for nitrogen assimilation and transport (6, 144). Two enzymes involved in these processes are discussed here.

PHOSPHOENOLPYRUVATE (PEP) CARBOXYLASE   Nodule PEP carboxylase has been purified from soybean, lupine, and alfalfa (103, 145). Analysis of PEP carboxylase isozymes by native-PAGE identified one activity band in roots and two bands in nodules of alfalfa, pea, and soybean. In common bean, two bands were observed in roots and three in nodules. Thus the additional PEP carboxylase bands observed could be regarded as nodule-specific components (31). Alfalfa ineffective nodules express only 4% of the nodule PEP carboxylase activity found in effective nodules (40).

SUCROSE SYNTHASE (SS)   SS has been purified and characterized from soybean nodules (94). A nodule-specific cDNA clone encoding the SS subunit has been isolated. Partial DNA sequence analysis showed 73% homology at the amino acid level with one of the maize sucrose synthase genes *(Sh)* (132). Incubation of purified SS from soybean nodules with free heme promotes an inactivation of the enzyme through dissociation into monomers, suggesting that the availability of free heme may regulate the activity of SS and the flow of carbon to nodule and bacteroid metabolism (132).

Synthesis of Nitrogen Transporters in Nodules

Nodulated legumes transport preferentially either amides or ureides to the shoot (6). Various plant enzymes that participate in the synthetic pathways of these compounds have been described (120). Only two examples are considered here.

ASPARTATE AMINOTRANSFERASE (AAT)   In legumes, AAT has been characterized from nodules of various plants (55, 106). Multiple nodule AAT isoenzymes located in different cell compartments have been described. One of these isoenzymes is regarded as nodule specific in alfalfa (55) and nodule enhanced in lupine (106). A considerable increase in AAT activity and polypeptide concentration occurred early (7–10 days) in both effective and ineffective alfalfa nodules (40). However, this activity declines earlier in Fix$^-$ nodules (55).

URICASE   A nodule-specific uricase (uricase II) has been identified and characterized from soybean and common bean nodules (97, 110). Immunogold-labeling analysis in soybean nodules showed that the uricase II enzyme is localized in peroxisomes of uninfected cells of the central tissue (97). Initial expression of uricase II transcript in soybean and common bean nodules is not dependent on active nitrogen fixation (75, 110). Incubation of soybean callus tissue and root cultures at varying oxygen levels demonstrated that low oxygen concentration (4–5%) induces uricase II activity and synthesis (75). Recently, it was found that uricase-II mRNA is expressed (about 10%) in empty nodules and accumulated in delayed nodulation in common bean (101). This result, together with data on Ngm-26 of soybean (95; see the section on PBM nodulins), suggests that there are distinct types of late nodulin genes, regulated by particular developmental and environmental conditions.

Peribacteroid Membrane (PBM) Nodulins

PBMs enclose the endosymbiont in the host cells of legume nodules. After endocytosis, the PBMs increase concomitantly with bacterial multiplication and differentiation, creating the interface between the two organisms and playing a major role in the control of metabolite exchange (26). In addition, it has been suggested that the PBM has a decisive role in preventing the phytoalexin production response by the host cell (139, 153). PBMs are derived originally from the plasmalemma, but their chemical composition suggests that the ER and the dictyosome also contribute to PBM biogenesis (16, 91, 146).

Nodulins associated with the PBM have been characterized in soybean (45). At least three nodulins (Ngm-24, Ngm-26, and Ngm-23) have been convincingly located in association with this membrane. Ngm-24 appears to be located in the inner surface of the PBM facing the peribacteroidal space; this orientation makes unlikely its function as a transmembrane transport protein (69). Analysis of the structure of Ngm-26 indicates that it is an integral protein facing both the cytoplasmic and peribacteroidal compartments, probably functioning in the exchange of metabolites between the plant and *Rhizobium* (44). Homology of Ngm-26 with other intrinsic membrane proteins (e.g. glycerol facilitator from *E. coli* and tonoplast intrinsic protein (TIP) from protein bodies of seeds) supports such a role in exchange (8, 66). This nodulin was recently shown to be phosphorylated by a Ca^{2+}-dependent protein kinase (150). We note with interest that Ca^{2+}-dependent protein kinase activities have been found in PBMs (73, 153). A third nodulin, Ngm-23, has also been assigned to the PBM. The function of this protein is not known, though its genetic regulation has interesting features, mentioned below (87; see the section on regulatory elements for nodulin expression).

PBM nodulin genes are differentially expressed, as evidenced by *Bradyrhizobium* mutants that uncouple peribacteroid membrane formation and PBM nodulin gene expression (90). In particular, *B. japonicum* mutant T8-1, defective in the endocytosis process, induces nodules almost completely devoid of intracellular bacteria where the PBM is not formed. In these nodules, the gene encoding Ngm-24 was expressed at very low levels; the gene encoding Ngm-26, on the other hand, was expressed at wild-type levels (44, 95).

Using monoclonal antibodies, Brewin et al (16, 146) have analyzed the molecular composition of bacteroid and PBM of pea. They have identified glycoproteins and glycolipids present in the infection-thread matrix, on the lumenal surface of the PBM, in the ER, and in dictyosomes.

Enhanced Nodule Proteins

Nodule-stimulated proteins were also reported in a number of legume species. Not considered "true" nodulins, they include polypeptides of a wide range of M_r whose expression is prominently or transiently augmented in nodules of pea (53), soybean (51, 72, 78), common bean (18), alfalfa (39, 83), cowpea

(135), and *Sesbania* (29). Although these proteins have not been characterized, the possibility that they play a role in the nodulation process deserves further attention.

NODULIN HOMOLOGIES AND GENE FAMILIES

Structural Nodulins

That many early nodulin genes encode cell wall structural proteins is inferred from the comparison of their derived amino acid sequence with the primary structure of several cell wall (hydroxy)proline-rich glycoproteins (HGRPs) (7, 20, 47, 141). In soybean and pea, *ENOD2* amino acid sequences contain alternated repeats of Pro-Pro-His-Glu-Lys and Pro-Pro-Glu-Tyr-Gln (46, 47, 141). This structure suggests that the amino acid composition of these pentapeptides could be the main requirement for the function of the *ENOD2* protein (N-75). Proteins of this type differ from the extensins, which are formed by (Hyp)$_4$-Ser pentapeptide repeats (20). The early nodulin *ENOD2* is a developmental gene marker, which permits us to distinguish nodule-like structures from tumors or secondary roots (60, 136). Other early nodulin gene products that may be involved in nodule morphogenesis are *GmENOD13* from soybean (47) and Nms-30 (34) from alfalfa. *GmENOD13* is homologous to *ENOD2* at the C-terminal part. Other early nodulins described in pea are *PsENOD5, PsENOD3*, and *PsENOD14* (114). The amino acid composition of *PsENOD5* resembles that of the arabinogalactan proteins (AGPs) (114). The central tissue and the cortex of soybean nodules contain high levels of AGP (19). *PsENOD3* and *PsENOD14* are 60% homologous and encode for ≈6-kD products. The arrangement of the four cysteines found in these nodulins suggests that they can bind a metal ion (114). This feature is common to other gene families (see below), though there is no significant homology between them (114). The presence of a putative signal peptide at the N-terminus of these proteins is in agreement with the idea that all have an extracellular location.

Nodulin Gene Families in Determinate Nodules

In soybean a well-characterized late nodulin gene family has been reported whose function is not understood (62, 112, 122). It consists of six or seven members with extensive stretches of sequence similarity within the coding and flanking regions. Examples of these genes include *GmN-20, GmN-22*, (112), *GmN-23* (87), *GmN-26b, GmN-27* (62), and *GmN-44* (or *E27*) (122). This nodulin gene family shows some interesting features: (*a*) Along with the Lbs, these are the most abundantly transcribed genes in soybean nodules; (*b*) the gene products have two domains that are arranged in paired Cys-X$_7$-Cys motifs, resembling zinc-finger sequences (13, 112); (c) all products also exhibit a conserved region that encodes for a putative signal peptide. In spite of their similarities, these proteins are associated with different cellular compartments.

In common bean, abundant nodule-specific transcripts that encode for a group of 30-kD products (Npv-30) have been reported (18). Sequence analysis of two cDNAs and a genomic clone revealed high homology with the former nodulin gene family from soybean (F. Campos, in preparation). The Npv-30 protein also contains two motifs resembling zinc fingers and a putative signal peptide contained in the N-terminal domain. The above data indicate that these nodulin families might have a common function, most likely as metal ion carriers in infected cells (112). In soybean, this ion transport would flow from the cytoplasm (Ngm-27 contribution) to the PBM (Ngm-23) and into the peribacteroidal space (Ngm-44) (112).

Immunological approaches to nodulin classification have identified closer relationships between legumes with similar nodule morphology and physiology (determinate vs meristematic type; see Figure 1) than between species related by conventional taxonomical criteria (127).

Function and Evolution of Plant Hemoglobins

Leghemoglobins (Lbs), the most abundant nodulins, function as oxygen carriers, facilitating O_2 diffusion to bacteroids (3, 154). Lbs are products of a multigene family whose expression is strongly activated prior to nitrogen fixation. In soybean, four Lbs are induced at slightly different points during nodule formation, suggesting a developmental control of these genes (65). Different Lb products seem to be associated with distinct oxygen affinities that may be required for nodule function (3, 28). Nonlegumes also contain hemoglobin-like proteins in their symbiotic structures (4), and hemoglobin could function as an oxygen-sensing mechanism in growing root tips (2). Finally, the widespread existence of hemoglobin genes in organisms of all kingdoms has led to the proposal of a common evolutionary origin of these genes (5).

REGULATORY ELEMENTS FOR NODULIN EXPRESSION

Regulated plant genes are activated by either developmental or environmental stimuli (11). Nodulin genes appear to be induced directly or indirectly by plant and bacterial signals in a cell- or organ-specific manner. Study of the regulation of nodulin gene expression requires a close examination of the promoter and 5' upstream flanking sequences. Such study has been enabled by use of *Lotus corniculatus* transformed with *Agrobacterium rhizogenes*

(104). Recently *Medicago truncatula* has been developed as another species for studying the molecular genetics of symbiosis (9). Experimental strategies have been reviewed by de Bruijn et al (28). Studies of transgenic plants containing chimeric genes have included soybeans Lbs (64, 65) and *GmN-23* (67); *Sesbania rostrata* Lbs (131) and *SrENOD2* (28); common bean *gln-γ* and *gln-β* (43) and P*arasponia* hemoglobin (15). Sequences present in the 5' upstream region of some late nodulins have been characterized. Some of these regions share consensus motifs and various functional elements conferring enhanced organ-specificity or reduced activation of gene expression (28, 65, 67, 131). Furthermore, in common bean-*gln-γ* (43a) the soybean *GmN-23* and *lbc3* 5' regions, A-T rich DNA elements with the ability to bind *trans*-acting factors have been found. These factors are present in nodule extracts but also in roots (43a, 64) and leaves (64). Biochemical analysis and DNA binding studies of some of these factors indicate functional relationships with regulatory nuclear proteins (human HMG I) (64), suggesting that chromatin structure may be an important controlling mechanism for organ-specific expression. It has also been reported that one of these nodulin genes *(ENOD2 from S. rostrata)* can be regulated by hormones and environmental conditions (cytokinins, anaerobiosis) (28), and it has been demonstrated that *lb* regulatory sequences are functional between legumes (67) and nonlegumes (15). Recently, a DNA-binding factor of bacterial origin that could regulate nodulin gene expression has been reported (152).

PLANT GENETICS AND NODULIN GENES

Legume Mutants in the Nodulation Process

Classical genetic studies have described a number of loci in legumes that control *Rhizobium* root hair infection, nodule formation, and nitrogen fixation (reviewed in 76, 109, 143). At least 45 mutations in 8 different species conditioning 5 major phenotypes have been identified (143): (*a*) non-nodulation, (*b*) ineffective (producing tumor-like nodules), (*c*) ineffective (producing early senescence), (*d*) supernodulation/nitrate tolerant nodulation (Nts); additionally, (*e*) nodulation in the absence of *Rhizobium* (Nar) has been reported (136). In alfalfa, non-nodulating and ineffective plant mutants have been characterized (40, 143). In peas, at least 17 genes affecting nodule formation or encoding nodule-specific proteins have been located on the chromosome linkage map (151). Almost 50% of the nodulation-impaired mutants mapped near the *lb* cluster on chromosome 1 (151). Of particular interest is *sym-13,* which produces a nonfixing nodule and maps near the loci encoding nodule-enhanced GS (151). The *sym-5* locus is a mutational hotspot, represented by seven independently derived mutant lines with decreased nodulation. The recessive *sym-5* mutants have a temperature-sensitive phenotype. An altered polypeptide in *sym-5* mutants may be related to this phenotype (43).

Recently, the presence of symbiosis-specific polypeptides (mycorrhizins) has been reported in the vesicular-arbuscular mycorrhiza-soybean interaction (156). Remarkably, nodulation-impared pea mutants are also defective in mycorrhizal infection (Myc⁻) (37, 151). The fact that the same loci control simultaneously both symbioses in this legume suggests common blockages at early stages of the interactions (48). In soybean, at least eight mutations affecting nodulation and nitrogen fixation have been reported (54, 143). Several supernodulating and nitrate-tolerant *(nts)* soybean mutants have been isolated (54). Non-nodulating mutants (single recessive mutations) have also been reported. These mutations condition nodulation with all *B. japonicum* strains and are root-controlled traits. It has been argued that pseudoinfections may precede and potentiate actual infection sites. Normal nitrogen-fixing nodules were obtained occasionally at high bacterial titers, suggesting that this mutant may have a higher response threshold to a bacterial signal (109). The possibility was explored that flavonoids exuded by the mutant plants were responsible for the non-nodulating phenotype, but no significant differences in *nod* gene inducibility between wild-type and mutant plants were found (129).

One clue to identifying some of the factors producing non-nodulating plants may be found in a recent report on root hair development in *Arabidopsis thaliana*. This work identifies single recessive mutations affecting proper initiation and normal elongation of root hairs; it suggests that some of these genes may be required for the synthesis both of structural elements such as the root hair cytoskeleton and of some calcium regulators. Furthermore, the abnormal root hair phenotypes observed with these mutants are similar to the root hair deformations induced by *Rhizobium* in compatible hosts (115).

CONCLUDING REMARKS

The information accumulated so far indicates that with a few initial signals *Rhizobium* can induce morphological and metabolic events in both leguminous and nonleguminous plants. Indeed the field is moving rapidly, and the chemical nature of some *Rhizobium* signals has been determined. These molecules are involved in plant organogenesis and early nodulin gene expression. The study of the symbiotic process is also uncovering essential and novel aspects of plant development and metabolism. The identification and analysis of additional plant mutations affecting the symbiotic process could provide new and unexpected insights into the plant genes involved in this process. Some of these mutant genes may also participate in other plant-microbe interactions, as exemplified in the mycorrhiza-legume association, where Nod⁻ plant mutants are infected by neither *Rhizobium* nor mycorrhiza. Finally, studies of the metabolic aspects of nodulation have been limited to the description of the main biochemical events involved, but further progress requires that we understand the molecular factors that activate and regulate nodulation. Particularly, an integrative view of the carbon and nitrogen flows in symbiosis is needed.

ACKNOWLEDGMENTS

We thank our colleagues for providing us with reprints and preprints during the preparation of this manuscript. The research work of this group was partially supported by grant NAS (BNF-MX-87-77), H.P. was supported by Centro de Investigacion sobre Enfermedades Infecciosas del Instituto Nacional de Salud Publica. We thank particularly Carmen, Dalia, Lourdes, and Micheline for their help and patience.

Literature Cited

1. Anderson, M. P., Vance, C. P., Heichel, G. H., Miller, S. S. 1989. Purification and characterization of NADH-glutamate synthase from alfalfa root nodules. *Plant Physiol.* 90:351–58
2. Appleby, C. A., Bogusz, D., Dennis, E. S., Peacock, W. J. 1988. A role for haemoglobin in all plant roots? *Plant Cell Environ.* 11:359–67
3. Appleby, C. A. 1984. Leghemoglobin and *Rhizobium* respiration. *Annu. Rev. Plant Physiol.* 35:443–78
4. Appleby, C. A., Bogusz, D., Dennis, E. S., Fleming, A. I., Landsmann, J. 1988. The vertical evolution of plant hemoglobin genes. In *Nitrogen Fixation: Hundred Years After*, ed. H. Bothe, F. J. de Bruijn, W. E. Newton, pp. 623–28. Stuttgart: Gustav Fischer
5. Appleby, C. A., Dennis, E. S., Peacock, W. J. 1990. A primaeval origin for plant and animal haemoglobins? *Aust. J. Syst. Bot.* 3:81–90
6. Atkins, C. A. 1987. Metabolism and translocation of fixed nitrogen in the nodulated legume. *Plant Soil* 100:157–69
7. Averyhart-Fullard, V., Datta, K., Marcus, A. 1988. A hydroxyproline-rich protein in the cell wall. *Proc. Natl. Acad. Sci. USA* 85:1083–85
8. Baker, M. E., Saier, M. H. 1990. A common ancestor for bovine lens fiber major intrinsic protein, soybean nodulin-26 protein and *E. coli* glycerol facilitator. *Cell* 60:185–86
9. Barker, D. G., Bianchi, S., Blondon, F., Datte, Y., Duc, G., et al. 1990. *Medicago truncatula,* a model plant for studying the molecular genetics of the *Rhizobium*-legume symbiosis. *Plant Mol. Biol. Rep.* 8:40–49
10. Bauer, W. D. 1981. Infection of legumes by rhizobia. *Annu. Rev. Plant Physiol.* 32:407–49
11. Benfey, P. N., Chua, N. H. 1989. Regulated genes in transgenic plants. *Science* 244:174–81
12. Bennett, M. J., Lightfoot, D. A., Cullimore, J. V. 1989. cDNA sequence and differential expression of the gene encoding the glutamine synthetase polypeptide of *Phaseolus vulgaris* L. *Plant Mol. Biol.* 12:553–65
13. Berg, J. M. 1990. Zinc finger domains: hypothesis and current knowledge. *Annu. Rev. Biophys. Biophys. Chem.* 19:405–21
14. Bisseling, T., Franssen, H., Govers, F., Horvath, B., Moerman, M., et al. 1990. Early nodulin gene expression during the pea-*Rhizobium* symbiosis. In *Advances in Molecular Genetics of Plant-Microbe Interactions*, ed. H. Hennecke, D. P. S. Verma, 1:300–3. Dordrecht: Kluwer Academic
15. Bogusz, D., Llewellyn, D. J., Craig, S., Dennis, E. S., Appleby, C. A., et al. 1990. Non-legume hemoglobin genes retain organ-specific expression in heterologous transgenic plants. *Plant Cell* 2:633–41
16. Brewin, N. J., Kannenberg, E. L., Perotto, S., Wood, E. A. 1990. Surface interactions of plant and bacterial mem-

branes in the *Rhizobium*-legume symbiosis. See Ref. 14, pp. 325–30

17. Calvert, H. E., Pence, M. K., Pierce, M., Malik, N. S. A., Bauer, W. D. 1984. Anatomical analysis of the development and distribution of *Rhizobium* infections in soybean roots. *Can. J. Bot.* 62:2375–84

18. Campos, F., Padilla, J., Vázquez, M., Ortega, J. L., Enriquez, C., Sánchez, F. 1987. Expression of nodule-specific genes in *Phaseolus vulgaris* L. *Plant Mol. Biol.* 9:521–32

19. Cassab, G. I. 1986. Arabinogalactan proteins during the development of soybean root nodules. *Planta* 168:441–46

20. Cassab, G. I., Varner, J. 1988. Cell wall proteins. *Annu. Rev. Plant Physiol. Plant Mol. Biol.* 39:351–53

21. Chen, F. L., Cullimore, J. V. 1988. Two isoenzymes of NADH-dependent glutamate synthase in root nodules of *Phaseolus vulgaris* L.: purification, properties and activity changes during nodule development. *Plant Physiol.* 88:1411–17

22. Chen, F., Bennett, M. J., Cullimore, J. V. 1990. Effect of the nitrogen supply on the activities of isoenzymes of NADH-dependent glutamate synthase and glutamine synthase in root nodules of *Phaseolus vulgaris* L. *J. Exp. Bot.* 41:1215–22

23. Cock, J. M., Mould, R. M., Bennett, M. J., Cullimore, J. V. 1990. Expression of glutamine synthetase genes in roots and nodules of *Phaseolus vulgaris* following changes in the ammonium supply and infection with various *Rhizobium* mutants. *Plant Mol. Biol.* 14:549–60

24. Cocking, E. C. 1990. Nodulation of non-legumes by *Rhizobium*. See Ref. 61, pp. 813–23

25. Dakora, F. D., Atkins, C. A. 1989. Diffusion of oxygen in relation to structure and function of legume root nodules. *Aust. J. Plant Physiol.* 16:131–40

26. Day, D. A., Price, D., Udvardi, M. K. 1989. Membrane interface of the *Bradyrhizobium japonicum-Glycine max* symbiosis: peribacteroid units from soybean nodules. *Aust. J. Plant Physiol.* 16:69–84

27. Dazzo, F. B., Truchet, G. L. 1983. Interaction of lectins and their saccharide receptors in the *Rhizobium*-legume symbiosis. *J. Membr. Biol.* 73:1–16

28. de Bruijn, F. J., Szabados, L., Schell, J. 1990. Chimeric genes and transgenic plants to study the regulation of genes involved in symbiotic plant-microbe interactions (nodulin genes). *Dev. Genet.* 11:182–96

29. de Lajudie, P., Huguet, T. 1988. Plant gene expression during effective and ineffective nodule development of the tropical stem-nodulated legume *Sesbania rostrata. Plant Mol. Biol.* 10:537–48

30. Delauney, A. J., Verma, D. P. S. 1988. Cloned nodulin genes for symbiotic nitrogen fixation. *Plant Mol. Biol. Rep.* 6:279–85

31. Deroche, M.-E., Carrayol, E. 1989. Some properties of legume nodule phosphoenolpyruvate carboxylase. *Plant Physiol. Biochem.* 27:379–86

32. Díaz, C. L., Melchers, L. S., Hooykaas, P. J. J., Lugtenberg, B. J. J., Kijne, J. W. 1989. Root lectin as determinant of host-plant specificity in the *Rhizobium*-legume symbiosis. *Nature* 338:579–81

33. Dicker, J. W., Turian, G. 1990. Calcium deficiencies and apical hyper-branching in wild-type and the "frost" and "spray" morphological mutants of *Neurospora crassa. J. Gen. Microbiol.* 136:1413–20

34. Dickstein, R., Bisseling, T., Reinhold, V. N., Ausubel, F. M. 1988. Expression of nodule-specific genes in alfalfa root nodules blocked at an early stage of nodule development. *Genes Dev.* 2:677–87

35. Dixon, R. A., Lamb, C. J. 1990. Molecular communication in interactions between plants and microbial pathogens. *Annu. Rev. Plant Physiol. Plant Mol. Biol.* 41:339–67

36. Djordjevic, M. A., Gabriel, D. W., Rolfe, B. G. 1987. *Rhizobium*—the refined parasite of legumes. *Annu. Rev. Phytopathol.* 25:145–68

37. Duc, G., Trouvelot, A., Gianinazzi-Pearson, V., Gianinazzi, S. 1989. First report on non-mycorrhizal plant mutants (Myc) obtained in pea (*Pisum sativum* L.) and fababean (*Vicia faba* L.). *Plant Sci.* 60:215–22

38. Dudley, M. E., Jacobs, T. W., Long, S. R. 1987. Microscopic studies of cell divisions induced in alfalfa roots by *Rhizobium meliloti. Planta* 171:289–301

39. Dunn, K., Dickstein, R., Feinbaum, R., Burnett, B. K., Peterman, T. K., et al. 1988. Developmental regulation of nodule-specific genes in alfalfa root nodules. *Mol. Plant-Microbe Interact.* 1:66–74

40. Egli, M. A., Griffith, S. M., Miller, S. S., Anderson, M. P., Vance, C. P. 1989. Nitrogen assimilating enzyme activities and enzyme protein during de-

velopment and senescence at effective and plant gene controlled ineffective alfalfa nodules. *Plant Physiol.* 91:898–904

41. Fearn, J. C., LaRue, T. A. 1990. An altered constitutive peptide in *sym* 5 mutants of *Pisum sativum* L. *Plant Mol. Biol.* 14:207–16

42. Forde, B. G., Cullimore, J. V. 1989. The molecular biology of glutamine synthetase in higher plants. In *Oxford Surveys of Plant Molecular and Cell Biology*, ed. B. J. Miflin, 6:247–96. Oxford: Clarendon

43. Forde, B. G., Day, H. M., Turton, J. F., Wen-jun, S., Cullimore, J. V., et al. 1989. Two glutamine synthetase genes from *Phaseolus vulgaris* L. display contrasting developmental and spatial patterns of expression in transgenic *Lotus corniculatus* plants. *Plant Cell.* 1:391–401

43a. Forde, B. G., Freeman, J., Oliver, J. E., Pineda, M. 1990. Nuclear factors interact with conserved A/T-rich elements upstream of a nodule-enhanced glutamine synthetase gene from french bean. *Plant Cell* 2:925–39

44. Fortin, M. G., Morrison, N. A., Verma, D. P. S. 1987. Nodulin-26, a peribacteroid membrane nodulin is expressed independently of the development of the peribacteroid compartment. *Nucleic Acids Res.* 15:813–24

45. Fortin, M. G., Zelechowska, M., Verma, D. P. S. 1985. Specific targeting of the membrane nodulins to the bacteroid enclosing compartment in soybean nodules. *EMBO J.* 4:3041–46

46. Franssen, H. J., Nap, J. P., Gloudemans, T., Stiekema, W., Van Dam, H., et al. 1987. Characterization of complementary DNA for nodulin-75 of soybean: a gene product involved in early stage of root nodule development. *Proc. Natl. Acad. Sci. USA* 84:4495–99

47. Franssen, H. J., Scheres, B., van der Wiel, C., Bisseling, T. 1988. Characterization of soybean (hydroxy)–proline-rich nodulins. In *Molecular Genetics of Plant-Microbe Interactions 1988*, ed. R. Palacios, D. P. S. Verma, pp. 321–26. St. Paul, MN: APS Press

48. Gianinazzi-Pearson, V., Gianinazzi, S., Guillemin, J. P., Trouvelot, A., Duc, G. 1990. Genetic and cellular analysis of resistence to VA mycorrhizal fungi in pea plants. See Ref. 14, pp. 336–42

49. Gloudemans, T., Bhuvaneswari, T. V., Moerman, M., Van Brussel, T., Van Kammen, A., et al. 1989. Involvement of *Rhizobium leguminosarum* nodulation genes in gene expression in

pea root hairs. *Plant Mol. Biol.* 12:157–67

50. Gloudemans, T., Bisseling, T. 1989. Plant gene expression in early stages *Rhizobium*-legume symbiosis. *Plant Sci.* 65:1–14

51. Gloudemans, T., De Vries, S. C., Bussink, H. J., Malik, N. S. A., Franssen, H. J., et al. 1987. Nodulin gene expression during soybean (*Glycine max*) nodule development. *Plant Mol. Biol.* 8:395–403

52. Govers, F., Gloudemans, T., Moerman, M., Van Kammen, A., Bisseling, T. 1986. *Rhizobium* nod genes are involved in inducing an early nodulin gene. *Nature* 323:5645–66

53. Govers, F., Nap, J. P., Van Kammern, A., Bisseling, T. 1987. Nodulins in the developing root nodule. *Plant Physiol. Biochem.* 25:309–22

54. Gresshoff, P. M., Mathews, A., Krotzky, A., Olsson, J. E., Carroll, B. J. 1988. Supernodulation and nonnodulation mutants of soybean. See Ref. 47, pp. 364–69

55. Griffith, S. M., Vance, C. P. 1989. Aspartate aminotransferase in alfalfa root nodules: purification and partial characterization. *Plant Physiol.* 90:1622–29

56. Groat, R. G., Schrader, L. E. 1982. Isolation and immunochemical characterization of plant glutamine synthetase in alfalfa (*Medicago sativa* L.) nodules. *Plant Physiol.* 70:1759–61

57. Halverston, L. J., Stacey, G. 1986. Signal exchange in plant-microbe interactions. *Microbiol. Rev.* 50:193–225

58. Hauber, I., Herth, W., Reiss, H.-D. 1984. Calmodulin in tip-growing plant cells, visualized by fluorescing calmodulin-binding phenothiazines. *Planta* 162:33–39

59. Hirel, B., Bouet, C., King, B., Layzell, D., Jacobs, F., et al. 1987. Glutamine synthetase genes are regulated by ammonia provided externally or by symbiotic nitrogen fixation. *EMBO J.* 6:1167–71

60. Hirsch, A. M., Bhuvaneswari, T. V., Torrey, J. G., Bisseling, T. 1989. Early nodulin genes are induced in alfalfa root outgrowths elicited by auxin transport inhibitors. *Proc. Natl. Acad. Sci. USA* 86:1244–48

61. Hollingsworth, R. I., Squartini, A., Philip-Hollingsworth, S., Dazzo, F. B. 1990. Had and noi signals from *Rhizobium trifolii*. In *Nitrogen Fixation: Achievements and Objectives*, ed. P. M. Gresshoff, L. E. Roth, G. Stacey, W. E. Newton, p. 260. New York: Chapman & Hall

62. Jacobs, F. A., Zhang, M., Fortin, M.

G., Verma, D. P. S. 1987. Several nodulins of soybean share structural domains but differ in their subcellular locations. *Nucleic Acids Res.* 15:1271–80

63. Jacobs, M., Rubery, P. H. 1988. Naturally occurring auxin transport regulators. *Science* 241:346–49

64. Jacobsen, K., Laursen, N. B., Ostergaard, J. E., Marcker, A., Poulsen, C., et al. 1990. HMGI-like proteins from leaf and nodule nuclei interact with different A-T motifs in soybean nodulin promoters. *Plant Cell* 2:85–94

65. Jensen, E. Ø., Stougaard, J., Jorgensen, J. E., Sandal, N., de Bruijn, F. J., et al. 1988. Regulation of nodule-specific plant genes. See Ref. 4, pp. 605–9

66. Johnson, K. D., Hofte, H., Chrispeels, M. J. 1990. An intrinsic tonoplast protein of protein storage vacuoles in seeds is structurally related to a bacterial solute transporter (GlpF). *Plant Cell.* 2:525–32

67. Jorgensen, J.-E., Stougaard, J., Marcker, A., Marcker, K. A. 1988. Root nodule specific gene regulation: analysis of the soybean nodulin N-23 gene promoter in heterologous symbiotic systems. *Nucleic Acids Res.* 16:39–50

68. Kachar, B., Reese, T. 1988. The mechanism of cytoplasmic streaming in characean algal cells: sliding of endoplasmic reticulum along actin filaments. *J. Cell Biol.* 106:1545–52

69. Katinakis, P., Verma, D. P. S. 1985. Nodulin-24 of soybean codes for a peptide of the peribacteroid membrane and was generated by tandem duplication of a sequence resembling an insertion element. *Proc. Natl. Acad. Sci. USA* 82:4157–61

70. Kijne, J. W., Diaz, C. L., de Pater, B. S., Smit, G., Bakhuizen, R., et al. 1990. Surface interactions between *Rhizobia* and legume root hairs. See Ref. 61. In press

71. Kohno, T., Shimmen, T. 1988. Accelerated sliding of pollen tube organelles along *Characeae* actin bundles regulated by Ca^{2+}. *J. Cell Biol.* 106:1539–43

72. Kouchi, H., Tsukamoto, M., Tajima, S. 1989. Differential expression of nodule-specific (nodulin) genes in the infected, uninfected and cortical cells of soybean *(Glycine max)* root nodules. *Plant Physiol.* 135:608–17

73. Krishnan, H. B., Pueppke, S. G. 1989. In vitro phosphorylation of soluble proteins from soybean, *Glycine max,* root nodules: inhibition of protein kinase activity by zinc. *Symbiosis* 7:127–38

74. Lara, M., Porta, H., Padilla, J., Folch, J., Sánchez, F. 1984. Heterogeneity of glutamine synthetase polypeptides in *Phaseolus vulgaris* L. *Plant Physiol.* 76:1019–23

75. Larsen, K., Jochimsen, B. U. 1986. Expression of nodule-specific uricase in soybean callus tissue is regulated by oxygen. *EMBO J.* 5:15–19

76. LaRue, T. A., Kneen, B. E., Gardside, E. 1985. Plant mutants in symbiotic nitrogen fixation. In *Analysis of Plant Genes Involved in the Legume-Rhizobium Symbiosis,* ed. R. Marcelin, pp. 39–48. Paris: OECD Publications

77. Legocki, A. B., Boron, L., Szczglowski, K. 1990. Yellow lupin genes coding for the nodule-specific form of glutamine synthetase and (hydroxy)proline-rich protein. Presented 8th Int. Congr. Nit. Fix., Knoxville, Abstr. G-24

78. Legocki, R. P., Verma, D. P. S. 1980. Identification of "nodule-specific" host proteins (nodulins) involved in the development of *Rhizobium*-legume symbiosis. *Cell* 20:153–63

79. Lerouge, P., Roche, P., Faucher, C., Maillet, F., Truchet, G., et al. 1990. Symbiotic host-specificity of *Rhizobium meliloti* is determined by a sulphated and acylated glucosamine oligosaccharide signal. *Nature* 344:781–84

80. Lloyd, C. W. 1989. The plant cytoskeleton. *Curr. Opin. Cell Biol.* 1:30–35

81. Long, S. R. 1989. *Rhizobium* genetics. *Annu. Rev. Genet.* 23:483–506

82. Long, S. R. 1989. *Rhizobium*-legume nodulation: life together in the underground. *Cell* 56:203–14

83. Lullien, V., Barker, D. G., de Lajudie, P., Huguet, T. 1987. Plant gene expression in effective and ineffective root nodules of alfalfa *(Medicago sativa)*. *Plant Mol. Biol.* 9:469–78

84. Lynn, D. G., Chang, M. 1990. Phenolic signals in cohabitation: implications for plant development. *Annu. Rev. Plant Physiol. Plant Mol. Biol.* 41:497–526

85. Marmé, D. 1989. The role of calcium and calmodulin in signal transduction. In *Second Messengers in Plant Growth and Development,* ed. W. F. Boss, D. J. Morré, pp. 57–80. New York: Alan R. Liss

86. Martínez, E., Romero, D., Palacios, R. 1990. The *Rhizobium* genome. *Crit. Rev. Plant. Sci.* 9:59–93

87. Mauro, V. P., Nguyen, T., Katinakis, P., Verma, D. P. S. 1985. Primary structure of the soybean nodulin-23 gene and potential regulatory elements in the 5'-flanking regions of nodulin and leghemoglobin genes. *Nucleic Acids Res.* 13:239–49

88. Means, A. R., Chafouleas, J. G. 1982. Regulation by and of calmodulin in

mammalian cells. *Cold Spring Harbor Symp.* 46:903–8

89. Mellor, R. B. 1989. Bacteroids in the *Rhizobium*-legume symbiosis inhabit a plant internal lytic compartment: implications for other microbial endosymbioses. *J. Exp. Bot.* 40:831–39

90. Mellor, R. B., Garbers, C., Werner, D. 1989. Peribacteroid membrane nodulin gene induction by *Bradyrhizobium japonicum* mutants. *Plant Mol. Biol.* 12:307–16

91. Mellor, R. B., Werner, D. 1987. Peribacteroid membrane biogenesis in mature legume root nodules. *Symbiosis* 3:75–100

92. Miflin, B. J., Lea, P. J. 1980. Ammonia assimilation. In *The Biochemistry of Plants. A Comprehensive Treatise,* ed. B. J. Miflin, 5:169–202. New York: Academic

93. Moerman, M., Nap, J. P., Govers, F., Schilperoort, R., Van Kammen, A., et al. 1987. *Rhizobium* nod genes are involved in the induction of two early nodulin genes in *Vicia sativa* root nodules. *Plant Mol. Biol.* 9;171–79

94. Morell, M., Copeland, L. 1985. Sucrose synthase of soybean nodules. *Plant Physiol.* 78:149–54

95. Morrison, N., Verma, D. P. S. 1987. A block in the endocytosis of *Rhizobium* allows cellular differentiation in nodules but affects the expression of some peribacteroidal membrane nodulins. *Plant Mol. Biol.* 9:185–96

96. Nap, J. P., Bisseling, T. 1990. Nodulin function and nodulin gene regulation in root nodule development. In *The Molecular Biology of Symbiotic Nitrogen Fixation,* ed. P. M. Gresshoff, pp. 181–229. Boca Raton, FL: CRC Press

97. Nguyen, T., Zelechowska, M. G., Foster, V., Bergmann, H., Verma, D. P. S. 1985. Primary structure of the soybean nodulin-35 gene encoding nodule specific uricase II localized in peroxisomes of uninfected cells of soybean. *Proc. Natl. Acad. Sci. USA* 82:5040–44

98. Norris, J. H., Macol, L. A., Hirsch, A. M. 1988. Nodulin gene expression in effective alfalfa nodules and in nodules arrested at three different stages of development. *Plant Physiol.* 88:321–28

99. Ortega, J. L., Blanco, L., Lara, M. 1990. Effect of high CO_2 concentration in nitrogen and carbon metabolism of the *Phaseolus vulgaris* root-nodules. Presented at 5th Int. Symp. Mol. Genet. Plant-Microbe Intract., Interlaken. Abstr. p. 154

100. Padilla, J. E., Campos, F., Conde, V., Lara, M., Sánchez, F. 1987. Nodule-specific glutamine synthetase is expressed before the onset of nitrogen fixation in *Phaseolus vulgaris* L. *Plant Mol. Biol.* 9:65–74

101. Padilla, J. E., Sánchez, F. 1990. Nodulin expression and organogenesis in mutant-induced nodules of common bean, (*Phaseolus vulgaris* L.). Presented at 5th Int. Symp. Mol. Genet. Plant-Microbe Intract., Interlaken. Abstr. p. 136

102. Peters, N. K., Verma, D. P. S. 1990. Phenolic compounds as regulators of gene expression in plant-microbe interactions. *Mol. Plant-Microbe. Interact.* 3:4–8

103. Peterson, J. B., Evans, H. J. 1979. Phosphoenolpyruvate carboxylase from soybean nodule cytosol. Evidence for isozymes and kinetics of the most active component. *Biochim. Biophys. Acta* 567:445–52

104. Petit, A., Stougaard, J., Kuhle, A., Marcker, K. A., Tempe, J. 1987. Transformation and regeneration of the legume *Lotus corniculatus:* a system for molecular studies of symbiotic nitrogen fixation. *Mol. Gen. Genet.* 207:245–50

105. Reiss, H. D., Herth, W. 1979. Calcium gradients in tip growing plant cells visualized by chlorotetracycline fluorescence. *Planta* 146:615–21

106. Reynolds, P. H. S., Boland, M. J., Farnden, K. J. F. 1981. Enzymes of nitrogen metabolism in legume nodules: partial purification and properties of the aspartate aminotransferases from lupin nodules. *Arch. Biochem. Biophys.* 209: 524–33

107. Röhm, M., Werner, D. 1987. Isolation of root hairs from seedlings of *Pisum sativum.* Identification of root hair specific proteins by in situ labelling. *Physiol. Plant.* 69:129–36

108. Rolfe, B. G. 1988. Flavones and isoflavones as inducing substances of legume nodulation. *Biofactors* 1:3–10

109. Rolfe, B. G., Gresshoff, P. M. 1988. Genetic analysis of legume nodule initiation. *Annu. Rev. Plant Physiol. Plant Mol. Biol.* 39:297–319

110. Sánchez, F., Campos, F., Padilla, J., Bonneville, J.-M., Enríquez, C., et al. 1987. Purification, cDNA cloning, and developmental expression of the nodule-specific uricase from *Phaseolus vulgaris* L. *Plant Physiol.* 1987:1143–47

111. Sánchez, F., Quinto, C., Vázquez, M., De las Peñas, A., Cevallos, M. A., et al. 1988. The symbiotic association between *Phaseolus vulgaris* and *Rhizobium phaseoli.* See Ref. 47, pp. 370–75

112. Sandal, N. N., Bojsen, K., Marcker,

K. A. 1987. A small family of nodule specific genes from soybean. *Nucleic Acids Res.* 15:1507–19

113. Scheres, B., van de Wiel, C., Zalenski, A., Horvath, B., Spaink, H., et al. 1990. The ENOD12 product is involved in the infection process during the pea-*Rhizobium* interaction. *Cell* 60:281–94

114. Scheres, B., van Engelen, F., van der Knaap, E., van de Wiel, C., van Kammen, A., et al. 1990. Sequential induction of nodulin gene expression in the developing pea nodule. *Plant Cell* 2:687–700

115. Schiefelbein, J. W., Somerville, C. 1990. Genetic control of root hair development in *Arabidopsis thaliana*. *Plant Cell* 2:235–43

116. Schmidt, J., John, M., Wieneke, U., Stacey, G., Schell, J. 1990. Studies on the function of *Rhizobium meliloti* nodulation genes. See Ref. 14, pp. 150–55

117. Schmidt, J., Wingender, R., John, M., Wieneke, U., Schell, J. 1988. *Rhizobium meliloti* nodA and nodB genes involved in generating compounds that stimulate mitosis of plant cells. *Proc. Natl. Acad. Sci. USA* 85:8578–82

118. Schmit, A.-C., Lambert, A.-M. 1990. Microinjected fluorescent phalloidin *in vivo* reveals the F-actin dynamics and assembly in higher plant mitotic cells. *Plant Cell* 2:129–38

119. Schnepf, E. 1986. Cellular polarity. *Annu. Rev. Plant Physiol.* 37:23–47

120. Schubert, K. R. 1986. Products of biological nitrogen fixation in higher plants: synthesis, transport and metabolism. *Annu. Rev. Plant Physiol.* 37:539–74

121. Sengupta-Gopalan, C., Pitas, J. W. 1986. Expression of nodule-specific glutamine synthetase genes during nodule development in soybeans. *Plant Mol. Biol.* 7:189–99

122. Sengupta-Gopalan, C., Pitas, J. W., Thompson, D. V., Hoffman, L. M. 1986. Expression of host genes during root nodule development in soybeans. *Mol. Gen. Genet.* 203:410–20

123. Sethi, R. S., Reporter, M. 1981. Calcium localization pattern in clover root hair cells associated with infection processes: studies with auromycin. *Protoplasma* 105:321–25

124. Sharma, S., Signer, E. R. 1990. Temporal and spatial regulation of the symbiotic genes of *Rhizobium meliloti* in plant revealed by transposon Tn5-*gusA*. *Genes Dev.* 2:344–56

125. Sprent, J. I. 1989. Which steps are essential for the formation of functional legume nodules? *New Phytol.* 11:129–53

126. Staiger, C. J., Schliwa, M. 1987. Actin localization and function in higher plants. *Protoplasma* 141:1–12

127. Stegink, S. J., Vaughn, K. C. 1990. Immunotaxonomy of nodule-specific proteins. *Cytobios* 61:7–20

128. Strittmatter, G., Chia, T.-F., Trinh, T. H., Katagiri, F., Kuhlemeier, C., Chua, N. H. 1989. Characterization of nodule-specific cDNA clones from *Sesbania rostrata* and expression of the corresponding genes during initial stages of stem nodules and root nodules formation. *Mol. Plant Microbe Interact.* 2:122–27

129. Sutherland, T. D., Bassam, B. J., Schuller, L. J., Gresshoff, P. M. 1990. Early nodulation signals of the wild type and symbiotic mutants of soybean *(Glycine max)*. *Mol. Plant-Microbe Interact.* 3:122–28

130. Suzuki, A., Vidal, J., Nguyen, J., Gadal, P. 1984. Occurrence of ferredoxin-dependent glutamate synthase in plant cell fraction of soybean root nodule *(Glycine max)*. *FEBS Lett.* 173:204–8

131. Szabados, L., Ratet, P., Grunenberg, B., de Bruijn, F. J. 1990. Functional analysis of the *Sesbania rostrata* leghemoglobin glb3 gene 5' upstream region in transgenic *Lotus corniculatus* and *Nicotiana tabacum* plants. *Plant Cell* 2:973–86

132. Thummler, F., Verma, D. P. S. 1987. Nodulin-100 of soybean is the subunit of sucrose synthase regulated by the availability of free heme in nodules. *J. Biol. Chem.* 262:14730–36

133. Tingey, S. V., Walker, E. L., Coruzzi, G. M. 1987. Glutamine synthetase genes of pea encode distinct polypeptides which are differentially expressed in leaves, roots and nodules. *EMBO J.* 6:1–9

134. Tran Thanh Van, K., Toubart, P., Cousson, A., Darvill, A. G., Gollin, D. J., et al. 1985. Manipulation of the morphogenetic pathway of tobacco explants by oligosaccharins. *Nature* 314:615–17

135. Trese, A. T., Pueppke, S. G. 1990. Modulation of host gene expression during initiation and early growth of nodules in cowpea, *Vigna unguiculata* (L.) Walp. *Plant Physiol.* 92:946–53

136. Truchet, G., Barker, D. G., Camut, S., de Billy, F., Vasse, J., Huguet, T. 1989. Alfalfa nodulation in the absence of *Rhizobium*. *Mol. Gen. Genet.* 219:65–68

137. Turgeon, G., Bauer, W. D. 1982. Electron microscopy of infection thread formation in soybean. *Plant Physiol.* 69:156

138. van Batenburg, F. H. D., Jonker, R., Kijne, J. W. 1986. *Rhizobium* induces marked root hair curling by redirection of tip growth: a computer simulation. *Physiol. Plant.* 66:476–80

139. van de Wiel, C., Nap, J. P., van Lammeren, A., Bisseling, T. 1988. Histological evidence that a defense of the host plant interferes with nodulin gene expression in *Vicia sativa* root nodules induced by an *Agrobacterium* transconjugant. *J. Plant Physiol.* 132:446–52

140. van de Wiel, C., Norris, J. I. I., Bochenek, B., Dickstein, R., Bisseling, T., et al. 1990. Nodulin gene expression and ENOD2 localization in effective, nitrogen-fixing and ineffective, bacteria-free nodules of alfalfa. *Plant Cell* 2:1009–17

141. van de Wiel, C., Scheres, B., Franssen, H., van Lierop M.-J., van Lammeren, A., et al. 1990. The early nodulin transcript ENOD2 is located in the nodule parenchyma (inner cortex) of pea and soybean root nodules. *EMBO J.* 9:1–7

142. Van Kammen, A. 1984. Suggested nomenclature for plant genes involved in nodulation and symbiosis. *Plant Mol. Biol. Rep.* 2:43–45

143. Vance, C. P., Egli, M. A., Griffith, S. M., Miller, S. S. 1988. Plant regulated aspects of nodulation and N2 fixation. *Plant Cell Environ.* 11:413–27

144. Vance, C. P., Giffith, S. M. 1988. The molecular biology of N metabolism. In *Advanced Plant Physiology and Molecular Biology,* ed. D. H. Turpin, D. T. Dennis. Essex: Longman Scientific

145. Vance, C. P., Stade, S. 1984. Alfalfa root nodule carbon dioxide fixation. II. Partial purification and characterization of root nodule phosphoenolpyruvate carboxylase. *Plant Physiol.* 75:261–64

146. VandenBosch, K. A., Bradley, D. J., Knox, J. P., Perotto, S., Butcher, G. W., Brewin, N. J. 1989. Common components of the infection thread matrix and the intercellular space identified by immunocytochemical analysis of pea nodules and uninfected roots. *EMBO J.* 8:335–42

147. Verma, D. P. S., Delauney, A. J. 1988. Root nodule symbiosis: nodulins and nodulin genes. In *Temporal and Spatial Regulation of Plant Genes,* ed. D. P. S. Verma, R. Goldberg, pp. 169–99. Berlin: Springer-Verlag

148. Vincent, J. M. 1980. Factors controlling the *Rhizobium* symbiosis. In *Nitrogen Fixation,* ed. W. E. Newton, W. H. Orme-Johnson, 2:103–22. Baltimore: University Park Press

149. Wacker, I., Schnepf, E. 1990. Effects of nifedipine, verapamil, and diltiazem on tip growth in *Funaria hygrometrica.* *Planta* 180:492–501

150. Weaver, C. D., Stacey, G., Roberts, D. M. 1990. Phosphorylation of nodulin-26 by a Ca-dependent protein kinase. See Ref. 61, p. 774

151. Weeden, N. F., Kneen, B. E., LaRue, T. A. 1990. Map position of *sym* genes involved in nodulation and nitrogen fixation in *Pisum sativum.* See Ref. 61, p. 323

152. Welters, P., Felix, G., Schell, J., de Bruijn, F. J. 1990. Specific interactions of rhizobial proteins (*trans*-acting factors) with distinct DNA sequences in the promoter region of *Sesbania rostrata* leghemoglobin *(lb)* genes. See Ref. 61, p. 775

153. Werner, D., Bassarab, S., Humbeck, C., Kape, R., Kinnback, A., et al. 1988. Nodule proteins and compartments. See Ref. 4, pp. 507–15

154. Wittenberg, J. B., Wittenberg, B. A. 1990. Mechanisms of cytoplasmic hemoglobin and myoglobin function. *Annu. Rev. Biophys. Biophys. Chem.* 19:217–41

155. Wood, S. M., Newcomb, W. 1989. Nodule morphogenesis: the early infection of alfalfa (*Medicago sativa*) root hairs by *Rhizobium meliloti.* *Can. J. Bot.* 67:3108–22

156. Wyss, P., Mellor, R. B., Wiemken, A. 1990. Vesicular-arbuscular mycorrhizas of wild-type soybean and nonnodulating mutants with *Glomus mosseae* contain symbiosis-specific polypeptides (mycorrhizins), immunologically cross-reactive with nodulins. *Planta* 182:22–26

157. Yokohama, K., Kaji, H., Nishimura, K., Miyaji, M. 1990. The role of microfilaments and microtubules in apical growth and dimorphism of *Candida albicans.* *J. Gen. Microb.* 136:1067–75

158. Zaat, S. A. J., Van Brussel, A. A. N., Tak, T., Pees, E., Lugtenberg, B. J. J. 1987. Flavonoids induce *Rhizobium leguminosarum* to produce *nodABC* gene-related factors that cause thick, short roots and root hair responses on common vetch. *J. Bacteriol.* 169:3388–91

Annu. Rev. Plant Physiol. Plant Mol. Biol. 1991. 42:529–51

MOLECULAR GENETIC APPROACHES TO PLANT HORMONE BIOLOGY

Harry Klee

Monsanto Company, 700 Chesterfield Village Parkway, Chesterfield, Missouri 63198

Mark Estelle

Department of Biology, Indiana University, Bloomington, Indiana 47405

KEY WORDS: mutant, plant development, transgenic, phytohormone, mechanism of hormone action

CONTENTS

INTRODUCTION

Physiological studies conducted over the last half century have established a role for plant hormones in virtually every aspect of development. Although the techniques that have been used historically are imprecise, the picture that

1040-2519/91/0601-0529$02.00

has emerged is remarkably accurate, if incomplete. Classical approaches to the understanding of phytohormone action have been heavily weighted toward application and observation. Most of the pioneering work on hormone action used approaches that involve 1. excision of an organ and replacement of it with some combination of hormones, 2. excision of an organ or tissue and growth in vitro, or 3. exogenous application of a hormone or inhibitor. For example, some of the earliest experiments consisted of removing a shoot apex, replacing it with an agar block containing plant extract, and observing the effects on growth. While the conclusions of much of this work are entirely correct, what we can learn in this manner is severely limited. Exogenous application of any biological material is subject to limitations of uptake, transport, sequestration, and metabolism. Further, it is difficult to quantitate the amount of active material within the target tissue. For these reasons, it has been difficult to establish a direct relationship between a hormone and a particular developmental process. Rather, what has emerged is a picture of a complex interaction among the various hormones. Here we present a summary of the contributions that molecular biology and genetics have made to our understanding of hormone action, with emphasis on genes and ultimately on the use of those genes in transgenic organisms. Molecular genetic techniques are having a major impact on our understanding of the biology and biochemistry of phytohormones. This is happening at a time when equally impressive advances in analytical capacity are occurring (32). The combined effect is that remarkable progress has been made in a very short time.

The development of molecular approaches to studying plant hormone action has been accompanied by an increase in the use of mutants to study hormone processes in plants (43, 78). The benefits of this kind of analysis are several. First, the phenotypic characterization of hormone mutants can provide information on the role of hormones in plant physiology and development. Second, mutants can be used to study the biochemistry of hormone biosynthesis and hormone action. Finally, in some plant species mutants can be used to isolate and characterize the genes required for hormone processes.

Two classes of hormone mutants have been identified either among breeding populations or by direct screening of mutagenized populations. The first class consists of mutants that are hormone deficient. Generally a mutant is considered hormone deficient if biochemical analysis shows a reduction in hormone level and exogenous application of hormone restores a wild-type phenotype to the mutant plant. These mutants have been used extensively to analyze hormone biosynthetic pathways and to study the physiological role of plant hormones.

The second class of hormone mutants consists of those deficient in one or more hormone responses. Mutants with altered hormone responses have been identified in a number of plant species (85). The biochemical lesion has not been defined for any of these mutants. However, recent genetic and physiological studies of hormone response mutants have provided new information on the way hormones may act, and interact, to regulate development.

The ultimate goal of genetic analysis is to obtain the gene affected by a mutation. However, it has not been necessary to wait for the cloning of mutant genes to initiate an entirely novel approach to hormone biology. There are numerous examples of bacteria and fungi that synthesize or metabolize phytohormones (63). In some cases, the genes encoding these activities have been cloned. Having a gene that directs synthesis of a hormone or a rate-limiting precursor opens up the possibility of increasing its level in a controlled manner in vivo. With the large number of transcriptional promoters currently available, synthesis can be targeted to specific tissue types or regulated by an external stimulus such as heat shock. Availability of the plant genes also permits us to shut them off by using antisense RNA techniques. Thus, levels of a hormone can be increased or decreased in very precise and reproducible ways. Alteration of the steady-state level of a hormone permits a direct test of the biological effects of the hormone on a tissue. Increasing the rate of synthesis can also be useful in elucidating the pathways of hormone metabolism. If the concentration of a cytokinin is suddenly increased by 10 or 100 times, how does a plant compensate? Yet another major benefit of having the genes that direct hormone synthesis is that one has molecular probes to define the site(s) and regulatory controls of synthesis. Despite many years of research, we still do not know precisely where or when hormones are made in the plant. Gene probes will allow unambiguous determination of the answer to this question. The transgenic approach to hormone biology has begun to yield interesting data. As more genes become available, our potential increases for understanding hormone synthesis and action, permitting us to manipulate plant growth and development for agronomic and horticultural purposes.

GENETIC ANALYSIS OF HORMONE PROCESSES

Hormone Biosynthesis

The most extensively characterized hormone deficient mutants are the gibberellic acid (GA) deficient dwarf mutants of maize and pea (94). In a series of pioneering studies, Phinney and colleagues used the dwarf mutants of maize (*dwarf-1, dwarf-2, dwarf-3, dwarf-5,* and *anther ear-1*) to characterize the GA biosynthetic pathway (74). By determining the location of the block in the biosynthetic pathway for each mutant, these researchers demonstrated that GA_1 is the active GA in stem elongation. Similar results have been obtained with the dwarf mutants of pea (94). GA deficient mutants have also been isolated in *Arabidopsis (ga1, ga2, and ga3)* (48) and tomato *(ga-1* and *ga-2)* (49). The biochemical defect in each of these mutants has not been defined precisely, but the available evidence indicates that the GA biosynthetic pathway in these species is similar to that of maize and pea (39).

Largely because of the existence of the GA-deficient mutants, we now know a great deal about GA biosynthesis. However, a number of important questions remain to be answered. What is the cellular location of the pathway? Is GA_1 synthesized in the responding organ or is it transported from a different part of the plant? We also know little about regulation of the pathway and the role that changes in GA concentration may play in stem elongation. Grafting experiments with pea suggest that GA_1 is synthesized primarily in the young, expanding apical tissue and cannot be transported from other parts of the plant (79). In contrast, grafting studies in maize indicate that both GA_1 and precursors can be transported through the plant (41). Definitive answers to these questions will require the isolation and characterization of the genes involved in GA biosynthesis. Phinney et al (73) have taken the first step towards this goal in maize, by isolating new GA-responding dwarfs using the transposon *Mu-1* as a mutagen. Also, the *Arabidopsis GA₁* gene has been isolated using a subtractive hybridization procedure (T.-P. Sung and F. Ausubel, personal communication). We can expect to learn a lot about the regulation of GA biosynthesis in the next few years.

Recent studies on abscisic acid (ABA) biosynthesis indicate that ABA is probably derived from a carotenoid such as xanthophyll. This hypothesis is supported by studies of the ABA-deficient mutants of tomato, *flacca (fla)* and *sitiens (sit)* (64). Conversion of [^{13}C] xanthophyll to ABA in *fla* and *sit* plants was dramatically reduced relative to wild-type plants, indicating that the block in ABA biosynthesis in these mutants lies between xanthophyll and ABA (71). An ABA-deficient mutant of *Arabidopsis (aba)* has also been isolated, but little is known about the nature of the biochemical lesion (46).

Mutants with defined defects in auxin, cytokinin, or ethylene biosynthesis have not been identified in any plant species. Recently several auxin auxotrophic variants of *Nicotiana plumbaginifolia* were isolated in tissue culture (8; P. King, personal communication). However, subsequent studies indicated that the auxotrophic phenotype in these mutants is not due to an auxin deficiency (P. King, personal communication). Similarly, a temperature-sensitive mutant of *Hyoscyamus muticus* that requires exogenous auxin to grow at high temperature is not auxin deficient (68). A cytokinin-deficient mutant of tomato has also been described (90), but the cause of this deficiency has not been determined.

What could explain the lack of auxin, cytokinin, and ethylene biosynthetic mutants? It is possible that there are several genes for most or all of the steps in biosynthesis of these hormones. Alternatively, mutants with low levels of auxin or cytokinin may be inviable. The auxin auxotrophic mutants in *Nicotiana* and *Hyoscyamus* described above both die in the absence of auxin, suggesting that auxin is essential for viability (8, 68). In contrast, it seems unlikely that an ethylene-deficient mutant would be inviable. Mutants of *Arabidopsis* (*etr, ein*) that are completely insensitive to ethylene are almost normal in appearance (7, 24). It may be that ethylene-deficient mutants have not been identified because, unlike the GA-deficient dwarfs or the ABA-deficient wilty mutants, an ethylene-deficient mutant does not have an obvious phenotype.

Physiological Roles of Plant Hormones

SEED DEVELOPMENT AND GERMINATION   In a series of elegant studies, Karssen, Koornneef, and colleagues have used ABA and GA mutants of *Arabidopsis* to study the hormonal regulation of seed dormancy and germination (39). The behavior of the ABA-deficient mutant *aba* and the ABA resistant mutants *abi1*, *abi2*, and *abi3* provided the first clear evidence that ABA is required for induction of seed dormancy. All four mutants exhibit reduced seed dormancy. In addition, by examining the progeny of reciprocal crosses between the *aba* mutant and wild-type plants, Karssen et al (38) showed that induction of dormancy in the developing seed requires ABA synthesis in the embryo and endosperm. ABA produced in maternal tissue or supplied exogenously does not induce dormancy in seeds homozygous for the *aba* mutation. Apparently maternal ABA cannot be transported into the developing embryo.

The GA-deficient mutants of *Arabidopsis*, *ga1*, *ga2*, and *ga3*, all require exogenous GA to germinate, indicating that GA is required for germination (48). Furthermore, dormancy is established in the *ga1* mutant at the same time during seed development as in wild type, suggesting that GA is not involved in regulating the establishment of dormancy. Based on these results, Karssen & Lacka (40) propose that ABA and GA act at different times during development of the seed. ABA acts early to establish dormancy, and GA is required later to stimulate germination. Other experiments suggest that dormancy-breaking conditions may exert their effect by increasing the GA sensitivity of the embryo, thereby permitting germination to occur (40).

Studies involving isolated embryos from a number of species indicate that ABA has a role in embryogenesis outside of dormancy induction (111). For this reason it was initially surprising that the *aba* and *abi* mutations do not have more dramatic effects on seed development. Koornneef et al (45) suggest that the ABA mutants are leaky and retain enough ABA or ABA response to permit normal development of the embryo. To test this hypothesis, they constructed the ABA-deficient, ABA-resistant double mutant, *aba/aba; abi3/abi3*. Seed development in the double mutant was dramatically affected in a number of ways, including a decrease in water loss, a decrease in accumulation of seed storage proteins, and a loss of desiccation tolerance.

These results indicate that ABA, albeit in very small amounts, is required for normal embryogenesis. Furthermore, because the levels of ABA in the seed are apparently in excess of the amount of ABA required for normal embryogenesis, changes in ABA sensitivity may be more important than changes in hormone concentration at this stage in development.

GRAVITROPISM   According to the Cholodny & Went hypothesis, gravitropic growth is the result of differential growth on the two sides of a responding organ (19). It has been suggested that in roots, differential growth is due to the asymmetric distribution of either indole-3-acetic acid (IAA) or ABA. The ABA-deficient mutants of *Arabidopsis* (46) and tomato (97) exhibit normal root gravitropism, indicating that ABA is not required for this process. In contrast, the phenotype of auxin-resistant mutants in several species indicates that auxin plays an important role in root gravitropism. The *diageotropica* (*dgt*) mutant of tomato was originally identified because of a defect in shoot gravitropism and was subsequently shown to be auxin resistant (42, 112). Both the shoot and root of *dgt* plants display unsupported horizontal or diageotropic growth. Recent biochemical studies show that shoots of *dgt* plants have reduced levels of an auxin-binding protein (30; see below).

In *Arabidopsis*, the auxin-resistant mutants *axr1*, *axr2*, and *aux1*, all display defects in root gravitropism. In the recessive *axr1* mutants, the gravitropic response is intact but significantly slower than in wild type (53). The dominant *axr2* mutant is much more severely affected (107). The roots of this mutant are almost completely agravitropic, and the shoots are also unable to orient properly. Both the *axr1* and *axr2* mutants have a number of morphological defects in addition to defects in gravitropism, suggesting that the wild-type genes are required for many auxin-regulated growth processes (53, 107). The *aux1* mutation is much more specific. The only defect in *aux1* mutants, apart from hormone resistance in the root, is the loss of root gravitropism (56, 62, 75). The aerial part of the plant is normal in appearance and growth behavior. This phenotype suggests that the role of the *AUX1* gene is confined to the hormonal regulation of gravitropism. It is possible that there is an auxin response pathway that functions specifically to regulate root gravitropism.

PLANT STATURE   Studies on GA-deficient mutants have clearly demonstrated the importance of gibberellic acid in determining plant stature (see 94 for a recent review). In pea, GA deficiency affects different cell layers in different ways. In the epidermal layer of GA-deficient mutants there is a decrease in cell length but little difference in cell number, while in the outer layer of the cortex the opposite is true (79). These results suggest that the role of GA in stem growth is to stimulate both cell division and cell elongation, at least in pea.

Auxin is also thought to play a critical role in stem elongation. In classic experiments with excised segments of maize coleoptiles or pea internodes, auxin was shown to stimulate cell elongation within several minutes (35). However it has been difficult to demonstrate a role for auxin in stem elongation in the intact plant (25). The phenotype of several auxin-resistant mutants of *Arabidopsis*, suggest that auxin is involved in stem elongation. The *dwf*, *axr1*, and *axr2* mutants all have shortened internodes (53, 61, 107). In the *axr1* mutant this phenotype appears to be due to a reduction in cell number per internode (53), while in the *axr2* mutant shortened internodes are clearly due to a dramatic reduction in cell elongation (C. Lincoln and M. Estelle, unpublished). However, because the *axr1* and *axr2* mutants both have a pleiotropic phenotype (see below), it is difficult to link changes in stem elongation directly with a defect in auxin action.

STRESS RESPONSES   ABA is thought to play an important role in a plant's response to a number of stress conditions, including drought and cold (111). Two recent studies of ABA-deficient mutants provide excellent examples of the use of mutants to study the physiological role of hormones. Heino et al (27) have shown that ABA is required for cold acclimation. These workers found that the *aba* mutant of *Arabidopsis* does not cold acclimate properly but that acclimation could be restored by application of ABA. In addition, certain ABA-induced proteins, previously thought to be involved in cold acclimation, accumulate to normal levels in cold-treated *aba* plants. This result indicates that these proteins will accumulate in response to either cold treatment or ABA and that their presence is not sufficient to produce cold acclimation.

Another study of ABA-deficient mutants by Pena-Cortes et al (72a) has suggested a new role for ABA. When tomato or potato plants are attacked by an insect pest, they produce two proteinase inhibitors called PI-I and PI-II. Accumulation of these inhibitors is induced by wounding and occurs systemically. Pena-Cortes et al found that wounding also causes a systemic increase in ABA levels and that treatment of wild-type plants with ABA causes PI accumulation in the absence of wounding. In ABA-deficient mutants of tomato or potato, wound-induced accumulation of PI proteins does not occur, indicating that ABA is required for this response.

Because many plant stresses stimulate ethylene biosynthesis, ethylene has often been called the stress hormone (108). Ethylene has also been shown to cause an increase in the expression of many defense-related genes (9, 15). For these reasons, ethylene may play a central role in many plant defense responses. However, in most cases, a causal relationship between ethylene action and plant defense has not been demonstrated. Several ethylene-resistant mutants in *Arabidopsis* have now been described (7, 24). These mutants should be useful in studies of the role of ethylene in stress and defense responses.

Mechanisms of Hormone Action

The molecular mechanisms of action for each of the five major plant hormones remain poorly understood. It is generally believed that hormone action involves hormone binding to a protein receptor (105). A protein that may be an auxin receptor has been identified in corn and several other species (29; see below). However, receptors for the other hormones have not been identified (105). Similarly, the biochemical events that follow formation of a hormone-receptor complex are largely unknown. A number of investigators have attempted to determine whether second messengers characterized in animal systems play a role in plant signal transduction (23). Although some experiments indicate that protein phosphorylation, phosphotidylinositol metabolism, and changes in Ca^{2+} concentration (16, 76, 84) may be involved in various aspects of hormone action, the evidence is not yet definitive. We do not review this extensive literature here. Instead, we describe what has been learned about hormone action from the study of hormone-resistant mutants.

ETHYLENE   Three ethylene-insensitive mutants of *Arabidopsis* (*etr*, *ein1-1*, and *ein2-1*) have now been described (7, 24). Because the *etr* and *ein1-1* mutation are both dominant and map to the same chromosomal region, they are probably allelic. The *ein2* mutation is recessive and unlinked to *etr* and *ein1*. A wide range of ethylene responses have been characterized in the *etr* mutant, including inhibition of hypocotyl elongation, stimulation of germination, induction of chlorophyll loss, and accumulation of peroxidases (7). For every response, the *etr* mutant was completely insensitive to ethylene. Furthermore, the partitioning of ^{14}C ethylene in mutant leaves was less than that in wild-type leaves. Taken together, these results suggest that the *etr* gene may code for an ethylene receptor. It is interesting to speculate on how a dominant mutation results in an apparent loss of a function—in this case, ethylene response. Bleecker et al (7) suggest a number of possibilities. The ETR protein may function as a multimer. The presence of defective subunits in a large proportion of the multimers may reduce function sufficiently to cause the resistant phenotype. Alternatively, the ETR protein may act to suppress ethylene response in the absence of ethylene. Ethylene binding to the ETR protein may relieve the suppression. According to this model, the ETR protein would be unable to bind ethylene, resulting in constitutive suppression. A project to clone the *etr* gene by chromosome walking is under way in the laboratories of E. Meyerowitz and A. Bleecker (89). Recent transformation results indicate that the gene has been localized to a single cosmid clone (89; C. Chang and A. Bleecker, personal communication) suggesting that molecular information on the nature of this interesting gene may be forthcoming.

ABA   Like the other major plant hormones, ABA appears to regulate a diverse set of physiological and developmental processes in plants (111). How hormones act to regulate so many different processes is central to our understanding of plant hormone action. In the case of ABA, the hormone may act through one response pathway or via a number of separate or partially overlapping pathways. The ABA-insensitive mutants (*abi1*, *abi2*, and *abi3*) were isolated by selecting for germination in the presence of exogenous ABA (47). All three mutants have decreased dormancy, but only *abi1* and *abi2* are susceptible to wilting. These results suggested that *abi1* and *abi2* affect all ABA responses, while the *abi3* mutation is specific to seed responses. Finkelstein & Somerville (21) examined additional ABA-inducible responses in the *abi* mutants, including inhibition of seedling growth, proline accumulation, ABA-regulated protein synthesis, and seed storage protein accumulation. They found that the *abi1* and *abi2* mutations affected all of the responses except storage protein accumulation. The *abi3* mutation decreased seed storage protein accumulation and also conferred some resistance to ABA growth inhibition in the seedling. The two classes of mutants, the *abi1*, *abi2* class and the *abi3* class, affect different but overlapping sets of responses. The behavior of double mutants was also informative. The *abi1 abi3* double mutant was much less sensitive to exogenous ABA than either of the single mutants while the *abi1 abi2* double mutant displayed no increased sensitivity. These results suggest that *abi1* and *abi2* define one ABA response pathway, while the *abi3* gene lies on a second pathway. However, Finkelstein & Somerville (21) note that if the *abi* mutations are leaky, their results could be explained by a single response pathway. They favor the two-pathway model because the three mutations affect different responses. Interestingly, Finkelstein & Somerville (21) were unable to recover a plant homozygous for both the *abi2* and *abi3* mutations. Recent studies indicate that pollen that carries both mutations does not develop to maturity (R. Finkelstein, personal communication). This is the first report that ABA may be involved in pollen development. Perhaps ABA is required for desiccation tolerance during pollen development.

Studies on the *viviparous-1* (*vp1*) mutants of maize have also begun to reveal the complexity of hormone response pathways. The *vp1* mutants are defective in many aspects of seed maturation, including developmental arrest and anthocyanin biosynthesis (13, 81). The mature plant is normal in appearance, suggesting that the wild-type *vp1* gene functions only in the seed. Robichaud & Sussex (82) have shown that the viviparous phenotype in the *vp1* mutant probably results from reduced ABA sensitivity in the embryo. However, the other aspects of the *vp1* phenotype do not appear to be related to ABA action, since the ABA-deficient mutants *vp2*, *vp7*, and *vp9*, have normal anthocyanin biosynthesis (81). McCarty et al (57) have cloned the *vp1* gene by transposon tagging using the Robertson's Mutator element, thus allowing the first molecular analysis of a hormone-response mutant. Their studies confirm that the *vp1* gene is expressed in the embryo and endosperm only, and also show that the gene is not expressed in *vp1* mutants. In addition, it was found that expression of *C1*, a regulatory gene for the anthocyanin, pathway is blocked in *vp1* mutants. The C1 protein has partial amino acid similarity to the *myb* oncogene product (72), suggesting that *C1* is a transcriptional regulator. McCarty et al (57) postulate that VP1 protein acts to regulate a number of signal transduction pathways including anthocyanin biosynthesis through *C1*, as well as an ABA response pathway.

AUXIN   Many physiological and molecular studies of auxin action have focused on early responses to auxin (35, 58, 98). When tobacco protoplasts are treated with auxin, a rapid hyperpolarization of the plasma membrane occurs (17). This hyperpolarization appears to require the action of the plasma membrane H^+-pumping ATPase (4). Because hyperpolarization occurs within a few seconds, it may represent a very early step in the chain of events leading to cellular response. To test the physiological relevance of membrane hyperpolarization, Ephritikhine et al (17) examined auxin response in the *Rac*⁻ mutant of tobacco. This auxin-resistant mutant was isolated in tissue culture. Plants regenerated from the resistant line were also resistant, and genetic analysis showed that resistance is due to a single dominant mutation. Protoplasts prepared from mutant leaves required a 10-fold higher concentration of auxin to achieve the same hyperpolarization as wild-type protoplasts. This result shows that membrane hyperpolarization is related to auxin response. It also indicates that the *Rac*⁻ mutation affects a very early step in auxin signal transduction.

The auxin-resistant mutant of tomato, *dgt*, has also been used to examine the physiological relevance of an in vitro result. The *dgt* mutant was originally identified because of its diageotropic growth behavior and later found to be resistant to auxin (42, 112). Hicks et al (30, 31) have used a tritium-labeled azido derivative of IAA to detect two auxin-binding proteins of M_r 42 and 44 in membrane preparations from wild-type tomato and zucchini. Competition experiments suggest that these proteins may be analogous to the IAA uptake carrier characterized in plasma membrane vesicles from zucchini. When the azido auxin is used to examine membrane preparations from *dgt* tomatoes, these proteins are absent from the shoot but present in the root. This result indicates that the auxin-binding proteins function in some aspect of auxin action. However, it is not clear that the *dgt* locus codes for these proteins. It is possible that the 42- and 44-kDa proteins are produced by separate genes in the root and shoot. Alternatively the *dgt* gene may regulate the expression of the auxin-binding protein genes, either directly or indirectly.

MUTANTS AND HORMONE INTERACTION   There are many examples in the literature of interactions between hormones (52). For example, it is clear from tissue culture studies (87), as well as from the experiments with transgenic plants discussed below, that the ratio of auxin to cytokinin is a critical factor in many developmental processes. Other types of interactions have been described. Raskin & Kende (77) have demonstrated that in deep-water rice, ethylene acts to increase the GA sensitivity of stem tissue, resulting in a stimulation of stem elongation. At present, we know nothing about the molecular nature of these interactions. However, the phenotypes of several hormone-resistant mutants of *Arabidopsis* and *Nicotiana* suggest that some signal transduction components may be shared by several hormones. The *aux1*, *axr1*, and *axr2* mutants of *Arabidopsis* were originally isolated by screening for resistance to auxin (18, 53, 54, 56, 75, 107). Subsequently, these mutants were found to be resistant to at least one other plant hormone, indicating that a single mutation can disrupt response to several hormones. The dominant *axr2* mutant has the most striking phenotype. Its roots are resistant to the growth-inhibiting effects not only of auxin but also of ethylene, ABA, and cytokinin (107; A. K. Wilson and M. Estelle, unpublished). The *aux1* mutants are resistant to auxin and ethylene only (75), while the *axr1* mutants are resistant to auxin and cytokinin (C. Lincoln and M. Estelle, unpublished). A reduction in hormone uptake probably does not explain these phenotypes because each mutant is resistant to a specific subset of hormones. There are at least two other ways of explaining multiple hormone resistance. The *AUX1*, *AXR1*, and *AXR2* genes may encode proteins that function in transducing more than one hormone signal. In animal systems, second messengers such as Ca^{2+} and inositol 1,4,5-trisphosphate are required for transduction of many cellular signals. The *Arabidopsis* genes may be required for the action of such a second messenger. Alternatively, sensitivity to one hormone may be regulated by the action of a second hormone. For example, the level of an ethylene receptor may be positively regulated by auxin. If this is the case, an auxin-resistant mutant would also be ethylene resistant because ethylene sensitivity is under auxin control. In fact, experiments with the *aux1* mutant show that ethylene sensitivity increases in roots exposed to low concentrations of auxin (F. B. Pickett and M. Estelle, unpublished).

Multiple hormone resistance is not confined to *Arabidopsis*. The *iba1* mutant of *Nicotiana plumbaginifolia* was recovered in a screen for mutants that germinate on medium containing high levels of auxin (6). Using germination as their assay, these workers found that the mutant was also resistant to ABA and paclobutrazol, an inhibitor of GA biosynthesis. Other aspects of the phenotype include early germination, a very low level of auxin resistance in mesophyl protoplasts, and a tendency to wilt. Bitoun et al (6) propose that early germination and the wilty phenotype are due to a change in the ABA/GA ratio in the mutant plants and suggest that auxin resistance is a secondary effect of this change. However, resistance to ABA, similar to that found in the *abi* mutants of *Arabidopsis* (47), might also explain these results. It will be interesting to determine if ABA-deficient or -resistant mutants in other species also display auxin resistance. Preliminary evidence indicates that the *abi3* mutant of *Arabidopsis* is not resistant to auxin (A. K. Wilson and M. Estelle, unpublished). In any case the phenotype of the *ibal* mutant provides further evidence that hormone signals are integrated during plant development.

Another type of hormone interaction involves regulation of hormone levels by a second hormone. For example, auxin has been shown to stimulate the synthesis of ethylene in many plant tissues (108). Other studies indicate that GA may regulate the level of IAA (36, 50). Law & Davies (51) performed IAA determinations on several pea lines and found that the amount of IAA in GA-deficient dwarfs was less than half that of wild type. They also show that peas with the genotype *le/le;cry^s/cry^s* had higher IAA levels than wild type. This gene combination produces plants that have the "slender" phenotype; they resemble GA-deficient dwarfs treated with a saturating dose of GA. It has been suggested that the slender phenotype is due to constitutive activation of a GA-regulated response pathway required for stem elongation (94). Law & Davies (51) hypothesize that GA may act to stimulate stem elongation, in part by increasing IAA levels. Thus the slender phenotype may result from overproduction of auxin. However, this hypothesis is not consistent with the observation that treatment of GA-deficient plants with auxin does not stimulate stem elongation. We still have much to learn about the interaction of GA and IAA during stem elongation. GA may affect IAA levels by stimulating biosynthesis or by inhibiting catabolism. In addition, it is not known if the relatively small changes in IAA level are sufficient to produce changes in growth. If auxin acts through changes in sensitivity, as suggested by Trewavas (101), a two-fold increase in auxin level may have no effect on stem elongation.

Molecular Cloning of Mutant Genes

Several approaches to gene isolation have been developed. In maize, transposons such as *Ac*, *Spm*, and *Mu* have been used to tag and clone many genes (14). The recent demonstration of *Ac* transposition in transgenic tobacco (3, 37), tomato (110), and *Arabidopsis* (104) suggests that this technique may soon be available in these species as well.

In *Arabidopsis*, the small genome size and small repetitive DNA fraction have allowed the development of additional approaches to gene isolation (60, 89). For example, several groups are attempting to isolate genes by chromosome walking from a closely linked restriction fragment length polymorphism (RFLP) (89). To facilitate this approach, approximately 200 RFLPs have been identified between different ecotypes of *Arabidopsis*. Linkage maps of these RFLPs have been generated and aligned with existing maps of visible markers (10, 67). The major advantage of this strategy is that it can be applied to any gene identified by mutation. An alternative approach to gene isolation in *Arabidopsis* is to use the T-DNA of *Agrobacterium tumefaciens* as a tag (20). This approach involves screening for mutants among progeny families of transgenic plants. If a mutant phenotype cosegregates with the inserted T-DNA, the mutation is probably a consequence of the insertion. It is then a straightforward matter to isolate the mutant gene by screening for clones that include the T-DNA. Several genes have been isolated in this way (20, 28, 109). Once a mutant has been identified, it is easy to isolate the gene. However, because each transgenic plant carries only one or two T-DNA inserts, large numbers of transgenic families must be screened to be assured of an insert in any particular gene.

Another promising approach to gene isolation has recently been described by Straus & Ausubel (95). This technique involves a series of subtractive hybridization steps using DNA from wild-type plants and mutant plants that are homozygous for a deletion in the gene of interest. The GA_1 gene of *Arabidopsis* has been cloned with this technique (T.-P. Sung and F. Ausubel, personal communication), and the technique may also work in plants with more complex genomes. The development of this approach to gene cloning has stimulated interest in using mutagens known to cause deletions in other organisms—e.g. butane, psoralin (11), and γ irradiation (66).

GENES AND TRANSGENIC PLANTS USED TO STUDY HORMONE ACTION

Genes Controlling Hormone Synthesis and Metabolism

AGROBACTERIUM   Synthesis of phytohormones is essential for the success of *Agrobacterium tumefaciens* and *A. rhizogenes* as plant pathogens. Genes that direct synthesis of auxin and cytokinin are located within the region of DNA transferred to the plant during transformation (T-DNA). These genes are active in transformed plant tissue, and the excess production of hormones leads to the characteristic crown gall tumor (reviewed in 106). To the best of our knowledge, the *Agrobacterium* genes have no homology to the endogenous plant genes, and numerous attempts to use the genes to obtain their plant counterparts have failed.

iaaM and iaaH   IAA is synthesized in a two-step process in crown gall tumors. The product of the *iaaM* gene, tryptophan monooxygenase, converts tryptophan to indole-3-acetamide (IAM). This intermediate is then converted to IAA by indoleacetamide hydrolase (IAAH). IAM is not a normal intermediate in plant auxin biosynthesis, and expression of both *iaaM* and *iaaH* is normally necessary for synthesis of IAA (34). Expression of *iaaM* alone leads to accumulation of IAM but does not affect IAA levels in transformed tissue (103). An increase in the level of IAA can also be achieved by overexpressing *IAAM* alone if a strong transcriptional promoter such as the cauliflower mosaic virus (CaMV) 19S or 35S promoter is used (44). In this circumstance, conversion of IAM to IAA must occur via hydrolysis, either chemically or by a nonspecific amidohydrolase. Expression of a *19S/IAAM* gene in petunia or tobacco can lead to a 10-fold increase in free IAA. Expression of *IAAH* alone, with either its own promoter or a stronger one, has no effect in plants.

ipt   The product of the *Agrobacterium* cytokinin synthesis gene, isopentenyl transferase (IPT), condenses isopentenyl pyrophosphate and AMP to produce isopentenyl AMP (iPMP). Synthesis of iPMP is probably the rate-limiting reaction in cytokinin biosynthesis since it is rapidly converted by the plant to a series of more biologically active cytokinins, most notably zeatin derivatives (59). Introduction of this gene into plants has proven difficult. Even under the control of weak transcriptional promoters, transgenic shoots do not form roots. Transformed tobacco and potato shoots have been successfully grafted. These shoots have elevated levels of cytokinin and abnormal phenotypes (69). Fertile transgenic plants have been obtained when the gene is placed under the control of regulated promoters such as the hsp70 or tissue-specific promoters (59). Transgenic tobacco plants containing an *hsp70/IPT* gene have increased cytokinins even in the absence of heat shock. Heat shock results in increases of as much as 100-fold in different cytokinins.

In *A. tumefaciens* a second cytokinin biosynthesis gene, *tzs*, is homologous to *ipt* and carries out the same reaction. This gene is located outside of the T-DNA and is active in the bacterium (5). The role of *tzs* is unknown at this time but the gene may be involved in bacterial host range.

In addition to the known auxin and cytokinin biosynthetic genes, there are a number of other genes located within the *Agrobacterium* T-DNA that, when expressed in plants, greatly affect morphology. One gene in the *A. tumefaciens* T-DNA, *tml*, may act through alteration of hormone signal transduction, but not by directly synthesizing active hormone (91, 100). Other genes known to affect plant morphology include the *rolA*, *rolB*, and *rolC* from *A. rhizogenes* (86, 93). While the actual functions of these genes remain a mystery, there is suggestive evidence that they act by altering the normal hormone signal transduction pathway, making transformed cells more sensitive to auxin (92).

IAA-LYSINE SYNTHETASE   Microorganisms are excellent sources of genes affecting phytohormone metabolism. Another example of an organism that has provided useful genes is *Pseudomonas syringae* pv. *savastanoi*. This plant pathogen produces galls on olive and oleander (65). Like *A. tumefaciens*, it does so by synthesizing IAA, and its auxin synthetic genes are related to those of *A. tumefaciens*. Besides being able to synthesize IAA, *P. savastanoi* contains a gene that further modifies it. The *iaaL* gene encodes IAA-lysine synthetase (IAAL), which conjugates IAA to lysine (22, 80). Auxin-amino acid conjugates are much less biologically active and are believed to be storage forms of the hormone (12). Constitutive overproduction of iaaL in tobacco, under the control of the CaMV 35S promoter, leads to upwards of a 20-fold reduction in free IAA and concomitant increases in IAA-lysine (82a).The transgenic plants exhibit morphological effects suggestive of auxin deprivation.

ETHYLENE BIOSYNTHESIS   Until recently, all of the progress in isolating phytohormone metabolic genes has been limited to bacteria. Despite a great deal of effort, the plant enzymes have proven recalcitrant to purification, being present in vanishingly small quantities in most cases. Within the past year, several groups have isolated genes involved in ethylene synthesis. Ethylene is synthesized from S-adenosylmethionine via the intermediate, 1-aminocyclopropane-1-carboxylic acid (ACC). The gene for ACC synthase has been cloned from tomato and squash (83, 102). It is likely that the gene encoding the final enzyme in the pathway has also been isolated. This gene was identified by its ability, in the antisense orientation, to inhibit ethylene synthesis (26). It has been referred to in the literature as the ethylene forming enzyme (EFE) and now appears to be an ACC oxidase. The isolation of this gene clearly demonstrates the power of molecular approaches to hormone biology. Not only had the enzyme never been purified; its mechanism of action was unproven. Amino acid homology between this enzyme and flavone-3-hydroxylase from *Antirrhinum majus* suggests that the enzyme acts by oxidation of ACC. Both ACC synthase and ACC oxidase should be useful tools for controlling synthesis of ethylene.

AUXIN BINDING PROTEINS   A great deal of effort has gone into purification of potential auxin receptors. Recently, genes encoding a family of auxin binding proteins (ABPs) have been isolated from maize and *Arabidopsis* (29, 33, 99). Whether these proteins are receptors remains to be determined. The proteins were originally identified as having a high affinity for auxins. Both affinity data and antibody experiments are suggestive of a receptor role for these proteins. However, the predicted protein sequences have provided puzzling information about their identities. At least one member of a receptor family should be located on the outer surface of the cell. The amino acid sequences of all of the members contain potential transit peptides, but each also terminates with the carboxyl-terminal tetrapeptide KDEL sequence. This sequence is thought to be an anchor that prevents export from the endoplasmic reticulum. Experiments with transgenic plants containing natural and modified versions of the genes should prove illuminating.

What We Have Learned from Transgenic Plants

AUXIN AND CYTOKININ   What happens when hormone metabolizing genes are introduced into plants? Perhaps the most remarkable aspect of the introduction of genes that alter hormone concentrations is the fact that fertile plants can be recovered. In tobacco, the range of IAA concentration between overproducers (iaaM) and underproducers (iaaL) is 200-fold. With cytokinins, the picture is more complicated because these compounds take multiple forms, but increases of 100–200-fold are tolerated. While hormones produce a number of major effects on different developmental processes, it is fair to say that a large number of cells and organs do not respond to alterations in hormone levels. For example, overproduction of auxin or cytokinin in maturing petunia embryos has no detectable effect (44; H. Klee, unpublished). Such results indicate that perception as well as synthesis of hormones is developmentally controlled.

From transgenic plant experiments we have learned of a complex interaction among cytokinins, auxins, and ethylene. For example, overproduction of IAA with the *A. tumefaciens* genes has some major phenotypic effects on plants (discussed below). Significant increases in auxin stimulate ethylene production, and many "auxin" effects are, in reality, ethylene effects. Therefore, the best way to summarize the available data is to concentrate on tissues rather than the individual hormones.

APICAL DOMINANCE   That auxin and cytokinin are capable of controlling apical dominance is well established (96). High auxin levels suppress release of lateral buds from dormancy while cytokinin stimulates their growth. Thus, it is not surprising that plants overexpressing the *iaaM* gene exhibit almost complete apical dominance (44), and plants overexpressing the *ipt* gene exhibit reduced apical dominance (59, 88). What is more surprising is that the extreme apical dominance in auxin-overproducing plants can be overcome by either crossing them with cytokinin overproducing plants or exogenously applying cytokinin to a dormant lateral bud (H. Klee, unpublished). These results suggest that absolute levels of auxin and cytokinin are not the determinants of lateral growth. Rather, the ratio of auxin to cytokinin controls growth. Dormancy can be induced by raising the auxin level 10-fold but can be relieved by further increasing the cytokinin in the bud. If this is the case, then reducing the effective auxin level should be equivalent to increasing the cytokinin level. This prediction was verified when plants overexpressing the *iaaL* gene were produced (82a). These plants, containing 5- to 20-fold lower IAA levels than controls, exhibited a reduced apical dominance similar to the *ipt*-overproducing plants. Thus, lateral growth appears to be regulated by the ratio of these two hormones, and absolute levels appear to be secondary.

VASCULAR GROWTH   The effect of auxin on vascular differentiation is well established (1). Auxin is thought to affect xylem formation both quantitatively and qualitatively. The effect of cytokinins on the process is somewhat less clear, but it is thought that they promote vascular growth (1). Work with transgenic plants having altered levels of IAA generally confirm the direct relationship between auxin and the degree of differentiation of xylem (44, 82a). Auxin-overproducing plants contain many more xylem elements than control plants. The cells are, however, smaller. Conversely, plants with lowered IAA levels contain fewer xylem elements of generally larger size. These results support the idea that auxin stimulates cell division within the vascular cambium and that there is a direct relationship between auxin content and rate of cell division (2).Cell size is most likely affected because auxin stimulates secondary cell wall formation. The more rapidly the wall is synthesized, the less time there is for cell expansion.

The effects of cytokinins on xylem formation are less clear cut. In tissue culture systems, cytokinin is essential for vascular element formation. In transgenic *ipt* plants, xylem formation is clearly inhibited by increased cytokinins. This result is analogous to cytokinin's effect on apical dominance, where excess cytokinin is equivalent to decreased auxin. Where data gathered using transgenic plants conflicts with observations from tissue culture we can see the power of the transgenic approach to clarify the roles of hormones in complex developmental processes.

ROOT GROWTH   Cytokinin and auxin do not always appear to be antagonistic to one another. Roots seem adversely affected by any alteration in hormone level. All manipulations of auxin and cytokinins result in slower growth of the root system (44, 59). Increased auxin does stimulate adventitious root formation; but overall, plants so treated have less root mass than controls.

OTHER EFFECTS   Both auxin and cytokinin perturbations have other interesting effects on plant growth. Auxin alteration can have profound effects on leaf development: Overproduction can lead to epinastic growth, presumably owing to induction of ethylene, and leaves are generally smaller and narrower than those of controls. Underproduction, in contrast, leads to a very dramatic wrinkling of the leaf. Histological analysis suggests that the wrinkling results from incomplete development of the vascular system relative to the rest of the leaf.

Cytokinins also have a significant effect on potato tuberization (69). When a constitutively expressed *ipt* gene was introduced into potato plants, abnormal shoots were regenerated. While these shoots would not form roots, they could be grafted onto a normal root stock. They were much more likely than controls to form precocious tubers, confirming the role of cytokinin in tuber formation.

ACC OXIDASE   This gene was first isolated as a cDNA induced during ripening in tomato. The gene was introduced in its antisense orientation into tomato for the purpose of identifying a phenotype associated with a nonfunctional gene. The antisense gene reduced ethylene production in a gene dosage–dependent manner (26). ACC oxidase activity was reduced by 93% in homozygous plants.These plants will be invaluable to the process of elucidating the role of ethylene in a variety of developmental processes.

CONCLUSIONS

The classic experiments of Skoog & Miller (87) established a precedent for an antagonistic relationship between hormones. In their work, undifferentiated tobacco tissue could be induced to form roots, shoots, or undifferentiated callus depending on the ratio of auxin to cytokinin. Analysis of hormone mutants and transgenic plants with altered auxin and cytokinin levels has confirmed the complex interactions between hormones. For example, the multiple hormone resistances of the auxin-resistant *Arabidopsis* and *Nicotiana* mutants hints at a complex interrelationship. Many processes (e.g. apical dominance) are relatively insensitive to absolute levels of hormone and seem to be regulated by the auxin-to-cytokinin ratio. A further complication is the intimate relationship between auxin and ethylene. Many of the "auxin" effects observed in transgenic plants are really caused by ethylene. Similarly, seed dormancy and germination appear to be regulated by the antagonistic balance of ABA and GA. This interrelationship of the hormones suggests that they may act through common signal-transduction pathways. One can imagine a scenario where, for example, auxin stimulates some "receptor" and cytokinin acts to reverse the reaction. The metabolic and/or developmental fate of the cell is then determined by the degree of stimulation of that transduction pathway.More auxin would be equivalent to less cytokinin, and vice versa. An alternative interpretation that cannot be ruled out at this point is that one hormone may act by altering either the level of, or the plant's sensitivity to, a second hormone. It is known, for example, that auxin can influence the stability and metabolism of cytokinin (70). Thus, it will be critical to measure the levels of all hormones in mutants and transgenic plants to obtain a complete picture.

It will be interesting to extend the transgenic approach using genes for the synthesis and metabolism of other hormones. For example, the ability to shut off ethylene synthesis via antisense RNA can be coupled with auxin over-production. This will allow a clear distinction between auxin and ethylene effects. In a similar way, the effects of auxins, cytokinins, and ethylene on gibberellins can be determined.

It is likely that many of the genes encoding biosynthetic activities of the gibberellins and ABA will be cloned in the next several years. The genes for ethylene synthesis have already been cloned. The auxin and cytokinin synthesis genes remain elusive, and it is not obvious how these will be obtained. Fortunately, bacterial substitutes are available in these cases. We are thus likely to have available to us tools for manipulation of all of the phytohormones within the next 5–10 years, if not sooner. These tools will help us to discover the biological roles of hormone molecules and may allow us to manipulate their levels to the benefit of agriculture.

Literature Cited

1. Aloni, R. 1988. Vascular differentiation within the plant. In *Vascular Differentiation and Plant Growth Regulators*, ed. T. E. Timell, pp. 39-62. Berlin/Heidelberg: Springer-Verlag

2. Aloni, R., Zimmermann, M. H. 1983. The control of vessel size and density along the plant axis—a new hypothesis. *Differentiation* 24:203-8

3. Baker, B., Coupland, G., Federoff, N. V., Starlinger, P., Schell, J. 1987. Phenotypic assay for excision of the maize controlling element *Ac* in tobacco. *EMBO J.* 6:1547-54

4. Barbier-Brygoo, H., Ephritikhine, G., Klambt, D., Ghislain, M., Guern, J. 1989. Functional evidence for an auxin receptor at the plasmalemma of tobacco mesophyll protoplasts. *Proc. Natl. Acad. Sci. USA* 86:891-95

5. Beaty, J. S., Powell, G. K., Lica, L., Regier, D. A., MacDonald, E. M., et al. 1986. *Tzs*, a nopaline Ti plasmid gene from *Agrobacterium tumefaciens* associated with trans-zeatin biosynthesis. *Mol. Gen. Genet.* 203:274-80

6. Bitoun, R., Rousselin, P., Caboche, M. 1990. A pleiotropic mutation results in cross-resistance to auxin, abscisic acid and paclobutrazol. *Mol. Gen. Genet.* 220:234-39

7. Bleecker, A. B., Estelle, M. A., Somerville, C., Kende, H. 1988. Insensitivity to ethylene conferred by a dominant mutation in *Arabidopsis thaliana*. *Science* 241:1086-89

8. Blonstein, A. D., Vahala, T., Koornneef, M., King, P. J. 1988. Plants regenerated from auxin-auxotrophic variants are inviable. *Mol. Gen. Genet.* 215:58-64

9. Broglie, K. E. Gaynor, J. J., Broglie, R. M. 1986. Ethylene-regulated gene expression: molecular cloning of the genes encoding an endochitinase from *Phaseolus vulgaris*. *Proc. Natl. Acad. Sci. USA* 83:6820-24

10. Chang, C., Bowman, J. L., DeJohn, A. W., Lander, E. S., Meyerowitz, E. M. 1988. Restriction fragment length polymorphism linkage map for *Arabidopsis thaliana*. *Proc. Natl. Acad. Sci. USA*. 85:6856-60

11. Cimino, G. D., Gamper, H. B., Isaacs, S. T., Hearst, J. E. 1985. Psoralens as photoactive probes of nucleic acid structure and function: organic chemistry, photochemistry, and biochemistry. *Annu. Rev. Biochem.* 54:1151-94

12. Cohen, J. D., Bandurski, R. S. 1982. Chemistry and physiology of the bound auxins. *Annu. Rev. Plant Physiol* 33:403-30

13. Dooner, H. K. 1985. *Viviparous-1* mutation in maize conditions pleiotropic enzyme deficiencies in the aleurone. *Plant Physiol.* 77:486-88

14. Doring, H.-P., Starlinger, P. 1986.

Molecular genetics of transposable elements in plants. *Annu. Rev. Genet.* 20: 175-200

15. Ecker, J. R., Davis, R. W. 1987. Plant defense genes are regulated by ethylene. *Proc. Natl. Acad. Sci. USA* 84:5202-6

16. Einspahr, K. J., Thompson, G. A. Jr. 1990. Transmembrane signaling via phosphatidylinositol 4,5-bisphosphate hydrolysis in plants. *Plant Physiol.* 93:361-66

17. Ephritikhine, G., Barbier-Brygoo, H., Muller, J-F., Guern, J. 1987. Auxin effect on the transmembrane potential difference of wild-type and mutant tobacco protoplasts exhibiting a differential sensitivity to auxin. *Plant Physiol.* 83:801-4

18. Estelle, M. A., Somerville, C. R. 1987. Auxin-resistant mutants of *Arabidopsis* with an altered morphology. *Mol. Gen. Genet.* 206:200-6

19. Feldman, L. J. 1985. Root gravitropism. *Physiol. Plant.* 65:341-44

20. Feldmann, K. A., Marks, M. D., Christianson, M. L. Quatrano, R. S. 1989. A dwarf mutant of *Arabidopsis* generated by T-DNA insertion mutagenesis. *Science* 243:1351-54

21. Finkelstein, R. R., Somerville, C. R. 1990. Three classes of abscisic acid (ABA)-insensitive mutations of *Arabidopsis* define genes which control overlapping subsets of ABA responses. *Plant Physiol.* 94:1172-79

22. Glass, N. L., Kosuge, T. 1986. Cloning of the gene for indoleacetic acid-lysine synthetase from *Pseudomonas syringae* subsp. *savastanoi*. *J. Bacteriol.* 166: 598-602

23. Guern, J. 1987. Regulation from within: the hormone dilemma. *Ann. Bot.* 60:75-102

24. Guzman, P., Ecker, J. R. 1990. Exploiting the triple response of *Arabidopsis* to identify ethylene-related mutants. *Plant Cell* 2:513-23

25. Hall, J. L., Brummell, D. A., Gillespie, J. 1985. Does auxin stimulate the elongation of intact plant stems? *New Phytol.* 100:341

26. Hamilton, A. J., Lycett, G. W., Grierson, D. 1990. Antisense gene that inhibits synthesis of the hormone ethylene in transgenic plants. *Nature* 346:284-87

27. Heino, P., Sandeman, G., Lang, V., Nordin, K. and Palva, E. T. 1990. Abscisic acid deficiency prevents development of freezing tolerance in *Arabidopsis thaliana* (L.) Heynh. *Theor. Appl. Genet.* 79:801-6

28. Herman, P. L., Marks, M. D. 1989.

Trichome development in *Arabidopsis*. II. Isolation and complementation of the *GLABROUS1* gene. *Plant Cell* 1:1051-55

29. Hesse, T., Feldwisch, J., Balshuesemann, D., Bauw, G., Puype, M., et al. 1989. Molecular cloning and structural analysis of a gene from *Zea mays* L.coding for a putative receptor for the plant hormone auxin. *EMBO J.* 8:2453-62

30. Hicks, G. R., Rayle, D. L., Jones, A. M., Lomax, T. L. 1989. Specific photoaffinity labeling of two plasma membrane polypeptides with an azido auxin. *Proc. Natl. Acad. Sci. USA* 86: 4948-52

31. Hicks, G. R. Rayle, D. L., Lomax, T. L. 1989. The *diageotropica* mutant of tomato lacks high specific activity auxin binding sites. *Science* 245:52-54

32. Horgan, R. 1987. Instrumental methods of plant hormone analysis. In *Plant Hormones and Their Role in Plant Growth and Development*, ed. P. J. Davies, pp. 222-39. Boston: Martinus Nijhoff

33. Inohara, N., Shimomura, S., Fukui, T., Futai, M. 1989. Auxin-binding protein located in the endoplasmic reticulum of maize shoots: molecular cloning and complete primary structure. *Proc. Natl. Acad. Sci. USA* 86:3564-68

34. Inze, D., Follin, A., Van Lijsebettens, M., Simoens, C., Genetello, C., et al. 1984. Genetic analysis of the individual T-DNA genes of *Agrobacterium tumefaciens*; further evidence that two genes are involved in indole-3-acetic acid synthesis. *Mol. Gen. Genet.* 194:265-74

35. Jacobs, M., Ray, P. 1976. Rapid auxin-induced decrease in free space pH and its relationship to auxin-induced growth in maize and pea. *Plant Physiol.* 58: 203-9

36. Jindal, K. K., Hemberg, T. 1976. Influence of gibberellic acid on growth and endogenous auxin levels in epicotyl and hypocotyl tissue of normal and dwarf bean plants. *Physiol. Plant.* 38:78-82

37. Jones, J. D. G., Carland, F., Lim, E., Ralston, E., Dooner, H. K. 1990. Preferential transposition of the maize element *Activator* to linked chromosomal locations in tobacco. *Plant Cell* 2:701-7

38. Karssen, C. M., Brinkhorst-van der Swan, D. L. C., Breekland, A. E., Koornneef, M. 1983. Induction of dormancy during seed development by endogenous abscisic acid: studies on abscisic acid deficient genotypes of *Arabidopsis thaliana* (L.) Heynh. *Planta* 157:158-65

39. Karssen, C. M., Groot, S. P. C., Koorn-

neef, M. 1987. Hormone mutants and seed dormancy in *Arabidopsis* and tomato. *Soc. Exp. Biol. Semin. Ser.* 32:119-32

40. Karssen, C. M., Lacka, E. 1986. A revision of the hormone balance theory of seed dormancy: studies on gibberellin and/or abscisic acid-deficient mutants of *Arabidopsis thaliana*. In *Plant Growth Substances 1985*, ed. M. Bopp, pp. 315-23. Heidelberg/Berlin: Springer-Verlag

41. Katsumi, M., Foard, D. E., Phinney, B. O. 1983. Evidence for the translocation of gibberellin A_3 and gibberellin-like substances in grafts between normal and *dwarf-1* and *dwarf-5* seedlings of *Zea mays* L. *Plant Cell Physiol.* 24:379-88

42. Kelly, M. O., Bradford, K. J. 1986. Insensitivity of the diageotropica tomato mutant to auxin. *Plant Physiol.* 82:713-17

43. King, P. J. 1988. Plant hormone mutants. *Trends Genet.* 4:157-62

44. Klee, H. J., Horsch, R. B., Hinchee, M. A., Hein, M. B., Hoffmann, N. L. 1987. The effects of overproduction of two *Agrobacterium tumefaciens* T-DNA auxin biosynthetic gene products in transgenic petunia plants. *Genes Dev.* 1:86-96

45. Koornneef, M., Hanhart, C. J., Hilhorst, H. W. M., Karssen, C. M. 1989. In vivo inhibition of seed development and reserve protein accumulation in recombinants of abscisic acid biosynthesis and responsiveness mutants in *Arabidopsis thaliana*. *Plant Physiol.* 90:463-69

46. Koornneef, M., Jorna, M. L., Brinkhorst van der Swan, D. L. C., Karssen, C. M. 1982. The isolation of abscisic acid (ABA) deficient mutants by selection of induced revertants in non-germinating gibberellin sensitive lines of *Arabidopsis thaliana* (L.) Heynh. *Theor. Appl. Genet.* 61:385-93

47. Koornneef, M., Reuling, G., Karssen, C. M. 1984. The isolation and characterization of abscisic acid-insensitive mutants of *Arabidopsis thaliana*. *Physiol. Plant.* 61:377-83

48. Koornneef, M., van der Veen, J. H. 1980. Induction and analysis of gibberellin sensitive mutants in *Arabidopsis thaliana* (L.) Heynh. *Theor. Appl. Genet.* 58:257-63

49. Koornneef, M., van der Veen, J. H., Spruit, C. J. P., Karssen, C. M. 1981. The isolation and use of mutants with an altered germination behavior in *Arabidopsis thaliana* and tomato. In *Induced Mutation — a Tool in Plant Research*,

ed. P. Howard Kitto, pp. 227-32. Vienna:IAEA-SM. 251 pp.

50. Law, D. M., Hamilton, R. H. 1984. Effects of gibberellic acid on endogenous indole-3-acetic acid levels in dwarf pea. *Plant Physiol.* 75:255-56

51. Law, D. M., Davies, P. J. 1990. Comparative indole-3-acetic acid levels in the slender pea and other pea phenotypes. *Plant Physiol.* 93:1539-43

52. Leopold, A. C., Nooden, L. D. 1984. Hormonal regulatory systems in plants. In *Encyclopedia of Plant Physiology*, Vol. 10. *Hormonal Regulation of Development II*, ed. T. K. Scott, pp. 4-22. New York: Springer-Verlag

53. Lincoln, C., Britton, J. H., Estelle, M. 1990. Growth and development of the *axr1* mutants of *Arabidopsis*. *Plant Cell.* 2:1071-80

54. Lincoln, C. Estelle, M. 1990. The *axr1* mutation of *Arabidopsis* is expressed in both roots and shoots. *Proc. Iowa Acad. Sci.* In press

55. MacMillan, J., Phinney, B. O. 1987. Biochemical genetics and the regulation of stem elongation by gibberellins. In *Physiology of Cell Expansion During Plant Growth*, ed. D. J. Cosgrove, D. P. Knievel, pp. 156-71. Rockville: Am. Soc. Plant Physiol.

56. Maher, E. P., Martindale, S. J. B. 1980. Mutants of *Arabidopsis* with altered responses to auxins and gravity. *Biochem. Genet.* 18:1041-53

57. McCarty, D. R., Carson, C. B., Stinard, P. S., Robertson, D. S. 1989. Molecular analysis of *viviparous-1*: an abscisic acid-insensitive mutant of maize. *Plant Cell* 1:523-32

58. McClure, B. A., Guilfoyle, T. 1987. Characterization of a class of small auxin-inducible soybean polyadenylated RNAs. *Plant Mol. Biol.* 9:611-24

59. Medford, J. I., Horgan, R., El-Sawi, Z., Klee, H. J. 1989. Alterations of endogenous cytokinins in transgenic plants using a chimeric isopentenyl transferase gene. *Plant Cell* 4:403-13

60. Meyerowitz, E. M. 1987. *Arabidopsis thaliana*. *Annu. Rev. Genet.* 21:93-112

61. Mirza J. I., Maher, E. P. 1980. More 2,4-D resistant mutants. *Arabidopsis Inf. Serv.* 17:103-7

62. Mirza, J. L., Olsen, G. M., Iversen, T. H., Maher, E. P., 1984. The growth and gravitropic responses of wild-type and auxin-resistant mutants of *Arabidopsis thaliana*. *Physiol. Plant.* 60:516-22

63. Morris, R. O. 1987. Genes specifying auxin and cytokinin biosynthesis in prokaryotes. See Ref. 32, pp. 636-655

64. Neill, S. J., Horgan, R. 1985. Abscisic

acid production and water relations in wilty tomato mutants subjected to water deficiency. *J. Exp. Bot.* 36:1222-31

65. Nester, E. W., Kosuge, T. 1981. Plasmids specifying plant hyperplasias. *Annu. Rev. Microbiol.* 35:531-65

66. Neuffer, M. G. 1982. Mutant Induction in maize. In *Maize For Biological Research*, ed. W. F. Sheridan, pp. 61-64. Grand Forks, ND: Univ. North Dakota Press

67. Nam, G.-H., Giraudat, J., den Boer, B., Moonan, F., Loos, W. D. B., et al. 1989. *Plant Cell* 1:699-705

68. Oetiker, J., Gebhardt, C., King, P. J. 1990. *Planta* 180:220-28

69. Ooms, G., Lenton, J. R. 1985. T-DNA genes to study plant development: precocious tuberisation and enhanced cytokinins in *A. tumefaciens* transformed potato. *Plant Mol. Bio.* 5:205-12

70. Palni, L. M., Burch, L., Horgan, R. 1988. The effect of auxin concentration on cytokinin stability and metabolism. *Planta* 174:231-34

71. Parry, A. D., Neill, S. J., Horgan, R. 1988. Xanthoxin levels and metabolism in wild-type and wilty mutants of tomato. *Planta* 173:397-404

72. Paz-Ares, J., Wienand, U., Peterson, P. A., Saedler, H. 1987. The regulatory locus c1 of *Zea mays* encodes a protein with homology to myb proto-oncogene products and with structural similarities to transcriptional activators. *EMBO J.* 6:3553-58

72a. Pena-Cortes, H., Sanchez-Serrano, J., Mertens, R., Prat, S. 1990. Abscisic acid is involved in the wound induced expression of the proteinase inhibitor II gene in potato and tomato. *Proc. Natl. Acad. Sci. USA* 86:9851-55

73. Phinney, B. O., Freeling, M., Robertson, D. S., Spray, C. R., Silverthorne, J. 1986. Dwarf mutant of maize-the gibberellin pathway and its molecular future. In *Plant Growth Substances*, ed. M. Bopp, pp. 55-64. Berlin/Heidelberg: Springer-Verlag

74. Phinney, B. O., Spray, C. R. 1982. Chemical genetics and the gibberellin pathway in *Zea mays* L. In *Plant Growth Substance 1982*, ed. P. F. Waering, pp. 101-10. New York: Academic

75. Pickett, F. B., Wilson, A. K., Estelle, M. 1990. The aux1 mutation of *Arabidopsis* confers both auxin and ethylene resistance. *Plant Physiol.* 94:1462-66

76. Poovaiah, B. W., Reddy, A. S. N., McFadden, J. J. 1987. Calcium messenger system: role of protein phosphorylation and inositol bisphospholipids. *Physiol. Plant.* 69:569-73

77. Raskin, I., Kende, H. 1984. Role of gibberellin in the growth response of submerged deep water rice. *Plant Physiol.* 76:947-50

78. Reid, J. B. 1990. Phytohormone mutants in plant research. *J. Plant Growth Regul.* 9:97-111

79. Reid, J. B., Murfet, I. C., Potts, W. C. 1983. Internode length in *Pisum*. II. Additional information on the relationship and action of loci Le, La, Cry, Na and Lm. *J. Exp. Bot.* 134:349-64

80. Roberto, F. F., Klee, H., White, F., Nordeen, R., Kosuge, T. 1990. Expression and fine structure of the gene encoding N^ϵ-(indole-3-acetyl)-L-lysine synthetase from *Pseudomonas savastanoi*. *Proc. Natl. Acad. Sci. USA* 87:5797-801

81. Robertson, D. S. 1955. The genetics of vivipary in maize. *Genetics* 40:745-60

82. Robichaud, C. S., Sussex, I. M. 1986. The response of *viviparous-1* and wild-type embryos of *Zea mays* to culture in the presence of abscisic acid. *J.Plant Physiol.* 126:235-42

82a. Romano, C., Hein, M., Klee, H. 1991. Inactivation of auxin in tobacco transformed with the indoleacetic acid-lysine synthetase gene of *Pseudomonas*. *Genes Dev.* In press

83. Sato, T., Theologis, A. 1989. Cloning the mRNA encoding 1-aminocyclopropane-1-carboxylate synthase, the key enzyme for ethylene synthesis in plants. *Proc. Natl. Acad. Sci. USA* 86:6621-25

84. Schroeder, J. I., Hagiwara, S. 1989. Cytosolic calcium regulates ion channels in the plasma membrane of *Vicia faba* guard cells. *Nature* 338:427-30

85. Scott, I. M. 1990. Plant hormone response mutants. *Physiol. Plant.* 78:147-52

86. Sinkar, V. P., Pythoud, F., White, F. F., Nester, E. W., Gordon, M. P. 1988. rolA locus of the Ri plasmid directs developmental abnormalities in transgenic tobacco plants. *Genes Dev.* 2:688-97

87. Skoog, F., Miller, C. O. 1957. Chemical regulation of growth and organ formation in plant tissues cultured in vitro. *Symp. Soc. Exp. Biol.* 11:188-231

88. Smigocki, A., Owens, L. 1989. Cytokinin-to-auxin ratios and morphology of shoots and tissues transformed by a chimeric isopentenyl transferase gene. *Plant Physiol.* 91:808-11

89. Somerville, C. 1989. *Arabidopsis* blooms. *Plant Cell* 1:1131-35

90. Sossountzov, L., Malkiney, R., Scotta, B., Sabbagh, I., Habricot, Y., et al. 1988. Immuncytochemical localization

of cytokinins in *Craigella* tomato and sideshootless mutant. *Planta* 175:291-304

91. Spanier, K., Schell, J., Schreier, P. H. 1989. A functional analysis of T-DNA gene 6b: the fine tuning of cytokinin effects on shoot development. *Mol. Gen. Genet.* 219:209-16

92. Spano, L., Mariotti, D., Cardarelli, M., Branca, C., Costantino, P. 1988. Morphogenesis and auxin sensitivity of transgenic tobacco with different complements of Ri T-DNA. *Plant Physiol.* 87:479-83

93. Spena, A., Schmulling, T., Koncz, C., Schell, J. 1987. Independent and synergistic activity of *rolA*, *B* and *C* loci in stimulating abnormal growth in plants. *EMBO J.* 6:3891-99

94. Stoddart, J. L. 1987. Genetic and hormonal regulation of stature. In *Developmental Mutants in Higher Plants*, ed. H. Thomas, D. Grierson, 32:155-80. Cambridge: Cambridge Univ. Press

95. Straus, D., Ausubel, F. M. 1990. Genomic subtraction for cloning DNA corresponding to deletion mutations. *Proc. Natl. Acad. Sci. USA* 87:1889-93

96. Tamas, I. A. 1987. Hormonal regulation of apical dominance. See Ref. 32, pp. 393-410

97. Taylor, I. B. 1987. ABA deficient tomato mutants. In *Developmental Mutants in Higher Plants*, ed. H. Thomas, D. Grierson, pp. 197-221. Cambridge: Cambridge Univ. Press

98. Theologis, A. 1986. Rapid gene regulation by auxin. *Annu. Rev. Plant Physiol.* 37:407-38

99. Tillmann, U., Viola, G., Kayser, B., Siemeister, G., Hesse, T., et al. 1989. cDNA clones of the auxin binding protein from corn coleoptiles. Isolation and characterization by immunogold methods. *EMBO J.* 8:2463-67

100. Tinland, B., Huss, B., Paulus, F., Bonnard, G., Otten, L. 1989. *Agrobacterium tumefaciens 6b* genes are strain-specific and affect the activity of auxin as well as cytokinin genes. *Mol. Gen. Genet.* 219:217-24

101. Trewavas, A. 1981. How do plant growth substances work. *Plant, Cell Environ.* 4:202-28

102. Van Der Straeten, D., Van Wiemeersch, L., Goodman, H. M., Van Montagu, M. 1990. Cloning and sequence of two different cDNAs encoding 1-aminocyclopropane-1-carboxylate synthase in tomato. *Proc. Natl. Acad. Sci. USA* 87:4859-63

103. Van Onckelen, H., Rudelsheim, P., Inze, D., Follin, A., Messens, E., et al. 1985. Tobacco plants transformed with the *Agrobacterium* gene 1 contain high amounts of indole-3-acetamide. *FEBS Lett.* 181:373-76

104. Van Sluys, M. A., Tempe, J., Fedoroff, N. 1987. Studies on the introduction and mobility of the maize *Activator* element in *Arabidopsis thaliana* and *Daucus carota*. *EMBO J.* 6:3881-89

105. Venis, M. 1985. *Hormone Sites in Plants*. New York/London: Longmans

106. Weiler, E., Schroder, J. 1987. Hormone genes and crown gall disease. *Trends Biol. Sci.* 12:271-75

107. Wilson, A. K., Pickett, F. B., Turner, J. C., Estelle, M. 1990. A dominant mutation in *Arabidopsis* confers resistance to auxin, ethylene and abscisic acid. *Mol. Gen. Genet.* 222:377-83

108. Yang, S. F., Hoffman, N. E. 1984. Ethylene biosynthesis and its regulation in higher plants. *Annu. Rev. Plant Physiol.* 35:155-89

109. Yanofsky, M. F., Ma, H., Bowman, J. L., Drews, G. N., Feldmann, K. A.,Meyerowitz, E. M. 1990. The protein encoded by the *Arabidopsis* homeotic gene agamous resembles transcription factors. *Nature* 346:35-39

110. Yoder, J. I. 1990. Rapid proliferation of the maize transposable element *Activator* in transgenic tomato. *Plant Cell* 2:723-30

111. Zeevaart, J. A. D., Creelman, R. A. 1988. Metabolism and physiology of abscisic acid. *Annu. Rev. Plant Physiol. Plant Mol. Biol.* 39:439-73

112. Zobel, R. W. 1974. Control of morphogenesis in the ethylene-requiring tomato mutant, *diageotropica*. *Can. J. Bot.* 52:735-41

—

Annu. Rev. Plant Physiol. Plant Mol. Biol. 1991. 42:553–78

FUNCTIONAL ASPECTS OF THE LICHEN SYMBIOSIS*

Rosmarie Honegger

Institute of Plant Biology, University of Zürich, CH-8008 Zürich, Switzerland

KEY WORDS: mycobiont-photobiont interface, carbohydrate transfer, water relations, gas exchange, polyol metabolism

CONTENTS

INTRODUCTION

Lichens are the symbiotic phenotype of nutritionally specialized fungi that live as ecologically obligate biotrophs in symbiosis with algal and/or cyano-bacterial photobionts. These are extracellularly located endosymbionts of lichen thalli. Species names of lichens refer to the fungal partner. About 8% of terrestrial ecosystems are lichen dominated (82), dominance being defined not only by a high "lichen quotient" (number of lichen species per higher plant

---

*Dedicated to Professor Sir David Smith, who has given friendly encouragement and stimulating advice to a whole generation of lichen physiologists

species), but also by a quantitative abundance that leads to a high percentage (up to 100%) of ground cover by lichens. These lichen-dominated ecosystems are the sites where vascular plants are at their physiological limits while the poikilohydrous lichen thalli gain a distinct ecological advantage from their capacity to survive extreme cold, heat, and drought stress unharmed in a state of dormancy. Various aspects of lichen ecophysiology have been examined in field and laboratory experiments (reviewed in 69). In recent times lichenologists began to achieve a better understanding of the biology of lichens by combining structural and physiological investigations. The present review focuses on aspects of lichen biology where the interpretation of physiological data has been facilitated by anatomical and ultrastructural studies. It also indicates fruitful topics for further experimental study.

Peculiarities of the Lichen Symbiosis

The lichen symbiosis is extremely diverse with regard to the taxonomic affiliation of the partners involved and the morphological and anatomical features of the symbiotic phenotype (Table 1; 40, 56, 60, 66). Besides this diversity there is also an amazing amount of analogy within lichens. Morphologically similar symbiotic phenotypes are formed by unrelated fungal taxa in association with diverse photobionts (e.g. foliose, dorsiventrally organized thalli; Figures 1b–c), but there are distinct differences with regard to functional aspects such as water uptake and translocation (e.g. 79), mycobiont-photobiont interactions, and mobilization of carbohydrates in analogous forms (see below) and even within closely related species (81).

Lichens are generally considered to be mutualistic symbioses—i.e. symbiotic systems in which the biological fitness of both partners is increased in the symbiotic state (84, 106). This is certainly true for highly evolved taxa with an internally stratified thallus. However, the biology of crustose species (the majority of lichens; Table 1; Figure 1a) has not been adequately investigated. Many of these more primitive symbiotic systems are likely to be more or less mild forms of fungal parasitism on algae or cyanobacteria. The present review focuses on some taxa of advanced lichens with morphologically complex, internally stratified (heteromerous) thalli. From a functional point of view these heteromerous lichens are comparable with ectomycorrhizae. In both symbioses a photoautotroph is provided with water and mineral nutrients by a fungus that, for its part, gains photosynthates from its life partner. However, ectomycorrhizal fungi contact and ensheath comparably large, multicellular host structures that are donors but not producers of photosynthates. It is the photoautotrophic partner that competes for space above ground, thereby securing appropriate illumination of its photosynthetically active parts. Lichen-forming fungi, on the other hand, associate with a population of minute photoautotrophic microorganisms that must be kept

Table 1   Taxonomic diversity of lichen mycobionts and photobionts. Data from 12, 28, 31, 32, 42, 43, 58, 60, 113

LICHEN MYCOBIONTS approx. 21% of all fungi

Ascomycotina :
    46% of Ascomycotina (approx. 13,250 spp.) are lichenized
    98 % of lichen-forming fungi are ascomycetes
        11 out of 46 orders include lichen-forming representatives
        5 orders are exclusively lichenized
        7 are mainly lichenized
        4 are mainly non-lichenized

Basidiomycotina :
    0.3% of Basidiomycotina (approx. 50 spp.) are lichenized
    0.4% of lichen-forming fungi are basidiomycetes
        Hymenomycetes, Holobasidiomycetidae:
        2 out of 9 (mainly non-lichenized) orders contain lichen-forming representatives

Deuteromycotina :
    1.2% of Deuteromycotina (approx. 200 spp.) are lichenized
    1.6% of lichen-forming fungi are deuteromycetes

Specificity, Selectivity

Approx. 85% of lichen mycobionts are symbiotic with green algae
        10% with cyanobacteria
        3-4% simultaneously with both, green algae and cyanobacteria

Many lichen mycobionts are **moderately specific**, i.e. accept several spp. of one genus as photobionts. >40% of lichen mycobionts associate with unicellular green algae of the genus *Trebouxia* (Pleurastrales, Ulvophyceae). These mycobionts are highly **selective** with regard to their photobiont: compatible spp. are seldom found in the aposymbiotic state outside lichen thalli or their symbiotic vegetative propagules, and the most common unicellular aerophilic algae are not acceptable partners.

Symbiotic phenotypes

Approx. 55% of lichen-forming fungi form homoeomerous (non-stratified) thalli
        20% form either placodioid or squamulose, and
        25% form foliose or fruticose, internally stratified (heteromerous) thalli

---

LICHEN PHOTOBIONTS (so far less than 200 spp. identified)

Note: in less than 2% of all lichen spp. has the photobiont ever been identified at the species level; fairly often not even the generic affiliation is known. As many photobionts change their shape and fine structure quite drastically in the lichen symbiosis they have to be isolated and cultured under defined condition for identification. The range of compatible photobiont spp. per mycobiont is unknown.

Chlorophyta : unicellular and filamentous spp.
        Charophyceae :   1 genus
        Chlorophyceae   10 genera, 1 order
        Ulvophyceae   11 genera, 5 orders

Xanthophyta :   1sp.
Phaeophyta :   1sp.
Cyanobacteria : unicellular or filamentous colonies of heterocystous and non-heterocystous, N_2-fixing taxa.
        13 genera, 4 orders

NON-STRATIFIED (HOMOEOMEROUS) THALLI

e.g. crustose

Graphis scripta

numerous spp. of the

Arthoniales	Opegraphales
Caliciales	Ostropales
Dothideales	Pertusariales
Graphidales	Pyrenulales
Gyalectales	Teloschistales
Lecanorales	Verrucariales
Leotiales	

a

INTERNALLY STRATIFIED (HETEROMEROUS) THALLI

e.g. foliose (dorsiventrally organized)

with upper cortex only

Peltigera venosa

examples: numerous spp. of the genera

Peltigera, Solorina(Peltigerales)

Cladonia (Lecanorales)

b

with upper and lower cortex

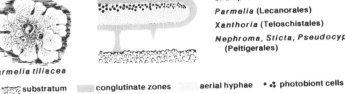

Parmelia tiliacea

examples: numerous spp. of the genera

Parmelia (Lecanorales)

Xanthoria (Teloschistales)

Nephroma, Sticta, Pseudocyphellaria
(Peltigerales)

c

substratum   conglutinate zones   aerial hyphae   • ∴ photobiont cells

Figure 1 Examples illustrating a small part of the morphological and anatomical diversity in the lichen symbiosis.

photosynthetically active. Lichen mycobionts are therefore obliged to secure adequate illumination, to facilitate the gas exchange of their photobiont cell population, and to compete for space above ground (59). The scarcity of some of the most commonly acquired green algal photobionts in natural habitats, and spatial limitations at the sites where aerophilic unicellular green algae and cynobacteria normally occur (uppermost layers or surface of soil, bark, rock, etc), may have been among the factors that imposed a strong evolutionary pressure on lichen-forming fungi to rise above the substratum by forming three-dimensional vegetative structures in which the photobiont cells are housed, carried along by growth processes, maintained, and controlled (62). Rising above the substratum subjects most lichen mycobionts to changes in temperature and moisture content more drastic than those a mycorrhizal fungus experiences. Among the numerous exceptions are lichen taxa that have adapted to stable, moist, and shady habitats such as evergreen rainforests of temperate climates (36). Owing to their unique morphological features, complex lichen thalli are, even in the most recent literature (e.g. 30), often referred to as plants. Such categorization is misleading, not only because plants and fungi are distinct kingdoms, but mainly because lichen thalli are not individuals but genetically heterogenous consortia with an unknown number of participants. Several fungal genotypes, originating from different sexual or asexual propagules, and several genotypes of the photobiont may and often do participate in the formation of a single thallus (65, 66, 83).

CARBOHYDRATE TRANSFER

Lichens were the first symbioses of fungi and photoautotrophic partners in which the mobile carbohydrates have been chemically identified. In their pioneering work, Smith and coworkers (reviewed in 45, 104–106) designed the so-called "inhibition technique" or "isotope trapping technique" on the basis of the following assumptions: 1. The apoplastic space of both partners in the lichen symbiosis is easily accessible, especially in thallus fragments that have been horizontally dissected along the photobiont layer. 2. No specialized, tight connections exist between the partners; mobile carbohydrates are released into the apoplastic space, whence they are taken up by the neighboring hyphae of the mycobiont. 3. The nutritional requirement of the fungal partner can be satisfied by incubation of dissected thallus fragments on a carbohydrate-containing liquid medium. Further uptake will be blocked or inhibited. (Best inhibition was achieved when the particular carbohydrate was offered that moves from the photobiont to the mycobiont.) Photosynthates produced in a subsequent $^{14}CO_2$ pulse are released into the apoplastic space, whence they leak out into the incubation medium that covers the cells of both partners in the photobiont layer. Radioactively labeled leakage products can be analyzed after recovery from the incubation medium.

A large series of taxonomically diverse lichens with different taxa of green algal or cyanobacterial photobionts have been investigated with the inhibition technique. Additional experiments include 1. incubation of complete thallus fragments on a $^{14}CO_2$-containing liquid medium, dissection after several periods of time, and analysis of labeled metabolites in the fungal portion (medullary layer); 2. inhibition of active release by means of dinitrophenol, arsenate, or FCCP (109), and of the fungal metabolism of cyanobacterial lichens by means of digitonin (105); 3. autoradiography of cryostat sections prior to and after extraction of soluble carbohydrates (109); 4. comparison of the carbohydrate production and release (qualitative and quantitative) of symbiotic and cultured photobionts (37, 48, 97); and 5. attempts to stimulate the carbohydrate release in cultured photobionts either by lowering the pH or by adding either thallus homogenates or mycobiont-derived secondary metabolites (summaries: 45, 105, 106).

The following results were obtained: 1. Only one type of molecule was recovered from the incubation medium, its chemical nature depending on the taxonomic identity of the photobiont. Cyanobionts produce an extracellular glucan that is hydrolyzed by a mycobiont-derived glucanase to glucose, the latter being recovered in the incubation medium (44). Green algal photobionts release a genus-specific, acyclic polyhydric alcohol (polyol), namely either ribitol *(Trebouxia, Coccomyxa, Myrmecia),* sorbitol *(Hyalococcus, Sticho-coccus),* or erythritol *(Trentepohlia)* (48, 94). 2. All photobionts reduce the production and release of soluble carbohydrates quite drastically and within short periods upon isolation from the thallus. In the isolated state they synthesize more insoluble carbohydrates than in the symbiotic state (37, 97). 3. Neither pH changes nor the addition of thallus homogenates or mycobiont-derived secondary metabolites stimulates the production and release of soluble carbohydrates in photobionts kept on agar or in liquid media (for summaries see 45, 105, 106). However, the release of substantial amounts of ribitol and sucrose was reported in aposymbiotic cultures of *Trebouxia* spp. kept on moist filter paper (88). 4. There are significant quantitative differences with regard to the release of soluble carbohydrates between different taxa of photobionts, even within closely related, anatomically very similar lichen species (94). 5. Mycobionts convert the mobile carbohydrate into mannitol, irrespective of their taxonomic affiliation (105). The mechanisms of carbohydrate release are not yet understood.

Two sets of data indicated that inhibition-technique experiments probably did not allow the recovery of the whole fraction of mobile compounds moving from photobiont to mycobiont: First, the meager and slow translocation rates [2–4% of total fixed C within 3 hr (94)] in heteromerous Lecanorales and Teloschistales with *Trebouxia* photobiont are unlikely to satisfy the nutritional requirements of the quantitatively predominant mycobiont. The ecologically most successful species of extreme environments (arctic-alpine, antarctic, and desert ecosystems) belong to this group of lichens (reviewed in 69). Second, the rates of translocation in closely related *Peltigera* species with either cyanobacterial *(Nostoc* sp.), or green algal *(Coccomyxa* sp.) photobionts (94) are quantitatively quite different. Both types of lichens grow equally fast, and photosymbiodemes have been reported (8, 110). Photosymbiodemes are different phenotypes, often found intermixed with each other as "chimerae," presumably formed by the same mycobiont in symbiosis with either a cynobacterial or green algal photobiont (67). Because the fungal material involved in individual lichen thalli is probably genetically heterogeneous, one cannot be certain that the same fungal genotype is capable of producing different phenotypes in association with either photobiont. Translocation of 20–40% of total fixed C within 3 hr was measured in *Peltigera* spp. with *Nostoc* cyanobionts, but only 10–15% within 3 hr in *Peltigera aphthosa* with *Coc-*

comyxa photobiont (94). The photobiont cell number per thallus area was not significantly different in the two types of symbiosis.

Comparative light microscopy (LM) investigations of the mycobiont-photobiont interface in different types of lichens had already revealed significantly different modes of mycobiont-photobiont interactions in distantly related lichen taxa (93, 111). Therefore a detailed analysis of the mycobiont-photobiont interface, with special regard to the possible routes of translocation of photosynthates, water, and mineral nutrients between the partners in different types of highly evolved lichens, seemed desirable. Such studies had to focus not only on structural aspects, but also on the chemical composition of the cell wall of the photobiont.

STRUCTURAL AND FUNCTIONAL ASPECTS OF THE MYCOBIONT-PHOTOBIONT INTERFACE

The photobiont cells of morphologically complex foliose and fruticose lichens are not casually dispersed within the thallus but are positioned by the fungal partner at the periphery of the gas-filled thalline interior underneath the conglutinate cortical layer (Figures 1*b–c*, 5), an optimal situation with regard to gas exchange and illumination. The mycobiont-photobiont interface of such lichens fulfils a triple function. It is the site of carbohydrate mobilization; and the passive water flow, enriched with dissolved mineral nutrients (18) and presumably also with fungal metabolites, reaches the photobiont cells at the contact site. Moreover, the mycobiont shifts the photobiont cells by means of intercalary growth processes over short distances within the photobiont layer (38, 51, 55, 57). There are no free photobiont cells in heteromerous thalli. All such cells are connected to the mycobiont, the relationship depending on the taxonomic identity and on structural peculiarities of the partners involved.

The photobiont layer of heteromerous thalli appears surprisingly uniform with regard to photobiont cell numbers per thallus area. However, the photobiont cell population is not homogenous with regard to cell size and metabolic activity (25, 46, 47, 54, 55, 80). Both smallest green algal photobiont cell sizes (as a result of regular autospore formation) and highest photosynthetic activity were found in growing thalline areas [e.g. thallus margins in species with marginal growth (25, 46, 47)]. Nongrowing zones, on the other hand, contained few or even no autosporangia; instead, they exhibited a high proportion of outsized photobiont cells that have exceeded the size required for cell division or sporulation, respectively, without forming autospores (25, 46). Shortage of mineral nutrients and the quantitatively significant export of photosynthates to the fungal partner are likely to be among the factors that regulate cell turnover of the endosymbiotic photobiont cell population (46, 47), but neither explains the high percentage [>50% in *Parmelia sulcata*

(25)] of outsized photobiont cells in nongrowing thalline areas. A mycobiont-derived, possibly reversible inhibitory factor was postulated (55), its chemical nature being as yet unknown. Mycobiont-derived secondary metabolites may be involved in regulatory processes, at least in taxa where crystalline phenolic secondary products have been observed at the mycobiont-photobiont interface [e.g. Parmeliaceae (53, 55)].

Mycobiont-Cyanobiont Relationships in the Peltigera-Nostoc Symbiosis (>30 spp.)

Colonies of the heterocyst-bearing, filamentous genus *Nostoc* are enclosed within a hygrophilic, gelatinous sheath composed of fibrillar polysaccharides [presumably glucans (27)] embedded within an amorphous matrix (49). Thin-walled intragelatinous fungal protrusions invade this gelatinous sheath, presumably by means of hydrolytic enzymes (Figures 2*a,b*). Intragelatinous protrusions are found close to the cyanobacterial cells, which are, however, never attacked by the mycobiont. Neither the outer membrane nor the murein sacculus of the *Nostoc* cells is penetrated by the fungal partner (6, 49). Blebbing and the release of small vesicles by the cyanobacterial outer membrane have been observed in distantly related symbiotic and nonsymbiotic cyanobacteria (6, 15, 16, 49, 91). In the *Nostoc* cyanobiont of *Peltigera* spp. these vesicles are released not only in the vicinity of the intragelatinous protrusions, but over the whole cell surface as well. They are likely involved in sheath synthesis (6, 49).

The thin-walled intragelatinous protrusions are lateral outgrowths of the aerial hyphae of the uppermost part of the medullary layer. These aerial hyphae are covered by a mainly proteinaceous, hydrophobic cell wall surface layer with distinct rodlet pattern (49), a common feature of aerial hyphae of asco- and basidiomycetes and imperfect fungi (summarized in 50, 59). Intragelatinous protrusions lack this particular wall surface layer; they are not aerial hyphae. Water and dissolved mineral nutrients are taken up by the cortical layer, the mechanisms of uptake and translocation being poorly understood in the Peltigeraceae. Goebel (33, 34) postulated rapid water uptake by the very large cortical cells, which he supposed to be dead and empty. His observations have not been tested experimentally. Cortical cells of submarginal thallus areas (near the growing edge) appeared viable but highly vacuolate in transmission-electron-microscopic (TEM) studies (49; R. Honegger, unpublished results); older parts of the thallus have not been examined. The route of passive water flow in the transition zone between the conglutinate cortical layer and the gas-filled thalline interior is unknown, but it can be excluded that the hydrophobic surfaces of the aerial hyphae are getting wet (33, 34; R. Honegger, unpublished results). Rapid water translocation towards the photobiont cells is likely to occur within the apoplastic space of the fungal cell wall underneath the hydrophobic wall surface layer.

Peltigera - Nostoc symbiosis: intragelatinous fungal protrusions

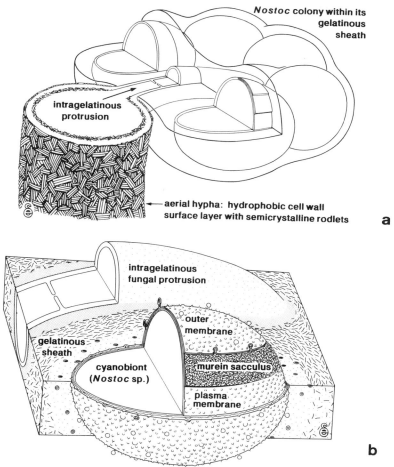

Figure 2 Diagrams illustrating the mycobiont-photobiont interface in *Peltigera* spp. with *Nostoc* cyanobiont. Ultrastructural data from 6, 42, 51; diagrams designed by Sybille Erni.

Most *Peltigera* species grow in moist habitats. Although tolerating desiccation, they are adapted to continuously moist conditions, an advantageous property with regard to laboratory culturing (107, 108). Although substantial amounts of water [~ 95% of thallus dry weight at 97% relative humidity (rh)] can be taken up by the mycobiont of *Peltigera* spp. from vapor, *Nostoc* photobionts start CO_2 fixation only after wetting with liquid water (75, 78). This feature was also observed in the *Nostoc* photobionts of diverse heteromerous Peltigerales and heteromerous and homoeomerous Lecanorales, and in nonsymbiotic *Nostoc* spp. (73–75, 78). By measuring 77 K fluorescence emission and excitation spectra of drying and rehydrating *Peltigera* thalli with *Nostoc* cyanobiont, Bilger and coworkers (5) demonstrated that the energy transfer from the phycobiline pigments to photosystem II is interrupted by desiccation, presumably owing to a functional detachment of the phycobilisomes, and can be restored only when rehydration occurs with liquid water. The same physiological properties were observed in the nonsymbiotic *Nostoc commune* (5), and in the *Nostoc* cyanobiont of the gelatinous, homoeomerous *Collema flaccidum* (102).

It seems easy to correlate structural and physiological data in the *Peltigera-Nostoc* symbiosis. Inhibition-technique experiments were carried out at water saturation when both partners were metabolically active. After horizontal dissection of thallus fragments and incubation on liquid media the *Nostoc* colonies became soaked and even covered by the incubation fluid. Symbiotic and nonsymbiotic *Nostoc* spp. and other cyanobacteria with gelatinous sheaths invest a high proportion of photosynthetically fixed C in the synthesis of sheath polysaccharides (summarized in 23) which, in the symbiotic system, are readily hydrolyzed by enzymes [probably glucanases (44)] released by the intragelatinous fungal protrusions. Glucose as the product of this hydrolytic activity was leaking out into the incubation medium, whence it could be recovered and analyzed.

Mycobiont-Photobiont Relationships in the Peltigeracae-Coccomyxa symbiosis (~ 15 spp.)

The chlorococcalean genus *Coccomyxa* comprises relatively common and widespread nonsymbiotic species and photobionts of various nonrelated, small groups of lichen-forming asco- and basidiomycetes (58, 63, 111, 113). *Coccomyxa* cell walls are tripartite (Figures 3*a,b*): An amorphous innermost layer of quite unequal thickness (probably partly as a result of artificial swelling during preparative procedures for conventional freeze-etching and ultrathin sectioning for TEM) borders upon a thin (30 ± 10 nm), electron-dense layer of regular thickness mainly composed of short cellulose fibers (10, 11, 63, 92). A thin (15 ± 5 nm) trilaminar outermost wall layer of membrane-like appearance revealed particularly interesting structural and chemical properties. It is brittle owing to a thin layer of sporopollenin, an enzymatically nondegradable biopolymer located in the electron-transparent central part of the trilaminar layer. This particular wall component remains intact after exhaustive extraction with a range of organic solvents, saponification under alkaline conditions, and incubation in 85% phosphoric acid at 50–55°C for 13–30 days (10, 11). The infrared (IR) spectra of *Coccomyxa* sporopollenin were almost identical to those of the hydrocarbon-rich, chlorophycean *Botryococcus braunii* (3); both of these spectra differed from that of *Lycopodium* sporopollenin, which is used by most investigators as reference material, at the wave number of 720 cm^{-1} (3, 10, 11). Recent ^{13}C NMR studies on sporopollenin biosynthesis indicated that green algal sporopollenins are derived from fatty acids, while *Lycopodium* sporopollenin seems to polymerize from carotenoids (39). Sporopollenin-like components whose IR spectra were almost identical with those of *Lycopodium* sporopollenin have been reported in a wide range of symbiotic and nonsymbiotic algae subjected, either as entire cells or as nonextracted wall fragments, to acetolysis—a widely used but chemically less than optimal preparative procedure (summarized in 11). During acetolysis the ubiquitous carotenoids (cytoplasmic or wall bound) polymerize to a resistant material, even in cell wall samples that contain no sporopollenin [e.g. cultured green algal lichen photobionts of the genera *Trebouxia, Myrmecia,* and *Pseudochlorella,* or *Hibiscus* leaf cell walls (10, 11)]. [Reports on sporopollenin in the *Myrmecia* photobiont of *Baeomyces rufus* (63) were based on a mis-identification of the photobiont, which has been described as an *Elliptochoris* sp. (112). *Elliptochloris* was the only genus besides *Coccomyxa* among the chlorophycean lichen photobionts so far investigated that contained sporopollenin (10, 11).]

The sporopollenin-containing outermost wall layer plays a key role in the symbiotic relationship between peltigeracean mycobionts and their *Coccomyxa* photobionts. Simple wall-to-wall apposition and very tight adhesion of the hydrophobic cell wall surface layer of the mycobiont to the trilaminar sheath of the *Coccomyxa* cell wall were observed in freeze-fracturing preparations and ultrathin sections (Figures 3a,b; 50, 51, 58, 63). During autospore formation the amorphous inner layer and part of the cellulosic middle layer are enzymatically degraded, but the trilaminar sheath persists as a nondegradable mother cell wall. Stacks of mother cell walls were observed within the algal layer of the thallus, and substantial numbers of mother cell walls with characteristic convolute margins (Figure 3a) were isolated from liquid media of aposymbiotically cultured *Coccomyxa* spp. (10, 11, 63).

As in the *Peltigera-Nostoc* symbiosis the modes of water uptake and translocation from the cortical layer to the thalline interior are not known in the Peltigeraceae with endosymbiotic *Coccomyxa* spp. Moreover, it is unknown how carbohydrates are translocated from *Coccomyxa* cells to the tightly adhering mycobiont hyphae. The sporopollenin-containing outermost wall layer of *Coccomyxa* cells is permeable for relatively small molecules but not for compounds with higher molecular weight (e.g. aniline blue, M_r 738; 10, 11).

It is particularly interesting that *Coccomyxa* cells, in contrast to cyanobacterial photobionts, start photosynthesis at thalline moisture contents around 70% of thallus dry weight, which can be achieved, at least under laboratory conditions, by the use of water vapor at about 97% relative

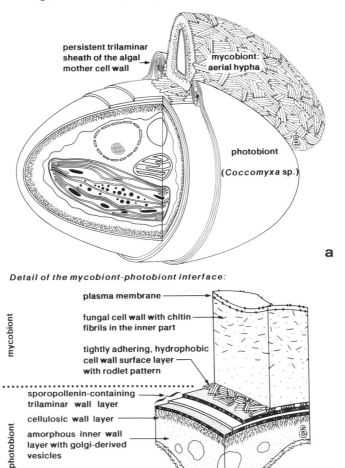

Peltigera - *Coccomyxa* symbiosis: wall - to - wall apposition

persistent trilaminar
sheath of the algal
mother cell wall

mycobiont:
aerial hypha

photobiont

(*Coccomyxa* sp.)

a

Detail of the mycobiont-photobiont interface:

plasma membrane

fungal cell wall with chitin
fibrils in the inner part

tightly adhering, hydrophobic
cell wall surface layer
with rodlet pattern

mycobiont

sporopollenin-containing
trilaminar wall layer

cellulosic wall layer

amorphous inner wall
layer with golgi-derived
vesicles

plasma
membrane

photobiont

b

Figure 3 Diagrams illustrating the mycobiont-photobiont interface in *Peltigera* and *Solorina* spp. with *Coccomyxa* photobiont. Note: algal cell wall dimensions in Figure 3*a* are not to scale. Data from 10, 11, 49–51, 58, 63; diagrams designed by Sibylle Erni.

humidity (75). The same physiological properties were observed in other Peltigerales with either cyanobacterial or green algal photobionts (73–75). Liquid water is required by cyanobionts, but not by green algal photobionts, for the restoration of the photosynthetic apparatus after a drought-stress event, but photosynthetic maxima were attained in water-saturated "wet" thalli [~ 250% moisture content per thallus dry weight (75, 98)]. The massive cortical layer of water-saturated thalli is glass-like and highly translucent (21). The energy transfer, as concluded from chlorophyll fluorescence, is highest in photobionts of water-saturated thalli (68).

Symbiotic *Coccomyxa* photobionts produce substantial amounts of ribitol, part of which was released into the incubation medium in either inhibition-technique experiments (94) or immediately after isolation from the thallus (37). When kept in liquid media, isolated *Coccomyxa* cells reduce drastically the production and release of ribitol. Instead, sucrose is produced as soluble carbohydrate, and distinctly more ethanol-insoluble compounds are synthesized than in the symbiotic state (37).

Only 10–15% of total fixed C was released as ribitol by symbiotic *Coccomyxa* cells of *Peltigera* spp. in inhibition-technique experiments (94). This is presumably not the whole fraction of mobile carbohydrates, but we can only speculate about the reasons for these meager translocation rates. The mycobiont-photobiont interface in the Peltigeraceae-*Coccomyxa* symbiosis is more hydrophobic than in the *Peltigera-Nostoc* system. It is likely that in inhibition-technique experiments only a portion of the *Coccomyxa* cells have been wetted and immersed; the mobile fraction of the other cells was not leaking out into the incubation medium. A comparative investigation of the carbohydrate metabolism of Peltigerales with either cyanobacterial or green algal photobionts with ^{13}C NMR techniques, as already performed in the *Xanthoria-Trebouxia* symbiosis (86), might be particularly interesting.

Mycobiont-Photobiont Relationships in Foliose Parmeliaceae with Trebouxia Photobionts (~ 750 spp.)

The pleurastralean genus *Trebouxia* De Puymaly [Ulvophyceae (103); taxonomy according to (32)] comprises about 24 species; 2 are nonsymbiotic, and the rest are photobionts of various groups of distantly related lichen-forming ascomycetes. More than 40% of all lichen mycobionts and the majority of highly evolved foliose and fruticose taxa associate with *Trebouxia* spp. all of which are seldom found in the aposymbiotic state in most habitats that can support lichen growth. The morphologically complex foliose and fruticose thalli of the Parmeliaceae are the product of a long cohabitation of lecanoralean ascomycetes with *Trebouxia* spp. That this relationship is very successful may be concluded from the ecological breadth of the partners and from the structural and taxonomic diversity of the fungal exhabitant (40, 41). In lichens the symbiotic way of life has triggered considerable evolutionary innovation in the fungal partner while the endosymbiotic photobionts have remained simple (62). A high percentage of Parmeliaceae have reduced the sexual reproductive cycle and disperse efficiently by means of symbiotic vegetative propagules; thus they elegantly overcome problems related to the acquisition of compatible *Trebouxia* cells, problems that must be solved at each reproductive cycle by sexually reproducing lichen mycobionts (90). Parmeliaceae (like most Lecanorales) and their *Trebouxia* photobionts are adapted to regular wetting and drying cycles; many tolerate extreme climatic conditions. A large number can be stored at $-20°C$ in the desiccated state (24, 89), but none can be kept or cultured under continuously wet conditions.

Trebouxia cells are characterized by their large, central chloroplast which in most species carries a central pyrenoid (29, 32). *Trebouxia* cell walls are composed mainly of a fibrillar inner part and an amorphous outer layer with acidic polysaccharides and proteins (13, 14, 50). Species-specific modifications of the outer wall layer were detected in aposymbiotically cultured *Trebouxia* isolates (28). The fibrillar elements are cellulosic in some taxa, but yield, upon hydrolyzation, glucose and mannose in others (10, 71). Following autospore formation the autosporangial wall is degraded, presumably (as concluded from aposymbiotically cultured *Trebouxia* isolates) by means of alga-derived hydrolytic enzymes (50). No resistant residue was obtained after exhaustive extraction, saponification under alkaline conditions, and subsequent phosphoric acid treatment (85%, 50–55°C, 13–30 days) of isolated cell walls (10, 11). However, acetolysis of insufficiently extracted cell wall preparations yielded small amounts of resistant material with infrared spectra comparable to those of *Lycopodium* sporopollenin (presumably artificial polymers of cell wall bound carotenoids); these must be regarded as method-dependent preparative artifacts (10, 11) rather than as sporopollenins (70, 71).

A high percentage of *Trebouxia* cells of numerous primitive crustose lichens are invaded by intracellular fungal haustoria (51, 52, 93, 111). Parmeliacean mycobionts do not normally produce intracellular protrusions but contact their *Trebouxia* photobionts by means of intraparietal haustoria [formerly termed "intramembraneous haustoria" (111)], a peculiarity of highly evolved lecanoralean mycobionts and a product of a highly coordinated developmental process in both partners (Figures 4a,b; 50–52). The structurally complex mycobiont-photobiont interface of Parmeliaceae was recognized by earlier light microscopists, many of whom, even in our century and long after the discovery of the dual nature of lichens by Schwendener (101), could not believe that these *Trebouxia* cells are not asexually produced spores ("gonidia") of the contacting fungal hyphae—i.e. the haustorial complex (e.g. 20). Parmeliacean mycobionts contact juvenile *Trebouxia* cells at a very early developmental stage when the latter are still enclosed within the degrading mother cell wall, and long before they have attained their full size (Figure 4b; 26, 38, 51, 52, 54). During maturation the developing *Trebouxia* cells are shifted over short distances by means of intercalary growth processes of the fungal partner within the haustorial complex (Figure 4b; 51).

The fungal elements involved in haustorium formation are aerial hyphae of the gas-filled thalline interior. They have a thin, highly hydrophobic wall surface layer built up by proteins, lipids, and mycobiont-derived, partly

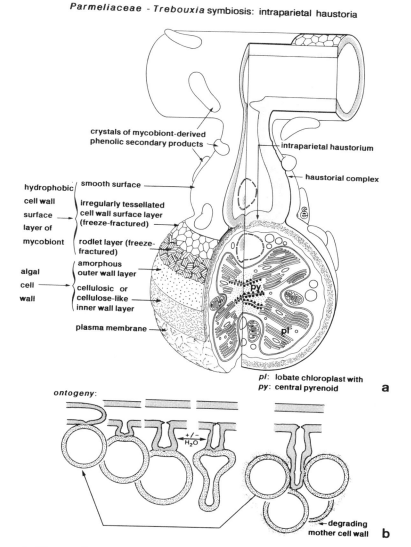

Parmeliaceae - Trebouxia symbiosis: intraparietal haustoria

crystals of mycobiont-derived
phenolic secondary products

intraparietal haustorium

haustorial complex

hydrophobic
cell wall
surface
layer of
mycobiont

smooth surface

irregularly tessellated
cell wall surface layer
(freeze-fractured)

rodlet layer (freeze-
fractured)

amorphous
outer wall layer

algal
cell
wall

cellulosic or
cellulose-like
inner wall layer

plasma membrane

pl: lobate chloroplast with
py: central pyrenoid

a

ontogeny:

+/−
H₂O

degrading
mother cell wall

b

Figure 4   Diagrams illustrating the mycobiont-photobiont interface in the Parmeliaceae and their *Trebouxia* photobionts. Data from 9, 50–53, 55, 56; Figure 4a designed by Sibylle Erni.

phenolic secondary compounds. These components crystallize on and within this cell wall surface layer, thus enhancing its hydrophobicity. In freeze-fracture preparations this hydrophobic wall surface layer has a characteristic, irregularly tessellated pattern in all Lecanorales so far examined (Figure 4a;

26, 50, 51, 53, 56, 58). At the immediate contact site between the growing hyphal tip and the young, not fully developed *Trebouxia* autospores of *Cladonia* spp. (Lecanorales), the hydrophobic cell wall surface layer was found to be composed of a rodlet layer. This layer is soon covered by an amorphous, probably lipidic and phenolic material that obscures the rodlets, each little unit in the mosaic of the irregularly tessellated surface layer corresponding to a bundle of rodlets (50, 51, 58). The hydrophobic cell wall surface layer spreads from the mycobiont over the wall surface of the *Trebouxia* photobiont at the very first contact, thus sealing the photobiont cell with a thin layer of water-repellent material (Figure 4a; 50, 53, 58). The significance of this sealing process with regard to the translocation of solutes in the apoplastic continuum between the symbionts is obvious. Water and dissolved nutrients are passively taken up by the gelatinous extracellular material of the conglutinate cortical layer, then passively translocated towards the algal cells within the apoplastic space of the fungal cell wall; they finally reach the photobiont cells at the haustorial complex (Figure 5). Fungal and algal cell wall surfaces of the gas-filled thalline interior remain dry even at water saturation [as concluded from low-temperature scanning electron microscopy (LTSEM) and histochemical studies (9, 56)], an optimal situation with regard to the gas exchange of the photobiont.

Parmeliaceae and other Lecanorales with *Trebouxia* photobiont are capable of photosynthetically fixing CO_2 at low thalline moisture contents; liquid water is not required for the restoration of the photosynthetic apparatus after drought stress (72, 74). Apparent CO_2 uptake from the thalline exterior was measured in some species at 20% moisture content per thallus dry weight, and temperature-dependent optima at $60 \pm 20\%$ (72, 74) can be achieved, at least in laboratory experiments, after several hours' incubation at 97% relative humidity (64, 72, 74, 98). Nevertheless, highest rates of energy transfer, as determined by chlorophyll fluorescence emission and excitation spectra at 77 K, were recorded in water-saturated thalli after spraying with liquid water (5).

The very low amounts and slow rates of translocation of carbohydrates [2–4% of total fixed C as ribitol within 3 hr (94) as measured with the inhibition technique] must be seen in correlation (a) with the complex haustorial apparatus, which is likely the site of active uptake, and (b) mainly with the hydrophobicity of the wall surfaces of both symbionts, which protects the cells from getting wet. Solutes released into the apoplastic space of the cell wall do not leak from the sealed algal cells unless the interface has been damaged during dissection. Therefore it seems reasonable to assume that only a minor fraction of the mobile carbohydrates has been recovered in isotope-trapping experiments. Ecophysiologists refer to the possibility that water saturation, as applied in inhibition-technique experiments, might be a less than optimal situation with regard to photosynthesis in this type of lichen (see

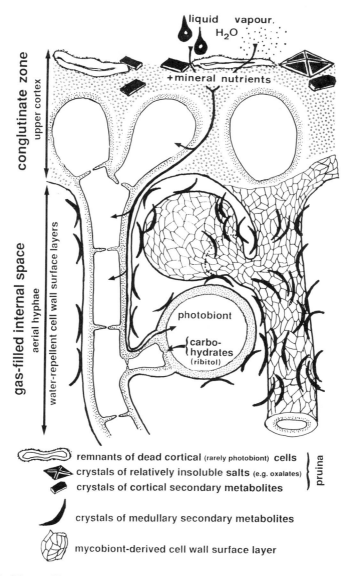

conglutinate zone
upper cortex

gas-filled internal space
aerial hyphae

water-repellent cell wall surface layers

liquid   vapour.
H₂O

+mineral nutrients

photobiont

{carbo-
hydrates
(ribitol)

remnants of dead cortical (rarely photobiont) cells
crystals of relatively insoluble salts (e.g. oxalates)   } pruina
crystals of cortical secondary metabolites

crystals of medullary secondary metabolites

mycobiont-derived cell wall surface layer

Figure 5 Diagram illustrating functional aspects in Parmeliaceae with *Trebouxia* photobionts. Comparable, horizontally dissected thallus fragments were used in inhibition-technique experiments (94). There is an apoplastic continuum between the fungal partner and the *Trebouxia* cells. After 59.

below). The presumed decline both of photosynthetic activity at high thalline water contents and of green algal and fungal polyol metabolism at different levels of thallus hydration is of central interest in lichen physiology.

CO_2 UPTAKE AND PHOTOSYNTHESIS IN HETEROMEROUS LICHENS

All foliose and most of the fruticose Lecanorales and Teloschistales form a peripheral conglutinate cortical layer that covers the entire thallus surface (Figure 1c). Cortical layers on the upper and lower thallus surfaces are also formed by some representatives of the Peltigerales, while others (e.g. the genera *Peltigera* and *Solorina*) have an upper cortex only (Figure 1b). The dimensions of the cortex may vary within different ecotypes, but its basic anatomical features are species specific (e.g. 79, 81) and have often been used as taxonomic markers (e.g. 7). The multifarious functions of cortical layers are far from being understood. The cortex is the site of water uptake, its anatomy and chemical composition influencing the imbibition rate (81). It provides mechanical stability to the whole thallus, being relatively soft and cartilaginous when wet and less elastic or even brittle when dry. Light absorption and transmission are influenced by the state of hydration of the cortex, but also by its thickness and by such contents as insoluble mineral complexes (e.g. oxalates of Ca, Cu, Mg, Mn) and mycobiont-derived, crystalline secondary products that may absorb, reflect, or transform light, or form insoluble complexes with metal ions (summarized in 61). The onset of cortex formation in early ontogenetic stages, possibly as a result of the establishment of a successful nutritional relationship between compatible partners, marks an important developmental step in the expression of the symbiotic phenotype of the fungal genotype (1, 2, 57, 62, 90, 100, 107, 108).

The cortical layer plays a key role in gas exchange. Irregularly thick cortical layers often have small, pore-like discontinuities, while the massive, regularly thick cortex of other species is interrupted by regularly formed aeration pores (pseudocyphellae) built up by groups of nonconglutinate, loose aerial hyphae with hydrophobic cell wall surfaces (summarized in 61, 66). When thalline water content is low the cortical layer allows CO_2 diffusion; the swollen extracellular gelatinous matrix of the fully hydrated cortex blocks CO_2 diffusion, as concluded from measurements of CO_2 uptake from the thalline exterior at different levels of thalline hydration in a wide range of heteromerous lichens (reviewed in 69, 76). In Lecanorales and Teloschistales, highest CO_2 uptake from the thalline exterior was recorded at thalline water contents below 80% of thallus dry weight, while a temperature-dependent decline (most prominent at around 20°C and smallest around 0°C) was recorded at water saturation [$\sim$ 150% moisture content per thallus dw (69, 76,

77)]. On the assumption of linear correlations between CO_2 uptake from the thalline exterior and photosynthetic activity of the endosymbiotic photobiont cell population, this "water depression" of CO_2 uptake is usually interpreted as a decline in photosynthesis at high thalline water contents. However, the carboxylation capacity of *Trebouxia* photobionts seems not to be affected by high water contents (77). Light transmission through the cortical layer is best in the fully hydrated state (21). Based upon fluorescence emission and excitation spectra at 77 K (ratio of relative fluorescence yields, ϕ_{480}/ϕ_{438}), the energy transfer between the light-harvesting pigment protein complex and photosystem II is highest in water-saturated thalli. [Measurements were made in both laboratory (5) and field experiments (99, and B. Schröter, personal communication).] Are *Trebouxia* photobionts condemned to cease photosynthesis under these highly favorable (at least theoretically) conditions because of a shortage in CO_2?

The endosymbiotic algal cell population of heteromerous lichens amounts to less than 20% of the thalline biomass [10–15% in *Parmelia sulcata* (25)]. A large proportion of the thalline volume [30–50% in *P. sulcata* (25)] is gas-filled intercellular space in the medullary and algal layers. It seems reasonable to postulate highest metabolic activity of the fungal partner at water saturation. If this assumption is correct one might expect enough mycobiont-derived respiratory CO_2 within the gas-filled thalline interior to secure the photosynthetic activity of the green algal endosymbiont. It is at least imaginable that CO_2 uptake from the thalline exterior is not linearly linked to the photosynthetic activity of the endosymbiotic photobiont cell population over all levels of thallus hydration. Vascular plant modelling of CO_2 exchange is unlikely to be applicable in a system where the photoautotroph is hidden within more or less compact peripheral structures of a quantitatively predominant C-heterotroph whose consumptive behavior is largely unknown. Fully hydrated heteromerous lichens may resemble tiny bottle gardens with mycorrhizal plants inside whose external CO_2 level remains unchanged while the photoautotrophic inhabitants are photosynthetically active. In this context it might be interesting to examine 1. the CO_2 pressure within the thalline interior of different types of heteromerous lichens (with only upper, or with upper and lower cortex, respectively) at different levels of thallus hydration and illumination, and 2. the temperature optima of aposymbiotically cultured lichen myco- and photobionts.

POLYOL METABOLISM AND WATER RELATIONS OF LICHEN MYCOBIONTS AND PHOTOBIONTS

The biological role of polyols in lichens deserves further investigation because they may play a more diverse and important physiological role than previously assumed (22). Polyols have been referred to as storage products, and contents of up to 10% of thallus dry weight have been reported in the symbiotic phenotype (19, 35, 85, 87). However, polyols can certainly not be regarded as safe storage products, because they are washed from thalli in substantial amounts in the rewetting phase after drought stress (19, 87). Polyols are more likely to be stress metabolites with important physiological functions in the desiccation tolerance of lichen mycobionts and green algal photobionts, especially as protecting agents of enzyme systems during stress events, and as turgor regulators ("osmoregulators") during the wetting and drying cycles (4, 22). The severity of the regularly returning drought stress events may often have been underestimated as a result of relatively imprecise concepts regarding thalline water relations.

Until recently it was believed that the relatively low cellular water contents of mycobiont and photobiont cells in lichen thalli would not be strongly affected during the regular wetting and drying cycles (64). Water was assumed to be taken up and kept by capillary forces between the mycobiont and photobiont cells of the thalline interior. This capillary water was supposed to be lost by evaporation during drying, followed by the water that had been stored by the gelatinous matrix of the conglutinate cortex (64). However, as has been shown in ultrastructural and histochemical studies (9, 33, 34, 51–53, 55, 56), the wall surfaces of fungal and green algal cells of the thalline interior are more or less strongly hydrophobic, depending on the taxonomic identity of the partners involved (see previous paragraphs). There is no capillary water on the hydrophobic wall surfaces of the thalline interior. Drying means, first, the loss of water from the apoplastic space of the cell walls of both partners, and second, partial loss of cellular water, leading to drastic but reversible cell shrinkage, as observed in LTSEM (9). This drastic cell shrinkage can only be visualized after physical fixation by rapid freezing, but not with conventional preparative procedures (chemical fixation) for electron microscopy.

It is important to note that the water flow that moves passively back and forth within the apoplastic continuum during the regularly occurring de- and rehydration processes translocates not only dissolved mineral nutrients but also a whole range of passively (during drought stress) and actively released metabolites from both mycobiont and photobiont. At least over short distances, mycobiont-derived secondary metabolites are translocated that are released in a yet unknown form (glycosides?) and crystallize within and on the hydrophobic wall surface layer of aerial hyphae of lecanoralean mycobionts and on their *Trebouxia* photobionts (53). Some of the soluble carbohydrates of the apoplastic space are washed out by rain (natural or artificial) and thus can be analyzed. In the rehydration phase after drought-stress events, carbohydrates in the range of 0.2–1.7% of thallus dry weight, corresponding to 7.5–24% of the total alcohol-soluble carbohydrates of the thallus, were lost from *Peltigera* spp. with either *Nostoc* cyanobiont, or *Coccomyxa* photobiont, and from 3 lecanoralean species (2 Parmeliaceae) with *Trebouxia* photobiont; the ethanol-soluble fraction was almost perfectly paralleled in the carbohydrates detected in the leakage fluids (87). These leakage fluids contained mannitol and smaller amounts of trehalose, both presumably of fungal origin, and either ribitol from green algal or glucose from cyanobacterial photobionts besides short-chain oligosaccharides of unknown origin (87). To what extent these leakage losses occur in the natural habitat and how they affect the carbon budget of the symbiotic system remain to be investigated, but in laboratory experiments leakages provide an interesting opportunity to study the types of molecules present in the apoplastic continuum of lichen thalli.

There is no contradiction between data gathered from leakage-fluid analyses and those obtained with the inhibition technique. Leakage fluids recovered from the thalline surface are likely to contain distinctly more mycobiont-derived compounds than those originating from the immediate vicinity of the photobiont cells. In ^{13}C NMR studies in *Xanthoria calcicola* with *Trebouxia* photobiont [a foliose Teloschistales with a mycobiont-photobiont interface structurally and functionally comparable to that in Parmeliaceae (57)], ribitol was found to move from the photobiont to the mycobiont, where it was readily converted into arabitol and mannitol (86), but obviously in higher rates than previously assumed on the basis of inhibition-technique experiments (94–96).

The drastic changes in cellular water contents experienced, often in combination with heat or cold stress, by both partners of desiccation-tolerant lichen species, all of which recover rapidly and within short periods after a stress event, necessitate 1. a strong conservation of metabolic systems during the stress period, and 2. mechanisms for rapid repair and efficient restitution of membrane systems, especially of the plasma membrane (4). The considerable amounts of carbohydrates (polyols and others) recovered from leakage fluids may result from passive release after stress-induced membrane alterations.

By isolating green algal photobionts from lichen thalli and subsequent culturing under continuously moist conditions, or even in liquid media enriched with glucose (e.g. 37), these algal cells are allowed to adapt to conditions extremely different from those in the symbiotic state. It is not surprising that their carbon metabolism is no longer the same as in the lichen thallus [less polyol, more sucrose, and more insoluble carbohydrates are produced compared to the symbiotic state (37, 97)]. Studies on the metabolic activities at different levels of thallus hydration in a parmeliacean *Ramalina* sp. with the *Trebouxia* photobiont indicate that biochemical events associated with sugar alcohol metabolism operate at lower water contents than those associated with the production of either sugar, sugar phosphates, or lipids (17). It might be particularly interesting to investigate the carbohydrate metabolism and fine structure of aposymbiotic green algal photobionts cultured under various conditions, including fluctuating moisture contents and regular drought stress events.

CONCLUDING REMARKS

Slow-growing organisms of little or no economic value (but nevertheless of high ecological importance) such as lichens, many of which have successfully adapted to life under harsh conditions, are not of central interest in a scientific world where rapid plant growth and high crop yields are more and more influencing the choice of research topics and research organisms. Lichenologists, and even more so lichen physiologists, are minority groups among mycologists and botanists. Accordingly, less is known about the physiological aspects of lichens than about those of other symbiotic systems, such as the *Rhizobium*-legume relationship. Neither the classical nor the molecular genetics of lichen mycobionts or photobionts has been investigated. Culturing of the symbiotic system under defined, axenic conditions is still not routinely performed, the factors triggering the expression of the symbiotic phenotype in the fungal partner being largely unknown. Nevertheless, lichens do have an advantage in that their mycobiont-photobiont interface is easily accessible for ultrastructural and physiological studies of possible modes and routes of solute translocation between the symbionts (104). Certainly lichens can not serve as model systems for investigating fluxes between photoautotrophs and fungal cells in general. Besides considerable differences in life strategies there is too much diversity within the kinds of mycobiont-photobiont relationships in different types of lichens, all of which, on the other hand, differ significantly from the various types of interactions in mycorrhizae and plant pathogenic associations (59). Nevertheless, comparisons may be fruitful for designing new experimental approaches in either one of these systems.

ACKNOWLEDGMENTS

I am very grateful to Sibylle Erni for skillfully designing Figures 2–4a. My sincere thanks are due to Drs. B. Eller, L. Kappen, and B. Schröter for stimulating discussions.

Literature Cited

1. Ahmadjian, V., Jacobs, J. B. 1982. Artificial reestablishment of lichens. III. Synthetic development of *Usnea strigosa. J. Hattori Bot. Lab.* 52:393–99
2. Ahmadjian, V., Jacobs, J. B. 1985. Artificial reestablishment of lichens. IV. Comparison between natural and synthetic thalli of *Usnea strigosa. Lichenologist* 17:149–65
3. Berkaloff, C., Casadevall, E., Largeau, C., Metzger, P., Peracca, S., Virlet, J. 1983. The resistant polymer of the walls of the hydrocarbon-rich alga *Botryococcus braunii. Phytochemistry* 22:389–97
4. Bewley, J. D., Krochko, J. E. 1982. Desiccation tolerance. In *Physiological Plant Ecology II. Water Relations and Carbon Assimilation/Encycl. Plant Physiol.*, New Ser., ed. O. L. Lange, P. S. Nobel, C. B. Osmond, H. Ziegler, 12B:325–78. Berlin: Springer-Verlag

5. Bilger, W., Rimke, S., Schreiber, U., Lange, O. L. 1989. Inhibition of energy-transfer to photosystem II in lichens by dehydration: different properties of reversibility with green and blue-green phycobionts. *J. Plant Physiol.* 134:261–68

6. Boissière, M.-C. 1982. Cytochemical ultrastructure of *Peltigera canina*: some features related to its symbiosis. *Lichenologist* 14:1–28

7. Bowler, P. A. 1981. Cortical diversity in the Ramalinaceae. *Can. J. Bot.* 59:437–52

8. Brodo, I. M., Richardson, D. H. S. 1978. Chimeroid associations in the genus *Peltigera*. *Lichenologist* 10:157–70

8a. Brown, D. H., ed. 1985. *Lichen Physiology and Cell Biology*. New York: Plenum

8b. Brown, D. H., Hawksworth, D. L., Bailey, R. H., eds. 1976. *Lichenology: Progress and Problems*. London: Academic

9. Brown, D. H., Rapsch, S., Beckett, A., Ascaso, C. 1987. The effect of desiccation on cell shape in the lichen *Parmelia sulcata* Taylor. *New Phytol.* 105:295–99

10. Brunner, U. 1985. *Ultrastrukturelle und chemische Untersuchungen an Flechtenphycobionten aus 7 Gattungen der Chlorophyceae (Chlorophytina) unter besonderer Berücksichtigung sporopollenin-ähnlicher Biopolymere*. Dissertation, Univ. Zürich 143 pp.

11. Brunner, U., Honegger, R. 1985. Chemical and ultrastructural studies on the distribution of sporopollenin-like biopolymers in 6 genera of lichen phycobionts. *Can. J. Bot.* 63:2221–30

12. Bubrick, P., Frensdorff, A., Galun, M. 1985. Selectivity in the lichen symbiosis. See Ref. 8a, pp. 319–34

13. Bubrick, P., Galun, M. 1980. Proteins from the lichen *Xanthoria parietina* which bind to phycobiont cell walls. Correlation between binding patterns and cell wall cytochemistry. *Protoplasma* 104:167–73

14. Bubrick, P., Galun, M. 1980. Symbiosis in lichens: differences in cell wall properties of freshly isolated and cultured phycobionts. *FEMS Microbiol. Lett.* 7:311–13

15. Büdel, B., Rhiel, E. 1987. Studies on the ultrastructure of some cyanolichen haustoria. *Protoplasma* 139:145–52

16. Butler, R. D., Allsopp, T. 1972. Ultrastructural investigations in the Stigonemataceae. *Arch. Mikrobiol.* 82:283–99

17. Cowan, D. A., Green, T. G. A., Wilson, A. T. 1979. Lichen metabolism. 1. The use of tritium labelled water in studies of anhydrobiotic metabolism in *Ramalina celastri* and *Peltigera polydactyla*. *New Phytol.* 82:489–503

18. Crittenden, P. D. 1989. Nitrogen relations of mat-forming lichens. In *Nitrogen, Phosphorus and Sulphur Utilization by Fungi*, ed. L. Boddy, R. Marchant, D. J. Read, pp. 243–68. Cambridge: Cambridge Univ. Press

19. Dudley, S. A., Lechowicz, M. J. 1987. Losses of polyol through leaching in subarctic lichens. *Plant Physiol.* 83:813–15

20. Elfving, F. 1913. Untersuchungen über die Flechtengonidien. *Acta Soc. Sci. Fenn.* 44:1–71

21. Ertl, L. 1951. Ueber die Lichtverhältnisse in Laubflechten. *Planta* 39:245–70

22. Farrar, J. F. 1988. Physiological buffering. See Ref. 30, Vol. 2, pp. 101–105

23. Fay, P. 1983. *The Blue-Greens*. London: Edward Arnold

24. Feige, G. B., Jensen, M. 1987. Photosynthetic properties of lichens stored at −25°C for several years. *Bibl. Lichenol.* 25:319–23

25. Fiechter, E. 1990. *Thallusdifferenzierung und intrathalline Sekundärstoffverteilung bei Parmeliaceae (Lecanorales, lichenisierte Ascomyceten)*. Inauguraldissertation, Univ. Zürich. 104 pp.

26. Fiechter, E., Honegger, R. 1987. Seasonal variations in the fine structure of *Hypogymnia physodes* (lichenised Ascomycetes) and its *Trebouxia* photobiont collected near the city of Zürich. *Plant Syst. Evol.* 158:249–63

27. Frey-Wyssling, A., Stecher, H. 1954. Ueber den Feinbau des Nostoc-Schleimes. *Z. Zellforsch. Mikrosk. Anat.* 39:515–19

28. Friedl, T. 1989. Systematik und Biologie von *Trebouxia* (Microthamniales, Chlorophyta) als Phycobiont der Parmeliaceae (lichenisierte Ascomyceten). Inauguraldissertation, Univ. Bayreuth. 218 pp.

29. Friedl, T. 1989. Comparative ultrastructure of pyrenoids in *Trebouxia* (Microthamniales, Chlorophyta). *Plant Syst. Evol.* 165:145–59

30. Galun, M., ed. 1988. *Handbook of Lichenology*, Vols. I–III. Boca Raton: CRC Press

31. Galun, M., Bubrick, P. 1984. Physiological interactions between the partners of the lichen symbiosis. In *Cellular Interactions. Encyclopedia of Plant Physiology*, ed. H. F. Linskens, J. Heslop-Harrison, pp. 362–401. Berlin: Springer

32. Gärtner, G. 1985. Die Gattung *Trebouxia* Puymaly (Chlorellales, Chlorophyceae). *Arch. Hydrobiol. Suppl.* 71:495–548

33. Goebel, K. von. 1926. Die Wasseraufnahme der Flechten. *Ber. Dtsch. Bot. Ges.* 44:158–61

34. Goebel, K. von. 1926. Morphologische und biologische Studien. Ein Beitrag zur Biologie der Flechten. *Ann. jard. Bot. Buitenzorg* 36:1–83

35. Gorin, P. J. A., Baron, M., Iacomini, M. 1988. Storage products in lichens. See Ref. 30, Vol. 3, pp. 9–23

36. Green, T. G. A., Lange, O. L. 1990. Ecophysiological adaptations of the *Pseudocyphellaria* lichens to south temperate rain forests. *4th Intl. Mycol. Congr. Regensburg: Abstr.*, p. 121

37. Green, T. G. A., Smith, D. C. 1974. Lichen physiology. XIV. Differences between lichen algae in symbiosis and in isolation. *New Phytol.* 73:753–66

38. Greenhalgh, G. N., Anglesea, D. 1979. The distribution of algal cells in lichen thalli. *Lichenologist* 11:2183–92

39. Guilford, W. J., Schneider, D. M., Labovitz, J., Opella, S. J. 1988. High resolution solid state ^{13}C NMR spectroscopy of sporopollenins from different plant taxa. *Plant Physiol.* 86:134–36

40. Hawksworth, D. L. 1988. The variety of fungal-algal symbioses, their evolutionary significance, and the nature of lichens. *Bot. J. Linn. Soc.* 96:3–20

41. Hawksworth, D. L. 1988. Coevolution of fungi with algae and cyanobacteria in lichen symbioses. In *Coevolution of Fungi with Plants and Animals*, ed. K. A. Pirozynski, D. L. Hawksworth, pp. 125–48. London: Academic

42. Hawksworth, D. L. 1988. The fungal partner. See Ref. 30, Vol. 1, pp. 35–38

43. Hawksworth, D. L., Sutton, B. C., Ainsworth, D. C. 1983. *Ainsworth and Bisby's Dictionary of the Fungi*. Kew: Commonwealth Mycol. Inst. 7th ed.

44. Hill, D. J. 1972. The movement of carbohydrate from the alga to the fungus in the lichen *Peltigera polydactyla*. *New Phytol.* 71:31–39

45. Hill, D. J. 1976. The physiology of lichen symbiosis. See Ref. 8b, pp. 457–97

46. Hill, D. J. 1985. Changes in photobiont dimensions and numbers during codevelopment of lichen symbionts. See Ref. 8a, pp. 303–17

47. Hill, D. J. 1989. The control of the cell cycle in microbial symbionts. *New Phytol.* 112:175–84

48. Hill, D. J., Ahmadjian, V. 1972. Relationship between carbohydrate movement and the symbiosis in lichens with green algae. *Planta* 103:267–77

49. Honegger, R. 1982. Cytological aspects of the triple symbiosis in *Peltigera aphthosa*. *J. Hattori Bot. Lab.* 52:379–91

50. Honegger, R. 1984. Cytological aspects of the mycobiont-phycobiont relationship in lichens. Haustorial types, phycobiont cell wall types, and the ultrastructure of the cell wall surface layers in some cultured and symbiotic myco- and phycobionts. *Lichenologist* 16:111–27

51. Honegger, R. 1985. Fine structure of different types of symbiotic relationships in lichens. See Ref. 8a, pp. 287–302

52. Honegger, R. 1986. Ultrastructural studies in lichens. I. Haustorial types and their frequencies in a range of lichens with trebouxioid phycobionts. *New Phytol.* 103:785–95

53. Honegger, R. 1986. Ultrastructural studies in lichens. II. Mycobiont and photobiont cell wall surface layers and adhering crystalline lichen products in four Parmeliaceae. *New Phytol.* 103:797–808

54. Honegger, R. 1987. Isidium formation and the development of juvenile thalli in *Parmelia pastillifera* (Lecanorales, lichenized ascomycetes). *Bot. Helv.* 97:147–52

55. Honegger, R. 1987. Questions about pattern formation in the algal layer of lichens with stratified (heteromerous) thalli. *Bibl. Lichenol.* 25:59–71

56. Honegger, R. 1988. The functional morphology of cell-to-cell interactions in lichens. In *Cell to Cell Signals in Plant Animal and Microbial Symbiosis*, ed. S. Scannerini, D. C. Smith, P. Bonfante, V. Gianinazzi-Pearson, NATO ASI Ser. H, 17:39–53. Berlin: Springer

57. Honegger, R. 1990. Mycobiont-photobiont interactions in adult thalli and in axenically resynthesized prethallus stages of *Xanthoria parietina* (Teloschistales, lichenized Ascomycetes). *Bibl. Lichenol.* 38:191–208

58. Honegger, R. 1990. Surface interactions in lichens. In *Experimental Phycology. Cell Walls and Surfaces, Reproduction, Photosynthesis*, ed. W. Wiessner, D. G. Robinson, R. C. Starr, pp. 40–54. Berlin: Springer

59. Honegger, R. 1990. Haustoria and hydrophobic cell wall surface layers in lichens. In *Electron Microscopy Applied in Plant Pathology*, ed. K. Mendgen, D. E. Lesemann, pp. 277–90. Heidelberg: Springer

60. Honegger, R. 1991. Lichens: structural

features. In *Algal Symbioses*, ed. W. Reisser. Bristol: Biopress. In press

61. Honegger, R. 1991. Developmental biology of ascomycetous lichens. In *Developmental Biology of Ascomycetes*, ed. N. D. Read, D. Moore. London: Academic. In press

62. Honegger, R. 1991. Symbiosis and fungal evolution: symbiosis and morphogenesis. In *Evolution and Speciation: Symbiosis as a source of Evolutionary Innovation*, ed. L. Margulis, R. Fester. Cambridge, MA: MIT Press. In press

63. Honegger, R., Brunner, U. 1981. Sporopollenin in the cell wall of *Coccomyxa* and *Myrmecia* phycobionts of various lichens: an ultrastructural and chemical investigation. *Can. J. Bot.* 59:2713–34

64. Jahns, H. M. 1984. Morphology, reproduction and water relations—a system of morphogenetic interactions in *Parmelia saxatilis*. *Nova Hedwigia Beih.* 79:715–37

65. Jahns, H. M. 1988. The establishment, individuality and growth of lichen thalli. *Bot. J. Linn. Soc.* 96:21–29

66. Jahns, H. M. 1988. The lichen thallus. See Ref. 30, Vol. 1, pp. 95–143

67. James, P. W., Henssen, A. 1976. The morphological and taxonomic significance of cephalodia. See Ref. 8b, pp. 27–77

68. Jensen, M., Feige, G. B. 1987. The effect of desiccation and light on the 77K chlorophyll fluorescence properties of the lichen *Peltigera aphthosa*. *Bibl. Lichenol.* 25:325–30

69. Kappen, L. 1988. Ecophysiological relationships in different climatic regions. See Ref. 30, Vol. 2, pp. 37–100

70. König, J., Peveling, E. 1980. Vorkommen von Sporopollenin in der Zellwand des Phycobionten *Trebouxia*. *Z. Pflanzenphysiol.* 98:459–64

71. König, J., Peveling, E. 1984. Cell walls of the phycobionts *Trebouxia* and *Pseudotrebouxia*: constituents and their location. *Lichenologist* 16:129–44

72. Lange, O. L. 1980. Moisture content and CO_2 exchange of lichens. I. Influence of temperature on moisture-dependent net photosynthesis and dark respiration in *Ramalina maciformis*. *Oecologia* 45:82–87

73. Lange, O. L., Green, T. G. A., Ziegler, H. 1988. Water status related photosynthesis and carbon isotope discrimination in species of the lichen genus *Pseudocyphellaria* with green or blue-green photobionts and in photosymbiodemes. *Oecologia* 75:494–501

74. Lange, O. L., Kilian, E. 1985. Reaktivierung der Photosynthese trockener Flechten durch Wasserdampfaufnahme aus dem Luftraum: artspezifisch unterschiedliches Verhalten. *Flora* 176:7–23

75. Lange, O. L., Kilian, E., Ziegler, H. 1986. Water vapour uptake and photosynthesis of lichens: performance differences in species with green and blue-green algae as phycobionts. *Oecologia* 71:104–10

76. Lange, O. L., Matthes, U. 1981. Moisture-dependent CO_2 exchange of lichens. *Photosynthetica* 15:555–74

77. Lange, O. L., Tenhunen, J. D. 1981. Moisture content and CO_2 exchange of lichens. II. Depression of net photosynthesis in *Ramalina maciformis* at high water content is caused by increased thallus carbon dioxide diffusion resistance. *Oecologia* 51:426–29

78. Lange, O. L., Ziegler, H. 1986. Different limiting processes of photosynthesis in lichens. In *Biological Control of Photosynthesis*, ed. R. Marcelle, H. Clijsters, M. Van Poucke, pp. 147–61. Dordrecht: Martinus Nijhoff

79. Larson, D. W. 1981. Differential wetting in some lichens and mosses: the role of morphology. *Bryologist* 84:1–15

80. Larson, D. W. 1983. The pattern of production within individual *Umbilicaria* lichen thalli. *New Phytol.* 94:409–19

81. Larson, D. W. 1984. Habitat overlap/niche segregation in two *Umbilicaria* lichens: a possible mechanism. *Oecologia* 62:118–25

82. Larson, D. W. 1987. The absorption and release of water by lichens. *Bibl. Lichenol.* 25:351–60

83. Larson, D. W., Carey, C. K. 1986. Phenotypic variation within "individual" lichen thalli. *Am. J. Bot.* 73:214–23

84. Lewis, D. H. 1987. Evolutionary aspects of mutualistic associations between fungi and photosynthetic organisms. In *Evolutionary Biology of the Fungi*, ed. C. M. Brasier, A. D. M. Rayner, D. Moore, pp. 161–78. Cambridge: Cambridge Univ. Press

85. Lewis, D. H., Smith, D. C. 1967. Sugar alcohols (polyols) in fungi and green plants. I. Distribution, physiology and metabolism. *New Phytol.* 66:143–84

86. Lines, C. E. M., Ratcliffe, R. G., Rees, T. A. V., Southon, T. E. 1989. A ^{13}C NMR study of photosynthate transport and metabolism in the lichen *Xanthoria calcicola* Oxner. *New Phytol.* 111:447–56

87. MacFarlane, J. D., Kershaw, K. A. 1985. Some aspects of carbohydrate

metabolism in lichens. See Ref. 8a, pp. 1–8

88. Maruo, B., Hattori, T., Takahashi, H. 1965. Excretion of ribitol and sucrose by green algae into the culture medium. *Agric. Biol. Chem.* 29:1084–89

89. Nash, T. H., Kappen, L., Lösch, R., Matthes-Sears, U., Larson, D. W. 1987. Cold resistance of lichens. *Bibl. Lichenol.* 25:313–17

90. Ott, S. 1987. Reproductive strategies in lichens. *Bibl. Lichenol.* 25:81–93

91. Peveling, E. 1973. Vesicles in the phycobiont sheath as possible transfer structures between the symbionts in the lichen *Lichina pygmaea*. *New Phytol.* 72:343–45

92. Peveling, E., Galun, M. 1976. Electron microscopical studies on the phycobiont *Coccomyxa* Schmidle. *New Phytol.* 77:713–18

93. Plessl, A. 1963. Ueber die Beziehungen von Pilz und Alge im Flechtenthallus. *Oesterr. Bot. Z.* 110:194–269

94. Richardson, D. H. S., Hill, D. J., Smith, D. C. 1968. Lichen physiology. XI. The role of the alga in determining the pattern of carbohydrate movement between lichen symbionts. *New Phytol.* 67:469–86

95. Richardson, D. H. S., Smith, D. C. 1966. The physiology of symbiosis in *Xanthoria aureola* (Ach.)Erichs. *Lichenologist* 3:202–6

96. Richardson, D. H. S., Smith, D. C. 1968. Lichen physiology. IX. Carbohydrate movement from the *Trebouxia* symbiont of *Xanthoria aureola* to the fungus. *New Phytol.* 67:61–68

97. Richardson, D. H. S., Smith, D. C. 1968. Lichen physiology. X. The isolated alga and fungal symbionts of *Xanthoria aureola*. *New Phytol.* 67:69–77

98. Rundel, P. W. 1988. Water relations. See Ref. 30, Vol. 2, pp. 17–36

99. Schröter, B., Kappen, L. 1990. Primary production and water relations of *Usnea antarctica* Du Rietz in the maritime Antarctic. *4th Intl. Mycol. Congr. Regensburg: Abstr.*, p. 155

100. Schuster, G., Ott, S., Jahns, H. M. 1985. Artificial cultures of lichens in the natural environment. *Lichenologist* 17:247–53

101. Schwendener, S. 1867. Ueber die wahre Natur der Flechtengonidien. *Verh. Schweiz. Naturforsch. Ges.* 51:88–90

102. Sigfridsson, B. 1980. Some effects of humidity on the light reaction of photosynthesis in the lichens *Cladonia impexa* and *Collema flaccidum*. *Physiol. Plant.* 49:320–26

103. Sluiman, H. J., Lokhorst, G. M. 1988. The ultrastructure of cellular division (autosporogenesis) in the coccoid green alga, *Trebouxia aggregata*, revealed by rapid freeze fixation and freeze substitution. *Protoplasma* 144:149–59

104. Smith, D. C. 1978. What can lichens tell us about real fungi? *Mycologia* 70:915–34

105. Smith, D. C. 1980. Mechanisms of nutrient movement between lichen symbionts. In *Cellular Interactions in Symbiosis and Parasitism*, ed. C. B. Cook, P. W. Pappas, E. D. Rudolph, pp. 197–227. Columbus: Ohio State Univ. Press

106. Smith, D. C., Douglas, A. 1987. *The Biology of Symbiosis*, pp. 1–302. London: Edward Arnold

107. Stocker-Wörgötter, E., Türk, R. 1988. Culture of the cyanobacterial lichen *Peltigera didactyla* from soredia under laboratory conditions. *Lichenologist* 20:369–75

108. Stocker-Wörgötter, E., Türk, R. 1990. Thallus formation of the cyanolichen *Peltigera didactyla* from soredia under laboratory conditions. *Bot. Acta* 103:315–21

109. Tapper, R. 1981. Direct measurements of translocation of carbohydrate in the lichen, *Cladonia convoluta*, by quantitative autoradiography. *New Phytol.* 89:429–39

110. Toensberg, T., Holtan-Hartwig, J. 1983. Phycotype pairs in *Nephroma*, *Peltigera* and *Lobaria* in Norway. *Nord. J. Bot.* 3:681–88

111. Tschermak, E. 1941. Untersuchungen über die Beziehungen von Pilz und Alge im Flechtenthallus. *Oesterr. Bot. Z.* 90:233–307

112. Tschermak-Woess, E. 1985. *Elliptochloris bilobata*, kein ganz seltener Phycobiont. *Herzogia* 7:105–16

113. Tschermak-Woess, E. 1988. The algal partner. See Ref. 30, Vol. 1, pp. 39–92

Annu. Rev. Plant Physiology Plant Mol. Biol. 1991. 42:579–620

THE ROLES OF HEAT SHOCK PROTEINS IN PLANTS

Elizabeth Vierling

Department of Biochemistry, University of Arizona, Tucson, Arizona 85721

KEY WORDS: heat shock response, molecular chaperone, HSP70, thermotolerance, stress
              proteins

CONTENTS

1040-2519/91/0601-0579$02.00

INTRODUCTION

Environmental conditions that change temperature, light environment, water status, or hormone balance lead to altered gene expression in plants. At the molecular level, one of the best-characterized environmental responses is the response to high temperature or heat shock. Ten years ago it was shown that when seedlings are shifted to temperatures five or more degrees above optimal growing temperatures, synthesis of most normal proteins and mRNAs is repressed, and transcription and translation of a small set of "heat shock proteins" (HSPs) is initiated (7, 88, 118, 121, 127 for review). The heat shock response is not unique to plants. It was first discovered in *Drosophila* in the 1960s and has been described in a wide range of organisms including *Escherichia coli, Saccharomyces cerevisiae*, and humans (100, 115, 126). A comparison of the response in different organisms has shown that it is highly conserved in two important ways: (*a*) the molecular mechanism of gene induction by heat shows many similarities among diverse eukaryotes; and (*b*) the major HSPs are highly homologous among eukaryotes, and in several cases, homologous proteins have been identified in prokaryotes as well. The evolutionary conservation of the heat shock response argues strongly that the production of HSPs is a fundamental and essential process.

A basic question with respect to environmentally regulated changes in gene expression is whether they are required for the organism's survival. Plants develop tolerance to normally lethal temperatures if they are first subjected to certain treatments at high but nonlethal temperatures. In plants, as well as in other organisms, considerable evidence suggests that HSP production is an essential component of this short-term development of thermotolerance (88, 89, 100, 118, 127). All plant species tested, including the green alga *Chlamydomonas*, produce HSPs in response to elevated temperatures (118). Plant species adapted to temperate environments, including crop plants such as soybean, pea, maize, and wheat, begin to synthesize HSPs when tissue temperatures exceed 32–33° C. HSP synthesis increases with increasing temperature, and the temperature of maximum HSP synthesis is positively correlated with each species' optimum growth temperature. This pattern of HSP production suggests that the response is finely tuned to the physiology of the organism and further supports its biological relevance.

Here I summarize recent progress toward understanding HSP function in plants. A number of reviews should be consulted for more detailed information concerning other organisms (16, 101, 115, 126, 153) and for early studies of the plant heat shock response (7, 118, 121, 127). Several recent reviews also provide excellent summaries of the mechanism of HSP gene activation (63, 115, 126).

HEAT SHOCK PROTEINS DEFINED

Several classes of HSPs have been described in eukaryotes, including plants. They are designated by their approximate molecular weights in kDa as HSP110, HSP90, HSP70, HSP60, and low molecular weight (LMW) HSPs (15–30 kDa) (121, 126). Ubiquitin, a small protein involved in ATP-dependent, intracellular proteolysis (77), is also referred to as an HSP (101, 121, 126). These proteins all fit the criterion that they are heat induced in a majority of cell types in a wide range of organisms. HSP90, HSP70, and HSP60 have also been studied in *E. coli*, and a protein with LMW HSP characteristics has been described in the prokaryote *Mycobacterium* (119). Given the extensive conservation of the HSPs, much of what is learned about their function in any organism can be extrapolated to other organisms. Consequently, information from many different sources is contibuting to our understanding of HSP function in plants.

The importance of HSPs clearly extends beyond their potential role in protection from high-temperature stress. Although HSPs (with the exception of ubiquitin) were first characterized because their expression increased in response to elevated temperature, some HSPs are found at significant levels in normal, nonstressed cells or are produced at particular stages of the cell cycle or during development in the absence of stress (16, 100, 101, 115). Additionally, certain normal cellular proteins are homologous to HSPs and do not exhibit increased expression in response to high temperature. Another important observation is that different proteins of the HSP70, HSP60, and LMW HSP classes are present in various cellular compartments, including semi-autonomous organelles—the mitochondria and chloroplasts (3, 35, 48, 98, 107, 113). The similarity of these proteins is known from cross-reactivity of specific antibodies and in many cases from direct sequence analysis of cloned genes. Thus, HSPs are actually members of multi-gene superfamilies in which not all members are regulated by heat. The term HSP cognate (abbreviated HSC) has been applied to those HSP family members that are expressed in the absence of heat stress. Here I use the term HSP cognate (or HSC) interchangeably with HSP homolog.

Heat is not the only stress treatment that leads to elevated expression of many HSPs. Ethanol, arsenite, heavy metals, amino acid analogues, glucose starvation, calcium ionophores, and a number of other treatments affect the synthesis of some, or all, HSPs in different organisms (19, 36, 40, 46, 121, 127, 129, 179,). Consequently, HSPs have also been referred to more generally as "stress proteins." However, it should be noted that only arsenite has been found to elicit a response involving a high level of expression of all HSPs in plants (36, 118). Clearly, HSPs are not just general stress proteins produced in response to any severe cellular perturbation. In plants, most HSPs are not synthesized in response to water stress (unless accompanied by heat stress) (90), anaerobic stress (151), cold stress (64, 65, 132, 182), or salt stress (36, 68). These data indicate that production of the full complement of HSPs occurs in response to specific changes not common to all stresses.

It is not yet possible to define precisely how HSPs contribute to an organism's ability to survive high temperatures. Determining the biochemical activities of individual HSPs is a necessary step before the mechanism of their proposed protective effects can be understood. Studies of how HSPs participate in basic cellular processes in the absence of heat stress are also providing insight into HSP function during stress. An emerging principle is that the role of HSPs involves stabilization of proteins in a particular state of folding (47, 101, 115, 126, 134, 136, 148, 153). Accumulating evidence indicates that through this mechanism, HSP90, HSP70, and HSP60 facilitate a wide diversity of important processes including protein folding, transport of proteins across membranes, assembly of oligomeric proteins, and modulation of receptor activities. All of these functions require the alteration or maintenance of specific polypeptide conformations. Based on these activities HSP90, HSP70, and HSP60 have been termed "molecular chaperones" or "polypeptide chain binding proteins" (47, 115, 136, 148). Only limited information is available concerning HSP110 in any organism. Recent studies of a yeast gene encoding a protein in this size class (HSP104) suggest this HSP may be critical for thermotolerance (152). LMW HSP function remains enigmatic but should prove particularly interesting in plants, where this HSP superfamily is unusually complex. Although it is firmly established that ubiquitin functions in intracellular proteolysis (77), its role in the heat stress response is not completely clear (115, 121). Studies of HSPs are providing exciting insights into the physiology of both normal and stressed plant cells.

HSP70: CHARACTERISTICS AND FUNCTIONS

HSP70 Diversity

HSP70 was one of the first eukaryotic genes cloned and has been extensively studied in many organisms (100, 101, 115). In addition to heat-regulated HSP70 genes, eukaryotes contain related genes that do not all exhibit increased expression during heat stress. The diversity of HSP70 genes is partly accounted for by the presence of distinct HSP70 homologs in the cytoplasm, in the lumen of the endoplasmic reticulum (ER) (116, 125, 147), and in the matrix of mitochondria (3, 35, 48, 98, 113). The HSP70 homologs located in the ER are also called "binding protein" (BiP) or "glucose regulated protein" (GRP). In plants HSP70s are also found in chloroplasts (3, 107). The prokaryotic homolog of HSP70 is the DnaK protein, which is present under normal growth conditions and is induced by high temperature (8, 55). The ubiquity of this family of proteins suggests that HSP70s perform critical, fundamental cellular functions.

Genetic analysis in *S. cerevisiae* has provided the most direct evidence that HSP70s are essential proteins of multiple cellular compartments (34, 101). This simple eukaryote has eight HSP70 genes which comprise four complementation groups or gene families, with nucleotide identities ranging from 50 to 96% [the *SSA* family — four genes; the *SSB* family — two genes; and the *SSC* and *KAR2* families, each with one gene (34)]. The *SSA* proteins are primarily cytoplasmic, the *SSC* protein is mitochondrial (35), and the *KAR2* protein is found in the ER (125, 147). The intracellular location of the *SSB* proteins has not been determined. The regulation of the different genes is complex; some show increased expression during heat stress, and others show little change. The *SSA* family, *SSC* and *KAR2*, are essential genes; deletion mutants are nonviable. When only two of the four *SSA* genes are mutated, cells are still viable but are unable to grow at elevated temperatures. It is interesting that deletion of the two *SSB* genes results in cold-sensitive growth. The number of HSP70 genes and the variations in their regulation are at least as complex, if not more complex, in higher eukaryotes.

Cytoplasmic HSP70 Genes and Proteins

In plants as in other eukaryotes, HSC70 is found constitutively in the cytoplasm of all tissues, and additional HSP70 is produced during high-temperature stress (24, 100, 120, 121). Genes encoding HSP70 homologs have been isolated from maize (146), petunia (179), *Arabidopsis* (180), tomato (45, A. Bennett, personal communication), soybean (145), pea (43), and *Chlamydomonas* (174). The DNA and derived amino acid sequences have been determined for eight of these genes. Seven of the characterized genes are most similar to the cytoplasmic HSP70 genes of yeast (*SSA* family) and humans (Table 1). This fact, combined with the absence of appropriate amino-terminal signal sequences required for localization to the ER, mitochondria, or chloroplasts indicates that these genes encode cytoplasmic forms of HSP70. The remaining plant gene, tomato BiP in Table 1, is an ER protein (see below). The sequence relationships between the different plant HSP70s also support the assignment of tomato BiP to a separate HSP70 class. An overall comparison of the seven cytoplasmic HSP70s shows that they are 75.0% identical and 91.0% similar at the amino acid level. When tomato BiP is added to the comparison the identity and similarity values drop to 54.9% and 82.5%, respectively. In general, there is greater similarity among eukaryotic cytoplasmic HSP70 homologs from different species than there is between HSP70 homologs in different cellular compartments of a single species. These data suggest an ancient divergence of these different HSP70

Table 1 Relationship of plant HSP70 proteins to other eukaryotic and prokaryotic HSP70s

Protein	Protein characteristics			Percent amino acid identity (similarity)[a] to:			
	Reference (Accession #)[b]	Molecular weight[c]	pI[d]	*S. cerevisiae* SSA1	Human HSP70	Human GRP78	*E. coli* DnaK
L. esculentum HSC70-1	e (NA)	71,286	4.98	71.4 (83.6)	75.7 (85.7)	63.6 (78.3)	49.1 (67.2)
L. esculentum HSC70-2	e (NA)	70,707	4.91	71.6 (83.1)	75.9 (86.2)	63.3 (78.9)	49.5 (68.1)
P. sativum HSP70	f (NA)	71,166	5.00	72.3 (84.3)	74.0 (84.7)	64.2 (79.0)	48.7 (68.4)
P. sativum HSC70	f (NA)	71,003	4.83	71.7 (83.6)	75.7 (86.5)	63.4 (78.7)	49.4 (67.5)
P. hybrida HSP70	179 (X06932)	71,226	4.93	72.0 (82.5)	76.4 (86.2)	63.6 (77.7)	48.9 (66.8)
G. max HSP70	145 (NA)	70,963	5.29	71.8 (83.9)	74.1 (84.2)	63.3 (78.5)	49.0 (67.3)
Z. mays HSP70	146 (X03658)	70,473	5.06	71.7 (83.0)	75.6 (85.6)	62.8 (77.7)	47.5 (65.7)
L. esculentum BiP[h]	i (NA)	70,063	4.89	64.1 (78.5)	63.1 (78.3)	70.3 (83.0)	50.9 (71.4)
S. cerevisiae SSA1	161 (X12926)	69,766	4.82	– –			
Human HSP70	79 (A29160)	69,867	5.32	74.7 (86.4)	– –		
Human GRP78	167 (A29821)	70,261	4.81	65.7 (79.4)	62.9 (77.3)	– –	
E. coli DnaK	8 (K01298)	69,114	4.67	50.0 (67.5)	49.2 (66.1)	51.2 (69.5)	–

[a] Percentage identity and similarity were calculated using the GAP computer program (set with default parameters) of the Wisconsin GCG sequence analysis software
[b] Accession numbers are from the EMBL, NBRF, or Genbank data bases.
[c] Molecular weight determined from the derived amino acid sequence
[d] pI calculated from the derived amino acid sequence
[e] N. Duck, personal communication
[f] L. Lauzon and E. Vierling, unpublished
[h] Molecular weight, pI, and all sequence comparisons were calculated without the amino-terminal signal sequence.
[i] A. Bennett, personal communication.
NA—Not available

forms. The comparisons in Table 1 also emphasize the high degree of HSP70 conservation between eukaryotes and prokaryotes. Eukaryotic HSP70s show close to 50% identity and greater than 65% similarity with the *E. coli* DnaK protein, making the HSP70s among the most highly evolutionarily conserved proteins known.

The HSP70 genes that have been isolated from plants appear to represent only a small number of the total genes encoding cytoplasmic HSP70s in any one plant species. For example, Wu et al (180) suggest there are as many as 12 HSP70 genes in *Arabidopsis*. The stringency of nucleic acid hybridization used in these experiments may not have detected more highly diverged HSP70 homologs, such as those encoding the organelle HSP70s. Therefore, the total complexity of the HSP70 superfamily may be higher, even in a species with a very small genome.

The significance of the presence of multiple genes encoding cytoplasmic HSP70s is not known, and the differences in their regulation and structure are only beginning to be characterized. There is emerging evidence that certain HSP70 genes are regulated in a tissue specific fashion or during precise developmental stages in plants. A tomato HSC70 cDNA described by Duck et al (45) exhibits a tissue specific pattern of constitutive expression and little or no change in mRNA level during heat treatment. Using in situ hybridization techniques, expression of the HSC70 mRNA was localized to the vascular system of the ovary, the inner integument of developing seeds, and the lateral root tips. Some expression was also observed in the transmitting tissue of the style, in the tapetum of immature anthers, and in 4 mm leaves. This complex expression pattern may reflect the activity of more than one HSC70 gene, a possibility that could not be distinguished because of the relaxed stringency of the in situ hybridizations. DeRocher et al (43) isolated two cDNAs encoding HSP70 homologs from pea and found that they have very different patterns of expression. One cDNA (pea HSC70) detects an mRNA that is expressed at high levels in many tissues in the absence of stress, including leaves, roots, and developing seeds. The other HSP70 cDNA hybridizes to mRNA from heat-stressed leaves and roots, but not to control mRNA from these tissues (24, 43). This latter cDNA also detects an mRNA that accumulates late in seed development and is present in mature seeds. Whether similar changes in HSP70 expression during seed development occur in other plant species is unknown.

Some HSP70 homologs may also respond to stresses that do not typically induce most HSPs. A two- to four-fold increase in HSC70 mRNA and a corresponding increase in a specific HSC70 protein have been observed during cold acclimation in several plant species (64, 122; C. Guy, personal communication). This is an interesting parallel to the *SSB* gene products in yeast, which also increase in abundance at low temperature (34). However, the cold-increased HSP70 accumulation has not been correlated with cold-tolerance; a similar response was observed in both cold-tolerant and -intolerant species (C. Guy, personal communication).

The number and homology of HSP70 genes have complicated attempts to examine the regulation of HSP70 expression. The presence of significant levels of HSC70 RNA and protein in most normal cells adds further difficulty to the assessment of stress-induced changes. Careful experiments with gene-specific DNA probes, coupled with two-dimensional gel analysis of proteins, are necessary for critical evaluation of HSP70 regulation. Further studies will undoubtedly reveal additional complexity in the expression patterns of the HSP70 gene family.

Cytoplasmic HSP70 Functions

Studies of HSP70 (or HSC70) biochemistry, along with analysis of *S. cerevisiae* mutants, have led to a general model of how HSP70s act at the molecular level. All HSP70 homologs bind ATP, can be isolated by ATP affinity chromatography, and exhibit a weak ATPase activity (52, 53, 136, 177). Pelham speculated in 1986 (134) that HSP70s and homologs played a role in ATP-dependent protein folding and assembly, a hypothesis that has received considerable support from studies of HSP70s in stressed and unstressed cells (136). The HSC70s have been implicated in several functions involving changes in protein conformation. Rothman and others have demonstrated that mammalian HSC70 catalyzes the ATP-dependent disassembly of clathrin-coated vesicles in vitro (23, 42, 148). A model for this reaction has been proposed in which HSC70 binds to a specific conformational form of clathrin light chains and stimulates uncoating accompanied by the hydrolysis of ATP (42). In yeast, the cytoplasmic *SSA* proteins appear to facilitate transport of proteins across both the ER membrane and the mitochondrial membranes (27, 44). Viable *SSA* mutants accumulate precursor proteins destined for these two compartments. Furthermore, ER and mitochondrial protein transport systems reconstituted in vitro show enhanced transport in the presence of added HSP70 or HSC70 (27). In this context it is proposed (27, 44, 136) that HSP70 homologs unfold precursor proteins or otherwise maintain them in a form suitable for transport across membranes. An even more general function has been suggested by Beckmann et al (11). These workers report that HSC70 binds to a large percentage of all nascent polypeptides. They suggest that HSP70 and homologs facilitate correct folding of many proteins. Although the function of HSC70s in plants has not been directly addressed, their involvement in protein transport processes is under investigation in several laboratories.

The function of HSP70 during stress is believed to be similar to that of the constitutive homologs (136). To date, no protein sequence characteristics have been identified as being specific to either the heat-induced or constitutive HSP70s (101, 115). Thus, there are no obvious structural differences in the heat-induced HSP70 homologs to indicate unique heat-stress-related functions. The total amount of HSP70 produced during heat stress is also typically less than the amount already present in the cell (16). It has been proposed that HSP70 binds to disassociated or denatured proteins produced during heat stress, facilitating protein refolding or reassembly accompanied by the hydrolysis of ATP (134, 136).

During high-temperature stress the additional HSP70 that is produced localizes primarily to the nucleolus and then redistributes to the cytoplasm during recovery. This phenomenon has been observed in all higher eukaryotes (100), including plants (112, 120, 121, 128). A 17-amino-acid segment of human HSP70, when fused to a heterologous protein, will act as a nuclear targeting signal (41). Whether this segment targets HSP70 to the nucleus is not known, and the mechanism of the reversible nuclear migration is not understood. Presumably either a change in binding of HSP70 to other cellular components or an alteration of HSP70 conformation is involved in controlling intracellular localization (41, 112). The HSP70 found associated with nucleoli during heat stress can be dissociated with ATP, but not with nonhydrolyzable ATP analogues (136). Pelham suggests that the protein protects preribosomes from denaturation during stress (134, 136). Pre-existing HSC70 also localizes to the nucleus during stress (120, 121). Therefore, the functional distinction between HSP70 and HSC70 is unclear.

The most highly conserved domains of the HSP70s, including the prokaryotic homolog (8), are found in the amino-terminal half of the protein (approximately amino acids 1–350). Biochemical studies and X-ray crystallography demonstrated that this half of the protein binds ATP (52). The structure of the ATP binding fold is similar to that of the the ATP binding domains of hexokinase and actin (52, 84), an exciting observation that should aid in understanding the mechanism of HSP70 ATPase activity.

Interaction of HSP70 with its substrates is poorly understood. It has been proposed that the carboxyl-terminal half of HSP70 determines substrate specificity (23, 52). This conclusion is based on loss of clathrin binding by a 44-kDa amino-terminal fragment of HSC70 (23), and on the fact that the carboxyl-terminal half contains the most divergent domains of the protein. Flynn et al (53) reported that HSP70s bind specifically to a number of short, synthetic peptides and that release of the peptides requires hydrolysis of ATP. The peptide binding region of HSP70 was not localized in these experiments. Pelham's original proposal suggested HSP70s bind exposed hydrophobic protein surfaces, preventing their improper association with other polypeptides or polypeptide segments (134). Binding studies with hydrophobic peptide substrates have not yet been reported. Significant work remains to be done to establish how HSP70s recognize their proposed diverse substrates and how they can distinguish correctly folded from unfolded or improperly folded proteins.

HSP70 in the Endoplasmic Reticulum

An HSP70 homolog localized to the lumen of the ER has been studied in mammals and is known to be homologous to the *S. cerevisiae KAR2* protein (116, 125, 136, 147). This HSP70 homolog was initially characterized in mammalian cells as a polypeptide complexed with unassembled immunoglobulin heavy chains in the ER (hence BiP, for "binding protein"; 116). An ER protein that exhibited increased expression during glucose starvation (GRP for "glucose regulated protein") was studied independently and found to be an HSP70 homolog identical to BiP (116). BiP/GRP shares extensive sequence homology with cytoplasmic HSP70s, along with the ability to bind ATP and short peptides (53). This HSP70 homolog is present at significant levels in many cells (comprising an estimated 0.4% of microsomal protein in some mammalian tissues) and does not show a high level of heat induction. Characterization of BiP/GRP activity in mammalian cells has demonstrated a much broader and more general function than originally assigned to this protein. It is hypothesized that, like HSP70, BiP/GRP stabilizes newly synthesized proteins in the ER until they fold or assemble into their correct structure (56, 135). Recognition that the essential *KAR2* gene from yeast was a BiP/GRP homolog extended the occurrence of this member of the HSP70 family to lower eukaryotes.

As mentioned above (see Table 1), an HSP70 homolog has been identified in tomato that has the characteristics expected for BiP/GRP (A. Bennett, personal communication). The tomato BiP is more similar to other eukaryotic BiP/GRPs than to the plant cytoplasmic HSP70s, and it has an amino-terminal signal sequence and carboxyl-terminal ER-retention signal, as would be expected for an ER protein. Cooper & Ho (33) reported a 72-kDa HSP was enriched in ER membranes from heat-stressed corn roots, and an HSC70 has been isolated from microsomes of developing wheat grains (G. Galili, personal communication). These proteins are most probably also BiP homologs. BiP is likely to be found as an abundant component of all plant cells actively engaged in protein synthesis involving ER-bound ribosomes. In this regard, seed storage protein synthesis is one essential process in which BiP may participate. The involvement of BiP in this process is now under investigation (G. Galili, personal communication).

Mitochondrial and Chloroplast HSP70s

The presence of HSP70 homologs in semi-autonomous organelles was first established when the essential, nuclear *SSC1* gene of yeast was found to encode a mitochondrial protein (35). *SSC1* shows greater similarity to the DnaK protein, which is the *E. coli* HSP70 homolog, than to eukaryotic cytoplasmic or ER HSP70s. Subsequently, a similar mitochondrial protein has been identified in *Trypanosoma cruzi* (48), mammalian cells (98, 113), and *Euglena* (3). The *T. cruzi* protein deduced from the DNA sequence is homologous to DnaK and *SSC1*. The mammalian and *Euglena* proteins have been related to DnaK by cross-reactivity to specific anti-DnaK antibodies. Undoubtedly, a similar protein is present in higher plant mitochondria. Neumann et al (121) reported that a 68-kDa HSP was localized to pea mitochondria. Preliminary characterization of a cDNA corresponding to the 68-kDa protein has suggested it is DnaK-like (D. Neumann, personal communication). In all species, the mitochondrial proteins are found in the absence of heat stress, as would be expected of an essential protein.

DnaK homologs have now been identified in chloroplasts of pea (107), maize (3), spinach (3), *Arabidopsis* (E. Vierling, unpublished), and *Euglena* (3). These proteins are localized to the chloroplast stroma and show greater cross-reactivity with anti-DnaK antibodies than with antibodies against eukaryotic HSP70s. They are present in the absence of temperature stress, and little change in protein level has been observed after heat stress (107). They are apparently nucleus-encoded proteins, as there is no evidence for the existence of a DnaK-like gene on any of the sequenced chloroplast genomes. The presence of prokaryotic-type HSP70s in chloroplasts and mitochondria is consistent with the hypothesis of an endosymbiotic origin of these organelles.

It would not be surprising to find that the organelle DnaK-like proteins perform functions similar to *E. coli* DnaK. *E. coli* DnaK mutants are impaired for growth at all temperatures and are completely unable to grow at elevated temperatures (55). Existing data support the hypothesis that DnaK participates in a range of functions that, like eukaryotic HSP70 functions, involve protein folding or assembly (54, 55, 160). Experiments with *SSC1* mutants in yeast further suggest the interesting possibility that the organelle homologs are essential for protein import into organelles (85). Temperature-sensitive *SSC1* mutants accumulate processed mitochondrial precursors that appear unable to complete entry into the organelle. Thus there is evidence that HSC70s found both outside (the *SSA* gene products) and inside (the *SSC* gene product) mitochondria participate in protein transport. By analogy, one would predict that HSC70 also participates in protein transport into chloroplasts.

Experiments with antibodies that recognize eukaryotic HSP70s have identified two additional HSP70 homologs that appear to be localized to chloroplasts (107). One of these proteins is in the chloroplast stroma. The other is associated with the envelope, probably between the inner and outer envelope membranes. Further investigations of the functions of these proteins will be difficult because similar, abundant HSP70 homologs are present in other parts of the cell.

HSP60

The HSP60s were the first HSPs to be termed "molecular chaperones," a role supported by both genetic and biochemical studies (47, 55, 66, 76). The eukaryotic HSP60 homologs are nucleus-encoded proteins found in mitochondria (66) and chloroplasts (76). They are abundant components of both these organelles even in the absence of heat stress. The chloroplast homolog was originally identified as the ribulose bisphosphate carboxylase (Rubisco) binding protein (9, 149, 150). No homologous proteins have been described in any other eukaryotic cellular compartment.

The prokaryotic HSP60 homolog is the *E. coli* GroEL gene product (55, 76). GroEL is present in normal cells and exhibits significant heat induction. In its native state the 57-kDa GroEL monomer forms a homo-oligomer of 14 subunits arranged in 2 hollow-core, stacked rings of 7 subunits. GroEL null mutants are inviable at any temperature. Viable GroEL mutants were originally identified as defective in lambda phage assembly. Studies of additional mutant alleles suggest GroEL participates in many cellular processes, consistent with the hypothesis that it facilitates protein folding or assembly (55, 168). GroEL and its eukaryotic homologs, mitochondrial HSP60 and the ribulose bisphosphate binding protein, are also referred to as chaperonin 60s (103).

In *E. coli* a protein encoded by the same operon as GroEL, the GroES protein (10 kDa), appears to interact with GroEL. Certain GroES mutants act as suppressors of GroEL defects (55). Recently a small protein (9 kDa) has been identified in mammalian mitochondria and pea chloroplasts that is most likely the organelle homolog of GroES (103). Thus, the mitochondrial and chloroplast chaperonin systems appear to parallel directly the *E. coli* system.

Mitochondrial HSP60

Mitochondrial HSP60 was initially identified in *Tetrahymena* and subsequently in *S. cerevisiae*, *Xenopus*, and maize, as an HSP that was antigenically related to the *E. coli* GroEL protein (111, 141). The isolated protein was also found to have a native structure similar to that of GroEL (111, 141). Isolation and sequencing of the *S. cerevisiae* HSP60 gene provided direct evidence of its homology to GroEL (143). *S. cerevisiae* HSP60 is 54% identical to *E. coli* GroEL at the amino acid level, and the regions of identity are evenly distributed along the protein (143). The HSP60 gene was subsequently shown to be the defective locus in a conditional-lethal yeast mutant that exhibited impaired assembly of mitochondrial proteins (26). In an HSP60 mutant incubated at the restrictive temperature, the β-subunit of the F_1 ATPase was correctly imported and processed but was not assembled into the membrane-bound ATPase complex (26). In the same mutant, two other mitochondrial proteins, the Rieske Fe/S protein and cytochrome b_2, were imported into the mitochondria but were incompletely processed and not properly localized to the intermembrane space.

A basic model of the interaction of HSP60 with newly imported, un-assembled proteins has been developed from further studies of yeast (67, 130). After import, proteins associate with the high-molecular weight HSP60 complex in a partially unfolded state. Association of HSP60 with unfolded proteins has been observed by co-immunoprecipitation and co-migration on native gels or gel-filtration columns. As is the case for HSP70, the manner in which HSP60 recognizes and binds to unfolded proteins is unclear. Folding is dependent on the presence of ATP and is proposed to occur on the surface of the HSP60 complex (130). Release of the folded (or partially folded) protein occurs in a second step which may require another mitochondrial protein component and additional ATP. The second component required for release is being sought in biochemical experiments and by analyzing other mutants. It is very possible that the other component is the homolog of *E. coli* GroES, but this has not yet been demonstrated in yeast. Investigation of additional conditional alleles of HSP60 should aid in dissecting its mechanism of action.

Mitochondrial HSP60 has been identified in many different plant species, and maize HSP60 has been extensively characterized (140, 141). As with yeast HSP60, maize HSP60 protein levels increase two- to three-fold during a four hour heat stress of maize seedlings (141). HSP60 also makes up a larger percentage of total protein during imbibition and early seedling growth as compared to 4 day-old or mature seedlings. Perhaps more HSP60 is needed during germination or other periods of active mitochondrial division and development.

Experiments with maize HSP60 suggest that, in addition to its role in folding nucleus-encoded, imported proteins, this molecular chaperone may also assist in correct assembly of proteins synthesized inside the mitochondria. By several criteria maize HSP60 was found to associate with two newly synthesized mitochondrial proteins, one of which was identified as the α-subunit of F_1 ATPase (140). However, it has not yet been shown that the HSP60/α-subunit complex is a productive intermediate in the normal assembly pathway of ATPase.

Chloroplast Chaperonin 60

The nucleus-encoded, GroEL (or HSP60) homolog in chloroplasts is the Rubisco subunit binding protein, also known as chloroplast chaperonin 60. It is proposed to be involved in assembly of the Rubisco holoenzyme (149). The plant Rubisco holoenzyme is composed of 16 subunits of two types: eight chloroplast-encoded large subunits and eight nucleus-encoded small subunits. Studies of Rubisco assembly in isolated chloroplasts suggested that newly synthesized large subunits were complexed with another protein prior to their incorporation into holoenzyme (9, 150). The isolated Rubisco subunit binding protein from pea was found to be a 720-kDa complex composed of two subunits of approximately 61 kDa (α) and 60 kDa (β) (75). Sequence analysis of cDNAs from wheat, *Ricinus communis*, and *Brassica napus* showed that the amino acid sequence of the α-subunit is 46% identical to *E. coli* GroEL (76, 108) and 43% identical to *S. cerevisiae* HSP60 (143). Interestingly, the amino aicd sequence of the β-subunit, determined from a *Brassica napus* cDNA, is also 46% identical to *E. coli* GroEL, but only 49% identical to the plant α-subunits (108). It is not known if the α- and β-subunits assemble together to form the native oligomeric structure, or if there are distinct α- and β-subunit homo-oligomers. There is no evidence for two HSP60 subunit types in mitochondria or prokaryotes.

The function of chloroplast chaperonin 60 in Rubisco assembly has received support from studies demonstrating that *E. coli* GroEL and GroES facilitate assembly of cyanobacterial or *Rhodospirillum rubrum* Rubisco. Functional prokaryotic Rubisco can be expressed and assembled in *E. coli* in vivo, dependent on the expression of GroEL and GroES (59). It will be interesting if similar experiments can be done to assemble active eukaryotic Rubisco in the presence of the α- and β-subunits of chaperonin 60 expressed in *E. coli*. Goloubinoff et al (58) also developed an in vitro system to assay assembly of the dimeric *R. rubrum* Rubisco. In their system, reconstitution of active enzyme from denatured *R. rubrum* subunits required the addition of GroEL, GroES, and MgATP. The mammalian mitochondiral and pea chloroplast GroES homologs have been identified by their ability to substitute functionally for GroES in this assay (103). The GroES component appears essential for release of the folded Rubisco subunit from the surface of the chaperonin 60. In the current model of assembly, the chaperonin functions primarily in proper folding of the individual subunits, and may or may not facilitate holoenzyme assembly.

While there is evidence that the *E. coli* and mitochondrial HSP60s interact with many different proteins, little information is available concerning the role of the chloroplast protein in functions other than Rubisco assembly. Lubben et al (102) reported that several proteins imported into chloroplasts in vitro can be found stably associated with the chloroplast chaperonin. It remains to be proven that such interactions occur in vivo as part of normal folding or assembly pathways.

Although the chloroplast chaperonin is a member of the HSP60 family of HSPs, no information concerning the expression of this protein during heat stress has been published. The protein is an abundant chloroplast component in the absence of stress, and it would be surprising if it increased more than two- to three-fold under stress conditions. Thus, the significance of this protein to plant stress responses is currently unknown.

LMW HEAT SHOCK PROTEINS

The LMW HSP Gene Superfamily

All plants synthesize multiple LMW HSPs ranging in size from approximately 17 to 28 kDa. The LMW HSPs are not expressed at detectable levels in leaves of plants grown at optimal temperatures, but in many species they are among the most abundant proteins induced by heat stress (24, 104, 118). LMW HSP genes have been cloned from soybean (36, 38, 40, 117, 138, 142, 154, 155), pea (71, 95, 97, 131, 172), *Arabidopsis* (25, 73, 166), carrot (184), petunia (25), *Chenopodium* (95), wheat (110), lily (18), maize (60, 121, 124), and *Chlamydomonas* (61). Analysis of these genes indicates that most of the LMW HSPs belong to four multi-gene families. Two of these gene families most likely encode cytoplasmic proteins, one encodes a chloroplast-localized protein, and one appears to encode an endomembrane protein (74). Thus, the LMW HSPs in plants comprise a gene superfamily encoding proteins specific to different cellular compartments. This fact is an interesting parallel to what is observed for the HSP70 proteins, although unlike the case of the HSP70s, no constitutive LMW HSP homologs have yet been identified.

Compared to higher plants, other eukaryotes contain far fewer LMW HSP genes. For example, *Drosophila* has four major LMW HSP genes, and only one LMW HSP gene has been identified in yeast and mammals (101). There is also no evidence for organelle-localized LMW HSPs outside the plant kingdom.

Evidence that most plant LMW HSPs belong to four gene families is summarized in Table 2. The table shows the percentage of amino acid identity and similarity between LMW HSPs representative of all available sequences, and a proposed classification for the LMW HSP genes is presented. The comparisons show that there is greater identity among certain genes from different species than there is among different genes of the same species. Three groups consist of genes from four plant species with representatives of both dicots and monocots. Class I cytoplasmic LMW HSPs include pea HSP18.1, soybean HSP17.5-E, *Arabidopsis* HSP17.6, and wheat HSPC5-8. The similarity among these genes ranges from 80.1 to 92.9% (identity 68.2–85.1%); the highest value is from the comparison of pea and soybean, plants from the same taxonomic family. In the other comparisons of these genes the identity and similarity values are all strikingly lower, even when genes of the same species are considered. For example, pea HSP18.1 is only 59.7% similar to pea HSP17.7, 72.5% similar to pea HSP22.7, and 52.0% similar to pea HSP21; and the identity values for these comparisons are all below 50%. It is not surprising that comparisons between the corresponding nucleic acid sequences of the LMW HSPs yield even lower percentages. Thus, genes from different families (or classes) cross-hybridize only at very low stringency, if at all. Together these data suggest that the LMW HSP

Table 2 Amino acid sequence relationship between LMW HSPs from plants

| | | | | Percent amino acid identity (similarity) to:[a] | | | | | | | | | | | | | | | | |
| | | | | Cytoplasmic class I | | | | Cytoplasmic class II | | | | Nuclear-encoded chloroplast-localized | | | | Endomembrane localized | | Unclassified | |
Protein	Reference (Accession #)[b]	Mol. wt.[c] (kd)	pI[c]	*P. sativum* HSP18.1	*G. max* HSP17.5-E	*A. thaliana* HSP17.6	*T. aestivum* C5-8	*P. sativum* HSP17.7	*G. max* HSP17.9-D	*Z. mays* HSP18.3	*L. longiflorum* pEMPR-C6	*P. sativum* HSP21	*P. hybrida* HSP22	*A. thaliana* HSP21	*Z. mays* HSP26	*P. sativum* HSP 22.7	*G. max* HSP22	*C. reinhardtii* HSP22	*G. max* HSP26
Class I:																			
P. sativum HSP18.1	121 (M3389)	18.1	5.96	—															
G. max HSP17.5-E	37 (M11395)	17.5	6.31	**85.1** **(92.9)**	—														
A. thaliana HSP17.6	73 (X16076)	17.6	5.24	**72.0** **(84.7)**	**71.4** **(85.1)**	—													
T. aestivum C5-8	110 (X13431)	16.9	5.95	**71.6** **(84.5)**	**72.3** **(84.5)**	**68.2** **(80.1)**	—												
Class II:																			
P. sativum HSP17.7	97 (M33901)	17.7	6.82	38.3 (59.7)	37.5 (59.2)	38.4 (59.6)	40.5 (62.2)	—											
G. max HSP17.9-D	142 (X07159)	17.9	6.29	37.1 (59.6)	38.4 (56.3)	34.0 (56.4)	34.7 (60.0)	**78.3** **(92.4)**	—										
Z. mays HSP18.3	60 (NA)	17.8	5.19	36.5 (59.0)	37.9 (60.7)	34.0 (58.8)	39.6 (61.7)	**62.4** **(84.1)**	**64.8** **(84.9)**	—									
L. longiflorum pEMPR-C6	18 (NA)	17.6	8.98	40.9 (62.0)	37.7 (55.1)	35.5 (57.2)	36.2 (58.0)	**53.3** **(71.5)**	**50.4** **(71.2)**	**54.7** **(73.4)**	—								

Chloroplast-localized:[d]

Protein	Ref	Accession[b]	MW[c]	pI[c]																
P. sativum HSP21	172	(X07187)	26.2	6.77	30.3 (52.0)	29.7 (56.8)	27.9 (57.1)	31.0 (58.6)	28.3 (53.1)	28.3 (57.2)	32.5 (58.4)	25.9 (47.4)	—							
P. hybrida HSP22	25	(X54103)	26.8	8.33	32.9 (52.6)	36.4 (57.3)	38.0 (61.3)	34.5 (57.9)	30.8 (54.8)	30.6 (57.1)	33.8 (60.6)	31.1 (54.1)	**67.6 (84.1)**	—						
A. thaliana HSP21	25	(X54102)	25.6	10.20	35.1 (57.6)	37.4 (57.8)	32.9 (56.4)	34.0 (59.0)	30.6 (54.4)	31.0 (60.0)	36.1 (61.1)	34.6 (57.7)	**65.2 (80.3)**	**61.2 (81.5)**	—					
Z. mays HSP26	124	(NA)	26.4	8.41	32.9 (57.7)	36.4 (58.7)	29.5 (57.7)	33.3 (57.8)	31.5 (57.5)	30.7 (54.9)	33.6 (59.2)	34.1 (54.8)	**63.9 (80.9)**	**55.9 (79.3)**	**57.1 (77.7)**	—				

Endomembrane:

Protein	Ref	Accession[b]	MW[c]	pI[c]																
P. sativum HSP22.7	[e]	(M33898)	22.7	7.26	48.3 (72.5)	47.1 (70.6)	42.6 (65.2)	44.7 (67.3)	35.3 (57.1)	32.7 (59.1)	33.7 (57.7)	30.7 (53.6)	31.1 (58.5)	30.6 (55.4)	27.8 (55.1)	30.6 (55.4)	—			
G. max HSP22	[f]	(NA)	22.0	6.69	53.0 (71.8)	50.3 (69.9)	45.2 (66.5)	52.3 (70.0)	36.9 (63.7)	34.0 (62.3)	36.6 (56.7)	35.5 (58.5)	31.2 (60.6)	31.5 (54.9)	31.2 (54.9)	28.7 (56.3)	**76.2 (89.4)**	—		

Unclassified:

Protein	Ref	Accession[b]	MW[c]	pI[c]																
C. reinhardtii HSP22	61	(X15053)	16.7	6.53	34.6 (51.0)	33.3 (51.0)	40.1 (57.2)	35.3 (54.0)	33.5 (49.7)	32.9 (51.6)	35.1 (58.9)	30.2 (51.1)	23.8 (57.8)	25.7 (55.4)	27.5 (56.2)	30.6 (56.5)	23.4 (55.4)	24.0 (53.8)	—	
G. max HSP26	40	(NA)	26.0	5.31	21.7 (45.4)	25.2 (42.2)	23.0 (47.3)	15.2 (39.1)	19.1 (45.9)	18.9 (49.7)	14.6 (45.1)	22.0 (48.0)	17.7 (39.4)	12.3 (36.9)	14.0 (39.7)	11.7 (38.3)	22.0 (42.9)	14.1 (42.2)	12.5 (43.8)	—

[a] Percentage identity and similarity were calculated using the GAP program (set with default parameters) of the Wisconsin GCG DNA sequence analysis software.

[b] Accession number for the EMBL, Gen Bank or NBRF data base.

[c] Molecular weight and pI were determined from the derived amino acid sequence.

[d] Percentage similarity and identity calculated without the transit peptide

[e] K. Helm and E. Vierling, unpublished

[f] R. Nagao, personal communication

NA—not available

Comparisons between proteins in the same class have been indicated in boldface.

families arose from ancient gene duplications that occurred before the divergence of monocots and dicots.

The same trend in amino acid sequence comparisons holds for the class II cytoplasmic HSPs identified in Table 2 as pea HSP17.7, soybean HSP17.9-D, maize HSP18.3, and lily EMPR-C6. The pea and soybean genes in this group have previously been referred to as class VI LMW HSPs (97, 142, 154). This classification was derived from early cloning studies and was based on cross-hybridization of soybean cDNAs (154, 155). In several cases the hybridization differences resulted from the fact that the cDNAs were small and comprised primarily of the divergent 3' noncoding regions. When the complete sequences were obtained for soybean cDNAs originally classified as class I, II, and VII LMW HSPs, the encoded proteins were at least 90% similar; they should be considered members of the same gene family (38, 117, 142). Class III and V HSP cDNAs from soybean hybridized to minor RNA species that may not have been heat inducible, class V mRNAs were greater than 4 kb in size, and class III and V proteins were never identified in hybrid-selection/translation experiments (154). These clones are no longer available, and their relationship to sequenced LMW HSP genes is unknown (R. Nagao, personal communication). Because the prior soybean classifications are no longer valid, it appears appropriate to redesignate class VI genes as class II cytoplasmic LMW HSPs.

Lily EMPR-C6, although most similar to class II HSPs, is less closely related than the remaining class II HSPs are to each other. Lily EMPR-C6 is also unusual, in that it was initally identified as an abundant transcript expressed during meiosis (18) (see below). Only in subsequent studies was it shown to be induced by heat (B. Bouchard, personal communication). Characterization of additional LMW HSP genes from other species may identify proteins more similar to lily EMPR-C6 and may thus define a new class of LMW HSPs. However, at present EMPR-C6 is included in the class II category.

The class I and II genes are typically encoded by multigene families (18, 60, 117, 118, 154; L. Zimmerman, personal communication), and class I and II genes listed in Table 2 represent only one member of the corresponding family. As mentioned above, several members of the soybean class I genes have been sequenced; the proteins are greater than 90% similar (38, 117, 142). This family is comprised of 10–13 different polypeptides in this species (154, 155). Multiple class I genes have also been sequenced from *Arabidopsis* (73, 166), pea (71), and carrot (L. Zimmerman, personal communication). They are also approximately 90% similar at the amino acid level. There are fewer class I polypeptides in these species than in soybean. Three class II genes, which show at least 90% amino acid identity, have been sequenced from maize (60). The class II gene families from soybean and pea are less complex than the class I families, but multiple members have not yet been sequenced.

The third group of LMW HSPs for which several sequences are available are nuclear genes that encode chloroplast-localized HSPs. Again, the inter-species similarity is high within this HSP class, while their similarity to other LMW HSP classes is lower (Table 2).

Less information is available regarding the remaining genes in Table 2. One group, represented by pea HSP22.7 and soybean HSP22, appears to encode proteins that localize to the endomembrane system (74; see below). Soybean HSP22 was previously designated a class IV cDNA. I propose that this group be referred to as the endomembrane LMW HSPs, without a numerical designation unless additional families of endomembrane proteins are discovered. HSP22 from *Chlamydomonas* is not easily classified by this type of comparison, as it shows the greatest identity to class I cytoplasmic HSPs but higher overall similarity to the chloroplast HSPs. Finally, soybean HSP26, which is a cytoplasmic protein, shows less than 50% similarity to all other sequenced LMW HSPs. Additional criteria suggest it should not be considered to belong to the same gene superfamily as the other LMW HSPs (see below). Future studies may uncover further complexity in the plant LMW HSPs, as already discussed concerning lily EMPR-C6. However, hybrid-selection/translation experiments with the available soybean and pea clones suggest that most highly induced mRNAs in these species are represented by genes that have already been cloned (88, 118; E. Vierling, unpublished). The proposed classification adequately describes the characterized higher plant genes and allows for the addition of new classes as they are established.

LMW HSP Structure

In general, within the above-defined LMW HSP classes, there is much higher sequence conservation in the carboxyl-terminal portion of the proteins. This fact is illustrated by the amino acid sequence alignment of class I cytoplasmic HSPs from several species presented in Figure 1. In the carboxyl-terminal two thirds of the protein there are 73 identical of 106 amino acids. In contrast, the amino-terminal one third shows significantly less identity, with only 15 identical of 52 amino acids. As expected, overall identity is higher when only the dicot sequences are compared, but the same disparity in conservation of the amino- vs the carboxyl-terminus remains.

Similar multi-species comparisons of class II cytoplasmic HSPs from pea, soybean, and maize show that the sequences align with minimal gaps and that the greatest divergence occurs in the amino-terminal one third of these proteins. The lily EMPR-C6 protein contains a 15 amino acid insertion following the conserved carboxyl-terminal domain and is correspondingly

```
P. sativum   HSP18.1  :   MSLIPSFFSGRRSNVFDPFSLDVWDPLKDFPFSNSSPSASFPRENPAFVSTRV
G. max       HSP17.5-E :   MSLIPGFFGGRRSNVFDPFSLDMWDPFKDFHVPTSSVSA----ENSAFVSTRV
A. thaliana  HSP17.6  :   MSLIPSIFGGRRTNVFDPFSLDVFDPFEGFLTP-SGLANAPAMDVAAFTNAKV
T. aestivum  C5-8     :   MSIV------RRSNVFDPFADLWADPFDTFRSIVPAISGGSS-ETAAFANARV
                          **..      ** ******   **. *        ** .*
```

```
HSP18.1    DWKETPEAHVFKADLPGLKKEEVKVEVEDDRVLQISGERSVEKEDKNDEWHRVERSSGKFLRRFRLP
HSP17.5-E  DWKETPEAHVFKADIPGLKKEEVKVQIEDDRVLQISGERNVEKEDKNDTWHRVERSSGKFTRRFRLP
HSP17.6    DWRETPEAHVFKADLPGLRKEEVKVEVEDGNILQISGERSNENEEKNDKWHRVERSSGKFTRRFRLP
C5-8       DWKETPEAHVFKVDLPGVKKEEVKVEVEDGNVLVVSGERSREKEDKNDKWHRVERSSGKFVRRFRLP
           **.*********  *.**..******..**   .*  .****   *  *.*** ***********  ******
```

```
HSP18.1    ENAKMDKVKASMENGVLTVTVPKEEIKKAEVKSIEISG
HSP17.5-E  ENAKVNEVKASMENGVLTVTVPKEEVKKPDVKAIEISG
HSP17.6    ENAKMEEIKASMENGVLSVTVPKVPEKKPEVKSIDISG
C5-8       EDAKVEEVKAGLENGVLTVTVPKAEVKKPEVKAIEISG
           *.**.   .**.*****.***** **.** *.***
```

Figure 1 Amino acid sequence comparison among Class I cytoplasmic LMW HSPs from several plant species. The multiple sequence alignment was performed as in reference 25. Asterisks indicate residues conserved in all four sequences, periods indicate positions of conservative replacements. Identical residues are also indicated in bold. Dashes indicate gaps. See Table 2 for references to individual sequences.

shorter at the amino terminus. With the exception of this insertion, the lily protein shows good alignment with the other class II HSPs.

The sequence divergence between the LMW HSP classes is illustrated by the comparison of representative members from the class I, II, endomembrane, and chloroplast LMW HSPs from pea (Figure 2). There is only limited homology among these LMW HSPs, in direct contrast to the similarity seen in Figure 1 among class I LMW HSPs from different species. In the comparison of the different HSP classes (Figure 2), consensus region I is 27 amino acids long (aa 116–143 of HSP18.1) with 9 identical and 7 conservative replacements. The conserved Pro and Gly-Val-Leu sequence in this region are also present in all LMW HSPs of other eukaryotes (101). Consensus region II is 29 amino acids long (aa 66–95 of HSP18.1) with 9 identical and 8 conservative replacements. The 77 amino acid region encompassed by these two domains may have a more conserved structure than is evidenced by this comparison. As noted by many investigators, hydropathy analysis indicates that the poorly conserved sequence between consensus I and II is part of a highly hydrophilic domain present in all LMW HSPs (38, 101, 117, 142). A corresponding 76 amino acid domain is the most highly conserved region seen in a comparison of all four *Drosophila* LMW HSPs, and this same region is homologous to the α-crystallin structural proteins of the eye lens (81). Both the LMW HSPs and the α-crystallins form high-molecular-weight (10–20S) complexes in vivo, and it has been suggested that the domain shared by these proteins, in particular the consensus region I segment, is essential for formation of these complexes (81, 101).

As shown in Table 2, soybean HSP26 is not significantly homologous to any of the four LMW HSP families compared in Figure 1. It contains neither

```
HSP18.1 (Class I)
HSP17.7 (Class II)
HSP22.7 (endomemb)
HSP21   (chlplst)   MAQSVSLSTIASPILSQKPGSSVKSTPPCMASFPLRRQLPRLGLRNVRAQAGGDGDNKDN                                                                           MSLK


HSP18.1                                      MSLIPSFFSGRRSNVFDPFSLDVWDPLKDFPFSNSSPSASFPRENP
HSP17.7                                      MDFRLMDLDSPLFNTLHHIMDLTDDTTEKNLNAPTRTYVRDA
HSP22.7   PLNMLLVPFLLLILAADFPLKAKASLLP-FIDSPNTLLSDLWSDRFPDP---FRVLEQIPGRNIGGGEI
HSP21     SVEVHRVNKDDQGTAVERKPRRSSIDISPFGLLDPWSPMRSMRQMLDTMDRIFEDAITIPYGVEKHEPS
                                       *

HSP18.1   AFVST-RVDWKETPEAHVFKA--DLPGLKKEEVKVEVEDDRVLQISGERSVEKEDKNDEWHRVERSSGK
HSP17.7   KAMAATPADVKEHPNSYVFMV--DMPGVKSGDIKVQVEDENVLLISGERKREEEKEGVKYLKMERRIGK
HSP22.7   ITLSHARVDWKETPEGHVIMV--DVPGLKKDDIKIEVEENRVLRVSGERKKEEDKKGDHWHRVERSYGK
HSP21     ------RVPWEIKDEEHEIRMRFDMPGVSKEDVKVSVEDDVLVIKSDHR--EENGGEDCWSRK--SYSC
                               *.**.    ..*. **. .. .   *  *  *
                                        CONSENSUS II
                                          29 aa
HSP18.1   FLRRFRLPENAKMDKVKASMENGVLTVTVPK---EEIKKAEVKSIEISG
HSP17.7   LMRKFVLPENANIEAISAISQDGVLTVTVNKLPPPEPKKPKTIQVKVA
HSP22.7   FWRQFKLPQNVDLDSVKAKMENGVLTLTLHKLSHDKIKGPRMVSIVEEDDKPSKIVNDELK
HSP21     YDTRLKLPDNCEKEKVKAELKDGVLYITIPKT------KIERTVIDVQIQ
          .  **.*   .  . .*** .*. *
                  CONSENSUS I
                     27 aa
```

Figure 2 Comparison of amino acid sequences among LMW HSPs of *P. sativum* from different LMW HSP classes. See Figure 1's legend for details of the comparison. Consensus regions I and II as discussed in the text are underlined.

the invariant Pro and Gly-Val-Leu residues near the carboxyl-terminus nor the conserved residues of consensus region II (40). Optimal alignment with any other LMW HSP requires insertion of multiple gaps. These data suggest that HSP26 is not evolutionarily related to the major abundant LMW HSPs and is probably not a member of the same gene superfamily. This conclusion is also supported by the fact that soybean HSP26 is unique in being strongly induced by many stresses other than heat. Most other LMW HSPs are either not induced or only moderately induced by these other stresses (36). Thus HSP26 is most likely structurally and functionally distinct from the other LMW HSPs.

Plastid LMW HSPs

The localization of a LMW HSP to the chloroplast has now been firmly established in diverse plant species (24, 25, 57, 92, 124, 165, 170–172). The LMW chloroplast HSP is nucleus encoded, and in higher plants it is synthesized as a precursor, which is processed to its mature size during import into the chloroplast. This was the first HSP identified in chloroplasts. In *Arabidopsis* it is encoded by a single gene (E. Vierling, unpublished), and in soybean and pea it is encoded by a small gene family (172). The existence of other, distinct families of LMW chloroplast HSPs has not yet been demonstrated.

The LMW chloroplast HSPs from four divergent plant species are com-

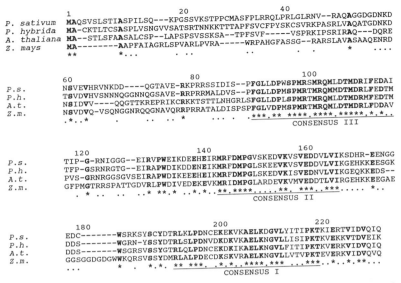

```
              1                20                40
P. sativum   MAQSVSLSTIASPILSQ---KPGSSVKSTPPCMASFPLRRQLPRLGLRNV--RAQAGGDGDNKD
P. hybrida   MA-CKTLTCSASPLVSNGVVSATSRTNNKKTTTAPFSVCFPYSKCSVRKPASRLVAQATGDNKD
A. thaliana  MA--STLSFAASALCSP--LAPSPSVSSKSA--TPFSVF-----VSPRKIPSRIRAQ---DQRE
Z. mays      MA-------AAPFAIAGRLSPVARLPVRA-----WRPAHGFASSG--RARSLAVASAAQENRD
             **          *...       ..              ..      .    *      ....
```

```
             60                80                100
P.s.   NSVEVHRVNKDD----QGTAVE-RKPRRSSIDIS--PFGLLDPWSPMRSMRQMLDTMDRIFEDAI
P.h.   TSVDVHVSNNNQGGNNQGSAVE-RRPPRRMALDVS--PFGLLDPMSPMRTMRQMMDTMDRLFEDTM
A.t.   NSIDVV----QQGTTKREPRIKCRKKTSTTLNHGRLSFGLLDPLSPMRTMRQMLDTMDRMFEDTM
Z.m.   NSVDVQ-VSQNGGNRQQGNAVQRRPRRATALDISPSPFGLVDPMSPMRTMRQMLDTMDRLFDDAV
       .*..*      .    .. ... *    ... .***.** ****.****.*****.*,*..
                                           CONSENSUS III
```

```
             120               140               160
P.s.   TIP-G-RNIGGG--EIRVPWEIKDEEHEIRMRFDMPGVSKEDVKVSVEDDVLVIKSDHR-EENGG
P.h.   TFP-GSRNRGTG--EIRAPWDIKDDENEIKMRFDMPGLSKEEVKVSVEDDVLVIKGEHKKEESGK
A.t.   PVS-GRNRGGSGVSEIRAPWDIKEEEHEIKMRFDMPGLSKEDVKISVEDNVLVIKGEQKKEDS--
Z.m.   GFPMGTRRSPATTGDVRLPWDIVEDEKEVKMRIDMPGLARDEVKVMVEDDTLVIRGEHKKEEGAE
       . *  ..  ..   ..* **.*  ..*.*...**.****..**. ***..*** ..... *..
                                    CONSENSUS II
```

```
             180               200               220
P.s.   EDC-------WSRKSYSCYDTRLKLPDNCEKEKVKAELKDGVLYITIPKTKIERTVIDVQIQ
P.h.   DDS-------WGRN-YSSYDTRLSLPDNVDKDKVKAELKNGVLLISIPKTKVEKKVTDVEIK
A.t.   DDS-------WSGRSVSSYGTRLQLPDNCEKDKIKAELKNGVLFITIPKTKVERKVIDVQIQ
Z.m.   GGSGGDGDGWWKQRSVSSYDMRLALPDECDKSKVRAELKNGVLLVTVPKTEVERKVIDVQVQ
       ...        *  .  *.*..** ***..**.. ****.***..... ***..*..*.**...
                              CONSENSUS I
```

Figure 3 Comparison of amino acid sequences among LMW chloroplast HSPs from four plant species. Details of the comparison as in Figure 1. Consensus regions I, II, and III as described in the text are underlined. Sequence numbers refer to the *P. sativum* sequence.

pared in Figure 3. The amino-terminal 45–50 residues show no sequence conservation between species. This region corresponds to the transit peptide that is removed during import into the chloroplast (86). Within the sequence corresponding to the mature proteins the three most highly conserved regions are indicated as consensus regions I, II, and III. Regions I and II delimit that segment of these proteins that shows homology to other LMW HSPs (see Figure 2). Consensus region III contains a highly conserved domain unique to the chloroplast HSPs. This domain, corresponding to amino acids 90–117 of the complete pea sequence (including the transit peptide), is predicted to form a Met-rich, amphipathic α-helix (25). In the helix predicted for the pea, soybean, petunia, *Arabidopsis*, and maize proteins, 100% of the residues on the hydrophilic face are absolutely conserved. On the hydrophobic face, four of six positions contain invariant Met residues. This remarkable domain, which is more conserved than any other part of the protein, is undoubtedly important for chloroplast HSP function. Recently a similar domain, called a "methionine bristle," has been described in a 54-kDa protein component of signal recognition particle (13). The methionine bristle domain has been proposed either to bind the signal sequence of nascent polypeptides (13) or to be an RNA binding domain necessary for association of the 54-kDa protein with the 7S RNA of signal recognition particle (146a). The function of the chloroplast LMW HSP Met-rich domain remains to be determined.

Nucleus-encoded *Chlamydomonas* LMW HSP22 can be isolated in association with thylakoid membranes and may be a chloroplast protein (61). However, the protein has not been unequivocally localized to the chloroplast (61, 95). Unlike other nucleus-encoded chloroplast proteins it does not have an amino-terminal transit peptide. It contains the characteristic carboxyl-terminal amino acid sequences that identify it as a LMW HSP, but it lacks the unique domain found in the higher plant LMW chloroplast HSPs. The latter fact suggests that the *Chlamydomonas* protein may have a somewhat different function in comparison to the characterized higher plant chloroplast HSP.

Endomembrane LMW HSPs

Cell-fractionation studies of heat-shocked plant tissues have indicated that certain LMW HSPs are enriched in the ER (33, 163). Several lines of evidence substantiate these observations and support the conclusion that soybean HSP22 and pea HSP22.7 (Table 2 and Figure 2) are endomembrane-localized proteins. The amino-terminal 25–30 residues of pea HSP22.7, presented in Figure 2, are similar to a consensus signal peptide. The amino-terminal 8 residues carry a net positive charge, and there is a 14-residue hydrophobic core followed by a 6–7 residue polar region typically adjacent to the cleavage site (176). The cleavage site of HSP22.7 is unknown, but based on the "-3,-1 rule" of von Heijne (175), it can be predicted to be between residues 28 and 29. The hydropathy profile of the first 29 residues of HSP22.7 also conforms to that of other eukaryotic proteins transported through the ER. The amino terminus of soybean HSP22 contains a comparable sequence (R. Nagao, personal communication). Helm & Vierling (74) demonstrated that the in vitro translation product of RNA transcribed from the pea HSP22.7 cDNA is transported and processed by isolated microsomes. Furthermore, HSP22.7-specific antibodies react with a microsomal protein that is present in heat stressed roots and not found in control roots (74). The protein is resistant to protease treatment of the microsomal fraction, indicating it is inside the microsomal vesicles. There is no data to suggest that this protein is transported to other organelles or secreted. However, the carboxyl-terminal four residues of pea HSP22.7 (DELK) and soybean HSP22 (KQEL) are not identical to any of the carboxyl-terminal tetrapeptides identified as ER-retention signals (135). Thus, further experiments are required to confirm that this protein is localized only in the ER, and to define the mechanism of its retention in the ER.

It is interesting that ER-localized LMW HSPs have not been found in other eukaryotes. Perhaps they are present but are not heat regulated. Because of the limited identity among the LMW HSP classes, ER-homologs are not likely to be detected by antibodies or DNA probes that are specific for the cytoplasmic LMW HSPs.

LMW HSP Function

The function of LMW HSPs remains unknown. Genetic approaches, which have been critical for probing the function of HSP70, have not helped establish a role for the LMW HSPs. Despite evidence that a LMW HSP homolog is present in prokaryotes [i.e. *Mycobacterium* (119)], no comparable protein has been characterized in *E. coli*. In *Saccharomyces cerevisiae* only a single LMW HSP gene has been identified (137). Yeast mutants in which the LMW HSP gene was deleted, disrupted, or overexpressed showed little detectable change in phenotype, even under severe stress conditions (137, 164). Based on these data, Susek & Lindquist (164) suggest that LMW HSPs are products of "selfish" or ancient viral DNA and that they have no function during heat stress. The complexity of the LMW HSPs in plants, along with the conservation of distinct LMW HSP classes, argues against this explanation. It is possible that yeast contains additional, unidentified LMW HSP genes that compensated for the missing function in the null mutants. Alternatively, LMW HSPs may perform functions that are dispensable in less complex or nonplant eukaryotes.

Biochemical experiments have also failed to define a function for LMW HSPs. In many organisms the cytoplasmic HSPs can be detected in 10–20S particles or in aggregates larger than one megadalton (4–6, 31, 101, 121, 126, 129). The larger LMW HSP aggregates, referred to as "heat shock granules," form reversibly from the smaller particles, depending on the duration and severity of stress conditions (4–6, 31, 129). The formation of heat shock granules may explain many of the discrepancies in studies of LMW HSP localization. LMW HSPs have been reported to be in the nucleus, bound to membranes, associated with ribosomes or associated with prosomes (7, 99, 101, 118). Some of these associations most likely resulted from copurification of heat shock granules with different cell fractions (101, 105, 126). For example, LMW HSP aggregates have now successfully been separated from ribosomes and prosomes (43a, 129). Furthermore, electron microscopic studies of LMW HSPs in plants indicate heat shock granules accumulate in the perinuclear region rather than within the nucleus (121, 129). It is also clear that some reports of LMW HSP association with membranes resulted from the presence of the specific chloroplast and endomembrane LMW HSPs. It would appear that under most conditions class I and II LMW HSPs in plants are located in the cytoplasm.

Studies of the LMW HSP complexes will undoubtedly be important in elucidating the function of these ubiquitous proteins. The composition of the 10–20S particles and heat shock granules has been investigated in several organisms. In tomato, heat shock granules appear to contain LMW cytoplasmic HSPs, HSP70, other unidentified proteins, and an RNA component (129). Nover et al (129) reported that the RNA component was selectively composed of mRNAs present prior to heat stress and devoid of HSP mRNA. They suggest that heat shock granules function to protect and store normal cellular mRNAs during stress. However, in chicken, *Drosophila*, and mammals the particles are composed primarily of LMW HSPs, and the presence of other proteins or RNA has not been confirmed (5, 6, 31). During heat stress normal cellular mRNAs are present, but not on polysomes; thus differential partitioning of the control and heat shock mRNAs is not surprising. Whether HSP granules actively sequester specific mRNAs, or cofractionate with the stored mRNA population, remains to be determined.

Because heat shock granules accumulate only when cells having very high concentrations of HSPs are subjected to severe heat shock (121), it is possible that LMW HSPs are not found as heat shock granules under most conditions. This possibility suggests that the active form of the LMW HSPs could be the 10–20S particle. Alternatively, the 10–20S particles could be nonfunctional, being made only as insurance against more severe stress, at which time they would become functional heat shock granules. However, the LMW HSPs are expressed in the absence of heat stress during defined developmental stages in *Drosophila* (4), and only the 10–20S LMW HSP particles are formed. These data support the idea that this form is functional. Collier et al (31) proposed that the 10–20S LMW HSP particles have an enzymatic function and that the larger aggregates are an inactive "stored" form of the enzyme. The biochemistry of the 10–20S particles deserves further investigation.

Researchers have also speculated, based on the homology of LWM HSPs to the α-crystallin structural proteins of the eye lens, that the LMW HSPs maintain the structural integrity of heat-stressed cells (101). However, the crystallins are structural as a result of their very high concentration in the lens (<50% of soluble protein). There is no evidence that LMW HSPs accumulate to much over 1% of plant cell protein even under severe stress conditions (24, 104, 118). Furthermore, other families of lens crystallins, such as ϵ and τ crystallins, have been found to be homologous to specific glycolytic enzymes, suggesting that these structural proteins actually evolved from enzymes (138). Thus, based on the crystallin homology, an enzymatic role for the LMW HSPs is as plausible as a structural role.

How the LMW chloroplast HSP functions is also enigmatic. Kloppstech and colleagues (57, 92) have proposed that the chloroplast HSP functions to protect or repair photosystem II during stress. Photosystem II is one of the more heat-sensitive components of the chloroplast (14). The hypothesis that it may be involved in protecting photosynthetic partial reactions is based on observations that the pea chloroplast HSP was bound to thylakoid membranes. However, thylakoid localization occurred only at temperatures above 38° C at relatively high light intensities. Below 38° C the protein showed no strong association with membranes (57). Unfortunately, no assay is available to demonstrate the function of the observed membrane association.

Using a variety of stress conditions, including temperatures of 38° C or 40° C at moderate light intensities, Chen et al (24) did not find a significant proportion of the pea chloroplast HSP associated with thylakoids. Greater than 80% of the protein was present in the soluble chloroplast fraction (24). The chloroplast HSP was also produced at significant levels in root tissues, indicating it may function in all types of plastids, not only photosynthetic organelles (24). Chen et al (24) proposed that if the chloroplast HSP protects photosynthesis, it does so as a consequence of general structural or enzymatic properties rather than through direct interaction with photosystem II. The presence of the protein in nonphotosynthetic plastids and its homology to cytoplasmic and endomembrane LWM HSPs support this hypothesis and suggest chloroplast HSP performs functions in addition to protecting photosynthesis.

Other evidence for a role of chloroplast HSPs in protecting photosynthesis has come from studies of *Chlamydomonas*. In *Chlamydomonas* a pretreatment at high temperature in the dark ameliorated photosystem II damage during subsequent heat treatments in the light (158). This effect was attributed to HSP22 localization to, and protection of, photosystem II. However, recent studies with antibodies against HSP22 showed that it was not synthesized in the dark (82), suggesting that the observed protection is not directly due to this HSP. As already mentioned, the *Chlamydomonas* protein lacks a highly conserved domain present in higher plant chloroplast HSPs, and it may interact with different chloroplast components.

OTHER HSPS

HSP90 and HSP110

The HSP90 class of proteins ranges in size from approximately 80 to 94 kDa. In addition to cytoplasmic forms of HSP90, vertebrates have an ER-localized homolog (also called GRP94) that is expressed in response to glucose starvation (101). An *E. coli* homolog, the product of the htpG gene, shows 40% amino acid identity to eukaryotic HSP90s (101). Genes encoding proteins with homology to HSP90 have only recently been isolated from plants. DNA sequence analysis of maize (121), *Brassica* (121) and *Arabidopsis* HSP90 (32; Y. Komeda, personal communication) genes shows that the proteins are approximately 70% identical to HSP90 from other eukaryotes. *Arabidopsis* has at least two HSP90 genes, one of which contains three introns (Y. Komeda, personal communication). HSP90 mRNA level is strongly induced during heat stress; however, at very high temperatures the appearance of a higher-molecular-weight transcript suggests that the mRNA is not being completely spliced. The HSP90 homolog in *Drosophila* (HSP82) also contains an intron, and HSP82 transcripts are not efficiently spliced during severe stress (183). Consequently, maximal HSP82 expression occurs at more moderate stress temperatures in *Drosophila*, and the same may be true in plants.

There have been no specific studies of the HSP90 protein in plants. It is easily identified in total protein profiles of in vivo labeled plant tissues, and it is recovered in the soluble fraction of lysed cells (33, 99, 121). In other eukaryotes, HSP90 is an abundant component of nonstressed cells, and it increases during stress. *S. cerevisiae* has two genes in this HSP family, HSP82 and HSC82. Disruption of both genes is lethal, and mutations in either gene impair growth at elevated temperatures (17). HSP90 has been extensively studied in mammalian cells where it can be isolated in association with steroid hormone receptors (101). A current model of HSP90 function proposes that it maintains these receptors (and possibly other proteins) in a specific conformation until the cell receives appropriate receptor-activating signals (101, 148, 153). Thus, HSP90 can also be considered as a type of molecular chaperone that acts to modify or maintain the conformation of proteins to which it binds. Recent experiments in which the mammalian glucocorticoid receptor was expressed in yeast suggest that the HSP82-receptor interaction is actually required for receptor activity (138a). Considerable further work is needed to confirm this model and to identify HSP90 substrates in plants.

Most eukaryotes, including plants, synthesize a high-molecular-weight HSP of approximately 110 kDa, which also appears to be a soluble cellular component (33, 101). No HSP of this size class has been identified in *Drosophila*. Borkovich et al (17) suggested that a homologous protein is present but constitutive in this organism. In plants, synthesis of HSP110 is more transient than that of other HSPs; abundant synthesis is primarily limited to the first hour of heat stress (118). The first eukaryotic gene encoding an HSP of this size range has been isolated from *S. cerevisiae* (HSP104) (152). The yeast HSP104 gene is not essential for growth at normal or high temperatures, but deletion mutants are incapable of developing thermotolerance (152). Although HSP104 appears necessary for thermotolerance, it is not likely to be sufficient. Undoubtedly other components are also required. Nonetheless, HSP104 is the first HSP for which genetic evidence supports a role in thermotolerance. It will be interesting to learn the extent of homology between HSP104 and HSPs of this size class in higher eukaryotes.

Ubiquitin

Heat stress increases ubiquitin transcription in many eukaryotes (22, 30, 101, 121). Ubiquitin is a highly conserved, 76-amino-acid protein which is covalently attached to other proteins, marking them for degradation (77). Ubiquitin-protein conjugation can also be reversible, and may serve regulatory roles distinct from protein turnover (77). Increased production of ubiquitin during heat stress may reflect an increased demand for removal of proteins damaged by the stress.

In plants, ubiquitin is encoded by a multi-gene family, and response of these genes to heat stress is complex (22, 30). In both *Arabidopsis* and maize the level of some ubiquitin gene transcripts increased during stress treatments, while levels of others decreased or remained unchanged. The significance of these different responses and their effects on the overall level of ubiquitin-mediated protein degradation during stress are unknown. Wheat roots treated at 42° C showed a 30% decrease in free ubiquitin and a corresponding increase in ubiquitin conjugated to other proteins (51). The resolution of these experiments did not permit identification of ubiquitin substrates unique to stressed tissues. It is not surprising that protein degradation increases at temperatures 15 or more degrees above normal growth temperatures. Whether ubiquitin activity exerts control over the production of other HSPs by removing inactive proteins remains an open question.

Additional Heat-Regulated Proteins

Numerous proteins that exhibit increased synthesis or elevated mRNA levels during heat stress have been identified in addition to those described above. This is to be expected given the dramatic nature of increasing cell temperature 5–15° C.In general, heat induction of other proteins is restricted to particular organisms or cell types, and their level of induction is significantly lower than that of HSP70 or the LMW HSPs. Nonetheless, some of these changes may prove to be important, evolutionarily conserved responses to heat stress, and deserve further investigation. For example, mRNA levels of certain enzymes of the glycolytic pathway increase during heat stress (101), including a specific glyceraldehyde-3-phosphate dehydrogenase gene from maize (151). Heat shock induction of a low-molecular-weight protein kinase in sugarcane cells has also been reported (114). These changes may be part of a complex re-equilibration of cellular metabolism at elevated temperatures.

As already discussed, HSPs homologous to major classes of HSPs have been identified in chloroplasts and mitochondria. These proteins are all nucleus encoded, synthesized in the cytoplasm, and transported into the organelle. Several investigators have reported that heat stress also induces expression of specific organelle-encoded proteins (94, 159). However, other researchers have provided data that conflict with these observations (123, 171). Because no genes with homology to HSPs have been identified on either the mitochondrial or chloroplast genomes, changes in organelle gene expression would not involve HSPs of the major classes defined here.

Cell-fractionation studies have indicated that certain nucleus-encoded LMW HSPs may be found in plant mitochondria (28, 33, 99), but evidence that these are distinct mitochondrial LMW HSPs is lacking. LMW HSPs reported to be associated with mitochondria are most likely LMW HSPs from plastid, endomembrane, or heat shock granule contamination of the mitochondrial preparations. There has been no direct demonstration that heat-induced proteins are transported into mitochondria, or that mitochondria-associated HSPs are resistant to protease digestion and therefore localized within the organelle. However, given the presence of LMW HSPs in three plant cell compartments, it is possible that future studies will identify specific mitochondrially targeted HSPs.

RELEVANCE OF HSPS TO PLANT GROWTH AND DEVELOPMENT

Expression of HSPs in the Natural Environment

Many studies of HSP expression have been performed in the laboratory, often subjecting organisms to nonphysiological, abrupt increases in temperature. Under these conditions HSP gene transcripts accumulate within the first five minutes of heat stress (118, 145). Further work has shown that induction of HSP synthesis does not require an abrupt stress. HSP expression also occurs when plants experience a gradual increase in temperature, more typical of the natural environment (2, 21, 24, 90, 118, 170). Unfortunately, few studies have actually examined HSP expression in the field. Kimpel & Key (90) showed that field-grown soybeans accumulate LMW HSP mRNAs on hot days. HSP mRNA expression was higher in water-stressed plants than in well-watered plants, probably due to differences in leaf temperature. Leaf protein profiles suggested that LMW HSPs also accumulated (J. Kimpel, personal communication). In other experiments, cotton grown under recurring water stress exhibited elevated leaf temperatures and accumulated proteins with molecular weights corresponding to those of HSPs (21). Both of these studies, although limited to examining plants suffering from severe stress, established that HSP expression does occur in the natural environment.

LMW HSP expression has also been studied in the laboratory using intact plants subjected to temperature stress treatments designed to mimic temperature changes in a field environment (24, 170). Nine-day-old pea plants were subjected to increasing morning temperatures, high midday temperatures, and decreasing afternoon temperatures under conditions of high humidity to limit transpirational cooling. Levels of LMW chloroplast and class I cytoplasmic HSPs were then determined immunologically (24; A. DeRocher, unpublished). Accumulation of the LMW HSPs was shown to be directly proportional to the midday leaf temperature (24; A. DeRocher, unpublished). The class I cytoplasmic HSPs were detectable even when the maximum leaf temperature was only 32° C, suggesting that HSPs may frequently be expressed even under mild stress conditions. Plants were monitored during a week of recovery following the stress, and the half-life of the HSPs was determined to be approximately 52 hr and 37 hr for the chloroplast and cytoplasmic HSP, respectively. The persistence of these HSPs suggests that their participation in recovery processes could be as important as their possible role during the stress period itself.

Production of HSPs may occur frequently in plant structures not effectively cooled by transpiration. Reproductive structures that have limited numbers of stomates, or are too large for rapid heat exchange, will be at higher temperatures than surrounding leaves. For example, pea pods have been found to reach extremely high temperatures during later stages of development in the field (69). HSP expression is also likely to occur at other stages of plant development during which cooling mechanisms are limited, such as during seedling emergence. Continuous high temperatures during seed maturation are certainly detrimental in many instances (87). However, seed temperatures high enough to induce HSPs may occur under optimal growth conditions for many species. Commercially produced pea seeds that have not been treated to post-harvest heat drying, and that exhibit maximal germination potential, have high levels of HSP mRNAs in both axis and cotyledon tissues (173). These data indicate that HSPs are expressed under the standard conditions employed in pea seed production. HSP mRNA accumulation in this context may not result solely from elevated seed temperature during development. HSP mRNAs have been detected in pea and wheat seeds that developed under temperatures below the temperature of HSP induction in vegetative tissues (see below; 43, 70, 71). More observations of HSP expression in the natural environment may reveal that production of HSPs is an integral part of many phases of plant life history, even for individuals growing in optimal environments.

HSP Expression During Plant Development

Two questions have been addressed with regard to HSP expression during plant development: 1. can HSP synthesis be induced at all stages of development, and 2. are HSPs developmentally regulated in the absence of heat stress? Concerning the first question, it has been established that induction of HSP synthesis in response to high-temperature stress occurs in most tissues that are transcriptionally active. The transcription of HSP genes and HSP synthesis have been shown to occur in all vegetative tissues, in mid-maturation seeds (2, 7, 29), in the aleurone of imbibed seeds (20, 163), and in germinating embryos (70, 72, 78). Two developmental stages in which full activation of the heat shock response does not occur are very early embryo development (pre-torpedo stage) (139, 184) and pollen germination (157, 169, 181). In both these cases failure to accumulate HSP mRNA in response to heat stress cannot be completely accounted for by reduced overall transcriptional activity. Despite failure to activate HSP genes, early embryos synthesize a significant number of HSPs in response to heat stress. Apparently the embryos already contain HSP mRNAs that are not actively translated unless the embryo is heat stressed (139, 184). Both germinating pollen (181) and globular embryos (184) are very heat sensitive. Whether this is in part a result of their inability to mount a full heat shock response is speculative.

The expression of HSP mRNA and protein during development in the absence of heat stress has been studied in a number of eukaryotes (16). As already mentioned, there is clearly tissue and developmental specificity in expression of certain HSP70 homologs in plants (43, 45). LMW HSP mRNAs are found not only early in embryogenesis, but class I LMW HSP mRNAs are also found in fully developed seeds of pea and wheat (70–72, 173). In contrast to the situation in early embryo development, accumulation of LMW HSP mRNAs during late seed development is accompanied by accumulation of the corresponding proteins (70, 71). It could be argued that LMW HSPs are expressed in seeds as a safeguard against possible stress, and that these HSPs have no specific functions in seed development. However, the fact that heat shock elicits the rapid production of HSP mRNAs in developing seeds (2, 29) makes this possibility seem unlikely. Early imbibing wheat embryos are very thermotolerant (1). HSPs accumulated during seed development might contribute to this tolerance, but this possibility has not been tested. Further studies of additional plant species, along with examination of the localization of LMW HSPs, should aid in formulating hypotheses concerning HSP function in development, and should also contribute to an understanding of HSP function during stress.

Although germinating pollen has been reported incapable of HSP synthesis, it is interesting that class II LMW HSP mRNAs are expressed during specific stages of microsporogenesis in *Lilium* and maize in the absence of stress (18; B. Bouchard, personal communication). As mentioned previously, the lily EMPR-C6 cDNA was originally isolated as an abundant meiotic prophase-specific transcript (18). These results further emphasize the potential significance of HSPs in normal plant growth and development. Whether other HSPs are also produced during meiosis in plants has not been examined.

Role of HSPs in Thermotolerance

Many organisms, including plants, can survive otherwise lethal high-temperature treatments if they are first subjected to a pretreatment at nonlethal high temperatures. This phenomenon is called acquired thermotolerance. Extensive experiments have shown that pretreatments which lead to acquisition of thermotolerance are conditions under which HSPs are synthesized (89, 100, 118). Using etiolated soybean seedlings, Lin and colleagues have shown not only that the rate of synthesis of LMW HSPs is correlated with the development of thermotolerance (99), but also that LMW HSP accumulation parallels acquired thermotolerance (C.-Y. Lin personal communication). These and other correlative data strongly suggest, but do not demonstrate definitively, that the accumulation of HSPs is important for protection from thermal killing.

The problem in drawing conclusions from correlative data is well-illustrated by experiments with *S. cerevisiae*. Although HSP expression correlates with the development of thermotolerance in this organism, HSP90, HSP26, and HSP70 mutants all show the ability to express acquired thermotolerance (101). As discussed above, HSP104 is the only yeast gene that has been demonstrated to be essential for thermotolerance (152).

Research using mammalian and *Drosophila* cells has tried to determine the role of HSPs in thermotolerance by altering only HSP concentrations, rather than using pretreatments that can have a variety of different effects. In mammalian cells, inactivating HSP70 by microinjection of antibodies (144) or specifically blocking HSP70 gene transcription (83) made cells more thermosensitive. Landry et al (96) showed that overproduction of LMW HSPs in transformed mammalian cells improved their short-term thermotolerance. When only HSP70 expression was increased or decreased in *Drosophila* tissue culture cells, thermotolerance was correspondingly increased or decreased (S. Lindquist, personal communication). These studies provide additional support for the hypothesis that HSPs contribute to thermotolerance. Experiments using transgenic technology to alter specifically the expression of HSPs have not yet been tried in plants.

For a number of plant species, heritable differences in resistance to high temperatures at various life stages have been documented (15). There is no clear evidence that HSPs contribute to these genetically determined differences in heat tolerance, although recently investigations of this possibility have been initiated (49, 50, 78, 93, 106, 133). Studies of this type are complicated by differences in the genetic background of the cultivars used and by the number and complexity of the HSP genes. For example, LMW HSP heterogeneity between cultivars has been documented in carrot (80), wheat (93), maize (182) and barley (106), and significant differences are also seen between species in a single genus (49, 50). Because of this natural polymorphism in HSP characteristics, correlating qualitative differences in HSPs with differences in heat tolerance can be misleading unless appropriate genetic tests are performed. In cases where a genetic analysis has been attempted, correlations between HSPs and thermotolerance have not been found. For example, several unique HSPs were identified in a thermotolerant cotton line (49). When backcrossed repeatedly to a thermosensitive line of cotton, the unique HSPs did not segregate with the thermotolerance phenotype. Extensive differences in the profile of LMW HSPs were also observed between *Lycopersicon esculentum* and *L. pennellii* (50), but the interspecific cross showed no linkage of HSPs to heat tolerance. These results emphasize the importance of genetic experiments for evaluating the contribution of HSPs to thermotolerance.

SUMMARY AND FUTURE PROSPECTS

Molecular analysis indicates that the major classes of HSPs synthesized by plants are homologous to HSPs of other eukaryotes. Several proteins with homology to HSPs, or in some cases HSPs themselves, are also components of unstressed plant cells. The discovery that different proteins of the major HSP families are found in more than one cellular compartment further indicates HSPs play a role in basic biochemical processes. The HSPs and their homologs must perform many essential functions in both normal and stressed cells, but these functions are only beginning to be understood. The current hypothesis that HSP60, HSP70, and HSP90 function to alter the conformation or assembly of other protein structures provides an exciting starting point for further investigations. Detailed analysis of the molecular mechanisms of these processes, and determination of the protein substrates involved, will be important to an understanding of how these processes may protect or allow recovery of the heat-stressed cell.

Certain aspects of the heat shock response are unique to plants. Because other eukaryotes lack chloroplasts, one obvious difference in the plant heat shock response is the presence of HSP70, HSP60, and LMW HSPs in these organelles. Although these HSPs are similar to their counterparts in other eukaryotes or prokaryotes, their significance to chloroplast structure and function require investigation. The abundance and diversity of the LMW HSPs are also unique to plants. Apparent structural homology indicates that the cytoplasmic, chloroplast, and endomembrane LMW HSPs may perform basically the same functions, involving different substrates. Whatever this function is, it is required in three plant cell compartments. Perhaps these proteins act as molecular chaperones, analogous to the HSP60 and HSP70 homologs also found in the several cell compartments. However, in contrast to HSP70 and HSP60, which are present both constitutively and at elevated temperatures, only certain LMW HSPs are present in the absence of stress, and then only during restricted periods of development. Other than these limited exceptions, LMW HSP function appears restricted to stress conditions.

Understanding the function of HSPs in plants will require much additional work. More complete characterization of the HSP genes along with further studies of the intracellular localization and biochemical properties of HSPs will be required. The use of gene-specific cloned DNA probes and highly specific antibodies will be critical to obtaining information about each of the classes of HSPs. Genetic approaches that could complement information gained from molecular and biochemical work include identification of specific HSP mutants and creation of mutant transgenic plants.

An important and fascinating problem in heat shock research that has not been discussed in this review concerns how high temperature is perceived by the cell and transduced to activate HSP expression. All evidence indicates that the production of HSPs is a cell-autonomous phenomenon. The heat shock response is not elicited by signals transmitted between cells or plant organs. Gene transcription is a major control point in HSP expression, and considerable progress has been made toward elucidating how HSP gene transcription is induced in plants (19, 63, 91). *Cis*-acting DNA elements ("heat shock elements" or HSEs) that are necessary for heat-induced transcription have been identified in the 5' promoter regions of plant HSP genes (10, 12, 39, 62, 63, 156). The plant HSE is the same as the element found in HSP genes of other eukaryotes (115). Specific *trans*-acting protein factors that bind to HSEs have recently been cloned from plants (185, L. Nover, personal communication) and found to have a DNA binding domain similar to that of the yeast (162, 178) and *Drosophila* (30a) HSP gene transcription factors. Other protein factors that bind in the HSP gene promoter region have also been identified (37). Determining what role these factors play in the signal transduction chain will require detailed analysis of how they act at HSP gene promoters. Continued investigation of HSP gene transcription in plants promises to provide information about the structure of plant gene promoters and insight into mechanisms by which plants sense environmental conditions.

Can plant tolerance to high temperatures be increased by changing the expression of HSPs? There is still no clear answer to this question. It would be interesting to determine if HSPs map to quantitative trait loci linked to heat tolerance, but at present appropriately characterized plant material is not available. Breeding for traits that contribute to maintaining plant structures at optimal growth temperatures, even when ambient temperatures are much higher (15), is one approach to generating thermotolerant plants that obviously does not involve HSPs. It seems unlikely that simply increasing the amount of a single HSP would increase an organism's optimum growth temperature. However, even an increase of one or two degrees in the maximum temperature at which plants will survive could be sufficient to allow higher crop productivity in seasons with great temperature extremes. Manipulating HSPs for plant improvement will only be possible when the mechanism of HSP induction and the roles of individual HSPs are better understood.

ACKNOWLEDGMENTS

I thank Drs. J. Kimple, J. Cushman, and M. Mishkind for their helpful criticism of this manuscript, and many colleagues for communicating unpublished information. Special thanks to the members of my laboratory, Q.

Chen, A. DeRocher, L. Lauzon, and Dr. K. Helm for their research contributions cited here, and for help with preparation of figures and tables. E. V. also acknowledges Dr. J. L. Key for initial insights into the heat shock response in plants. The laboratory has been supported by grants from the NSF, USDA, and NIH, and by Arizona Hatch Project Funds.

Literature Cited

1. Abernathy, R. H., Theil, D. S., Petersen, N. S., Helm, K. 1989. Thermotolerance is developmentally dependent in germinating wheat seed. *Plant Physiol.* 89:569-76
2. Altschuler, M., Mascarenhas, J. P. 1985. Transcription and translation of heat shock and normal proteins in seedlings and developing seeds of soybean exposed to a gradual temperature increase. *Plant Mol. Biol.* 5:291-97
3. Amir-Shapira, D., Leustek, T., Dalie, B., Weissbach, H., Brot, N. 1990. Hsp70 proteins, similar to *Escherichia coli* DnaK, in chloroplasts and mitochondria of *Euglena gracilis*. *Proc. Natl. Acad. Sci. USA* 87:1749-52
4. Arrigo, A.-P. 1987. Cellular localization of HSP23 during *Drosophila* development and following subsequent heat shock. *Dev. Biol.* 22:39-48
5. Arrigo, A.-P., Suhan, J. P., Welch, W. J. 1988. Dynamic changes in the structure and intracellular locale of the mammalian low-molecular-weight heat shock proteins. *Mol. Cell. Biol.* 8:5059-71
6. Arrigo, A.-P., Welch, W. J. 1987. Characterization and purification of the small 28,000-dalton mammalian heat shock protein. *J. Biol. Chem.* 262:15359-69
7. Atkinson, B. G., Walden, D. B., eds. 1985. *Changes in Eukaryotic Gene Expression in Response to Environmental Stress*. Orlando: Academic. 379 pp.
8. Bardwell, J. C. A., Craig, E. A. 1984. Major heat shock gene of *Drosophila* and *Escherichia coli* heat-inducible *dnaK* gene are homologous. *Proc. Natl. Acad. Sci. USA* 81:848-52
9. Barraclough, R., Ellis, R. J. 1980. Assembly of newly synthesized large subunits into ribulose bisphosphate carboxylase in isolated pea chloroplasts. *Biochim. Biophys. Acta* 606:19-31
10. Baumann, G., Raschke, E., Bevan, M., Schöffl, F. 1987. Functional analysis of sequences of a soybean heat shock gene in transgenic tobacco plants. *EMBO J.* 6:1161-66
11. Beckmann, R. P., Mizzen, L. A., Welch, W. J. 1990. Interaction of Hsp70 with newly synthesized proteins: Implications for protein folding and assembly. *Science* 248:850-54
12. Benfey, P. N., Chua, N. H. 1989. Regulated genes in transgenic plants. *Science* 244:174-81
13. Bernstein, H. D., Poritz, M. A., Strub, K., Hoben, P. J., Brenner, S., Walter, P. 1989. Model for signal sequence recognition from amino-acid sequences of 54K subunit of signal recognition particle. *Nature* 340:482-86
14. Berry, J. A., Bjorkman, O. 1980. Photosynthetic response and adaptation to temperatures in higher plants. *Annu. Rev. Plant Physiol.* 31:491-543
15. Blum, A. 1988. *Plant Breeding for Stress Environments*. Boca Raton, FL: CRC Press
16. Bond, U., Schlesinger, M. J. 1987. Heat-shock proteins and development. *Adv. Genet.* 24:1-28
17. Borkovich, K. A., Farrelly, F. W., Finkelstein, D. B., Taulien, J., Lindquist, S. L. 1989. hsp82 is an essential protein that is required in higher concentrations for growth of cells at high temperatures. *Mol. Cell. Biol.* 9:3919-30
18. Bouchard, R. A. 1990. Characterization of expressed meiotic prophase repeat transcript clones of *Lilium*: meiosis-specific expression, relatedness, and affinities to small heat shock protein genes. *Genome* 33:68-79
19. Brodl, M. R. 1990. Biochemistry of heat shock responses in plants. In *Environmental Injury to Plants*, ed. F. Katterman, pp. 113-35. NY: Academic
20. Brodl, M. R., Belanger, F. C., Ho, T.-H. D. 1990. Heat shock proteins are not required for the degradation of α-amylase mRNA and the delamellation of endoplasmic reticulum in heat-stressed barley aleurone cells. *Plant Physiol.* 92:1133-41
21. Burke, J. J., Hatfield, J. L., Klein, R. R., Mullet, J. E. 1985. Accumulation of

heat shock proteins in field-grown cotton. *Plant Physiol.* 78:394-98

22. Burke, T. J., Callis, J., Vierstra, R. D. 1988. Characterization of a polyubiquitin gene from *Arabidopsis thaliana*. *Mol. Gen. Genet.* 213:435-43

23. Chappell, T. G., Konforti, B. B., Schmid, S. L., Rothman, J. E. 1987. The ATPase core of a clathrin uncoating protein. *J. Biol. Chem.* 262:746-51

24. Chen, Q., Lauzon, L., DeRocher, A., Vierling, E. 1990. Accumulation, stability, and localization of a major chloroplast heat shock protein. *J. Cell Biol.* 110:1873-83

25. Chen, Q., Vierling, E. 1991. Analysis of conserved domains identifies a unique structural feature of a chloroplast heat shock protein. *Mol. Gen. Genet.* In press

26. Cheng, M. Y., Hartl, F.-U., Martin, J., Pollock, R. A., Kalousek, F., et al. 1989. Mitochondrial heat-shock protein hsp60 is essential for assembly of proteins imported into yeast mitochondria. *Nature* 337:620-25

27. Chirico, W. J., Waters, M. G., Blobel, G. 1988. 70K heat shock related proteins stimulate protein translocation into microsomes. *Nature* 332:805-10

28. Chou, M., Chen, Y.-M., Lin, C.-Y. 1989. Thermotolerance of isolated mitochondria associated with heat shock proteins. *Plant Physiol.* 89:617-21

29. Chrispeels, M. J., Greenwood, J. S. 1987. Heat stress enhances phytohemagglutinin synthesis but inhibits its transport out of the endoplasmic reticulum. *Plant Physiol.* 83:778-84

30. Christensen, A. H., Quail, P. H. 1989. Sequence analysis and transcriptional regulation by heat shock of polyubiquitin transcripts from maize. *Plant Mol. Biol.* 12:619-32

30a. Clos, J., Westwood, J. T., Becker, P. B., Wilson, W., Lambert, K., Wu, C. 1990. Molecular cloning and expression of a hexameric *Drosophila* heat shock factor subject to negative regulation. *Cell* 63:1085-97

31. Collier, N. C., Heuser, M. A., Levy, M. A., Schlesinger, M. J. 1988. Ultrastructural and biochemical analysis of the stress granula in chicken embryo fibroblasts. *J. Cell Biol.* 106:1131-39

32. Conner, T. W., LaFayette, P. R., Nagao, R. T., Key, J. L. 1990. Sequence and expression of a HSP83 from *Arabidopsis thaliana*. *Plant Physiol.* 94:1689-95

33. Cooper, P., Ho, T.-H. D. 1987. Intracellular localization of heat shock proteins in maize. *Plant Physiol.* 84:1197-203

34. Craig, E. A. 1990. Regulation and function of the HSP70 multigene family of *Saccharomyces cerevisiae*. See Ref. 115, pp. 301-21

35. Craig, E. A., Kramer, J., Shilling, J., Werner-Washburne, M., Holmes, S., et al. 1989. SSC1, an essential member of the yeast HSP70 multigene family encodes a mitochondrial protein. *Mol. Cell. Biol.* 9:3000-8

36. Czarnecka, E., Edelman, L., Schöffl, F., Key, J. L. 1984. Comparative analysis of physical stress responses in soybean seedlings using cloned heat shock cDNAs. *Plant Mol. Biol.* 3:45-58

37. Czarnecka, E., Fox, P. C., Gurley, W. B. 1990. In vitro interaction of nuclear proteins with the promoter of soybean heat shock gene *GmHSP17.5-E*. *Plant Physiol.* 94:935-43

38. Czarnecka, E., Gurley, W. B., Nagao, R. T., Mosquera, L. A., Key, J. L. 1985. DNA sequence and transcript mapping of a soybean gene encoding a small heat shock protein. *Proc. Natl. Acad. Sci. USA* 82:3726-30

39. Czarnecka, E., Key, J. L., Gurley, W. B. 1989. Regulatory domains of the *Gmhsp17.5-E* heat shock promoter of soybean: a mutational analysis. *Mol. Cell. Biol.* 9:3457-63

40. Czarnecka, E., Nagao, R. T., Key, J. L., Gurley, W. B. 1988. Characterization of *Gmhsp26-A*, a stress gene encoding a divergent heat shock protein of soybean: heavy-metal-induced inhibition of intron processing. *Mol. Cell. Biol.* 8:1113-22

41. Dang, C. V., Lee, W. M. F. 1989. Nuclear and nucleolar targeting sequences of c-erb-A, C-myb, N-myc, p53, HSP70, and HIV tat proteins. *J. Biol. Chem.* 264:18019-23

42. DeLuca-Flaherty, C., McKay, D. B., Parham, P., Hill, B. L. 1990. Uncoating protein (hsc70) binds a conformationally labile domain of clathrin light chain LCa to stimulate ATP hydrolysis. *Cell* 62:875-87

43. DeRocher, A., Lauzon, L., Vierling, E. 1990. *HSP70* expression during seed development. *J. Cell. Biochem.* 14E:298 (Abstr.)

43a. de Sa, C. M., Rollet, E., de Sa, M.-F. G., Tanguay, R. M., Best-Belpomme, M., et al. 1989. Prosomes and heat shock complexes in *Drosophila melanogaster* cells. *Mol. Cell. Biol.* 9:2672-81

44. Deshaies, R. J., Koch, B. D., Werner-Washburne, M., Craig, E. A., Schek-

man, R. 1988. A subfamily of stress proteins facilitates translocation of secretory and mitochondrial precursor polypeptides. *Nature* 332:800-5

45. Duck, N., McCormick, S., Winter, J. 1989. Heat shock protein Hsp70 cognate expression in vegetative and reproductive organs of *Lycopersicon esculentum*. *Proc. Natl. Acad. Sci. USA* 86:3674-78

46. Edelman, L., Czarnecka, E., Key, J. L. 1988. Induction and accumulation of heat shock-specific poly(A+)RNAs and proteins in soybean seedlings during arsenite and cadmium treatments. *Plant Physiol.* 86:1048-56

47. Ellis, J. 1987. Proteins as molecular chaperones. *Nature* 328:378-79

48. Engman, D. M., Kirchhoff, L. V., Donelson, J. E. 1989. Molecular cloning of *mtp70*, a mitochondrial member of the *hsp70* family. *Mol. Cell. Biol.* 9:5163-68

— 49. Fender, S. E., O'Connell, M. A. 1989. Heat shock protein expression in thermotolerant and thermosensitive lines of cotton. *Plant Cell Rep.* 8:37-40

50. Fender, S. E., O'Connell, M. A. 1990. Expression of the heat shock response in a tomato interspecific hybrid is not intermediate between the two parental responses. *Plant Physiol.* 93:1140-46

51. Ferguson, D. L., Guikema, J. A., Paulsen, G. M. 1990. Ubiquitin pool modulation and protein degradation in wheat roots during high temperature stress. *Plant Physiol.* 92:740- 46

52. Flaherty, K. M., DeLuca-Flaherty, C., McKay, D. B. 1990. Three-dimensional structure of the ATPase fragment of a 70K heat-shock cognate protein. *Nature* 346:623-28

53. Flynn, G. C., Chappell, T. G., Rothman, J. E. 1989. Peptide binding and release by proteins implicated as catalysts of protein assembly. *Science* 245:385-90

54. Gaitanaris, G. A., Papavassiliou, A. G., Rubock, P., Silverstein, S. J., Gottesman, M. E. 1990. Renaturation of lambda repressor requires heat shock proteins. *Cell* 61:1013-20

55. Georgopoulos, C., Ang, D., Liberek, K., Zylicz, M. 1990. Properties of the *Escherichia coli* heat shock proteins and their role in bacteriophage lambda growth. See Ref. 115, pp. 191-221

56. Gething, M. J., Sambrook, J. 1990. Transport and assembly processes in the endoplasmic reticulum. *Sem. Cell Biol.* 1:65-72

57. Glaczinski, H., Kloppstech, K. 1988. Temperature-dependent binding to the thylakoid membranes of nuclear-coded chloroplast heat-shock proteins. *Eur. J. Biochem.* 173:579-83

58. Goloubinoff, P., Christeller, J. T., Gatenby, A. A., Lorimer, G. H. 1989. Reconstitution of active dimeric ribulose bisphosphate carboxylase from an unfolded state depends on two chaperonin proteins and Mg-ATP. *Nature* 342:884-88

59. Goloubinoff, P., Gatenby, A. A., Lorimer, G. H. 1989. GroE heat-shock proteins promote assembly of foreign prokaryotic ribulose bisphosphate carboxylase oligomers in *E. coli*. *Nature* 337:44-47

60. Goping, I. S., Frappier, J. R. H., Walden, D. B., Atkinson, B. G. 1991. Sequence, identification, and characterization of cDNAs encoding two different members of the 18 kDa heat shock protein family of *Zea mays* L. *Plant Mol. Biol.* In press

61. Grimm, B., Ish-Shalom, D., Even, D., Glaczinski, H., Ottersbach, P., et al. 1989. The nuclear-encoded 22-kDa heat-shock protein of *Chlamydomonas*. *Eur. J. Biochem.* 182:539-46

62. Gurley, W. B., Czarnecka, E., Nagao, R. T., Key, J. L. 1986. Upstream sequences required for efficient expression of a soybean heat shock gene. *Mol. Cell. Biol.* 6:559-65

63. Gurley, W. B., Key, J. L. 1991. Transcriptional regulation of the heat shock response: a plant perspective. *Biochemistry*. In press

64. Guy, C. L. 1990. Cold acclimation and freezing stress tolerance: role of protein metabolism. *Annu. Rev. Plant Physiol. Plant Mol. Biol.* 41:187-223

65. Hajela, R. K., Horvath, D. P., Gilmour, S. J., Thomashow, M. F. 1990. Molecular cloning and expression of cor (cold-regulated) genes in *Arabidopsis thaliana*. *Plant Physiol.* 93:1246-52

66. Hallberg, R. L. 1990. A mitochondrial chaperonin: genetic, biochemical, and molecular characteristics. *Sem. Cell. Biol.* 1:37-45

67. Hartl, F.-U., Neupert, W. 1990. Protein sorting to mitochondria: evolutionary conservations of folding and assembly. *Science* 247:930-38

68. Harrington, H. M., Alm, D. M. 1988. Interaction of heat shock and salt shock in cultured tobacco cells. *Plant Physiol.* 88:618-25

69. Hawthorne, L. R., Kerr, L. B., Campbell, W. F. 1966. Relation between temperature of developing pods and seeds and scalded seeds in garden pea. *Am. Soc. Hortic. Sci.* 88:437-40

70. Helm, K. W., Abernathy, R. H. 1990.

Heat shock protein and their mRNAs in dry and early imbibing embryos of wheat. *Plant Physiol.* 93:1626-33

71. Helm, K. W., DeRocher, A., Lauzon, L., Vierling, E. 1990. Expression of low molecular weight heat shock proteins during seed development. *J. Cell. Biochem.* 14E:301 (Abstr.)

72. Helm, K. W., Petersen, N. S., Abernathy, R. H. 1989. Heat shock response of germinating embryos of wheat. *Plant Physiol.* 90:598-605

73. Helm, K. W., Vierling, E. 1989. An *Arabidopsis* cDNA clone encoding a low molecular weight heat shock protein. *Nucleic Acids Res.* 17:7995

74. Helm, K. W., Vierling, E. 1990. A member of the eukaryotic superfamily of small heat shock proteins is located in the endomembrane system of *Pisum sativum. J. Cell Biol.* 111:69a

75. Hemmingsen, S. M., Ellis, R. J. 1986. Purification and properties of ribulosebisphosphate carboxylase large subunit binding protein. *Plant Physiol.* 80:269-76

76. Hemmingsen, S. M., Woolford, C., van der Vies, S. M., Tilly, K., Dennis, D. T. 1988. Homologous plant and bacterial proteins chaperone oligomeric protein assembly. *Nature* 333:330-34

77. Hershko, A. 1988. Ubiquitin-mediated protein degradation. *J. Biol. Chem.* 263:15237-40

78. Howarth, C. 1989. Heat shock proteins in *Sorghum bicolor* and *Pennisetum americanum.* I. Genotypic and developmental variation during seed germination. *Plant Cell Environ.* 12:471-77

79. Hunt, C., Morimoto, R. I. 1985. Conserved features of eukaryotic hsp70 genes revealed by comparison with the nucleotide sequence of human hsp70. *Proc. Natl. Acad. Sci. USA* 82:6455-59

80. Hwang, C. H., Zimmerman, J. L. 1989. The heat shock response of carrot: protein variations between cultured cell lines. *Plant Physiol.* 91:552-58

81. Ingolia, T. D., Craig, E. A. 1982. Four small *Drosophila* heat shock proteins are related to each other and to mammalian α-crystallins. *Proc. Natl. Acad. Sci. USA* 79:2360-64

82. Ish-Shalom, D., Kloppstech, K., Ohad, I. 1990. Light regulation of the 22 kd heat shock gene transcription and its translation product accumulation in *Chlamydomonas reinhardtii. EMBO J.* 9:2657-61

83. Johnston, R. N., Kucey, B. L. 1988. Competitive inhibition of hsp70 gene

expression causes thermosensitivity. *Science* 242:1551-54

84. Kabsch, W., Mannherz, H. G., Suck, D., Pai, E. F., Holmes, K. C. 1990. Atomic structure of the actin:DNase I complex. *Nature* 347:37-44

85. Kang, P.-J., Ostermann, J., Schilling, J., Neupert, W., Craig, E. A., et al. 1990. Hsp70 in the mitochondrial matrix is required for translocation and folding of precursor proteins. *Nature* 348:137-43

86. Keegstra, K., Olsen, L. J., Theg, S. M. 1989. Chloroplastic precursors and their transport across the envelope membranes. *Annu. Rev. Plant Physiol. Plant Mol. Biol.* 40:471-501

87. Keigl, P. J., Mullen, R. E. 1986. Changes in soybean seed quality from high temperature during seed fill and maturation. *Crop Sci.* 26:1212-16

88. Key, J. L., Kimpel, J., Vierling, E., Lin, C.-Y., Nagao, R. T., et al. 1985. Physiological and molecular analyses of the heat shock response in plants. See Ref. 7, pp.327-48

89. Kimpel, J. A., Key, J. L. 1985. Heat shock in plants. *Trends Biochem. Sci.* 10:353-57

90. Kimpel, J. A., Key, J. L. 1985. Presence of heat shock mRNAs in field grown soybeans. *Plant Physiol.* 79:672-78

91. Kimpel, J. A., Nagao, R. T., Goekjian, V., Key, J. L. 1990. Regulation of the heat shock response in soybean seedlings. *Plant Physiol.* 94:988-95

92. Kloppstech, K., Meyer, G., Schuster, G., Ohad, I. 1985. Synthesis, transport and localization of a nuclear coded 22-kd heat-shock protein in the chloroplast membranes of peas and *Chlamydomonas reinhardi. EMBO J.* 4:1902-9

93. Krishnan, M., Nguyen, H. T., Burke, J. J. 1989. Heat shock protein synthesis and thermal tolerance in wheat. *Plant Physiol.* 90:140-45

94. Krishnasamy, S., Mannan, R. M., Krishnan, M., Gnanam, A. 1988. Heat shock response of the chloroplast genome in *Vigna sinensis. J. Biol. Chem.* 263:5104-9

95. Kruse, E., Kloppstech, K. 1991. Heat shock proteins in plants — an approach to understand the function of plastid HSP. *Topics Photosynth.* 11. In press

96. Landry, J., Chretien, P., Lambert, H., Hickey, E., Weber, L. A. 1989. Heat shock resistance conferred by expression of the human HSP27 gene in rodent cells. *J. Cell Biol.* 109:7-15

97. Lauzon, L., Helm, K. W., Vierling, E.

Evidence that HSP production is an essential component of thermotolerance [handwritten annotation]

1990. A cDNA clone from *Pisum sativum* encoding a low molecular weight heat shock protein. *Nucleic Acids Res.* 18:4274

98. Leustek, T., Dalie, B., Smir-Shapira, D., Brot, N., Weissbach, H. 1989. A member of the Hsp70 family is localized in mitochondria and resembles *Escherichia coli* DnaK. *Proc. Natl. Acad. Sci. USA* 86:7805-8

99. Lin, C.-Y., Roberts, J. K., Key, J. L. 1984. Acquisition of thermotolerance in soybean seedlings. *Plant Physiol.* 74:152-60

100. Lindquist, S. 1986. The heat shock response. *Annu. Rev Biochem.* 45:39-72

101. Lindquist, S., Craig, E. A. 1988. The heat shock proteins. *Annu. Rev Genet.* 22:631-77

102. Lubben, T. H., Donaldson, G. K., Viitanen, P. V., Gatenby, A. A. 1989. Several proteins imported into chloroplasts form stable complexes with the GroEL-related chloroplast molecular chaperone. *Plant Cell* 1:1223-30

103. Lubben, T. H., Gatenby, A. A., Donaldson, G. K., Lorimer, G. H., Viitanen, P. V. 1990. Identification of a groES-like chaperonin in mitochondria that facilitates protein folding. *Proc. Natl. Acad. Sci. USA* 87:7683-87

104. Mansfield, M. A., Key, J. L. 1987. Synthesis of the low molecular weight heat shock proteins in plants. *Plant Physiol.* 84:1007-17

105. Mansfield, M. A., Key, J. L. 1988. Cytoplasmic distribution of heat shock proteins in soybean. *Plant Physiol.* 86:1240-46

106. Marmiroli, N., Lorenzoni, C., Stanca, A. M., Terzi, V. 1989. Preliminary studies of the inheritance of temperature stress proteins in barley (*Hordeum vulgare* L.). *Plant Sci.* 62:147-56

107. Marshall, J. S., DeRocher, A. E., Keegstra, K., Vierling, E. 1990. Identification of heat shock protein hsp70 homologues in chloroplasts. *Proc. Natl. Acad. Sci. USA* 87:374-78

108. Martel, R., Cloney, L. P., Pelcher, L. E., Hemmingsen, S. M. 1990. Unique composition of plastid chaperone-60: α and β polypeptide-encoding genes are highly divergent. *Gene.* 94:181-87

109. Deleted in proof

110. McElwain, E., Spiker, S. 1989. A wheat cDNA clone which is homologous to the 17 kd heat shock protein gene family of soybean. *Nucleic Acid Res.* 17:1764

111. McMullin, T., Hallberg, R. L. 1988. A highly conserved mitochondrial protein is structurally related to the protein encoded by the *E. coli groEL* gene. *Mol. Cell. Biol.* 8:371-80

112. Milarski, K. L., Morimoto, R. I. 1989. Mutational analysis of the human HSP70 protein: distinct domains for nucleolar localization and adenosine triphosphate binding. *J. Cell Biol.* 109:1947-62

113. Mizzen, L. A., Chang, C., Garrels, J. I., Welch, W. J. 1989. Identification, characterization, and purification of two mammalian stress proteins present in mitochondria, grp75, a member of the hsp 70 family and hsp58, a homolog of the bacterial groEL protein. *J. Biol. Chem.* 264:20664-75

114. Moisyadi, S., Harrington, H. M. 1990. Functional characterization of a low molecular weight heat shock protein in cultured sugarcane cells. *Plant Physiol.* 93:88 (Abstr.)

115. Morimoto, R. I., Tissieres, A., Georgopoulos, C., eds. 1990. *Stress Proteins in Biology and Medicine.* New York: Cold Spring Harbor Laboratory Press. 450 pp.

116. Munro, S., Pelham, H. R. B. 1986. An HSP70-like protein in the ER: identity with the 78kd glucose-regulated protein and immunoglobulin heavy chain binding protein. *Cell* 46:291-300

117. Nagao, R. T., Czarnecka, E., Gurley, W. B., Key, J. L. 1985. Genes for low-molecular-weight heat shock proteins of soybeans: sequence analysis of a multigene family. *Mol. Cell. Biol.* 5:3417-28

118. Nagao, R. T., Kimpel, J. A., Vierling, E., Key, J. L. 1986. The heat shock response: a comparative analysis. In *Oxford Surveys of Plant Molecular & Cell Biology*, ed. B. J. Miflin, 3:384-438. Oxford: Oxford Univ. Press

119. Nerland, A. H., Mustafa, A. S., Sweetser, D., Godal, T., Young, R. A. 1988. A protein antigen of *Mycobacterium leprae* is related to a family of small heat shock proteins. *J. Bact.* 170:5919-21

120. Neumann, D., zur Nieden, U., Manteuffel, R., Walter, G., Scharf, K.-D., Nover, L. 1987. Intracellular localization of heat-shock proteins in tomato cell cultures. *Eur. J. Cell. Biol.* 43:71-81

121. Neumann, D., Nover, L., Parthier, B., Rieger, R., Scharf, K.-D., et al. 1989. Heat shock and other stress response systems of plants. *Biol. Zentralbl.* 108:1-156

122. Neven, L. G., Haskell, D. W., Guy, C. L. 1990. A heat shock cognate comes out in the cold. *Cryobiology* 27:661

123. Nieto-Sotelo, J., Ho, T.-H. D. 1987. Absence of heat shock protein synthesis in isolated mitochondria and plastids

from maize. *J. Biol. Chem.* 262:12288-92

124. Nieto-Sotelo, J., Vierling, E., Ho, T.-H. D. 1990. Cloning, sequence analysis and expression of a cDNA encoding a plastid localized heat shock protein in maize. *Plant Physiol.* 93:1321-28

125. Normington, K., Kohno, K., Kozutsumi, Y., Gething, M.-J., Sambrook, J. 1989. *S. cerevisiae* encodes an essential protein homologous in sequence and function to mammalian BiP. *Cell* 57:1223-36

126. Nover, L. 1990. *Heat Shock Response.* Boca Raton: CRC Press. In press

127. Nover, L., Hellmund, D., Neumann, D., Scharf, K.-D., Serfling, E. 1984. The heat shock response of eukaryotic cells. *Biol. Zentralbl.* 103:357-435

128. Nover, L., Munsche, D., Neumann, D., Ohme, K., Scharf, K.-D. 1986. Control of ribosome biosynthesis in plant cell cultures under heat-shock conditions. *Eur. J. Biochem.* 160:297-304

129. Nover, L., Scharf, K.-D., Neumann, D. 1989. Cytoplasmic heat shock granules are formed from precursor particles and are associated with a specific set of mRNAs. *Mol. Cell. Biol.* 9:1298-308

130. Ostermann, J., Horwich, A. L., Neupert, W., Hartl, F.-U. 1989. Protein folding in mitochondria requires complex formation with hsp60 and ATP hydrolysis. *Nature* 341:125-30

131. Otto, B., Grimm, B., Ottersbach, P., Kloppstech, K. 1988. Circadian control of the accumulation of mRNAs for light- and heat-inducible chloroplast proteins in pea (*Pisum sativum* L.). *Plant Physiol.* 88:21-25

132. Ougham, H. J. 1987. Gene expression during leaf development in *Lolium temulentum*: patterns of protein synthesis in response to heat-shock and cold-shock. *Physiol. Plant.* 70:479-84

133. Ougham, H., Stoddart, J. L. 1986. Synthesis of heat-shock protein and acquisition of thermotolerance in high-temperature tolerant and high temperature susceptible lines of *Sorghum. Plant Sci.* 44:163-67

134. Pelham, H. R. B. 1986. Speculations on the functions of the major heat shock and glucose regulated proteins. *Cell* 46:959-61

135. Pelham, H. R. B. 1989. Control of protein exit from the endoplasmic reticulum. *Annu. Rev. Cell Biol.* 5:1-23

136. Pelham, H. R. B. 1990. Functions of the HSP70 protein family: an overview. See Ref. 115, pp. 287-99

137. Petko, L., Lindquist, S. 1986. Hsp26 is not required for growth at high temperatures, not for thermotolerance, spore development, or germination. *Cell* 45:885-94

138. Piatigorsky, J., Wistow, G. J. 1989. Enzyme/crystallins: gene sharing as an evolutionary strategy. *Cell* 51:197-99

138a. Picard, D., Khursheed, B., Garabedian, M. J., Fortin, M. G., Lindquist, S., Yamamoto, K. R. 1990. Reduced levels of hsp90 compromise steroid receptor action in vivo. *Nature* 348:166-68

139. Pitto, L., LoSchiavo, F., Giuliano, G., Terzi, M. 1983. Analysis of heat-shock protein pattern during somatic embryogenesis of carrot. *Plant Mol. Biol.* 2:231-37

140. Prasad, T. K., Hack, E., Hallberg, R. L. 1990. Function of the maize mitochondrial chaperonin hsp60: specific association between hsp60 and newly synthesized F1-ATPase alpha subunits. *Mol. Cell. Biol.* 10:3679-986

141. Prasad, T. K., Hallberg, R. L. 1989. Identification and metabolic characterization of the *Zea mays* mitochondrial homolog of the *Escherichia coli* groEL protein. *Plant Mol. Biol.* 12:609-18

142. Raschke, E., Baumann, G., Schoffl, F. 1988. Nucleotide sequence analysis of soybean small heat shock genes belonging to two different multigene families. *J. Mol. Biol.* 199:549-57

143. Reading, D. S., Hallberg, R. L., Myers, A. M. 1989. Characterization of the yeast *HSP60* gene coding for a mitochondrial assembly factor. *Nature* 337:655-59

144. Riabowol, K. T., Mizzen, L. A., Welch, W. J. 1988. Heat shock is lethal to fibro-blasts microinjected with antibodies against hsp70. *Science* 242:433-36

145. Roberts, J. K., Key, J. L. 1990. Isolation and characterization of a soybean HSP70 gene. *Plant Mol. Biol.* In press

146. Rochester, D. E., Winter, J. A., Shah, D. M. 1986. The structure and expression of maize genes encoding the major heat shock protein, hsp70. *EMBO J.* 5:451-58

146a. Romisch, K., Webb, J., Lingelbach, K., Gausepohl, H., Dobberstein, B. 1990. The 54-kD protein of signal recognition particle contains a methionine-rich RNA binding domain. *J. Cell Biol.* 111:1793-1802

147. Rose, M. D., Misra, L. M., Vogel, J. P. 1989. *KAR2*, a karyogamy gene, is the yeast homolog of the mammalian *BiP/GRP78* gene. *Cell* 57:1211-21

148. Rothman, J. E. 1989. Polypeptide chain binding proteins: catalysts of protein folding and related processes in cells. *Cell* 59:591-601

149. Roy, H. 1989. Rubisco assembly: a model system for studying the mechanism of chaperonin action. *Plant Cell* 1:1035-42

150. Roy, H., Bloom, M., Milos, P., Monroe, M. 1982. Studies on the assembly of large subunits of ribulose bisphosphate carboxylase in isolated pea chloroplasts. *J. Cell Biol.* 94:20-27

151. Russell, D. A., Sachs, M. A. 1989. Differential expression and sequence analysis of the maize glyceraldehyde-3-phosphate dehydrogenase gene family. *Plant Cell.* 1:793-803

152. Sanchez, Y., Lindquist, S. L. 1990. HSP104 required for induced thermotolerance. *Science* 248:1112-15

153. Schlesinger, M. J. 1990. Heat shock proteins. *J. Biol. Chem.* 265:12111-14

154. Schöffl, F., Key, J. L. 1982. An analysis of mRNAs for a group of heat shock proteins of soybean using cloned cDNAs. *J. Mol. Appl. Genet.* 1:301-14

155. Schöffl, F., Key, J. L 1983. Identification of a multigene family for small heat shock proteins in soybean and physical characterization of one individual coding region. *Plant Mol. Biol.* 2:269-78

156. Schöffl, F., Rieping, M., Baumann, G., Bevan, M., Angermuller, S. 1989. The function of plant heat shock promoter elements in the regulated expression of chimaeric genes in transgenic tobacco. *Mol. Gen. Genet.* 217:246-53

157. Schrauwen, J. A. M., Reijnen, W. H., DeLeeuw, H. C. G. M., VanHerpen, M. M. A. 1986. Response of pollen to heat stress. *Acta Bot. Neerl.* 35:321-27

158. Schuster, G., Even, D., Kloppstech, K., Ohad, I. 1988. Evidence for protection by heat-shock proteins against photoinhibition during heat-shock. *EMBO J.* 7:1-6

159. Sinibaldi, R. M., Turpen, T. 1985. A heat shock protein is encoded within mitochondria of higher plants. *J. Biol. Chem.* 260:15382-85

160. Skowyra, D., Georgopoulos, C., Zylicz, M. 1990. The *E. coli* dnaK gene product, the hsp70 homolog, can reactivate heat-inactivated RNA polymerase in an ATP hydrolysis-dependent manner. *Cell* 62:939-44

161. Slater, M. R., Craig, E. A. 1989. The *SSA1* and *SSA2* genes of the yeast *Saccharomyces cerevisiae*. *Nucleic Acids Res.* 17:805-6

162. Sorger, P. K., Pelham, H. R. B. 1988. Yeast heat shock factor is an essential DNA binding protein that exhibits temperature-dependent phosphorylation. *Cell* 54:855-64

163. Sticher, L., Biswas, A. K., Bush, D. S., Jones, R. L. 1990. Heat shock inhibits α-amylase synthesis in barley aleurone without inhibiting the activity of endoplasmic reticulum marker enzymes. *Plant Physiol.* 92:506-13

164. Susek, R. E., Lindquist, S. L. 1989. hsp26 of *Saccharomyces cerevisiae* is related to the superfamily of small heat shock proteins but is without a demonstrable function. *Mol. Cell. Biol.* 9:5265-71

165. Süss, K.-H., Yordanov, I. T. 1986. Biosynthetic cause of in vivo acquired thermotolerance of photosynthetic light reactions and metabolic responses of chloroplasts to heat stress. *Plant Physiol.* 81:192-99

166. Takahashi, T., Komeda, Y. 1989. Characterization of two genes encoding small heat-shock proteins in *Arabidopsis thaliana*. *Mol. Gen. Genet.* 219:365-72

167. Ting, J., Lee, A. S. 1988. Human gene encoding the 78,000 dalton glucose regulated protein and its pseudogene's structure, conservation and regulation. *DNA* 7:275-86

168. van Dyk, T. K., Gatenby, A. A., LaRossa, R. A. 1989. Demonstration by genetic suppression of interaction of GroE products with many proteins. *Nature* 342:451-53

169. van Herpen, M. M. A., Reijnen, W. H., Schrauwen, J. A. M., de Groot, P. F. M., Jager, J. W. H., Wullems, G. J. 1989. Heat shock proteins and survival of germinating pollen of *Lilium longiflorum* and *Nicotiana tabacum*. *J. Plant Physiol.*

170. Vierling, E., Harris, L. M., Chen, Q. 1989. The major heat shock protein in chloroplasts show antigenic conservation among diverse higher plant species. *Mol. Cell. Biol.* 9:461-68

171. Vierling, E., Mishkind, M. L., Schmidt, G. W., Key, J. L. 1986. Specific heat shock proteins are transported into chloroplasts. *Proc. Natl. Acad. Sci. USA* 83:361-65

172. Vierling, E., Nagao, R. T., DeRocher, A. E., Harris, L. M. 1988. A heat shock protein localized to chloroplasts is a member of a eukaryotic superfamily of heat shock proteins. *EMBO J.* 7:575-81

173. Vierling, E., Sun, A. 1987. Developmental expression of heat shock proteins in higher plants. In *Environ-*

mental Stress in Plants, ed. J. Cherry, pp. 343-54. Berlin: Springer-Verlag

174. von Gromoff, E. D., Treier, U., Beck, C. F. 1989. Three light-induced heat shock genes of *Chlamydomonas reinhardii*. *Mol. Cell. Biol.* 9:3911-18

175. von Heijne, G. 1983. Patterns of amino acids near signal-sequence cleavage sites. *Eur. J. Biochem.* 133:17-21

176. von Heijne, G. 1985. Signal sequences: the limits of variation. *J. Mol. Biol.* 184:99-105

177. Welch, W. J., Feramisco, J. R. 1985. Rapid purification of mammalian 70,000-dalton stress proteins: affinity of the proteins for nucleotides. *Mol. Cell. Biol.* 5:1229-37

178. Wiederrecht, G., Seto, D., Parker, C. S. 1988. s Isolation of the gene encoding the *S. cerevisiae* heat shock transcription factor. *Cell* 54:841-53

179. Winter, J., Wright, R., Duck, N., Gasser, C., Fraley, R., Shah, D. 1988. The inhibition of petunia hsp70 mRNA processing during CdCl2 stress. *Mol. Gen. Genet.* 211:315-19

180. Wu, C. H., Caspar, T., Browse, J., Lindquist, S., Somerville, C. 1988. Characterization of an *HSP70* cognate gene family in *Arabidopsis*. *Plant Physiol.* 88:731-40

181. Xiao, C.-M., Mascarenhas, J. P. 1985. High temperature-induced thermotolerance in pollen tubes of *Tradescantia* and heat shock proteins. *Plant Physiol.* 78:887-90

182. Yacob, R. K., Filion, W. G. 1986. Temperature stress in maize: a comparison of several cultivars. *Can. J. Genet. Cytol.* 28:1125-31

183. Yost, H. J., Petersen, R. B., Lindquist, S. 1990. RNA metabolism: strategies for regulation in the heat shock response. *Trends Genet.* 6:223-27

184. Zimmerman, J. L., Apuya, N., Darwish, K., O'Carroll, C. 1989. Novel regulation of heat shock genes during carrotsomatic embryo development. *Plant Cell* 1:1137-46

185. Scharf, K.-D., Rose, S., Zott, W., Schöffl, F., Nover, L. 1990. Three tomato genes code for heat stress transcription factors with a region of remarkable homology to the DNA-binding domain of the yeast HSF. *EMBO J.* 9:4495-4501

Annu. Rev. Plant Physiol. Plant Mol. Biol. 1991. 42:621–49

MOLECULAR STUDIES ON THE DIFFERENTIATION OF FLORAL ORGANS

Charles S. Gasser

Department of Biochemistry and Biophysics, University of California, Davis, California 95616

KEY WORDS: anther, pistil, flower development, pollen, tissue-specific genes

CONTENTS

1040-2519/91/0601-0621$02.00

INTRODUCTION

The flowers of angiosperms have fascinated both humanists and naturalists for as long as records have been kept. The fact that flowers are the source of fruits and seeds that make up the bulk of the world's food supply has given near-religious significance to flowering in many cultures. Early taxonomists recognized that structural similarities between flowers of plants with very different growth habits provided logical phylogenetic groupings. Recently molecular techniques have been applied to the study of flower formation. One aspect of this research is the examination of the final stages of the development of floral organs. The initial goals of such investigation are the identification of genes predominantly expressed in the different organs and tissues of the flower. Clones of these genes may then serve as tools with which to study the regulation of the differentiation process and illuminate the functions of the differentiated tissues. In this review I describe the progress that has been made in the identification and characterization of such genes and discuss the ways these studies can complement other approaches to the study of flower development.

The typical flower consists of two sterile and two fertile sets of organs. The outer organs, the sepals, are commonly leaf-like and photosynthetic. The next set of organs, the petals, are also leaf-like and often contain differentiated pigmented plastids, the chromoplasts, responsible for their striking colors. The next set of organs, the androecium, is made up of the stamens, which are responsible for the formation and dissemination of the microgametophytes (pollen grains). The innermost organs, the carpels, are collectively referred to as the gynoecium and are the location of production of the ovules, which contain the megagametophytes. When one or more carpels are fused into a single structure they are often referred to as a pistil. Significant alterations in this general scheme of flower formation have been produced by the pressures of natural selection. Whole sets of organs are absent in many flowers, and in others, organ fusion or modification can make it difficult to classify a floral organ precisely.

Flowering is an attractive system for the study of development in plants. During the flowering process a series of unique events occur that typify all of the aspects of plant development: organogenesis, differential cell division, cellular differentiation, and alterations in gene expression. While flowering is a single continuous process, for purposes of discussion it is commonly divided into two phases, floral initiation, and floral development. Floral development can be further subdivided into two steps: the initiation and commitment of lateral organ primordia on the floral meristem, and the development of the floral organs into their final physical forms, with the differentiation of specific tissues within each organ (Figure 1).

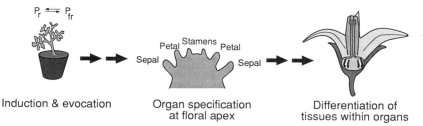

$P_r \rightleftharpoons P_{fr}$

Sepal Petal Stamens Petal Sepal

Induction & evocation

Organ specification at floral apex

Differentiation of tissues within organs

Figure 1 Examples of decision points in flowering. Discrete decisions must be made by tissues or individual cells during the formation of flowers. The first decision is whether or not to flower, represented in the left panel of the figure by the interconversion of the two forms of phytochrome, a common mediator of floral induction. The cells of the apex are then directed to undergo evocation, resulting in the conversion of the vegetative meristem to a floral meristem. The meristem may produce branches, but eventually the branches terminate in apexes for single flowers. As illustrated in the center panel, a series of organ primordia are produced at the floral apex. Early in the formation of the primordia, decisions must be made as to which floral organs the primordia will form (*center*). Genetic changes can result in fundamental alterations in this set of decisions (see text). Finally, as the organs begin to take on their characteristic forms each cell must be correctly directed to differentiate biochemically and anatomically so that all of the differentiated tissues in the flower form in proper spatial relationship to each other.

Floral Initiation

The environmental and physiological stimuli responsible for the initiation of flowering have been studied extensively in a number of plant species. A similarly large body of information exists concerning the other aspect of floral initiation, the reorganization of the meristem, called evocation. A recent review by G. Bernier in this series (5) integrates this information and describes the current models for the regulation of these processes. These studies indicate wide variations in the requirements for induction in different angiosperms, and it is difficult to formulate a single mechanism of induction that encompasses these differences. The reader is referred to the aforementioned review for Dr. Bernier's insightful analysis.

Lateral Organ Initiation and Specification

After conversion of the vegetative meristem into a floral meristem, the floral apex initiates formation of a single flower, or of an inflorescence, by production of a series of branches, each of which will terminate in an apex for a single flower. The floral apexes produce lateral organ primordia in a pattern that is usually totally different from the vegetative phyllotaxic pattern. The conservation of this pattern within a species demonstrates that it is primarily genetically determined, although biophysical phenomena may be involved in the manifestation of the genetic program (22).

Very soon after their initiation, the primordia begin to differentiate into the characteristic floral organs. Which organ each primordium will become is under strict genetic control. Significant progress in understanding the genetic control of these determinations has been made through the analysis of mutants with altered patterns of floral organs. The most extensively studied of these are the homeotic mutants in which relatively normal organs appear in the wrong locations. These mutants are described in more detail in the article by E. Coen in this issue. I discuss this work further only as it directly relates to the primary subject of this review.

Floral Organ Differentiation

Following specification of the fates of the primordia, each primordium undergoes a series of programmed cell divisions and cell expansions that produce the final form of the floral organ. During this process the cells differentiate anatomically and biochemically, taking on the properties that allow the various tissues of the floral organs to carry out their individual functions. Examples of differentiated tissues include the stigma, transmitting tissue, placenta, and ovules within the carpels, and the endothecium, tapetum, and micospores in the stamens. The differentiation of the pollen grains must be considered part of this process because the production of viable pollen is one of the most critical processes in the development of a mature flower. The specialization of these tissues is affected by unique proteins and secondary metabolites produced by protein-mediated reactions. These proteins are products of genes expressed preferentially or exclusively in specific differentiating tissues. In formulating a complete theory on how flowers are formed and function, we must understand both these final developmental steps and the two earlier regulatory events described above.

A logical approach to understanding floral tissue differentiation is to identify the genes expressed in the various tissues, to characterize their regulation, and to determine the nature of their products. Recently a number of laboratories have initiated such studies. Here I describe how such genes have been identified, outline the progress in determining the functions and mechanisms of regulation of these genes, and discuss some of the ramifications of this work.

I begin with a comment on the concept of organ- and tissue-specificity. In many cases genes discussed in this review have only been detected in a single organ or tissue of the plant. The methods used in these studies are various and include northern blots with total RNA, northern blots with poly-A^+ RNA, in situ hybridization, and various immunodetection methods. These methods can vary in sensitivity over several logs depending upon how they are applied. While it is useful shorthand to refer to a gene as "specific" to a given organ or tissue when it is not detected in other locations, the certainty implied by this term suggests that it should be used with caution. The gene might be demonstrated to be nonspecific by the application of more sensitive methods than those used in initial studies. While one could set an arbitrary quantity differential as a criterion of specificity (for example, consider a gene specific to a location if it occurs there in quantities 2000 times greater than in any other location) problems with the terminology would still remain. Specificity implies that the gene is not expressed in any other location at any time of development. Demonstration of this exclusivity would be experimentally challenging. While detection and hybridization can be correctly referred to as "specific," for the purpose of this review I use the weaker term "predominant" when referring to gene expression.

METHODS OF ISOLATING CLONES OF ORGAN-PREDOMINANT GENES

Protein-Related Methods

One way to identify genes that are predominantly expressed in a floral organ is first to identify proteins, the ultimate products of gene expression, that are detectable only in those organs. Two general approaches have been taken: characterization of a protein of known function that is expected to have a flower-predominant expression pattern, and identification of proteins with the desired specificity whose functions are not known. The first approach is usually not used expressly for the purpose of isolating a flower-specific gene. Rather, during the study of a floral process one may be able to identify a protein involved in the process. Further characterization of the process may be facilitated by the isolation of the protein and eventual isolation of the gene. The chalcone synthase genes that mediate an important step in the phenylpropanoid pathway are a good example of this progression. Genes for this enzyme were isolated because of a general interest in phenylpropanoid metabolism in plants, and have subsequently been studied in flowers (see below).

The second protein-based approach involves the identification of proteins of unknown function based simply on their specificity in flowers. Such proteins can be identified by comparison of proteins from floral or vegetative organs on one-dimensional, or two-dimensional gels (see, for example, 19, 29). Identification of proteins by comparative gel electrophoresis allows for an assessment of the differences between the organs but does not necessarily facilitate the isolation of the proteins or the related genes.

Eliezer Lifschitz and his colleagues in the Department of Biology at the Technion-Israel Institute of Technology have used a novel method of identifying proteins that are flower-specific (35). They utilized a series of standard protein purification steps—centrifugation, ion exchange chromatography, and

HPLC hydrophobic separation—to identify and compare approximately 500 major proteins present in leaves and floral tissues. This method appears to allow for recognition of as many proteins as two-dimensional gel analysis, while having the advantage that it also results in partial purification of each of the proteins. Antibodies were raised to several proteins that were predominantly present in flowers or floral meristems. These antibodies have been used to characterize the expression patterns of the proteins in flowers and to screen cDNA expression libraries to isolate clones of the mRNAs encoding the proteins (35).

Evans et al have reported on the production of monoclonal antibodies to flower-specific antigens (17). Hybridomas were prepared from mice that had been immunized with extracts isolated from a mixture of tobacco flowers at several stages of development. The antibodies were then screened for organ specificity and for independence of their epitopes. Using this method they were able to isolate 13 lines producing antibodies that reacted primarily with pistil extracts, and 1 line producing antibodies that were specific to an anther antigen. The epitopes recognized by antibodies produced in this manner can be proteins, free carbohydrates, or bound carbohydrates such as those in glycoproteins or glycolipids. In fact, the authors conclude that most of their antibodies recognize carbohydrate epitopes. The carbohydrate-specific antibodies are useful in identifying biochemically distinct cells within flowers but do not facilitate the isolation of genes that are predominantly expressed in flower parts. The antibodies that recognize protein epitopes could be used to isolate flower-specific cDNA clones from expression libraries as described above.

Nucleic Acid Methods

Direct screening of nucleic acids has been used to isolate clones of flower-specific genes. The simpler of these methods relies on differential hybridization of two or more probes to individual recombinant clones. Screening is accomplished by plating out a cDNA library made from floral tissues and making duplicate filter replicas of the plated clones. The filters are then hybridized with a labeled probe made from floral RNA and with a probe made from RNA from a vegetative organ, or from a mixture of vegetative organs. The intensity of the signal produced by each clone on the filters is roughly proportional to the steady-state level of mRNA from the corresponding gene in that tissue. Clones that produce strong signals with the floral probe, and weak or no signals with the vegetative probe, derive from genes with enhanced expression in floral tissues. Examples of the successful use of this procedure include the isolation of cDNA clones for genes expressed in pistils (18, 21), stamens (21, 33a, 53), and pollen (9, 25, 60). This method was also used in the initial isolation of genes encoding pistil-specific self-incompatibility-associated proteins from *Brassica* (45) and *Nicotiana* (2).

A variation on this method has been used to directly isolate genomic clones of differentially expressed genes from *Arabidopsis thaliana* (52). It is possible to do this in *Arabidopsis* because it lacks the large quantities of repetitive DNA present in most higher plants that can complicate screening of genomic libraries. By hybridizing replicate filters of cosmid clones arrayed according to the pattern of 96-well microtiter plates Simoens et al were able to isolate a number of clones for genes that show light regulation, or preferential expression in stems, calli, or inflorescences. This method can accelerate the screening for clones by eliminating the need for the construction of a cDNA library from the organ of interest. The construction of such a cDNA library does, however, often provide an enrichment because of the higher levels of differentially expressed messengers in the starting RNA.

All differential screening methods are limited in sensitivity by the presence of strong signals from clones representing very highly expressed genes. In long exposures, the signals from these clones can mask the signals of clones representing weakly expressed genes. Despite these limitations, clones representing approximately 0.05% of the total message have been isolated by this method (18).

The second nucleic acid–based method for isolation of organ- or tissue-predominant sequences relies on producing a population of RNA or cDNA from which the commonly expressed sequences have been removed. This method is technically more demanding, but has the potential to identify organ-predominant sequences present at very low levels. In practice, there are several alternative methods for performing the "subtractive hybridization." One method (14) that has been used in other applications is first to produce single-stranded cDNA from the tissue of interest (the "subtracting population"). This cDNA is then hybridized in solution to an excess of RNA from a second tissue and passed over a hydroxylapatite column. The sequences in common form duplexes that are retained on the column while unhybridized cDNA and RNA flow through. The unbound material is then treated with base to hydrolyze the RNA. The remaining material is cDNA greatly enriched for sequences present in the tissue of interest and absent from the second tissue. This cDNA can then be used to synthesize a probe to screen an existing library or to prepare a library enriched for the desired sequences. An additional variation in this procedure includes the use of biotinylated nucleotides on the subtracting population to allow the substitution of affinity columns, which bind biotin, for the hydroxylapatite columns (16). A combination of subtractive hybridization and differential screening have been used by Sommer et al (54) to isolate the *Def* gene of *Antirrhinum majus,* a gene that regulates the placement of organs in flowers.

More recently, novel methods have been developed that allow for amplification of sequences prior to performing the subtractive hybridization. These methods rely either on cloning (48) or on the polymerase chain reaction

(16, 56) to produce multiple copies of an entire RNA population prior to the subtractive hybridization; these extra steps facilitate the isolation of rare sequences. These methods have not yet been applied to the isolation of flower-predominant clones, and readers are referred to the cited publications for more detailed methods.

IDENTIFIED FLOWER-PREDOMINANT GENES

The methods described in the previous section have been applied to a number of different plant systems. Genes that have interesting expression patterns have been isolated from petals, stamens, and carpels. Because sepals are structurally very similar to leaves, there have been no reports of special efforts to isolate specific genes from these floral organs. In this section I review the isolation and characterization of some of the genes described to date.

Petal-Predominant Genes

Early experiments indicated that the total set of genes expressed in petals is nearly identical to the set expressed in leaves (30). The often-striking coloration of the petals of many plants is their most noticeable difference from leaves. Considerable research has focused on the enzymes involved in synthesis of the pigments that color petals. Most floral pigments are products of the flavonoid pathway, a branch of the general phenylpropanoid pathway that is also responsible for the synthesis of precursors to lignin and precursors to defense-related compounds (26). The first committed step in flavonoid synthesis is mediated by the enzyme chalcone synthase (CHS), which catalyzes the condensation of 4-coumaroyl-CoA and malonyl-CoA to form naringenin chalcone. Clones encoding this enzyme, and the enzyme for the next step in the pathway, chalcone isomerase (CHI), have been isolated from a number of species (compiled in 43). The initial clones were isolated by transposon tagging (64) or by immunological methods (63) with subsequent genes isolated using heterologous probes. Since the function of these genes in flowers was known prior to their isolation, research has focused on the specificity and mechanism of their regulation. I discuss this work below in the section on regulation of floral gene expression.

Stamen-Predominant Genes

Development of the higher-plant stamen represents an intimate interaction between two generations: the sporophyte consisting of the microsporocytes, filaments, and anther walls (including the epidermis, the endothecium, the middle layers, and the tapetum); and the gametophyte, or mature pollen grain. This interesting interaction and the importance of pollen in the reproductive cycle have led to the careful examination of gene expression during stamen development and microsporogenesis. The development of pollen grains has been especially well characterized and is described in an extensive review by J. P. Mascarenhas in last year's *Annual Review of Plant Physiology and Plant Molecular Biology* (37). These studies indicate that there is substantial overlap (60–65%) between genes expressed in pollen and those expressed in vegetative tissues (37). Pollen-predominant mRNAs fall into high, medium, and low abundance classes, but a relatively high proportion are in the high-abundance fraction (37). I present here a summary and update on the properties of cloned genes that have been identified as preferentially expressed in pollen and stamens.

Stinson et al (55) used the differential screening method to isolate cDNA clones of genes preferentially expressed in pollen from *Tradescantia paludosa* and *Zea mays* (maize). The clones were found to derive from genes present in single or low copy number in the genomes of these organisms. On northern blots containing equal amounts of total RNA these clones had strong signals in pollen RNA, and no visible signals in RNA from vegetative organs or from female reproductive structures. The steady-state levels of RNA during pollen development were also determined for three of the clones. For all of these clones the level of RNA continuously increased from an undetectable level in microspores to maximal accumulation in mature pollen grains. Hybridizing RNA decreased dramatically in pollen germinated in vitro. Ungerminated pollen contains a significant amount of stored mRNA that is translated upon germination (reviewed in 37). The pattern of accumulation and depletion of RNA hybridizing to the three characterized clones suggests that they represent such stored messengers. Further studies on these and other pollen-derived clones will aid in determining the nature of the proteins necessary for the first steps following pollen germination.

A fourth clone, Zmc13, which was isolated from maize pollen by Stinson et al (55), has been more fully characterized by Hanson et al (25). In contrast to the three clones described above, mRNA from the Zmc13 gene not only accumulates in maturing pollen but also maintains a high steady-state level during pollen tube germination. The sequence of the cDNA indicates that the Zmc13 mRNA could encode a protein of approximately 18,000 daltons with a putative signal peptide sequence. Twell et al (58) have independently isolated a cDNA clone, LAT52, for a single-copy gene from tomato that encodes a protein with 32% sequence identity to the Zmc13 protein. The LAT52 protein sequence contains a putative signal peptide region similar to that observed in the maize clone. The timing of expression of this gene in tomato pollen appears to be similar to the observed pattern in maize. In addition, northern blot analysis detected expression in tomato petals. In situ hybridization of tomato anther tissue with this clone demonstrated that mRNA from this gene accumulated both in the pollen grains and in a region of the anther endothecium and epidermis (60). Expression in the anther wall was primarily on the abaxial side and was confined longitudinally to the region adjacent to the locule of the anther. The amount of hybridizing RNA in anther walls appeared to be roughly equivalent to the amount in pollen grains (60). The functions of the Zmc13 and LAT52 gene products are not known, and they do not share detectable similarity with any sequences in the current DNA and protein databases (25, 58).

The LAT52 cDNA clone was isolated in a general screen for clones that are preferentially expressed in mature stamens of tomato (39). Four additional clones that do not cross-hybridize with LAT52 were also isolated (39). All four share LAT52's pattern of in situ hybridization on mature tomato stamens, with strong signals in pollen grains and in a portion of the abaxial wall of the stamen (60). Their expression differs only in the timing of accumulation of RNA in the stamens. DNA sequence analysis of two of these clones indicated that they encode proteins which share 54% amino acid identity even though they were not found to cross-hybridize (65). Both predicted proteins contain hydrophobic amino-terminal regions that may be signal peptides. They also share significant sequence identity with pectate lyases from several plant pathogenic bacteria of the genus *Erwinia* (65). Pectate lyases are secreted by these bacteria during infection to digest the walls of plant cells (34). The presence of putative signal peptides in the predicted protein products of the plant genes indicates that they also may be secreted by pollen grains or pollen tubes. Wing et al (65) hypothesize that the proteins may be produced by growing pollen tubes to aid in penetration of the stylar transmitting tissue. They also suggest that degradation of pectic polysaccharides by these enzymes could produce precursors for the synthesis of wall components for growing pollen tubes. The two genes were shown to be genetically linked and are located on chromosome 3 (65).

Brown & Crouch (9) have used a differential screening procedure to isolate cDNA clones that are preferentially expressed in the anthers of *Oenothera organensis*. One set of cross-hybridizing cDNA clones isolated in this screen was extensively characterized (9). Southern blotting experiments show that this set of cDNA clones derive from a small gene family with 6–8 members. Restriction enzyme maps of the cDNA clones show that they fall into six classes, indicating that several of the genes (or alleles) of the family are expressed in pollen. The accumulation of mRNA corresponding to these clones shows kinetics similar to that of the clones from maize and tomato described above: They are first detectable after the tetrad stage of pollen development and reach maximum levels in mature pollen. By means of sequencing, these clones were shown to encode proteins 54% identical to tomato fruit polygalacturonase (PG) (9, 23). This degree of similarity makes it likely that the encoded proteins have PG activity. In tomato fruits PG is synthesized as a precursor protein with a hydrophobic sequence that directs the enzyme to extracellular spaces where it digests cell wall components during ripening (15, 23). The cDNA clones isolated from *Oenothera* pollen were truncated at the 5'-ends, so it was not possible to determine if they contained similar signal sequences (9). A portion of one of the cDNA clones was expressed in *E. coli* as a fusion protein with β-galactosidase (9). Antibodies raised to this protein fusion detected proteins of 40,000–45,000 daltons on western (i.e. protein) blots of pollen grain extracts. The timing of accumulation of the proteins in pollen roughly paralleled the accumulation of mRNA described above. The protein was also detected in pollen tubes by both western blotting and in situ localization experiments. The protein was not detected in other floral or vegetative organs.

Genes that are maximally expressed in parts of stamens other than the pollen grains have also been identified. Goldberg (21) and Koltunow et al (33a) have reported on the isolation of a large number of cDNA clones for genes that are predominantly expressed in tobacco anthers. The identified genes fall roughly into two classes: some showing detectable levels of RNA throughout anther development with maximal RNA levels just prior to anthesis, and the others showing maximal levels of RNA relatively early in anther development (stages 1–6) followed by decline to undetectable levels by anthesis. Genomic Southern blot hybridizations and the isolation of homologous but not identical clones showed that the anther-predominant genes are present in the tobacco genome in from one to five copies. Several of the cDNA clones have been extensively characterized (33a).

Clone TA56 is a member of the late-expressing class, and was found to encode a thiol endopeptidase (33a). In situ hybridizations to anther sections showed that the *TA56* gene was expressed in cells of the connective, between locules on one side of the filament, and in cells of the adjacent stomium, the site of rupture during anther dehiscence. While hybridizing RNA is present in these regions as early in development as the tetrad stage of meiosis (stage 1), levels of this RNA are maximal just prior to breakdown of the connective and stomium. On the basis of the timing and pattern of expression it was hypothesized that this endopeptidase may be directly involved in the process of degeneration of the connective and stomium which leads to anther dehiscence (33a). The sequence of a second clone, TA20, which showed similar timing of expression was not found to be significantly similar to any sequences in the databases. In situ hybridization showed that the spatial distribution of RNA hybridizing to TA20 was very different from that of TA56. TA20 RNA was detected at significant levels in all cells of the anther except for the vascular bundle, the tapetum, and the cells at the junction of the connective and the stomium (which had the highest levels of TA56 RNA). This gene is additionally expressed at lower levels in specific tissues of the pistil (33a; see below).

Three of the clones with peak expression earlier in anther development were similarly characterized. In in situ hybridization experiments RNA from the genes corresponding to these clones was detected exclusively in the tapetal cells that surround the locules of the anthers during pollen maturation. A genomic clone corresponding to one of these clones (TA29) was shown to encode a 321 amino acid polypeptide that contains 20% glycyl residues (51, 33a). A hydrophobic amino-terminal region of the protein has properties of a signal peptide that could direct the protein outside of the cell. Because one of the primary functions of the tapetum is to provide materials for the outer layers of pollen grains, this protein could be exported from the tapetum and become a part of the pollen wall. Glycine-rich proteins have been shown to be present in walls of other plant cells (12). The previously identified wall proteins do, however, differ significantly from the TA29 gene product in that they contain levels of glycine approaching 70%. A second tapetal gene, *TA32*, was found to encode a lipid transfer protein (33a). This protein could also function to transfer materials from the tapetum to the developing pollen grains. The third tapetal gene was not homologous to any previously described sequence. For all three of these genes the increase and decline in detectable RNA closely parallels the formation and degeneration of the tapetum (33a).

Smith et al (53) have isolated 11 classes of stamen-predominant cDNA clones from immature stamens of tomato. All but one of the classes are derived from genes present in single or low copy number in the tomato genome. Nine of the clones hybridized to immature stamen RNA with no detectable hybridization to RNA from mature stamens or other reproductive or vegetative organs. In situ hybridization experiments with three of the stamen-predominant clones showed that expression of the corresponding genes was confined to the cells of the tapetum. As seen for the tapetal clones described above, the timing of expression of these genes closely paralleled the appearance of the tapetal layer and its degeneration as the pollen grains matured (53).

The stamens and pollen of many plants are colored by the same pigments found in petals. The proteins of the phenylpropanoid pathway would, therefore, also be expected to be expressed in the stamens of these species. This expectation has been confirmed for some of the flavonoid synthesis genes that have been characterized in petals (32, 62).

Several additional reports on preliminary characterization of clones expressed in developing pollen or mature anthers have been published. Appels et al (3) have isolated putative meiosis-specific cDNA clones from *Lilium* microsporocytes, and Herdenberger et al (29) have isolated at least one cDNA clone from sunflower that is expressed preferentially in mature anthers.

Conclusions on Stamen-Predominant Genes

Several interesting observations can be made from the initial characterizations of genes with enhanced expression in stamens. All of the genes for which expression was detected only in stamens were found to express in tapetum, pollen, or anther wall. Because the pollen and tapetum are highly specialized cells in the stamens, isolation of clones for unique mRNAs from these tissues does not come as a surprise. In addition, the tapetum is known to be the most transcriptionally active tissue during early differentiation of the stamen (37), and pollen is known to accumulate mRNA actively as it matures (37).

The large amount of mRNA that accumulates during pollen maturation is stored in the dormant grain for translation during germination (37). The identified pollen-derived clones come primarily from genes whose transcripts show a pattern of accumulation consistent with their being members of this class of genes. Several of the identified genes also continue to produce mRNA in the germinating pollen tube. The protein products of mRNA that accumulate in mature pollen have been hypothesized to be enzymes and structural proteins necessary for initial germination of pollen tubes, and for penetration of pollen tubes into the style (37). The results outlined above show that two different genes of this class encode proteins that are homologs of enzymes for polysaccharide digestion (9, 65). The products of these genes could act soon after germination of the pollen tube to digest components of the extracellular material in the stigma and style, facilitating tube entry into and passage through the transmitting tissue. The genes encoding a homolog of pectate lyase include a putative signal peptide region that could direct the protein outside the cell (65) in a manner consistent with the proposed function. Owing to truncation of the cDNA clones it is not known if the pollen-predominant polygalacturonase homolog includes a signal peptide (9). Because the action of this enzyme in another location in the plant is to digest components of the cell wall (20), it is likely that it is transported outside of the pollen tubes. It is also possible that the primary function of these enzymes is to digest materials in the stigma and style to provide subunits for synthesis of the wall of the pollen tube. Recent evidence shows that the health of the male parent can have a significant impact on the success of the pollen the plant sheds (67). If stressed plants produce pollen with reduced stored mRNA for the polysaccharide digesting enzymes, then this pollen may be unable to germinate and grow as rapidly as pollen from unstressed plants, putting them at a competitive disadvantage.

Ursin et al (60) show that all five of the cDNA clones they isolated that express primarily in pollen are also expressed at readily detectable levels in a region of the anther wall. This includes the genes encoding proteins homologous to pectinolytic enzymes. One possible function of the accumulation of such enzymes in the walls of maturing anthers is digestion of cell wall components to aid in dehiscence and release of the pollen grains. However, dehiscence of tomato anthers occurs through the formation of longitudinal slits on the adaxial surface, and the visualized accumulation occurs along the abaxial surface of the anther. In contrast, location and enzymatic activity of the *TA56* gene (thiol endoprotease) described by Goldberg and coworkers (21, 33a) are consistent with the product of this gene participating in the dehiscence of tobacco anthers (33a). Similarly, the identification of tapetal-predominant gene products as putative cell wall proteins and lipid carrier proteins (33a) is consistent with the hypothesis that this tissue functions to provide materials for the maturation of pollen grains.

Carpel-Predominant Genes

The most extensively studied set of genes known to express primarily in the gynoecium are genes encoding self-incompatibility-associated glycoproteins (S-genes). S-genes have been cloned from plants with either the sporophytic or gametophytic systems of incompatibility. The genes for sporophytic incompatibility are expressed primarily in the stigmatic region (46), and genes for gametophytic incompatibility are expressed in the transmitting tissue of the style and in the tissues of the placenta through which pollen tubes grow (13). Neither the protein nor the RNA for these genes have been detected in pollen or germinating pollen tubes. It has recently been shown that the S-gene product in *Nicotiana alata* is homologous to known ribonucleases and has ribonuclease activity (38). Recent evidence indicates that differential degradation of RNA in elongating pollen tubes may be a critical part of the incompatible response in *Nicotiana alata* (38a). For an extensive discussion of this subject the reader is referred to the review by Nasrallah et al in this volume.

Outside of the work on self-incompatibility, research on genes with enhanced expression in the gynoecium has not been extensive. My research group reported on the isolation and initial characterization of several such genes (10, 18). Using a differential screening approach we isolated 23 cDNA clones for genes that express at much higher levels in tomato pistils than in tomato seedlings. The clones were found to fall into 11 noncross-hybridizing classes. Southern blotting experiments indicated that the clones derived from genes that were single copy, or members of small gene families. Northern blotting experiments showed that these genes were expressed predominantly in reproductive organs; but when 5 μg of poly A$^+$ RNA was used per lane, some hybridization was also detectable for most of the clones in one or more vegetative organs. Ten of the eleven clones also showed some detectable expression in stamens, although in most cases the level was considerably lower than in pistils (18; K. Budelier and C. Gasser, unpublished). In cases where the genes are members of small families, the RNA detected in stamens may derive from a family member not expressed in the pistil.

Three of these genes have been further characterized by in situ hybridization to examine the tissue-specificity of expression within the pistil. Clones 9617 and 9608 were shown to express in the periphery of the strands of the transmitting tissue in the style, and in the inner cell layers of the integument, respectively (18). mRNA hybridizing to a third clone, 9612, was found in the outer layers of the strands of transmitting tissue, but only in the upper two thirds of the style (10). RNA from each of the three genes identified by these clones is undetectable in immature pistils as little as three days prior to anthesis but reaches maximal levels of approximately 0.05–0.1 % of the mRNA at anthesis (10, 18). The sequence of the 9612 cDNA indicates that it encodes a 404-amino-acid protein of approximately M_r 44,000 (10). Like several of the the stamen-predominant proteins described above, the predicted 9612 protein includes a hydrophobic amino-terminal region that may represent a signal peptide to direct the protein outside of cells (10). The protein also includes two potential N-linked glycosylation sites. It is therefore possible that this protein represents one of the many glycoproteins present in the extracellular mucilage of the transmitting tissue (28).

The *TA20* gene of tobacco, which expresses in most cell layers of anthers (see above), was additionally shown to express in pistils (21, 33a). In situ hybridizations show that within the pistil, RNA from this gene is present in cells on the surface of the placenta, specific sites in the periphery of the ovaries, and parenchyma tissue subtending the stigma and surrounding the transmitting tract of the style (21, 33a). This gene, therefore, appears to be primarily expressed in relatively undifferentiated cells of both anthers and pistils in tobacco.

Pathogenesis-Related Genes

Several studies have shown that the products of genes whose expression has been associated with pathogenesis or wounding are also expressed during flowering in undamaged plants. Memelink et al (41) found that a chitinase gene was expressed at significant levels in mature tobacco flowers. The levels of chitinase RNA in roots were, however, higher than the levels in flowers. Lotan et al (36) used antibodies to pathogenesis-related proteins to show that proteins related to β-1,3-glucanases and endochitinases were present in the flowers of unstressed, uninfected plants. In particular, proteins that react strongly with antibodies to the β-1,3-glucanase class of proteins were detected in the style. In situ reactions with these antibodies localized these proteins to the outer layers of the transmitting tissue, a pattern very similar to that seen for one of the tomato pistil genes described above (36).

A set of cDNAs preferentially expressed in young floral buds produced in a tobacco explant system were isolated by Meeks-Wagner et al (40). Several of these cDNAs have been shown to encode pathogenesis-related proteins including chitinase, β-1,3-glucanase, osmotin, and extensin (47). Chitinase and osmotin RNA were also detected during flower development in intact plants (47). It has been suggested (36, 47) that the proteins encoded by these genes may be directly involved in important flower-specific processes such as pollen tube growth (47) or the vegetative-to-floral meristem conversion (47). Alternatively, these compounds may be necessary for protection of the critical floral structures from pathogen attack. Further study on the expression pattern and properties of these genes in flowers will be needed to clarify their role in flowering.

SPECIALIZATION OF TISSUES IS INDICATED BY TISSUE-PREDOMINANCE OF GENES

Consideration of the anatomy of floral organs has greatly helped us to understand their functions. For example, our current understanding of the process of fertilization is only possible because of detailed studies on the structures of the cells of transmitting tissue of the style, the placental surface, the funniculus, the integument, and the megagametophyte. Superimposed over the anatomical differentiation of the tissues are biochemical differences that result from the expression of tissue-predominant genes. A map of the "biochemical anatomy" of a flower could indicate previously unknown functional differentiations of tissues or of sets of cells within tissues. Such a map can aid in determination of the functions of genes, and can help in determining the pattern and number of regions of differential gene expression in a flower, for correlation with models of floral gene regulation.

As described for several examples in the previous section, in situ hybridization can be used to define the specific location of expression of a gene within a floral organ. Figure 2 (*left*) shows an example of a micrograph of a tomato ovule in a longitudinal section of a tomato pistil that was hybridized with the pistil-predominant cDNA clone pMON9608 (18). Hybridization is confined to two cell layers of the single integument, indicating the presence of mRNA from the 9608 gene in these cells. This indicates that these cells have a unique function that is facilitated or enabled by the 9608 gene product and that is different from adjacent cells that do not express this gene. Figure 2 (*right*) shows a section of a similar ovule that has been stained with safranin and fast green. The inner layers of the integument stain slightly darker than the surrounding cells, but the staining does not show the significant biochemical difference readily apparent in the in situ hybridization experiment.

Several additional novel biochemical compartments have been identified by these methods in floral organs. Five different pollen cDNA clones hybridize to a histologically indistinct region of tomato anther walls (60), defining it as a separate biochemical compartment. [Several novel compartments within

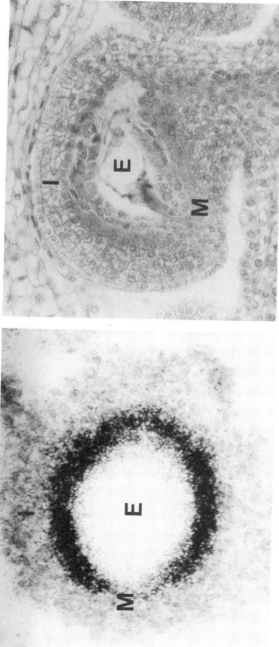

Figure 2 Biochemical differentiation of tissues. The left panel shows a section through a tomato ovule from a tomato flower at anthesis. The section was hybridized in situ with a radioactive probe made from a pistil-predominant cDNA clone (18), coated with photographic emulsion, exposed, and developed. The cells in which this gene is expressed are located by the dark autoradiographic grains over the innermost layers of the single integument. The right panel shows a section of a similar ovule that has been stained with fast green and safranin. The cells of several of the inner layers of the integument stain more intensely, indicating some difference from the cells of the outer layers. I: integument; M: micropyle; E: egg sac.

tobacco anthers were identified by the patterns of expression of the genes described by Kultunow et al (33a; see above).] Similarly, the tomato pistil cDNA clone pMON9612 (10, 18) hybridizes to cells in the upper two thirds of the transmitting tissue of the style and not to similar cells in the lower third of the style. This indicates a previously undescribed biochemical difference between these two parts of a single tissue that may have important biological implications. For example, this difference could play a role in a novel property of the transmitting tissue that was recently identified by Sanders & Lord (49). They demonstrated directed movement of latex beads through the style, indicating the existence of a previously unsuspected directional motive force for pollen tube growth inherent in the transmitting tissue. It is possible that the observed gradient of 9612 gene expression is part of the difference between the different levels in the style that drive this process. This hypothesis is completely speculative and is meant to serve only as an example of how newly identified biochemical differences between cell types within a tissue can be incorporated into descriptions of biological function.

Continuing analysis of additional floral clones will allow the construction of a more complete map of the zones of gene expression within the floral organs. These zones must be accounted for both in models of the regulation of gene expression during flowering and in descriptions of the functions of the differentiated tissues of flowers.

PROGRESS IN UNDERSTANDING REGULATION OF FLOWER-PREDOMINANT GENE EXPRESSION

Regulation of Genes Involved in Floral Pigment Synthesis

Owing to the easy identification of mutants and the early isolation of clones of structural genes, the regulation of floral pigment synthesis is the best-understood area of floral gene regulation. Studies on the regulation of these genes have been performed utilizing naturally existing mutations and in vitro modification and re-introduction of cloned genes. These studies are extensive, and a complete review and analysis of the results would require a full chapter in this volume. Here I present only representative examples of this work and attempt to summarize the most important conclusions.

The expression of chalcone synthase (*CHS*) genes in *Petunia hybrida* (petunia) flowers is a relatively well-studied example. Eight *CHS* genes have been identified in petunia, of which two express in flowers (32, 33). The 5'-flanking regions of the two flower-expressed genes (*CHSA* and *CHSJ*) have been isolated and attached to the *E. coli* β-glucuronidase (GUS) coding region. The chimeric gene constructs were introduced into petunias by *Agrobacterium*-mediated gene transfer, and the resulting plants have been analyzed for GUS activity (33). The pattern of GUS activity in the flowers correlated well with the known locations of *CHS* expression (which do not always correlate exactly with pigment formation) (33). Expression was seen in petals, stamens, and ovaries. These experiments demonstrate that the isolated genomic fragments contain promoter sequences that direct flower-predominant expression of the attached coding sequences. The *CHSA* promoter has been further characterized by systematic deletions, followed analysis of expression in transgenic plants (61). These experiments indicated that an 800-bp promoter fragment produced maximal expression in corollas, with progressively lower levels of expression produced by fragments of 530, 220, and 67 bp. However, even the 67-bp fragment accurately reproduced the petal-predominant expression pattern seen for the endogenous gene (61). Control of transcription of this gene is, therefore, mediated by a series of quantitative elements in the 5'-flanking region, and a short proximal region of the promoter fragment is sufficient to direct low-level organ-predominant expression. Analysis of a *CHS* promoter fragment from beans corroborates the finding that flower-predominant *CHS* expression is regulated at the level of transcription (50).

The promoter regions of phenylalanine ammonia-lyase (*PAL*) and chalcone isomerase (*CHI*) genes have been similarly studied. In one example, the two *CHI* genes found in petunia (63) were shown to be differentially expressed in flowers. Expression of the *CHI* B gene is detectable only in immature stamens, while the *CHI* A gene contains two distinct promoters, one of which is predominantly utilized in corollas, with the other being most active in mature stamens (62). Transcriptional control, once again, appears to be the most important determinant in expression of these genes. A similar conclusion can be drawn from studies on the *PAL* promoter from *Phaseolus vulgaris* (6). In these studies the bean *PAL* promoter was shown to regulate GUS expression according to the expected program in transgenic tobacco and potato plants (6).

Genetic tools have been used by Enrico Coen and coworkers to show that CHS enzymes in pea are controlled by a unique set of *trans*-acting factors in flowers (27). Mutations in the genes encoding these factors that dramatically alter CHS levels in petals have no effect on the expression of *CHS* genes in roots. This same lab has used similar methods to define three *cis*-acting transcriptional regulatory elements within the promoter region of a pigment synthesis gene, *Pal* (*pallida*, not phenylalanine ammonia lyase), of unknown function in *Antirrhinum majus* (1). Specific deletions of the promoter region of this gene that were produced by excision of a transposable element alter the nature and tissue-specificity of the response of this gene to a second *trans*-acting regulatory gene (1). These experiments suggest the existence of multiple transcription factor binding sites in the promoter regions of flower-predominant genes and imply that a complex set of interactions between the promoters and the regulatory factors is responsible for the pattern of expression of the genes (1, 27; see below).

While many of the pigment-related genes are additionally expressed in other locations in response to infection and other environmental insults, studies on their expression will continue to provide important information on the mechanism of regulation of flower-predominant gene expression.

Regulation of Stamen Genes

Studies have been initiated to determine the specific sequences responsible for expression of pollen-predominant genes. Twell et al (57, 59) have examined the 5'-flanking sequences of two stamen-predominant genes for transcriptional activity in transient and stable transformation experiments. In the first of these studies a test vector was constructed by attaching 0.6-kb fragment from the 5'-end of the *LAT52* gene (58; see above) to the GUS coding sequence. This construct was introduced into pollen, stamens, and leaves by particle-gun-mediated transformation (57). High levels of GUS expression were detected in tomato and tobacco pollen that had been bombarded with particles coated with this construct. No expression was detected in leaves or in the anther wall. Control experiments using a CaMV 35S promoter fused to the GUS coding sequence demonstrated that the particle gun method could produce productive transformation of all of these tissues. The specificity of expression in the pollen was, therefore, conferred by regulatory sequences within the *LAT52* 5'-region.

The *LAT52*/GUS transcriptional fusion described above, and a similar construction using a 1.4 kb 5'-flanking fragment from a second anther-predominant gene, *LAT59* (60), were additionally characterized by stable introduction into tomato, tobacco, and *Arabidopsis* (59). These experiments showed that the *LAT52* promoter fragment produced detectable expression only in the pollen of all three plants, confirming the accuracy of the transient-expression experiments described above. Activity was also detected in the tubes of germinating pollen. The *LAT59* construction was also expressed at very high levels in pollen, with some activity also detectable in roots and seeds (60- to 500-fold lower than in pollen). The only significant discrepancy between the expression pattern observed for the endogenous and introduced genes related to expression in the anther wall. Both the LAT52 and LAT59 cDNA clones hybridized to RNA present in the anther wall (59, 60). In transgenic plants no GUS activity was detected in this tissue with either of the promoters, and no GUS RNA was detectable in walls of anthers from the *LAT59*/GUS plants (59). This could indicate that additional regulatory sequences not contained in the tested promoter fragments are present in the endogenous genes. Alternatively, the hybridization detected with these probes could derive from related genes with sequences sufficiently diverged to prevent detection in the stringent hybridizations used to estimate the copy number of these genes (58).

The tobacco tapetal-predominant gene, *TA29*, was shown to be transcriptionally regulated by hybridization to RNA synthesized in vitro in isolated nuclei (33a). Introduction of a fusion between the 5'-flanking region of this gene and the GUS coding region into tobacco plants led to specific production of GUS protein and mRNA in the tapetal cells. This region was additionally shown to produce tapetal-specific expression correctly in transgenic oilseed rape plants (36a). Deletion analysis was used to show that a region extending from -207 to -85 relative to the start of transcription was sufficient to produce tapetal-specific expression when attached to a minimal promoter region (33a).

The tapetum-specific regulatory region was additionally used in experiments to evaluate the effect of premature destruction of the tapetum on anther development, and in a novel strategy for hybrid crop production. In these experiments the *TA29* regulatory region was fused to a coding sequence for diphtheria toxin (33a) or ribonuclease (36a) and introduced into plants. Tapetum-specific production of either the diphtheria toxin or ribonuclease resulted in destruction of the tapetum and complete male sterility. The anthers of these plants were otherwise normal, indicating that the tapetum is only necessary for pollen production, and not for the development of other anther structures. Readers are referred to the paper by Mariani et al (36a) for a discussion of how these transgenic plants could be used in a breeding program for production of hybrid seed.

The transformation experiments demonstrate that transcriptional control is sufficient to explain the stamen- and pollen-predominant expression patterns of these genes. The conservation of regulation of the chimeric genes in tobacco, oilseed rape, and *Arabidopsis* indicates that the factors controlling the regulation of these genes are conserved in different species.

Regulation of Pistil Genes

Budelier et al (10) have examined the functionality of the putative promoter of a tomato gene, referred to as *9612*, that is predominantly expressed in the upper part of the transmitting tissue of the styles of mature flowers. Transcripts from this gene also accumulate to detectable levels in immature stamens (50 times lower than the maximal level in pistils), declining to an undetectable level as the stamens mature. A 1.2-kb fragment of the 5'-flanking region of this gene was attached to the GUS coding sequence and introduced into tomato and tobacco plants by *Agrobacterium*-mediated gene transfer. In transgenic tomato plants the pattern of expression of the chimeric gene in pistils was essentially identical to the pattern of expression of the endogenous gene. Onset of expression of the chimeric gene in stamens was at a time similar to that of the endogenous gene, but the GUS protein was still present in mature pollen, several days after the endogenous transcript would have disappeared. In a manner consistent with the very low level of expression of the endogenous *9612* in other parts of the plant, GUS activity was not detectable outside of the tomato flowers (10).

When tobacco plants containing the *9612*/GUS construct were examined, the GUS activity was found to be confined to the flowers and the pattern of expression of the introduced chimeric gene in tobacco stamens was similar to that seen in tomato stamens. However, in contrast with the situation in tomatoes, the expression associated with the gynoecium occurred only at the base of, and in, the septum of the ovary. No GUS activity was detected in the style (10). In addition, the onset of expression occurred in very young buds, much earlier than in tomatoes. An endogenous homolog to *9612* that expresses at high levels in the style was shown to be present in tobacco.

Appropriate expression of the chimeric gene in tomato pistils and aberrant expression of the same construction in tobacco pistils indicate a lack of conservation of either the regulatory factors or the locations of the regulatory factors between these two members of a single family. By contrast, the anther- and pollen-predominant genes described above, and most other regulated plant genes (reviewed in 4), show conservation of regulation even across family boundaries.

Regulation of two different alleles of genes encoding self-incompatibility proteins (S-proteins) of *Brassica oleracea* has also been examined in transgenic plants (31, 44). In *B. oleracea*, transcripts and protein products of the S-genes are detected only in the papillar cells of the stigma. When transferred into the self-compatible species *Nicotiana tabacum*, expression of these genes was detected only in pistils, with the highest levels present in mature flowers. In situ localization of the protein products of the introduced genes showed that protein accumulated only in the transmitting tissue of the style and not in the stigma. The S13 allele accumulated in the upper region of the transmitting tissue of the style, and the S22 allele accumulated throughout the transmitting tissue and in the outer layers of the placenta. The region in which the introduced S13 allele is expressed is superficially similar to the region in which the chimeric *9612*/GUS gene (10; see above) is expressed when introduced into tobacco. The pattern of expression of the S22 allele is very similar to the pattern observed for the endogenous S-gene of the related species, *Nicotiana alata* (2).

The difference in regulation of these genes between *B. oleracea* and tobacco may be the result of differences between these two species, or the result of the use of fragments that do not contain all of the necessary regulatory determinants. Since experiments on the introduction of these genetic constructs into *B. oleracea* have not been reported, this second possibility cannot be ruled out. The similarity of the S23 expression pattern in *N.*

tabacum to the pattern of expression of the *N. alata* S-gene may indicate a conservation of regulatory determinants within S-genes of different species (31). This would be surprising, because the complete lack of similarity between the sporophytic (*B. napus*) and gametophytic (*N. alata*) S-gene products indicates that they are likely to have evolved independently (see the review by Nasrallah et al in this volume).

Models for Floral Gene Regulation

In the transformation experiments described above, information sufficient for appropriate spatial distribution of expression was found to be contained within the 5'-flanking regions of the genes. In most cases the 5'-flanking regions also directed expression correctly when introduced into heterologous species. This observation is consistent with previous research in plants (reviewed in 4) and animals (reviewed in 42). While other steps in the gene expression process (e.g., differential processing and transport of RNA from the nucleus, mRNA stability, translational efficiency, and protein stability) have clearly been shown to affect the level of expression of a given gene, there are only a few examples where these processes regulate developmental specificity of gene expression. If currently cloned plant genes are representative, then alterations in the rate of transcription regulate tissue-specific gene expression, with the other mechanisms acting to alter the final level of protein.

Control of gene transcription in animals and plants appears to be mediated by the binding of transcription factors to specific control sequences associated with the genes (for reviews see 7, 24, 42). Several lines of evidence indicate that similar mechanisms are active in the regulation of gene expression in flowers. For example, genetic evidence for transcription factors is provided by work on the *pal* gene of *A. majus* cited above in the section on pigment gene regulation (1). This work indicated that the *pal* gene is regulated by at least two independent factors that interact with the 5'-flanking sequences (1). Additional evidence of a role for transcription factors in flower development is provided by the observation that two genes that regulate the developmental fate of floral organ primorida, the *AG* gene of *Arabidopsis* (66) and the *Def* gene of *A. majus* (54), encode proteins that are homologous to DNA binding transcription factors.

Accepting the assertion that flower-predominant genes are regulated at the transcriptional level by binding of transcription factors, we can construct working models of the mechanism by which these factors control the spatial distribution of gene expression within flowers. In the conceptually simplest model each region of specific gene expression (an "expression zone") is defined by the presence of a single unique transcription factor that binds to and coordinately regulates genes specific to that zone. This model implies the existence of an extremely large number of transcription factors and is inconsistent with observations on several well-studied genes that have two or more transcription factor binding sites important in regulation. In a somewhat more elegant combinatorial model, pictured in Figure 3, several transcription factors act in concert to produce the complex pattern of expression zones present in the flower. All genes that express in flowers have binding sites for flower-specific factors, and the cells of each organ contain factors that bind to the regulatory sequences of genes expressed in that organ. Tissue specificity within an organ is produced by the binding of additional tissue-specific factors. A gene is expressed only when the appropriate set of factors have bound to the regulatory regions. The tissue-specific factors could be "reused" in the sense that the same factor could produce a different specificity when it is present in combination with different additional transcription factors. This is illustrated in Figure 3 by the utilization of the same factor in petals for mesophyll-specific expression and in pistils for stigma-specific expression. I illustrate only the most simple form of this model, which certainly does not represent the in vivo situation accurately. The model can be expanded to include factors that negatively regulate gene expression and additional molecules that bind to and modify the activity of factors that directly interact with genes.

The combinatorial model of gene regulation in flowers is consistent with the hypothesis of Lifschitz (35) that all cells in a flower are maintained in the "floral state" by a "floral program." In the simplified model the mediator of the floral state would be the hypothetical transcription factor present in all floral organs (Figure 3).

These hypotheses on the regulation of the final stages of floral development can be tested through experiments on the tissue-predominant genes described in previous sections. If the combinatorial model is correct, then we would expect genes that express in different regions of a single floral organ to have several transcription-factor binding sites in common. Because such sites consist of only a small number of bases and often have some degree of variability, direct identification by sequence comparison is probably not possible. The critical regions must first be defined by modification or deletion and re-introduction into plants. As described above, these experiments are already in progress in a number of labs. Careful comparison of regions found to be important in regulation will allow for determination of the similarities and differences between the genes expressed in different compartments.

As more is learned from studies of flower-predominant genes it will be important to integrate this information with results gained from other approaches. For example, the *AG* gene of *Arabidopsis* and the *Def* gene of *A. majus* have recently been isolated (54, 66). Mutations in these genes lead to homeotic alterations in the flowers of these plants (8, 54). On the basis of homology to factors that have been characterized in other systems, the products of these genes appear to be transcription factors (54, 66). The *AG*

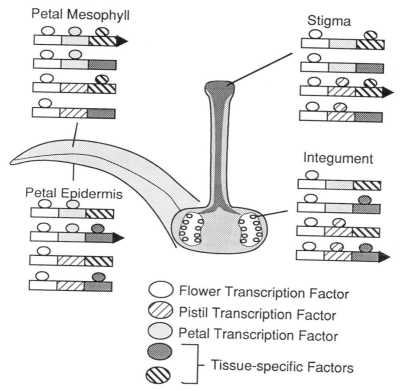

Figure 3 Combinatorial model of spatial regulation of gene transcription. In this simplified model the pattern of flower-predominant gene expression is regulated by factors that directly correlate with the structural organization of the flower. The bars represent the 5'-flanking regions of four genes that are expressed only in specific tissues of the flower. The different regions of the bars represent sequence motifs to which the similarly shaded factors can bind. Genes that are expressed predominantly in the flower contain a binding site for a "floral transcription factor" that is only produced in floral cells. Similarly, genes that express predominantly in the petals or pistil contain binding sites for additional petal or pistil factors that are only present in the respective organs. Tissue specificity is provided by the presence of a site for binding of factors that are only present in certain tissues of each organ. A given gene is only expressed when all three factors are present to activate transcription (indicated by the arrowheads in the figure). Note that the same tissue-specific factor can be used in different organs because it is the combination of factors, not any single factor, that determines which genes are expressed. Similarly, the pistil and petal factors could have other functions in other parts of the plant where the floral factor is absent.

gene product is essential for the formation of normal stamens and carpels (8). In situ hybridizations show that this gene is expressed at detectable levels only in the androecium and gynoecium (66), and that the transcript persists throughout stamen and carpel development (E. M. Meyerowitz, personal communication). The properties of *AG*, a putative transcription factor specific to a subset of the organs of flowers, are consistent with its product being one of the organ-specific regulatory factors predicted by the combinatorial model. The *A. majus Def* gene is expressed at varying levels throughout the flower at all stages of development (54) and is, therefore, more difficult to associate with our simplified model. The availability of the clone of this gene (54), and a newly isolated mutable allele (11), should help to clarify the role of *Def*. Experimentation on potential interactions of the *Def* and *AG* gene products with the promoter regions of the flower-predominant genes will be important in determining their roles in the differentiation of floral tissues. For a more detailed discussion of the homeotic genes see the the chapter in this volume by E. Coen.

CONCLUDING COMMENTS

The research outlined in this chapter represents only the initial steps in a quest to understand the formation and function of the differentiated cells and tissues in floral organs. Genes with flower-predominant expression patterns that appear to be transcribed at significant levels only in specific tissues have been isolated from petals, stamens, and carpels. These clones provide a starting point for characterization of the final stages of floral differentiation. In combination with work being done on the initiation of flowering and the determination of the pattern of floral organs, information gained from these and future studies will allow the formulation of progressively more refined models of this critical developmental process.

Literature Cited

1. Almeida, J., Carpenter, R., Robbins, T. P., Martin, C., Coen, E. S. 1989. Genetic interactions underlying flower patterns in *Antirrhinum majus*. *Genes Dev.* 3:1758-67

2. Anderson, M. A., Cornish, E. C., Mau, S.-L., Williams, E. G., Hoggart, R., et al. 1986. Cloning of cDNA for a stylar glycoprotein associated with expression of self-incompatibility in *Nicotiana alata*. *Nature* 321:38-44

3. Appels, R., Bouchard, R. A., Stern, H. 1982. cDNA clones from meiotic-specific poly(A)+ RNA in *Lilium*: Homology with sequences in wheat, rye and maize. *Chromosoma* 85:591-602

4. Benfey, P. N., Chua, N.-H. 1989. Regulated genes in transgenic plants. *Science* 244:174-81

5. Bernier, G. 1988. The control of floral evocation and morphogenesis. *Annu. Rev. Plant Physiol.* 39:175-219

6. Bevan, M., Shufflebottom, D., Edwards, K., Jefferson, R., Schuch, W. 1989. Tissue- and cell-specific activity of a phenylalanine ammonia-lyase promoter in transgenic plants. *EMBO J.* 8:1899-906

7. Biggin, M. D., Tijan, R. 1989. Transcription factors and the control of *Drosophila* development. *Trends Genet.* 5:377-83

8. Bowman, J. L., Smyth, D. R., Meyerowitz, E. M. 1989. Genes directing flower development in *Arabidopsis*. *Plant Cell* 1:37-52

9. Brown, S. M., Crouch, M. L. 1990. Characterization of a gene family abundantly expressed in *Oenothere organensis* pollen that shows sequence similarity to polygalacturonase. *Plant Cell* 2:263-74

10. Budelier, K. A., Smith, A. G., Gasser, C. S. 1990. Regulation of a stylar transmitting tissue-specific gene in wild-

type and transgenic tomato and tobacco. *Mol. Gen. Genet.* 224:183-92

11. Carpenter, R., Coen, E. S. 1990. Floral homeotic mutation produced by transposon-mutagenesis in *Antirrhinum majus. Genes Dev.* 4:1483-93

12. Condit, C. M., McLean, B. G., Meagher, R. B. 1990. Characterization of the expression of the petunia glycine-rich protein-1 gene product. *Plant Physiol.* 93:596-602

13. Cornish, E. C., Pettitt, J. M., Bonig, I., Clarke, A. E. 1987. Developmentally controlled expression of a gene associated with self-incompatibility in *Nicotiana alata. Nature* 326:99-102

14. Davis, M. A., Cohen, D. I., Nielsen, E. A., Steinmetz, M., Paul, W. E., Hood, L. 1984. Cell-type-specific cDNA probes and the murine I region: the localization and orientation of A-d alpha. *Proc. Natl. Acad. Sci. USA* 81:2194-98

15. DellaPenna, D., Alexander, D. C., Bennett, A. B. 1986. Molecular cloning of tomato fruit ripening polygalacturonase: analysis of polygalacturonase mRNA levels during ripening. *Proc. Natl. Acad. Sci. USA* 83:6420-24

16. Duguid, J. R., Dinauer, M. C. 1990. Library subtraction of in vitro cDNA libraries to identify differentially expressed genes in scrapie infection. *Nucleic Acids Res.* 9:2789-92

17. Evans, P. T., Holaway, B. L., Malmberg, R. L. 1988. Biochemical differentiation in the tobacco flower probed with monoclonal antibodies. *Planta* 175:259-69

18. Gasser, C. S., Budelier, K. A., Smith, A. G., Shah, D. M., Fraley, R. T. 1989. Isolation of tissue-specific cDNAs from tomato pistils. *Plant Cell* 1:15-24

19. Gasser, C. S., Smith, A. G., Budelier, K. A., Hinchee, M. A., McCormick, S., et al. 1988. Isolation of differentially expressed genes from tomato flowers. In *Plant Gene Research*, Vol. 5: *Temporal and Spatial Regulation of Plant Genes*, ed. D. P. S. Verma, R. B. Goldberg, pp. 83-96. New York: Springer-Verlag

20. Giovannoni, J. J., DellaPenna, D., Bennett, A. B., Fischer, R. L. 1989. Expression of a chimeric polygalacturonase gene in transgenic *rin* (ripening inhibitor) tomato fruit results in polyuronide degradation but not fruit softening. *Plant Cell* 1:53-63

21. Goldberg, R. B. 1988. Plants: novel developmental processes. *Science* 240:1460-67

22. Green, P. B. 1988. A theory for inflorescence development and flower formation based on morphological and biophysical analysis of *Escheveria. Planta* 175:153-69

23. Grierson, D., Tucker, G. A., Keen, J., Ray, J., Bird, C. R., Schuch, W. 1986. Sequencing and identification of a cDNA clone for tomato polygalacturonase. *Nucleic Acids Res.* 14:8595-603

24. Gruissem, W. 1990. Of fingers, zippers, and boxes. *Plant Cell* 3:827-28

25. Hanson, D. D., Hamilton, D. A., Travis, J. L., Bashe, D. M., Mascarenhas, J. P. 1989. Characterization of a pollen-specific cDNA clone from *Zea mays* and its expression. *Plant Cell* 1:173-79

26. Harborne, J. B. 1986. Nature, distribution and function of plant flavonoids. In *Plant Flavonoids in Biology and Medicine: Biochemical, Pharmacological and Structure-Activity Relationships*, ed. V. Cody, E. Middleton, J. B. Harborne, pp.25-42. New York: Liss

27. Harker, C. L., Ellis, T. H. N., Coen, E. S. 1990. Identification and genetic regulation of the chalcone synthase multigene family in pea. *Plant Cell* 2:185-94

28. Helsop-Harrison, J. 1983. Self-incompatibility: phenomenology and physiology. *Proc. R. Soc. London Ser. B* 218:371-95

29. Herdenberger, F., Evrard, J.-L., Kuntz, M., Tessier, L.-H., Klein, A., et al. 1990. Isolation of flower-specific cDNA clones from sunflower. *Plant Sci.* 69:111-22

30. Kamalay, J. C., Goldberg, R. B. 1980. Regulation of structural gene expression in tobacco. *Cell* 19:935-46

31. Kandasamy, M. K., Dwyer, K. G., Paolillo, D. J., Doney, R. C., Nasrallah, J. B., Nasrallah, M. E. 1990. *Brassica* S-proteins accumulate in the intercellular matrix along the path of pollen tubes in transgenic tobacco pistils. *Plant Cell* 2:39-49

32. Koes, R. E., Spelt, C. E., van den Elzen, P. J. M., Mol, J. N. M. 1989. Cloning and molecular characterization of the chalcone synthase multigene family of *Petunia hybrida. Gene* 81:245-57

33. Koes, R. E., van Blokland, R., Quattrocchio, F., van Tunen, A. J., Mol, J. N. M. 1990. Chalcone synthase promoters in petunia are active in pigmented and unpigmented cell types. *Plant Cell* 2:379-92

33a. Koltunow, A. M., Truettner, J., Cox, K. H., Wallroth, M., Goldberg, R. B. 1990. Different temporal and spatial gene expression patterns occur during anther development. *Plant Cell* 2:1201-24

34. Kotoujansky, A. 1987. Molecular genet-

ics of pathogenesis by soft-rot *Erwinias*. *Annu. Rev. Phytopathol.* 25:405-30

35. Lifschitz, E. 1988. Molecular markers for the floral program. *Flowering Newsl.* 6:16-20

36. Lotan, T., Ori, N., Fluhr, R. 1989. Pathogensis-related proteins are developmentally regulated in tomato flowers. *Plant Cell* 1:881-87

36a. Mariani, C., De Beuckeleer, M., Truettner, J., Leemans, J., Goldberg, R. B. 1990. Induction of male sterility in plants by a chimaeric ribonuclease gene. *Nature* 347:737-41

37. Mascarenhas, J. P. 1990. Gene activity during pollen development. *Annu. Rev. Plant Physiol. Plant Mol. Biol.* 41:317-38

38. McClure, B. A., Haring, V., Ebert, P. R., Anderson, M. A., Simpson, R. J., et al. 1989. Style self-incompatibility gene products of *Nicotiana alata* are ribonucleases. *Nature* 342:955-57

38a. McClure, B. A., Gray, J. E., Anderson, M. A., Clarke, A. E. 1990. Self-incompatibility in *Nicotiana alata* involves degradation of pollen RNA. *Nature* 347:757-80

39. McCormick, S., Smith, A., Gasser, C., Sachs, K., Hinchee, M., et al. 1987. Identification of genes specifically expressed in reproductive organs of tomato. In *Tomato Biotechnology*, ed. D. J. Niven, R. A. Jones, pp. 255-65. New York: Alan R. Liss

40. Meeks-Wagner, D. R., Dennis, E. S., Tran Thanh Van, K., Peacock, W. J. 1989. Tobacco genes expressed during in vitro floral initiation and their expression during normal plant growth. *Plant Cell* 1:25-35

41. Memelink, J., Hoge, J. H. C., Schilperoort, R. A. 1987. Cytokinin stress changes the developmental regulation of several defense-related genes in tobacco. *EMBO J.* 6:3579-83

42. Mitchell, P. J., Tijan, R. 1990. Transcriptional regulation in mammalian cells by sequence-sepcific DNA binding proteins. *Science* 245:371-78

43. Mol, J. N. M., Stuitje, A. R., Gerats, A. G. M., Koes, R. E. 1988. Cloned genes of phenylpropanoid metabolism in plants. *Plant Mol. Biol. Rep.* 6:274-78

44. Moore, H. M., Nasrallah, J. B. 1990. A *Brassica* self-incompatibility gene is expressed in the stylar transmitting tissue of transgenic tobacco. *Plant Cell* 2:29-38

45. Nasrallah, J. B., Kao, T.-H., Goldberg, M. L., Nasrallah, M. E. 1985. A cDNA clone encoding an S-locus-specific gly-coprotein from *Brassica napus*. *Nature* 318:263-67

46. Nasrallah, J. B., Yu, S.-M., Nasrallah, M. E. 1988. Self-incompatibility genes of *Brassica oleracea*: expression, isolation, and structure. *Proc. Natl. Acad. Sci. USA* 85:5551-55

47. Neale, A. D., Wahleithner, J. A., Lund, M., Bonnett, H. T., Kelly, A., et al. 1990. Chitinase, β-1,3-glucanase, osmotin and extensin are expressed in tobacco explants during flower formation. *Plant Cell* 2:673-84

48. Palazzolo, M. J., Hyde, D. R., Vijay-Raghavan, K., Mecklenburg, K., Benzer, S., Meyerowitz, E. 1989. Use of a new strategy to isolate and characterize 436 *Drosophila* cDNA clones corresponding to RNAs detected in adult heads but not in early embryos. *Neuron* 3:527-39

49. Sanders, L. C., Lord, E. M. 1989. Directed movement of latex particles in the gynoecia of three species of flowering plants. *Science* 243:1606-8

50. Schmid, J., Doerner, P. W., Clouse, S. D., Dixon, R. A., Lamb, C. J. 1990. Developmental and environmental regulation of a bean chalcone synthase promoter in transgenic tobacco. *Plant Cell* 2:619-31

51. Serunick, J., Truettner, J., Goldberg, R. B. 1990. The nucleotide sequence of an anther-specific gene. *Nucleic Acids Res.* 18:3403

52. Simoens, C. R., Peleman, J., Valvekens, D., Van Montagu, M., Inzé, D. 1988. Isolation of genes expressed in specific tissues of *Arabidopsis thaliana* by differential screening of a genomic library. *Gene* 67:1-11

53. Smith, A. G., Gasser, C. S., Budelier, K. A., Fraley, R. T. 1990. Identification and characterization of stamen- and tapetum-specific genes from tomato. *Mol. Gen. Genet.* 222:9-16

54. Sommer, H., Beltrán, J.-P., Huijner, P., Pape, H., Lönnig, W.-E., et al. 1990. *Deficiens*, a homeotic gene involved in the control of flower morphogenesis in *Antirrhinum majus*: the protein shows homology to transcription factors. *EMBO J.* 9:605-13

55. Stinson, J. R., Eisenberg, A. J., Willing, R. P., Pe, M. E., Hanson, D. D., Mascarenhas, J. P. 1987. Genes expressed in the male gametophyte of flowering plants and their isolation. *Plant Physiol.* 83:442-47

56. Timblin, C., Battey, J., Kuehl, W. M. 1990. Application for PCR technology to subtractive cDNA cloning: identification of genes expressed specifically in

murine plasmacytoma cells. *Nucleic Acids Res.* 6:1587-93

57. Twell, D., Klein, T. M., Fromm, M. E., McCormick, S. 1989. Transient expression of chimeric genes delivered into pollen by microprojectile bombardment. *Plant Physiol.* 91:1270-74

58. Twell, D., Wing, R., Yamaguchi, J., McCormick, S. 1989. Isolation and expression of an anther-specific gene from tomato. *Mol. Gen. Genet.* 217:240-45

59. Twell, D., Yamaguchi, J., McCormick, S. 1990. Pollen-specific gene expression in transgenic plants: coordinate regulation of two different tomato gene promoters during microsporogenesis. *Development* 109:705-13

60. Ursin, V. M., Yamaguchi, J., McCormick, S. 1989. Gametophytic and sporophytic expression of anther-specific genes in developing tomato anthers. *Plant Cell* 1:727-36

61. van der Meer, I. M., Spelt, C. E., Mol, J. N. M., Stuitje, A. R. 1990. Promoter analysis of the chalcone synthase (chsA) gene of *Petunia hybrida*: a 67 bp promoter region directs flower-specific expression. *Plant Mol. Biol.* 15:95-109

62. van Tunen, A. J., Hartman, S. A., Mur, L. A., Mol, J. N. M. 1989. Regulation of chalcone flavone isomerase (CHI) gene expression in *Petunia hybrida*: the use of alternative promters in corolla, anthers and pollen. *Plant Mol. Biol.* 12:539-51

63. van Tunen, A. J., Koes, R. E., Spelt, C. E., van der Krol, A. R., Stuitje, A. R., Mol, J. N. M. 1988. Cloning of two chalcone flavone isomerase genes from *Petunia hybrida*: coordinate, light-regulated and differential expression of flavonoid genes. *EMBO J.* 7:1257-63

64. Wienand, U., Sommer, H., Schwarz, Z., Shepard, N., Saedler, H., et al. 1982. A general method to identify plant structural genes among genomic DNA clones using tansposable element induced mutations. *Mol. Gen. Genet.* 187:195-201

65. Wing, R. A., Yamaguchi, J., Larabell, S. K., Ursin, V. M., McCormick, S. 1989. Molecular and genetic characterization of two pollen-expressed genes that have sequence similarity to pectate lyases of the plant pathogen *Erwinia. Plant Mol. Biol.* 14:17-28

66. Yanofsky, M. F., Ma, H., Bowman, J. L., Drews, G. N., Feldmann, K. A., Meyerowitz, E. M. 1990. The protein encoded by the *Arabidopsis* homeotic gene *agamous* resembles transription factors. *Nature* 346:35-39

67. Young, H. J., Stanton, M. L. 1990. Influence of environmental quality on pollen competitive ability in wild radish. *Science* 248:1631-33

Annu. Rev. Plant Physiol. Mol. Bio. 1991. 42:651–74

OLIGOSACCHARIDE SIGNALS IN PLANTS: A CURRENT ASSESSMENT*

Clarence A. Ryan and Edward E. Farmer

Institute of Biological Chemistry, Washington State University, Pullman, Washington 99194-6340

KEY WORDS: developmental signals, defense gene signals, signal transduction, oligouronide signaling, oligosaccharide elicitors

CONTENTS

*Abbreviations: ACC, 1-aminocyclopropane-1-carboxylic acid; AVG, (aminoethoxyvinyl) glycine; CMC, carboxymethyl cellulose; DP, degree of polymerization; FITC, fluorescein isothiocyanate; FPLC, fast protein liquid chromatography; PG, polygalacturonase; PGA, polygalacturonic acid; PGIP, polygalacturonase inhibitor protein; PL, pectate lyase; PME, pectin methyl esterase; PR; pathogenesis related; RG, rhamnogalacturonan; TLC, thin cell layer.

1040-2519/91/0601-0651$02.00

INTRODUCTION

Understanding the biochemical basis of the signaling mechanisms that regulate the expression of plant genes is currently a major objective in plant biology. Vigorous research activity seeks to elucidate the mechanisms of action of traditional plant hormones and growth regulators, and to identify other signaling molecules that initiate gene expression during plant stress and plant development. Among the recent candidates for consideration as signaling molecules are defined oligosaccharides (32, 92), polypeptides (95; G. Pearce, D. Strydom, S. Johnson, C. A. Ryan, unpublished), and smaller molecules such as arachidonic acid (14), abscisic acid (85), jasmonic acid (4, 45), Ca^{2+} (86, 115), and inositol phosphates (39, 81).

Among these molecules, oligosaccharides have been most extensively studied. They appear to have multiple roles in signaling transduction systems that regulate both plant defensive and developmental processes. In the 1970s, Albersheim and colleagues demonstrated (6, 32) that purified cell walls from certain pathogenic fungi were potent elicitors in plant tissues of the synthesis of the defensive antibiotics called phytoalexins. Elicitor activities of fungal cell walls were initially associated with the β-glucan polysaccharide fractions (6). More recently the inducing activities derived from the fungal cell wall β-glucans, generated by hydrolytic enzymes of plants, have been shown to be a property of a hepta-β-glucosyl fragment (98). This implied that signaling pathways for defensive genes in plants responded to pathogen attacks by sensing chemical signals—i.e. β-glucan fragments, derived from the pathogen cell walls during the infection process (6, 32).

In the late 1970s and early 1980s other polysaccharides were found to activate plant defensive responses. Oligogalacturonide fragments from the plant cell wall, generated by the action of PG and PL, were found to activate both phytoalexin and proteinase inhibitor synthesis (13, 32, 53, 80). Subsequently, many other plant defensive responses were found to be activated by oligouronides (92). Fragments of another fungal cell wall polysaccharide, chitin (and its deacetylated product chitosan), released during pathogen attacks were potent inducers of plant defense responses (52).

During the early stages of pathogenesis, pectin-degrading enzymes secreted by the pathogens depolymerize the plant cell wall and solubilize oligouronides that can act as signals that initiate localized defensive responses (32, 92). These oligomers appear to be among the earliest signals that can activate defense responses in plants. Among the many responses induced in various plants by oligouronides are the syntheses of β-glucanases and chitinases that can fragment fungal cell walls. These oligomers can also act as potent signaling molecules, thereby amplifying signals for localized defenses. In this scenario, at least three structurally different oligosaccharides can play major signaling roles.

During the 1980s, a developmental role for oligosaccharides was proposed (50, 108). Since then, oligosaccharides have been implicated as signals that regulate both physiological and developmental responses in plants. These responses include morphogenesis in tobacco thin-layer explants (108), ethylene production during fruit ripening (8, 17), and the inhibition of auxin-induced elongation in pea stem segments (114). Xyloglucan fragments (114) and PGA fragments (16) appear to have a role in regulating the latter response and are discussed in detail below.

Despite the increasing evidence that oligosaccharides, called "oligosaccharins" (2), can regulate both defensive and developmental processes in plants, acceptance of the signaling role of oligosaccharides has been cautious. In this chapter we summarize research progress over the past three years in selected systems where oligosaccharides have been found to act as signals or regulators for plant processes. Reports that oligosaccharides regulate plant defensive systems and plant development have steadily proliferated. Reports that fungal cell wall oligosaccharides and pectic fragments elicit defensive responses in plants are becoming commonplace, further substantiating that these poly- and oligosaccharides are active inducers of defensive genes in plants. Structures of the active β-glucan elicitors have been reported, and membrane receptors for these active oligomers have been identified. Newly reported effects of oligosaccharides on membrane-associated processes such as protein phosphorylation, ion fluxes, and action potentials have strengthened the hypothesis that fragments of fungal and plant cell wall oligosaccharides can play important regulatory roles in plants.

Here we review recent data that provide new insights into the structural and/or functional aspects of oligosaccharide signaling in plant defense, growth, and development. Not meant to be comprehensive, our review focuses on assessment of the role of oligosaccharides as signaling molecules.

OLIGOSACCHARIDE SIGNALS FOR DEFENSIVE RESPONSES

Pectic Fragments

Three recent studies described below have provided novel evidence supporting the roles of pectic fragments as regulators of defensive responses in plants. In suspension cultures of castor beans, the structures of oligouronides required to elicit the biosynthesis of lignin have been determined (19); at the whole-plant level, the induction of β-glucanase and chitinase in tobacco by pectic fragments has been demonstrated (18); and the cellular locations of pectic fragments in plant tissues during fungal infection have provided further evidence that these fragments can act as signaling molecules in response to pathogen attacks (11).

Lignin synthesis was detected in suspension-cultured castor bean cells within 3 hr following addition of pectic fragments (19); the maximum rates of lignin synthesis occurred between 4 and 10 hr. Fractions from a PG digest of pectin, comprising oligouronides of about average DP (7–13), contained active elicitors. In unelicited suspension cultures six extracellular peroxidases (two anionic and four cationic) were detectable; but in response to the addition of pectin fragments of average DP = 7 to the cultures, four activities were altered (one anionic and three cationic) and three new enzymes appeared. The synthesis of lignin in response to the oligouronides was as fast or faster than the elicitation of phytoalexins by fungal elicitors in other systems. Neither pectic fragments nor the hepta-glucoside elicitor from *Phytophthora megasperma* f. sp. *glycinea* induced lignification in wheat leaves. Two β-glucan elicitors—pachyman, a water-insoluble β-1,3-glucan from the fungus *Poria cocos,* and a mycelial wall fraction from *P. megasperma* f. sp. *glycinea*—did elicit lignin deposition in wheat leaves (9), indicating that a signaling system for the synthesis of lignin in response to β-glucans was present in wheat; but the oligosaccharide signals may differ from those that signal the defensive genes of dicots.

Secretion of pectinolytic enzymes by the bacterial pathogen *Erwinia carotovora* during infection of tobacco resulted in the induction of chitinase and β-glucanase activity by the plants (18). Excised young tobacco plants, supplied with citrus pectic polysaccharides through their cut petioles, exhibited a 12-fold increase in chitinase activity within 48 hr. The maximum rate of induction in leaves occurred at about 12–24 hr—a time course similar to that of the induction of proteinase inhibitor synthesis by pectic fragments (112). Chitinase and β-glucanase levels increased in response to pathogen and virus infections, both locally and systemically; the chitinase activity had previously been shown to increase in melon plants supplied with chitin fragments (87, 88). The similarities in the induction of chitinases, β-glucanases, and proteinase inhibitors by pectic fragments suggest that the signaling systems regulating the various genes in tobacco, both locally and systemically, may be related or the same. A subset of PR proteins (which includes chitinases and β-glucanases) in tobacco can also be induced by xylanase (see below). Thus tobacco may be a good species in which to study the differential regulation of PR protein subsets by various cell wall-degrading hydrolases and the fragments they release.

The secretion of pectinolytic enzymes and the subsequent generation of pectic fragments from the plant cell wall matrix have been investigated during fungal and bacterial pathogenesis and insect attack. The fungal pathogen *Fusarium oxysporum* f. sp. *radicis-lycopersici* was applied to roots of tomato plants, and the localization of pectic cell wall components was studied utilizing a polygalacturonate-binding lectin from *Aplysia delipans* complexed

to colloidal gold (11). Electron microscopy revealed the release of oligouronide fragments at strategic sites during pathogenesis. In tomato root tissues
infected with *F. oxysporum* f. sp. *radicis-lycopersici*, cell wall modification
was seen adjacent to fungal penetration channels. Release of pectic fragments
was observed, with some fragments entering the altered phloem cells. Heavy
labeling of the fragments in intercellular spaces and wall appositions at the
junction of fungal and plant cells supported their possible role as signals for
activation of plant cells defensive systems. The *Aplysia* polygalacturonate-
binding lectin (12) promises to be a useful tool for further investigation of the
structure, function, and metabolism of pectic substances in plants. Monoclonal antibodies that bind specific Ca^{2+} polygalacturonate complexes have recently been developed (70) and should also provide probes for the study of
pectic fragments as signaling molecules in plant tissues.

Salivary fluids from some species of greenbug (aphid) contain PME and
pectolytic enzymes that can release pectic fragments into the phloem and
sieve elements of host plants (21, 72). Sap-feeding insects cause a variety of
systemic responses; these may be associated with oligosaccharides released
by the interactions of salivary enzymes with plant cell walls and transported in
the phloem. The plant-aphid system appears to have considerable promise for
investigating the biochemical events in hosts that regulate defensive responses
triggered by phloem-feeding insects. Lectins and antibodies specific for pectic
oligosaccharides might be useful in investigating such interactions.

β-Glucans and Viral Resistance

A partially hydrolyzed, highly branched β-glucan elicitor from *P. megasperma* induced viral resistance in tissues of several tobacco species when
sprayed, injected, or inoculated before, during, or within 8 hr after inoculation of tobacco leaves with virus particles (65). The glucan was effective
against several virus taxa, including (TMV), alfalfa mosaic virus, and tomato
black ring virus. At 0.1–10 μg per ml, the glucan reduced virus accumulation
50–100%. The glucan apparently affected the host plant itself, resulting in the
inhibition of both local lesions and viral multiplication. The resistance mechanism is not known and does not appear to involve known defenses against
pathogens, such as phytoalexin, lignin, or callose synthesis.

Polygalacturonase Inhibitors as Signal Regulators

Proteins that are inhibitors of PG (polygalacturonase-inhibiting proteins, or
PGIPs) are found in the apoplasts of many plants (1, 23). After synthesis these
proteins, which are apparently targeted to their extracellular locations, may
have a role in regulating plant defense responses (1, 23). The PGIPs appear to
have broad specificities against fungal PG. Cervone et al (25) have shown that

a purified *Phaseolus vulgaris* PGIP inhibited endoPG from several fungal origins *(Aspergillus, Fusarium, Colletotrichum)* but did not inhibit a bacterial *(Erwinia carotovara)* endoPG. Tomato endoPG was not inhibited nor were exoPGs from sorghum or oak pollen, but an exoPG from corn pollen was inhibited. PGIP did not inhibit endoPL from several fungi.

When an excess of pure *Phaseolus* PGIP (apparent M_r of 41,000) was added to a digest of PGA and *A. niger* endoPG, the digestion was dramatically slowed but not totally inhibited (Figure 1) (26). In the absence of PGIP, PGA was digested to small elicitor oligomers below DP = 9 (Figure 1) that are inactive in eliciting most localized defensive responses (93). In the presence of PGIP, oligomers above DP = 9 in the digest persisted for 24 hr. Only after 48 hr was oligomer size reduced to below DP = 9. The decreased

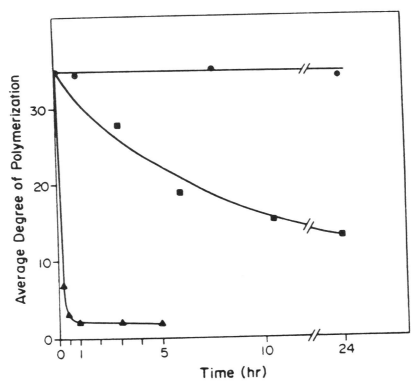

Figure 1 Depolymerization of polygalacturonic acid at pH 5 by fungal endopolygalacturonase in the presence or absence of *Phaseolus vulgaris* polygalacturonase inhibitor protein, PGIP. Polygalacturonic acid (10 mg ml^{-1}) alone, -●-; polygalacturonic acid (10 mg ml^{-1}) containing 10 pmol ml^{-1} endopolygalacturonase, -▲-; polygalacturonic acid (10 mg ml^{-1}) containing 10 pmol ml^{-1} endopolygalacturonase and 250 pmol ml^{-1} PGIP, -■-. From Cervone et al (26) with permission from the publisher.

rate of depolymerization of cell wall PGA clearly extended the effective half-lives of the elicitor oligouronides. Resistant plants and susceptible plants exhibited similar levels of the inhibitor (66), and therefore PGIPs are not considered to be determinants of resistance or susceptibility. The inhibitors appear to have a general role in reducing PG activities of attacking pathogens, a role that enhances the signaling capabilities of plants by increasing the period during which larger active oligouronide signals are present at the infection sites (26). The isolation of the cDNAs and/or genes that code for PGIPs may allow direct manipulation of the levels of PGIP expression in tissues, thereby enhancing the defense gene signaling capabilities of plants. It is also possible that PGIPs play roles in pectic fragment signaling phenomena other than signaling defense responses. Gene manipulation should reveal such responses.

Hypersensitivity and Necrosis

Hypersensitivity and necrosis are early events that occur in response to invading microorganisms. The hypersensitive response is defined as a rapid, localized necrosis—part of the mechanism leading to resistance (106). Mycelia of *Asperigillus niger* caused a necrotic reaction when applied to the surface of cowpea *(Vigna)* pods; this reaction was simulated by *A. niger* PG within 24 hr of application to cowpea pods (22, 24). If the enzyme was inactivated by either heat or specific antibodies, no induction of necrosis resulted. Di- and monogalacturonic acids were inactive as inducers of necrosis, but a mixture of oligouronides with DP $\geq$ 4 products of PG hydrolysis of polygalacturonic acid, produced necrotic lesions similar to those produced by the active enzyme. Oligosaccharides released from *Vigna* cell walls by purified *A. niger* PG also induced necrosis (22). Polygalacturonase from *Aspergillus,* which is not a plant pathogen, was much more active in releasing oligouronides from plant cell walls than the PG of pathogenic fungi and may therefore produce signal-active oligouronides rapidly.

Further evidence that killing factors may be associated with cell wall hydrolysis (54) was obtained with a mixture of commercial carbohydrases that killed suspension cultured wheat cells and tobacco cells. The enzymes acting on these cells apparently produced cell wall fragments that possessed the killing activities. Pectin lyases and xylanase, but not cellulase, killed rice cells, but no killing activities were found in filtrates produced by the action of these enzymes on the rice cells (61).

Recently the production of plant cell–killing activities by the rice blast *Magnaporthe grisea* has been investigated in detail (20, 34). An assay for plant cell death based on ^{14}C-leucine incorporation into plant cells was utilized. *M. grisea* produced heat-labile activities that killed suspension cultured rice, maize, and sycamore cells (34). When maize cell walls were treated with the *M. grisea* heat-labile killing activity, heat-stable killing activities were generated, indicating that the *M. grisea* factor was an enzyme(s) that fragmented the cell walls to produce the heat-stable killing activity (34).

The heat-labile killing activities from *M. grisea* were further analyzed by liquid chromatography and FPLC (20). PL, PME, and xylanase were all purified to homogeneity from *M. grisea;* but alone, or in combination, they did not kill plant cells. This complex situation now appears to be resolved. A recent report showed that *M. grisea* secreted an endoxylanase and an arabinosidase that act in concert to release arabinoxylans, which appear to be the killing factors derived from plant cell walls (63). Furthermore the arabinoxylan killing factors are ferulated, and removal of the esterified ferulic acids destroys more than 95% of the killing activities (63).

It was suggested (20) that the two enzymes, endoxylanase and arabinosidase, working in concert to produce killing oligosaccharides, may enhance the localization of the necrotic response to critical regions of infections, while three other types of oligosaccharides—oligouronides, β-glucans, or chitosans—can act as inducers of defensive responses.

Xylanase and Signal Production

Xylose-containing oligosaccharides may also be involved in other signaling events during pathogenesis. A cell wall–degrading mixture of enzymes of fungal origin, called cellulysin, induced ethylene production in tobacco leaf discs by inducing synthesis of ACC (3). No cellulase, pectinase, or PG activity was found in the ethylene-inducing proteins isolated from the enzyme mixture, but an active endoxylanase activity was detected (47). The endoxylanase activity was inseparable from ethylene-inducing activity, but all components of cellulysin having xylanase activity were not ethylene inducers, nor were the endoxylanase activities that were found in other cell wall–degrading mixtures of enzymes of fungal origin always associated with ethylene-inducing activities (47). The possibility that two enzymes act in concert to induce ethylene production has not been fully investigated.

Endoxylanase has also been implicated in the induction of PR proteins in tobacco leaves. Injection of purified *Trichoderma viride* endoxylanase into tobacco leaves induced the accumulation of a defined subset of PR proteins, but only near the areas of application (71). The response to xylanase was shown to be independent of light, unlike the induction of PR proteins by tobacco mosaic virus and α-aminobutyric acid (an inducer of PR proteins). It is not clear how the induction of PR proteins by endoxylanase relates to the induction of PR proteins (β-glucanase and chitinase) by oligouronide fragments, described above (18). The identification of the specific PR proteins induced by either endoxylanase or by oligouronides may resolve whether these two elicitors activate the synthesis of different subsets of these proteins.

An extracellular protein from *Phytophthora parasitica* that was found to be an elicitor of the phytoalexin capsidiol in tobacco callus cultures copurified with an endoxylanse activity (41). Although the enzyme activity was inseparable from elicitor activity, the possible role of the endoxylanase has not been established in this system.

Although xylanase activity was associated with the elicitation of phytoalexins, ethylene synthesis, and PR protein synthesis, the molecular basis for elicitation of these systems is still unknown. The effects are similar to those produced by pectic hydrolases and lyases, indicating that diverse signaling roles may exist for different plant cell wall degrading enzymes that depend upon the manner by which the cell wall is fragmented. It is possible that complex xyloglucan polymers act directly as signals that elicit ethylene, phytoalexins, and PR proteins in tobacco, although endoxylanases may also act indirectly by releasing pectic fragments closely associated with the xyloglucans to enhance signaling pathways. More research is needed to elucidate the biochemical basis for the actions of the various carbohydrases and their products in signaling defensive responses in plants.

OLIGOSACCHARIDES IN PLANT GROWTH AND DEVELOPMENT

Morphogenesis

The first reports of the effects of oligosaccharides as regulators of plant growth and development indicated that plant cell wall fragments, produced by partial acid hydrolysis of cell walls, influenced flowering and vegetative growth in *Lemna gibba* (50). Plant cell wall fragments were subsequently found to cause changes in the morphological characteristics of tobacco thin-layer cultured cells. Depending upon the composition of the medium, the fragments caused the cells to form vegetative shoots rather than flowers or vice versa (108). Attempts to reproduce these results were not fully successful, but a modified tobacco thin-cell layer (TCL) bioassay was developed in which pectic fragments influenced root and flower formation (38, 79). The fragments either inhibited root formation, enhanced rooting in a polar manner, or enhanced polar tissue enlargement leading to flower development, depending upon the auxin and kinetin levels in the media (38, 79). The effects were reproducible with pectic fragments obtained from either sycamore cell suspension cultures or cultured tobacco cells (38). The effects of the fragments cannot be explained by a mechanism in which they increase or decrease auxin or cytokinin concentrations, because changing these concentrations in the culture medium does not bring about the same effects as the pectic fragments (38).

The activities of the fragments were destroyed by incubating with endoPG but were stable under mild acid or base treatment (38). The cumulative evidence, while somewhat variable, confirmed that pectic fragments regulated organogenesis in the explants.

According to the TCL bioassay (V. Marfa and P. Albersheim, in preparation), the oligouronides present in the partially hydrolyzed plant cell wall preparations that caused flowers to form on tobacco explant possessed DPs greater than six. The larger, complex cell wall polysaccharides RGI and RGII were not active in the assay. When a series of purified oligogalacturonides generated by mild acid hydrolysis of citrus pectin were used, uronides below DP = 10 were inactive in the bioassays, whereas oligomers of DP = 10 and DP = 11 were moderately active, and those of DP = 12–14 were maximally active. The size range of active oligomers is strikingly similar to the range of oligouronides that elicit plant defensive responses against pathogens (93). Structure-activity relationships of oligouronides in plant defensive responses are further addressed below.

More information about the effects of oligouronide fragments on plant developmental programs is needed, including data on the origin and regulation of the developmental cues that might generate fragments in specific tissues or cells, the nature and location of enzymes involved in releasing such fragments, and the localization and/or transport of the fragments to target cells or tissues. If oligosaccharides are developmental cues, then major research efforts will be required to investigate the processes they regulate.

The hypothesis that oligosaccharides act as signaling molecules for developmental processes in plants gains support from a recent report of an oligosaccharide of prokaryotic origin that regulates morphogenic patterns in plant roots (67). The nitrogen-fixing symbiont *Rhizobium meliloti* produces an oligosaccharide, termed NodRm-1, that has recently been characterized as a tetra-β-1,4-D-glucosamine modified by both sulfation and a C-16 fatty acid (67). At low nanomolar concentrations, this oligosaccharide causes alfalfa cortical cells to dedifferentiate and initiate nodulation (leading to root hair deformation) in the absence of the bacterium. The molecule is host specific to alfalfa *(Medicago sativa)* and does not cause root hair deformation in vetch *(Vicia sativa* subsp. *nigra)* (67). *Rhizobium* strains lacking the ability to sulfate the glucosamine oligosaccharide produce a signal called NodRm-2 that causes root hair deformation in vetch but not in alfalfa (89). The action of oligosaccharide signals in nodulation may be mediated by root lectins (67).

Auxin Regulation

Two mechanisms have been suggested for auxin induction of plant cell elongation (57, 109), one resulting from H^+ secretion (62, 104), perhaps associated with the activation of hydrolases already present in the cell walls

(40, 77), and the second by activating de novo synthesis of cellulases (110) that loosen cell wall structures. Structural components of the xyloglucan polysaccharides of the primary cell walls can affect (regulate) auxin-stimulated growth in pea stem segments (114). A xyloglucan nonasaccharide (called XG9, Figure 2) inhibited auxin-induced elongation of pea stems, where smaller (XG7 and XG8) and larger (XG10) xyloglucans were totally inactive (76). The closely related XG7 lacked the terminal Fuc-Gal- groups and XG8 lacked Fuc-. XG10 possessed an extra galactose at the reducing terminus. Two smaller oligouronides, Fuc-Gal-Glc (obtained from human milk) and Fuc-Gal-Xyl-Glc-Glc, both possessed anti-auxin activity, with the trisaccharide nearly as active as XG9 (75). This suggested that a highly discriminatory receptor may be present that, in part, recognized the fucose portion of the molecule. High concentrations of XG9 (10^{-5} to 10^{-6} M) had diminishing anti-auxin activities (114), but higher concentrations of the pentasaccharides did not (76).

A possible mechanism for xyloglucan regulation of auxin activity may involve the differential regulation of cellulase activity during growth. Pea cellulase (40) and bean cellulase (77) were recently shown in vitro to be activated up to 10-fold by XG9 at 1–200 nM concentrations. Activation of the enzyme was only observed when xyloglucan polysaccharide was the substrate for the enzymes. No enzyme activation was detected when the soluble substrate CMC was employed for assays. The activity of XG9 may result from its ability to activate cellulases, thereby facilitating rapid growth of cells

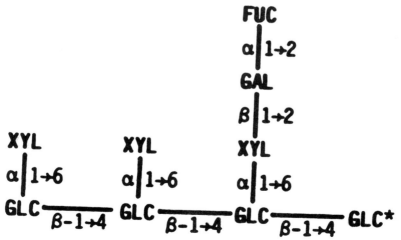

Figure 2   The structure of XG9, a naturally occurring anti-auxin. FUC = L-fucose, GAL = D-galactose, GLC = D-glucose, XYL = D-xylose. All residues are pyranose.

during cell elongation (77). Other xyloglucan oligomers were also activators of cellulase, suggesting that multiple products of degradation probably contribute to these effects.

XG7 and XG9 alternate along the length of the xyloglucan polysaccharide (57, 58), and both oligosaccharides can be released upon degradation by cellulase. Thus, a feedback mechanism might involve XG9, released by cellulase, as an activator of a latent cellulase activity specific for the midchain cleavage of xyloglucans (77).

In experiments designed to determine if XG9 was present in spinach cell cultures in vivo, and therefore a viable candidate as a regulator, the extracellular XG9 xyloglucan oligosaccharide was found to be present at a concentration of 430 nM (46). It is not known if XG9 is synthesized de novo, as an intermediate in the synthesis of cell walls, or as a degradation product of xyloglucan polysaccharide.

When radiolabeled XG9 (or XG9 acetylated at the galactose residue) was added to suspension cell cultures at 10 nM, little of the compound entered the cells and became degraded (10). Three fourths of the XG9 became bound to a soluble extracellular polymer, probably xyloglucan, via transglycosylation, indicating that the levels of XG9 in vivo may be modulated by this mechanism, thus helping to regulate the physiological process of cell elongation.

Ethylene Synthesis and Fruit Ripening

Ethylene production is essential to the regulation of tomato fruit ripening, activating the expression of specific nuclear genes at transcriptional and posttranscriptional levels (68). Recently, both PG and pectic fragments have been implicated as possible regulators of ethylene synthesis in ripening citrus fruit (7), tomato fruit (8, 17), and cell suspension cultures (107).

Pure tomato fruit cell wall polygalacturonase isoenzymes PG I and II, when infiltrated into preclimacteric green "cherry" tomato fruit together with PME, induced a several-hundred-fold increase in ethylene production over control fruit treated with a boiled enzyme (8). Both of the PGase enzymes were effective inducers of ethylene, PG I being more effective than PG II (8). Ethylene production increased for 17–20 hr following infiltration of the isoenzymes and decreased thereafter. All treated fruit turned red within one week of treatment, whereas control fruit remained green. PG II + PME also induced ethylene production and accelerated chlorophyll breakdown in two nonripening lines, Cornell III and *rin*, but fruit from another nonripening line, *nor*, did not respond (8). When infiltrated into unripe fruit, pectin fragments isolated from ripened tomatoes accelerated ethylene production and ripening. Mixtures of pectic fragments with DPs below 8 were inactive, but fragments above DP = 8 strongly induced both ethylene and CO_2 production. Products from the action of tomato PG on unripe pericarp did not induce ethylene or ripening, but the PG could rapidly have reduced the sizes of fragments to below the active range.

A nonripening tomato mutant, *rin*, exhibited reduced PG gene expression, polyuronide degradation, ethylene production, and lycopene accumulation during ripening (48, 105). Transformation of *rin* tomato plants with a tomato PG structural gene, driven by a tomato fruit promoter (the E8 promoter) that is fairly highly expressed in *rin* fruit, resulted in the expression of the PGase. Polyuronide degradation was observed in these *rin* fruit, but ethylene induction was not, suggesting that the degradation of cell wall polyuronides by PG was not sufficient to signal elevated rates of ethylene biosynthesis in tomato fruit (48). These results contrast with the results obtained by introducing pectic enzymes or pectic fragments into *rin* fruit by vacuum infiltration, where ethylene production is enhanced (8). It has been suggested that the location of the PG or pectic fragments during ripening may differ from the location when vacuum infiltrated (49). Although the roles of PG and pectic fragments in signaling ethylene induction during fruit ripening appear to be complex and are not yet understood, the recent evidence indicates that PG and pectic fragments can affect the fruit ripening processes. PG levels have been reduced in transgenic tomato plants by expressing PG antisense mRNA (99). These plants should be useful in the investigation of the relationships between PG and ethylene production.

In suspension cultures, an association among pectic fragments, ethylene production, and morphogenesis has been noted. In this case a galacturonic acid polymer of 18 kDa (100 α-1,4-galacturonic acid residues) was released from soybean suspension cultures when treated with colchicine to effect cell expansion and separation (59). This polysaccharide, when recovered and added back to suspension cultures, immediately enhanced cell expansion and separation. This property of the polymer was lost when digested with pectinase. AVG, a specific inhibitor of ethylene, severely inhibited cell expansion and separation induced by colchicine but did not inhibit the production of the polysaccharide by the suspension cultures. Thus the galactouronides may contribute to the morphogenic effects by regulating ethylene production.

SIGNAL TRANSDUCTION

Fungal Glucan Receptors

Considerable progress has been made in identifying high-affinity receptors for fungal β-glucan elicitors. Two laboratories have demonstrated that the heptaglucoside and larger fungal glucan elicitors strongly bind to plant membranes in a manner that suggests a receptor-mediated interaction.

Evidence that soybean membranes may possess high-affinity binding sites for 1,3-1,6-β-glucan elicitor fractions derived from the walls of the fungal pathogen *Phytophthora megasperma* f. sp. *glycinea* (94) was first reported by Ebel et al (29). The binding of glucans to soybean microsomal membranes was saturable and reversible, and it exhibited high affinity for several poly- and oligoglucans of fungal origin, with a preference for branched glucans of DP above 12 (29). The β-glucan elicitor occurred in oligomers of a range of sizes from a mild acid hydrolysate of *P. megasperma* cell wall polysaccharides. The glucan fraction likely contained as part of its structure the heptaglucoside elicitor derived from *P. megasperma* cell walls [isolated previously by Sharp et al (97, 98)], as well as other β-glucan fragments. Using ^{125}I-tyramine-glucan conjugates, a K_d of 37 nM was determined (29). Detergent-protein micelles of the soybean membranes prepared with three different detergents retained β-glucan-binding activities, with K_ds of 10–30 nM (31). The micelles eluted in sizing gels with M_r values of about 300,000–380,000.

Soybean membranes containing the high-affinity sites were assayed for binding with a series of β-glucan fragments having DPs from 5–22. Binding of individually purified glucans was assayed in competition with a radioactively labeled glucan of DP = 18 (30). As the DPs increased above 5, the competition for the membranes also increased, to a maximum DP of about 20. The heptaglucoside elicitor (97, 98) was obtained by chemical synthesis and was analyzed for binding characteristics in the same assay. The binding of this glucan was the highest of any assayed. A radioactively labeled heptaglucoside bound the membranes with a K_d of about 3 nM. A close correlation was found between elicitor activity of the various glucan preparations and membrane binding activity. The highest activities resided in the heptaglucoside, which is also the most active oligosaccharide elicitor of phytoalexins known (27, 32).

In a parallel and more detailed study Hahn et al (28) demonstrated that a tyramine conjugate radiolabeled with ^{125}I retained elicitor activity and exhibited saturable, reversible, single-site binding with an apparent K_d of 0.75 nM. This concentration is comparable to that required for elicitation of the phytoalexin glyceollin in soybean cotyledons and is a little lower than the K_d found by Cosio et al of 3 nM (30). Nonspecific binding accounted for less than 5% of the binding at saturation, and the association-dissociation was relatively rapid (28). The binding component was also found in membranes of roots, hypocotyls, and leaves.

The binding of ^{125}I-tyramine-hepta-β-glucan-1 to soybean root membranes was competitively inhibited by several small structurally related β-glucans (28). Both tyramine-hepta-β-glucoside-1 and heptaglucoside-1 were highly efficient competitors of the labeled glucoside for binding sites in the root membranes. Hepta-β-glucoside-5, which was 800 times less effective than heptaglucoside-1 as an elicitor of glyceollin in soybean cotyledons, was 760 times less active than heptaglucoside-1 in competing for membrane binding sites. Octa-β-glucoside-9, which has intermediate phytoalexin-inducing activity, was 100 times less active than heptaglucoside-1 in the competition assay. Thus the membrane-binding and phytoalexin-inducing activities of the β-glucosides are closely correlated (28).

The identification of receptors for glucan elicitor in membranes provides a major step toward the isolation and characterization of an oligosaccharide receptor. These and other studies, discussed below, strongly argue that the plasma membranes act as the site(s) of (a) initial binding of elicitor molecules and (b) the initiation of early events in the signal transduction pathways that lead to activation of phytoalexin synthesis and perhaps other defensive or developmental responses.

Many questions remain regarding the specificity of oligosaccharide signals. The heptaglucoside elicitor is inactive in many plants where the plant cell wall–derived carbohydrates are active. Are there numerous receptors to accommodate various other oligosaccharides that activate defensive and developmental responses? Do different receptors network into the same or different signaling circuits? How do the receptors convey information into these circuits? And what are the components that constitute intracellular signaling pathway(s)? The isolation of the β-glucan receptor should provide both an important advance in understanding the early events of the signaling pathways and a focal point for future research.

Pectic Fragments

Oligouronides that induced proteinase inhibitor synthesis in tomato leaves enhanced the in vitro phosphorylation of a 34-kDa protein (pp34) present in purified plasma membranes from tomato and potato leaves (44). The oligouronides strongly enhanced the phosphorylation of pp34 at concentrations near 3 μg ml^{-1} in the assay, but concentrations higher than about 300 μg ml^{-1} suppressed this effect. The phosphorylation of pp34 in response to oligouronides was shown to occur at a threonine residue(s), indicating the involvement of membrane-associated protein kinase of the serine/threonine type. When γ-^{35}S-ATP was used as substrate in place of γ-ATP^{32}, phosphorylation of pp34 was also strongly enhanced by uronides, but background phosphorylation was greatly reduced (43).

A series of oligouronides from 2–20 moieties in length was isolated from enzyme digests of PGA, and individual oligomer sizes were assayed to determine chain-length specificity in enhancing phosphorylation of pp34 (43). Only oligouronides of DP = 14 and above were active. Thus, the phosphorylation exhibited an oligomer specificity that resembled the oligouronide specificity found both with several localized defensive responses in various plant tissues and with developmental or morphological responses in tissue cultures (cf the section above on morphogenesis). Nearly all defensive responses tested have been elicited by oligogalacturonides of DP > 9. Oligogalacturonides having DPs of about 10 or above form intermolecular complexes (called egg-box structures) with Ca^{2+} as a bridging ion (reviewed in 64). Thus the intermolecular uronide complexes may interact with plant cells to initiate localized defensive or developmental responses (42, 43). This hypothesis also provides a testable explanation for the DP requirement of the oligogalacturonide-enhanced in vitro phosphorylation of pp34 (43). At present, however, there is no direct evidence that the phosphorylation of pp34 is an integral part of any oligouronide signaling system in plants.

The interaction of fungal cell wall–derived elicitors and plant cell wall pectic fragments with suspension cultured soybean cells was investigated by light microscopy (60). Using FITC-labeled fungal elicitor from *Verticillium* cell walls, the elicitor was found to be attached to the plant cell surface within 4 hr of addition to the cells. The fluorescein label was then internalized and appeared to concentrate in the cell vacuole. FITC-labeled PGA followed the same course of events as the fungal elicitor, but with most of the labeled elicitor internalized within 2 hr. ^{125}I-labeled PGA was taken into cells at a rate of 10^6 molecules/cell/sec for at least 2 hr. KCN blocked the process, suggesting that it was energy dependent (60). Labeled PGA was recovered from cells unmodified. Cooling cells to 4°C reduced the uptake by 88% of that at 23°C. Nonlabeled PGA reduced uptake of labeled PGA 10-fold, indicating that specific receptors may be involved. If internalization through endocytosis is part of the signal transduction mechanism leading to defense gene activation, it will be important to determine the fate of the oligomers once they are inside the cytoplasm of cells. Specific receptors for PGA have not been identified, but the internalization of PGA by endocytosis would be consistent with the presence of such receptors.

In vivo Phosphorylation

A possible role for protein phosphorylation in signaling the interaction of β-glucan elicitor was investigated with suspension cultured soybean cells (51). A *P. megasperma* β-glucan fraction changed in vivo protein phosphorylation patterns within 5 min after elicitor treatment. Several proteins exhibited either increases or decreases in phosphorylation, and the changes persisted for at least 15 min. Among proteins clearly showing altered in vitro phosphorylation in response to glucan elicitor were proteins of M_r 94, 78, and 69. Of about 35 phosphoproteins detected in vivo, 7 of the proteins labeled in vitro were similar, as judged by two-dimensional electrophoresis. One protein, pp69, exhibited decreased phosphorylation in vivo in response to glucan elicitor, but phosphorylation was enhanced in vitro at a serine residue(s) using different effectors (51). For example, the phosphorylation of this protein was somewhat enhanced by Ca^{2+} [previously shown to be an essential component of the phytoalexin response in soybeans (100)] but was more strongly enhanced by small-molecular-weight components (under 1000 M_r) present in extracts from soybean suspension cultures. This effector resembled 3-phosphoglycerate, but its identity has not yet been established.

Phosphorylation of several proteins in suspension cultures of parsley *(Petroselinum crispum)* were noted upon addition of a fungal protein elicitor of phytoalexins in these tissues (33). The elicitor activity and the phosphorylation-inducing activity were lost when the elicitor was treated with pronase, indicating that oligosaccharides were not the primary basis of the activity. The phosphorylations were reversible when elicitor was removed. Most of the phosphorylations were not observed in heat-shocked, UV-light-treated, or $HgCl_2$-treated cells. A 45-kDa polypeptide was phosphorylated within 1 min of elicitor addition. The rapid phosphorylation of a nuclear protein (26 kDa) was also observed. In all, about 16 polypeptides displayed altered in vivo phosphorylation after elicitor treatment of the parsley cells. Since the elicitor used was not fully characterized, it is not yet clear (*a*) what the relationship might be between the fungal protein phytoalexin elicitor(s) and the β-glucan elicitor(s) and/or (*b*) whether any of the phosphorylated proteins may also be phosphorylated in response to any glucan elicitors.

H_2O_2 Production and Lipid Peroxidation

The production of H_2O_2, and/or superoxide (O_2^-) occurs in healthy plants during lignification, a process that can itself be triggered by fungal attacks or by elicitors, including pectic fragments (see above). The O_2^- generated during the hypersensitive response of potato tuber tissue (36) and protoplasts (37) in response to hyphal wall components from *Phytophthora infestans* may be involved in activating resistance responses.

When added to suspension cultured soybean cells, an elicitor from *Verticillium dahliae* (containing both protein and carbohydrate components) caused the production of H_2O_2 within 5 min (5). H_2O_2 generation resulted in the destruction of compounds in the medium, such as indoleacetic acid and the fluorescent dyes oxonol VI, carbocyanine, and pyramine. H_2O_2 addition to soybean suspension cultures also resulted in phytoalexin (glyceollin) production, suggesting that H_2O_2 can play a role in signaling phytoalexin induction, perhaps by generating other signals.

The generation of H_2O_2 and O_2^- might be expected to have a number of consequences for plant cells, particularly for lipids which are especially vulnerable to peroxidation. Addition of a galactoglucomannan-containing elicitor preparation from *Colletotrichum lindemuthianum* to *Phaseolus* cotyledons and to suspension cultured *Phaseolus* cells produced compounds typical of lipid peroxidation—e.g. malondialdehyde and lipofuscin-like pigments (90). Bean phytoalexins were induced both by the *C. lindemuthianum* elicitor and by a superoxide-generating system (83). Lipid peroxidation and lipoxygenase induction (as well as electrolyte leakage) were also noted in tomato leaf tissue treated with elicitor preparations from *Cladosporium fulvum* (83).

The chemiluminescence of luminal was enhanced when soybean cells were treated with a crude glucan elicitor fraction from *Phytophthora megasperma* (69). This effect was related to superoxide production because it was inhibited when cells were treated in catalase and superoxide dismutase as well as the peroxidase inhibitors cyanide and salicylhydroxamic acid.

The generation of O_2^- and H_2O_2 in elicitor-treated plant cells is similar to processes in animals in response to infection. Foreign agents cause granulocytes in the blood or other fluids to generate superoxide (O_2^-), which is rapidly converted to H_2O_2 and other active oxygen species (55). These reactive oxygen species can cause membrane damage, including lipid peroxidation, and can cause localized responses such as inflammation, and, in some cases, mutational events leading to transformation and carcinogenesis.

Reports of lipid peroxidation and lipoxygenase induction may also be related to the recent observations that the lipid-derived, volatile compound methyl jasmonate can induce the synthesis of defensive protease inhibitors in several plant species (45). The wound-inducible proteinase inhibitor genes are also activated by oligogalacturonides (13) and a recently discovered 18-residue polypeptide (82). Jasmonic acid was proposed as an intermediate in the intracellular transduction mechanism of wound-inductibility of the proteinase inhibitor genes. Oligogalacturonide-induced synthesis of proteinase inhibitors is inhibited by aspirin (acetylsalicylic acid) (15, 35), which is known to inhibit lipoxygenases in mammalian cells. If aspirin has a similar action in plants it may result from inhibition of the production of jasmonic acid in wounded plants.

Ion Fluxes

A number of reports have related ion fluxes and membrane depolarization to the regulation of plant defensive responses (56, 74, 84, 91, 101–103, 111, 113, 115). Most studies have not used chemically characterized elicitors. However, the effects of two size ranges of pectic fragments (DP = 1–7 and DP = 10–20) were examined for their effects on membrane potential in tomato leaf cells (103). Both sets of oligogalacturonides at concentrations of 0.5–1.0 mg ml^{-1} caused rapid membrane depolarization, which was reversible upon removal of the oligogalacturonides. Monogalacturonic acid was inactive. Fungal elicitors isolated from cell walls of *Colletotrichum lagenarium* and *Phytophthora parasitica* had similar effects on membrane potential in *Cucumis* and *Nicotiana* species, respectively (84). Proton accumulation in the root-bathing medium was inhibited by fungal elicitors, an effect attributed to the direct inhibition of H$^+$-ATPase. Recent studies have further implicated

H^+-ATPase in oligosaccharide signaling. Treatment of *Arachis* (peanut) suspension cultures with a water-insoluble mycelial elicitor of stilbene synthesis isolated from *P. cambivora* resulted in the elicitation of enzymes of stilbene synthesis (101). Na_3VO_4 (a known plasma membrane H^+-ATPase inhibitor) mimicked the effect of the fungal elicitor. The authors hypothesized that changes in membrane potential, modulated by H^+-ATPase, may mediate signaling, leading to stilbene production. Vanadate was first shown to induce PAL induction in *Vigna* suspension cell cultures (56), and both vanadate and fusicoccin induced phytoalexin synthesis in soybean cotyledons, mimicking the effects of the glucan elicitors (74). The accumulation of proteinase inhibitors in tomato, induced by pectic fragments, was also inhibited by low concentrations of fusicoccin (35).

SUMMARY

Roles for oligosaccharides as signals for activating defensive and developmental processes have continued to be reported and supported in the recent literature. Studies of recognition events employing defined oligosaccharides indicate that oligosaccharides interact with cellular membrane receptors in a hormone-like manner to affect gene regulation. Recent studies concerning the regulation of signal production (e.g. studies of the roles PGIPs play in regulating the effective half-lives of oligouronides in infected tissue; and investigations of possible roles for cell wall fragments in morphogenesis, hypersensitivity, fruit ripening, and cell elongation) provide further evidence that oligosaccharides signal developmental responses. In plants, as in animals, our understanding of how complex carbohydrates may function in signaling and recognition systems is still in its infancy.

Literature Cited

1. Albersheim, P., Anderson, A. T. 1971. Proteins from plant cell walls inhibit polygalacturonase secreted by plant pathogens. *Proc. Natl. Acad. Sci. USA* 68:1815–19

2. Albersheim, P., Darvill, A. G., McNeil, M., Valent, B., Sharp, J. K., et al. 1983. Oligosaccharins, naturally occurring carbohydrates with biological regulatory functions. In *Structure and Function of Plant Genomes*, ed. O. Ciferri, L. Dure III, pp. 293–312. New York: Plenum

3. Anderson, J. D., Mattoo, A. K., Lieberman, M. 1982. Induction of ethylene biosynthesis in tobacco leaf discs by cell wall digesting enzymes. *Biochem. Biophys. Res. Commun.* 107:588–96

4. Anderson, J. M. 1989. Membrane-derived fatty acid as precursors to second messengers. In *Second Messengers in Plant Growth and Development*, ed. W. F. Boss, D. J. Morre, pp. 181–212. NY: Alan R. Liss

5. Apostol, I., Heinstein, P. F., Low, P. S. 1989. Rapid stimulation of an oxidative burst during elicitation of cultured plant cells. *Plant Physiol.* 90:109–16

6. Ayers, A. R., Valent, B. S., Ebel, J., Albersheim, P. 1976. Host-pathogen interactions. XI. Composition and structure of wall-released elicitor fractions. *Plant Physiol.* 57:766–74

7. Baldwin, E. A., Biggs, R. H. 1988. Cell-wall lysing enzymes and products of cell-wall digestion elicit ethylene in citrus. *Physiol. Plant.* 73:58–64

8. Baldwin, E. A., Pressey, R. 1988. Tomato polygalacturonase elicits ethylene production in tomato fruit. *J. Am. Soc. Hortic. Sci.* 113:92–95

9. Barber, M. S., Ride, J. P. 1988. A

quantitative assay for induced lignification in wounded wheat leaves and its use to survey potential elicitors of the response. *Physiol. Plant Pathol.* 32:185–97

10. Baydoun, E. A.-H., Fry, S. 1989. *In vivo* degradation and extracellular polymer-binding of xyloglucan nonasaccharide, a naturally-occurring antiauxin. *J. Plant Physiol.* 134:453–59

11. Benhamou, N., Chamberland, H., Pauze, E. J. 1990. Implication of pectic components in cell surface interactions between tomato root cells and *Fusarium oxysporum* f. sp. *radicis-lycopersici*. *Plant Physiol.* 92:995–1003

12. Benhamou, N., Gilboa-Garber, N., Trudel, J., Asselin, A. 1988. A new lectin-gold complex for ultrastructural localization of galacturonic acids. *J. Histochem. Cytochem.* 36:1403–11

13. Bishop, P. D., Makus, D. J., Pearce, G., Ryan, C. A. 1981. Proteinase inhibitor-inducing factor activity in tomato leaves residues in oligosaccharides enzymically released from cell walls. *Proc. Natl. Acad. Sci. USA* 78:3536–40

14. Bostock, R. M., Kuc, J., Lane, R. A. 1981. Eicosapentaenoic and arachidonic acids from *Phytophthora infestans* elicit fungitoxic sesquiterpenes in the potato. *Science* 212:67–69

15. Bowles, D. J. 1990. Defense-related proteins in higher plants. *Annu. Rev. Biochem.* 59:873–907

16. Branca, C., De Lorenzo, G., Cervone, F. 1988. Competitive inhibition of the auxin induced elongation α-D-oligogalacturonides in pea stem segments. *Physiol. Plant.* 72:499–504

17. Brecht, J. K., Huber, D. J. 1988. Products released from enzymically active cell wall stimulate ethylene production and ripening in preclimacteric tomato (*Lycopersicon esculentum* Mill.) fruit. *Plant Physiol.* 88:1037–41

18. Broekaert, W. F., Peumans, W. J. 1988. Pectic polysaccharides elicit chitinase accumulation in tobacco. *Physiol. Plant.* 74:740–44

19. Bruce, R. J., West, C. A. 1989. Elicitation of lignin biosynthesis and isoperoxidase activity by pectic fragments in suspension cultures of castor bean. *Plant Physiol.* 89:889–97

20. Bucheli, P., Doares, S. H., Albersheim, P., Darvill, A. 1990. Host-pathogen interactions. XXXVI. Partial purification of heat-labile molecules secreted by the rice blast pathogen that solubilize plant cell wall fragments (oligosaccharins) that kill plant cells. *Physiol. Mol. Plant Pathol.* 36:159–73

21. Campbell, B. C. 1986. Host-plant oligosaccharins in the honeydew of *Schizaphis graminum* (Rondani) (Insecta, Aphididae). *Experientia* 42:451–52

22. Cervone, F., De Lorenzo, G., Degra, L., Salvi, G. 1987. Elicitation of necrosis in *Vigna unguiculata* Walp. by homogeneous *Aspergillus niger* endopolygalacturonase and by α-D-galacturonate oligomers. *Plant Physiol.* 85:626–30

23. Cervone, F., De Lorenzo, G., Degra, L., Salvi, G., Bergami, M. 1987. Purification and characterization of a polygalacturonase inhibiting protein from *Phaseolus vulgaris* L. *Plant Physiol.* 85:631–37

24. Cervone, F., De Lorenzo, G., D'Ovidio, R., Hahne, M. G., Ito, Y., et al. 1989. Phytotoxic effects and phytoalexin-elicitor activity of microbial pectic enzymes. In *Phytotoxins and Plant Pathogenesis, NATO ASI Ser.*, ed. A. Graniti, R. D. Durbin, A. Ballio, H27:473–77. Berlin: Springer-Verlag

25. Cervone, F., De Lorenzo, G., Pressey, R., Darvill, A. G., Albersheim, P. 1989. Can *Phaseolus* PGIP inhibit pectic enzymes from microbes and plants? *Phytochemistry* 29:447–49

26. Cervone, F., Hahn, M. G., De Lorenzo, G., Darvill, A., Albersheim, P. 1989. Host pathogen interactions. XXXIII. A plant protein converts a fungal pathogenesis factor into an elicitor of plant defense responses. *Plant Physiol.* 90:542–48

27. Cheong, J.-J., Birberg, W., Fugedi, P., Pilotti, A., Garegg, P. J., et al. 1991. Structure-function relationships of oligo-β-glucoside elicitors of phytoalexin accumulation in plants. *Plant Cell.* 3:127–36

28. Cheong, J.-J., Hahn, M. 1991. A specific high-affinity binding site for the hepta-β-glucoside elicitor exists in soybean membranes. *Plant Cell.* 3:137–47

29. Cosio, E. G., Frey, T., Ebel, J. 1990. Solubilization and characteristics of the binding sites for fungal β-glucans from soybean cell membranes. *FEBS Lett.* 264:235–38

30. Cosio, E. G., Frey, T., Verduyn, R., van Boom, J., Ebel, J. 1990. High-affinity binding of a synthetic heptaglucoside and fungal glucan phytoalexin elicitors to soybean membranes. *FEBS Lett.* 271:223–26

31. Cosio, E. G., Popperl, H., Schmidt, W. E., Ebel, J. 1988. High-affinity binding of fungal β-glucan fragments to soybean (*Glycine max* L.) microsomal fractions

and protoplasts. *Eur. J. Biochem.* 175: 309–15

32. Darvill, A. G., Albersheim, P. 1984. Phytoalexins and their elicitors—a defense against microbial infection in plants. *Annu. Rev. Plant Physiol.* 35: 243–75

33. Dietrich, A., Mayer, J. E., Hahlbrock, K. 1990. Fungal elicitor triggers rapid, transient, and specific protein phosphorylation in parsley cell suspension cultures. *J. Biol. Chem.* 265:6360–68

34. Doares, S. H., Bucheli, P., Albersheim, P., Darvill, A. G. 1989. Host pathogen interactions. XXXIV. A heat labile activity secreted by a fungal phytopathogen releases fragments of plant cell walls that kill plant cells. *Mol. Plant Microbe Interact.* 2:346–53

35. Doherty, H. M., Selvendran, R. R., Bowles, D. J. 1988. The wound response of tomato plants can be inhibited by aspirin and related hydroxybenzoic acids. *Physiol Mol. Plant Pathol.* 33: 377–84

36. Doke, N. 1983. Involvement of superoxide generation in the hypersensitive response of potato tuber tissues to infection with an incompatible race of *Phytophthora infestans* and to the hyphal wall components. *Physiol Plant Pathol.* 23:345–57

37. Doke, N. 1983. Generation of superoxide anion by potato tuber protoplasts during the hypersensitive response to hyphal wall components of *Phytophthora infestans* and specific inhibition of the reaction by suppressors of hypersensitivity. *Physiol Plant Pathol.* 23: 359–67

38. Eberhard, S., Doubrava, N., Marfa, V., Mohnen, D., Southwick, A., et al. 1989. Pectic cell wall fragments regulate tobacco thin-cell-layer explant morophogenesis. *Plant Cell* 1:747–55

39. Einspahr, K. J., Thompson, G. A. Jr. 1990. Transmembrane signaling via phosphatidylinositol 4,5-bisphosphate hydrolysis in plants. *Plant Physiol.* 93:361–66

40. Farkas, V., Maclachlan, G. 1988 Stimulation of pea 1,4-β-glucanase activity by oligosaccharides derived from xyloglucan. *Carbohydr. Res.* 184:213–20

41. Farmer, E. E., Helgeson, J. P. 1987. An extracellular protein from *Phytophthora parasitica* var *nicotianae* is associated with stress metabolite accumulation in tobacco callus. *Plant Physiol.* 85:733–40

42. Farmer, E. E., Moloshok, T. D., Ryan,

C. A. 1990. In vitro phosphorylation in response to oligouronide elicitors: structural and biological relationships. *Curr. Top. Plant Biochem. Physiol.* 9:249–58

43. Farmer, E. E., Moloshok, T. D., Saxton, M. J., Ryan, C. A. 1991. Oligosaccharide signaling in plants: specificity of oligouronide-enhanced plasma membrane protein phosphorylation. *J. Biol. Chem.* 266:3140–45

44. Farmer, E. E., Pearce, G., Ryan, C. A. 1989. *In vitro* phosphorylation of plant plasma membrane proteins in response to the proteinase inhibitor inducing factor. *Proc. Natl. Acad. Sci. USA* 86: 1539–42

45. Farmer, E. E., Ryan, C. A. 1990. Interplant communication: Airborne methyl jasmonate induces synthesis of proteinase inhibitors in plant leaves. *Proc. Natl. Acad. Sci. USA* 87:7713–16

46. Fry, S. C. 1986. *In vivo* formation of xyloglucan nonasaccharides: a possible biologically active cell wall fragment. *Planta* 169:443–53

47. Fuchs, Y., Saxena, A., Gamble, H. R., Anderson, J. D. 1989. Ethylene biosynthesis-inducing protein from cellulysin is an endoxylanase. *Plant Physiol.* 89:138–43

48. Giovannoni, J. J., DellaPenna, D., Bennett, A. B., Fischer, R. L. 1989. Expression of a chimeric polygalacturonase gene in transgenic *rin* (ripening inhibitor) tomato fruit results in polyuronide degradation but not fruit softening. *Plant Cell* 1:53–63

49. Giovannoni, J. J., DellaPenna, D., Bennett, A. B., Fischer, R. L. 1990. Polygalacturonase and tomato fruit ripening. *Hortic. Rev.* In press

50. Gollin, D. J., Darvill, A. G., Albersheim, P. 1984. Plant cell wall fragments inhibit flowering and promote vegetative growth in *Lemna gibba*. *Biol. Cell* 51:275–80

51. Grab, D., Feger, M., Ebel, J. 1989. An endogenous factor from soybean (*Glycine max* L.) cell cultures activates phosphorylation of a protein which is dephosphorylated *in vivo* in elicitor-challenged cells. *Planta* 179:340–48

52. Hadwiger, L., Beckman, J. M. 1980. Chitosan as a component of pea/ *Fusarium solani* interactions. *Plant Physiol.* 67:170–73

53. Hahn, M. G., Darvill, A. G., Albersheim, P. 1981. Host-pathogen interactions. XIX. The endogenous elicitor, a fragment of a plant cell wall polysaccharide that elicits phytoalexin

accumulation in soybeans. *Plant Physiol.* 68:1161–69

54. Hahne, G., Lörz, H. 1988. Release of phytotoxin factors from plant cell walls during protoplast isolation. *J. Plant Physiol.* 132:345–50

55. Halliwell, B., Gutteridge, J. M. C. 1985. *Free Radicals in Biology and Medicine.* Oxford: Clarendon

56. Hattori, T., Ohta, Y. 1985. Induction of phenylalanine ammonia lyase activation and isoflavone glycoside accumulation in suspension cultured cells of red bean, *Vigna angularis,* by phtoalexin elicitors, vanadate and elevation of medium pH. *Plant Cell Physiol.* 26:1101–10

57. Hayashi, T. 1989. Xyloglucan in the primary cell wall. *Annu. Rev. Plant Physiol. Plant Mol. Biol.* 40:139–68

58. Hayashi, T., Maclachlan, G. 1984. Pea xyloglucan and cellulose. I. Macromolecular organization. *Plant Physiol.* 75:596–604

59. Hayashi, T., Yoshida, K. 1988. Cell expansion and single-cell separation induced by colchicine in suspension-cultured soybean cells. *Proc. Natl. Acad. Sci. USA* 85:2618–22

60. Horn, M. A., Heinstein, P. F., Low, P. S. 1989. Receptor-mediated endocytosis in plant cells. *Plant Cell* 1:1003–9

61. Ishii, S. 1988. Factors influencing protoplast viability of suspension-cultured rice cells during isolation process. *Plant Physiol.* 88:26–29

62. Jacobs, M., Ray, P. M. 1975. Promotion of xyloglucan metabolism by acid pH. *Plant Physiol.* 56:373–76

63. Kauffman, S., Doares, S. H., Albersheim, P., Darvill, A. 1990. Involvement of pathogen-secreted enzymes and plant cell wall derived xylan fragments in plant cell death in the rice blast system. *5th Int. Symp. Mol. Genet.* Plant-Microbe Interact., Interlaken. Abstr. SO106, p. 32

64. Kohn, R. 1985. Ion binding on polyuronates-alginate and pectin. *Pure Appl. Chem.* 42:371–97

65. Kopp, M., Rouster, J., Fritig, B., Darvill, A., Albersheim, P. 1989. Host-pathogen interaction. XXXII. A fungal glucan preparation protects *Nicotianae* against infection by virus. *Plant Physiol.* 90:208–16

66. Lafitte, C., Barthe, J. P., Montillet, J. L., Touze, A. 1984. Glycoprotein inhibitors of *Colletotrichum lindemuthianum* endopolygalacturonase in near isogenic lines of *Phaseolus vulgaris* resistant and susceptible to anthracnose. *Physiol. Plant Pathol.* 25:39–53

67. Lerouge, P., Roche, P., Faucher, C.,

Maillet, F., Truchet, G., et al. 1990. Symbiotic host-specificity of *Rhizobium meliloti* is determined by a sulphated and acylated glucosamine oligosaccharide signal. *Nature* 344:781–84

68. Lincoln, J. E., Fischer, R. L. 1988. Diverse mechanisms for the regulation of ethylene-inducible gene expression. *Mol. Gen. Genet.* 212:71–75

69. Lindner, W. A., Hoffmann, C., Grisebach, H. 1988. Rapid elicitor chemiluminescence in soybean cell suspension cultures. *Phytochemistry* 27:2501–3

70. Liners, F., Letesson, J. J., Didembourg, C., Van Cutsem, P. 1989. Monoclonal antibodies against pectin: Recognition of a conformation induced by calcium. *Plant Physiol.* 91:1419–24

71. Lotan, T., Fluhr, R. 1990. Xylanase, a novel elicitor of pathogenesis-related proteins in tobacco, uses a non-ethylene pathway for induction. *Plant Physiol.* 43:811–17

72. Ma, R., Reese, J. C., Black, W. C. IV, Bramel-Cox, P. 1990. Detection of pectinesterase and polygalacturonase from salivary secretions of living greenbugs, *Schizaphis graminum* (Homoptera: Aphididae). *J. Insect Physiol.* 36:507–12

73. Deleted in proof

74. Mayer, M. G., Ziegler, E. 1988. An elicitor from *Phytophthora megasperma* f. sp. *glycinea* influences the membrane potential of soybean cotyledonary cells. *Physiol. Mol. Plant Pathol.* 33:397–407

75. McDougall, G., Fry, S. C. 1989. Anti-auxin activity of xyloglucan oligosaccharides: the role of groups other than the terminal β-L-fucose residue. *J. Exp. Bot.* 40:233–38

76. McDougall, G., Fry, S. C. 1989. Structure-activity relationships for xyloglucan oligosaccharides with anti-auxin activity. *Plant Physiol.* 89:883–87

77. McDougall, G., Fry, S. C. 1990. Xyloglucan oligosaccharides promote growth and activate cellulase: evidence for a role of cellulase in cell expansion. *Plant Physiol.* 93:1042–48

78. McNeil, M., Darvill, A. G., Fry, S. C., Albersheim, P. 1984. Structure and function of the primary cell walls of plants. *Annu. Rev. Biochem.* 53:625–63

79. Mohnen, D., Eberhard, S., Marfa, V., Doubrava, N., Toubart, P., et al. 1990. Control of root, vegetative shoot, and flower morphogenesis in tobacco thin-cell-layer expants. *Development* 108:191–201

80. Moloshok, T., Ryan, C. A. 1989. Di- and trigalacturonic acid and 4,5-di-

and 4,5-trigalacturonic acids: inducers of proteinase inhibitor genes in plants. *Methods Enzymol.* 179:566–70

81. Morse, M. J., Satter, R. L., Crain, R. C., Cote, G. G. 1989. Signal transduction and phosphatidylinositol turnover in plants. *Physiol. Plant.* 76:118–21

82. Deleted in proof

83. Peever, T. L., Higgins, V. J. 1989. Electrolyte leakage, lipoxygenase, and lipid peroxidation induced in tomato leaf tissue by specific and nonspecific elicitors from *Cladosporium fulvum. Plant Physiol.* 90:867–75

84. Pelissier, B., Thibaud, J. B., Grignon, C., Esquerre-Tugaye, M. T. 1986. Cell surfaces in plant-microorganism interactions. VII. Elicitor preparations from two fungal pathogens depolarize plant membrane. *Plant Sci.* 46:103–9

85. Pena-Cortes, H., Sanchez-Serrano, J. J., Mertens, R., Willmitzer, L., Prat, S. 1989. Abscisic acid is involved in the wound-induced expression of the proteinase inhibitor II gene in potato and tomato. *Proc. Natl. Acad. Sci. USA* 86:9851–55

86. Poovaiah, B. W., Reddy, A. S. N., McFadden, J. J. 1987. Calcium messenger system: role of protein phosphorylation and inositol bisphospholipids. *Physiol. Plant.* 69:569–73

87. Roby, D., Gadelle, A., Toppan, A. 1987. Chitin oligosaccharides as elicitors of chitinase activity in melon plants. *Biochem. Biophys. Res. Commun.* 143: 885–92

88. Roby, D., Toppan, A., Esquerre-Tugaye, M.-T. 1987. Cell surfaces in plant micro-organism interactions. VIII. Increased proteinase inhibitor activity in melon plants in response to infection by *Colletotrichum lagenarium* or to treatment with an elicitor fraction from this fungus. *Physiol. Mol. Plant Pathol.* 39: 453–60

89. Roche, P., Lerouge, P., Faucher, C., Vasse, J., Prome, J. C., et al. 1990. *Rhizobium meliloti* extracellular Nod signals. *5th Int. Symp. Mol. Genet. Plant-Microbe Interact.,* Interlaken. Abstr. L12, p. 7

90. Rogers, K. R., Albert, F., Anderson, A. J. 1988. Lipid peroxidation is a consequence of elicitor activity. *Plant Physiol.* 86:547–53

91. Rogers, K. R., Anderson, A. J. 1987. The effect of extracellular components from *Colletotrichum lindemuthianum* on membrane transport in vesicles isolated from bean hypocotyl. *Plant Physiol.* 84:428–32

92. Ryan, C. A. 1987. Oligosaccharide

93. Ryan, C. A. 1988. Oligosaccharides as recognition signals for the expression of defensive genes in plants. *Biochemistry* 27:8879–82

94. Schmidt, W. E., Ebel, J. 1987. Specific binding of a fungal glucan phytoalexin elicitor to membrane fractions from soybean *Glycine max. Proc. Natl. Acad. Sci. USA* 84:4117–21

95. Schottens-Toma, I. M. J., DeWit, P. J. G. M. 1988. Purification and primary structure of a necrosis-inducing peptide from the apoplastic fluids of tomato infected with *Cladosporium fulvum* (syn. *Fulvia fulva*). *Physiol. Mol. Plant Pathol.* 33:59–67

96. Schroeder, J. I., Hedrich, R. 1989. Involvement of ion channels and active transport in osmoregulation and signaling of higher plant cells. *Trends Biol. Sci.* 14:187–92

97. Sharp, J. K., McNeil, M., Albersheim, P. 1986. The primary structures of one elicitor-active and seven elicitor-inactive hexa(β - D - glucopyranosyl)D - glucitols isolated from the mycelial walls of *Phytophthora megaperma* f. sp. *glycinea. J. Biol. Chem.* 259:11321–36

98. Sharp, J. K., Valent, B., Albersheim, P. 1986. Purification and partial characterization of a β-glucan fragment that elicits phytoalexin accumulation in soybean. *J. Biol. Chem.* 259:11312–20

99. Smith, C. J. S., Watson, C. F., Ray, J., Bird, C. R., Morris, P. C., et al. 1988. Antisense RNA inhibition of polygalacturonase gene expression in transgenic tomatoes. *Nature* 334:724–26

100. Staeb, M. R., Ebel, J. 1987. Effects of Ca^{2+} on induction by fungal elicitor in soybean cells. *Arch. Biochem. Biophys.* 257:416–23

101. Steffens, M., Ettl, F., Kranz, D., Kindl, H. 1989. Vanadate mimics effects of fungal cell wall in eliciting gene activation in plant cell cultures. *Planta* 177: 160–68

102. Strasser, H., Matern, U. 1986. Minimal time requirement for lasting elicitor effects in cultured parsely cells. *Z. Naturforsch.* 41c:222–27

103. Thain, J. F., Doherty, H. M., Bowles, D. J., Wildon, D. C. 1990. Oligosaccharides that induce proteinase inhibitor activity in tomato plants cause depolarization of tomato leaf cells. *Plant Cell. Environ.* 13:569–74

104. Theologis, A. 1986. Rapid gene regulation by auxin. *Annu. Rev. Plant Physiol.* 37:407–38

105. Tigchelaar, E. C., McGlasson, W. B.,

Buescher, R. W. 1978. Genetic regulation of tomato fruit ripening. *Hortic. Sci.* 13:508–13

106. Tomiyama, K. 1982. *Hypersensitive Cell Death: Its Significance and Physiology in Plant Infection: The Physiological and Biochemical Basis.*, ed. Y. Asada, W. Bushell, S. Ouchi, C. P. Vance, pp. 329–44. Berlin: Springer-Verlag

107. Tong, C. B., Labavitch, J. M., Yang, S. F. 1986. The induction of ethylene production from pear cell culture by cell wall fragments. *Plant Physiol.* 81:929–30

108. Tran Thanh Van, K., Toubart, P., Cousson, A., Darvill, A. G., Gollin, D. G., et al. 1985. Manipulation of the morphogenetic pathway of tobacco explants by oligosaccharides. *Nature* 314:615–17

109. Vanderhoef, L. N. 1979. Auxin regulated cell enlargement: Is there action at the level of gene expression? In *Genomic Organization and Expression in Plants, NATO ASI Ser. A*, ed. C. J. Leaver, 29:159–74. Edinburgh New York: Plenum

110. Verma, D. P. S., Maclachlan, G. A., Byrne, H., Ewings, D. 1975. Regulation and in vitro translation of message ribonucleic acid for cellulase from auxin treated pea epocotyls. *J. Biol. Chem.* 250:1019–26

111. Waldman, T., Kauss, H. 1990. Ca^{2+} influx, K^+ leakage and external alkalinization are early events in the induction of callose synthesis in suspension-cultured plant cells. *5th Int. Symp. Mol. Genet. Plant-Microbe Interact.*, Interlaken. Abstr. P378, p. 230

112. Walker-Simmons, M., Hadwiger, L., Ryan, C. A. 1983. Chitosans and pectic polysaccharides both induce the accumulation of the antifungal phytoalexin pisatin in pea pods and antinutrient proteinase inhibitors in tomato leaves. *Biochem. Biophys. Res. Commun.* 110: 194–99

113. Wildon, D. C., Doherty, H. M., Eagles, G., Bowles, D. J., Thain, J. F. 1989. Systemic responses arising from localized heat stimuli in tomato plants. *Ann. Bot.* 64:691–95

114. York, W. S., Darvill, A. G., Albersheim, P. 1984. Inhibition of 2-4-dichlorophenoxyacetic acid–stimulated elongation of pea stem segments by a xyloglucan oligosaccharide. *Plant Physiol.* 75:295–97

115. Young, D. H., Kauss, H. 1983. Release of calcium from suspension cultured *Glycine max* cells by chitosan, other polycations and polyamines in relation to effects on membrane permeability. *Plant Physiol.* 73:698–702

Annu. Rev. Plant Physiol. Plant Mol. Biol. 1991. 42:675-703

ROLE OF CELL WALL HYDROLASES IN FRUIT RIPENING

Robert L. Fischer

Department of Plant Biology, University of California, Berkeley, California 94720

Alan B. Bennett

Department of Vegetable Crops, Mann Laboratory, University of California, Davis, California 95616

KEY WORDS: polygalacturonase, cellulase, transgenic plant, regulation of gene expression, plant polysaccharides

CONTENTS

INTRODUCTION

Ripening is a complex process involving major transitions in fruit development and metabolism. Although ripening hallmarks the initiation of fruit

1040-2519/91/0601-0675$02.00

senescence, previous theories emphasizing the deteriorative aspects of ripening have been superseded by the view that ripening represents a well-coordinated process of organ differentiation that is genetically programmed (22). Because this developmental transition involves coordinated changes in a number of biochemical pathways, the regulation of ripening has been extensively studied as a model in plant development (22, 63, 64). Because ethylene regulates at least part of this developmental transition, ethylene biosynthesis, ethylene responses, and ethylene-regulated gene expression have been extensively studied in ripening fruit (20, 35, 64, 112, 114, 157, 169, 197).

In addition to other biochemical transitions, such as pigment biosynthesis and production of volatiles, softening accompanies the ripening of many fruit. It has been considered axiomatic that these textural changes result primarily from changes in cell wall structure (92). Fruit softening has been of interest to plant physiologists and biochemists because it has provided an opportunity to study the induction of specific cell wall hydrolases in relation to developmentally programmed changes in cell wall structure. In addition, the process of fruit softening is commercially important because it often dictates early harvest of fruit to avoid damage in subsequent handling, which can result in failure to develop optimum flavor and color. Ultimately, excessive softening and the associated enhancement in pathogen susceptibility limits the postharvest life of many fruit.

Aspects of fruit ripening and its regulation have been the subject of several reviews in this series (22, 107, 156, 197) and elsewhere (24, 63, 64, 65, 68, 92, 102, 150). This review focuses on one aspect of fruit ripening, the role of cell wall hydrolases, a topic last reviewed extensively in 1983 (92). Recently, a convergence in research focusing on genetic control of fruit ripening and cell wall biochemistry have led to rapid advances in understanding the regulation and function of specific cell wall hydrolases in ripening fruit. Here we emphasize current concepts in cell wall architecture, discuss alterations in that architecture that accompany ripening, and illustrate molecular genetic strategies that have begun to elucidate the function of specific cell wall hydrolases. The terrain covered in this review is, like the cell wall itself, tortuous. Our present understanding of these processes certainly fails to appreciate complexities that future studies will encounter and elucidate.

CELL WALL STRUCTURE AND RIPENING-ASSOCIATED CHANGES

A model of cell wall architecture is needed in order to interpret the physical consequences of changes in cell wall polymers that accompany fruit ripening. Such a model is also necessary if we are to predict the effects that particular cell wall hydrolases may have on the rheological properties of the cell wall. Unfortunately, the complexity and diversity of cell walls precludes the development of a precise three-dimensional model of universal applicability. Nevertheless, a number of tentative models have been advanced that seek to integrate chemical and enzymic analysis of specific cell wall components into a unified structure (54, 96, 110). Reviews of these models and of particular cell wall components are numerous (8, 29, 31, 52, 79, 122, 190).

The general polymeric composition of the primary cell wall is not controversial. The cell wall is approximately 30% cellulose, 30% hemicellulose, 35% pectin, and 5% protein in dicotyledonous plants (52), although pectin content may be substantially higher and protein content lower in fruit cell walls (102). All cell wall models have cellulose microfibrils embedded in a matrix of noncellulosic polysaccharides and protein. However, the specific interactions and cross-linkages between polymers have provided a fertile area of controversy. Similarly, in ripening fruit it is widely appreciated that cell wall structural changes occur, but the specific nature of the changes, their enzymic basis, and their contribution to changing rheological properties of the cell wall are disputed.

Cellulose

Cellulose is a linear β-(1→4) glucan that provides the mechanical strength of plant cell walls. Cellulose self-associates by intermolecular hydrogen bonding to form microfibrils of at least 36 glucan chains (29) and, as indicated below, becomes strongly associated with hemicellulose in the cell wall. Indeed, it has been suggested that the diameter of the cellulose microfibril may be determined, at least in part, by the binding of hemicellulose during cellulose synthesis, which prevents fasciation of small microfibrils into larger bundles (81). Thus, cellulose microfibrils exist as a complex with hemicellulose in situ — a complex established during cellulose synthesis and deposition. Because of this association, the physical properties of the microfibril are determined by intermolecular hydrogen bonding between cellulose and hemicellulose polymers.

Although one might anticipate that changes in cellulose structure are associated with softening of ripening fruit, this does not appear to be the case. Ultrastructural observations have documented the apparent dissolution of the cell wall fibrillar network in ripening avocado, pear, and apple (130, 133), and this dissolution could be reproduced by treatment of fruit tissues with fungal cellulase in vitro (15). It was suggested that this dissolution resulted from cellulolytic activity, but chemical analysis of cellulosic glucan levels indicating that cellulose levels remain constant or even increase slightly during ripening of pear and tomato fruit does not support this view (1, 72). It is possible that the observed ultrastructural changes resulted from cellulolytic activity that did not completely solubilize cell wall cellulose. Alternatively, the ultrastructural changes may have resulted from the degradation of a component of the noncellulosic matrix that resulted in loss of microfibrillar organization. In this regard it is notable that chemical extraction of hemicelluloses of onion cell walls resulted in loss of microfibrillar organization even though cellulose was presumably intact in these preparations (120).

Hemicellulose

In spite of the many areas of disagreement, all cell wall models posit that cellulose is associated with a monolayer of hemicellulose strongly hydrogen-bonded to the microfibril surface (29, 54, 96, 190). In dicotyledonous cell walls the principal hemicellulose is xyloglucan, a linear β-(1→4) glucosyl chain to which xylose and more complex side chains containing xylose, galactose, and fucose are attached to carbon 6 of glucosyl residues of the glucan backbone at regular intervals. In some cases xylose side chains are attached in a highly regular fashion at three consecutive glucose residues followed by an unsubstituted glucose (53, 79). This pattern of xylose substitution gives rise to repeating subunits of hepta and nonasaccharides, with the nonasaccharide containing a fucosyl-α-(1→2)-galactosyl-β-(1→2) xylosyl side chain attached in the third position of the three consecutively substituted glucose residues and the heptasaccharide being simply xylosylated in this position (79). It has been suggested that release of the xyloglucan nonasaccharide may antagonize the growth-promoting action of auxins elongating pea stems (118, 202).

The association of xyloglucan with cellulose microfibrils has been demonstrated in vitro to be spontaneous and highly specific, in that a 10-fold excess of β-(1→2) glucan, β-(1→3) glucan, β-(1→6) glucan, β-(1→3),(1→4) glucan, arabinogalactan, or pectin failed to compete with xyloglucan for binding to cellulose. In addition, fluorescein-labeled xyloglucan specifically labels hemicellulose-depleted cellulose microfibrils in situ, indicating that this interaction can occur spontaneously in the cell wall (81). However, the native xyloglucan/cellulose complex contains far higher quantities of xyloglucan than binds to cellulose in vitro, suggesting that the association of xyloglucan with cellulose in vivo must not only comprise a monolayer on the microfibril, but must also extend well into the cell wall matrix and perhaps span neighboring microfibrils (29, 53). This suggestion is supported by ^{3}H-fucose and lectin labeling, showing xyloglucan to be present both on and between cellulose microfibrils (80). An attempt to visualize cross-linkages in onion primary cell walls further indicated that after alkali extraction of cell wall hemicelluloses, cellulose microfibrils collapsed together and cross-linking fibers disappeared. This direct visualization of cell wall cross-linkages was interpreted to indicate that hemicelluloses, and specifically xyloglucan polymers, cross-link two or more cellulose microfibrils, and that these cross-linkages are responsible for holding cellulose microfibrils in a fixed position in the cell wall matrix (120).

Other hemicellulosic components of the dicot primary cell wall are far less abundant than xyloglucan, and their structure and function has not been studied as thoroughly. Xylans make up about 5% and 20% of the primary cell wall of dicots and grasses, respectively. In both dicotyledonous and graminaceous cell walls the xylans may function with xyloglucan in associating with and perhaps spanning cellulose microfibrils (29). Other hemicellulosic components include glucomannans and galactomannans. While these hemicellulosic components serve primarily as storage polysaccharides in seeds, some are found in cell walls, where they may have a structural function (8).

Ripening-associated modifications of hemicellulose structure have been documented in tomato, pepper, strawberry, and melons. Although only slight changes in the quantity of cell wall hemicelluloses occur, gel filtration chromatography indicates that the size of hemicellulose polymers decreases dramatically during ripening of these fruits (73, 93, 94, 116). Because hemicelluloses are a collection of diverse polymers, attempts have been made to identify the specific polysaccharide component responsible for the appearance of smaller hemicellulosic polymers. Examination of tomato fruit divided hemicelluloses into two fractions, HFI and HFII, extractable in 4 M and 8 M KOH, respectively (177). Only HFII, a fraction presumably bound tightly to cellulose microfibrils, showed an increase in small polysaccharide polymers during fruit ripening. Linkage analysis of this fraction indicated an increase in C4-linked and C4,6-linked mannosyl and C4-linked glucosyl residues with a corresponding decrease in C5-linked arabinosyl residues (177). Although the appearance of small hemicellulosic polymers has been interpreted as indicating that some components of hemicellulose undergo limited degradation (93), the latter study advanced an alternate interpretation: The molecular weight shift in hemicellulose may involve synthesis of small polymers enriched in mannosyl and glucosyl residues — perhaps glucomannan (71, 177). Ripening-associated changes in hemicellulose structure are likely to be important determinants of the textural changes in fruit, but the biochemical bases of hemicellulose turnover are largely uncharacterized.

Pectin

Pectins are a class of complex polysaccharides defined primarily by extractability in hot water, chelators, or dilute acid. These polysaccharides are often described in terms of "smooth" and "hairy" blocks which may reside as components of a single pectin polymer. The dominant feature of smooth blocks is a linear copolymer of α-(1→4)-linked galacturonic acid and its methyl ester. Inserted within this smooth homogalacturonan polymer are α-(1→2)-linked rhamnosyl residues whose spacing may be quite regular. It has been suggested that the rhamnosyl residues delineate homogalacturonan domains for methyl esterification or de-esterification, thus enabling Ca^{2+} cross-linking at regular intervals (52). Rhamnosyl residues within the homogalacturonan backbone also serve as attachment sites for arabinose- and galactose-rich side chains.

So-called "hairy" pectin blocks include rhamnogalacturonans I and II and are complex heteropolymers comprised of 12 different sugars. The backbone, rich in galacturonic acid and rhamnose, bears numerous side chains rich in arabinose and galactose but also containing fucose, methylfucose, methylxylose, apiose, glucuronic acid, aceric acid, keto-deoxy-octulosonic acid, and glucose (52). In general, the "hairy" rhamnogalacturonans are not susceptible to degradation by pectinase. Other pectins include arabinose- and galactose-rich polymers that may arise from the hairy rhamnogalacturonans' being released enzymically during extraction.

Pectins have generally been treated as polymers comprised of a continuum of molecular-size species. Recent analysis of tomato fruit cell wall pectins by high-performance size-exclusion chromatography, however, suggested that the apparent continuum of pectin molecular species could be resolved by curve-fitting techniques into five discrete macromolecular species (50). This interpretation suggests that the proposed pectin subunits function as an aggregate mosaic, with subunits associated by noncovalent interactions. This model presents the possibility that changes in cell wall integrity and pectin degradation may be brought about by changes in ionic strength of the extracellular matrix rather than by enzymic cleavage.

Ripening-associated changes in pectins have been extensively documented (92, 107). However, the structure of pectins, either in situ or after extraction, has been difficult to assess because of the tendency of the polymers to associate in various aggregation states. The simplest assay of pectin structure involves its extraction from the cell wall by chelators. The susceptibility of pectins to chelator extraction indicates that covalent cross-linkages to insoluble polymers have been cleaved and that the pectin polymer remains bound only by ionic bonds, presumably Ca^{2+} cross-linkages, to adjacent galacturonan polymers. During ripening of many fruit, notably tomato, apple, and pear, there is a dramatic increase in water- and chelator-soluble pectins. This observation suggests that pectin polymers are cleaved from covalent cross-linkages in the cell wall (92).

More informative assays of pectin structures include methods that assess the molecular size of the polymers. These methods rely on chelator extraction of pectins followed by size fractionation of the pectins by gel filtration (93, 161), by ultracentrifugation (160), or most recently by high-performance size-exclusion chromatography (50). Of paramount importance in these assays is the complete and rapid inactivation of endogenous polygalactur-onases by phenol-acetic acid–water treatment during cell wall isolation (93, 161). Failure to do this results in extensive pectin depolymerization, presum-ably owing to enhanced accessibility of extracted pectins to degradation by endogenous polygalacturonase (24, 154). By means of both gel filtration and ultracentrifugation, the average molecular size of pectins has been shown to decrease dramatically during fruit ripening suggesting that, in addition to cleavage of cross-linkages, the galacturonan backbone is degraded. Recent analysis of transgenic plants indicates that pectin degradation in tomato fruit, as assayed either by chelator solubility of pectin (60) or by changes in molecular size (43, 166), reflects the activity of a single enzyme, endo-polygalacturonase.

Other Cell Wall Components

Other neutral cell wall polymers apparently undergo extensive ripening-related alterations. Indeed, the loss of neutral sugars, especially galactose and arabinose, is quantitatively the largest ripening-associated change in cell wall composition in many fruit. For instance, 65% of the cell wall galactosyl residues are lost during tomato ripening (69, 72, 108, 192). In addition it has been suggested that galactose may promote ethylene production in ripening fruit, thus potentiating the ripening process (70, 98). Although cell wall arabinose solubilized during pear fruit ripening was identified as a component of pectin arabinan, indicating that arabinose loss occurred as a consequence of pectin degradation (1), other studies have shown that neutral sugar loss occurs independently of pectin degradation (101, 12, 72). Recent studies have further demonstrated that net loss of galactosyl residues occurs in two stages in tomato fruit, one that proceeds continuously throughout fruit development and another that is ripening-related and absent in the ripening-impaired tomato mutant, *rin* (97). The enzymic basis of cell wall galactosyl loss is uncertain, but it may result from the degradation of galactans by β-galactosidase II (136) and perhaps by a concomitant decrease in galactan synthesis (108).

CELL WALL HYDROLASES

The apparent changes in molecular size of cell wall polymers that accompany fruit ripening implicate the action of enzymes capable of degrading specific cell wall components. Consequently, the activity of ripening-induced cell wall hydrolases has been the subject of intense study for several decades (for previous reviews, see 92). Such studies have focused on the relatively few enzymes that accumulate to high levels in particular fruit. Most notable in this regard are the polygalacturonases and carboxymethylcellulases (Cx-cellulases), each of which accumulates to high levels in some, but not all, fruit. In recent years the use of molecular genetic approaches has provided the means to assess critically the function of these well-characterized enzymes in the ripening process. Here we summarize the biochemical characterization of the major cell wall hydrolases and review recent advances.

Polygalacturonase

As described above, large changes in pectin structure accompany the ripening of many fruit. These changes in structure have been attributed to the action of polygalacturonases, enzymes that catalyze the hydrolytic cleavage of α-(1→4) galacturonan linkages. Of the polygalacturonases that have been identified in fruit, both exo- and endo-polygalacturonases have been characterized. Polygalacturonase has been identified in tomato abscission zones (184) as well as in ripening fruit, and in pollen grains of diverse plants, including *Oenothera* and maize (27, 147). Pectate lyase catalyzes a similar cleavage of α-(1→4) galacturonan linkages but by a β-elimination rather than a hydrolytic reaction mechanism. Pectate lyases have not been identified in fruit but are present in microorganisms, and recently a mRNA having sequence similarity to fungal pectin lyase has been reported in tomato pollen (194).

EXO-POLYGALACTURONASE   A number of fruits contain both endo- and exo-polygalacturonases, including freestone peach, pear, cucumber, and papaya (141, 143, 144, 119, 32). In some fruits, most notably apple, only exo-polygalacturonase activity has been detected (102), raising the question of how pectin solubilization is achieved. However, recent evidence using more sensitive immunological and DNA probes suggests that even apple may contain detectable levels of an endo-polygalacturonase (H. Gradzina, personal communication). Similarly, clingstone peaches are reported to contain only the exo-polygalacturonase while freestone peaches contain both exo- and endo-polygalacturonase (141). The presence of the endo-polygalacturonase in freestone peaches is associated with a higher level of water-soluble pectins than in clingstone peaches (135). Exo-polygalacturonase activity was purified from a freestone peach variety and shown to consist of two enzymes that could be distinguished by cation exchange chromatography. These exo-polygalacturonases had a molecular mass of 66 kDa and, while present at low levels in developing fruit, increased dramatically in abundance during fruit ripening (45). This pattern of expression suggests that the exo-polygalacturonases may participate in the limited softening that occurs in clingstone peaches.

ENDO-POLYGALACTURONASE   Endo-polygalacturonase (referred to hereafter as polygalacturonase) activity has been identified in a number of ripening fruit and is correlated with increases in soluble pectins and softening that accompany ripening. It has often been suggested that polygalacturonase is primarily responsible for ripening-associated pectin degradation and fruit softening (16, 22, 92). The evidence supporting this contention includes: 1. the observation that in vitro degradation of isolated cell walls by polygalacturonase mimics the pectin degradation that occurs in vivo (36, 91, 173, 192), 2. a correlation between polygalacturonase levels and fruit softening (3, 25, 87, 170), and 3. the absence or deficiency of polygalacturonase in ripening-impaired tomato mutants that fail to soften (25, 145, 175).

Polygalacturonase has been studied in a number of fruit but has been analyzed at the molecular level most extensively in tomato. Tomato polygalacturonase cDNA and genomic clones have been characterized, and polygalacturonase gene expression in tomato plants has been analyzed. It has recently been established that, based upon cross-reactivity to antibody and DNA probes, polygalacturonase in peaches (111) and avocado (R. E. Christoffersen, personal communication) are structurally related to tomato polygalacturonase, suggesting that properties of this enzyme are widely conserved. Although we focus here on the properties of tomato polygalacturonase, the results are likely relevant to a number of fruit.

Polygalacturonase gene expression Polygalacturonase activity has long been known to increase dramatically during tomato fruit ripening (86). This increase in activity is associated with an increase in immunologically detectable protein (23, 181) and an increase in polygalacturonase mRNA (19, 40, 115, 164). Quantitative analysis of polygalacturonase mRNA levels indicated that polygalacturonase mRNA accumulated to extremely high levels in ripening tomato fruit, accounting for over 1% of the poly(A)$^+$ RNA (17, 42). Analysis of relative rates of polygalacturonase gene transcription by nuclear run-on assays indicated that the polygalacturonase gene becomes transcriptionally active at the onset of fruit ripening (44, 163). In these studies a close correlation in the timing of the first detectable polygalacturonase mRNA and first detectable transcriptional activity of the gene was observed, suggesting that the change in the rate of polygalacturonase gene transcription is an important determinant of the timing of polygalacturonase expression in tomato fruit. In addition, polygalacturonase gene transcription was reduced in ripening-impaired tomato fruit containing the *rin, nor* or *Nr* mutations (44), a finding consistent with previous observations that these mutant fruit fail to accumulate polygalacturonase mRNA (19, 42, 100, 113). Analysis of polygalacturonase gene transcription indicated that its relative rate of transcription was lower than that of other genes, in spite of high steady-state polygalacturonase mRNA levels. This result was interpreted as indicating that posttranscriptional processes, such as mRNA stability or transport, contribute to the high accumulation of polygalacturonase mRNA, relative to other transcripts (44, 163).

Spatial regulation of polygalacturonase gene expression and protein accumulation has also been examined. Whole fruit slices were used in a tissue blotting technique to immunolocalize polygalacturonase protein. These studies indicated that polygalacturonase accumulates first in the central columella and later in radial and outer pericarp walls, but not in locular tissue (174). This pattern of polygalacturonase accumulation parallels that of lycopene accumulation and suggests that these aspects of ripening are coordinated both temporally and spatially. Polygalacturonase mRNA was localized by in situ hybridization and found to be most prevalent in outer layers of the pericarp and in cell layers adjacent to vascular tissue in mid-ripe fruit (129).

Analysis of the polygalacturonase gene in tomato suggests that it is present in one copy per haploid genome (21, 42, 60, 100). The polygalacturonase gene has been cloned and sequenced, indicating a 7 kb transcription unit containing eight introns (21; Figure 1). Fusion of DNA fragments from 5' upstream regions of the polygalacturonase gene to the bacterial chloramphenicol acetyl transferase gene indicated that a 1.4 kb fragment directly upstream of the 5' end of the transcription unit confers ripening-specific expression (21).

Polygalacturonase isozymes and maturation   Tomato polygalacturonase is comprised of three isozymes, PG1, PG2A, and PG2B (4). In ripening tomato fruit PG1 accumulates first, followed by accumulation of PG2A and PG2B (23). The relationship of these isozymes to one another and to the polygalacturonase gene has been a major focus in understanding polygalacturonase function and especially the relative roles of specific isozymes in pectin degradation and fruit ripening. PG1 can be distinguished from PG2A and PG2B by its high heat stability (181) and high molecular weight (124, 142). PG1 appears to be a heterodimer composed of a subunit of PG2A or PG2B and a 41-kDa ancillary subunit (124; C. J. Brady, personal communication). A protease-sensitive, but heat-stable factor (termed the polygalacturonase converter because it converts PG2 to PG1) has been identified in extracts from vegetative as well as fruit tissue (182, 137). It is thought that the polygalacturonase converter is the 41-kDa ancillary subunit of PG1, but this possibility has not been demonstrated directly. Because the converter is found in a number of tissues that do not express polygalacturonase activity, and because under certain extraction conditions only PG2 is recovered, it has been suggested that PG1 is an artifact of extraction (138). In contrast, Knegt et al (104) have argued that PG1 is present in the cell wall in situ, and have proposed that the polygalacturonase converter "targets" polygalacturonase to its substrate in the cell wall.

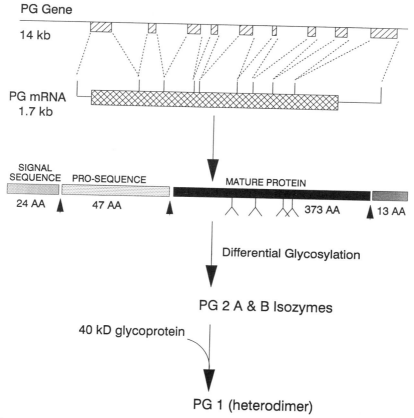

Figure 1   Schematic diagram of polygalacturonase expression in tomato fruit. Structure of the polygalacturonase gene is from Bird et al (21). Structure of the primary translation product deduced from sequences of the polygalacturonase cDNA with processing sites determined by comparison of the cDNA-deduced sequence with amino acid sequence from the mature polygalacturonase protein (67, 164) or from processing intermediates (41).

PG2A and PG2B apparently function as monomeric catalytic subunits. Based upon antibody cross-reactivity of PG2A and PG2B, it was suggested that these two proteins may differ only in glycosylation (4). Recently it has been shown that the amino and carboxy termini of PG2A and PG2B are identical, and that chemical deglycosylation renders the PG2A and PG2B enzymes indistinguishable by SDS polyacrylamide gel electrophoresis (41; C. J. Brady, personal communication). In addition, constitutive expression of a single tomato polygalacturonase cDNA in transgenic tobacco plants resulted in the accumulation of two polypeptides that were indistinguishable from tomato PG2A and PG2B, providing definitive evidence that both PG2 isozymes result from differential processing of the same gene product (128).

Analysis of the polygalacturonase cDNA sequence and deduced amino acid sequence identified several processing domains, indicating that polygalacturonase undergoes extensive co- and posttranslational processing (41, 67, 164; see Figure 1). Beginning at the amino terminus of the deduced amino acid sequence, a typical hydrophobic signal sequence is evident, which presumably directs cotranslational translocation of the nascent polypeptide into the endoplasmic reticulum, initiating transport of the protein to the cell wall. The signal sequence of polygalacturonase conforms precisely to the general structure of signal sequences (191, 18). Directly adjacent to the signal sequence is a pro-sequence whose boundary is defined by the signal sequence cleavage site at the amino terminus (41) and by the start of the mature protein at the carboxy terminus (164). The function of the pro-sequence is unknown, as are details of its removal during polygalacturonase maturation. For instance, it is not known if this domain is cleaved in single or multiple proteolytic steps, nor is it known where in the cell cleavage takes place. By analogy to pro-sequences in yeast, it has been speculated that the polygalacturonase pro-sequence may function to inactivate the protein during transport to the cell wall or to direct its final subcellular localization (41). Four potential glycosylation sites (Asn-X-Ser/Thr) are encoded by the polygalacturonase cDNA (67, 164) although it has not been determined if all of these sites are utilized. Finally, C-terminal proteolytic processing has been suggested by comparison of the amino acid sequence deduced from the cDNA with the C-terminal amino acid sequence of the mature protein (164). Although its significance has not been explored, this processing event is reminiscent of proteolytic processing events in a number of proteins that are transported through the endoplasmic reticulum and Golgi apparatus (56, 58, 148, 149). The posttranslational processing of polygalacturonase has been described in some detail, yet critical experiments that define the function of specific processing events or the function of domains removed during polygalacturonase maturation are needed to understand precisely how these processing events contribute to the accumulation of polygalacturonase.

Pectinmethylesterase

Polygalacturonase is more active in degrading demethylated than methylated pectin (146, 105, 162). Therefore, pectinmethylesterase, an enzyme that catalyzes demethylation of the C6 carboxyl group of galacturonosyl residues, may play an important role in determining the extent to which pectin is accessible to degradation by polygalacturonase. Indeed, it has been suggested that the increased susceptibility of tomato fruit cell walls to polygalacturonase action during ripening is due to the action of pectinmethylesterase (105).

Two isozymes of pectinmethylesterase have been identified in tomato fruit (140). Pectinmethylesterase activity is detectable throughout tomato fruit development, though the activity of one isozyme increases several fold during ripening (183). Cloning of a tomato fruit pectinmethylesterase cDNA clone allowed the characterization of mRNA levels that are highest in immature green fruit and then decline throughout maturation and ripening (149). This pattern of mRNA accumulation did not parallel the increase in pectin-methylesterase enzyme activity previously reported (161, 183). Three distinct tomato pectinmethylesterase genes have recently been isolated (A. Handa, personal communication), and it is likely that analysis of the expression of these genes will provide a more comprehensive understanding of the regulation of individual pectinmethylesterase isozymes in ripening fruit.

Cx-Cellulase

Cx-cellulases are a class of enzymes characterized by their ability to degrade carboxymethylcellulose (Cx-cellulose). Except in one unconfirmed case (168), plant Cx-cellulases have not been shown to degrade crystalline cellulose and should probably be referred to as endo-β-(1→4) glucanases to avoid potential confusion with some bacterial Cx-cellulases that do degrade crystalline cellulose (13). This raises a question of considerable importance: What is the in vivo substrate of plant Cx-cellulases?

The identification of endogenous Cx-cellulase substrates has been addressed by examining the activity of purified Cx-cellulases on pure polysaccharide polymers. Purified avocado fruit Cx-cellulase failed to degrade crystalline cellulose but was active against various carboxymethyl and hydroxyethyl celluloses, xyloglucan, and mixed β-(1→3),(1→4) glucans (78). Of the substrates tested, only xyloglucan is known to be present in avocado cell walls, suggesting that this may be the target of Cx-cellulase activity in ripening avocado. Cx-cellulase from elongating pea stems may also function in xyloglucan degradation in vivo (82). However, Cx-cellulase isolated from bean abscission zones did not have xyloglucanase activity (46). Thus, plant Cx-cellulases may comprise a group of related enzymes acting on distinct polysaccharide substrates in different tissues, though in several cases the endogenous substrate is proposed to be a β-(1→4)-linked hemicellulosic polysaccharide.

Ultrastructural studies implicated Cx-cellulase in cellulose degradation in that loss of fibrillar cellulose was associated with an increase in Cx-cellulase activity in ripening avocado and pear (14, 130, 133). It has, however, been observed recently that loss of cellulose microfibrillar organization detected at the ultrastructural level can result from chemical extraction of hemicellulose (120). This result may reconcile ultrastructural studies with the proposed in vivo activity of Cx-cellulase in that degradation of noncellulosic components of the cell wall by Cx-cellulase may result in the observed disorganization of fibrillar cellulose.

CX-CELLULASE GENE EXPRESSION IN RIPENING FRUIT   Cx-cellulase activity increases during ripening of tomato (76, 89, 131), strawberry (11), pear (15), peach (84), and avocado (5, 6, 130). Avocado Cx-cellulase activity is induced to an extremely high level, and this induction results from an accumulation of Cx-cellulase mRNA (34, 186). Avocado fruit Cx-cellulase is encoded by a small family of closely related genes (30). Two members of this family have been cloned, but only one gene, *cel1*, is expressed in ripening fruit. Hybridization of a bean abscission zone Cx-cellulase cDNA with avocado genomic DNA further indicates that the situation in avocado may be more complex with Cx-cellulases encoded by multiple gene families with relatively low sequence similarity between families (187).

Regulation of Cx-cellulase expression has also been studied in tomato fruit. Activity is high in very young fruit, decreases during maturation, and dramatically increases again during fruit ripening (7, 89, 134). Two cDNAs encoding tomato fruit Cx-cellulases have been identified (C. Lashbrook and A. B. Bennett, unpublished data). One corresponds to a mRNA that is abundant in immature fruit, is barely detectable in mature green fruit, and accumulates again during ripening. The other Cx-cellulase cDNA corresponds to a mRNA expressed only in ripening fruit. This indicates that in tomato, multiple Cx-cellulases are differentially expressed during fruit development, suggesting that they may have distinct physiological functions.

Other Cell Wall Hydrolases

Most research has focused on the relatively abundant cell wall hydrolases already discussed. However, the complexity of cell wall structural polysaccharides suggests that a large number of enzymes must ultimately be involved in its turnover. Enzymes reported to be present in ripening fruit tissue include a number of glycosidases (2, 193), endo-β-(1→4) mannanase (139), xylanase (195), α- and β-galactosidase (2, 136), and endo-β-(1→3) glucanase (85, 193). The role of an endo-β-(1→4) mannanase in fruit softening is suggested by its ripening-specific accumulation in tomato fruit (139), and its action could account for the observed increase in small hemicellulosic fragments rich in C4- and C4,6-linked mannosyl residues in ripening tomato (177). In addition, it has been proposed that β-galactosidase may be responsible for the solubilization of galactose observed during fruit ripening. This contention is supported by the resolution of three β-galactosidase isozymes in tomato, one of which (β-galactosidase II) is capable of degrading β-(1→4)-linked galactan (136). Unlike that of the other two β-galactosidase isozymes, β-galactosidase II activity increases during fruit ripening, an observation consistent with its potential role in galactose solubilization from a galactan polymer. Analysis of cell wall polysaccharide fragments indicates that the loss of galactose from the cell wall results from removal of β-(1→4)-linked galactose, a finding consistent with the proposal that β-galactosidase II functions as a β-(1→4) galactanase in vivo (159).

ANALYSIS OF CELL WALL HYDROLASE FUNCTION DURING FRUIT RIPENING

The studies reviewed above describe cell wall changes associated with fruit ripening and the major cell wall hydrolases present in ripening fruit. Experiments designed to elucidate cell wall hydrolase function have attempted to correlate specific enzyme activities with changes in cell wall structure and with discrete fruit ripening processes. In recent years, the ability to isolate and analyze the function of specific enzymes has improved dramatically. Initial experiments involved the use of inhibitors, environmental conditions, and mutations to determine the effect of modifying an array of cell wall hydrolase enzyme activities on fruit ripening. Currently, molecular genetic approaches are being used to determine the effect of modifying the activity of individual cell wall hydrolases on fruit ripening.

Physical, Genetic, and Biochemical Modification of Cell Wall Hydrolase Activity

In order to correlate cell wall hydrolase activities with ripening processes, a variety of methods were utilized that resulted in the overall inhibition of ripening. Cell wall hydrolase activities were measured in fruit where ripening had been inhibited by heat (127, 132, 203) or by treatment with compounds that reduced the biosynthesis and/or action of the fruit ripening hormone ethylene (37, 90, 112, 196). A genetic approach involved the analysis of cell wall hydrolytic activities in fruit from mutants such as *rin* (ripening inhibitor; 152), *nor* (non ripening; 176), and *Nr* (Neverripe; 88, 151). Plants bearing these mutations develop normally except that many aspects of fruit ripening are inhibited (reviewed in 66). For example, *rin* and *nor* fruit display reduced levels of softening, ethylene and CO_2 biosynthesis, lycopene pigment accumulation, and chlorophyll degradation (reviewed in 22).

Data from the experiments described above, for the most part, indicated that levels of polygalacturonase activity were positively correlated with fruit ripening and softening. Polygalacturonase enzyme activity (87), mRNA (40, 64, 112, 164), and gene transcription (44, 163) synchronously increased at the onset of ripening, although other fruit such as strawberry have been shown to soften in the absence of dramatic increases in polygalacturonase activity (103). Tomato fruit ripening mutants with delayed ripening exhibited decreased levels of polygalacturonase activity (175), mRNA (42, 66, 19), and gene transcription (44), whereas activities of several other cell wall enzymes (i.e. Cx-cellulase, pectinmethylesterase) appeared to be unaffected (28, 89,

175, 193). A rough correlation was demonstrated between levels of polygalacturonase activity and softening among different lines of cultivated tomatoes (3, 25, 170,175), although there are clear exceptions (178). Physical and chemical treatments that suppressed ripening likewise inhibited polygalacturonase gene expression (37, 90, 112, 127, 132, 196, 203). These results led to the hypothesis that polygalacturonase plays an essential role in the induction of fruit softening. However, because the methods employed all had multiple effects on fruit ripening and may influence a wide variety of enzyme activities, it was not possible to derive firm conclusions about the function of an individual cell wall hydrolase, such as polygalacturonase, from these experiments.

In an experiment designed to examine more specifically the function of a single enzyme, purified polygalacturonase protein was introduced into unripe wild-type and *rin* fruit by vacuum infiltration (10). Fruit receiving polygalacturonase responded by producing ethylene. By contrast, control fruit subjected to vacuum infiltration alone did not respond, leading to the suggestion that oligosaccharides produced by the action of polygalacturonase elicited the onset of ethylene biosynthesis. Others have shown that cell wall fragments stimulate ethylene biosynthesis when applied to cultured cells (179), tomato pericarp tissue (26), intact tomato fruit (71), or leaves (189). However, it is possible that vacuum infiltration introduced polygalacturonase into regions of the cell wall not normally accessed via the natural protein secretory pathway, where the action of the enzyme might be perceived by the plant as a wound, thus resulting in ethylene evolution (198). The possibility also exists that the polygalacturonase infiltrate contained other cell wall degrading enzymes, such as endoxylanases, that have been shown to elicit ethylene biosynthesis (55). Experiments with transgenic plants described below do not support the conclusion that the pectic fragments produced by the action of polygalacturonase lead to the production of ethylene (60, 166).

Molecular Genetic Modification of Cell Wall Hydrolase Activity in Tomato Fruit

Recent advances in molecular genetic techniques have made it possible to alter individual cell wall hydrolase activities. By genetically transforming plants with cloned cell wall hydrolase genes or antisense genes, it is now possible to increase or decrease specific enzyme activity levels in either wild-type or mutant fruit. As a result, the effect of altering a single enzyme activity on cell wall structure and fruit ripening can be assessed, and conclusions about the function of individual cell wall hydrolytic enzymes during fruit ripening can be drawn.

Among the fruits that have been intensively studied, tomato is perhaps the best model system for molecular genetic analysis of cell wall hydrolase function during fruit ripening. 1. The cultivated tomato plant is easy to grow and maintain, is self-pollinating, is amenable to clonal propagation, and has a relatively short generation time. 2. Tomato has been extensively analyzed genetically. Its haploid genome size is comparatively small for a higher plant, approximately 6.8×10^8 base pairs (57). Tomato has an extensive genetic map of over 300 morphological mutants, including mutations that affect fruit ripening (172). 3. Foreign DNA can be inserted into the tomato genome and fertile transgenic tomato plants can be regenerated using modified Ti plasmids from *Agrobacterium tumefaciens* (49, 117). 4. New genetic resources are being developed that may prove to be important in future experiments. A tomato restriction fragment polymorphism map is well developed (204). In addition, transposons from maize have been shown to function in the tomato genome (199–201). Thus, tomato is an excellent species for the molecular genetic analysis of specific cell wall hydrolases during fruit ripening.

COMPLEMENTATION OF THE *RIN* MUTATION WITH A CHIMERIC POLYGA-LACTURONASE GENE   To investigate the function of polygalacturonase enzyme activity, a molecular genetic strategy has been utilized to increase the polygalacturonase enzyme activity in the tomato fruit ripening mutant, *rin* (43, 60). The *rin* mutation inhibits many aspects of tomato fruit ripening including softening, polyuronide degradation, ethylene biosynthesis, lycopene accumulation, and polygalacturonase enzyme activity (60, 175). The lack of polygalacturonase enzyme activity in *rin* fruit does not result from a lesion in the polygalacturonase gene (21, 42); rather, it reflects a failure to initiate many ripening processes, including the activation of polygalacturonase gene transcription (44). *rin* fruit appear to be "fixed" at the unripe stage of fruit development and contain less than 1% of the wild-type level of polygalacturonase enzyme activity (60). Thus, induction of polygalacturonase gene expression in transgenic *rin* fruit complements the *rin* mutation specifically for polygalacturonase enzyme activity. One can thereby address whether the expression of the polygalacturonase gene is sufficient to bring about changes in cell wall structure, softening, and/or ripening in *rin* fruit.

In order for the polygalacturonase gene to be expressed in *rin* fruit, it was necessary to replace its promoter with one that would be active in the *rin* genetic background. To this end, a chimeric gene was constructed, designated *E8-PG*, consisting of the E8 promoter (38) joined to the polygalacturonase structural gene (60). The chimeric *E8-PG* gene was constructed so that the polygalacturonase enzyme produced is identical to native polygalacturonase at the level of amino acid sequence, processing, secretion, and enzyme activity. Although the function of the E8 gene is not known, its promoter was utilized because it allowed targeted expression of the polygalacturonase coding sequences throughout the *rin* fruit pericarp at the appropriate stage of development. The *E8-PG* gene was added to the *rin* genome via modified

Ti-plasmid vectors from *Agrobacterium tumefaciens*, and transgenic plants containing a single copy of the *E8-PG* gene were regenerated, designated *rin[E8/PG]* (59, 60).

Induction of polygalacturonase gene expression in rin fruit   Chimeric *E8-PG* gene expression was activated by exposing the *rin[E8/PG]* fruit to the ethylene analog propylene (121), resulting in the accumulation of polygalacturonase protein in the cell wall (60). The level of extractable polygalacturonase enzyme activity was approximately 60% of that found in wild-type fruit. By contrast, the level of extractable polygalacturonase enzyme activity in control air-treated *rin[E8-PG]* fruit, and in control propylene-treated *rin* fruit, was less than 1% of that found in wild-type fruit. The polygalacturonase isozymes produced by the propylene-treated *rin[E8-PG]* fruit were electrophoretically indistinguishable from the three native polygalacturonase isozymes found in wild-type fruit (PG1, PG2A, and PG2B) (43), proving that the three isozymes arise by differential processing of a single gene product, and providing a transgenic system for analyzing the physiological function of polygalacturonase.

Effect on cell wall structure   The expression of the polygalacturonase gene in transgenic *rin[E8/PG]* fruit proved to have a dramatic effect on cell wall structure. Cell walls of transgenic fruit were analyzed by the determination of chelator-soluble polyuronides extracted from acetone-insoluble solids, providing a measure of polyuronides that had been enzymically cleaved but remained ionically bound in the cell wall by calcium cross-linkage to other acidic cell wall polymers, and by the resolution of soluble polyuronides by gel filtration chromatography, providing a measure of polyuronide depolymerization. Control propylene-treated *rin* fruit and air-treated *rin[E8/PG]* fruit contained low levels of chelator-soluble polyuronides. Furthermore, gel filtration analysis of the chelator-soluble polyuronides indicated that they were in the high-molecular-weight class. By contrast, expression of the polygalacturonase gene in propylene-treated *rin[E8/PG]* fruit resulted in a large increase in the levels of chelator-soluble polyuronides, comparable to that found in ripening wild-type fruit (60). Furthermore, gel filtration analysis revealed that the chelator-soluble polyuronides had undergone depolymerization. That is, a significant fraction of the chelator-soluble polyuronides were in the low-molecular-weight class, 67% of the amount produced by wild-type fruit (43). Finally, the period of maximum polyuronide solubilization and depolymerization occurred coincident with the appearance of the PG1 isozyme, well before the appearance of the PG2A and PG2B isozymes. Taken together, these results indicate a primary role for polygalacturonase in polyuronide degradation, and suggest that the PG1 isozyme is sufficient to carry out both solubilization and depolymerization of polyuronides.

Effect on ripening   The physiological consequence of polygalacturonase-mediated polyuronide solubilization and depolymerization in transgenic *rin[E8/PG]* fruit proved to be limited (60, 61). That is, high levels of polygalacturonase enzyme activity and polyuronide degradation were not sufficient to increase softening, as measured by compressibility. Furthermore, neither ethylene production nor the accumulation of lycopene pigment increased. Results seen with fruit attached to the vine were similar to those seen with detached fruit. These findings indicate that polygalacturonase enzyme activity is sufficient for degrading polyuronide but is not sufficient for the induction of softening, ethylene production, or color development in *rin* tomato fruit.

SUPPRESSION OF POLYGALACTURONASE GENE EXPRESSION IN TOMATO FRUIT BY AN ANTISENSE GENE   Regulation of gene expression by the production of RNA sequences complementary to a given RNA occurs in nature in both prokaryotes (165) and eukaryotes (99). The production of antisense RNA has been utilized successfully to gain negative control of gene expression in a wide variety of biological systems (reviewed in 123), including higher plants (39, 47, 77, 83, 153, 154, 188). To investigate the function of polygalacturonase, the level of polygalacturonase activity has been reduced by producing antisense polygalacturonase RNA in fruit (163, 167). One can thereby address whether the expression of the polygalacturonase gene is necessary to bring about changes in cell wall structure, softening, and/or ripening in wild-type fruit.

To this end, antisense polygalacturonase genes were constructed using the cauliflower mosaic virus 35S promoter (74) to transcribe cloned polygalacturonase cDNA sequences in a reverse orientation relative to the endogenous gene. Sheehy et al (163) utilized 1600 base pairs spanning the polygalacturonase open reading frame whereas Smith et al (167) utilized 730 base pairs representing the 5'-end of the polygalacturonase mRNA. The antisense genes were added to the tomato genome via modified Ti-plasmid vectors from *Agrobacterium tumefaciens,* and transgenic plants were regenerated.

Antisense polygalacturonase gene expression in transgenic plants   Owing to the constitutive nature of the cauliflower mosaic virus 35S promoter, antisense polygalacturonase RNA was detected in all the major plant organs of transformed plants (163, 167). In ripe fruit from selected transgenic plants, the levels of polygalacturonase mRNA, protein, and enzyme activity were reduced to approximately 10% of that found in untransformed control ripe fruit (163, 167). Analysis of progeny from the primary transgenic plants demonstrated that reduced levels of polygalacturonase co-segregated with the antisense gene, verifying that the antisense gene is responsible for the suppression of polygalacturonase gene expression (158).

The mechanism by which the production of antisense RNA results in reduced levels of polygalacturonase gene expression is not yet clearly understood. Because the cytoplasmic polygalacturonase mRNA concentration was more severely reduced than the concentration of polygalacturonase transcription complexes extracted from isolated nuclei, it has been proposed that posttranscriptional mechanisms play an important role (163). No relationship between the amount of polygalacturonase antisense RNA and the reduction in polygalacturonase activity in fruit was observed (163, 166), although gene dosage clearly had an effect (166). That is, plants homozygous for the antisense gene displayed reduced levels of polygalacturonase mRNA, protein, and enzyme activity when compared to the hemizygous parent. Homozygous transgenic plants with less than 1% of wild-type levels of polygalacturonase enzyme, primarily in the form of the PG1 isoenzyme, have been identified (166), providing a system for analyzing the physiological function of polygalacturonase.

Effect on cell wall structure   The increase in chelator-soluble polyuronides associated with ripening was not affected by the reduction of polygalacturonase enzyme activity in transgenic fruit (166). That is, as ripening progressed both control fruit and fruit harboring the polygalacturonase antisense gene showed an equivalent increase in the level of cleaved polyuronides bound in the cell wall ionically by calcium cross-linkage to other acidic cell wall polymers. Thus, polygalacturonase activity above 1% of wild type levels is not necessary for polyuronide solubilization during fruit ripening. It remains to be seen whether or not further reduction of polygalacturonase will result in suppression of polyuronide solubilization. By contrast, determination of apparent molecular weights showed that depolymerization of chelator-soluble polyuronides was suppressed by expression of the antisense gene (166). Thus, levels of polygalacturonase activity above 1% of that found in wild-type fruit appear to be necessary for depolymerization of chelator-soluble polyuronides.

Effect on ripening and quality   The physiological consequence of suppressed polyuronide depolymerization on transgenic fruit ripening proved to be limited. Comparison of fruit containing 1% polygalacturonase with wild-type controls revealed no obvious differences in compressibility (166), ethylene production (166), or lycopene accumulation (163, 166). Thus, polygalacturonase-mediated depolymerization of polyuronides is not necessary for the fundamental developmental processes associated with tomato fruit ripening.

From a horticultural perspective, suppression of polygalacturonase gene expression may be beneficial. In field tests, ripe fruit with reduced levels of polygalacturonase (8% of wild-type) displayed a measurable improvement in storage life, in solids content, and in the consistency and viscosity of processed juice (106). Thus, after fruit have become ripe, polygalacturonase-mediated polyuronide depolymerization may play a role in reducing fruit integrity and processing quality.

FUTURE PROSPECTS

The molecular genetic procedures used to elucidate the role of polygalacturonase during tomato fruit ripening are general and can be applied to test the function of other cloned cell wall hydrolytic enzymes during ripening. Thus, it is currently possible to apply this strategy to analyze the function of Cx-cellulase (34, 185) and pectinmethylesterase genes (149) during fruit ripening. In addition, by genetically crossing the appropriate transgenic plants it will be possible to modify the activity of multiple cell wall hydrolytic enzymes and thus look for synergistic relationships between various enzyme activities. The ability to apply this technology to various fruits is expanding as the number of fruit-bearing plants that can be transformed and regenerated increases. This list currently includes apple (95), blackberry/raspberry (62), cucumber (180), eggplant (75), muskmelon (48), papaya (51), and strawberry (126). Finally, this strategy can also be applied to other physiological processes associated with cell wall hydrolysis, such as cell growth and organ abscission.

The results of the molecular genetic analysis of polygalacturonase function indicate that correlative data linking enzyme activity and fruit softening may not accurately predict enzyme function. The plant cell wall is enormously complex, and the number of different hydrolytic enzymes influencing its structural integrity is not known. Thus a detailed understanding of the function of hydrolytic enzymes encoded by genes that have been cloned (polygalacturonase and Cx-cellulase) may not provide a complete picture of how cell wall structure is controlled during ripening. Further biochemical analysis of other, less abundant, hydrolytic enzymes is needed. In addition, the multiple strategies that are now available for cloning mutagenized genes in dicotyledonous plants, such as transposon tagging (9), subtraction hybridization (171), and chromosome walking (125, 33), are applicable to tomato (172, 200). Thus, genetic approaches to the identification of important cell wall hydrolases may also prove effective in the future.

ACKNOWLEDGMENTS

We thank numerous colleagues for providing preprints of unpublished data and Dr. John Labavitch for critically reviewing this manuscript. Research in the authors' laboratories has been supported by grants from the USDA-Competitive Research Grants Office.

Literature Cited

1. Ahmed, A. E., Labavitch, J. M. 1980. Cell wall metabolism in ripening pear. I. Cell wall changes in ripening Bartlett pears. *Plant Physiol.* 65:1009-13
2. Ahmed, A. E., Labavitch, J. M. 1980. Cell wall metabolism in ripening pear. II. Changes in carbohydrate-degrading enzymes in ripening Bartlett pears. *Plant Physiol.* 65:1014-16
3. Ahrens, M. J., Huber, D. J. 1990. Physiology and firmness determination of ripening tomato fruit. *Physiol. Plant.* 78:8-14
4. Ali, Z. M., Brady, C. J. 1982. Purification and characterization of polygalacturonase of tomato fruits. *Aust. J. Plant Physiol.* 9:155-69
5. Awad, M. 1977. Variation in cellulase content of Fuerte avocado fruit after harvest. *Hortic. Sci.* 12:406
6. Awad, M., Young, R. E. 1979. Postharvest variation in cellulase, polygalacturonase and pectinmethylesterase in avocado fruits in relation to respiration and ethylene production. *Plant Physiol.* 64:306-8
7. Babbitt, J. K., Powers, M. J., Patterson, M. E. 1973. Effects of growth-regulators on cellulase, polygalacturonase, respiration, color, and texture of ripening tomatoes. *J. Am. Soc. Hortic. Sci.* 98:77-81
8. Bacic, A., Harris, P. J., Stone, B. A. 1988. Structure and function of plant cell walls. In *The Biochemistry of Plants: A Comprehensive Treatise*, ed. J. Preiss, 14:297-371. New York: Academic
9. Baker, B., Schell, J., Lörz, H., Fedoroff, N. 1986. Transposition of the maize controlling element "Activator" in tobacco. *Proc. Natl. Acad. Sci. USA* 83:4844-48
10. Baldwin, E., Pressey, R. 1988. Tomato polygalacturonase elicits ethylene production in tomato fruit. *J. Am. Soc. Hortic. Sci.* 113:92-95
11. Barnes, M. F., Patchett, B. J. 1976. Cell wall degrading enzymes and the softening of senescent strawberry fruit. *J. Food Sci.* 41:1392-95
12. Bartley, I. M. 1976. Changes in the glucans of ripening apples. *Phytochemistry* 13:2107-11
13. Beguin, P. 1990. Molecular biology of cellulose degradation. *Annu. Rev. Microbiol.* 44:219-48
14. Ben Arie, R., Kislev, N., Frenkel, C. 1979. Ultrastructural changes in the cell walls of ripening apple and pear fruit. *Plant Physiol.* 64:197-202
15. Ben Arie, R., Sonego, L., Frenkel, C. 1979. Changes in the pectic substance of ripening pears. *J. Am. Soc. Hortic. Sci.* 104:500-5
16. Bennett, A. B., DellaPenna, D. 1987. Polygalacturonase: its importance and regulation in ripening. In *Plant Senescence: Its Biochemistry and Physiology*, ed. W. W. Thompson, E. Nothnagel, R. Huffaker, pp. 98-107. Rockville, MD: Am. Soc. Plant Physiol.
17. Bennett, A. B., DellaPenna, D. 1987. Polygalacturonase gene expression in ripening tomato fruit. In *Tomato Biotechnology*, ed. D. Nevins, R. Jones, pp. 299-308. New York: Alan R. Liss
18. Bennett, A. B., Osteryoung, K. W. 1990. Protein transport and targeting within the endomembrane system of plants. In *Plant Biotechnology*. Vol. 1. *Plant Genetic Engineering*, ed. D. Grierson, pp. 200-39. Edinburgh: Blackie and Sons Limited
19. Biggs, M. S., Handa, A. K. 1988. Temporal regulation of polygalacturonase gene expression in fruits of normal, mutant, and heterozygous tomato genotypes. *Plant Physiol.* 89:117-25
20. Biggs, M. S., Harriman, R. W., Handa, A. K. 1986. Changes in gene expression during tomato fruit ripening. *Plant Physiol.* 81:395-403
21. Bird, C. R., Smith, C. J. S., Ray, J. A., Moureau, P., Bevan, M. W., et al. 1988. The tomato polygalacturonase gene and ripening-specific expression in transgenic plants. *Plant Mol. Biol.* 11:651-62
22. Brady, C. J. 1987. Fruit ripening. *Annu. Rev. Plant Physiol.* 38:155-78
23. Brady, C. J., MacAlpine, G., McGlasson, W. B., Ueda, Y. 1982. Polygalacturonase in tomato fruits and the induction of ripening. *Aust. J. Plant Physiol.* 9:171-78
24. Brady, C. J., McGlasson, W. B., Speirs, J. 1987. The biochemistry of fruit ripening. In *Tomato Biotechnology*, ed. D. Nevins, R. Jones, pp. 279-88. New York: Alan R. Liss
25. Brady, C. J., Meldrum, S. K., McGlasson, W. B., Ali, Z. M. 1983. Differential accumulation of the molecular forms of polygalacturonase in tomato mutants. *J. Food Biochem.* 7:7-14
26. Brecht, J. K., Huber, D. J. 1986. Stimulation of tomato fruit ripening by tomato cell-wall fragments. *Hortic. Sci.* 21:319
27. Brown, S. M., Crouch, M. L. 1990. Characterization of a gene family abun-

dantly expressed in *Oenothera organensis* pollen that shows sequence similarity to polygalacturonase. *Plant Cell* 2:263-74

28. Buescher, R. W., Tigchelaar, E. C. 1975. Pectinesterase, polygalacturonase, Cx-cellulase activities and softening of the *rin* tomato mutant. *Hortic. Sci.* 10:624-25

29. Carpita, N. C. 1987. The biochemistry of "growing" cell walls. In *Physiology of Cell Expansion During Plant Growth*, ed. D. J. Cosgrove, D. P. Knievel, pp 28-45. Rockville, MD: Am. Soc. Plant Physiol.

30. Cass, L. G., Kirven, K. A., Christoffersen, R. E. 1990. Isolation and characterization of a cellulase gene family member expressed during avocado fruit ripening. *Mol. Gen. Genet.* 223:76-86

31. Cassab, G. I., Varner, J. E. 1988. Cell wall proteins. *Annu. Rev. Plant Physiol. Plant Mol. Biol.* 39:321-53

32. Chan, H. T., Tam, S. Y. T. 1982. Partial separation and characterization of papaya endo- and exo-polygalacturonase. *J. Food Sci.* 47:1478-83

33. Chang, C., Bowman, J. L., DeJohn, A. W., Lander, E. S., Meyerowitz, E. M. 1988. Restriction fragment length polymorphism linkage map for *Arabidopsis thaliana*. *Proc. Natl. Acad. Sci. USA* 85:6856-60

34. Christoffersen, R. E., Tucker, M. L., Laties, G. G. 1984. Cellulase gene expression in ripening avocado fruit: the accumulation of cellulase mRNA and protein as demonstrated by cDNA hybridization and immunodetection. *Plant Mol. Biol.* 3:385-91

35. Christoffersen, R. E., Warm, E., Laties, G. G. 1982. Gene expression during fruit ripening in avocado. *Planta* 155:52-57

36. Crookes, P. R., Grierson, D. 1983. Ultrastructure of tomato fruit ripening and the role of polygalacturonase isoenzymes in cell wall degradation. *Plant Physiol.* 72:1088-93

37. Davies, K., Hobson, G. E., Grierson, D. 1988. Silver ions inhibit the ethylene stimulated production of ripening-related mRNAs in tomato. *Plant Cell Environ.* 11:729-38

38. Deikman, J., Fischer, R. L. 1988. Interaction of a DNA binding factor with the 5'-flanking region of an ethylene-responsive fruit ripening gene from tomato. *EMBO J.* 7:3315-20

39. Delauney, A. J., Tabaeizadeh, Z., Verma, D. P. S. 1988. A stable bifunctional antisense transcript inhibiting gene expression in transgenic plants. *Proc. Natl. Acad. Sci. USA* 85:4300-4

40. DellaPenna, D., Alexander, D. C., Bennett, A. B. 1986. Molecular cloning of tomato fruit polygalacturonase: analysis of polygalacturonase mRNA levels during ripening. *Proc. Natl. Acad. Sci. USA* 83:6420-24

41. DellaPenna, D., Bennett, A. B. 1988. In vitro synthesis and processing of tomato fruit polygalacturonase. *Plant Physiol.* 86:1057-63

42. DellaPenna, D., Kates, D. S., Bennett, A. B. 1987. Polygalacturonase gene expression in Rutgers, *rin*, *nor*, and *Nr* tomato fruits. *Plant Physiol.* 85:502-7

43. DellaPenna, D., Lashbrook, C. C., Toenjes, K., Giovannoni, J. J., Fischer, R. L., Bennett, A. B. 1990. Polygalacturonase isozymes and pectin depolymerization in transgenic rin tomato fruit. *Plant Physiol.* 94:1881-86

44. DellaPenna, D., Lincoln, J. E., Fischer, R. L., Bennett, A. B. 1989. Transcriptional analysis of polygalacturonase and other ripening associated genes in Rutgers, *rin*, *nor*, and *Nr* tomato fruit. *Plant Physiol.* 90:1372-77

45. Downs, C. G., Brady, C. J. 1990. Two forms of exopolygalacturonase increase as peach fruits ripen. *Plant Cell Environ.* 13:523-30

46. Durbin, M. L., Lewis, L. N. 1988. Cellulases in *Phaseolus vulgaris*. *Methods Enzymol.* 160:342-51

47. Ecker, J. R., Davis, R. W. 1986. Inhibition of gene expression in plant cells by expression of antisense RNA. *Proc. Natl. Acad. Sci. USA* 83:5372-76

48. Fang, G., Grumet, R. 1990. Agrobacterium tumefaciens mediated transformation and regeneration of muskmelon plants. *Plant Cell Rep.* 9:160-64

49. Fillatti, J. J., Kiser, J., Rose, B., Comai, L. 1987. Efficient transformation of tomato and the introduction and expression of a gene for herbicide tolerance. In *Tomato Biotechnology*, ed. D. Nevins, R. Jones, pp. 199-210. New York: Alan R. Liss

50. Fishman, M. L., Gross, K. C., Gillespie, D. T., Sondney, S. M. 1989. Macromolecular components of tomato fruit pectin. *Arch. Biochem. Biophys.* 274:179-91

51. Fitch, M. M. M., Manshardt, R. M., Gonsalves, D., Slightom, J. L., Sanford, J. C. 1990. Stable transformation of papaya via microprojectile bombardment. *Plant Cell Rep.* 9:189-94

52. Fry, S. C. 1988. *The Growing Plant Cell Wall: Chemical and Metabolic*

Analysis. New York: John Wiley & Sons. 332 pp.

53. Fry, S. C. 1989. The structure and functions of xyloglucan. *J. Exp. Bot.* 40:1-11

54. Fry, S. C., Miller, J. G. 1989. Towards a working model of the growing plant cell wall. In *The Biosynthesis and Biodegradation of Plant Cell Wall Polymers*, ed. N. G. Lewis, M. G. Paice, pp 36-46. Washington DC: Am. Chem. Soc.

55. Fuchs, Y., Anderson, J. D. 1987. Purification and characterization of ethylene inducing proteins from cellulysin. *Plant Physiol.* 84:732-36

56. Fujiyama, K., Takemura, H., Shibayama, S., Kobayashi, K., Choi, J.-K., et al. 1988. Structure of horseradish peroxidase isozyme C genes. *Eur. J. Biochem.* 173:681-87

57. Galbraith, D. W., Harkins, K. R., Maddox, J. M., Ayres, N. M., Sharma, D. P., Firoozabady, E. 1983. Rapid flow cytometric analysis of the cell cycle in intact plant tissues. *Science* 220:1049-51

58. Gausing, K. 1987. Thionin genes specifically expressed in barley leaves. *Planta* 717:241-46

59. Giovannoni, J. 1990. *The role of polygalacturonase in tomato fruit ripening.* PhD thesis. Univ. Calif., Berkeley. 191 pp.

60. Giovannoni, J. J., DellaPenna, D., Bennett, A. B., Fischer, R. L. 1989. Expression of a chimeric polygalacturonase gene in transgenic *rin* (ripening inhibitor) tomato fruit results in polyuronide degradation but not fruit softening. *Plant Cell* 1:53-63

61. Giovannini, J. J., DellaPenna, D., Bennett, A. B., Fischer, R. L. 1990. Polygalacturonase and tomato fruit ripening. *Hortic. Rev.* 11:217-27

62. Graham, J., McNicol, R. J., Kumar, A. 1990. Use of the GUS gene as a selectable marker for *Agrobacterium*-mediated transformation of *Rubus*. *Plant Cell, Tissue Organ Cult.* 20:35-39

63. Grierson, D. 1985. Gene expression in ripening tomato fruit. *Crit. Rev. Plant Sci.* 3:113-32

64. Grierson, D., Maunders, M. J., Slater, A., Ray, J., Bird, C. R., et al. 1986. Gene expression during tomato ripening. *Philos. Trans. R. Soc. Lond. Ser. B* 314:399-410

65. Grierson, D., Maunders, M. J., Holdsworth, M. J., Ray, J., Bird, C., et al. 1987. Expression and function of ripening genes. In *Tomato Biotechnology*, ed. D. J. Nevins, R. A. Jones, pp. 309-24. New York: Alan R. Liss

66. Grierson, D., Purton, M., Knapp, J., Bathgate, B. 1987. Tomato ripening mutants. In *Developmental Mutants in Higher Plants*, ed. H. Thomas, D. Grierson, pp. 73-94. Cambridge: Cambridge Univ. Press

67. Grierson, D., Tucker, G. A., Keen, J., Ray, J., Bird, C. R., Schuch, W. 1986. Sequencing and identification of a cDNA clone for tomato polygalacturonase. *Nucleic Acids Res.* 14:8595-603

68. Grierson, D., Tucker, G. A., Robertson, N. G. 1981. The molecular biology of ripening. In *Recent Advances in the Biochemistry of Fruit and Vegetables*, ed. J. Friend, M. J. C. Rhodes, pp. 149-60. London: Academic

69. Gross, K. C. 1984. Fractionation and partial characterization of cell walls from normal and non-ripening mutant tomatoes. *Physiol. Plant.* 62:25-32

70. Gross, K. C. 1985. Promotion of ethylene evolution and ripening of tomato fruit by galactose. *Plant Physiol.* 79:306-7

71. Gross, K. C. 1988. Cell wall dynamics. In *Biotechnology and Food Quality*, ed. S.-D. Kung, D. D. Bills, R. Quatrano, pp. 143-58. Boston: Butterworths

72. Gross, K. C., Wallner, S. J. 1979. Degradation of cell wall polysaccharides during tomato fruit ripening. *Plant Physiol.* 63:117-20

73. Gross, K. C., Watada, A. E., Kang, M. S., Kim, S. D., Kim, K. S., Lee, S. W. 1986. Biochemical changes associated with the ripening of hot pepper. *Physiol. Plant.* 66:31-36

74. Guilley, H., Dudley, R., Jonard, G., Balazs, E., Richards, K. 1982. Transcription of cauliflower mosaic virus DNA: detection of promoter sequences, and characterization of transcripts. *Cell* 30:763-73

75. Guri, A., Sink, K. C. 1988. *Agrobacterium* transformation of eggplant. *J. Plant Physiol.* 133:52-55

76. Hall, C. B. 1963. Cellulase in tomato fruits. *Nature* 200:1010-11

77. Hamilton, A. J., Lycett, G. W., Grierson, D. 1990. Antisense gene that inhibits synthesis of the hormone ethylene in transgenic plants. *Nature* 346:284-87

78. Hatfield, R., Nevins, D. J. 1986. Characterization of the hydrolytic activity of avocado cellulase. *Plant Cell Physiol.* 27:541-52

79. Hayashi, T. 1989. Xyloglucans in the primary cell wall. *Annu. Rev. Plant Physiol. Plant Mol. Biol.* 40:139-68

80. Hayashi, T., MacLachlan, G. 1984. Pea xyloglucan and cellulose. I. Macro-

molecular organization. *Plant Physiol.* 75:596-604

81. Hayashi, T., Marsden, M. P. F., Delmer, D. P. 1987. Pea xyloglucan and cellulose. V. Xyloglucan-cellulose interactions in vitro and in vivo. *Plant Physiol.* 83:384-89

82. Hayashi, T., Wong, Y. S., MacLachlan, G. A. 1984. Pea xyloglucan and cellulose. II. Hydrolysis by pea endo-1,4-β-glucanases. *Plant Physiol.* 75:605-10

83. Hemenway, C., Fang, R.-X., Kaniewski, W. K., Chua, N.-H., Tumer, N. E. 1988. Analysis of the mechanism of protection in transgenic plants expressing the potato virus coat protein or its antisense RNA. *EMBO J.* 7:1273-80

84. Hinton, D. M., Pressey, R. 1974. Cellulase activity in peaches during ripening. *J. Food Sci.* 39:783-85

85. Hinton, D. M., Pressey, R. 1980. Glucanases in fruits and vegetables. *J. Am. Hortic. Sci.* 105:499-502

86. Hobson, G. E. 1964. Polygalacturonase in normal and abnormal tomato fruit. *Biochem. J.* 92:324-32

87. Hobson, G. 1965. The firmness of tomato fruit in relation to polygalacturonase activity. *J. Hortic. Sci.* 40:66-72

88. Hobson, G. 1967. Effects of alleles at the "never-ripe" locus on ripening of tomato fruit. *Phytochemistry* 6:1337-41

89. Hobson, G. E. 1968. Cellulase activity during the maturation and ripening of tomato fruit. *J. Food Sci.* 33:588-92

90. Hobson, G. E., Nichols, R., Davies, J. N., Atkey, P. T. 1984. The inhibition of tomato fruit ripening by silver. *J. Plant Physiol.* 116:21-29

91. Huber, D. J. 1981. Polyuronide degradation and hemicellulose modifications in ripening tomato fruit. *J. Am. Soc. Hortic. Sci.* 108:405-9

92. Huber, D. J. 1983. The role of cell wall hydrolases in fruit softening. *Hortic. Rev.* 5:169-219

93. Huber, D. J. 1983. Polyuronide degradation and hemicellulose modifications in ripening tomato fruit. *J. Am. Soc. Hortic. Sci.* 108:405-9

94. Huber, D. J. 1984. Strawberry fruit softening: the potential roles of polyuronides and hemicelluloses. *J. Food Sci.* 49:1310-15

95. James, D. J., Passey, A. J., Barbara, D. J., Bevan, M.1989. Genetic transformation of apple (*Malus pumila* Mill.) using a disarmed Ti-binary vector. *Plant Cell Rep.* 7:658-61

96. Keegstra, K., Talmadge, K. W., Bauer, W. D., Albersheim, P. 1973. The structure of plant cell walls. III. A model of the walls of suspension-cultured sycamore cells based on the interconnections of the macromolecular components. *Plant Physiol.* 51:188-96

97. Kim, J., Solomos, T., Gross, K. C. 1991. Galactose metabolism and ethylene production during development and ripening of tomato (*Lycopersicon esculentum*) fruit. *Postharvest Biol. Technol.* In press

98. Kim, J., Gross, K. C., Solomos, T. 1987. Characterization of the stimulation of ethylene production by galactose in tomato (*Lycopersicon esculentum* Mill.) fruit. *Plant Physiol.* 85:804-7

99. Kimelman, D., Kirschner, M. W. 1989. An antisense mRNA directs the covalent modification of the transcript encoding fibroblast growth factor in *Xenopus* oocytes. *Cell* 59:687-96

100. Knapp, J., Moureau, P., Schuch, W., Grierson, D. 1989. Organization and expression of polygalacturonase and other ripening related genes in Ailsa Craig "Neverripe" and "Ripening inhibitor" tomato mutants. *Plant Mol. Biol.* 12:105-16

101. Knee, M. 1973. Polysaccharide changes in cell walls of ripening apple. *Phytochemistry* 12:1543-49

102. Knee, M., Bartley, I. M. 1981. Composition and metabolism of cell wall polysaccharides in ripening fruits. In *Recent Advances in the Biochemistry of Fruit and Vegetables*, ed. J. Friend, M. J. C. Rhodes, pp. 133-48. London: Academic. 275 pp.

103. Knee, M., Sargent, J., Osborne, D. 1977. Cell wall metabolism in developing strawberry fruits. *J. Exp. Bot.* 28:377-96

104. Knegt, E., Vermeer, E., Bruinsma, J. 1988. Conversion of the polygalacturonase isozymes from ripening tomato fruits. *Physiol. Plant.* 72:108-14

105. Koch, J. L., Nevins, D. J. 1989. Tomato fruit cell wall. I. Use of purified tomato polygalacturonase and pectinmethylesterase to identify developmental changes in pectins. *Plant Physiol.* 91:816-22

106. Kramer, M., Sanders, R. A., Sheehy, R. E., Melis, M., Kuehn, M., Hiatt, W. R. 1990. Field evaluation of tomatoes with reduced polygalacturonase by antisense RNA. In *Horticultural Biotechnology*, ed. A. B. Bennett, S. D. O'Neill, pp. 347-55. New York: Alan R. Liss

107. Labavitch, J. M. 1981. Cell wall turnover in plant development. *Annu. Rev. Plant. Physiol.* 32:385-406

108. Lackey, G. D., Gross, K. C., Wallner,

S. J. 1980. Loss of tomato cell wall galactan may involve reduced rate of synthesis. *Plant Physiol.* 66:532-33

109. Deleted in proof

110. Lamport, D. T. A. 1986. The primary cell wall: a new model. In *Cellulose: Structure, Modification and Hydrolysis*, ed. R. A. Young, R. M. Rowell, pp. 77-90. New York: John Wiley & Sons

111. Lee, E., Speirs, J., Gray, J., Brady, C. J. 1990. Homologies to the tomato endopolygalacturonase gene in the peach genome. *Plant Cell & Environ.* 13:513-21

112. Lincoln, J. E., Cordes, S., Read, E., Fischer, R. L. 1987. Regulation of gene expression by ethylene during *Lycopersion esculentum* (tomato) fruit development. *Proc. Natl. Acad. Sci. USA* 84:2793-97

113. Lincoln, J. E., Fischer, R. L. 1988. Regulation of gene expression by ethylene in wild-type and *rin* tomato (*Lycopersicon esculentum*) fruit. *Plant Physiol.* 88:370-74

114. Mansson, P. E., Hsu, D., Stalker, D. 1985. Characterization of fruit specific cDNAs from tomato. *Mol. Gen. Genet.* 200:356-61

115. Maunders, M. J., Holdsworth, M. J., Slater, A., Knapp, J. E., Bird, C. R., Schuch, W., Grierson, D. 1987. Ethylene stimulates the accumulation of ripening-related mRNAs in tomatoes. *Plant, Cell Environ.* 10:177-84

116. McCollum, T. G., Huber, D. J., Cantlife, D. J. 1989. Modification of polyuronides and hemicelluloses during muskmelon fruit softening. *Physiol. Plant.* 76:303-8

117. McCormick, S., Neidermeyer, J., Fry, J., Barnson, A., Horsch, R., Fraley, R. 1986. Leaf disc transformation of cultivated tomato (*L. esculentum*) using *Agrobacterium tumefaciens*. *Plant Cell Rep.* 5:81-84

118. McDougall, G. J., Fry, S. C. 1988. Inhibition of auxin-stimulated growth of pea stem segments by a specific nonasaccharide of xyloglucan. *Planta* 175:412-16

119. McFeeters, R. F., Bell, T. A., Fleming, H. P. 1980. An endopolygalacturonase in cucumber fruit. *J. Food Biochem.* 4:1-16

120. McMann, M. C., Wells, B., Roberts, K. 1990. Direct visualization of crosslinks in the primary plant cell wall. *J. Cell. Sci.* 96:323-34

121. McMurchie, E. J., McGlasson, W. B., Eaks, I. L. 1972. Treatment of fruit with propylene gives information about the biogenesis of ethylene. *Nature* 237:235-36

122. McNeil, M., Darvill, A. G., Fry, S., Albersheim, P. 1984. Structure and function of the primary cell walls of plants. *Annu. Rev. Biochem.* 53:625-63

123. Melton, D., ed. 1988. *Antisense RNA and DNA*. Cold Spring Harbor: Cold Spring Harbor Laboratory. 149 pp.

124. Moshrefi, M., Luh, B. S. 1983. Carbohydrate composition and electrophoretic properties of tomato polygalacturonase isoenzymes. *Eur. J. Biochem.* 135:511-14

125. Nam, H.-G., Giraudat, J., den Boer, B., Moonan, F., Loos, W. D. B., et al. 1989. Restriction fragment length polymorphism linkage map of *Arabidopsis thaliana*. *Plant Cell* 1:699-705

126. Nehra, N. S., Chibbar, R. N., Kartha, K. K., Datla, R. S. S., Crosby, W. L., Stushnoff, C. 1990. *Agrobacterium*-mediated transformation of strawberry calli and recovery of transgenic plants. *Plant Cell Rep.* 9:10-13

127. Ogura, N., Nakagawa, H., Takenhana, H. 1975. Effect of storage temperature of tomato fruits on changes of their polygalacturonase and pectinesterase activities accompanied with ripening. *J. Agric. Chem. Soc. Jpn.* 49:271-74

128. Osteryoung, K. W., Toenjes, K., Hall, B., Winkler, V., Bennett, A. B. 1990. Analysis of tomato polygalacturonase expression in transgenic tobacco. *Plant Cell* 2:1239-48

129. Pear, J., Ridge, N., Rasmussen, R., Rose, R., Houck, C. 1989. Isolation and characterization of a fruit specific cDNA and corresponding genomic clone from tomato. *Plant Mol. Biol.* 13:639-51

130. Pesis, E., Fuchs, Y., Zauberman, G. 1978. Cellulase activity and fruit softening in avocado. *Plant Physiol.* 61:416-19

131. Pharr, D. M., Dickinson, D. B. 1973. Partial characterization of Cx-cellulase and cellobiase from ripening tomato fruits. *Plant Physiol.* 52:577-83

132. Picton, S., Grierson, D. 1988. Inhibition of expression of tomato-ripening genes at high temperature. *Plant, Cell Environ.* 11:265-72

133. Platt-Aloia, K. A., Thomson, W. W., Young, R. E. 1980. Ultrastructural changes in the walls of ripening avocados: transmission, scanning and freeze fracture microscopy. *Bot. Gaz.* 141:366-73

134. Poovaiah, B. W., Nukaya, A. 1979. Polygalacturonase and cellulase enzymes in the normal Rutgers, and

mutant *rin* tomato fruits and their relationship to the respiratory climacteric. *Plant Physiol.* 64:534-37

135. Postlmayr, H. L., Luh, B. S., Leonard, S. J. 1966. Characterization of pectic changes in freestone and clingstone peaches during ripening and processing. *Food Tech.* 10:618-25

136. Pressey, R. 1983. β-galactosidases in ripening tomatoes. *Plant Physiol.* 71: 132-35

137. Pressey, R. 1984. Purification and characterization of tomato polygalacturonase converter. *Eur. J. Biochem.* 144: 217-21

138. Pressey, R. 1988. Reevaluation of the changes in polygalacturonases in tomatoes during ripening. *Planta* 174:39-43

139. Pressey, R. 1989. Endo-β-mannanase in tomato fruit. *Phytochemistry* 28:3277-80

140. Pressey, R., Avants, J. K. 1972. Multiple forms of pectinesterase in tomatoes. *Phytochemistry* 11:3139-42

141. Pressey, R., Avants, J. K. 1973. Separation and characterization of endopolygalacturonase and exopolygalacturonase from peaches. *Plant Physiol.* 52:252-56

142. Pressey, R., Avants, J. K. 1973. Two forms of polygalacturonase in tomatoes. *Biochim. Biophys. Acta* 309:363-69

143. Pressey, R., Avants, J. K. 1975. Cucumber polygalacturonase. *J. Food Sci.* 40:937-39

144. Pressey, R., Avants, J. K. 1976. Pear polygalacturonase: *Phytochemistry* 15: 1349-51

145. Pressey, R., Avants, J. K. 1982. Pectic enzymes in "Longkeeper" tomatoes. *Hortic. Sci.* 17:398-400

146. Pressey, R., Avants, J. K. 1982. Solubilization of cell walls by tomato polygalacturonases: effects of pectinesterases. *J. Food Biochem.* 6:57-74

147. Pressey, R., Reger, B. J. 1989. Polygalacturonase in pollen from corn and other grasses. *Plant Sci.* 59:57-62

148. Raikhel, N. V., Wilkens, T. A. 1987. Isolation and characterization of a cDNA encoding wheat germ agglutinin. *Proc. Natl. Acad. Sci. USA* 84:6745-49

149. Ray, J., Knapp, J. E., Grierson, D., Bird, C., Schuch, W. 1988. Identification and sequence determination of a cDNA clone for tomato pectin esterase. *Eur. J. Biochem.* 174:119-24

150. Rhodes, M. J. C. 1980. The maturation and ripening of fruits. In *Senescence in Plants*, ed. K. V. Thimann, pp. 157-205. Boca Raton, FL: CRC Press

151. Rick, C. 1956. New mutants. *Rep. Tomato Genet. Coop.* 6:22-23

152. Robinson, R., Tomes, M. 1968. Ripening inhibitor: a gene with multiple effects on ripening. *Rep. Tomato Genet. Coop.* 18:36-37

153. Rodermel, S., Abbott, M. S., Bogorad, L. 1988. Nuclear-organelle interactions: nuclear antisense gene inhibits ribulose bisphosphate carboxylase enzyme levels in transformed tobacco plants. *Cell* 55: 673-81

154. Rothstein, S. J., DiMaio, J., Strand, M., Rice, D. 1987. Stable and heritable inhibition of the expression of nopaline synthase in tobacco expressing antisense RNA. *Proc. Natl. Acad. Sci. USA* 84: 8439-43

155. Deleted in proof

156. Sacher, J. A. 1973. Senescence and postharvest physiology. *Annu. Rev. Plant Physiol.* 24:197-224

157. Saltveit, M. E., Bradford, K. J., Dilley, D. R. 1978. Siver ion inhibits ethylene synthesis and action in ripening fruits. *J. Am. Soc. Hortic. Sci.* 103:472-75

158. Schuch, W., Bird, C. R., Ray, J., Smith, C. J. S., Watson, C. F., et al. 1989. Control and manipulation of gene expression during tomato fruit ripening. *Plant Mol. Biol.* 13:303-11

159. Seymour, G. B., Colquhoun, I. J., Dupont, M. S., Parsley, K. R., Selvendran, R. R. 1990. Composition and structural features of cell wall polysaccharides from tomato fruits. *Phytochemistry* 29:725-31

160. Seymour, G. B., Harding, S. E. 1987. Analysis of the molecular size of tomato (*Lycopersicon esculentum* Mill.) fruit polyuronides by gel filtration and low-speed sedimentation equilibrium. *Biochem. J.* 245:463-66

161. Seymour, G. B., Harding, S. E., Taylor, A. J., Hobson, G. E., Tucker, G. A. 1987. Polyuronide solubilization during ripening of normal and mutant tomato fruit. *Phytochemistry* 26:1871-75

162. Seymour, G. B., Lasslett, Y., Tucker, G. A. 1987. Differential effects of pectolytic enzymes on tomato polyuronides in vivo and in vitro. *Phytochemistry.* 26:3137-39

163. Sheehy, R. E., Kramer, M. K., Hiatt, W. R. 1988. Reduction of polygalacturonase activity in tomato fruit by antisense RNA. *Proc. Natl. Acad. Sci. USA* 85:8805-9

164. Sheehy, R. E., Pearson, J., Brady, C. J., Hiatt, W. R. 1987. Molecular characterization of tomato fruit polygalacturonase. *Mol. Gen. Genet.* 208:30-36

165. Simons, R. W., Kleckner, N. 1988. Biological regulation by antisense RNA

in prokaryotes. *Annu. Rev. Genet.* 22: 567-600

166. Smith, C. J. S., Watson, C. F., Morris, P. C., Bird, C. R., Seymour, G. B., et al. 1990. Inheritance and effect on ripening of antisense polygalacturonase genes in transgenic tomatoes. *Plant Mol. Biol.* 14:369-79

167. Smith, C. J. S., Watson, C. F., Ray, J., Bird, C. R., Morris, P. C., et al. 1988. Antisense RNA inhibition of polygalacturonase gene expression in transgenic tomatoes. *Nature* 334:724-26

168. Sobotka, F. E., Stelzig, D. A. 1974. An apparant cellulase complex in tomato (*Lycopersicon esculentum*) fruit. *Plant Physiol.* 53:759-63

169. Speirs, J., Brady, C. J., Grierson, D., Lee, E. 1984. Changes in ribosome organization and mRNA abundance in ripening tomato fruit. *Aust. J. Plant Physiol.* 11:225-33

170. Speirs, J., Lee, E., Brady, C. J., Robertson, J., McGlasson, W. B. 1989. Endopolygalacturonase: messenger RNA, enzyme and softening in the ripening fruit of a range of tomato genotypes. *J. Plant Physiol.* 135:576-82

171. Straus, D., Ausubel, F. M. 1990. Genomic subtraction for cloning DNA corresponding to deletion mutations. *Proc. Natl. Acad. Sci. USA* 87:1889-93

172. Tanksley, S. D., Mutschler, M. A., Rick, C. M. 1987. Linkage map of the tomato (*Lycopersicon esculentum*) (2n=24). In *Genetic Maps*, ed. S. J. O'Brien, pp. 655-69. Cold Spring Harbor: Cold Spring Harbor Laboratory

173. Themmen, A. P. N., Tucker, G. A., Grierson, D. 1982. Degradation of isolated tomato cell walls by purified polygalacturonase in vitro. *Plant Physiol.* 69:122-24

174. Tieman, D., Handa, A. 1989. Immunocytolocalization of polygalacturonase in ripening tomato fruit. *Plant Physiol.* 90:17-20

175. Tigchelaar, E. C., McGlasson, W. B., Buescher, R. W. 1978. Genetic regulation of tomato fruit ripening. *Hortic. Sci.* 13:508-13

176. Tigchelaar, E. C., Tomes, M., Kerr, E., Barman, R. 1973. A new fruit ripening mutant, non-ripening (*nor*). *Rep. Tomato Genet. Coop.* 23:33

177. Tong, C. B. S., Gross, K. C. 1988. Glycosyl-linkage composition of tomato fruit cell wall hemicellulosic fractions during ripening. *Physiol. Plant.* 74:365-70

178. Tong, C. B. S., Gross, K. C. 1989.

Ripening characteristics of a tomato mutant, dark green. *J. Am. Soc. Hortic. Sci.* 114:635-38

179. Tong, C. B. S., Labavitch, J. M., Yang, S. F. 1986. The induction of ethylene production from pear cell culture by cell wall fragments. *Plant Physiol.* 81:929-30

180. Trulson, A. J., Simpson, R. B., Shahin, E. A. 1986. Transformation of cucumber (*Cucumis sativus* L.) plants with *Agrobacterium rhizogenes*. *Theor. Appl. Genet.* 73:11-15

181. Tucker, G. A., Robertson, N. G., Grierson, D. 1980. Changes in polygalacturonase isoenzymes during the 'ripening' of normal and mutant tomato fruit. *Eur. J. Biochem.* 112:119-24

182. Tucker, G. A., Robertson, N. G., Grierson, D. 1981. The conversion of tomato-fruit polygalacturonase isoenzyme 2 into isoenzyme 1 in vitro. *Eur. J. Biochem.* 115:87-90

183. Tucker, G. A., Robertson, N. G., Grierson, D. 1982. Purification and changes in activities of tomato pectinesterase isozymes. *J. Sci. Food Agric.* 33:396-400

184. Tucker, G. A., Schindler, C. B., Roberts, J. A. 1984. Flower abscission in mutant tomato plants. *Planta* 160: 164-67

185. Tucker, M. L., Durbin, M. L., Clegg, M. T., Lewis, L. N. 1987. Avocado cellulase: nucleotide sequence of a putative full length cDNA clone and evidence for a small gene family. *Plant Mol. Biol.* 9:197-203

186. Tucker, M. L., Laties, G. G. 1984. Interrelationship of gene expression, polysome prevalence, and respiration during ripening of ethylene and/or cyanide-treated avocado fruit. *Plant Physiol.* 74:307-15

187. Tucker, M. L., Milligan, S. 1991. Sequence analysis and comparison of avocado fruit and bean abscission cellulases. *Plant Physiol.* In press

188. van der Krol, A. R., Lenting, P. E., Veenstra, J., van der Meer, I. M., Does, R. E., et al. 1988. An anti-sense chalcone synthase gene in transgenic plants inhibits flower pigmentation. *Nature* 333:866-69

189. VanderMolen, G. E., Labavitch, J. M., Strand, L. L., DeVay, J. E. 1983. Pathogen-induced vascular gels: ethylene as a host intermediate. *Physiol. Plant* 59:573-80

190. Varner, J. E., Lin, L.-S. 1989. Plant cell wall architecture. *Cell* 56:231-39

191. von Heijne, G. 1985. Signal sequences:

the limits of variation. *J. Mol. Biol.* 184:99-105

192. Wallner, S. J., Bloom, H. L. 1977. Characteristics of tomato cell wall degradation in vitro. Implication for the study of fruit-softening enzymes. *Plant Physiol.* 60:207-10

193. Wallner, S. J., Walker, J. E. 1975. Glycosidases in cell wall-degrading extracts of ripening tomato fruits. *Plant Physiol.* 55:94-98

194. Wing, R. A., Yamaguchi, J., Larabell, S. K., Ursin, V. M., McCormick, S. 1990. Molecular and genetic characterization of two pollen-expressed genes that have sequence similarity to pectate lyases of the plant pathogen *Erwinia*. *Plant Mol. Biol.* 14:17-28

195. Yamaki, S., Kakiuchi, N. 1979. Changes in hemicellulose-degrading enzymes during development and ripening of Japanese pear fruit. *Plant & Cell Physiol.* 20:301-9

196. Yang, S. F. 1985. Biosynthesis and action of ethylene. *Hortic. Sci.* 20:41-45

197. Yang, S. F., Hoffman, N. E. 1984. Ethylene biosynthesis and its regulation in higher plants. *Annu. Rev. Plant Physiol.* 35:155-89

198. Yang, S. F., Pratt, H. K. 1978. The physiology of ethylene in wounded plant tissues. In *Biochemistry of Wounded*

Plant Storage Tissues, ed. G. Kahl, pp. 595-622. Berlin: Walter de Gruyter

199. Yoder, J. I. 1990. A genetic analysis of mutations recovered from tomato following *Agrobacterium*-mediated transformation with the maize transposable elements Activator and Dissociator. *Theor. Appl. Genet.* 79:657-62

200. Yoder, J. I. 1990. Rapid proliferation of the maize transposable element Activator in transgenic tomato. *Plant Cell* 2:723-30

201. Yoder, J. I., Palys, J., Alpert, D., Lassner, M. 1988. *Ac* transposition in transgenic tomato plants. *Mol. Gen. Genet.* 213:291-96

202. York, W. S., Darvill, A. G., Albersheim, P. 1984. Inhibition of 2,4-dichlorophenoxyacetic acid-stimulated elongation of pea stem segments by a xyloglucan oligosaccharide. *Plant Physiol.* 75:295-97

203. Yoshida, O., Nakagawa, H., Ogura, N., Sato, T. 1984. Effect of heat treatment on the development of polygalacturonase activity in tomato fruit during ripening. *Plant Cell Physiol.* 25:505-9

204. Young, N. D., Tanksley, S. D. 1989. Restriction fragment length polymorphism maps and the concept of graphical genotypes. *Theor. Appl. Genet.* 77:95-101

AUTHOR INDEX

T

Ta, T. C., 377, 382
Tabaeizadeh, Z., 693
Tague, B. W., 26, 32, 39, 41-43
Tahara, S., 227, 232, 233
Taiz, L., 117
Tajima, S., 374, 380, 382-84, 517
Tak, T., 508, 510
Takabe, T., 138, 140
Takahashi, H., 558
Takahashi, J. S., 355
Takahashi, N., 174, 482, 485
Takahashi, T., 593, 596
Takai, T., 158-61
Takayama, S., 404-6
Takehara, T., 94
Takemura, H., 686
Takenhana, H., 689, 690
Takimoto, A., 358, 359, 364, 365
Talmadge, K. W., 677, 678
Tam, S. Y. T., 682
Tamai, N., 322
Tamas, I. A., 544
Tanabe, T., 158-61
Tanaka, A., 445
Tanaka, I., 197
Tanchak, M. A., 36
Tang, D.-T., 193, 196
Tanguay, R. M., 602
Tani, T., 177
Tanksley, S. D., 425, 691, 695
Tanner, W., 38
Tantikanjana, T., 402, 405, 406
Tappel, A. L., 155
Tapper, R., 557
Tardieu, F., 67
Tasaka, Y., 470, 485
Taulien, J., 605
Tavladoraki, P., 363, 364
Taylor, A. J., 680, 681, 687
Taylor, I. B., 534
Taylor, I. E. P., 120
Taylor, J. P., 402, 413
Taylor, J. R. N., 24
Taylor, J. S., 63
Taylor, L. P., 213, 220
Taylor, P., 191, 194, 196, 198
Taylor, R., 107
Taylor, S. A., 56
Taylor, S. S., 290
Taylor, W., 354
Taylor, W. C., 425, 429, 432, 442
Tchang, F., 484
Teakle, G. R., 425
Teeter, M. M., 228-33
Telfer, A., 293, 295, 315, 322, 323
Tempe, J., 520, 540

Tenhunen, J. D., 67, 571
Tepfer, M., 118, 120
Terada, R., 213
Terao, T., 444
Terpstra, J., 482
Terzi, M., 608, 609
Terzi, V., 610
Tessier, L.-H., 625, 632
Tester, M., 430
Teucher, T., 470, 475
Thain, J. F., 668
Theg, S. M., 302, 600
Theil, E. C., 438
Theimer, R. R., 492
Theisen, T. W., 158, 161
Thellier, M., 104-6, 109
Themmen, A. P. N., 683
Theologis, A., 538, 543, 660
Theorell, H., 145, 149
Theriot, L. J., 230, 232
Theunis, C. H., 193, 194, 196, 197
Theuvenet, A. P. R., 105, 109
Thibaud, J.-B., 108, 110, 115, 116, 668
Thiel, D. S., 609
Thiel, G., 468
Thiele, B. J., 158, 164, 165
Thielen, A. P. G. M., 328
Thien, W., 432
Thienel, U., 363, 429
Thimann, K. V., 167, 168
This, P., 484
Thom, D., 104, 105
Thoma, S., 484
Thomas, A., 56, 60
Thomas, B., 357-60, 364-66
Thomas, E., 211
Thomas, H., 56, 60, 174, 175
Thomas, H. R., 395
Thomas, T. H., 66
Thomashow, M. F., 210, 582
Thome, U., 84
Thompson, A. G., 134
Thompson, D. M., 438
Thompson, D. V., 514, 518
Thompson, G. A. Jr., 468, 482, 536, 652
Thompson, J. E., 175, 176
Thompson, J. F., 152, 153, 159, 164, 169, 384, 385
Thompson, K. F., 402, 413
Thompson, R. D., 401
THOMPSON, W. F., 423-66; 425, 426, 429-32, 434, 435, 439, 446, 447
Thomson, A. B. R., 107
Thomson, W. W., 109, 120, 677, 687
Thornber, J. P., 283, 284, 442, 443
Thorne, S. W., 427

Thornley, W. R., 80, 81, 86, 93
Thorsness, M. K., 409, 410
Thummler, F., 516
Thunberg, E., 229, 230
Thurman, D. A., 136, 137, 139
Tieman, D., 684
Tigchelaar, E. C., 663, 683, 689
Tijan, R., 643
Tillmann, U., 31, 543
Tilly, K., 442, 590, 592
Tilney-Basset, R. A. E., 250
Tilton, V. R., 395
Timblin, C., 628
Timko, M. P., 435, 436
Timmerhaus, M., 334
Ting, J., 584
Tingey, S. V., 425, 514
Tinland, B., 542
Tissieres, A., 580, 587, 612
Tixier, M., 147, 170
Tjepkema, J. D., 378, 379
Tjian, R., 249
Tobin, E. M., 358, 364, 425, 426, 430, 434, 435, 437, 443, 444, 446
Toenjes, K., 681, 685, 691, 692
Toensberg, T., 558
Tokuhisa, J. G., 359, 426, 428
Tolbert, N. E., 481, 485
Tomasic, J., 80, 81
Tomes, M., 689
Tomiyama, K., 657
Tomizawa, K.-I., 426, 427, 433, 451
Tomos, A. D., 56, 60, 111, 119
Tong, C. B., 662
Tong, C. B. S., 679, 688, 690
Tooze, J., 32
Töpfer, R., 217
Toppan, A., 654
Toriyama, K., 213, 409, 410, 414
Tormanek, H., 492
Torrey, J. G., 510, 512, 518
Toubart, P., 510, 653, 659
Touchard, P., 109
Toulon, V., 108, 115
Touze, A., 657, 660
Toyoshima, K., 227, 232, 233
Tran Thanh Van, K., 510, 635, 653, 659
Traska, A., 134
Travis, J. L., 401, 626, 629, 630
Trebst, A., 336
Treier, U., 583
Treisman, R., 256
Trentham, D. R., 468
Trese, A. T., 518

SUBJECT INDEX

lichen, 560, 562
Transport
chloroplast envelope
see Chloroplast envelope,
metabolite translocators
secretory system
see Secretory system pro-
tein sorting
solutes, 104, 107
see also Electron transport
Transposon tagging, 628, 695
Traumatic acid, 173–79
Traumatin, 172–73, 179
Trebouxia photobionts
see Lichen symbiosis
Trehalose
SST activity barley, 90
Triacylglycerol
see Glycerolipid synthesis
Trifolium
leaf movement rhythm, 354
Triticum aestivum
sperm cells, 194–195
Triose phosphates, 130, 132–33
Trypsin, 292
Tryptophan, 166
monooxygenase, 541–42
Tunicamycin, 29, 33–34, 38
Tyramine, 664

U

Ubiquinol, 168
Ubiquitin
see Heat shock proteins
Unity of life
see Plant physiologist Erasmo
Marrè
Unstirred layers
apoplast, 107–8, 116, 118
Uricase
nodule, 516
Uronic acid, 104, 120

V

Vacuole
fructan
accumulation, 88, 91
metabolism, 91
sucrose transport, 95
see also Secretory system
storage proteins

Vascular growth
phytohormone action, 545
veg, 248
Verticillium dahliae, 667
Vicilin, 32–33, 43
Viral vectors
gene transfer, 212–13
Viscotoxin
see Thionins
Viscum album, 228

W

Water
availability measured by
plants, 60–62
root water status changes 60–
61, 64
soil drying and leaf water status, 56–57
transport
cell expansion, 119–20
Water free space, 105, 118
Water potentials, 57, 119
Water relations
see Lichen symbiosis
Western immunoblots
lipoxygenase, 148, 156–57
Wheat
lipoxygenase
growth, 170
senescence, 175
secretory system proteins, 24–
25
xylem sap, 65–66
Whorl
floral
see Floral homeotic genes
Wound response, 635
role
jasmonic acid, 668
lipoxygenase, 172–74, 176–
79
phytohormones, 535
see also Gene transfer to
plants

X

Xanthium pennsylvanicum
stomatal opening
circadian rhythm, 361
Xanthophyll, 334–35
abscisic acid, 532

Xenopus laevis
secretory system, 25–26, 31–
32
X-ray crystallography
heat shock proteins, 587
phaseolin structure, 34
Xylanase, 688
see also Oligosaccharide signals
Xylans, 679
Xylem
field studies, 67
ion imbalance as soil water
status indicator, 62
stomatal behavior and leaf
growth, 64–66
tomato cation exchange, 106
sap, 111
analysis unwatered plants,
63–65
maize sap, 65–66
pH and soil drying, 68–69
wheat sap, 65
water potentials, 57, 119
Xyloglucan, 653, 678–79, 687
auxin regulation, 661–62
Xyloglucanase, 687
Xylose, 678

Y

Yeast
cation uptake, 109
phaseolin, 26
secretory system
sorting signals, 39–43
see also *Saccharomyces cerevisiae*
Yield
nitrogen fixation, 376

Z

Zanoni, Giuseppina, 6–8, 11
Zea mays
sperm cells, 191–92, 194–95,
199–200
see also Corn
Zeaxanthin
fluorescence
quenching, 332,335
Zygomorphic flowers, 272–275

CUMULATIVE INDEXES

CONTRIBUTING AUTHORS, VOLUMES 34–42

Klee, H., 42:529–51
Kleinig, H., 40:39–59
Krause, G. H., 42:313–49
Kuhlemeier, C., 38:221–57
Kurkdjian, A., 40:271–303

L

Lamb, C. J., 41:339–67
Lara, M., 42:507–28
Lee, M., 39:413–37
Leong, S. A., 37:187–208
Letham, D. S., 34:163–97
Lewis, N. G., 41:455–97
Lin, W., 37:309–34
Lloyd, C. W., 38:119–39
Loewus, F. A., 34:137–61
Loewus, M. W., 34:137–61
Lucas, W. J., 34:71–104
Lucas, W. J., 41:369–419
Lumsden, P. J., 42:351–71
Lynn, D. G., 41:497–526

M

Møller, I. M., 37:309–34
Maliga, P., 35:519–42
Malmberg, R. L., 40:235–69
Mandava, N. B., 39:23–52
Mansfield, T. A., 41:55–75
Marrè, E., 42:1–20
Marx, G. A., 34:389–417
Mascarenhas, J. P., 41:317–38
Mazur, B. J., 40:441–70
Meeks, J. C., 40:193–210
Meins, F. Jr., 34:327–46
Melis, A., 38:11–45
Messing, J., 37:439–66
Mimura, T., 38:95–117
Morgan, J. M., 35:299–319
Morris, R. O., 37:509–38
Mullet, J. E., 39:475–502

N

Nakamoto, H., 36:255–86
Nasrallah, J. B., 42:393–422
Nasrallah, M. E., 42:393–422
Neilands, J. B., 37:187–208
Nester, E. W., 35:387–413
Neuburger, M., 40:371–414
Newton, K. J., 39:503–32
Nishio, T., 42:393–422

O

Oaks, A., 36:345–65
Ogren, W. L., 35:415–42
Olsen, L. J., 40:471–501

P

Padilla, J. E., 42:507–38
Palni, L. M. S., 34:163–97

Passioura, J. B., 39:245–65
Payne, P. I., 38:141–53
Pearcy, R. W., 41:421–53
Pérez, H., 42:507–28
Peters, G. A., 40:193–210
Pharis, R. P., 36:517–68
Phillips, R. L., 39:413–37
Pichersky, E., 40:415–39
Pickard, B. G., 36:55–75
Pollock, C. J., 42:77–101
Potrykus, I., 42:205–25
Powles, S. B., 35:15–44
Pradet, A., 34:199–224
Press, M. C., 41:127–51

R

Ranjeva, R., 38:73–93
Raymond, P., 34:199–224
Reinhold, L., 35:45–83
Rennenberg, H., 35:121–53
Robards, A. W., 41:369–419
Roberts, J. K. M., 35:375–86
Robinson, D., 39:53–99
Rogers, S., 38:467–86
Rolfe, B. G., 39:297–319
Russell, S. D., 42:189–204
Ryan, C. A., 42:651–74

S

Sachs, M. M., 37:363–76
Sánchez, F., 42:507–28
Sanders, D., 41:77–107
Satoh, K., 37:335–61
Scheel, D., 40:347–69
Schnepf, E., 37:23–47
Schroeder, J. I., 40:539–69
Schubert, K. R., 37:539–74
Schulze, E.-D., 37:247–74
Schwintzer, C. R., 37:209–32
Sentenac, H., 42:103–28
Serrano, R., 40:61–94
Shimmen, T., 38:95–117
Siedow, J. N., 42:145–88
Silk, W. K., 35:479–518
Silverthorne, J., 36:569–93
Smith, S. E., 39:221–44
Smith, T. A., 36:117–43
Snell, W. J., 36:287–315
Solomonson, L. P., 41:225–53
Somerville, C. R., 37:467–507
Somerville, C., 42:467–506
Sperry, J. S., 40:19–38
Spiker, S., 36:235–53
Steffens, J. C., 41:553–75
Steponkus, P. L., 35:543–84
Stewart, G. R., 41:127–51
Stitt, M., 41:153–85
Stocking, C. R., 35:1–14
Stone, B. A., 34:47–70
Strotmann, H., 35:97–120

Sweeney, B. M., 38:1–9
Sze, H., 36:175–208

T

Taiz, L., 35:585–657
Tang, P.-S., 34:1–19
Taylor, W. C., 40:211–33
Tazawa, M., 38:95–117
Theg, S. M., 38:347–89
Theg, S. M., 40:471–501
Theologis, A., 37:407–38
Thompson, W. F., 42:423–66
Thorne, J. H., 36:317–43
Ting, I. P., 36:595–622
Tjepkema, J. D., 37:209–32
Tobin, E. M., 36:569–93
Trelease, R. N., 35:321–47
Turgeon, R., 40:119–38
Tyree, M. T., 40:19–38

V

Vänngård, T., 39:379–411
van Huystee, R. B., 38:205–19
Vance, C. P., 42:373–92
Varner, J., 39:321–53
Vierling, E., 42:579–620

W

Wada, M., 40:169–91
Walbot, V., 36:367–96
Wayne, R. O., 36:397–439
Weil, C. F., 41:527–52
Weiler, E. W., 35:85–95
Weis, E., 42:313–49
Wessler, S. R., 41:527–52
White, M. J., 42:423–66
Whitfeld, P. R., 34:279–310
Wiemken, A., 37:137–64
Woodrow, I. E., 39:533–94

Y

Yamamoto, E., 41:455–97
Yang, S. F., 35:155–89
Yanofsky, M. F., 35:387–413
Yocum, C. F., 41:255–76

Z

Zaitlin, M., 38:291–315
Zeevaart, J. A. D., 39:439–73
Zeiger, E., 34:441–75
Zhang, J., 42:55–76
Ziegler, P., 40:95–117
Zurawski, G., 38:391–418

CHAPTER TITLES, VOLUMES 34–42

ANNUAL REVIEWS INC.
A NONPROFIT SCIENTIFIC PUBLISHER
4139 El Camino Way
P.O. Box 10139
Palo Alto, CA 94303-0897 • USA

ORDER FORM

ORDER TOLL FREE
1-800-523-8635
(except California)

FAX: 415-855-9815

Annual Reviews Inc. publications may be ordered directly from our office; through booksellers and subscription agents, worldwide; and through participating professional societies. **Prices subject to change without notice.**

ARI Federal I.D. #94-1156476

- **Individuals:** Prepayment required on new accounts by check or money order (in U.S. dollars, check drawn on U.S. bank) or charge to credit card — American Express, VISA, MasterCard.
- **Institutional buyers:** Please include purchase order.
- **Students:** $10.00 discount from retail price, per volume. Prepayment required. Proof of student status must be provided (photocopy of student I.D. or signature of department secretary is acceptable). Students must send orders direct to Annual Reviews. Orders received through bookstores and institutions requesting student rates will be returned. You may order at the Student Rate for a maximum of 3 years.
- **Professional Society Members:** Members of professional societies that have a contractual arrangement with Annual Reviews may order books through their society at a reduced rate. Check with your society for information.
- **Toll Free Telephone orders:** Call 1-800-523-8635 (except from California) for orders paid by credit card or purchase order and customer service calls only. California customers and all other business calls use 415-493-4400 (not toll free). Hours: 8:00 AM to 4:00 PM, Monday-Friday, Pacific Time. **Written confirmation** is required on purchase orders from universities before shipment.
- **FAX: 415-855-9815   Telex: 910-290-0275**
- **We do not ship on approval.**

Regular orders: Please list below the volumes you wish to order by volume number.

Standing orders: New volume in the series will be sent to you automatically each year upon publication. Cancellation may be made at any time. Please indicate volume number to begin standing order.

Prepublication orders: Volumes not yet published will be shipped in month and year indicated.

California orders: Add applicable sales tax. **Canada:** Add GST tax.

Postage paid (4th class bookrate/surface mail) **by Annual Reviews Inc.** UPS domestic ground service available (except Alaska and Hawaii) at $2.00 extra per book. Airmail postage or UPS air service also available at prevailing costs. UPS must have street address; P.O. Box, APO or FPO not acceptable.

ANNUAL REVIEWS SERIES	Prices postpaid, per volume USA & Canada / elsewhere		Regular Order Please Send Vol. Number:	Standing Order Begin With Vol. Number:
	Until 12-31-90	After 1-1-91		
Annual Review of ANTHROPOLOGY				
Vols. 1-16 (1972-1987)	$31.00/$35.00	$33.00/$38.00		
Vols. 17-18 (1988-1989)	$35.00/$39.00	$37.00/$42.00		
Vol. 19 (1990)	$39.00/$43.00	$41.00/$46.00		
Vol. 20 (avail. Oct. 1991)	$41.00/$46.00	$41.00/$46.00	Vol. _____	Vol(s). _____
Annual Review of ASTRONOMY AND ASTROPHYSICS				
Vols. 1, 5-14 (1963, 1967-1976)				
16-20 (1978-1982)	$31.00/$35.00	$33.00/$38.00		
Vols. 21-27 (1983-1989)	$47.00/$51.00	$49.00/$54.00		
Vol. 28 (1990)	$51.00/$55.00	$53.00/$58.00		
Vol. 29 (avail. Sept. 1991)	$53.00/$58.00	$53.00/$58.00	Vol. _____	Vol(s). _____
Annual Review of BIOCHEMISTRY				
Vols. 30-34, 36-56 (1961-1965, 1967-1987)	$33.00/$37.00	$35.00/$40.00		
Vols. 57-58 (1988-1989)	$35.00/$39.00	$37.00/$42.00		
Vol. 59 (1990)	$39.00/$44.00	$41.00/$47.00		
Vol. 60 (avail. July 1991)	$41.00/$47.00	$41.00/$47.00	Vol. _____	Vol(s). _____
Annual Review of BIOPHYSICS AND BIOPHYSICAL CHEMISTRY				
Vols. 1-11 (1972-1982)	$31.00/$35.00	$33.00/$38.00		
Vols. 12-18 (1983-1989)	$49.00/$53.00	$51.00/$56.00		
Vol. 19 (1990)	$53.00/$57.00	$55.00/$60.00		
Vol. 20 (avail. June 1991)	$55.00/$60.00	$55.00/$60.00	Vol. _____	Vol(s). _____

Annual Review of CELL BIOLOGY

Vols. 1-3	(1985-1987)	$31.00/$35.00	$33.00/$38.00		
Vols. 4-5	(1988-1989)	$35.00/$39.00	$37.00/$42.00		
Vol. 6	(1990)	$39.00/$43.00	$41.00/$46.00		
Vol. 7	(avail. Nov. 1991)	$41.00/$46.00	$41.00/$46.00	Vol(s). _____	Vol. _____

Annual Review of COMPUTER SCIENCE

Vols. 1-2	(1986-1987)	$39.00/$43.00	$41.00/$46.00		
Vols. 3-4	(1988, 1989-1990)	$45.00/$49.00	$47.00/$52.00	Vol(s). _____	Vol. _____

Series suspended until further notice. SPECIAL OFFER: Volumes 1-4 are available at the special promotional price of $100.00 USA & Canada / $115.00 elsewhere, when all 4 volumes are purchased at one time. Orders at the special price must be prepaid.

Annual Review of EARTH AND PLANETARY SCIENCES

Vols. 1-10	(1973-1982)	$31.00/$35.00	$33.00/$38.00		
Vols. 11-17	(1983-1989)	$49.00/$53.00	$51.00/$56.00		
Vol. 18	(1990)	$53.00/$57.00	$55.00/$60.00		
Vol. 19	(avail. May 1991)	$55.00/$60.00	$55.00/$60.00	Vol(s). _____	Vol. _____

Annual Review of ECOLOGY AND SYSTEMATICS

Vols. 2-18	(1971-1987)	$31.00/$35.00	$33.00/$38.00		
Vols. 19-20	(1988-1989)	$34.00/$38.00	$36.00/$41.00		
Vol. 21	(1990)	$38.00/$42.00	$40.00/$45.00		
Vol. 22	(avail. Nov. 1991)	$40.00/$45.00	$40.00/$45.00	Vol(s). _____	Vol. _____

Annual Review of ENERGY

Vols. 1-7	(1976-1982)	$31.00/$35.00	$33.00/$38.00		
Vols. 8-14	(1983-1989)	$58.00/$62.00	$60.00/$65.00		
Vol. 15	(1990)	$62.00/$66.00	$64.00/$69.00		
Vol. 16	(avail. Oct. 1991)	$64.00/$69.00	$64.00/$69.00	Vol(s). _____	Vol. _____

Annual Review of ENTOMOLOGY

Vols. 10-16, 18	(1965-1971, 1973)				
20-32	(1975-1987)	$31.00/$35.00	$33.00/$38.00		
Vols. 33-34	(1988-1989)	$34.00/$38.00	$36.00/$41.00		
Vol. 35	(1990)	$38.00/$42.00	$40.00/$45.00		
Vol. 36	(avail. Jan. 1991)	$40.00/$45.00	$40.00/$45.00	Vol(s). _____	Vol. _____

Annual Review of FLUID MECHANICS

Vols. 2-4, 7	(1970-1972, 1975)				
9-19	(1977-1987)	$32.00/$36.00	$34.00/$39.00		
Vols. 20-21	(1988-1989)	$34.00/$38.00	$36.00/$41.00		
Vol. 22	(1990)	$38.00/$42.00	$40.00/$45.00		
Vol. 23	(avail. Jan. 1991)	$40.00/$45.00	$40.00/$45.00	Vol(s). _____	Vol. _____

Annual Review of GENETICS

Vols. 1-21	(1967-1987)	$31.00/$35.00	$33.00/$38.00		
Vols. 22-23	(1988-1989)	$34.00/$38.00	$36.00/$41.00		
Vol. 24	(1990)	$38.00/$42.00	$40.00/$45.00		
Vol. 25	(avail. Dec. 1991)	$40.00/$45.00	$40.00/$45.00	Vol(s). _____	Vol. _____

Annual Review of IMMUNOLOGY

Vols. 1-5	(1983-1987)	$31.00/$35.00	$33.00/$38.00		
Vols. 6-7	(1988-1989)	$34.00/$38.00	$36.00/$41.00		
Vol. 8	(1990)	$38.00/$42.00	$40.00/$45.00		
Vol. 9	(avail. April 1991)	$41.00/$46.00	$41.00/$46.00	Vol(s). _____	Vol. _____

Annual Review of MATERIALS SCIENCE

Vols. 1, 3-12	(1971, 1973-1982)	$31.00/$35.00	$33.00/$38.00		
Vols. 13-19	(1983-1989)	$66.00/$70.00	$68.00/$73.00		
Vol. 20	(1990)	$70.00/$74.00	$72.00/$77.00		
Vol. 21	(avail. Aug. 1991)	$72.00/$77.00	$72.00/$77.00	Vol(s). _____	Vol. _____